Evolutionary Quantitative Genetics

Evolutionary Quantitative Genetics

STEVAN J. ARNOLD

Professor Emeritus, Oregon State University

OXFORD
UNIVERSITY PRESS

Great Clarendon Street, Oxford, OX2 6DP,
United Kingdom

Oxford University Press is a department of the University of Oxford.
It furthers the University's objective of excellence in research, scholarship,
and education by publishing worldwide. Oxford is a registered trade mark of
Oxford University Press in the UK and in certain other countries

Published in the United States of America by Oxford University Press
198 Madison Avenue, New York, NY 10016, United States of America

British Library Cataloguing in Publication Data
Data available

Library of Congress Control Number: 2022950957

ISBN 978–0–19–285938–9
ISBN 978–0–19–285939–6 (pbk.)

DOI: 10.1093/oso/9780192859389.001.0001

Printed and bound by
CPI Group (UK) Ltd, Croydon, CR0 4YY

To Russell Lande, pioneer and trailblazer

Preface

Over the last several decades, evolutionary quantitative genetics has emerged as the central discipline for understanding the evolution of quantitative traits. By quantitative traits we mean traits such as body mass, behavioural characteristics, and physiological measurements and many other characteristics that have continuous distributions and are affected by many genes. Evolutionary quantitative genetics has succeeded in modelling the evolution of such complex quantitative traits by taking a statistical approach to the elementary processes of selection, mutation, inheritance, and gene flow. Theory based on those elementary characterizations has succeeded in modelling trait evolution in a generation by generation format that can be extrapolated into the future.

The field of evolutionary quantitative urgently needs a succinct, up-to-date synthesis. A similar need for synthesis prevailed in the 1940s when multiple authors tackled the problem. In 1944 G. G. Simpson succeeded in his synthetic goal of reconciling major patterns of evolution revealed by palaeontology with an understanding of process gleaned from systematics and population genetics. Simpson's book was the fourth in a series of six books by Dobzhanksy, Mayr, Huxley, Schmalhausen, and Stebbins that established a new synthesis among the evolutionary disciplines. Simpson's contribution was especially significant because he used a powerful and novel version of Sewall Wright's adaptive landscape to conceptualize evolution over deep paleontological time. Almost forgotten today, many concepts of genetic saltation and vitalism were still current in the 1940s but were quickly eclipsed by the new synthesis. As a sign of Simpson's enduring legacy, we continue to use his concept of an adaptive landscape for phenotypes as a formal model that generates familiar evolutionary patterns on all timescales.

The need for a synthesis today is as dire as it was in the 1940s. Consider, for example, the study of phenotypic evolution. Evolutionary quantitative genetics has blossomed over the last 30 years, but it's relevance to many problems is still unappreciated by most evolutionary biologists. Astounding discoveries have been made in the evolution of development (evodevo), but that field remains a conceptual orphan. Likewise, progress in palaeontology has been largely independent of the rest of evolutionary biology. Part of the reason for these and other disconnects is that conceptual advances in evolutionary quantitative genetics have not been synthesized and made accessible. Students and researchers seeking an introduction are forced into the primary literature where they confront formidable technical hurdles (matrix algebra, advanced multivariate statistics, stochastic processes). In the absence of a succinct, accessible overview, investigators may not appreciate the forest even though they are familiar with particular trees. For all these reasons, phenotypic evolution urgently needs a synthesis.

The genesis of a book that would address the apparent need for a synthesis was sparked by the US National Science Foundation's OPUS program (Opportunities for Promoting Understanding through Synthesis), spearheaded by Mark Courtney. The OPUS guidelines suggested that the proposal identify 20 key papers, written by the grant writer, which would be the themes for 20 chapters in a synthetic work (book, video, etc). I was funded by the NSF in 2010 and set to work. One article on the list of 20 argued that evolutionary models with a moving adaptive peak had the best potential to explain adaptive radiations. That insight became the crux of the book's main argument.

The book is patterned after my teaching style in graduate courses, taught first at the University of Chicago and later at Oregon State University in Corvallis. The courses at Chicago were overview classes for beginning graduate students. The first course (aka LAW) was a collaboration with Russell Lande and Michael Wade. Later when Brian Charlesworth came to Chicago, we recruited him to the course and it became known as CLAW. Each of us also taught our own advanced course, so the intent of LAW and CLAW was to showcase the upcoming content of those individual courses. Once the class was underway, I noticed that the students were exasperated as each successive speaker used different notations and even styles of matrix algebra. I became the course secretary, who interrupted the speaker to point out that different names and symbols were being used for the same variable. When I moved to Corvallis, I brought two graduate courses with me. One course was patterned after CLAW, in the sense that its content was mainly theory. The other was my main graduate solo course at Chicago (Tools of the Trade), which was a survey of methodologies used in evolutionary biology. The upshot of these experiences was that I became familiar with both the theoretical and empirical content of evolutionary quantitative genetics and with ways to teach them.

My other pedagogical laboratory was the Evolutionary Quantitative Genetic Workshop, which I organized with Joseph Felsenstein. This workshop has run for a week each summer, beginning in 2011. From the onset, the workshop was intended as a bridge between quantitative genetics and comparative methods. We argued to funding agencies that a synthesis would benefit both of these fields, and we recruited faculty who could help us generate that synthesis. The workshop was a great success, and a synthesis unfolded over a period of years. Students benefited from a workshop taught from multiple perspectives, and the instructors experienced a master course in evolutionary biology. Our instructors and guest lecturers included: Patrick Carter, Marguerite Butler, Thomas Hansen, Luke Harmon, Paul Hohenlohe, Adam Jones, Michelle Lawing, Jonathan Losos, Trudy Mackay, Emilia Martins, Brian O'Meara, Patrick Phillips, Samantha Price, Liam Revell, Josef Uyeda, and Michael Whitlock. Examples of lectures, computer exercises, and preliminary drafts of chapters from my book are posted on the workshop website.

The organization and style of the book emerged from these teaching experiences and many years of research collaboration. The idea of the book is to use the theory of quantitative genetics to show conceptual continuity across time from generation-to-generation evolution to adaptive radiation in deep time. In the process, I endeavoured to resolve—as Simpson did in the 1940s—the apparent conceptual conflict between neontology and palaeontology. The main argument of the book is that the adaptive landscape concept can be used to understand both evolutionary process within lineages and the shaping of adaptive radiations. In particular, the book argues that models with a moving adaptive peak carry us further than any other conceptual approach yet devised.

The adaptive landscape concept turned out to be a major hurdle for students in the classroom, as well as an organizing principle for the book. The roots of the adaptive landscape idea are in technical accounts by Sewall Wright, George Gaylord Simpson, and Russell Lande. Perhaps because these accounts are difficult to follow, the mathematical versions of the adaptive landscape concept are unfamiliar to nearly all students, professors, and researchers. The mathematical theory for multiple quantitative traits evolving on a landscape is especially challenging. The book is designed to make the adaptive landscape, in all its mathematical splendour, understandable to a wider audience.

Acknowledgements

For inspiration and encouragement, I thank Marguerite Butler, Peter and Rosemary Grant, Joseph Felsenstein, Russell Lande, Johanna Schmitt, and Andy Sinauer.

For reviews of individual chapters, I am grateful to Reinhard Bürger, Patrick Carter, Paul Hohenlohe, Adam Jones, Russell Lande, Mark McPeek, Dolph Schluter, Monique Simon, Josef Uyeda, Bruce Walsh, Michael Whitlock, and Benjamin Wölfl. Of course, any errors that remain are mine alone.

I thank Ivan Phillipsen for his patience and artistic skill in the production of figures and A. Jesús Muñoz Pajares for programming the animations. The abstract expressionist art on the cover and chapter headers was created by my father, Richard E. Arnold.

Planning of the book was supported by the Opportunities for Promoting Understanding through Synthesis (OPUS) program of the US National Science Foundation. The US National Science Foundation also supported my research on garter snakes and my research with Lynne Houck on plethodontid salamanders throughout our careers.

The Evolutionary Quantitative Genetics Workshop was supported by the National Evolutionary Synthesis Center (NESCent), the National Institute for Mathematical and Biological Synthesis (NIMBios), Friday Harbor Laboratories, the Society for the Study of Evolution, the American Society of Naturalists, and the Society of Systematic Biologists.

I especially thank Lynne Houck for advice, encouragement, and patience during the long gestation of this book.

Stevan J. Arnold, Corvallis
July 2022

Contents

Introduction

Researchers and graduate students are the primary audience for this book. I have in mind a reader who seeks an overview of the theory for phenotypic evolution, as well as an understanding of its empirical basis. The book is intended as a tutorial for graduate and postdoctoral students, although it can also serve as an introduction and overview for researchers. This book assumes a background in population genetics and statistics, including analysis of variance and multiple regression. The text is supplemented with a website companion (https://phenotypicevolution.com/) that features animations and links to datasets.

This book argues that evolutionary models with a moving adaptive peak provide a powerful framework for understanding adaptive evolution and radiation. To construct that argument, the book builds from first principles to models of evolutionary processes and patterns. Along the way, the book summarizes and synthesizes diverse empirical literatures and integrates those findings into a conceptual framework provided by quantitative genetics. The literatures to be summarized and synthesized includes quantitative inheritance, mutation accumulation experiments, phenotypic integration, analysis of selection, responses to deliberate selection, and trait response to selection on all timescales. Evolutionary quantitative genetics is used as a conceptual framework to integrate all of these results.

I do not attempt a comprehensive treatment of theory in evolutionary quantitative genetics. That need has been met by two major reference books, Lynch and Walsh (1998) and Walsh and Lynch (2018). In contrast, the primary goal of this book is to understand and evaluate alternative models for adaptive radiation. To pursue that goal, I introduce and use the additive version of quantitative genetic theory, pioneered especially by Russell Lande, which is based on assumptions of prevalent polygeny, pervasive pleiotropy, Gaussian distribution of mutational effects at individual loci, additive inheritance, weak stabilizing selection, and persistent configuration of adaptive landscapes. This additive version has the advantage that its implications have been explored all the way from the mutation process to evolutionary patterns in deep time. Although additive theory holds centre stage, I mention and reference—but do not explore—departures from additivity (e.g., non-Gaussian distributions of allelic effects, dominance, epistasis, maternal effects, phenotypic plasticity).

Decisions about what to include in each chapter were dictated by two questions: 'Does inclusion help carry the main theme of the book, namely the evaluation of alternative models for adaptive radiation?' and 'Is quantitative genetic theory available for this topic?' Many important evolutionary phenomena were ignored because they failed to meet one or both of these criteria. As a consequence of implementing these two criteria, the book does not have something for everyone.

The particular strengths of the book are its organization around a central theme (the moving peak model of adaptive evolution and radiation) and an exposition of multivariate concepts that are essential for understanding the evolutionary consequences of

Evolutionary Quantitative Genetics. Stevan J. Arnold, Oxford University Press. © Stevan Arnold (2023).
DOI: 10.1093/oso/9780192859389.003.0001

interactions between traits. The centrality of trait interactions is emphasized throughout the book. A foundation for understanding how trait interactions generate selection is laid in the early chapters. That foundation is used to explore coevolution between males and females, as well as coevolution between interacting species. The book ends with an account of the bridge that is being built between quantitative genetics and community ecology.

The book is constructed as a tutorial that proceeds stepwise from process to pattern, from the elementary to the more complex. The backbone of the book is a linear conceptual argument from first principles and elementary processes up through the analysis of complex evolutionary patterns revealed on phylogenies, in the fossil record, and in ecological communities. The sequence of chapters begins with elementary processes (selection, mutation, inheritance) and builds up to chapters on trait differentiation in adaptive radiations. The final chapters provide a bridge to evolutionary ecology by showing how trait-mediated interactions between species affect coevolution and adaptive radiation.

The book is organized to enhance its use as a tutorial. Many of the chapters are grouped into pairs that treat the single character (univariate) case and then the multiple character (multivariate) case. This construction enables readers who are unfamiliar with matrix algebra to grasp central ideas and the flow of the main argument by focusing on the univariate chapters. Each chapter follows the same three-part sequence: (1) an introduction to theory, (2) a few illustrative empirical examples, (3) a summary of major empirical findings. Illustrative examples are of two kinds. Some study systems appear in many chapters (e.g., garter snake vertebral counts, *Drosophila* bristle numbers, the beaks of Darwin's finches) and provide empirical bridges between topics and chapters. Other examples appear as appropriate in only one or two chapters (e.g., morphometrics of *Anolis* lizards, defensive armour of sticklebacks). The intent of the examples is to make the nature of data apparent to readers and to provide connection to the empirical summaries. Two summaries of content are provided for each chapter: a bulleted list of major points (Conclusions) at the end of each chapter and a short narrative version of that list (Overview) at the start of each chapter. The aim of these two summaries is to point out the connection of each chapter to the main argument of the book.

Literature Cited

Lynch M. and B. Walsh. 1998. Genetics and Analysis of Quantitative Traits. Sinauer Associates, Inc., Sunderland, Massachusetts.

Walsh, B. and M. Lynch. 2018. Evolution and Selection of Quantitative Traits. Oxford University Press, New York.

CHAPTER 1

Selection on a Single Trait

Overview—Phenotypic selection can be measured by its effects on trait distributions within a generation. Our approach contrasts trait distributions before and after selection. This contrast is less intuitive than the comparison of traits in survivors and nonsurvivors, but it has an important statistical advantage. The difference in trait means before and after selection is equivalent to the covariance between a trait and fitness. Such selection differentials have been measured in a wide variety of natural populations and show that trait means are usually shifted by less than half a phenotypic standard deviation (mean about 0.6) within generations and that the modal value is close to zero. A similar perspective on trait variances shows that they usually contract by 0–50% within a generation or expand by 0–25% as a consequence of selection, with a mode close to zero.

In this chapter we will focus on simple descriptive characterizations of selection. Our account goes only a little way beyond older treatments of selection on quantitative traits in which correlations with other traits were ignored and no attempt was made to relate selection coefficients to equations for evolutionary change (Cook 1971; Endler 1986). In later chapters we will correct for the effects of selection on correlated characters, deduce modes of selection and fitness functions from changes in trait distributions, and use our measures of selection to model evolution.

In this and later chapters we employ a standard life cycle of events. (1) A cohort of zygotes (produced by the previous generation) expresses a phenotypic trait that does not change during ontogeny. We refer to the statistical characterization of the trait distribution at this life cycle stage as 'before selection'. (2) The cohort experiences some form of selection. We refer to the statistical characterization of the trait distribution after this selection event or episode as 'after selection' within the same generation as the before-selection trait distribution. (3) The cohort produces a new cohort of zygotes which comprises the next generation. In later chapters we will detail the occurrence of mutation, migration, and population regulation in this basic three-part life cycle.

1.1 Traits and Trait Distributions

The scale of trait measurement affects the study of selection and inheritance in subtle, as well as obvious ways. See Houle et al. 2011 for an extensive discussion of how measurement scale affects transformations, estimation error, and the relationship of the trait to fitness. Their discussion is illustrated with illuminating examples of pitfalls. Fitness itself is the ultimate phenotypic trait, so—not surprisingly—its measurement is both crucial

Evolutionary Quantitative Genetics. Stevan J. Arnold, Oxford University Press. © Stevan Arnold (2023).
DOI: 10.1093/oso/9780192859389.003.0002

and laden with implications and ramifications. Wagner (2010) provides a wide-ranging discussion of both empirical and theoretical implications of fitness measurement.

Many important traits show continuous distributions within populations, rather than discrete polymorphisms. Such traits are represented by multiple values rather than a few, so that the resulting distribution is continuous and often unimodal (Wright 1968). Although normal distributions are not universal, many traits approach such a distribution, or can be transformed so that they approach normality more or less closely (Wright 1968). In the following sections of this chapter, and in most of the chapters that follow, we will assume normality of trait distributions. This assumption is less restrictive than it may appear. In many theoretical situations the crucial assumptions are actually unimodality and symmetry rather than normality per se.

A few examples will illustrate the kinds of traits that are continuously distributed. The examples that follow were chosen because their statistical distributions are well known, and—more importantly—because they are the subjects of research from diverse points of view. Because this extensive backlog of information, we will use them as examples throughout this book.

Vertebral numbers in snakes (Figure 1.1) have been important characters in systematics since the time of Linnaeus because they often differentiate geographic races and closely related species, as well as higher taxa. Body and tail vertebrae are parts of a functional system that controls locomotion in snakes (Jayne 2020). Vertebral counts also serve as markers for the occupancy of different adaptive zones; as few as 100 vertebrae in fossorial species, but as many as 300 in arboreal species (Marx and Rabb 1972). In most snakes the vertebrae show a 1:1 correspondence with external scales (Figure 1.2), so counts can be

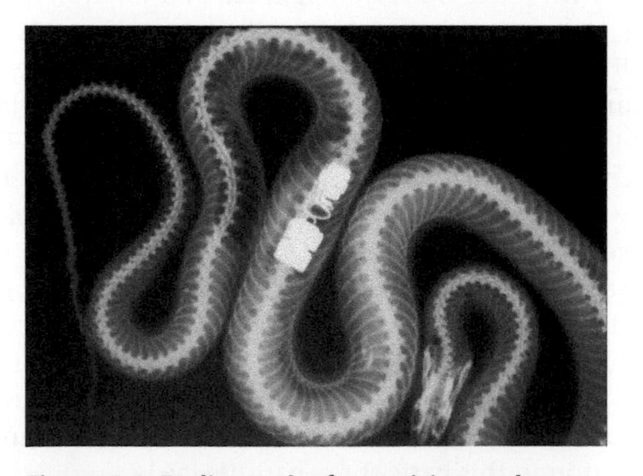

Figure 1.1 Radiograph of a natricine snake showing vertebrae in the body and tail. The distinction between the two kinds of vertebrae is not arbitrary. Ribs are attached to body vertebrae, but not to tail vertebrae. These two vertebral numbers can be assessed, without recourse to radiography, by counting ventral and subcaudal scales. The electronic object is a radiotransmitter used to study thermoregulation in free-ranging females during pregnancy.

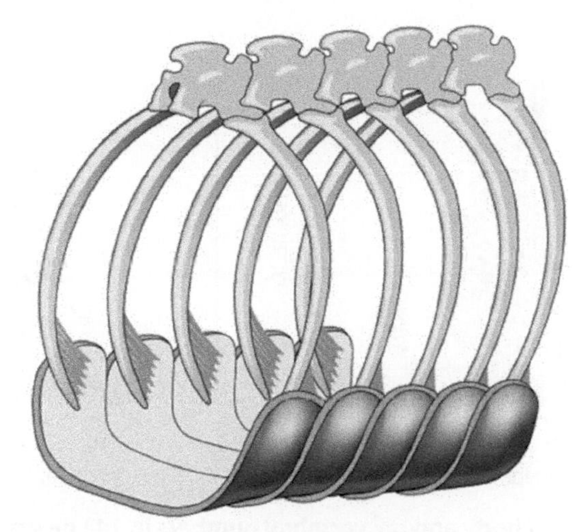

Figure 1.2 A five-vertebrae segment of the vertebral column of a natricine snake (*Natrix natrix*), showing functional connections between vertebrae, ribs, and ventral scales. Ribs articulate with the vertebral column, but muscles also connect the tips of the ribs of the ventral scales on the snake's ventral surface. Furthermore, a complicated system of muscles (not shown) connects between the ribs.

made using those scales (ventral and subcaudal) without recourse to radiography (Alexander and Gans 1966; Voris 1975). Furthermore, the transition from body vertebrae (with ribs) to tail vertebrae (with ribs) is marked by the anal scale, so counts on both body regions can be made in any specimen without a broken tail. In most snakes, both counts are sexually dimorphic, typically with more vertebrae in males. Counts from females in a single population of the garter snake *Thamnophis radix* are shown in Figure 1.3. Distributions of body and tail counts are generally unimodal and closely approximate normal or lognormal distributions (Kerfoot and Kluge 1971), as in these examples.

Counts of bristles on the thorax and abdomen of *Drosophila melanogaster* have been used in studies of inheritance and responses to deliberate selection since the 1940s (Mather 1941; Mather and Harrison 1949). Usually two kinds of counts are made: abdominal bristles (on the sternites located on the ventral surface of the abdomen) and sternopleural bristles (on the sternopleuron located laterally on the thorax) (Figure 1.4). The bristles are actually the moving parts of a mechanoreception system. When the bristles are moved they activate an electrical signal that is sent to the brain, keeping the fly aware of changes in its environment. Because the larger bristles (macrochaetae) on the sternopleuron are fewer in number and almost completely invariant, they are sometimes ignored so that the count is based only on the smaller, more numerous bristles (microchatae) (Clayton and Robertson 1957). Distributions of abdominal and sternopleural bristle numbers closely approach normal distributions (Figure 1.5).

As a final example, the dimensions of bird bills often reflect differences in food habits among species and so capture an essential feature of adaptive radiations (Schluter 2000). Diversification of bills is a pivotal feature of the adaptive radiation of the ground finches

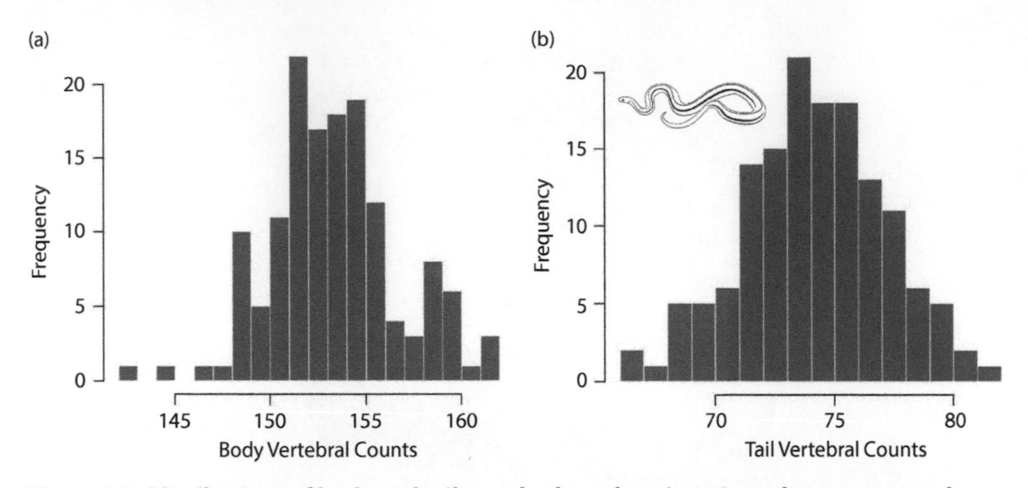

Figure 1.3 Distributions of body and tail vertebral numbers in 143 newborn garter snakes, *Thamnophis radix*. (a) Distribution of body vertebral counts. (b) Distribution of tail vertebral counts.

Data from Arnold and Bennett (1988).

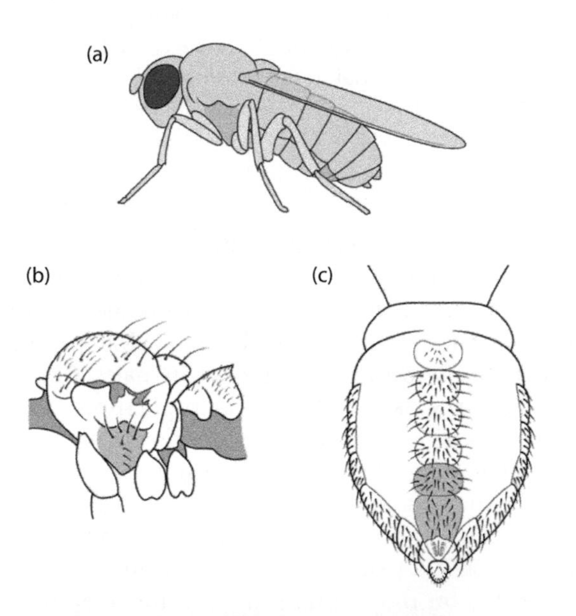

Figure 1.4 *Drosophila*, showing sites of important bristle counts. (a) Inset showing locations of bristle patches. (b) Sternopleuron location (coloured) on the thorax, showing eight sternopleural bristles (c) Abdominal tergites, showing abdominal bristles on the fourth and fifth tergites (coloured).

After Wheeler (1981).

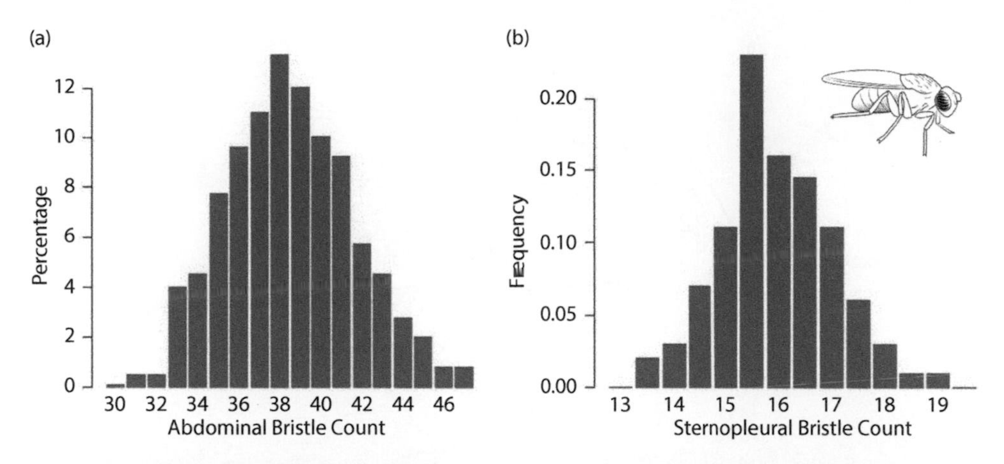

Figure 1.5 Histograms illustrating variation in *Drosophila melanogaster* bristle counts. (a) Abdominal bristle number (after Falconer 1989). (b) Sternopleuroal bristle number (i.e., sum of right and left sides). This histogram shows just the contributions to total variation from chromosome 2.

After Mackay and Lyman (2005).

(*Geospiza*) of the Galápagos, and for this reason many aspects of bills have been intensively studied (Lack 1947; Bowman 1961; Abbott et al. 1977; Grant 1986). The bills of *Geospiza* species differ greatly in size and shape and those differences reflect differences in mechanical properties (Bowman 1961). The bills of the largest species are powered by massive muscles that attach over large areas of the skull (Figure 1.6). The mechanical properties of this functional system enable the large species to crack and eat large seeds (Bowman 1961; Grant 1986). In genetic and ecological studies, bill measurements are made on individuals that have reached adult size so that there are no ontogenetic complications. Distributions of bill depth, for example, often approximate a normal distribution (Figure 1.7).

In this and the chapters that follow, we will assume that the trait does not change during ontogeny, as a result of age, growth, or experience. Some traits are naturally of this kind. Vertebral numbers in snakes and other vertebrates, for example, are determined relatively early in development and do not change during the postnatal ontogeny. Likewise, bristle numbers do not change once the fly ecloses from its pupal stage. In other cases ontogenetic invariance can be achieved by defining age-specific traits (e.g., size at age three years), as in the beak dimensions of *Geospiza*. A general solution to the issue of traits that vary with age, size, experience, environment, etc. can be achieved by treating them as *function-valued* or *infinite dimensional attributes* (Kirkpatrick and Heckman 1989; Gomulkiewicz and Kirkpatrick 1992; Kingsolver et al. 2001). In this approach, the phenotypic size of an individual is represented as a continuous function of age. The resulting theory closely follows the more simple theory for point-valued traits that is sketched here and in later chapters. In general, the main expressions remain the same except those involving traits values are transformed to continuous functions. In any case, the general point in trying to achieve size- and age-independence is that we want to define a phenotype that enables us to separate the effects of ontogeny from the effects of selection.

The choice of scale for a particular trait can be based on practical concerns as well as theoretical considerations (Houle et al. 2011). Homogeneity of variance among populations

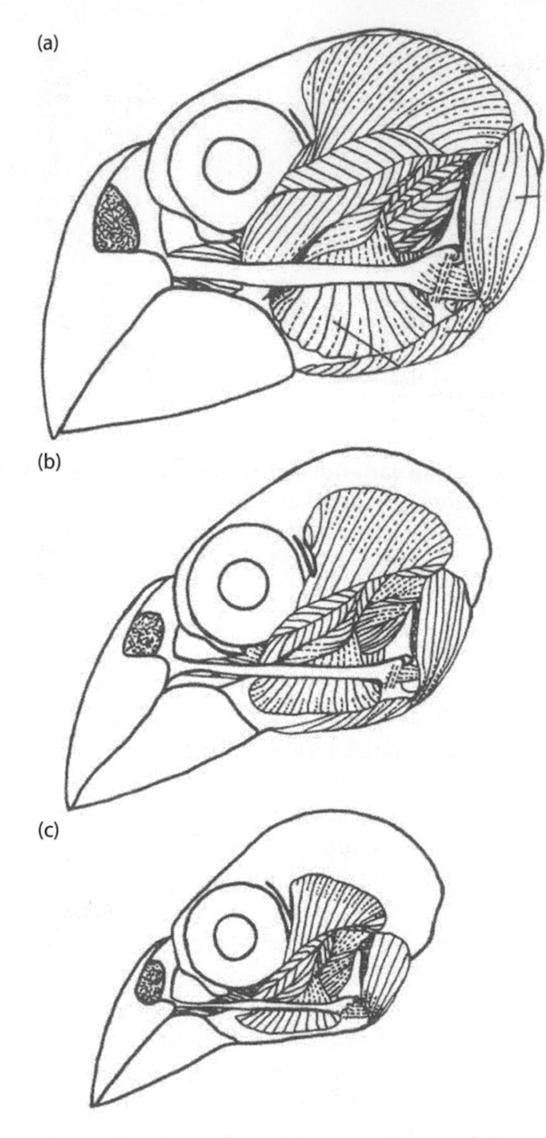

Figure 1.6 Superficial jaw musculature of three species of Galápagos finches. (a) *Geospiza magnirostris*. (b) *Geospiza fortis*. (c) *Geospiza fuliginosa*.

From Grant (1986) with permission.

or higher taxa is often desirable for then the evolution of the trait mean can be divorced from concerns about the evolution of trait variance. The logarithmic scale is often useful in attaining this kind of invariance and has useful properties in its own right (Wright 1968). On the other hand, transforming a trait with the sole goal of making its distribution approach normality is seldom useful. Most statistical tests assume that the distribution of errors is normal (not the trait distribution itself) and, in any case, are robust to even appreciable departures from normality.

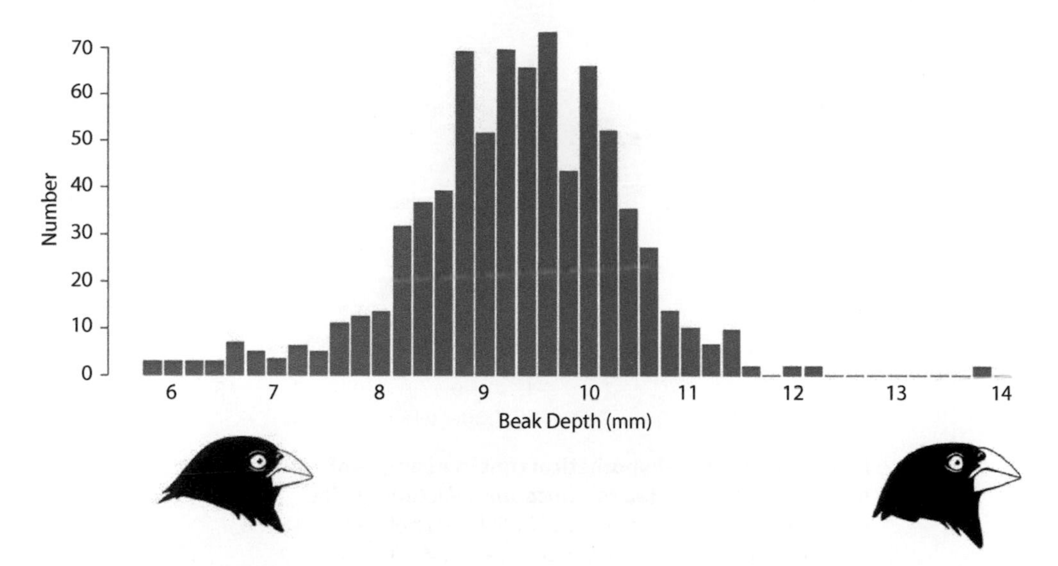

Figure 1.7 The distribution of beak depth of the medium ground finch (*Geospiza fortis*) on Daphne Major, Galápagos Islands, in 1976 before a drought (n = 751).

After Grant (1986).

1.2 Selection Changes the Trait Distribution

In the following account of how selection changes trait distributions, we have a particular, conventional life cycle in mind. Each generation in this life cycle begins at the zygote state and then the following steps ensue: (1) the zygotes mature and express their phenotypic traits, as well as a before-selection trait distribution, (2) selection acts, producing an after-selection distribution, (3) the surviving mature zygotes produce the next generation of progeny, which will inherit attributes from their parents. In this scheme, selection acts *within a generation* (the subject of this chapter) and has consequences *across generations* (the subject of Chapter 10, i.e., the response to selection). The theoretical results below are from Lande (1976) and Lande and Arnold (1983).

We are concerned here not with the agents of selection but with the statistical effects of those agents on a particular trait, z. Those statistical effects are evolutionarily important even though they fail to capture the personality of selection. Imagine the statistical distribution of the trait in a population before selection has acted (Figure 1.8a). We will call the continuous version of that distribution $p(z)$, a distribution function that might take any of a variety of forms. Later we will assume that the function is a normal distribution, but for the moment we will not make any assumptions about its form. Now imagine that as a consequence of selection some phenotypes increase in frequency, while others decrease (Figure 1.8b). Even though our trait is continuously distributed, we will represent that distribution with a histogram with a finite number of trait categories, because empirical cases will almost always be portrayed in that way because of relatively small sample sizes. We ascribe those changes in frequency to differences in fitness as a function of phenotype. The essence of selection is that all individuals with a particular phenotype, z, have

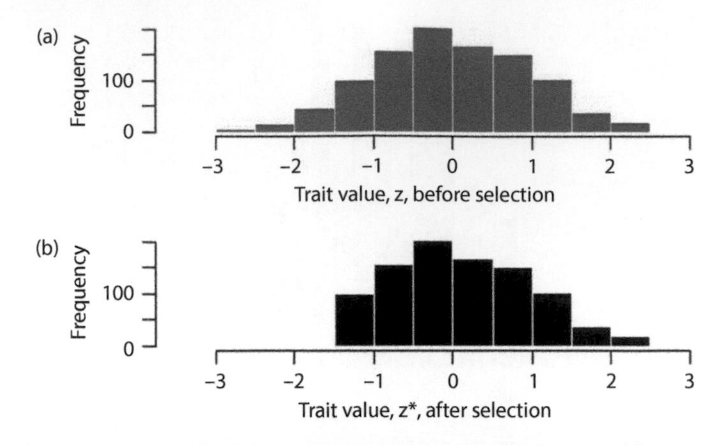

Figure 1.8 A single hypothetical trait in a sample of 1000 individuals is subjected to *truncation selection*. (a) The histogram of trait values for these individuals before selection is shown in blue (mean = – 0.02, variance = 0.96). (b) Only individuals with trait values greater than –1.5 (n = 921) survived selection. The trait distribution after selection is shown in black. Selection has shifted the trait mean and contracted its variance (mean = 0.11, variance = 0.75). Statistical values are $s = \bar{z}^* - \bar{z} = 0.13$; $(P^* - P)/P = -0.22$; $\left(P^* - P + s^2\right)/P = -0.21$.

an expected absolute fitness, which we call $W(z)$. To determine average absolute fitness in the population we need to weight each value of fitness by its frequency, in other words,

$$\bar{W} = \int p(z)\, W(z)\, dz. \tag{1.1}$$

The differences in fitness are crucial in determining how the frequency of individuals with phenotype z will be changed from $p(z)$ before selection to $p^*(z) = w(z)\,p(z)$ after selection, where $w(z) = W(z)/\bar{W}$ is relative fitness of an individual with phenotype z (Figure 1.9). Note that because mean relative fitness is

$$\bar{w} = \int p(z)\, w(z)\, dz = \int p(z)\, \frac{W(z)}{\bar{W}}\, dz = \int p^*(z)\, dz, \tag{1.2}$$

it equals 1. We will need the crucial function, $p^*(z)$, the frequency distribution after selection, to calculate various coefficients that can be used to characterize selection (Lande 1976; Lande and Arnold 1983).

To simplify particular theoretical results it will sometimes be useful to assume that the trait distribution before selection is normal. Under this assumption we have the following expression for the trait distribution before selection,

$$p(z) = \frac{1}{\sqrt{2\pi P}}\, exp\left\{\frac{-(z-\bar{z})^2}{2P}\right\}, \tag{1.3}$$

where P is the trait variance before selection and the $\sqrt{2\pi P}$ term is a normalization factor which ensures that the trait probabilities sum to one.

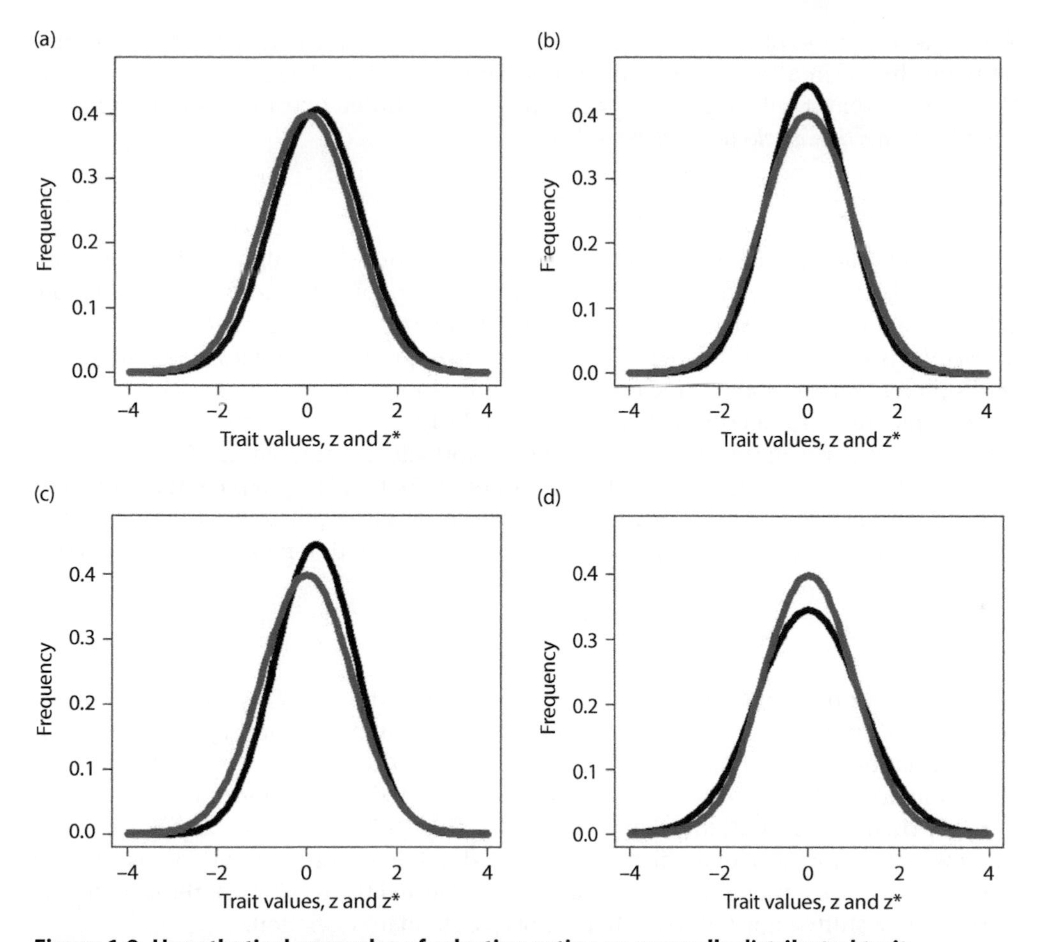

Figure 1.9 Hypothetical examples of selection acting on normally distributed trait distributions. Trait distributions are shown in blue before selection and in black after selection. (a) An upward shift in mean with little change in variance; $\bar{z} = 0.00$, $P = 1.00$, $\bar{z}^* = 0.02$, $P^* = 0.96$. (b) A contraction in variance with no shift in mean; $\bar{z} = 0.00$, $P = 1.00$, $\bar{z}^* = 0.02$, $P^* = 0.80$. (c) An upward shift in mean with a contraction in variance; $\bar{z} = 0.00$, $P = 1.00$, $\bar{z}^* = 0.20$, $P^* = 0.80$. (d) An expansion of variance with no shift in mean; $\bar{z} = 0.00$, $P = 1.00$, $\bar{z}^* = 0.02$, $P^* = 1.33$.

1.3 Shift in the Trait Mean, the Linear Selection Differential

A fundamental question is to ask whether selection changes the mean of the trait distribution. The mean before selection is, using the standard definition of the mean,

$$\bar{z} = \int p(z)\ z\,dz. \tag{1.4}$$

Using that same, familiar definition, the mean after selection must be

$$\bar{z}^* = \int p^*(z)\ z\,dz. \tag{1.5}$$

A natural way to express the effect of selection on the mean is to take the difference between the mean after selection and mean before selection. The difference is taken in this order so that it will be positive, when the mean is shifted upwards. This difference is called the *directional selection differential*,

$$s = \bar{z}^* - \bar{z}. \tag{1.6}$$

It is useful to measure the shift in mean given by s in units of within-population phenotypic standard deviation. If we let the variance in the population for trait z be P before selection, then the trait standard deviation is $\sqrt{P}$, and our *standardized directional selection differential* is $s' = s/\sqrt{P}$, which is sometimes known as i, the *selection intensity*. A standardized selection differential of 2 means that the trait mean has been shifted upwards by two phenotypic standard deviations (Lande and Arnold 1983).

Suppose we perform truncation selection on a normally distributed trait by saving only individuals in the top p proportion of the distribution for breeding, as is common in plant and animal breeding. For simplicity, let the trait mean be zero before selection and the trait variance be one. Then the selection intensity, i, is related to p in the following way,

$$i = s/\sqrt{P} = \frac{\varphi\left(z_{1-p}\right)}{p}, \tag{1.7}$$

where p is the proportion of individuals saved for breeding, $\varphi(x)$ is the unity normal or standard normal probability density function evaluated at x, and z_{1-p} is the probit transformation of $1-p$ (Walsh et al. 2018). The value of z_{1-p} can be obtained in R using the command pnorm $(1-p)$, and $\varphi(x)$ can be computed using dnorm. For example, if $p = 0.2$, then the truncation point is $z_{1-p} = 0.788$, $\phi(0.788) = 0.785$, and $i = 1.46$. In other words, if the mean of a normally distributed trait is zero and its variance is 1 before selection, and truncation selection is practised by selected the top 20% of the sample, the mean will be shifted upwards by 1.46 phenotypic standard deviations.

1.4 The Directional Selection Differential as a Covariance

The directional selection differential is an especially powerful descriptor of selection because it is a covariance as well as a shift in mean. Recall that the covariance between two variables, x and y, is defined as

$$Cov(x,y) = \int\int p(x,y)(x-\bar{x})(y-\bar{y})\,dxdy. \tag{1.8}$$

where $p(x,y)$ is the frequency of observations with values x and y. With a little rearrangement we can express this same equation as

$$Cov(x,y) = \int\int p(x,y)\,xy\,dxdy - \bar{x}\bar{y}. \tag{1.9}$$

Because relative fitness is a function of the phenotype, $w = w(z)$, the covariance between relative fitness and trait values is

$$Cov(w,z) = \int p(z)(w-1)(z-\bar{z})\,dz = \int p^*(z)\,z\,dz - \bar{z} = \bar{z}^* - \bar{z} = s. \tag{1.10}$$

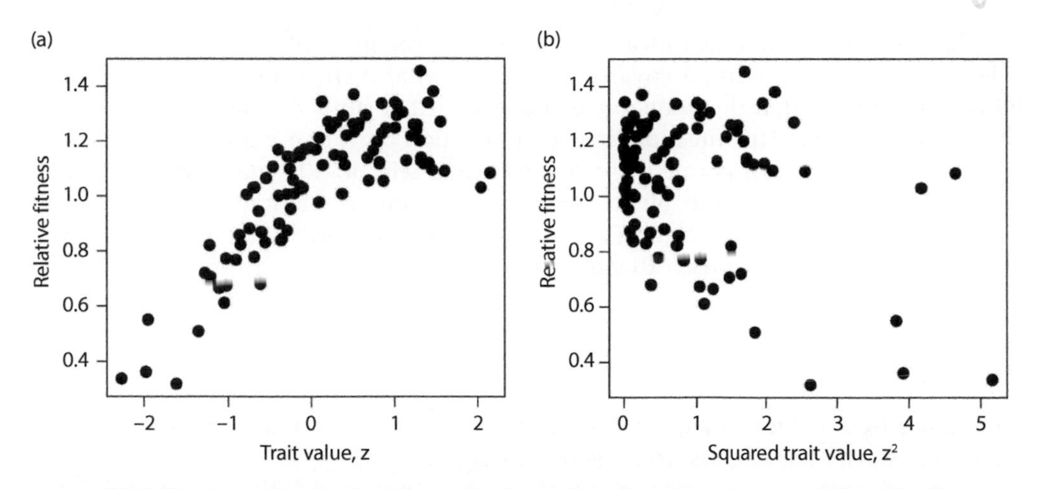

Figure 1.10 Scatterplots showing the equivalence between covariance and the selection differential, *s*. The sample of 100 individuals in these plots were drawn from the trait and fitness distributions, $p(z)$ and $w(z)$, used in Figure 1.9c. (a) Relative fitness as a function of trait value, z: $\bar{z} = -0.13$, $P = 0.98$, $\bar{w} = 0.97$, $Cov(w, z) = 0.21$, $r(w, z) = 0.81$. (b) Relative fitness as a function of squared trait values, z^2: mean $z^2 = 0.99$, $Var(z^2) = 0.98$, $Cov(w, z^2) = -0.17$, $r(w, z^2) = -0.46$.

In other words, the directional selection differential, s, is the covariance between relative fitness and trait values (Robertson 1966). This second definition of the directional selection differential will be of special importance in later sections. An example showing the equivalence of $Cov(w, z)$ and the directional selection differential is provided in Figure 1.10a, which shows relative fitness as a function of trait values. This sample was drawn from the trait and fitness distributions, $p(z)$ and $w(z)$, used in Figure 1.9c. As expected, the value of the covariance in Figure 1.10a (0.21) is very close in value to the shift in mean in Figure 1.9c (0.20).

1.5 Change in the Trait Variance, the Nonlinear Selection Differential

We can expect that selection might change the variance of a trait, just as it might shift the trait mean. Applying the same logic as before, the variance before selection is, using the standard expression for variance,

$$P = \int p(z)(z - \bar{z})^2 dz, \tag{1.11}$$

and after selection it is

$$P^* = \int p^*(z)(z - \bar{z})^2 dz. \tag{1.12}$$

By analogy with our treatment of the mean, we can measure the absolute effect of selection on the variance as $P^* - P$ and its proportional effect as $(P^* - P)/P$. For example, when this proportional measure is -0.5, the trait variance has been reduced by 50%.

A slightly more complicated measure of effects on variance will prove useful because it leads to equivalence with a covariance. The essential point behind this more complicated measure is that the same mode of selection that shifts the trait mean will also contract its variance. This effect of directional selection on variance is especially easy to characterize if the trait is normally distributed before selection. In that case, directional selection that shifts the mean by an amount s will contract the trait's variance by an amount s^2. Consequently, we can define a *nonlinear selection differential* so that it measures effects on variance from sources other than directional selection (e.g., from stabilizing and disruptive selection), viz,

$$C = P^* - P + s^2. \tag{1.13}$$

This same selection differential, not $P^* - P$, is equivalent to the covariance between relative fitness and squared deviations from the trait mean,

$$C = Cov\left(w, \tilde{z}^2\right), \tag{1.14}$$

where $\tilde{z} = z - \bar{z}$ (Lande and Arnold 1983). An example, showing the equivalence of $Cov\left(w, \tilde{z}^2\right)$ and the nonlinear selection differential is provided in Figure 1.10b, which shows relative fitness as a function of squared trait values. As expected, the covariance in Figure 1.10b (−0.17) is very close to the corrected change in variance, $C = P^* - P + s^2$, observed in the parent distributions shown in Figure 1.9c (−0.16), within the bounds of sampling error. As before, it is useful to standardize using the variance before selection to obtain a proportional measure of effects on variance, *a standardized nonlinear selection differential*, C/P. In the next section we will show that s and C reflect selection acting on correlated traits, as well as selection acting directly on the trait in question.

1.6 Estimates of Univariate Selection Differentials

Statistics associated with traits distributions before and after selection allow us to visualize the impact of selection. The following example illustrates how various univariate statistics contribute to our visualization of selection. In this example, crawling speed of newborn garter snakes (*Thamnophis radix*) was measured in the laboratory and related to counts of body and tail vertebrae (Arnold and Bennett 1988). Crawling speed is plausibly related to vertebral counts because larger vertebral counts promote the flexibility needed for snake locomotion (Kelley et al. 2003; Jayne 2020), Figure 1.11.

Figure 1.11 Diagram of a crawling snake showing how the body pushes against points in the environment. Contractions of musculature in particular segments of the body (black) produce the forces that move the snake forward.

From (Jayne 2020) with permission.

As is typical in snakes, males had several more tail vertebrae than females and a few more body vertebrae, so the sex difference in means was added to the counts of females, and the combined sample of 143 neonates was analysed. The trait sample before selection was standardized by subtracting the trait mean from each observation and dividing by the trait standard deviation, so that the means of each trait were zero and variances were equal to one. This standardization simplifies the interpretation of selection statistics.

The sample after selection, $p^*(z)$, in this example is not a sample of survivors, but is instead the trait distribution weighted by relative crawling speed (Figure 1.12). Because of the trait standardization, we can immediately see the changes in trait means and variances before we calculate the selection differentials (Table 1.1). We see that the body mean has been shifted upwards by slight amount (3.1% of standard deviation in the body vertebral count before selection). The tail mean has also been shifted upwards but to a smaller degree (0.4% of standard deviation in the body vertebral count before selection). The

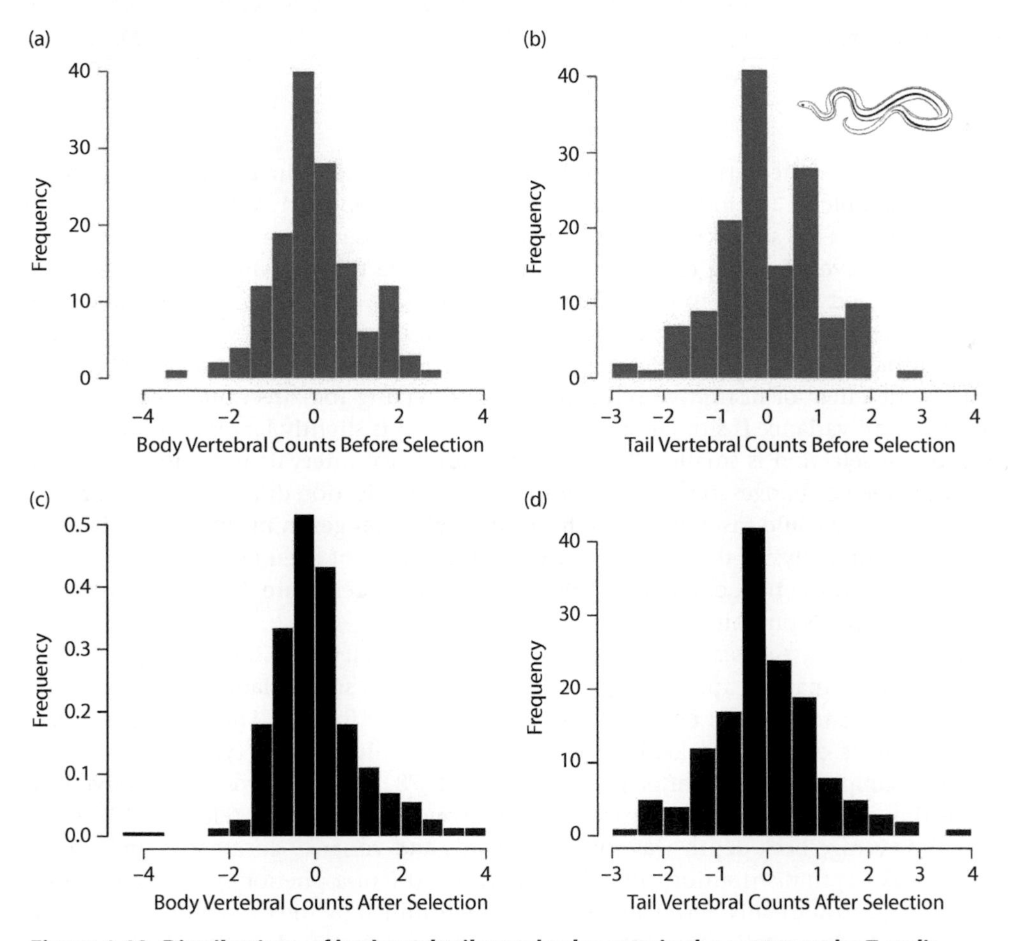

Figure 1.12 Distributions of body and tail vertebral counts in the garter snake *T. radix* before and after selection. The samples before selection (n = 143) are standardized to zero means. In the samples after selection, the vertebral count of each individual is weighted by relative crawling speed. (a) Distribution of body vertebral counts before selection. (b) Distribution of tail vertebral counts before selection. (c) Distribution of body vertebral counts after selection. (d) Distribution of tail vertebral counts after selection.

Table 1.1 Selection differentials describing effects of numbers of body and tail vertebrate on crawling speed in newborn garter snakes (*Thamnophis radix*) (n = 143). Standardized measures are denoted with a prime ('). Bootstrap estimates of 95% confidence limits in parentheses.

	Body vertebrae		Tail vertebrae	
	Mean	Variance	Mean	Variance
Before selection	153.414	11.437	74.257	9.094
Before selection'	0.000	1.000	0.000	1.000
After selection'	0.031	1.122	0.004	1.142
		$P_1^* - P_1$		$P_2^* - P_2$
Change in variance		0.122 (−0.062, 0.342)		0.142 (−0.015, 0.327)
	s_1	C_1	s_2	C_2
Selection differential	0.031 (−0.032, 0.096)	0.123 (−0.062, 0.328)	0.004 (−0.058, 0.063)	0.142 (−0.015, 0.327)
Correction term, s_i^2		0.001		0.00002

95% confidence intervals for these and other statistics were estimated by bootstrapping over the sample of 143 individuals and are reported in Table 1.1. We see that the confidence intervals for s_1 and s_2 do not overlap zero, so those estimates are not significant at the 0.05 level. Turning to the trait variance, we see that crawling performance has caused the body variance to expand by 12.2% ($P_1^* - P_1$). The effect of directional selection on the trait variance will be to decrease the variance by 0.1%, $s_1^2 = 0.001$, so when we add that amount to $P_1^* - P_1$ we obtain a selection differential of 12.3%, which represents the effect of disruptive selection, after correcting for directional selection. The effect on tail variance is similar, except the expansion is slightly larger and the effect of directional selection is smaller. Bootstrapped confidence intervals indicate that neither the uncorrected changes in variance nor the nonlinear selection differentials are different from zero. One could easily conclude from the slight changes in means and variances—and from our analysis—that the trait distributions have not been changed by selection. We shall see in the next chapter that this conclusion is premature, because we have not considered effects on trait covariance.

Ground finches (*Geospiza fortis*) present an especially well understood example of selection. Intensive mark-recapture studies on Dapne Major, a small island in the Galápagos, provided measurements of beak depth on 751 adult birds in 1976. Shortly after this field work, a severe drought markedly changed the availability of seed types on the island. When a sample was taken after the drought in 1978, only 90 birds had survived. An inspection of the beak depth distribution before and after selection (Figure 1.13) reveals that the average beak depth was shifted upwards and variance contracted. Indeed, the mean beak depth distribution shifted upwards by 60% of a phenotypic standard deviation ($s = 0.60$). Both tails of the distribution were trimmed by the drought condition, but especially the lower tail. Observational studies showed that birds with deep beaks were better able to open larger seeds that increased in frequency during the drought (Grant 1986).

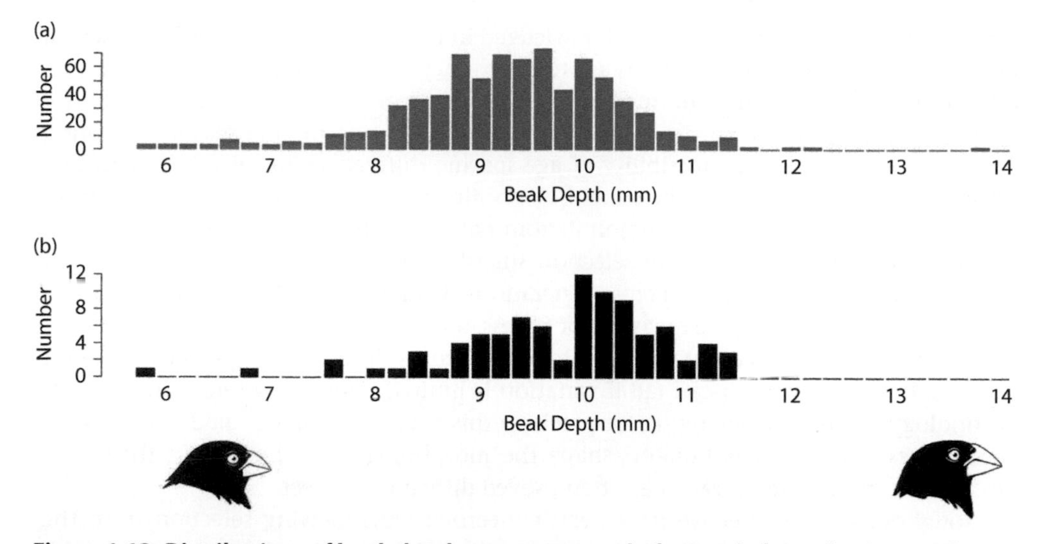

Figure 1.13 Distributions of beak depth measurements before and after selection on the Daphne Major population of *Geospiza fortis*. (a) The distribution in 1976, before selection (n = 751). (b) The distribution in 1978, after a drought killed many birds (n = 90 survivors).
Data from Boag and Grant (1984); after Grant (1986).

1.7 Technical Issues in Estimating and Interpreting Selection Differentials

Different kinds of data are used to infer selection and measure its impact. It is useful to recognize two broad categories of samples (Lande and Arnold 1983). In *longitudinal samples*, a set of individuals is followed through time. Phenotypes are measured before selection, as well as after selection, and a particular value for fitness can be assigned to each individual. Because fitness values are attached to individuals, the covariance forms for selection differentials (1.10, 1.14) can be used. The significance of fitness assignments is apparent when we consider the contrasting case of *cross-sectional samples*. In a cross-sectional study, one sample is taken before selection and another is taken after selection, but individuals are not tracked through time. As a consequence, more assumptions must be made to interpret selection differentials. In particular, one must assume that the sample before selection is representative of the statistical population that gave rise to the sample after selection. Although this assumption is straightforward in some cases, it can be tortuous if the study begins with a sample of survivors and the probable population before selection must be reconstructed (Blanckenhorn et al. 1999).

The difference between longitudinal and cross-sectional samples also affects estimation of standard errors. In the case of longitudinal data, estimation is straight-forward. The two key selection differentials are covariances which can be converted to correlations with well-characterized sampling properties, assuming a normal distribution of errors or by using nonparametric correlations. In the case of cross-sectional data, however, one must use the difference formulas (1.6, 1.13) to estimate selection differentials, and standard errors must be estimated by a re-sampling procedure (e.g., bootstrapping or jack-knifing).

Studies of selection are nearly always based on particular periods or episodes rather than lifetimes of exposure to selection. Because this restriction is universally recognized by investigators, it may not always be acknowledged in print. For example, studies of sexual selection often use mating success as a fitness currency. The selection that is measured is distinct but it is usually not summed up over a lifetime of episodes. Instead, a snapshot of selection is taken at a particular place and time (e.g., one mating season), ignoring differences in age and the possibility of age-specific differences in selection. A similar restricted focus is often taken in studies of viability selection. Such restrictions are so common that they become a common denominator in comparisons across studies of a particular kind. The restriction to selection snapshots will make a difference when we consider responses to selection across generations (Chapters 10–11), for then the focus will necessarily be on lifetime measures of fitness and selection.

The use of standing, natural variation to assess fitness differences is powerful when it succeeds, but the approach can fail if variation is limited. Measuring selection on floral morphology has been challenging for precisely this reason (Fenster et al. 2004). Despite abundant evidence that pollinators shape the morphologies of the flowers they visit, selection on specific floral traits has often proved difficult to detect.

Throughout this chapter we have been concerned with viewing selection from the standpoint of a single trait. The univariate measures of selection that we have considered (s, $P^* - P$, and C) are all useful, but they share a common ambiguity. Each of these indices reflects the effects of selection on correlated traits as well as on the trait in question. In the next chapter we will consider techniques for separating these two kinds of effects.

1.8 Surveys of Selection Differentials

Endler (1986) surveyed about 30 studies of approximately 24 species published between 1904 and 1985 that measured selection in natural or experimental populations exposed to selection. Those studies encompassed a wide range of organisms (plants, invertebrates, vertebrates) and traits (mostly linear measurements but some counts). Endler's survey indicates that selection typically changes trait means and variances by rather small amounts. The modal values for standardized change in mean, $(\bar{z}^* - \bar{z})/\sqrt{P}$, and standardized change in variance, $(P^* - P)/P$, are very close to zero (Figure 1.14a, b). Note that in Figure 1.14a, the positive and negative changes in the mean are grouped together, so that absolute values, $\left|(\bar{z}^* - \bar{z})/\sqrt{P}\right|$, are shown, since we are interested in the overall picture of how strong selection might be. In general, selection shifts the trait mean by less than half a phenotypic standard deviation (Figure 1.14a). Likewise, selection generally causes a less than 50% change in trait variance (Figure 1.14b). Contraction of variance is more common than expansion of variance; 68% of the values in Figure 1.14b are negative. The distribution of changes in trait mean and variance are portrayed in Figs. 1.14 c–d, where for purposes of illustration the traits are assumed to be normally distributed before and after selection. Notice that the trait mean can be shifted by more than a standard deviation and the variance can change by more than 100%, but instances of such dramatic changes are relatively rare.

The consequences of correcting the change in variance for the effect of directional selection are shown in Figure 1.15. The overall effect is, as Endler (1986) noted, extremely slight. Since the modal value of directional selection is close to zero, it is not surprising

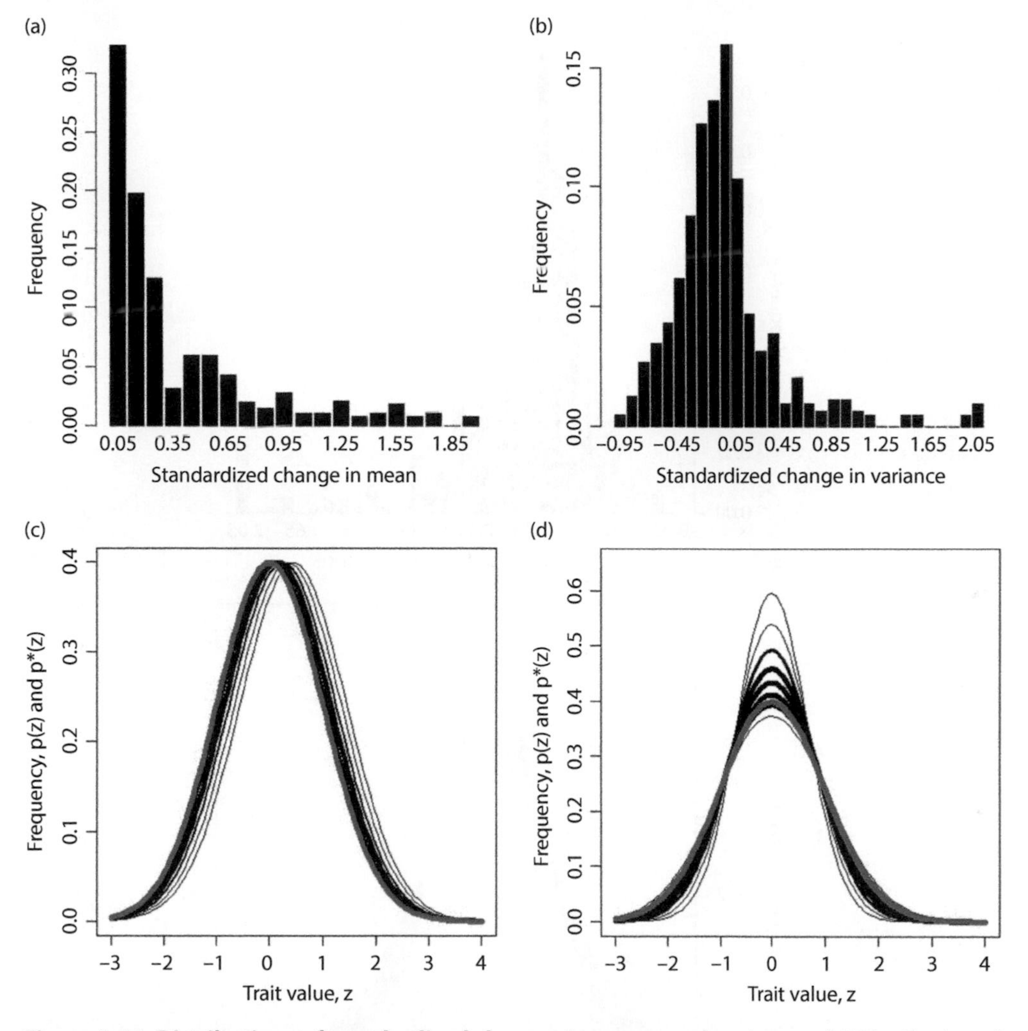

Figure 1.14 Distributions of standardized changes in means and variances. (a) Distribution of estimates of the standardized change in mean, $(\bar{z}^* - \bar{z})/\sqrt{P}$, n = 262. After Endler (1986). (b) Distribution of estimates of the standardized change in variance, $(P^* - P)/P$, $n = 330$. After Endler (1986). The red vertical line shows the transition from negative to positive change in variance. (c) Selection differential, *s*, illustrated as normal distributions after selection with mean shifted towards higher values. Line widths represent frequency in Endler's histogram; before selection (blue, with unit standard deviation and zero mean) and after selection (black). (d) Standardized change in variance, $(P^* - P)/P$, illustrated as normal distributions after selection with variance contracted or expanded. Line widths represent bin frequencies in Endler's histogram.

that correcting for this generally weak selection usually makes a small contribution to change in trait variance. The correction does, however, have the effect of making corrected expansions of variance nearly as common as contractions; only 52% of the values are negative in Figure 1.15.

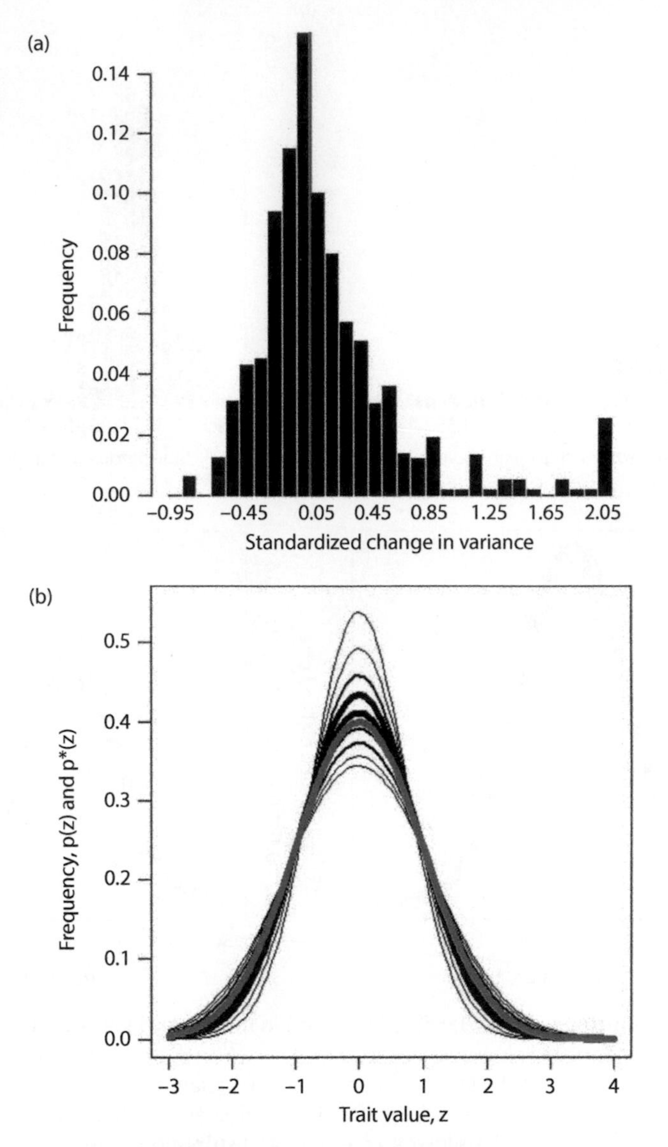

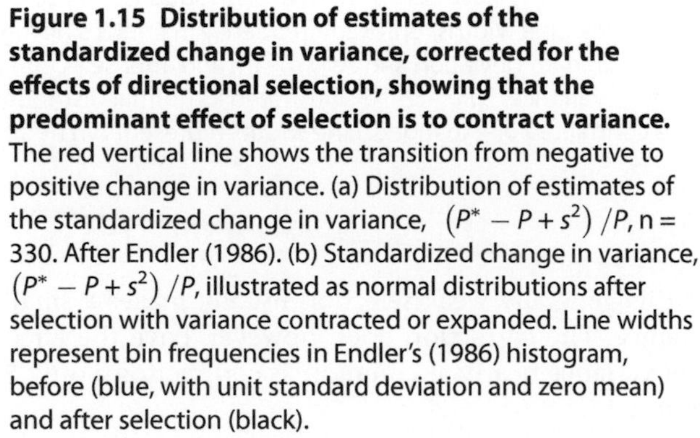

Figure 1.15 Distribution of estimates of the standardized change in variance, corrected for the effects of directional selection, showing that the predominant effect of selection is to contract variance. The red vertical line shows the transition from negative to positive change in variance. (a) Distribution of estimates of the standardized change in variance, $\left(P^* - P + s^2\right)/P$, n = 330. After Endler (1986). (b) Standardized change in variance, $\left(P^* - P + s^2\right)/P$, illustrated as normal distributions after selection with variance contracted or expanded. Line widths represent bin frequencies in Endler's (1986) histogram, before (blue, with unit standard deviation and zero mean) and after selection (black).

1.9 Conclusions

- Selection can be measured by the effects it has on trait distributions within a generation.
- The shift in the trait mean caused by selection is known as the linear selection differential. This shift can be measured by comparing the means of the trait distribution before and after selection.
- The shift in the trait variance caused by selection is known as the nonlinear selection differential. This shift can be measured by comparing the variances of the trait distribution before and after selection.
- These shifts in trait means and variances are generally small in natural populations. Both kinds of selection differential have modal values close to zero but with distributional tails that extend to large values.

Literature Cited

Abbott, I., L. K. Abbott, and P. R. Grant. 1977. Comparative ecology of Galápagos ground finches (*Geospiza* Gould): evaluation of the importance of floristic diversity and interspecific competition. Ecol. Monogr. 47:151–184.

Alexander, A. A., and C. Gans. 1966. The pattern of dermal-vertebral correlation in snakes and amphisbaenians. Zool Meded. 41:171–190.

Arnold, S. J., and A. F. Bennett. 1988. Behavioural variation in natural populations. V. Morphological correlates of locomotion in the garter snake (*Thamnophis radix*). Biol. J. Linn. Soc. 34:175–190.

Blanckenhorn, W. U., M. Reuter, P. I. Ward, and A. D. Barbour. 1999. Correcting for sampling bias in quantitative measures of selection when fitness is discrete. Evolution 53: 286–291.

Boag, P. T., and P. R. Grant. 1984. Darwin's finches (*Geospiza*) on Isla Daphne Major, Galápagos: breeding and feeding ecology in a climatically variable environment. Ecol. Monogr. 54: 463–489.

Bowman, R. I. 1961. Morphological differentiation and adaptation in Galápagos finches. University Calif. Publ. Zool. 58:1–302.

Clayton, G. A., and R. Robertson. 1957. An experimental check on quantitative genetical theory. II. The long-term effects of selection. J. Genet. 55:152–170.

Cook, L. M. 1971. Coefficients of Natural Selection. Hutchinson, London.

Endler, J. A. 1986. Natural Selection in the Wild. Princeton University Press, Princeton.

Falconer, D. S. 1989. Introduction to Quantitative Genetics. 3rd ed. Longman scientific & technical, J. Wiley & son, Essex.

Fenster, C. B., W. S. Armbruster, P. Wilson, M. R. Dudash, and J. D. Thomson. 2004. Pollination syndromes and floral specialization. Annu. Rev. Ecol. Evol. Syst. 35:375–403.

Gomulkiewicz, R., and M. Kirkpatrick. 1992. Quantitative genetics and the evolution of reaction norms. Evolution 46:390–411.

Grant, P. R. 1986. Ecology and Evolution of Darwin's Finches. Princeton University Press, Princeton.

Houle, D., C. Pélabon, G. P. Wagner, and T. F. Hansen. 2011. Measurement and meaning in biology. Q. Rev. Biol. 86:3–34.

Jayne, B. C. 2020. What defines different modes of snake locomotion? Integr. Comp. Biol. 60: 156–170.

Kelley, K., S. Arnold, and J. Gladstone. 2003. The effects of substrate and vertebral number on locomotion in the garter snake *Thamnophis elegans*. Funct. Ecol. 11:189–198.

Kerfoot, W. C., and A. G. Kluge. 1971. Impact of the lognormal distribution on studies of phenotypic variation and evolutionary rates. Syst. Zool. 20:459–464.

Kingsolver, J. G., R. Gomulkiewicz, and P. A. Carter. 2001. Variation, selection and evolution of function-valued traits. pp. 87–104 *in* A. P. Hendry and M. T. Kinnison, eds. Microevolution Rate, Pattern, Process. Springer Netherlands, Dordrecht.

Kirkpatrick, M., and N. Heckman. 1989. A quantitative genetic model for growth, shape, reaction norms, and other infinite-dimensional characters. J. Math. Biol. 27:429–450.

Lack, D. 1947. Darwin's Finches. Cambridge University Press, Cambridge.

Lande, R. 1976. Natural selection and random genetic drift in phenotypic evolution. Evolution 30:314–334.

Lande, R., and S. J. Arnold. 1983. The measurement of selection on correlated characters. Evolution 37:1210–1226.

Mackay, T. F. C., and R. F. Lyman. 2005. *Drosophila* bristles and the nature of quantitative genetic variation. Philos. Trans. R. Soc. B Biol. Sci. 360:1513–1527.

Marx, H., and G. B. Rabb. 1972. Phyletic analysis of fifty characters of advanced snakes. Fieldiana, Zoology 63:321.

Mather, K. 1941. Variation and selection of polygenic characters. J. Genet. 41:159–193.

Mather, K., and B. J. Harrison. 1949. The manifold effect of selection. Heredity 3:1–52.

Robertson, A. 1966. A mathematical model of the culling process in dairy cattle. Anim. Prod. 8:93–108.

Schluter, D. 2000. The Ecology of Adaptive Radiation. Oxford University Press, Oxford.

Voris, H. K. 1975. Dermal scale-vertebra relationships in sea snakes (Hydrophidae). Copeia 1975:746–755.

Wagner, G. P. 2010. The measurement theory of fitness. Evolution 64:1358–1376.

Walsh, B., M. Lynch, and M. Lynch. 2018. Evolution and Selection of Quantitative Traits. Oxford University Press, New York.

Wheeler, M. R. 1981. The Drosophilidae: a taxonomic overview. pp. 1–97 *in* M. Ashburner, H. L. Carson & J. N. Thompson Jr., eds. The Genetics and Biology of *Drosophila*. Vol. 3a. Academic Press, London.

Wright, S. 1968. Evolution and the Genetics of Populations: A Treatise. University of Chicago Press, Chicago.

CHAPTER 2

Selection on Multiple Traits

Overview—The phenotypic effect of selection on multiple traits can be assessed by its effects on multivariate trait distributions. As in the case of a single trait, the fundamental approach is to compare the first and second central moments of trait distributions before and after selection. Such a multivariate comparison of moments represents a major statistical improvement over trait-by-trait comparisons. By taking a multivariate approach we may be able to identify which traits are the actual targets of selection. Analysis of selection in natural systems reveals that the effects of selection on actual targets are often obscured by correlations between traits. In addition to identifying targets of selection, multivariate analysis also gives us a way to quantify functional interactions between traits. Measuring the strength of such interactions is especially important because most traits belong to one or more functional complexes.

Animal and plant breeders sometimes select on a single trait with the goal of improving their stocks. In the natural world, however, selection inevitably acts simultaneously on many traits. In this section we will introduce matrix algebra tools that will enable us to deal with this multivariate aspect of selection. In particular, we will move beyond the ambiguity of selection differentials. The parameters s and C are ambiguous because the shifts that they quantify could represent effects of selection acting on correlated traits as well as on the trait in question. Matrix algebra will help us to disentangle those direct and indirect effects, and it will help us measure how strongly traits interact in functional complexes. The theoretical results that follow are from Lande and Arnold (1983) unless noted otherwise.

2.1 Selection Changes the Multivariate Trait Distribution

Before we consider selection, we need to imagine the distribution of multiple traits before selection has happened. To visualize this distribution, picture a cloud of trait values in three-dimensional space. If more than three traits are involved, so that the cloud hangs in n-dimensional space, a standard convention is to depict those dimensions two or three at a time. An example of an actual two-dimensional trait distribution is shown in Figure 2.1. We will return to this example in Section 2.7, where we will discuss the effects of γ-coefficients on the covariance between two traits.

To consider how selection might affect such a bivariate distribution, it will be useful to consider the case of a hypothetical trait distribution that is subjected to truncation section. We will assume that the trait distribution is multivariate normal before selection, even though some of the results that follow do not depend on this assumption. An example is provided in Figure 2.2a, which shows a sample from normal distribution of just two

Evolutionary Quantitative Genetics. Stevan J. Arnold, Oxford University Press. © Stevan Arnold (2023).
DOI: 10.1093/oso/9780192859389.003.0003

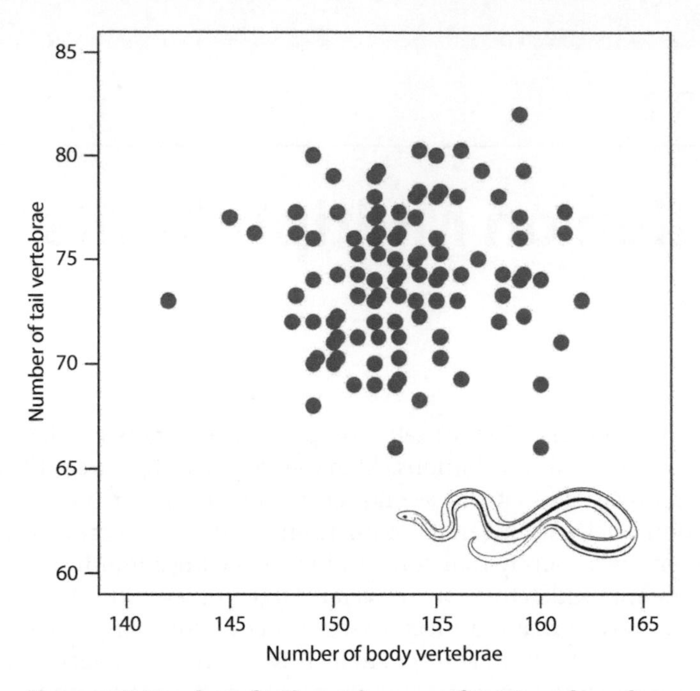

Figure 2.1 Number of tail vertebrae as a function of number of body vertebrae in a sample of 143 newborn garter snakes (*Thamnophis radix*). The two counts are not demonstrably correlated in this sample ($r = 0.07$).

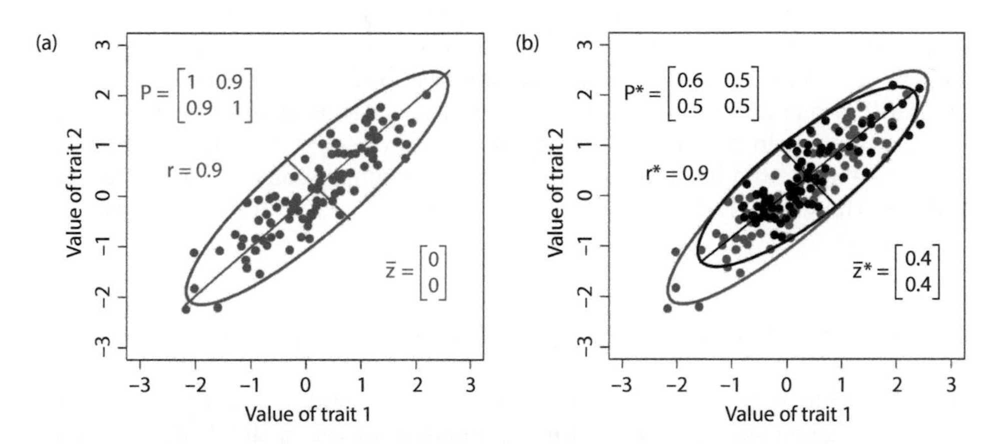

Figure 2.2 Two traits in a hypothetical sample of 100 individuals are subjected to truncation selection. (a) A scatterplot of trait values for these individuals before selection is shown in blue, along with the sample's 95% confidence ellipse (the contour line within which 95% of the observations are expected to occur). (b) Only individuals with trait values greater than – 1.0 for both traits survived selection (n = 81). The scatterplot of the sample after selection is shown in black along with its confidence ellipse. Selection has shifted the trait means upwards and contracted both variances, the covariance, and the correlation, although the latter effect is slight.

hypothetical traits, z_1 and z_2. We impose truncation selection, so that only individuals with $z_1 > -1$ and $z_2 > -1$ survive. The sample after selection is shown in Figure 2.2b. Calculation of selection measures (discussed below) confirm the impression that selection has shifted the bivariate mean and reduced dispersion in the sample.

Multivariate selection can change the trait distribution in a variety of ways that we might not have expected from a simple univariate view. In particular, bivariate selection can change trait covariances and correlations, as well as means and variances. Only contractions of variance and covariance are shown in Figure 2.3, but expansions can occur as well. As we shall see in a later section, the hypothetical selection regimes used to produce Figure 2.3 were, for the purposes of illustration, stronger than we would expect in nature.

To get a better feel for the bivariate normal distribution, consider the views in Figure 2.4, which show large samples from distributions with no correlation (Figure 2.4a) and a strong positive correlation (Figure 2.4b). With no correlation and equal variances for the two traits, the distribution is a symmetric hill. Positive correlation converts the distribution into a symmetric, ridge-shaped hill.

The points on the three-dimensional surfaces shown in Figure 2.4 are $p(\mathbf{z})$, the probabilities of observing a particular phenotype as a function of particular traits values, z_1 and z_2. We now want to consider the formula for those probabilities for the case of a bivariate normal distribution. For convenience we can represent those two values as a column vector, which we will call boldface z, $\mathbf{z}$ (see Appendix 1 in Arnold 1994 for basic conventions and rules of matrix algebra). In the present case, the phenotype is represented by just two traits, but in general it might be represented by m traits, so that $\mathbf{z}$ might be a very tall vector. The expression for a normal probability distribution in this general, m-trait case, is

$$p(\mathbf{z}) = \sqrt{(2\pi)^{-m}\left|\boldsymbol{P}^{-1}\right|}\, exp\left\{-\frac{1}{2}(\mathbf{z}-\bar{\mathbf{z}})^{\mathrm{T}}\boldsymbol{P}^{-1}(\mathbf{z}-\bar{\mathbf{z}})\right\}, \tag{2.1}$$

where $\boldsymbol{P}^{-1}$ is the inverse of the m x m variance-covariance matrix $\boldsymbol{P}$, || denotes determinant, $\bar{\mathbf{z}}$ is the column vector of means with m elements, and the superscript T denotes transpose, in this case, the conversion of a column vector into a row vector (see Lande 1979 for additional details of multivariate notation). As in the univariate case, the square root term is a normalization factor that insures that the probabilities sum to one.

2.2 The Linear Selection Differential, *s*, a Vector

We need to consider the effects of selection on the multivariate distribution, $p(\mathbf{z})$. Recall from Section 1.1 that relative fitness is the variable that translates the distribution before selection, $p(\mathbf{z})$, into the distribution after selection, $p^*(\mathbf{z})$. In our multivariate world, absolute fitness, $W(\mathbf{z})$, and relative fitness, $w(\mathbf{z})$, are functions of a multi-trait phenotype, $\mathbf{z}$. Indeed, we can substitute $\mathbf{z}$ for z, and a vector of trait means $\bar{\mathbf{z}}$ for $\bar{z}$ in expressions (1.4–1.6) and those same expressions apply, without an assumption about the multivariate distribution of $\mathbf{z}$. We pause to consider the selection differential, which is now an m-element column vector. In the two-trait case it is

$$\mathbf{s} = Cov(w, \mathbf{z}) = \begin{bmatrix} Cov(w, z_1) \\ Cov(w, z_2) \end{bmatrix} = \bar{\mathbf{z}} - \bar{\mathbf{z}}^* = \begin{bmatrix} \bar{z}_1 - \bar{z}_1^* \\ \bar{z}_2 - \bar{z}_2^* \end{bmatrix} = \begin{bmatrix} s_1 \\ s_2 \end{bmatrix}. \tag{2.2}$$

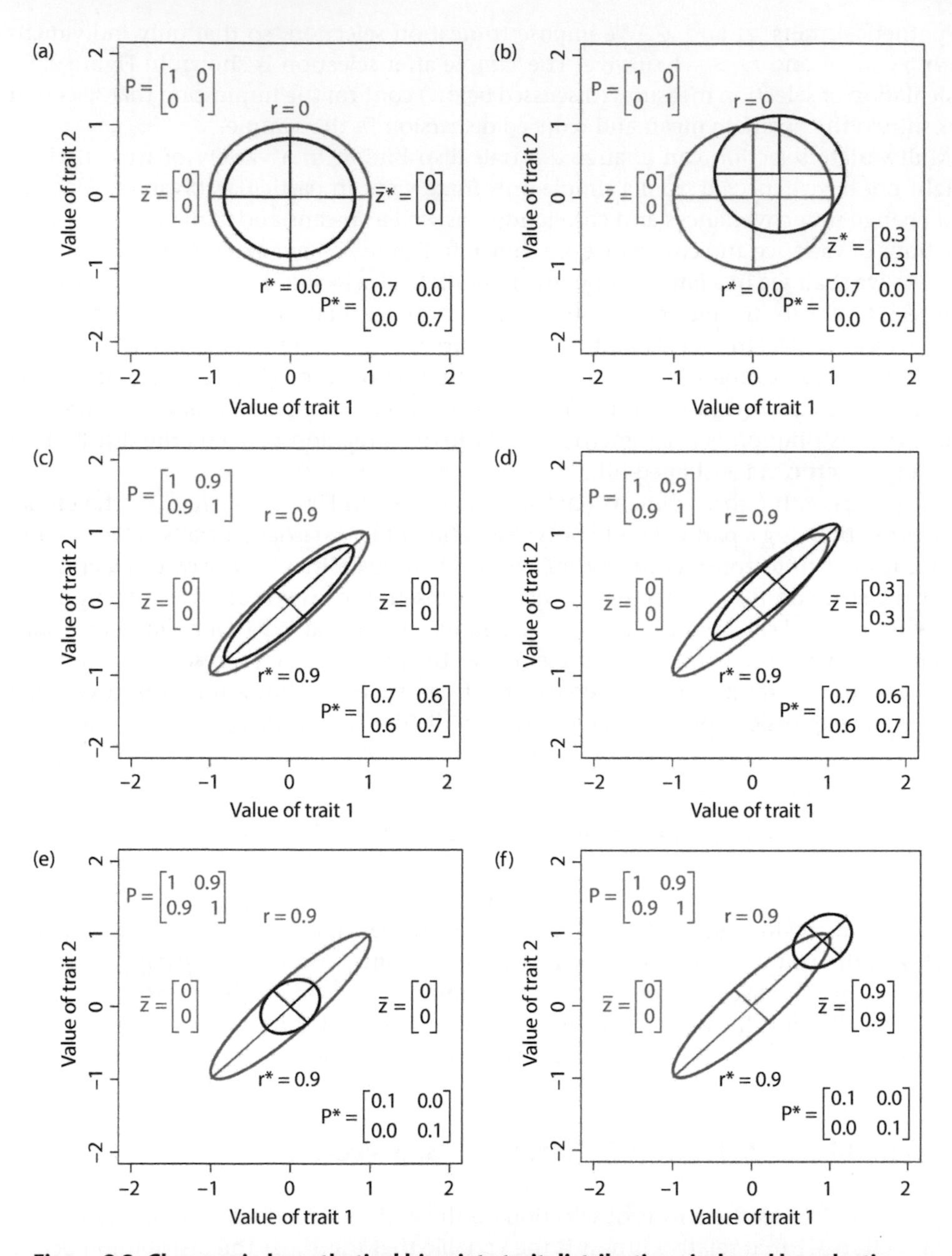

Figure 2.3 Changes in hypothetical bivariate trait distributions induced by selection. All of the distributions are normal before and after selection; 95% confidence ellipses are shown before selection (blue) and after selection (black). The position of bivariate means is shown with crosses. (a) Contractions in both variances with no shift in mean. (b) Contractions in both variances with an increase in the bivariate mean, $\bar{\boldsymbol{z}} = (0\ 0)^{\mathrm{T}}$, where the superscript T denotes transposition that converts a row vector into a column vector. (c) Contractions in both variances and covariance with no shift in mean. (d) Contractions in both variances and covariance with an upward shift in mean. (e) Substantial contractions in both variances and covariance with no shift in mean. (f) Substantial contractions in both variances and covariance with an upward shift in mean.

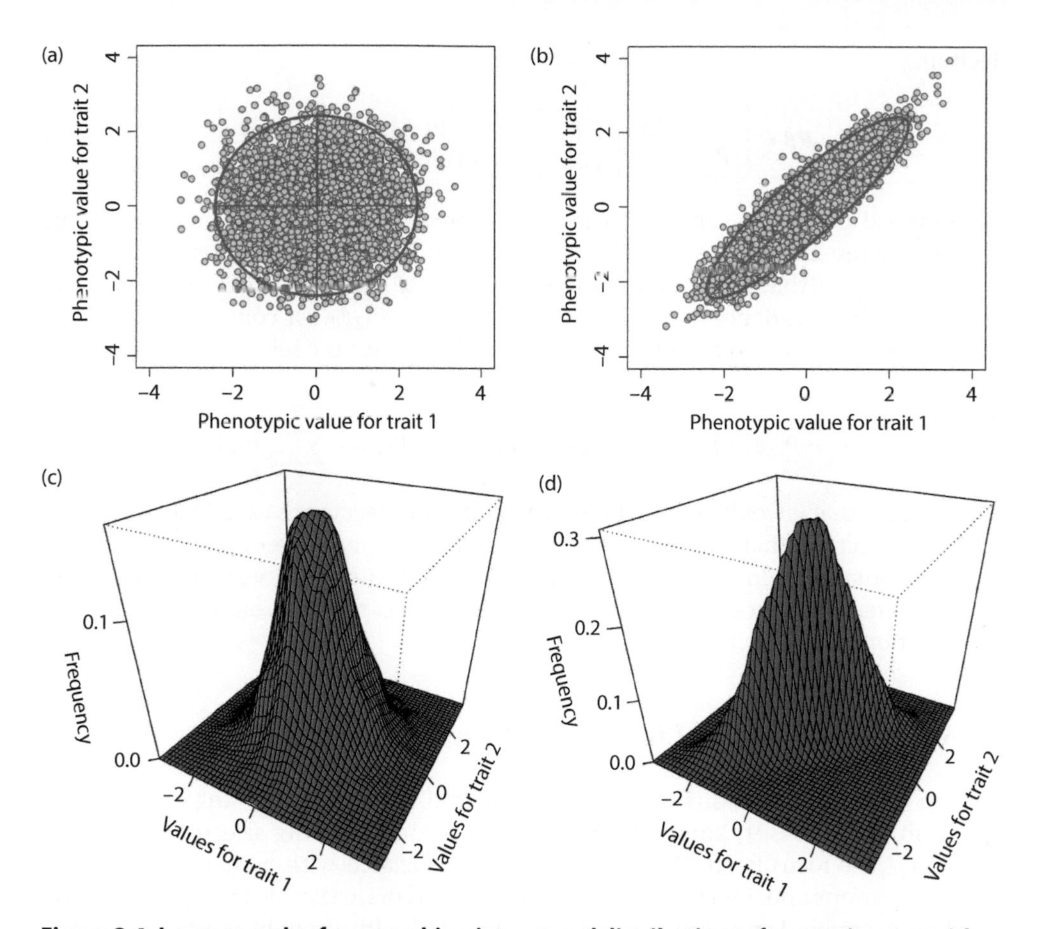

Figure 2.4 Large samples from two bivariate normal distributions of two traits, one with no trait correlation and the other with a strong positive correlation. (a) Above—a two-dimensional scatterplot of the individuals sampled from a distribution with no correlation: ($\bar{z}_1 = 0.0$, $\bar{z}_2 = 0.0$, $P_{11} = 1.0$, $P_{22} = 1.0, P_{12} = 1.0$, $r = 0.0$). Below—a three-dimensional view of the probability distribution with no correlation. (b) Above—a two-dimensional scatter plot of the individuals sampled from a distribution with a strong positive correlation: ($\bar{z}_1 = 0.0$, $\bar{z}_2 = 0.0$, $P_{11} = 1.0$, $P_{22} = 0.9, P_{12} = 1.0$, $r = 0.9$). Below—a three-dimensional view of the probability distribution with a strong positive correlation.

2.3 The Linear Selection Gradient, $\boldsymbol{\beta}$, a Vector

We now consider a new measure of selection, one that will account for correlations among the traits. That new measure is $\boldsymbol{\beta}$, the *directional selection gradient*, an m-element column vector. In general and in the two-trait case it is

$$\boldsymbol{\beta} \equiv \boldsymbol{P}^{-1}\boldsymbol{s} = \begin{bmatrix} \beta_1 \\ \beta_2 \end{bmatrix}, \tag{2.3}$$

where β_1 and β_2 are the selection gradients for traits 1 and 2, respectively. We can rearrange this last expression in a way that disentangles the direct and indirect effects of directional selection,

$$\boldsymbol{s} = \boldsymbol{P}\boldsymbol{\beta} = \begin{bmatrix} P_{11} & P_{12} \\ P_{12} & P_{22} \end{bmatrix} \begin{bmatrix} \beta_1 \\ \beta_2 \end{bmatrix} = \begin{bmatrix} P_{11}\beta_1 + P_{12}\beta_2 \\ P_{12}\beta_1 + P_{22}\beta_2 \end{bmatrix} = \begin{bmatrix} s_1 \\ s_2 \end{bmatrix}. \tag{2.4}$$

This portrayal of $\boldsymbol{s}$ tells us that the selection differential for a trait is composed of one term that represents the direct effect of selection on that trait on its mean, e.g., $P_{11}\beta_1$, and another term that represents the indirect effect of selection on another trait, acting through the covariance between the two traits, e.g., $P_{12}\beta_2$. Of course, in the general case, with m traits, the indirect effects can be very numerous; the selection differential for trait 1 is

$$s_1 = P_{11}\beta_1 + P_{12}\beta_2 + P_{13}\beta_3 + \cdots + P_{1m}\beta_m = P_{11}\beta_1 + \sum_{i=2}^{n} P_{1i}\beta_i. \tag{2.5}$$

Indeed, we see that it would be possible for the sum of indirect terms in (2.5) to overwhelm the direct effect, so that the selection differential, s_1, could have opposite sign to the selection gradient, β_1! In other words, when we consider the actual value for a particular selection differential, for example s_1, that value may reflect selection on other traits rather than selection on the trait in question.

2.4 An Estimate of the Linear Selection Gradient, β

Selection on beak dimensions in a Galápagos finch illustrates a surprising case in which a selection gradient is strikingly different from the corresponding selection differential (Table 2.1). The focus is on directional selection associated with a change in environmental conditions and, indeed, this form of selection is dramatically stronger than in the preceding example. For example, selection associated with drought conditions shifted average body weight by 62% of the standard deviation in this trait before selection. Although this and the other selection differentials suggest that heavier birds with larger beak dimensions were favoured by selection, the directional selection gradients tell a different story. They suggest that while selection favoured heavier birds, it also favoured birds with deep, narrow beaks. The picture of selection revealed by the selection gradients is consistent with ecological observations during the drought. Hard seeds became disproportionally common during the drought, and only large birds could crack these

Table 2.1 Directional selection on body size and beak dimension in a Galápagos finch (*Geospiza fortis*) arising from drought conditions on Daphne Major. The sample size before selection is 640; after selection the sample is 96. Standardized selection differentials and gradients are shown. An asterisk indicates significance at the 0.05 level. The t-test is used to compare the means of samples before and after selection. From Price et al. (1988) with permission.

Trait	Selection differential, s_i	Selection gradient, β_i ± s.e.
Weight	0.62*	0.51 ± 0.14*
Beak length	0.49*	0.17 ± 0.18
Beak depth	0.60*	0.79 ± 0.23*
Beak width	0.49*	−0.47 ± 0.21*

seeds. Birds with narrow beaks are apparently favoured by selection because birds with this morphology could twist and open the woody seed pods of *Tribulus cistoides* (Price et al. 1984).

How can the selection gradient for beak width have a different sign than its selection differential? We see from expression 2.5 that a shift in the mean, s_i, is composed of contributions from correlated traits. If those contributions ($P_{1i}\beta_i$) have positive signs, they can overwhelm the direct effect of selection on beak width ($P_{11}\beta_1$), which has a negative sign. Calculating $\boldsymbol{\beta}$ can help untangle the direct and indirect effects and reveal the actual targets of selection. Finally, this example illustrates the value of standardized selection coefficients. In the case of body weight, $\beta_i = 0.51$ means that if direct selection increased body weight by a before-selection standard deviation, relative fitness would be increased by 51%. Decreasing beak width by a standard deviation would increase relative fitness by 47%.

2.5 The Nonlinear Selection Differential, *C*, a Matrix

In general, selection can change the variances and covariances of the traits, as well as shift their means. Change in second central moments (variance and covariance) is considerably more complicated than in the univariate case, so we will need to consider it in detail. Focusing on the two-trait case for concreteness, the phenotypic variance-covariance matrix before selection is

$$\boldsymbol{P} = \int p(\mathbf{z})\,(\mathbf{z} - \bar{\mathbf{z}})^{\mathrm{T}}\,(\mathbf{z} - \bar{\mathbf{z}})\,d\mathbf{z} = \begin{bmatrix} P_{11} & P_{12} \\ P_{12} & P_{22} \end{bmatrix}, \tag{2.6}$$

where P_{11} and P_{22} are the variances of z_1 and z_2, and P_{12} is their covariance before selection. Selection may change any or all of the elements in $\boldsymbol{P}$, so that after selection the matrix becomes

$$\boldsymbol{P}^* = \int p^*(\mathbf{z})\,(\mathbf{z} - \bar{\mathbf{z}})^{\mathrm{T}}\,(\mathbf{z} - \bar{\mathbf{z}})\,d\mathbf{z} = \begin{bmatrix} P_{11}^* & P_{12}^* \\ P_{12}^* & P_{22}^* \end{bmatrix}. \tag{2.7}$$

Following the argument in Section 1.4, we will want to employ a correction factor for the effects of directional selection on variances and covariances. Here and henceforth we assume multivariate normality of the trait distribution before selection. In the two-trait case, that correction factor is

$$\mathbf{s}\mathbf{s}^{\mathrm{T}} = \begin{bmatrix} s_1^2 & s_1 s_2 \\ s_1 s_2 & s_2^2 \end{bmatrix}, \tag{2.8}$$

where the diagonal terms are effects on the trait variance (which will always be positive) and the off-diagonal term is the effect on the trait covariance (which may be positive or negative). In other words, we want to correct for the fact that directional selection on trait i will reduce its variance by an amount s_i^2, regardless of the sign of s_i, while directional selection on traits i and j will change their covariance by an amount $s_i s_j$. In the two-trait case, the nonlinear stabilizing selection differential is

$$\boldsymbol{C} = \boldsymbol{P}^* - \boldsymbol{P} + \mathbf{s}\mathbf{s}^{\mathrm{T}} = \begin{bmatrix} P_{11}^* - P_{11} + s_1^2 & P_{12}^* - P_{12} + s_1 s_2 \\ P_{12}^* - P_{12} + s_1 s_2 & P_{22}^* - P_{22} + s_2^2 \end{bmatrix} = \begin{bmatrix} Cov\left(w, \tilde{z}_1^2\right) & Cov\left(w, \tilde{z}_1 \tilde{z}_2\right) \\ Cov\left(w, \tilde{z}_1 \tilde{z}_2\right) & Cov\left(w, \tilde{z}_2^2\right) \end{bmatrix}. \tag{2.9}$$

Directional (linear) selection shifts the value of the trait mean (2.3); nonlinear (= stabilizing and disruptive) selection shifts the values of the quadratic variables $\tilde{z}_1^2$, $\tilde{z}_2^2$, and $\tilde{z}_1\tilde{z}_2$, where $\tilde{z} = z - \bar{z}$ (2.9). Notice that by the nature of the correction factor $\mathbf{ss}^{\mathrm{T}}$ (2.8), the nonlinear selection differential for a particular trait will be $C_{ii} = P_{ii}^* - P_{ii} + s_i^2$, regardless of how many traits are under selection. In other words, the diagonal elements in $\boldsymbol{C}$ correct for the effects of directional selection on the trait in question and not for effects exerted through correlations with other traits.

2.6 The Nonlinear Selection Gradient, γ, a Matrix

Just as we can solve for a vector, $\boldsymbol{\beta}$, that corrects for trait correlations in measuring directional selection, we can by similar operations obtain a set of selection coefficients that correct for trait correlations in measuring nonlinear selection. Those coefficients constitute the *nonlinear selection gradient*, $\boldsymbol{\gamma}$, an m x m symmetric matrix, defined here in the general case and illustrated in the two-trait case,

$$\boldsymbol{\gamma} \equiv \boldsymbol{P}^{-1}\boldsymbol{C}\boldsymbol{P}^{-1} = \begin{bmatrix} \gamma_{11} & \gamma_{12} \\ \gamma_{12} & \gamma_{22} \end{bmatrix}, \tag{2.10}$$

where γ_{11} and γ_{22} represent, respectively, the direct effects of stabilizing/disruptive selection on the variances of traits 1 and 2, and γ_{12} represents the direct effect of *correlational selection* on the covariance of traits 1 and 2, i.e., P_{12}. This result assumes that the trait distribution before selection is multivariate normal. To see what constitutes direct and indirect effects, we can rearrange the 2 × 2 version of (2.10) to obtain

$$\begin{aligned} \boldsymbol{C} = \boldsymbol{P}\boldsymbol{\gamma}\boldsymbol{P} &= \begin{bmatrix} P_{11} & P_{12} \\ P_{12} & P_{22} \end{bmatrix} \begin{bmatrix} \gamma_{11} & \gamma_{12} \\ \gamma_{12} & \gamma_{22} \end{bmatrix} \begin{bmatrix} P_{11} & P_{12} \\ P_{12} & P_{22} \end{bmatrix} \\ &= \begin{bmatrix} P_{11}^2\gamma_{11} + 2P_{11}P_{12}\gamma_{12} + P_{12}^2\gamma_{22} & P_{11}P_{12}\gamma_{11} + P_{12}^2\gamma_{12} + P_{11}P_{22}\gamma_{12} + P_{12}P_{22}\gamma_{22} \\ P_{11}P_{12}\gamma_{11} + P_{12}^2\gamma_{12} + P_{11}P_{22}\gamma_{12} + P_{12}P_{22}\gamma_{22} & P_{12}^2\gamma_{11} + 2P_{12}P_{22}\gamma_{12} + P_{22}^2\gamma_{22} \end{bmatrix} \\ &= \begin{bmatrix} C_{11} & C_{12} \\ C_{12} & C_{22} \end{bmatrix}. \end{aligned} \tag{2.11}$$

The $P_{11}^2\gamma_{11}$, $P_{22}^2\gamma_{22}$, and $P_{12}^2\gamma_{12}$ terms represent, respectively, direct effects on $\tilde{z}_1^2$, $\tilde{z}_2^2$, and $\tilde{z}_1\tilde{z}_2$. All of the other terms represent indirect effects mediated through covariances between the quadratic variables, $\tilde{z}_1^2$, $\tilde{z}_2^2$, and $\tilde{z}_1\tilde{z}_2$.

2.7 An Example of the Nonlinear Selection Differential, C, and the Nonlinear Selection Gradient, γ

Returning to the example of crawling speed in newborn garter snakes introduced in Chapter 1, we now consider how accounting for trait covariance affects our concept of selection (Table 2.2). In the first place, we have a new nonlinear selection differential to consider, C_{12}, which describes the total effect of nonlinear selection on the trait covariance, P_{12}. In the sample after selection, trait covariance has increased by 19.2%. Bootstrapping reveals that in only 1 boot sample out of 1000 did C_{12} take a value less than zero, so the positive change in covariance is highly significant (P = 0.001). Virtually all of the 19.2% expansion is due to nonlinear selection; directional selection contributes only a 0.8% decrease in covariance.

The selection gradients, which account for trait covariance, give much the same picture of selection as the selection differentials. Note, however, that the expressions we used for the selection gradients (2.3 and 2.10) assume that the trait distribution before selection is multivariate normal. For this reason, the regression relationships discussed in Chapter 4 are the recommended route for estimating the selection gradients, because they do not make this distributional assumption. Although most gradients are nonsignificant, the nonlinear selection gradient describing the effect of the trait product, z_1z_2, on

Table 2.2 Selection differentials and gradients describing effects of numbers of body and tail vertebrate on crawling speed in newborn garter snakes (*Thamnophis radix*) (n = 143). Selection gradients were estimated using expressions 2.3 and 2.10, which assume that the trait distribution before selection is multivariate normal. [a] The standard errors of $\boldsymbol{\beta}$ and $\boldsymbol{\gamma}$ were estimated from the standard deviations of corresponding bootstrap distributions (n = 1000 samples with replacement). [b] Bootstrap estimates of 95% confidence limits are shown in parentheses. Significance levels: $*$, P < 0.05; $**$, P < 0.01; $***$, P $\leq$ 0.001. Eigenvalues of the γ-matrix corresponding to the leading and minor eigenvector are denoted λ_1 and λ_2, respectively. 'Eigenslope' is the slope of the leading eigenvector.

	Body vertebrae		Tail vertebrae			
	mean	variance	mean	variance	covariance	correlation
Before selection	153.414	11.437	74.257	9.094	0.743	0.073
Before selection'	0.000	1.000	0.000	1.000	0.073	0.073
After selection'	0.031	1.122	0.004	1.142	0.265	0.234**
	s_1	C_{11}	s_2	C_{22}	C_{12}	
selection differential	0.031 (− 0.032, 0.096)	0.123 (− 0.062, 0.348)	0.004 (− 0.058, 0.063)	0.142 (− 0.015, 0.327)	0.192*** (0.074, 0.328)	
correction term, ss^T		0.001		0.00002	0.0001	
	β_1	γ_{11}	β_2	γ_{22}	γ_{12}	Eigenslope
selection gradient[a]	0.031 (− 0.033, 0.095)	0.097 (− 0.087, 0.301)	0.002 (− 0.065, 0.059)	0.116 (0.028, 0.292)	0.176** (0.066, 0.288)	1.055** (0.380, 2.156)
standard error [b]	0.033	0.098	0.031	0.082	0.059	λ_1 = 0.282 (0.136, 0.489) λ_2 = −0.069 (−0.221, 0.088)

crawling speed is positive and highly significant ($P < 0.001$). The positive value for this coefficient, γ_{12}, indicates that selection directly increases trait covariance. In other words, comparing samples before and after selection we see a change in trait covariance (from 0.073 to 0.265), but that change largely reflects direct effects on covariance rather than indirect effect via trait means or variances. The direct effect of selection is to increase the correlation and hence integration of the two traits.

2.8 The Canonical Form of the γ-matrix

The picture of multivariate nonlinear selection presented by $\boldsymbol{\gamma}$ may be easier to visualize if we use a different coordinate system. A natural alternative to the original trait axes is one in which the γ-matrix takes a diagonal form in which all the off-diagonal terms, which describe correlational selection, are zero. This diagonal form of the γ-matrix is known as its *canonical form* (Phillips and Arnold 1989)

$$\boldsymbol{\Lambda} = \mathbf{M}^{\mathrm{T}}\boldsymbol{\gamma}\mathbf{M}, \tag{2.12a}$$

where $\boldsymbol{\Lambda}$ is a matrix with the eigenvalues of $\boldsymbol{\gamma}$, λ_i, on its diagonal and zeros as its off-diagonal elements. $\mathbf{M}$ is a matrix whose columns are the eigenvectors of $\boldsymbol{\gamma}$ (normalized to unit length). The new axes, the eigenvectors, are a rotation of the original axes, and like those axes, they are orthogonal to one another.

2.9 An Example of the Canonical Form of the γ-matrix

A useful way to visualize the picture presented by the γ-matrix is to rotate trait axes so that $\boldsymbol{\gamma}$ takes a so-called canonical form (Phillips and Arnold 1989). Using the example in Table 2.2 to illustrate (2.12a), we find that the canonical form of the γ-matrix is

$$\begin{aligned}\boldsymbol{\Lambda} = \mathbf{M}^{\mathrm{T}}\boldsymbol{\gamma}\mathbf{M} &= \begin{bmatrix} 0.282 & 0 \\ 0 & -0.069 \end{bmatrix} \\ &= \begin{bmatrix} 0.688 & -0.726 \\ 0.726 & 0.688 \end{bmatrix}^{\mathrm{T}} \begin{bmatrix} 0.097 & 0.176 \\ 0.176 & 0.116 \end{bmatrix} \begin{bmatrix} 0.688 & -0.726 \\ 0.726 & 0.688 \end{bmatrix}. \end{aligned} \tag{2.12b}$$

In this canonical form, we see that $\boldsymbol{\gamma}$ has been transformed into a new diagonal matrix $\boldsymbol{\Lambda}$. The coefficients of the elements of $\boldsymbol{\Lambda}$ indicate that selection increases the variance along the first eigenvector ($\Lambda_{11} = \lambda_1 = 0.282$) and decreases the variance along the second eigenvector ($\Lambda_{22} = \lambda_2 = -0.069$). The off-diagonal elements of $\boldsymbol{\Lambda}$ are zero, so no direct effects act on trait covariance. The Λ-coefficients (eigenvalues) corresponding to the new major and minor axes are, respectively, 0.282 (0.95 CI = 0.136, 0.489) and – 0.069 (0.95 CI = –0.221, 0.088). In other words, selection on the major axis is disruptive and highly significant, while selection on the minor axis might be either stabilizing (as it is in the point estimate) or disruptive, but nonsignificant. The major axis (leading eigenvector) is inclined at about a 45 degree angle in trait space with a slope of 1.055 (0.95 CI = 0.380, 2.156); the minor axis is perpendicular to the major axis. The leading eigenvector of $\boldsymbol{\gamma}$ is important to us because we will show that it is the trait dimension of greatest functional integration (see next section), i.e., the trait direction in which trait covariance is most affected by selection exerted through crawling speed. We will return to the issue of visualization in Chapter 4 when we consider the topic of multivariate selection surfaces.

2.10 Technical Issues in Estimating and Interpreting Selection Gradients

Several cautions, common to all multivariate statistical analyses, should be kept in mind in estimating $\boldsymbol{\beta}$ and $\boldsymbol{\gamma}$ and interpreting those estimates. As we will see in Chapter 4, the key estimation formulas for $\boldsymbol{\beta}$ and $\boldsymbol{\gamma}$ (2.3 and 2.10) are equivalent to formulas for sets of multiple regression coefficients. Consequently, the issues we need to consider are usually discussed in the context of estimation by multiple regression (Lande and Arnold 1983; Mitchell-Olds and Shaw 1987; Wade and Kalisz 1990). We will view them in that light in Chapter 4.

Not including correlated traits in the study can lead to biased estimates of $\boldsymbol{\beta}$ and $\boldsymbol{\gamma}$. In particular, we need to consider the possibility that traits under directional or stabilizing selection are correlated with the measured traits but are not included in the study. This circumstance can cause us to over or underestimate our selection gradients, depending on the sign and magnitude of selection on the unmeasured trait and the pattern of correlation with the measured traits. In other words, what we interpret as direct effects of selection on our traits can be affected by traits that are excluded from the analysis for one reason or another. In practice, most investigators live with this limitation on interpretation because traits are included in the analysis precisely because they are likely to be under selection. In other words, prior information is brought to bear in choosing traits that partially mitigates the problem of influence from unmeasured traits. Nevertheless, the possibility of this kind of complication can never be completely eliminated and should be borne in mind in interpreting results.

Unmeasured environmental variables can produce an illusion of selection. This problem is related to the one just discussed, but here we are concerned with environmental effects that produce correlations between fitness and traits. Mitchell-Olds and Shaw (1987) discuss a telling, hypothetical example in which growing conditions vary spatially and cause some plants to be both large and fecund and others to be small and barren. If we fail to include growing conditions in our analysis (e.g., as covariates), we might erroneously conclude that plant size is under strong selection. As in the case of correlated, unmeasured traits, biased estimates have lead us to a false conclusion. One solution is the realization that correlational studies may not reveal causal relationships. In multivariate statistical analysis, we attempt to correct for correlations, but we may not succeed. For this reason, it is always wise to do a companion experimental study that manipulates traits of interest and gets us closer to an inference of causality (Wade and Kalisz 1990).

It is possible to make a strong inference of causality within the framework of correlational study. The argument is vividly portrayed in court cases that challenged the claim of tobacco companies that the link between smoking and lung cancer was not causal. Attorneys for cancer victims argued that three conditions help make a strong case for causality: plausibility, strength of correlation, and prevalence of correlation across replicate studies. All of these conditions can be considered in deciding whether a particular selection gradient represents a causal effect of phenotype on fitness.

The expressions presented in this chapter for estimating $\boldsymbol{\beta}$ and $\boldsymbol{\gamma}$ are useful when a population is sampled before and after selection (i.e., for the case of *cross-sectional data*). Those estimations require making assumptions about the two samples since they may have no individual in common (Lande and Arnold 1983). A key assumption then is that the sample before selection closely resembles the actual set of individuals after selection before they were exposed to selection. Alternatively, individuals may be followed through time so that their individual fitness values are assessed, bypassing these assumptions. In the case of such *longitudinal data*, a multiple regression approach can be used to estimate selection coefficients. That regression approach is described in Chapter 4. The notorious *factor of 1/2 problem* surrounding the estimates of $\boldsymbol{\gamma}$ in the literature before 2008 will also be discussed in Chapter 4 (Stinchcombe et al. 2008).

2.11 Surveys of Selection Gradients

The point of the surveys in this section is that the $\boldsymbol{\beta}$ and $\boldsymbol{\gamma}$ estimates correct for the effects of selection on multiple traits, unlike the surveys in Chapter 1 which showed shifts in mean and variance that reflect both direct and indirect effects. The advantage is that comparisons between $\boldsymbol{s}$ and $\boldsymbol{\beta}$ and between $\boldsymbol{C}$ and $\boldsymbol{\gamma}$ allow us to see how much the aggregated indirect effects contribute to the overall distribution of selection coefficients.

The surveys summarized here were compiled by Hoekstra et al. (2001), Kingsolver et al. (2001), and Stinchcombe et al. (2008). Kingsolver et al. (2001) and Hoekstra et al. (2001) updated Endler's (1986) survey, using similar criteria and obtaining a sample about an order of magnitude larger. This more recent survey of 62 studies (63 species) focused on selection acting on natural populations in natural circumstances. Like Endler's sample, updated survey included a wide range of organisms (plants, invertebrates, vertebrates) and an even wider range of traits (most were morphological measurements and counts, but some behaviour and life history traits were included). The sample consists of studies published between 1984 and 1997 and so slightly overlaps with Endler's sample. Stinchcombe et al. (2008) uncovered errors in the estimation of $\boldsymbol{\gamma}$ that permeate the literature prior to 2008. They resampled the studies compiled by Kingsolver et al. (2001) and corrected those errors in $\boldsymbol{\gamma}$ and added many additional studies. That new database was used to make the histogram shown in Figure 2.6g.

From a statistical point of view, distributions of selection differentials and gradients are extremely similar (Figures 2.5a and 2.5c). We get the same overall picture of directional selection, for example, whether we look at the differential, $\boldsymbol{s}$, or the gradient, $\boldsymbol{\beta}$. In both cases we see a distribution that is negative exponential in appearance with a modal value close to zero and with the vast majority of estimates in the range 0 to 0.5. In qualitative terms, directional selection tends to be weak, rarely shifting mean by more than half a within-population phenotypic standard deviation. The similarity between the distributions of $\boldsymbol{s}$ and $\boldsymbol{\beta}$ suggest that the indirect effects of selection, arising from phenotypic correlations between traits, make at most a minor contribution to the overall picture, although they may greatly affect any particular estimate of $\boldsymbol{\beta}$.

Turning to the distributions of nonlinear selection differentials and gradients, we again see much the same distributional picture (Figure 2.6). Both selection coefficients show leptokurtic (sharply peaked, thin tailed) distributions that are almost symmetrically centred about zero (Figures 2.6a, 2.6c). The selection differential C_{ii} is slightly biased towards

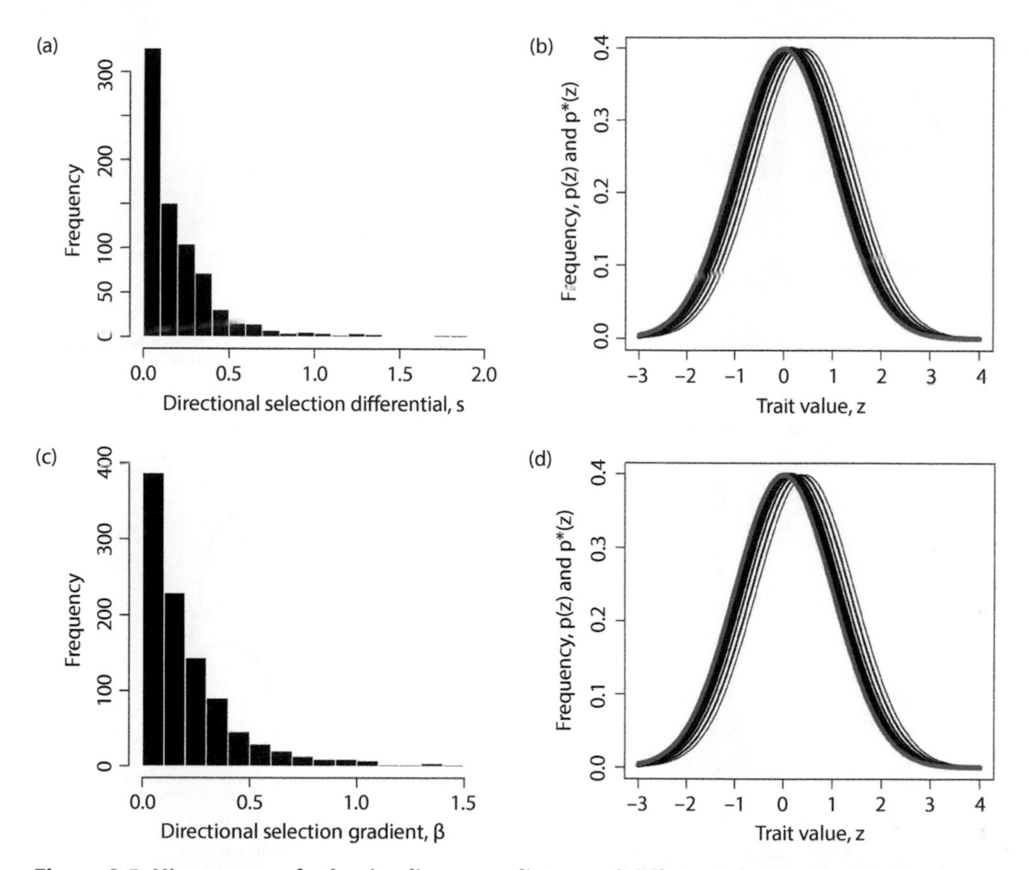

Figure 2.5 Histograms of selection linear gradients and differentials paired with frequency distributions that portray the magnitude of effects on means. (a) Histogram of the absolute values of directional selection differential estimates, s_i; 746 values from Kingsolver et al (2001) database. Three values greater than $|2|$ are not included. (b) Shifts in mean corresponding to the directional selection differentials in Figure 2.5a. Shifts corresponding to the five bins on the rightmost side of the distribution (0.05–0.45) are shown in black and account for 92% of the observations. (c) Histogram of the absolute values of directional selection gradient estimates, β_i; 992 values from Kingsolver et al (2001). Three values for s_i greater than $|2|$ were not included. (d) Shifts in mean corresponding to the directional selection gradients in Figure 2.5c. Shifts corresponding to the five bins on the rightmost side of the distribution (0.05–0.45) are shown in black and account for 90% of the observations.

negative values (stabilizing selection), while the gradient γ_{ii} has a slight bias in the other direction, towards disruptive selection. The overall effect of indirect effects appears to inflate the absolute values of C_{ii}, so that its distribution is less leptokurtic than that of γ_{ii}. Judging from the distribution of γ_{ii}, it is unusual for direct nonlinear selection to increase or decrease trait variance by more than 40%. A comparison of the distribution of γ_{ii} compiled by Kingsolver et al. (2001) with the one compiled by Stinchcombe et al. (2008) yields a surprising result. The two distributions are virtually identical! Although individual estimates of γ_{ii} may be inaccurate by a considerable amount prior to 2008, in the aggregate the errors appear to cancel out.

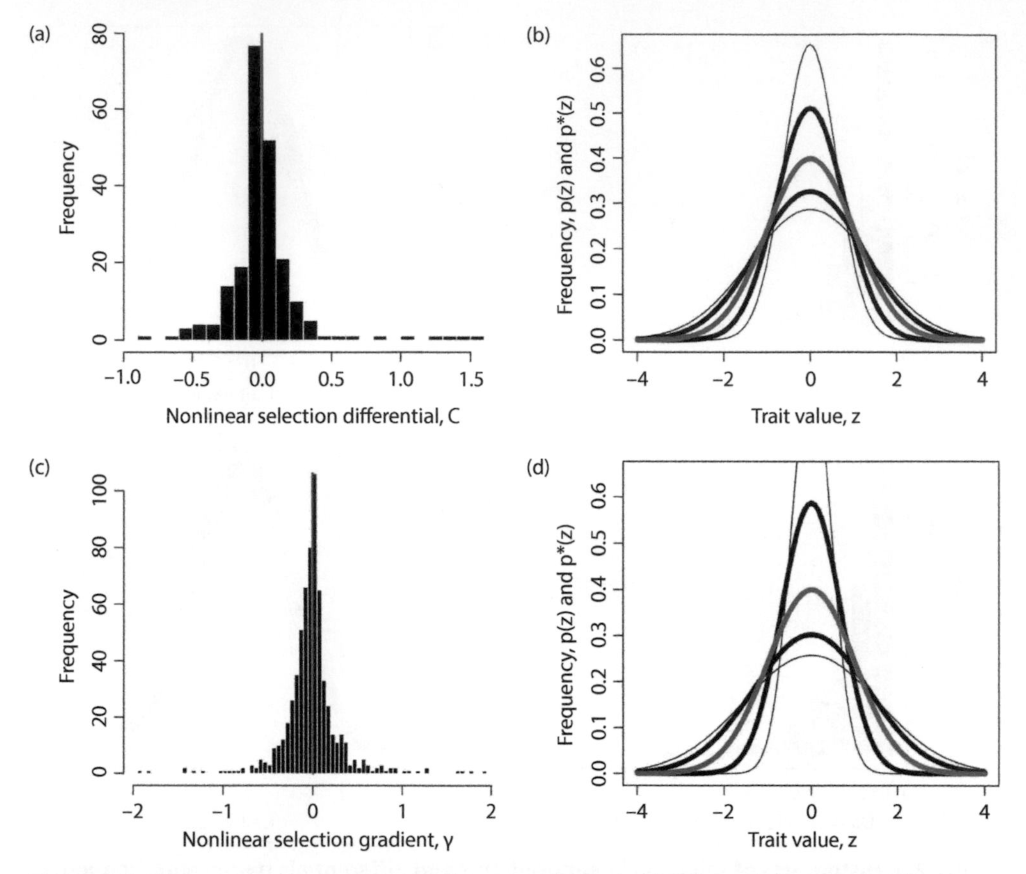

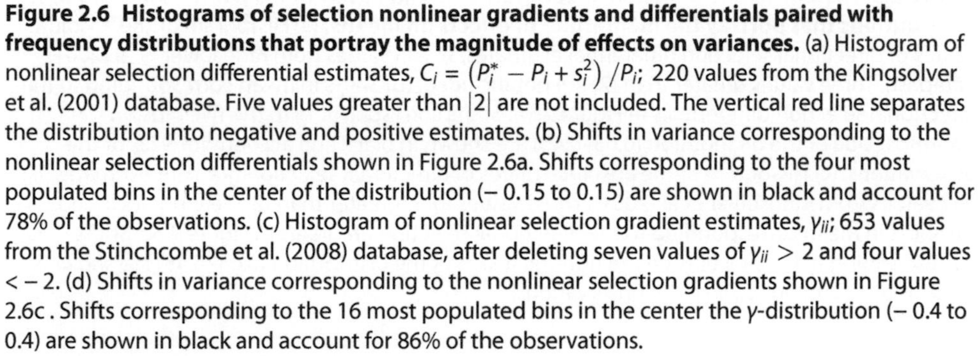

Figure 2.6 Histograms of selection nonlinear gradients and differentials paired with frequency distributions that portray the magnitude of effects on variances. (a) Histogram of nonlinear selection differential estimates, $C_i = \left(P_i^* - P_i + s_i^2\right) / P_i$; 220 values from the Kingsolver et al. (2001) database. Five values greater than $|2|$ are not included. The vertical red line separates the distribution into negative and positive estimates. (b) Shifts in variance corresponding to the nonlinear selection differentials shown in Figure 2.6a. Shifts corresponding to the four most populated bins in the center of the distribution (− 0.15 to 0.15) are shown in black and account for 78% of the observations. (c) Histogram of nonlinear selection gradient estimates, γ_{ii}; 653 values from the Stinchcombe et al. (2008) database, after deleting seven values of $\gamma_{ii} > 2$ and four values < -2. (d) Shifts in variance corresponding to the nonlinear selection gradients shown in Figure 2.6c . Shifts corresponding to the 16 most populated bins in the center the γ-distribution (− 0.4 to 0.4) are shown in black and account for 86% of the observations.

2.12 Conclusions

- Change in multivariate distribution caused by selection can be assessed by comparing vectors of means and matrices of variances and covariance before and after selection within a generation. These comparisons are the multivariate analogues of univariate selection differentials.
- Linear and nonlinear selection gradients offer an improvement over selection differentials. In particular, these two kinds of gradients correct for the effects of indirect selection that acts through trait correlations.

- Just as principal components enable us to visualize the structure of the multivariate phenotypic distribution (i.e., the *P*-matrix), a similar analysis enables us to visualize the structure of the γ-matrix. This analysis produces the canonical form of the γ-matrix.
- The estimation of selection gradients in particular cases reveals that correlations between traits can propagate indirect selection effects that in turn obscure the actual targets of selection.
- Surveys of linear selection gradients reveal that modal value of this form of selection is close to zero and that the mean is generally shifted by less than half a phenotypic standard deviation. Likewise, the modal value of nonlinear selection is close to zero, with trait variances increasing or decreasing by less than 40%. The distribution of γ_{ii} is slightly biased towards negative values (i.e., towards stabilizing rather than disruptive selection).
- Estimation of correlational selection (e.g., the off-diagonal elements of the γ-matrix) has been neglected in empirical studies. This neglect is unfortunate, because—as we shall see in Chapter 7—detecting this type of selection provides evidence that trait interactions affect fitness and so play a crucial role in functional complexes.

Literature Cited

Arnold, S. J. 1994. Multivariate inheritance and evolution: a review of concepts. pp. 17–48 *in* C. R. B. Boake (ed.), Quantitative Genetic Studies of Behavioral Evolution. University of Chicago Press, Chicago.

Endler, J. A. 1986. Natural Selection in the Wild. Princeton University Press, Princeton.

Hoekstra, H. E., J. M. Hoekstra, D. Berrigan, S. N. Vignieri, A. Hoang, C. E. Hill, P. Beerli, and J. G. Kingsolver. 2001. Strength and tempo of directional selection in the wild. Proc. Natl. Acad. Sci. USA 98:9157–9160.

Kingsolver, J. G., H. E. Hoekstra, J. M. Hoekstra, D. Berrigan, S. N. Vignieri, C. E. Hill, A. Hoang, P. Gibert, and P. Beerli. 2001. The strength of phenotypic selection in natural populations. American Naturalist 157:245–261.

Lande, R., and S. J. Arnold. 1983. The measurement of selection on correlated characters. Evolution 37:1210–1226.

Mitchell-Olds, T., and R. G. Shaw. 1987. Regression analysis of natural selection: statistical inference and biological interpretation. Evolution 41:1149–1161.

Phillips, P. C., and S. J. Arnold. 1989. Visualizing multivariate selection. Evolution 43:1209–1222.

Price, T., Grant, P. R., Gibbs, H. L., and Boag, P. T. 1984. Recurrent patterns of natural selection in a population of Darwin's finches. Nature 309:787–789.

Price, T., M. Kirkpatrick, S. J. Arnold, and others. 1988. Directional selection and the evolution of breeding date in birds. Science 240:798–799.

Stinchcombe, J. R., A. F. Agrawal, P. A. Hohenlohe, S. J. Arnold, and M. W. Blows. 2008. Estimating nonlinear selection gradients using quadratic regression coefficients: Double or nothing? Evolution 62:2435–2440.

Wade, M. J., and S. Kalisz. 1990. The causes of natural selection. Evolution 44:1947–1955.

CHAPTER 3

The Selection Surface and Adaptive Landscape for a Single Trait

Overview—Selection on a single trait can be visualized as a curve, the ISS, that relates trait values to fitness. The fundamental properties of this curve (slope and curvature) are intimately related to the changes in trait mean and variance that are induced by selection within a generation. For example, a steep selection curve dramatically shifts the mean. A ∩-shaped curve reduces trait variance within a generation. Although we generally will not know the shape of the ISS, we can deduce its properties and approximate its shape with various functions. To make predictions about how the trait will change from one generation to the next, we need a related curve, one that is averaged over the trait distribution. This curve is called the adaptive landscape (AL).

3.1 The Individual Selection Surface, ISS

The idea that selection is some kind of function is implicit in the names used to describe selection; e.g., directional, stabilizing, truncation, etc. In this chapter we will make this idea explicit in a way that helps us visualize selection in its many guises. Imagine selection described by some continuous function. In particular, imagine relative fitness of individuals of phenotype z as a function of trait values. An overall positive slope implies directional selection favouring higher values (Figure 3.1a), while an overall negative slope implies directional selection for smaller values. Downward curvature that straddles the phenotypic mean implies stabilizing selection in the sense that selection encourages the mean to reside in a specified range (Figure 3.1b, c). Upward curvature has the opposite effect, the sign of disruptive selection (Figure 3.1d). We will refer to the function in question as $w(z)$ and call it the *individual selection surface*, the ISS.

Our concern with the ISS is local in the sense that we will focus on its shape in the region of trait values in which individuals are likely to be observed within our focal population. For example, consider the issue of the average slope of the ISS. To calculate that average slope, we would need to take the slope or first derivative of the function at a particular value of z, $\partial w(z) / \partial z$ (Figure 3.2), weight that slope by the frequency of individuals at that point, $p(z)$, repeat these operations over the entire range of occupied values of z, and

Evolutionary Quantitative Genetics. Stevan J. Arnold, Oxford University Press. © Stevan Arnold (2023).
DOI: 10.1093/oso/9780192859389.003.0004

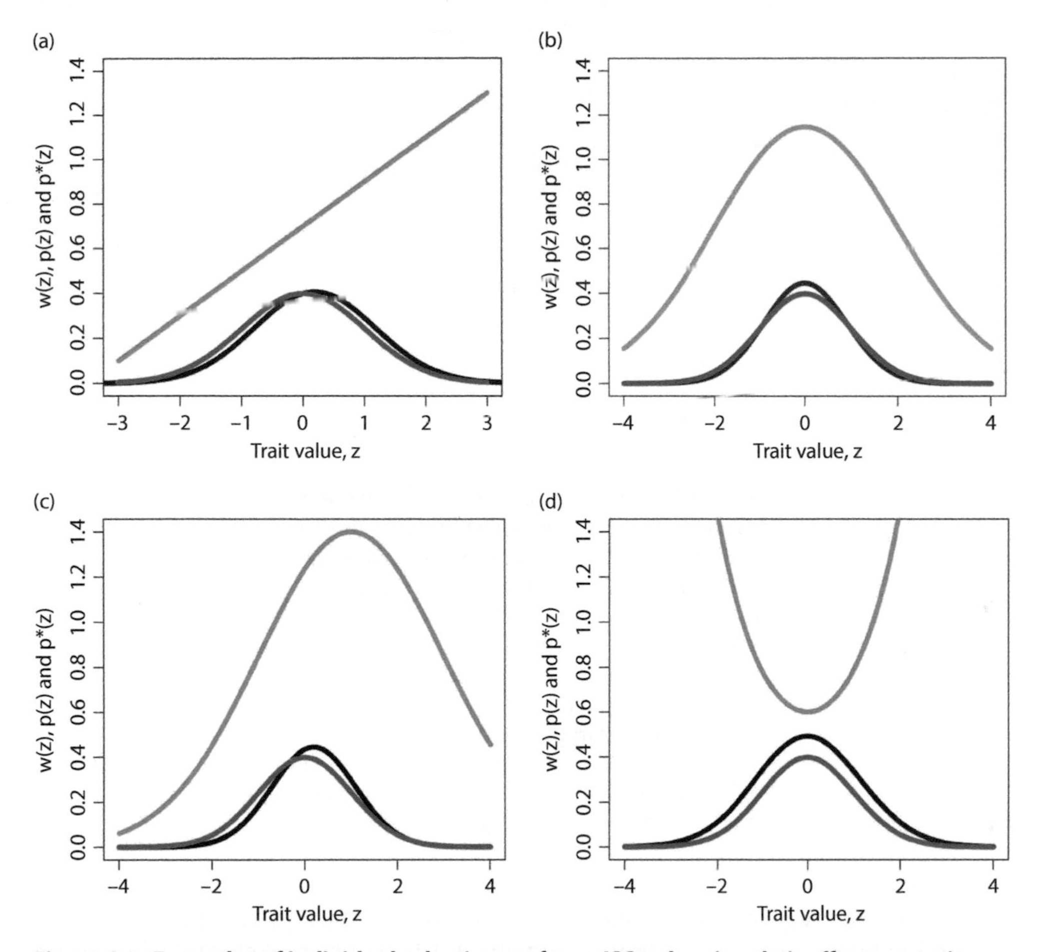

Figure 3.1 Examples of individual selection surfaces, ISSs, showing their effects on trait distributions. The statistics of the trait distributions before selection (blue), $p(z)$, and after selection (black), $p^*(z)$, are given in Figure 1.9. The ISS, $w(z)$, is shown as an orange curve. (a) Directional selection. The ISS is a linear function, $w(z) = \alpha + \beta z$, $\alpha = 1$, $\beta = 0.2$. The orange line has been shifted down 0.3 units for graphic effect. (b) Stabilizing selection. The ISS is a Gaussian function with $\theta = 0$, $\omega = 4$. (c) Directional and stabilizing selection. The ISS is a Gaussian function with $\theta = 1$, $\omega = 4$. (d) Disruptive selection. The ISS is a Gaussian function with $\theta = 0$, $\omega = -4$. The orange curve has been shifted down 0.8 units for graphic effect.

then add up all those weighted slopes, $p(z)\,\partial w(z)/\partial z$. This average slope turns out to be a familiar commodity, the directional selection gradient,

$$\beta = \int p(z)\,\frac{\partial w(z)}{\partial z}\,dz \tag{3.1}$$

(Lande and Arnold 1983). By a similar set of operations, we can evaluate the average curvature of the ISS, $\partial^2 w(z)/\partial z^2$, at each point and calculate the average of those curvatures (Figure 3.3). This average is equivalent to the nonlinear selection gradient,

$$\gamma = \int p(z)\,\frac{\partial^2 w(z)}{\partial z^2}\,dz. \tag{3.2}$$

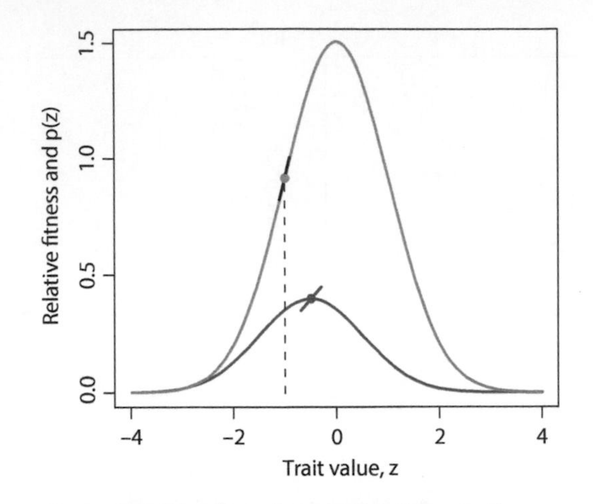

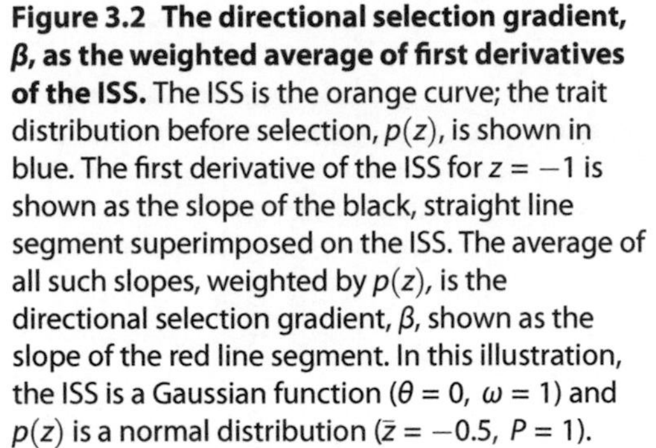
Figure 3.2 The directional selection gradient, β, as the weighted average of first derivatives of the ISS. The ISS is the orange curve; the trait distribution before selection, $p(z)$, is shown in blue. The first derivative of the ISS for $z = -1$ is shown as the slope of the black, straight line segment superimposed on the ISS. The average of all such slopes, weighted by $p(z)$, is the directional selection gradient, β, shown as the slope of the red line segment. In this illustration, the ISS is a Gaussian function ($\theta = 0$, $\omega = 1$) and $p(z)$ is a normal distribution ($\bar{z} = -0.5$, $P = 1$).

Surprisingly, these equivalencies hold whatever the form of the ISS so long as it is continuous and differentiable, and the trait is normally distributed before selection (Lande and Arnold 1983).

3.2 Linear and Quadratic Approximations to the ISS

Unless we are prophets, as we pretended to be in Figures 3.2 and 3.3, the ISS is not revealed to us directly, so we must deduce the properties of the ISS from data on relative fitness, $w(z)$, as a function of trait values, z. Suppose, for example, instead of an ISS revelation, we have the data plot shown in Figure 3.4. We can bypass the problem of the actual shape of the ISS and try to estimate the selection gradients, β and γ. It turns out that these gradients can be estimated by the familiar statistical procedures of linear and quadratic regression (Lande and Arnold 1983). Before fitting the regressions, it is useful to standardize the trait before selection so that it has a zero mean and a standard deviation of one. Fitness should be standardized in a different way, so that it has a mean of one. These standardizations yield selection gradients in readily interpretable units. Having accomplished these standardizations, we first estimate β by fitting a linear approximation to the ISS,

$$w(z) = \alpha + \beta z + \varepsilon, \tag{3.3}$$

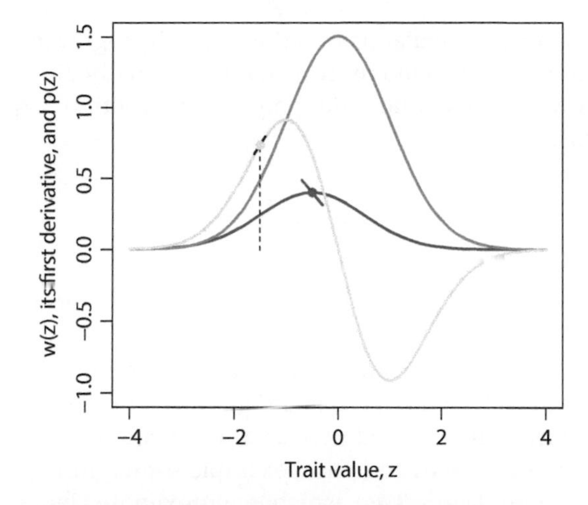

Figure 3.3 The nonlinear selection gradient, γ, as the weighted average of second derivatives of the ISS. The ISS is the orange curve, $w(z)$; the trait distribution before selection, $p(z)$, is shown in blue. The first derivative of the ISS is the yellow curve. A first derivative of this yellow curve (which is the second derivative of the ISS) at $z = -1.5$ is shown as a black line segment superimposed on the yellow curve. The average of all such slopes, weighted by the trait distribution, is the nonlinear selection gradient, $\gamma = -0.4375$, shown as the slope of the red line segment. In this illustration, the ISS is a Gaussian function ($\theta = 0,\ \omega = 1$) and $p(z)$ is a normal distribution ($\bar{z} = -0.5,\ P = 1$).

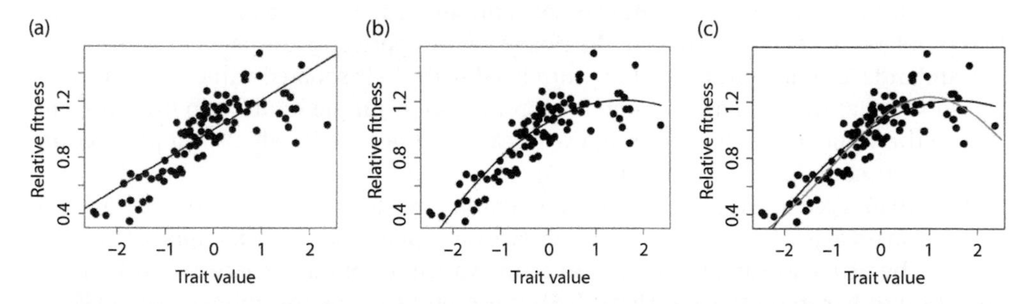

Figure 3.4 Linear and quadratic approximations to the ISS in a hypothetical example. Relative fitness, $w(z)$, is a function of trait value, z. (a) A linear fit to the data (Figure 1.10) using (3.3), $\alpha = 1,\ \beta = 0.21 \pm 0.02$ s.e. (b) A quadratic fit using (3.4), $\alpha = 1,\ \beta = 0.20 \pm 0.01$ s.e., $\gamma = -0.12 \pm 0.02$ s.e. (c) The actual ISS is a Gaussian function (3.8), shown in orange, $\theta = 1,\ \omega = 4$. The data points were generated by taking a random sample of trait values, z, from a normal distribution ($\bar{z} = 0,\ P = 1$). Those trait values were used in conjunction with the specified Gaussian function to produce corresponding, expected values of $w(z)$. Values of ε were drawn from a normal distribution (mean = 0, variance = 0.01) and added to expected value of w to produce the points in the figure (n = 100).

where ε is the deviation of a particular observation from the regression line (Figure 3.4a). The usual assumption in statistical inference is that ε is normally distributed with a mean of zero. To estimate γ, we use a second-order polynomial function, a quadratic function, to approximate the ISS,

$$w(z) = \alpha + \beta z + \frac{1}{2}\gamma z^2 + \varepsilon \tag{3.4}$$

(Figure 3.4b). The factor of $\frac{1}{2}$ is included so that γ will be the second derivative of the function, a measure of curvature.

In general, neither the linear (3.3) nor the quadratic approximation (3.4) is meant to actually imitate the asymmetry or the bumps and grinds that might be present in the ISS. Linear and quadratic functions are used because they enable us to estimate parameters of selection. For example, in the hypothetical example shown in Figure 3.4, the ISS was actually a Gaussian function (Figure 3.4c), which is approximated by a quadratic function. Despite the relatively poor fits of the linear and quadratic functions to this ISS, those fits do provide correct estimates of β and γ. Nevertheless, the actual shape of the ISS is also of interest. A variety of procedures and functions might be used to more accurately describe the shape of the ISS. We will consider one of these in the next section.

A few words about the actual steps in performing these two regressions may be useful. For the standardisation of trait values before selection, one simply subtracts the raw trait mean from each value and divides each of those new values by the raw trait standard deviation, producing a desired new trait with zero mean and standard deviation of one. The raw values of fitness are divided by the raw fitness mean to produce a new fitness variable (relative fitness) with a mean of one. If one wished to draw or express the two regressions, (3.3) or (3.4), one ignores the estimated value of α and sets its value to one. Why is this? The value of ε on the regression curve is zero, so it is dropped from the equation. Finally, to draw or express (3.4), one takes the value of β estimated from (3.3) and the value of γ estimated from (3.4). This last point may seem mysterious, but it is a practical solution to a bias problem. In general, if the trait distribution is not perfectly normal before selection, the standardized trait values, z, will be correlated with their squared values, z^2. This correlation will bias the estimate of β obtained from (3.4). A simple solution to this problem is to use the value of β estimated from the linear regression (3.3), which will produce an unbiased estimate (Lande and Arnold 1983).

Kalisz (1986) used linear and quadratic approximations to estimate β- and γ-coefficients for selection on germination timing in a winter annual plant. Just two examples are shown here from her extensive analysis of how selection varied in time and space in a single local population of her study species (Figure 3.5). It is useful to draw the approximated ISS, as we see in these examples, in which directional selection is relatively strong and statistically significant but curvature is very slight. The later feature is difficult to appreciate from just the point estimates of the γ-coefficients (shown in the figure caption). In other particular episodes of viability and fecundity selection, directional selection typically favoured early germination, whereas the ISS was either convex or concave (i.e., γ either negative or positive). Nevertheless, Kalisz (1986) argued that an intermediate optimum date for germination might vary from year to year. In Section 3.6, we will discuss the Chevin et al. (2015) approach to direct testing of a similar hypothesis in another species.

In passing, we note that in the univariate case, selection gradients and selection differentials are equivalent when fitness and trait values are standardized. Under that standardization scheme, relative fitness has a mean of one, and the trait values have a

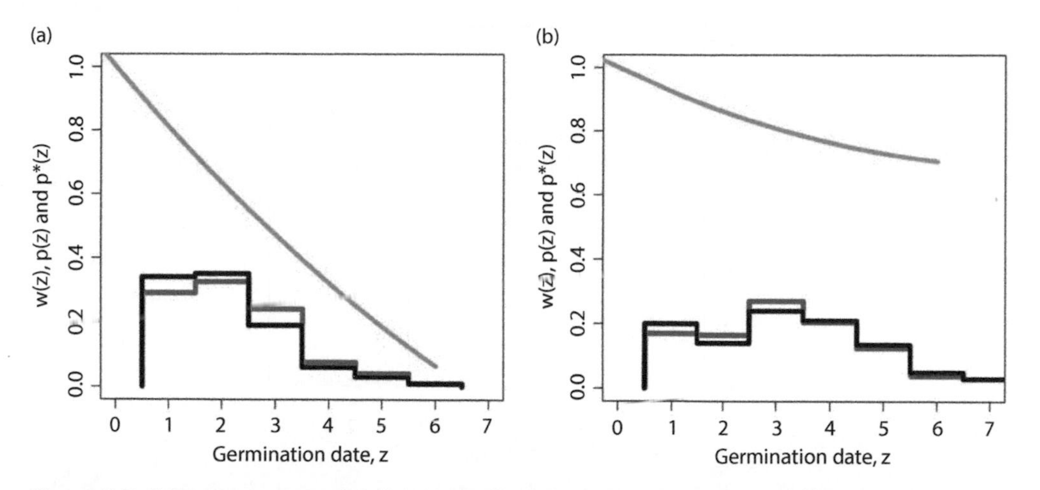

Figure 3.5 Selection on germination timing in Spring Blue-eyed Mary (*Collinsia verna*) in two study years after Kalisz (1986). Distributions of germination dates are shown before (blue) and after selection (black). The scores for germination date represent census periods lasting seven or 10 days. Quadratic approximations to the ISS are shown in orange. (a) For the 1982 sample (n = 2580 individuals that germinated and bore fruit), the ISS approximation is $w(z) = 1 + \beta z + \frac{1}{2}\gamma z^2$, where $\beta = -0.198^{***}$ and $\gamma = 0.014$ *ns* (*** denotes $P < 0.0001$ and *ns* denotes $P > 0.05$). Each of the six census periods lasted 10 days. (b) For the 1983 sample (n = 2,440 individuals that germinated and bore fruit), the ISS approximation is $w(z) = 1 + \beta z + \frac{1}{2}\gamma z^2$, where $\beta = -0.079^{*}$ and $\gamma = 0.010$ *ns* (* denotes $P < 0.05$). Each of the seven census periods lasted seven days.

mean of zero, with $P = 1$ (Section 1.4). With this standardization, we see from (2.2 and 2.9) that $\beta = s$ and $\gamma = C$, but these relationships will not hold in the multivariate case.

3.3 Cubic Spline Approximation to the ISS

The power of the quadratic approximations just described is that we can use them to estimate the selection coefficients β and γ. It is a remarkable fact that this estimation is legitimate and accurate even if the ISS is not quadratic. In some situations, however, quadratic approximation may lead us astray even though it does its job of estimating β and γ. One such case is shown in Figure 3.6. In this hypothetical example, truncation selection acts on the population. The actual shape of the ISS is step function with a step at a phenotypic value of +1. Quadratic regression will do a poor job of approximating the ISS for it will yield a disruptive curve with an inflexion point near –1. In general, when the shape of the ISS is of interest, regression techniques more sophisticated than quadratic regression can be used to approximate its shape. Schluter (1988) introduced the use of regression with cubic splines to describe fitness functions, and many subsequent studies have successfully used this approach to approximate the ISS.

An example of a cubic spline approximation to an unknown ISS is shown in Figure 3.7. Here survivorship of song sparrows (*Melospiza melodia*) is shown as a function of a linear combination of morphological measurements (PC2). As in the preceding example, fitness is binary. Measurements in the sample after selection are clustered near the middle of the range in trait values. The cubic spline approximation appropriately shows a convex function with an inflexion near the trait mean (stabilizing selection). The minor curves

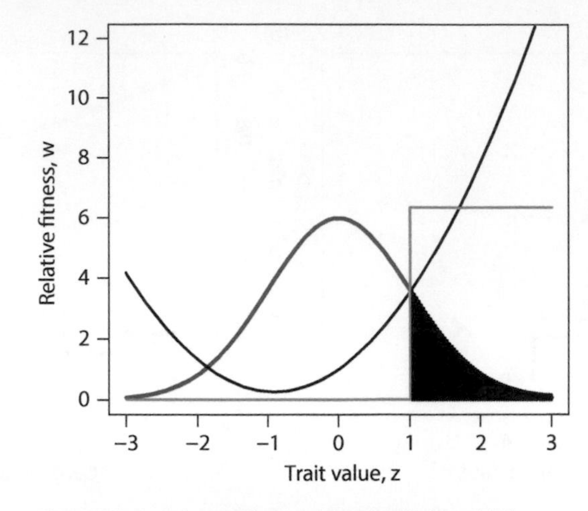

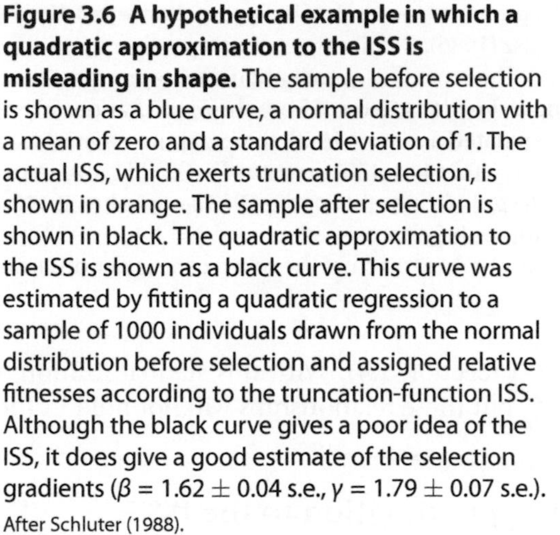
Figure 3.6 A hypothetical example in which a quadratic approximation to the ISS is misleading in shape. The sample before selection is shown as a blue curve, a normal distribution with a mean of zero and a standard deviation of 1. The actual ISS, which exerts truncation selection, is shown in orange. The sample after selection is shown in black. The quadratic approximation to the ISS is shown as a black curve. This curve was estimated by fitting a quadratic regression to a sample of 1000 individuals drawn from the normal distribution before selection and assigned relative fitnesses according to the truncation-function ISS. Although the black curve gives a poor idea of the ISS, it does give a good estimate of the selection gradients ($\beta = 1.62 \pm 0.04$ s.e., $\gamma = 1.79 \pm 0.07$ s.e.).
After Schluter (1988).

in the fitted function were not anticipated by the authors and are probably of no interest to readers. In this sense, a simple quadratic approximation, which would be a smooth convex function, would do the job just as well, while estimating β and γ. In some situation, however, minor curves or non-quadratic shapes might have biological significance.

3.4 The Adaptive Landscape, AL

Another fitness function, distinct from the ISS, will be important to us. This function is known as the *adaptive landscape*, the AL. In the AL, average fitness of the population, $\bar{W}$, is a function of its average trait value, $\bar{z}$. For our focal population, we evaluate the AL at a single point, the trait mean, $\bar{z}$. That narrow perspective on the AL, however, is sufficient to tell us the slope and curvature of the AL at that point. In particular,

$$\beta = \frac{\partial \bar{W}}{\bar{W} \partial \bar{z}} = \frac{\partial \, ln \bar{W}}{\partial \bar{z}} \tag{3.5}$$

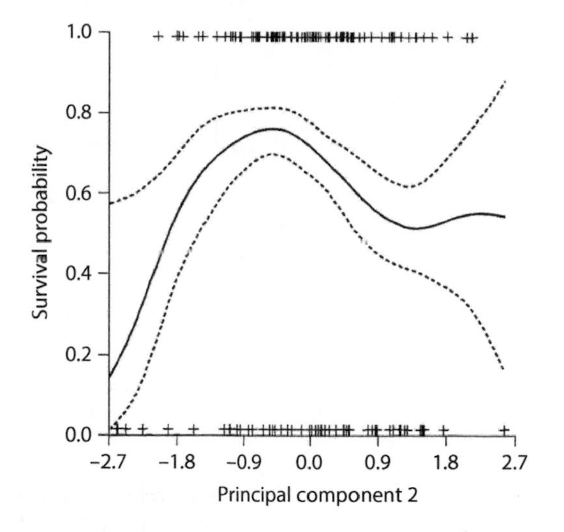

Figure 3.7 An example of a cubic spline approximation to an unknown ISS. The solid black curve is a cubic spline of overwinter survival of male song sparrows (n =152) as a function of morphological measurements (Principal component 2). The dashed curves show ± 1 standard error of the function, estimated by bootstrapping. The crosses at the top and bottom show the absolute fitness and trait values of males that did and did not survive winters. Fitness is not normalized by dividing by its mean, as in other ISSs that we have depicted.

From Schluter (1988) with permission.

and

$$\gamma - \beta^2 = \frac{\partial^2 \bar{W}}{\bar{W}\partial \bar{z}^2} = \frac{\partial^2 \, ln\bar{W}}{\partial \bar{z}^2}, \tag{3.6}$$

where $ln\bar{W}$ means the natural logarithm of average absolute fitness in the population (Lande 1979; Lande and Arnold 1983). The directional selection gradient, β, as the slope of the AL evaluated at the trait mean is portrayed in Figure 3.8.

Just as we can imagine that the ISS is a function that extends beyond the occupied region of phenotypes for our population, the same might be so for the AL. We also need to imagine that hypothetical probability distributions for z, so that we can average the ISS over those hypothetical distributions to visualize the corresponding AL. An easy way to generate these hypothetical distributions of z is to keep the variance, P, constant and simply shift the trait mean before selection, $\bar{z}$. The extension is especially simple if we assume that $p(z)$ is a normal distribution, and the ISS is a Gaussian function (Haldane 1954; Lande 1976) so that

$$p(z) = (1/\sqrt{2\pi P}) \exp\left\{\frac{-(z-\bar{z})^2}{2P}\right\} \tag{3.7}$$

and

$$w(z) = \exp\left\{\frac{-(z-\theta)^2}{2\omega}\right\}, \tag{3.8}$$

where θ is the optimum and $\sqrt{\omega}$ is the width of the function, analogous to a standard deviation (ω is analogous to a variance). If $p(z)$ keeps its shape (constant variance) while translating its mean, averaging the ISS over this translated function yields a Gaussian-shaped AL,

$$\bar{W} = \exp\left\{\frac{-(\bar{z}-\theta)^2}{2\,(\omega+P)}\right\} = \exp\left\{\frac{-(\bar{z}-\theta)^2}{2\Omega}\right\}, \tag{3.9}$$

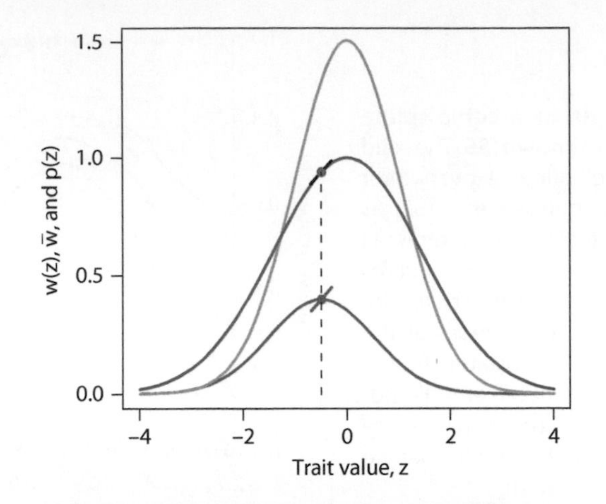

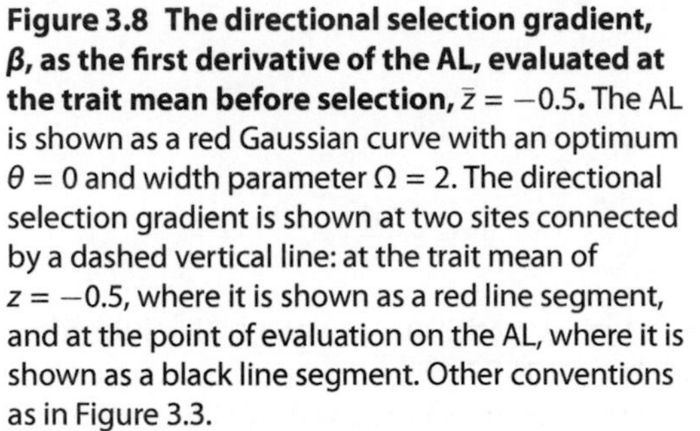

Figure 3.8 The directional selection gradient, β, as the first derivative of the AL, evaluated at the trait mean before selection, $\bar{z} = -0.5$. The AL is shown as a red Gaussian curve with an optimum $\theta = 0$ and width parameter $\Omega = 2$. The directional selection gradient is shown at two sites connected by a dashed vertical line: at the trait mean of $z = -0.5$, where it is shown as a red line segment, and at the point of evaluation on the AL, where it is shown as a black line segment. Other conventions as in Figure 3.3.

with the same optimum as the ISS, θ, but a larger 'variance', $\Omega = \omega + P$ (Figure 3.9). The first derivative or slope of the AL, evaluated at the population mean, is

$$\frac{\partial\, ln\bar{W}}{\partial \bar{z}} = \Omega^{-1}\,(\theta - \bar{z}) = \beta \tag{3.10a}$$

and its curvature is

$$\frac{\partial^2\, ln\bar{W}}{\partial \bar{z}^2} = -\Omega^{-1} = \gamma - \beta^2 \tag{3.10b}$$

(Lande 1979; Phillips and Arnold 1989; Jones et al. 2003). Rearranging (3.10b), we obtain a general expression for translating between γ- and ω-coefficients in the univariate case,

$$\gamma = \beta^2 - \Omega^{-1}. \tag{3.10c}$$

A useful approximation for converting between γ and ω under Gaussian assumptions is

$$\gamma \approx -\frac{1}{\omega}, \tag{3.11}$$

an expression that works best when γ is close to zero or with a trait value at or close to the optimum when $\gamma > -1$ (Figure 3.10).

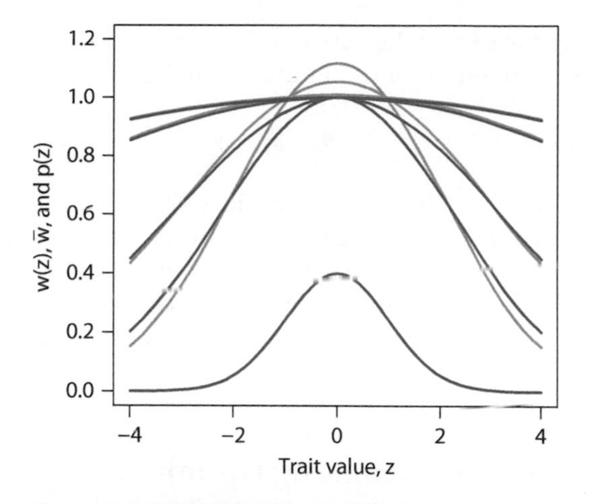

Figure 3.9 Gaussian ISSs and their corresponding Gaussian ALs for a range of values of ω. The blue curve shows a normal trait distribution before selection, $p(z)$, with a mean of 0 and a variance $P = 1$. The wide orange curves show the individual selection surfaces for $\omega = 99$ (at the top), 49, 9, and 4 (at the bottom). The narrow red curves show the corresponding adaptive landscapes with width parameters $\Omega = \omega + P = 100, 50, 10$, and 5.

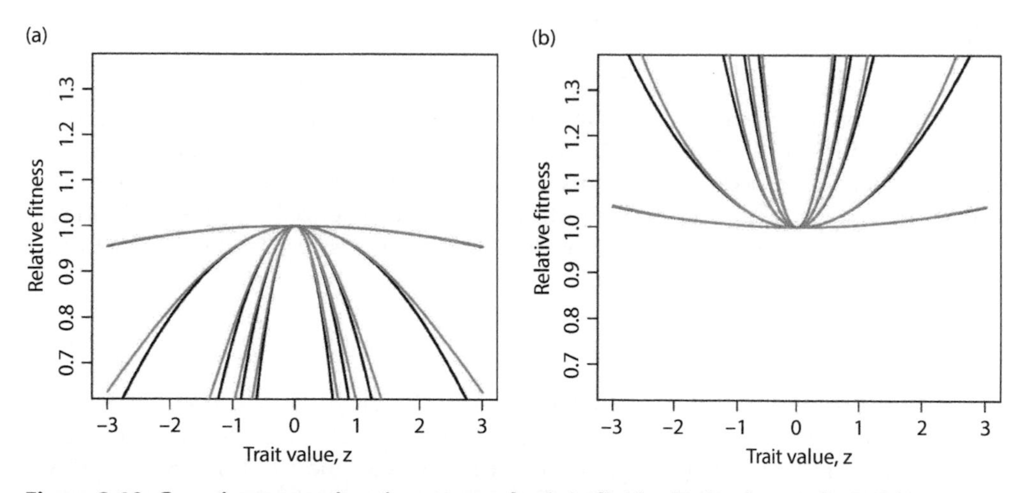

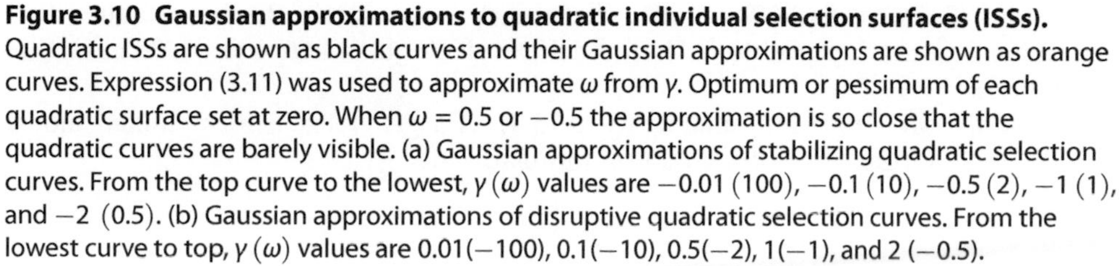

Figure 3.10 Gaussian approximations to quadratic individual selection surfaces (ISSs). Quadratic ISSs are shown as black curves and their Gaussian approximations are shown as orange curves. Expression (3.11) was used to approximate ω from γ. Optimum or pessimum of each quadratic surface set at zero. When $\omega = 0.5$ or -0.5 the approximation is so close that the quadratic curves are barely visible. (a) Gaussian approximations of stabilizing quadratic selection curves. From the top curve to the lowest, $\gamma\,(\omega)$ values are -0.01 (100), -0.1 (10), -0.5 (2), -1 (1), and -2 (0.5). (b) Gaussian approximations of disruptive quadratic selection curves. From the lowest curve to top, $\gamma\,(\omega)$ values are 0.01(-100), 0.1(-10), 0.5(-2), 1(-1), and 2 (-0.5).

In general, by taking the second derivative of (3.4) and setting it equal to zero, we find that the distance to the optimum (or pessimum) from the trait mean is

$$(\bar{z} - \theta) = \beta / - \gamma \tag{3.12}$$

(Mitchell-Olds and Shaw 1987; Phillips and Arnold 1989).

The Gaussian form of the ISS (3.8) is also useful in theoretical work because it allows us to immediately solve for the mean and variance after selection. If a trait is normally distributed before selection and subjected to Gaussian selection (3.8), it will be normally distributed after selection with mean, $\bar{z}^* = (\bar{z}\omega + \theta P)/\Omega$, and variance, $P^* = (\omega P)/\Omega$ (Lande 1981).

3.5 Gaussian ISS With a Fluctuating Optimum

The main rationale for approximating the ISS with a Gaussian function is that the resulting surface provides a direct bridge to the AL and hence to the analysis and modelling of adaptive radiations. An excellent example of this approach is provided by an analysis of selection on breeding date by (Chevin et al. 2015). To apply their model one needs time series data in which the ingredients for estimating an ISS (i.e., samples of individual values for fitness and a trait) are available for each of a series of years or generations. The basic idea is to fit a model of stabilizing selection in which the position of the fitness optimum changes through time. As we shall see in later chapters, this feature of a moving optimum is shared by an important class of models for adaptive radiation in deep evolutionary time. For the moment, however, we will focus on modelling a moving Gaussian peak for a few dozen generations.

Using a Gaussian model for the ISS (3.8), Chevin et al. (2015) modelled the selection on laying date, z, that arises from movement of an intermediate optimum. As the optimum, θ_t, moves through time, its position is a function of an observed environmental variable, x_t, and a stochastic process that governs the movement. Each generation this stochastic process contributes a value ϵ_t and its distribution is characterized by an autocovariance function $Cov(\epsilon_t, \epsilon_{t+h}) = \alpha^{|h|}\sigma_\epsilon^2$ and a mean of zero. In other words, our model for the position of the optimum at generation t is

$$\theta_t = A + Bx_t + \epsilon_t, \tag{3.13}$$

where A is the regression intercept, and B is the regression slope. Furthermore, the parameters used to describe the movement of the optimum (B, σ_ϵ^2 and α) are related to parameters of stabilizing and directional selection under the Gaussian model of the ISS and AL via expressions (3.10a) and (3.10b). Notice that the optimum might fluctuate about either a stationary or moving position. The model can distinguish between these and other possibilities.

The estimation expression, analogous to (3.4), is a generalized linear mixed model (GLMM) with a logarithmic link function for relative fitness,

$$ln w_t(z) = \mu_t + \beta_z z + \zeta_t z + \frac{1}{2}\gamma_{zz} z^2 + \gamma_{xz} x_t z, \tag{3.14}$$

where μ_t is a fixed effect of time; β_z is a fixed, linear effect of z; ζ_t is a random effect of z (with autocovariance function $Cov(\zeta_t, \zeta_{t+h}) = \alpha^{|h|}\sigma_\zeta^2$); γ_{zz} is a quadratic effect of z; and γ_{xz}

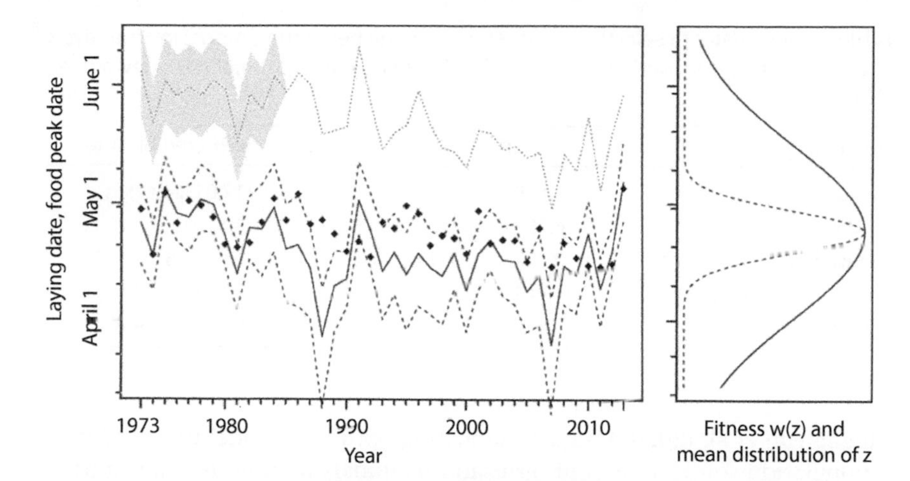

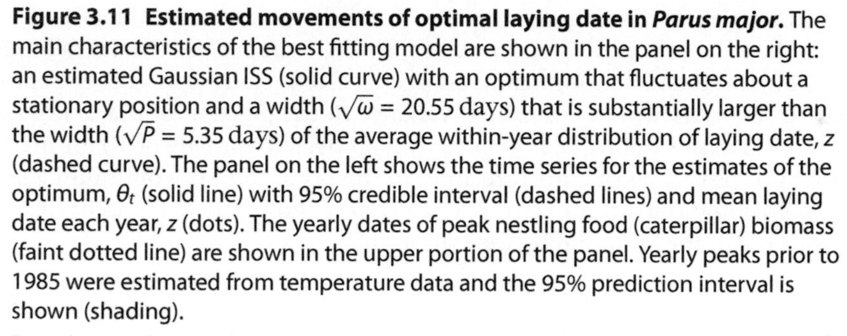

Figure 3.11 Estimated movements of optimal laying date in *Parus major*. The main characteristics of the best fitting model are shown in the panel on the right: an estimated Gaussian ISS (solid curve) with an optimum that fluctuates about a stationary position and a width ($\sqrt{\omega}$ = 20.55 days) that is substantially larger than the width ($\sqrt{P}$ = 5.35 days) of the average within-year distribution of laying date, z (dashed curve). The panel on the left shows the time series for the estimates of the optimum, θ_t (solid line) with 95% credible interval (dashed lines) and mean laying date each year, z (dots). The yearly dates of peak nestling food (caterpillar) biomass (faint dotted line) are shown in the upper portion of the panel. Yearly peaks prior to 1985 were estimated from temperature data and the 95% prediction interval is shown (shading).

From Chevin et al. (2015) with permission.

is an interaction between x_t and z. The process parameters of interest (ω, θ_t, B, σ^2_ϵ, $w_{max, t}$) are functions of the estimated parameters (γ_{zz}, β_z, γ_{xz}, ζ_t, μ_t), as detailed by Chevin et al. (2015). For example, one solves for ω using the estimated value of γ_{zz} and expression (3.10).

Using this approach, Chevin et al. (2015) analysed an extensive time series for fecundity as a function of laying date in a passerine bird (*Parus major*) (Visser et al. 2006; Reed et al. 2013a, b) (Figure 3.11). Their analysis yielded a well-supported characterization of selection acting on laying date (Table 3.1).

The best fitting model included a temperature covariate (meant to represent the availability of food for nestlings), a stationary optimum, and a first-order autocorrelation for the random effect of z. In other words, the analysis does not support the impression that might be gained from Figure 3.11 of a long term temporal trend in the optimum. Instead, the evidence is for no long-term trend (i.e., stationary θ) and a time series in which the position of the optimum is correlated only between adjacent years.

Instead of following the approach used in many empirical studies of how selection gradients vary in time and space, Chevin et al. (2015) did not attempt to estimate β for each time interval (or location in space). Instead they used the entire data time series to estimate the parameters that characterize models of fluctuating selection (e.g., Bull 1987; Charlesworth 1993; Lande and Shannon 1996). As a consequence, their approach builds a much needed empirical bridge to these models and to models that apply in deep evolutionary time.

Table 3.1 Estimates of selection parameters from the best-fitting model in an analysis of a breeding date-fecundity time series in the Great Tit (*Parus major*) by Chevin et al. (2015).

Parameter	Posterior mean ± s.e.	95% credible interval
$\sqrt{\omega}$ **(days)**	20.55 ± 1.7	(17.85, 24.42)
σ_ϵ **(days)**	6.75 ± 1.66	(4.4, 10.53)
α **(autocorrelation)**	0.3029 ± 0.2419	(−0.2176, 0.7113)
A **(intercept, April day)**	19.43 ± 1.95	(14.97, 22.89)
B **(slope, days/°C)**	−5.01 ± 1.09	(−7.38, −2.93)

De Villemereuil et al. (2020) tested a fluctuating optimum model for breeding date in 39 wild populations of 21 species of birds and mammals with results that mirrored the finding of Chevin et al. (2015) in *Parus major*. In this larger dataset, a Gaussian ISS with a stabilizing shape but fluctuating optimum was strongly supported. The overall estimate of ω was 6.22 for birds and 4.94 for mammals ($P = 1$). The overall position of the optimum (θ_t) suggested selection for early breeding. Variance in the position of the optimum (σ_θ^2) varied between datasets from < 0.25 to > 9, with an overall estimate of 10 for mammals and 3.6 for birds. Estimates of autocorrelation in the position of the optimum generally could not be distinguished from zero. The study also showed that plasticity tended to enhance tracking of the optimum by the trait mean.

3.6 Standardizing the Directional Selection Gradient by the Trait Mean

In some circumstances it is useful to standardize the selection gradients by the trait mean rather than by the trait variance (Hereford et al. 2004). Recalling that the directional selection differential is the equivalent to the covariance between relative fitness and trait values (1.10), the univariate selection gradient is $\beta = Cov\,(w, z)\,/P$. We standardize β, so that

$$\beta' = \sqrt{P}\beta. \tag{3.15}$$

β' is variance standardized in the sense that it gives the change in relative fitness for a change in the trait of one phenotypic standard deviation before selection (Lande and Arnold 1983). Alternatively, we can standardize by the trait mean to obtain

$$\beta_\mu = \bar{z}\,\beta = \bar{z}\,Cov\,(w, z)\,/P, \tag{3.16}$$

which gives the change in relative fitness for a proportional change in the trait mean (Hereford et al. 2004). This mean-standardized gradient is a natural way to measure selection on a logarithmic trait scale. Furthermore, as the authors argue, β_μ is a readily interpretable measure of the strength of selection because a value of one is equivalent to selection on relative fitness itself (and hence an upper bound on β_μ).

3.7 Surveys of Selection Gradients and Their Implications for the Shape of Selection Surfaces and Adaptive Landscapes

3.7.1 *Shape of the ISS deduced from surveys of selection gradients*

The shape of the ISS in nature emerges from the two surveys summarized in Chapter 2, Kingsolver et al. (2001) and Stinchcombe et al. (2008). The most frequent result from empirical studies is an almost flat curve with almost no slope and no curvature. We reach this conclusion by noting that the most frequent estimate of β is very close to zero (Figure 3.12a). This fact is poorly reflected in Figure 3.12b, in which the widest ISS is for a bin category centred at $\beta = 0.05$. Also note that the histogram in Figure 3.12a is for the absolute value of β. If we had plotted the estimated values, in which the signs are arbitrary functions of measurement scale, the plot would be approximately symmetrical about zero. Turning to γ, we see such an approximately symmetrical histogram in Figure 3.12c, which implies that the most frequent γ estimate is very close to zero, which implies a nearly flat curve that is bowed slightly upward or downward. Curves corresponding to four bin categories (centred at $\gamma = 0.3,\ 0.1, -0.1, -0.3$) are shown in Figure 3.12d. In other words, the most frequent result in empirical studies is an individual selection surface that is nearly flat. It is unusual to estimate a slope steeper than the thinnest orange line in Figure 3.12b or more curved than the thinnest curves in Figure 3.12d.

3.7.2 *Surveys of mean-standardized selection gradients*

Surveys of mean-standardized selection gradients provide another valuable perspective. Hereford et al. (2004) conducted a survey of 240 univariate studies and computed both variance- and mean-standardized βs and presented their results on both arithmetic and log 10 scales (Figure 3.13). The results for mean-standardized β on the arithmetic scale (Figure 3.13a) resemble the histogram of the larger sample assembled by Kingsolver et al. (2001). The distribution has a median value of β' close to zero (0.19), a 19% change in relative fitness when the trait value is changed by one phenotypic standard deviation, a modal category even closer to zero and tail extending to larger values (> 1.0). Viewing the same distribution on a log scale (Figure 3.13b), we see more clearly that only the larger values attain statistical significance. The distribution of mean-standardized gradients, β_μ, also has a mode close to zero on the arithmetic scale, with a median of 1.45 and a long tail extending to larger values. A correction for bias changes the median estimate to 0.48. In other words, the median value of β_μ represents a trait change of nearly half the value of the trait mean. On the log scale we see that those larger values form a distinct cluster, as they do in the case of β on the log scale, and only the largest values tend to be statistically significant.

3.7.3 *Distance to an optimum deduced from surveys of selection gradients*

We can calculate the distance from the trait mean to the optimum of the ISS (which are equivalent to distances from the optimum of the AL), assuming a convex ISS, using (3.12). Taking the median values of those distances, we find that most trait means are within one trait phenotypic standard deviation of the optimum and that it is unusual for the trait mean to be more than five standard deviations from the optimum (Figure 3.14a). Likewise,

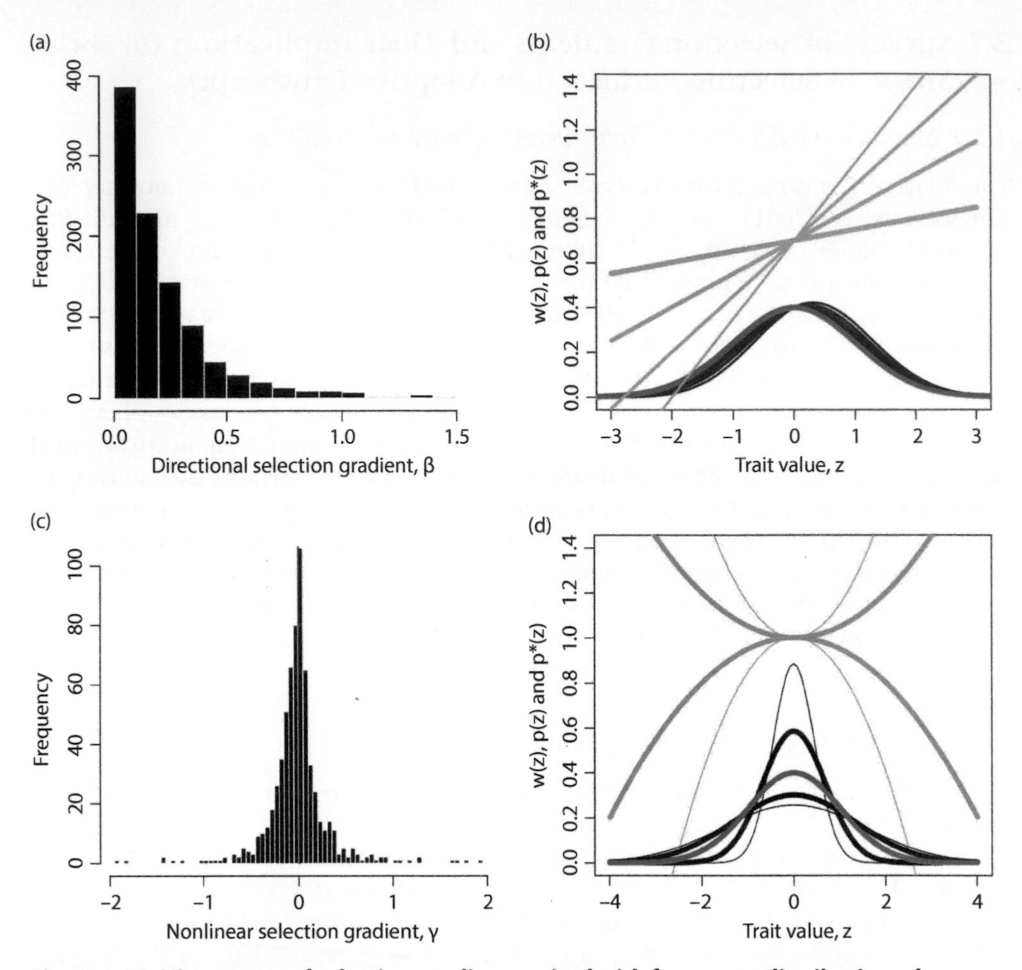

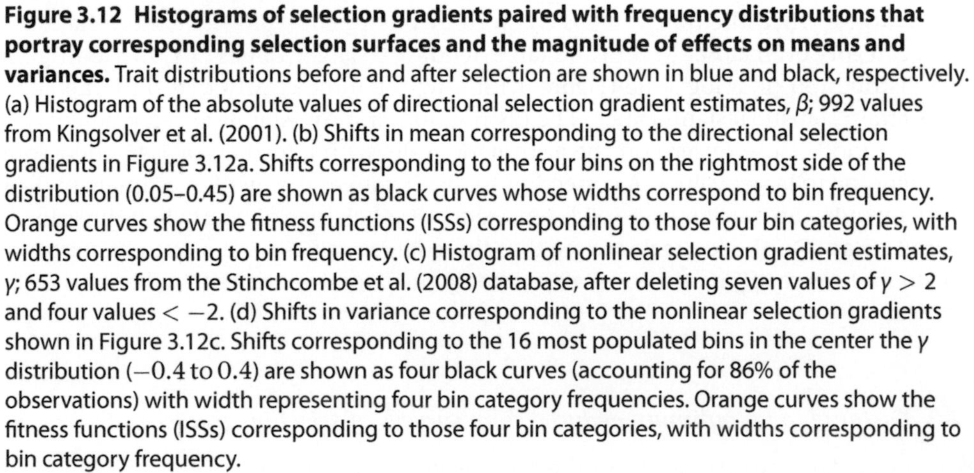

Figure 3.12 Histograms of selection gradients paired with frequency distributions that portray corresponding selection surfaces and the magnitude of effects on means and variances. Trait distributions before and after selection are shown in blue and black, respectively. (a) Histogram of the absolute values of directional selection gradient estimates, β; 992 values from Kingsolver et al. (2001). (b) Shifts in mean corresponding to the directional selection gradients in Figure 3.12a. Shifts corresponding to the four bins on the rightmost side of the distribution (0.05–0.45) are shown as black curves whose widths correspond to bin frequency. Orange curves show the fitness functions (ISSs) corresponding to those four bin categories, with widths corresponding to bin frequency. (c) Histogram of nonlinear selection gradient estimates, γ; 653 values from the Stinchcombe et al. (2008) database, after deleting seven values of $\gamma > 2$ and four values < -2. (d) Shifts in variance corresponding to the nonlinear selection gradients shown in Figure 3.12c. Shifts corresponding to the 16 most populated bins in the center the γ distribution (-0.4 to 0.4) are shown as four black curves (accounting for 86% of the observations) with width representing four bin category frequencies. Orange curves show the fitness functions (ISSs) corresponding to those four bin categories, with widths corresponding to bin category frequency.

a survey of local adaptation, as revealed by transplant studies, revealed that the native population advantage in relative fitness was 45% Hereford (2009). Furthermore, fluctuation in the optimum, drift in the mean, and developmental instability can all contribute to deviation of the trait mean from an adaptive peak (Hansen et al. 2008).

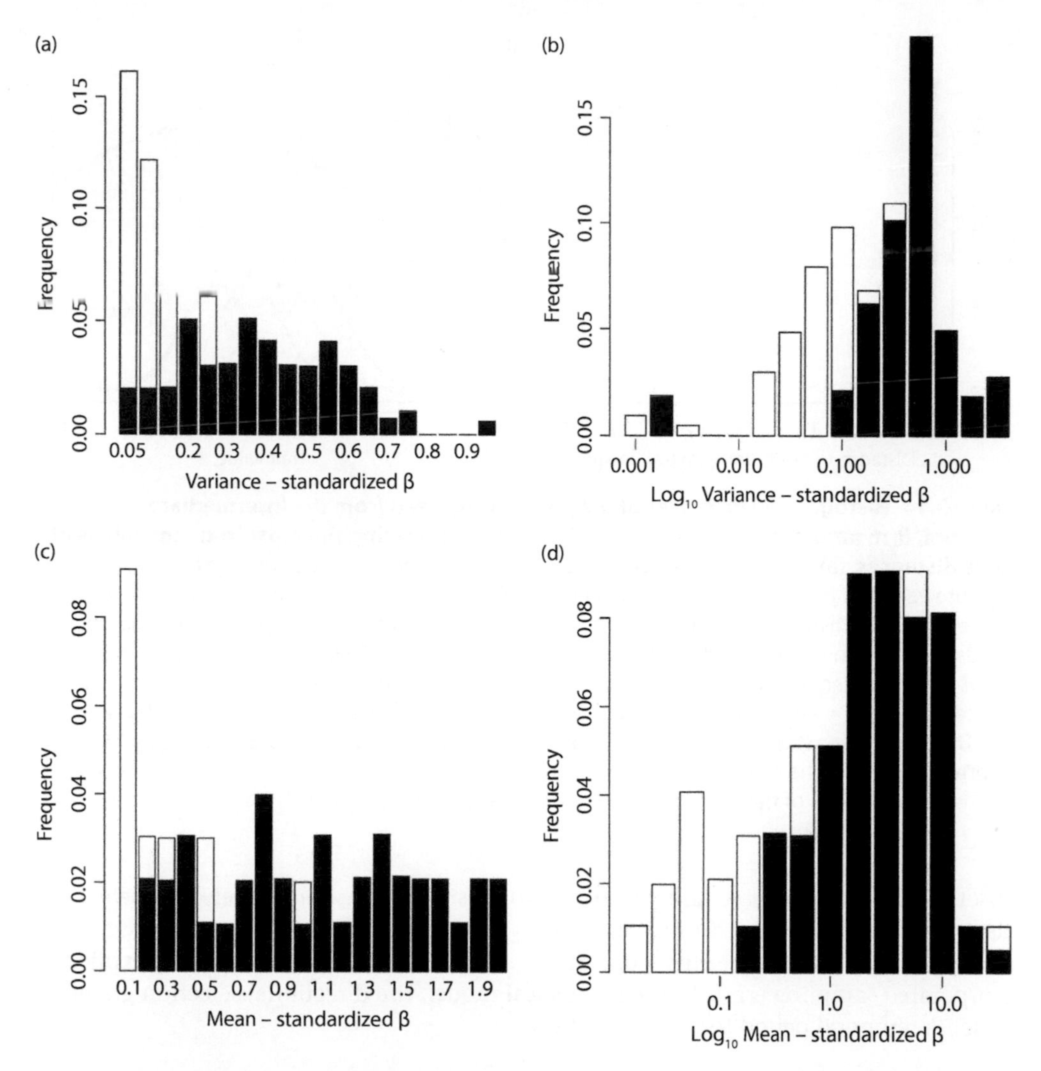

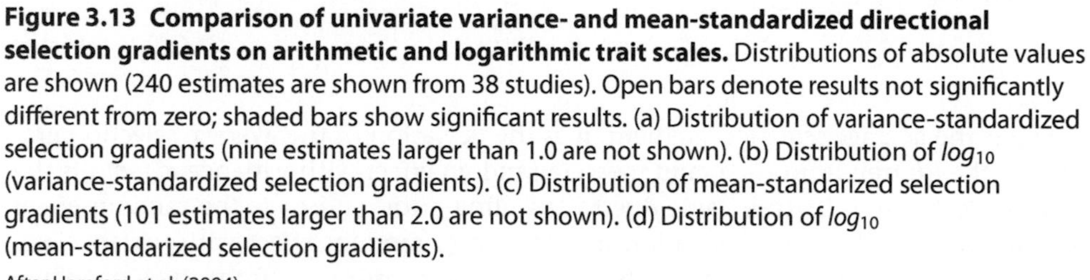

Figure 3.13 Comparison of univariate variance- and mean-standardized directional selection gradients on arithmetic and logarithmic trait scales. Distributions of absolute values are shown (240 estimates are shown from 38 studies). Open bars denote results not significantly different from zero; shaded bars show significant results. (a) Distribution of variance-standardized selection gradients (nine estimates larger than 1.0 are not shown). (b) Distribution of log_{10} (variance-standardized selection gradients). (c) Distribution of mean-standarized selection gradients (101 estimates larger than 2.0 are not shown). (d) Distribution of log_{10} (mean-standarized selection gradients).

After Hereford et al. (2004).

3.7.4 *Temporal variation in selection gradients*

Several dozen empirical studies have explored the important topic of how selection varies through time (e.g., Kalisz 1986; Grant and Grant 2002). Siepielski et al. (2009) compiled a database of selection estimates from 89 studies in which selection had been studied through time. One of their primary conclusions was that the strength of selection often varied considerably from year to year. Morrissey and Hadfield (2012) reanalysed a large

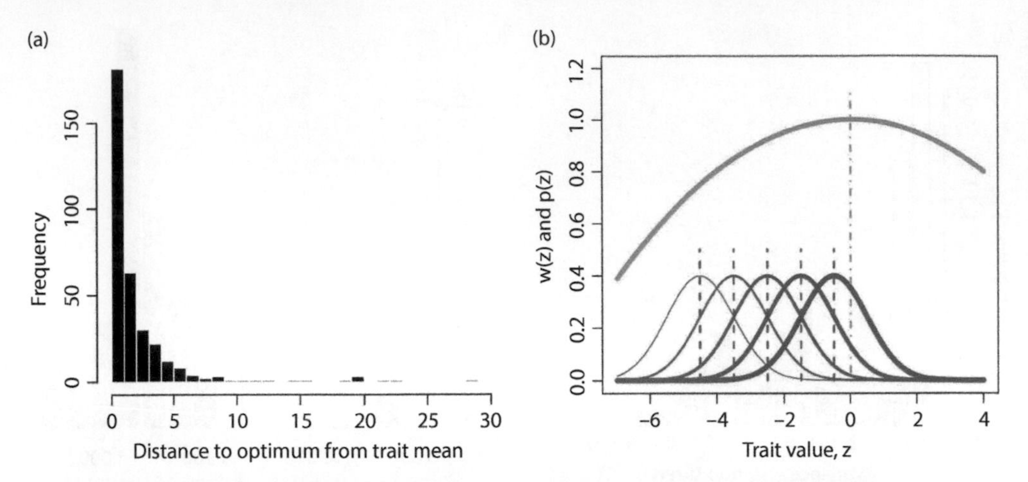

Figure 3.14 Histogram of distances of the phenotypic mean from the intermediate optimum, θ, in natural populations, with blue curves portraying the most frequent values of those distances. (a) Values were calculated using (3.12) with paired values of negative γ and absolute values of β from the Stinchcombe et al. (2008) database (n = 339, dropping two values greater than 50). Distance to the optimum is measured in units of within-population phenotypic standard deviation before selection. The median distance value is 0.906. (b) The orange curve shows the most frequent value of γ with a negative value in the histogram shown in Figure 3.12c (bin centered at $\gamma = -0.025$ with a count of 80). The optimum, θ, is shown with a vertical dash-dot line. The blue curves show $p(z)$ with trait means at distances of −0.5, −1.5, −2.5, −3.5 and −4.5 from θ, corresponding to the first five bins in the histogram with relative frequencies portrayed with line widths. The positions of trait means are shown with vertical dashed lines.

subset of the same database and reached the opposite conclusion, namely that selection was remarkably consistent from year to year. How is this possible? The Morrissey and Hadfield (2012) study benefited from the use of a meta-analysis model that explicitly incorporated sampling error. In their statistical model, the directional selection gradient in the *jth* temporal period for the *ith* trait was

$$\beta_{i,j} = \mu + u_i + m_{i,j} + e_{i,j}, \tag{3.17}$$

where μ is the average selection gradient, u_i is the deviation of the average selection gradient for the *ith* trait from the grand mean (a random effect with variance σ_μ^2), $m_{i,j}$ is the deviation of $\beta_{i,j}$ from its true value due to sampling error, and $e_{i,j}$ is the random effect contribution of within-study variation in selection (a random effect with variance σ_e^2). The variance of the two random effects (σ_e^2 and σ_μ^2) were estimated from the data, while the variance in sampling errors was estimated from standard errors (*SE*) published in the studies ($\sigma_{m_{i,j}}^2 = SE_{i,j}^2$). The first two random effect variances in turn provide an overall estimate of the repeatability of directional selection ($\sigma_\mu^2 / \left[\sigma_\mu^2 + \sigma_e^2\right]$), which was 87.6% with a 95% credible interval of 82% − 91%. In other words, only 12.4% of the variation in directional selection resided within studies (across time), while 87.6% resided among studies. It would be informative to extend this analysis and determine how much variation in β could be attributed to fluctuation in the trait optimum, as in the Chevin et al. (2015) analysis.

3.8 The Ecology of Selection

The simplicity of the function describing stabilizing selection (probably the commonest mode of selection) gives conceptual power but it should not hide the fact that many ecological factors and features of biological design are responsible for the particular location of the optimum and downward curvature at the two ends of the function (Travis 1989). If we consider the length of the tail in a hypothetical bird, for example, it seems likely that a tail that is too short will fail to provide aerodynamic lift at take off, while a tail that is too long will interfere with manoeuvrability and landing. In other words, different kinds of performance and selective agents are likely to cause downward curvature at the two ends of the ISS. The ISS is complex in the sense that it summarizes the interaction of a set of traits with an environment of numerous selective agents and contexts.

3.9 Technical Issues in Estimating and Interpreting Selection Surfaces

No one mode of selection analysis is likely to satisfy all the expectations of the investigator or reader. Selection gradients, for example, provide useful measures of the intensity of selection in the same form that appears in response to selection equations, but they may fail to describe the actual shape of the ISS. Consequently, a quadratic approximation (Section 3.2) to the ISS might be used to estimate β and γ, but a cubic spline (Section 3.3) or some other averaging function may be needed to accurately describe the shape of the ISS.

The ISS is a multivariate beast whose overall appearance cannot generally be appreciated by viewing it one trait at a time. Even in the simple case in which only two traits are under selection, each of the two univariate views of selection may be misleading. Plotting fitness as a function of a shape trait may show a flat trend line because strong selection for increasing values of shape prevails in small individuals, while strong selection for decreasing values of shape prevails in large individuals. In this case, averaging selection across one trait (shape) masks the true mode of selection acting on another trait (size). In general, we must be wary that a univariate view or coefficient accurately describes selection. Usually it will not. In the next chapter we will consider truly multivariate visions of the selection beast that are likely to give a more accurate characterization of selection.

3.10 Conclusions

- Selection on a single trait can be visualized as a curve (ISS) that relates the scores for a particular phenotypic trait to the fitness of individuals.
- This curve is mathematically related to the changes in trait mean and variance that are caused by selection.
- Key properties of the ISS can be deduced by approximating this curve with various functions.
- The selection gradients, β and γ, can be estimated with regression functions.
- The ISS is related to the adaptive landscape (AL) in which fitness and trait values are averaged over the phenotypic distribution of the trait.
- The average slope and curvature of the AL are functions of the linear and nonlinear selection gradients.

- Surveys of linear and nonlinear selection gradients suggest that trait means are generally in the neighbourhood of an optimum of the ISS and hence close to an adaptive peak.
- The AL is important because it is the function that can shape the long-term evolution of the trait mean.

Literature Cited

Bull, J. J. 1987. Evolution of phenotypic variance. Evolution 41:303–315.

Charlesworth, B. 1993. Directional selection and the evolution of sex and recombination. Genet. Res. 61:205–224.

Chevin, L.-M., M. E. Visser, and J. Tufto. 2015. Estimating the variation, autocorrelation, and environmental sensitivity of phenotypic selection. Evolution 69:2319–2332.

de Villemereuil, P., A. Charmantier, D. Arlt, P. Bize, P. Brekke, L. Brouwer, A. Cockburn, S. D. Côté, F. S. Dobson, S. R. Evans, M. Festa-Bianchet, M. Gamelon, S. Hamel, J. Hegelbach, K. Jerstad, B. Kempenaers, L. E. B. Kruuk, J. Kumpula, T. Kvalnes, A. G. McAdam, S. E. McFarlane, M. B. Morrissey, T. Pärt, J. M. Pemberton, A. Qvarnström, O. W. Røstad, J. Schroeder, J. C. Senar, B. C. Sheldon, M. van de Pol, M. E. Visser, N. T. Wheelwright, J. Tufto, and L.-M. Chevin. 2020. Fluctuating optimum and temporally variable selection on breeding date in birds and mammals. Proc. Natl. Acad. Sci. USA 117:31969–31978.

Grant, P. R., and B. R. Grant. 2002. Unpredictable evolution in a 30-year study of Darwin's Finches. Science 296:707–711.

Haldane, J. B. S. 1954. The measurement of natural selection. Proc IX Intl Cong Genet 1:480–487.

Hansen, T. F., J. Pienaar, and S. H. Orzack. 2008. A comparative method for studying adaptation to a randomly evolving environment. Evolution 62:1965–1977.

Hereford, J. 2009. A Quantitative survey of local adaptation and fitness trade-offs. Am. Nat. 173:579–588.

Hereford, J., T. F. Hansen, and D. Houle. 2004. Comparing strengths of directional selection: How strong is strong? Evolution 58:2133–2143.

Jones, A. G., S. J. Arnold, and R. Bürger. 2003. Stability of the G-matrix in a population experiencing pleiotropic mutation, stabilizing selection, and genetic drift. Evolution 57:1747–1760.

Kalisz, S. 1986. Variable selection on the timing of germination in *Collinsia verna* (Scrophulariaceae). Evolution 40:479–491.

Kingsolver, J. G., H. E. Hoekstra, J. M. Hoekstra, D. Berrigan, S. N. Vignieri, C. E. Hill, A. Hoang, P. Gibert, and P. Beerli. 2001. The strength of phenotypic selection in natural populations. Am. Nat. 157:245–261.

Lande, R. 1976. Natural selection and random genetic drift in phenotypic evolution. Evolution 30:314–334.

Lande, R. 1979. Quantitative genetic analysis of multivariate evolution, applied to brain: body size allometry. Evolution 33:402–416.

Lande, R. 1981. Models of speciation by sexual selection on polygenic traits. Proc. Natl. Acad. Sci. USA 78:3721–3725.

Lande, R., and S. J. Arnold. 1983. The measurement of selection on correlated characters. Evolution 37:1210–1226.

Lande, R., and S. Shannon. 1996. The role of genetic variation in adaptation and population persistence in a changing environment. Evolution 50:434–437.

Mitchell-Olds, T., and R. G. Shaw. 1987. Regression analysis of natural selection: Statistical inference and biological interpretation. Evolution 41:1149–1161.

Morrissey, M. B., and J. D. Hadfield. 2012. Directional selection in temporally replicated studies is remarkably consistent. Evolution 66:435–442.

Phillips, P. C., and S. J. Arnold. 1989. Visualizing multivariate selection. Evolution 43:1209–1222.

Reed, T. E., V. Grøtan, S. Jenouvrier, B.-E. Sæther, and M. E. Visser. 2013a. Population growth in a wild bird is buffered against phenological mismatch. Science 340:488–491.

Reed, T. E., S. Jenouvrier, and M. E. Visser. 2013b. Phenological mismatch strongly affects individual fitness but not population demography in a woodland passerine. J. Anim. Ecol. 82:131–144.

Schluter, D. 1988. Estimating the form of natural selection on a quantitative trait. Evolution 42:849–861.

Siepielski, A. M., J. D. DiBattista, and S. M. Carlson. 2009. It's about time: The temporal dynamics of phenotypic selection in the wild. Ecol. Lett. 12:1261–1276.

Stinchcombe, J. R., A. F. Agrawal, P. A. Hohenlohe, S. J. Arnold, and M. W. Blows. 2008. Estimating nonlinear selection gradients using quadratic regression coefficients: Double or nothing? Evolution 62:2435–2440.

Travis, J. 1989. The role of optimizing selection in natural populations. Annu. Rev. Ecol. Evol. Syst. 20:279–296.

Visser, M. E., L. J. M. Holleman, and P. Gienapp. 2006. Shifts in caterpillar biomass phenology due to climate change and its impact on the breeding biology of an insectivorous bird. Oecologia 147:164–172.

CHAPTER 4

The Selection Surface and Adaptive Landscape for Multiple Traits

Overview—Selection is often viewed and analyzed as a single-trait phenomenon or multiple traits are treated as isolated, single traits. This univariate focus is unfortunate because it misses crucial aspects of selection that are only revealed in a multivariate treatment of the problem. Crucial new aspects that are uncovered in multivariate selection analysis allow us to: (1) distinguish between direct and indirect targets of selection, (2) diagnose the correct shape of the selection surface and hence the shape of the adaptive landscape, (3) estimate the force of directional selection in units that can be used to assess evolutionary responses of trait means to selection, and (4) estimate the force of nonlinear selection in units that can are used to assess immediate effects on genetic variance and covariance, as well as long-term contributions to evolutionary patterns, such as stasis. However, all of these selling points are subject to provisos and limitations. Nevertheless, the case for pursuing multivariate rather than univariate selection analysis is compelling.

Natural selection acts simultaneously on many traits. An important consequence of this fact of nature is that we need a multivariate conceptualization of selection to deal with selection and evolution in the natural world. Textbooks in evolutionary biology have been slow to embrace these multivariate inevitabilities. Instead, depictions of selection remain locked on a vision of selection in which selection acts on single traits and is purely directional, a vision developed in the 1940–1960 world of plant and animal breeding. That world included the origin of selection indices, a multivariate tool for handling directional selection on multiple traits (Hazel 1943), but the conceptualization of multivariate stabilizing selection is a relatively recent development (Lande 1979; Lande and Arnold 1983; Phillips and Arnold 1989).

Many important features of the effects of multivariate selection on trait distributions can be appreciated by considering bivariate selection that is stabilizing. The selection surface corresponding to this form of selection is convex (Figure 4.1) with the optimum situated near the bivariate mean of the trait distribution before selection. Fitness falls off in all directions away from the optimum.

Another convention that will prove useful is the representation of the eigenvalues and eigenvectors of a bivariate stabilizing selection surface with a single elliptical contour line and a pair of orthogonal axes. In Figure 4.2 we show this type of representation for the selection surface depicted in Figure 4.1. To understand these surface portrayals, consider a familiar bivariate normal distribution of phenotypes, $p(\mathbf{z})$. The 95% confidence

Evolutionary Quantitative Genetics. Stevan J. Arnold, Oxford University Press. © Stevan Arnold (2023).
DOI: 10.1093/oso/9780192859389.003.0005

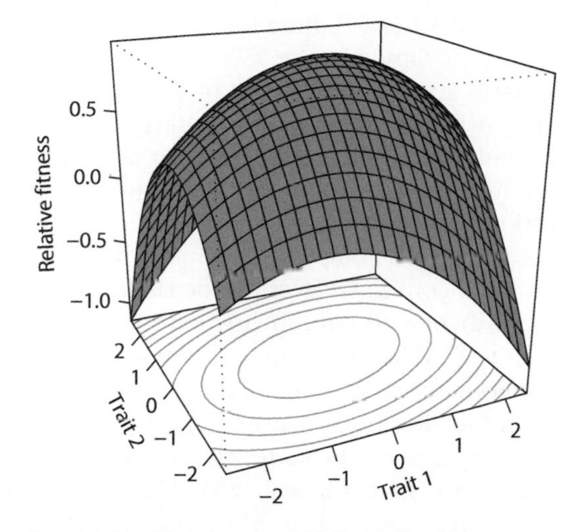

Figure 4.1 Bivariate stabilizing selection represented as a convex selection surface. Relative fitness, $w(\boldsymbol{z})$, corresponds to points on the surface as a function of the values of two traits z_1 and z_2. A contour representation of the surface is projected onto the $z_1 \times z_2$ plane with contours at increments of 0.2 in relative fitness. This surface is quadratic with $\alpha = 1$, $\beta_1 = 0.22$, $\beta_2 = -0.08$, $\gamma_{11} = -0.31$, $\gamma_{22} = -0.18$, and $\gamma_{12} = 0.07$.

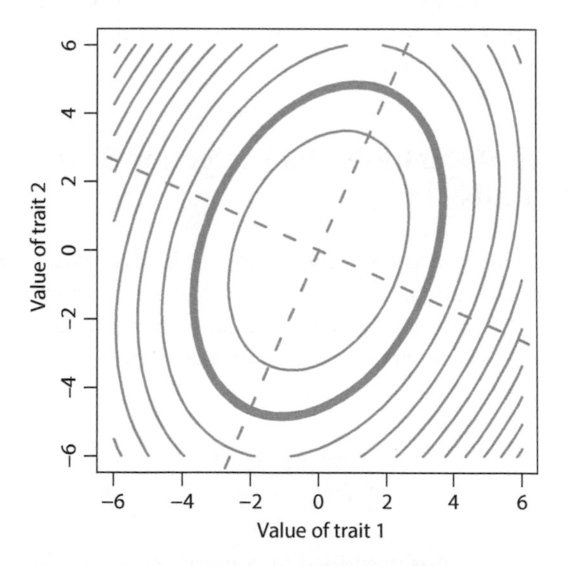

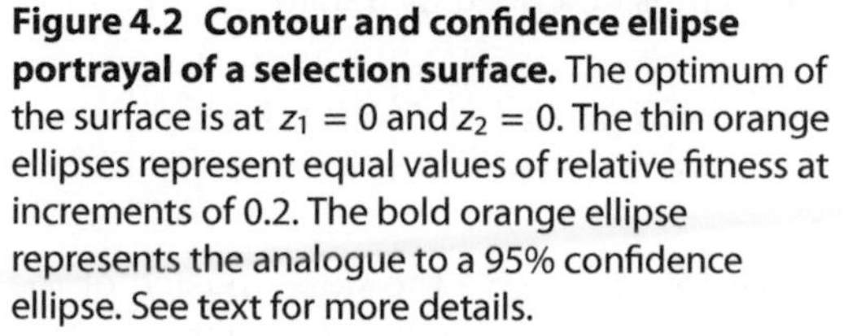
Figure 4.2 Contour and confidence ellipse portrayal of a selection surface. The optimum of the surface is at $z_1 = 0$ and $z_2 = 0$. The thin orange ellipses represent equal values of relative fitness at increments of 0.2. The bold orange ellipse represents the analogue to a 95% confidence ellipse. See text for more details.

ellipse for the bivariate mean lies $1.96\sqrt{P_{ii}}$ from the mean along each of its principal axes or eigenvectors (see Figure 2.2). Similarly, the 95% confidence ellipse for $w(\mathbf{z})$ that is a Gaussian function is situated $1.96\sqrt{\lambda_i}$ along each of the two eigenvectors, $i = 1$ and 2, from the stationary point (where the eigenvectors intersect). This confidence ellipse is shown in Figure 4.2 as a bold orange ellipse. The smaller this ellipse, the stronger is the stabilizing selection. The further the optimum of the surface is from the bivariate mean, the stronger is the directional aspect of selection.

Other examples of stabilizing selection surfaces are shown in Figure 4.3 using the confidence ellipse portrayal. Such convex surfaces can be circular (Figure 4.3a, b), elliptical with a positive (Figure 4.3c, d), or a negative inclination (Figure 4.3e, f). Many other configurations are possible as well. When the optima of these concave surfaces are located away from the bivariate mean before selection, directional selection is imposed and the bivariate mean is pulled in that direction (Figure 4.3b, c, f). In the rest of this chapter we will elaborate on these concepts of multivariate selection and make them more precise. In the process we will see that the selection depicted in Figure 4.3 is substantially stronger than we expect to see in nature.

In Figure 4.3, we specified selection, imposed it on trait distributions and observed the consequences within a generation. We now want to consider the problem of deducing the shape of the selection surface from actual data. As we pointed out in Chapter 1, those data might take one of two forms: (1) longitudinal data with a sample of individuals together with their trait measurements before selection and an estimate of each individual's fitness, or (2) cross-sectional data with one sample of individuals and their trait measurements before selection and a second sample, with measurements, taken after selection. In either case, we want to deduce the properties of the ISS from the data. The properties we have in mind are the average slopes and curvatures of the ISS, properties that we wish to mathematically connect to actual observations. We will need to begin by defining the properties of multivariate slope and curvature.

4.1 Key Properties of the Individual Selection Surface, ISS, for Multiple Traits

In general, we want to think of the relative fitness of individuals, $w(\mathbf{z})$, as a continuous function or surface of the values of two or more traits, $\mathbf{z}$. This surface might have a complicated shape, but we will be concerned only with simple quadratic shapes. If we consider a particular point on a two-trait version of this surface (Figure 4.1), the slope at that point is a vector,

$$\begin{bmatrix} \partial w(\mathbf{z})/\partial z_1 \\ \partial w(\mathbf{z})/\partial z_2 \end{bmatrix}, \tag{4.1a}$$

and the curvature at that point is described by a matrix,

$$\begin{bmatrix} \partial^2 w(\mathbf{z})/\partial z_1^2 & \partial^2 w(\mathbf{z})/\partial z_1 \partial z_2 \\ \partial^2 w(\mathbf{z})/\partial z_1 \partial z_2 & \partial^2 w(\mathbf{z})/\partial z_1^2 \end{bmatrix}. \tag{4.1b}$$

The two elements in the vector (4.1a) give the slope in each of the two trait directions, z_1 and z_2, with a positive sign indicating that fitness increases with trait values, a negative

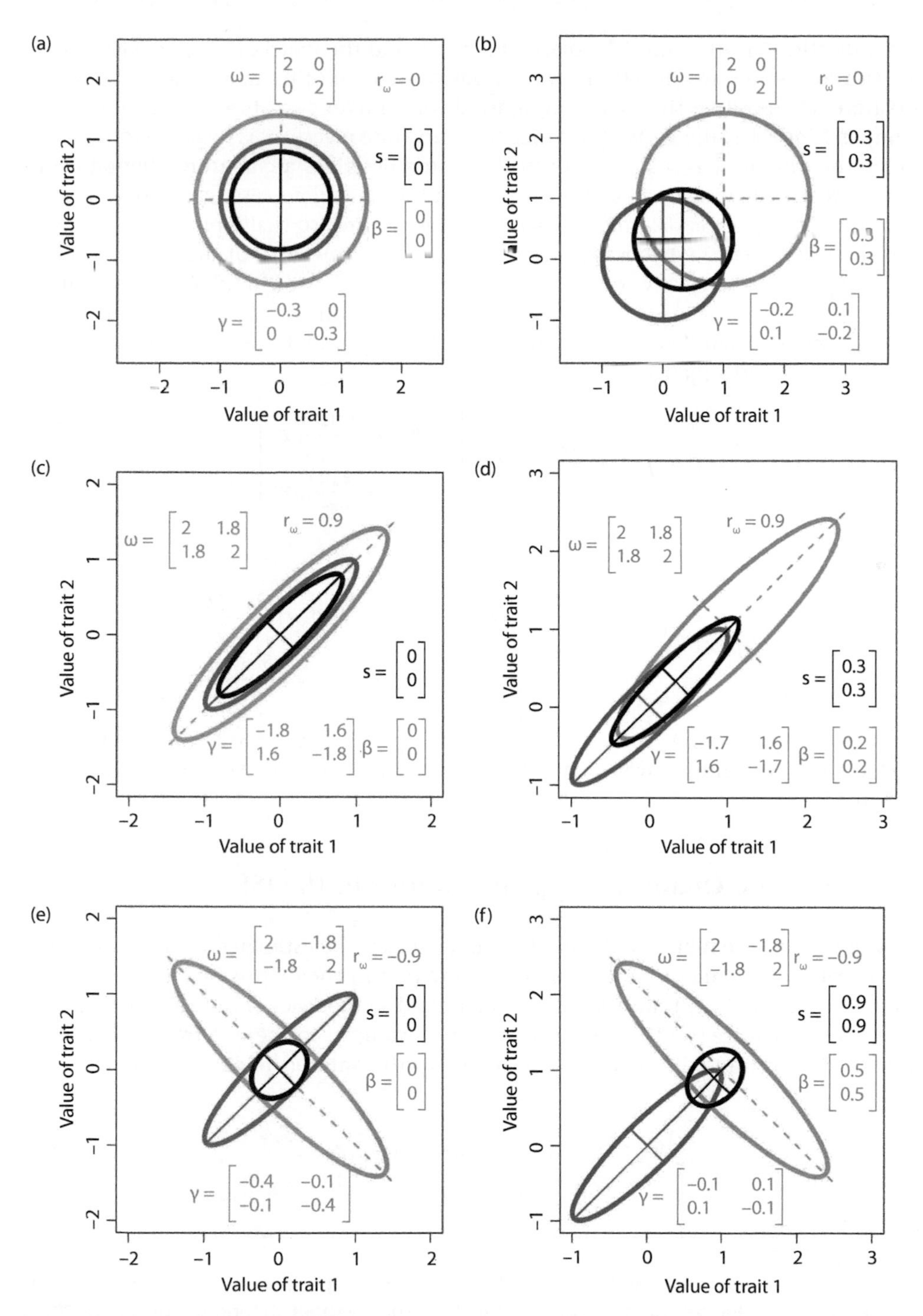

Figure 4.3 Various forms of Gaussian selection and their effects on bivariate trait distributions. Ellipses comparable to 95% confidence ellipses are shown for Gaussian ISSs (orange). Phenotypic trait distributions are bivariate normal before (blue 95% confidence ellipses) and after selection (black 95% confidence ellipses). Bivariate means, variances, and covariances, before and after selection, are given in the caption for Figure 2.3. Parameters for the ISSs are the same in Figure 2.3. The relationship of the γ-matrix to the ω-matrix will be discussed in Section 4.4. In the left-hand panel, the trait distribution experiences nonlinear selection, but no directional selection. In the right-hand

(Continued)

sign indicating the opposite. The diagonal elements in the matrix (4.1b) give the curvature in those same two directions, with negative sign denoting downward (stabilizing) curvature and positive sign denoting upward (disruptive) curvature. In our illustrated example, Figure 4.1, the signs of two diagonal curvature measures is negative (downward curvature). The off-diagonal element in the matrix describes a curvature phenomenon, correlational selection, that has no analogue in the univariate case. A positive sign for this element, $\partial^2 w(\mathbf{z}) / \partial z_1 \partial z_2$, means that the surface is tilted so that it promotes a positive correlation between the two traits, while a negative sign means that the surface is tilted so that it promotes a negative correlation. In the illustrated case $\partial^2 w(\mathbf{z}) / \partial z_1 \partial z_2$ is positive.

As in the univariate case, the average slope and curvature of the multivariate ISS are equivalent to our familiar selection gradients, $\boldsymbol{\beta}$ and γ, which are now, respectively, a column vector and a matrix:

$$\boldsymbol{\beta} = \int p(\mathbf{z}) \frac{\partial w(\mathbf{z})}{\partial \mathbf{z}} d\mathbf{z} = \begin{bmatrix} \int p(\mathbf{z}) \dfrac{\partial w(\mathbf{z})}{\partial z_1} d\mathbf{z} \\ \int p(\mathbf{z}) \dfrac{\partial w(\mathbf{z})}{\partial z_2} d\mathbf{z} \end{bmatrix} \tag{4.2a}$$

and

$$\gamma = \int p(\mathbf{z}) \frac{\partial^2 w(\mathbf{z})}{\partial \mathbf{z}^2} d\mathbf{z} = \begin{bmatrix} \int p(\mathbf{z}) \dfrac{\partial^2 w(\mathbf{z})}{\partial {z_1}^2} d\mathbf{z} & \int p(\mathbf{z}) \dfrac{\partial^2 w(\mathbf{z})}{\partial z_1 \partial z_2} d \\ \int p(\mathbf{z}) \dfrac{\partial^2 w(\mathbf{z})}{\partial z_1 \partial z_2} d\mathbf{z} & \int p(\mathbf{z}) \dfrac{\partial^2 w(\mathbf{z})}{\partial {z_2}^2} d\mathbf{z} \end{bmatrix} \tag{4.2b}$$

(Lande and Arnold 1983). These are the same averaging functions as in (3.1, 3.2), except that now the averaging is over each of two elements in $\boldsymbol{\beta}$ and over each of three distinct elements in γ.

4.2 Linear and Quadratic Approximations to the ISS

The elements in the multivariate selection gradients can be estimated by approximating the ISS with linear and quadratic surfaces. (Note that in the discussions that follow, we assume that each of the traits has been standardized so that each mean is zero, and $w(\mathbf{z})$ has been standardized so that its mean is 1). For example, in the two-trait case, we can estimate $\boldsymbol{\beta}$ by fitting a linear regression model to the data on relative fitness and trait values,

$$w(\mathbf{z}) = \alpha + \boldsymbol{\beta}^{\mathrm{T}} \mathbf{z} + \varepsilon = \alpha + \beta_1 z_1 + \beta_2 z_2 + \varepsilon, \tag{4.3}$$

Figure 4.3 *(Continued)*
panel, directional selection has been added by shifting the position of the optimum, θ. (a) Symmetrical stabilizing selection contracts the trait variance without shifting the mean. (b) When the optimum is displaced from the trait mean in both directions, the trait distribution after selection shifts towards the optimum. (c) Positive correlational selection is reflected in the positive inclination of the *ISS*-ellipse. When this ellipse is aligned with the *P*-ellipse (blue), the inclination of the *P**-ellipse (black) is not changed by selection. (d) Displacement of the optimum from the trait mean, combined with aligned γ- and *P*-matrices, shifts the trait distribution towards the optimum. (e) Nonalignment of the γ- and *P*-matrices causes a dramatic change in trait covariance (compare *P*- and *P**-ellipse). (f) Nonalignment of the γ- and *P*-matrices, combined with a displaced optimum, shifts the trait distribution towards the optimum.

where T denotes transpose (i.e., $\boldsymbol{\beta}^T$ is the row vector version of $\boldsymbol{\beta}$). The fitted surface is a plane, with $\alpha = 1$ describing its elevation, β_1 and β_2 its inclination, and ε representing the departure of individual data points from the regression surface in the vertical dimension. We estimate the elements $\boldsymbol{\gamma}$ in fitting a curvilinear regression model that corresponds to a quadratic surface,

$$w(\mathbf{z}) = \alpha + \boldsymbol{\beta}^T\mathbf{z} + \frac{1}{2}\mathbf{z}^T\boldsymbol{\gamma}\mathbf{z} + \varepsilon = \alpha + \beta_1 z_1 + \beta_2 z_2 + \frac{1}{2}\gamma_{11} z_1^2 + \frac{1}{2}\gamma_{22} z_2^2 + \gamma_{12} z_1 z_2 + \varepsilon, \tag{4.4}$$

where z_1^2, z_2^2, and $z_1 z_2$ are the squares and products of trait values, the so-called quadratic variables. (Note that these are the same as $\tilde{z}_1^2$, $\tilde{z}_2^2$, and $\tilde{z}_1\tilde{z}_2$ in Section 2.5, but here we have deleted the tildes for simplicity.) As in the univariate case, γ_{11} and γ_{22} are *nonlinear selection gradients*, describing downward curvature (*stabilizing selection*) when their signs are negative and upward curvature (*disruptive selection*) when their signs are positive. And, as in the univariate case, the factors of $\frac{1}{2}$ are present so that the stabilizing selection gradients are second derivatives (Stinchcombe et al. 2008). A new kind of coefficient is represented by γ_{12}. This *correlational selection gradient* describes orientation of the ISS in the z_1 by z_2 dimensions, with a positive sign corresponding to an upward tilt and a negative sign corresponding to a downward tilt.

Note that (4.4) provides a statistical model that can be used to estimate the elements of $\boldsymbol{\beta}$ and $\boldsymbol{\gamma}$. The requisite data are measures of trait values and relative fitness for each individual in a sample. Such data are *longitudinal* in the sense that ordinarily individuals must be literally or figuratively followed through time to estimate relative fitness. Fitting the quadratic regression model (4.4) with least squares is a standard problem in multivariate statistics than can be accomplished in many statistics packages. Extracting actual estimates of $\boldsymbol{\beta}$ and $\boldsymbol{\gamma}$, however, requires some attention to details, which are discussed below.

The rationale for treating the estimation of $\boldsymbol{\beta}$ as a regression problem (4.4) is very strong. Because $\boldsymbol{s}$ is a vector of covariances between relative fitness and traits (2.2), comparison of $\boldsymbol{P}^{-1}\boldsymbol{s}$ with the definition of partial regression coefficients (Kendall and Stuart 1973) reveals that $\boldsymbol{\beta}$ is a vector of partial regression coefficients (Lande and Arnold 1983). For example, the directional selection gradient for the first of m traits, β_1, is the partial regression of relative fitness on z_1, holding all the other traits constant. By analogy and assuming multivariate normality of the trait distribution, $\boldsymbol{\gamma}$ is a matrix of partial regression coefficients, where each element represents a partial regression of relative fitness on quadratic variables, holding all the other quadratic variables constant (2.8 and 2.9).

Since all the elements of $\boldsymbol{\beta}$ and $\boldsymbol{\gamma}$ are contained in one quadratic regression model, one is tempted to use just (4.4) to estimate all the elements in these gradients. One should resist this temptation! If the distributions of the traits are not perfectly symmetrical, as they are in the multivariate normal case, skewness in the distributions will cause the regular trait values, z_1 and z_2, to be correlated with the quadratic variables, z_1^2, z_2^2, and $z_1 z_2$. These correlations will distort the estimates of $\boldsymbol{\beta}$. A simple way to avoid this problem (the heartbreak of multivariate skewness) is to use (4.3) to estimate $\boldsymbol{\beta}$ and (4.4) to estimate $\boldsymbol{\gamma}$. Another, more complicated solution involving orthogonal polynomials is discussed by Lande and Arnold (1983).

Despite the simplicity of the quadratic approximation to the ISS (4.4), it can be used to represent a large variety of selection possibilities. In the two-trait case, for example, quadratic surfaces may be hill (Figure 4.4a), a saddle (Figure 4.4b), a nearly level ridge (Figure 4.4c), a rising ridge (Figure 4.4d), as well as other possibilities. Figure 4.4 also

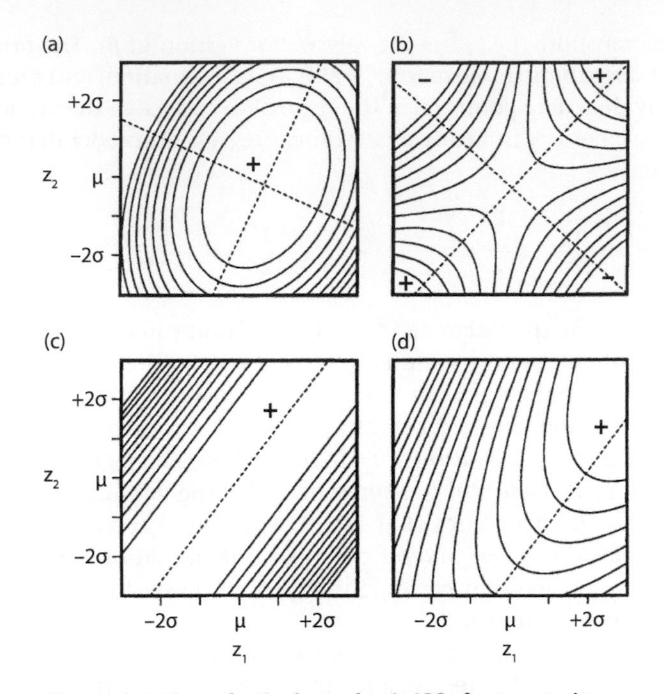

Figure 4.4 Hypothetical quadratic ISSs for two traits. Peaks are denoted with a + sign and depressions with a − sign. Dotted lines represent the canonical (principal) axes of the surface. Trait means are denoted with μ and phenotypic standard deviation with σ. Despite the differences in the appearance of these surfaces, the values of the elements in their γ-matrices are only slightly different. See Phillips and Arnold (1989) for details.

From Phillips and Arnold (1989).

illustrates the point that it is generally difficult to visualize the quadratic surface that is specified by even a simple 2 × 2 matrix of $\boldsymbol{\gamma}$ values. All four surfaces illustrated in Figure 4.4 have the same value and sign (negative) for their stabilizing selection coefficients, γ_{11} and γ_{22}. All of the correlational selection coefficients, γ_{12}, have the same sign (positive), but differ in magnitude from surface to surface. The directional selection gradient, $\boldsymbol{\beta}$, is the direction of steepest uphill slope from the bivariate mean and that direction differs in Figure 4.4d from that in the other three surfaces. If the bivariate mean is located at a stationary point on the surface (the intersection of the two eigenvectors), as in Figures 4.4a and b, both of the elements in $\boldsymbol{\beta}$ are zero, i.e., there is no directional selection.

4.3 Canonical Analysis of the Quadratic Approximation to the ISS

Figure 4.4 illustrates the important point that we cannot deduce the shape of the fitted quadratic surface from a simple inspection of the γ-matrix. The four surfaces illustrated in Figure 4.4 differ radically in shape yet their γ-matrices are extremely similar. One can determine the shape of the surface by plotting expression (2.9) or, more elegantly, by conducting a canonical analysis of the γ-matrix (Section 2.65) that yields its eigenvalues,

λ, and eigenvectors. Recalling our earlier discussion (Section 2.8), the eigenvectors are a rotation of the original trait axes which have the property that the first eigenvector is in the direction that has the greatest curvature, the second eigenvector is in an orthogonal direction with the next greatest curvature, and so on. The eigenvectors of the surfaces illustrated in Figure 4.4 are shown as dotted lines. These eigenvectors and their associated eigenvalues are calculated directly from γ-matrix (Phillips and Arnold 1989).

Using the general form of (4.4), we rotate the axes of the original trait space so that $\gamma = \mathbf{M}\Lambda\mathbf{M}^{\mathrm{T}}$, where $\mathbf{M}$ is an orthogonal matrix whose columns are the eigenvectors of γ normalized to unit length, and Λ is a matrix with the eigenvalues of γ on its diagonal and zeros on all the off-diagonals. In the two-trait case, eigenvector-matrix $\mathbf{M}$ is

$$\mathbf{M} = \begin{bmatrix} \boldsymbol{m}_1 & \boldsymbol{m}_2 \end{bmatrix} = \begin{bmatrix} m_{11} & m_{21} \\ m_{12} & m_{22} \end{bmatrix}, \tag{4.5a}$$

where $\boldsymbol{m}_1$ and $\boldsymbol{m}_2$ are, respectively, the leading and trailing eigenvectors, whose elements are given by the matrix in (4.5a). The eigenvalue-matrix Λ describes the multivariate curvature of γ in a simple and convenient way,

$$\Lambda = \mathbf{M}^{\mathrm{T}}\gamma\mathbf{M} \tag{4.5b}$$

(Phillips and Arnold 1989). In the two-trait case, the canonical form of γ is

$$\Lambda = \begin{bmatrix} \lambda_1 & 0 \\ 0 & \lambda_2 \end{bmatrix}. \tag{4.5c}$$

Using this rotation, our new trait variables are $\boldsymbol{y} = \mathbf{M}^{\mathrm{T}}\mathbf{z}$, with directional selection gradient $\boldsymbol{b} = \mathbf{M}^{\mathrm{T}}\boldsymbol{\beta}$. Our quadratic approximation of the ISS becomes

$$w(\boldsymbol{y}) = \alpha + \boldsymbol{b}^{\mathrm{T}}\boldsymbol{y} + \frac{1}{2}\boldsymbol{y}^{\mathrm{T}}\Lambda\boldsymbol{y}. \tag{4.5d}$$

Using the γ-matrix one can determine the *stationary point*, $\mathbf{z}_0$, on the fitted quadratic surface, which may be a fitness maximum, minimum, or saddle point. This stationary point is a column vector,

$$\mathbf{z}_0 = -\gamma^{-1}\boldsymbol{\beta}, \tag{4.5e}$$

and the value of relative fitness at the stationary point is

$$w_0 = \alpha + \frac{1}{2}\boldsymbol{\beta}^{\mathrm{T}}\mathbf{z}_0. \tag{4.6a}$$

An additional, useful transformation is to shift the trait origins (zero points) to the stationary point, again with the rotation of axes described above. The new trait axes are

$$\boldsymbol{y} = \mathbf{M}^{\mathrm{T}}(\mathbf{z} - \mathbf{z}_0), \tag{4.6b}$$

and our new quadratic approximation of the ISS is

$$w(\boldsymbol{y}) = w_0 + \frac{1}{2}\boldsymbol{y}^{\mathrm{T}}\Lambda\boldsymbol{y}, \tag{4.6c}$$

which conveniently lacks terms for linear selection (Phillips and Arnold 1989).

By calculating the eigenvalues of the selection surface, one can deduce its properties. If all the eigenvalues of $\boldsymbol{\gamma}$ are negative the surface is a dome (convex); if they are all positive the surface is a bowl (concave). Mixed signs indicate a saddle-shaped surface, with a stationary point at the saddle. See Phillips and Arnold (1989) for further discussion.

When the eigenvalues of $\boldsymbol{\gamma}$ are of mixed sign, it will often be useful to order them using their absolute values, $|\lambda_i|$, that is to say, by greatest curvature, whether concave or convex (Blows and Brooks 2003). The contrasting convention used by most computational algorithms is to rank eigenvalues by their raw values, so that all eigenvalues with positive sign receive higher rank than all eigenvalues with negative sign, irrespective of the magnitude of curvature. The distinction becomes important when we wish to determine the direction on the surface that has the least effect on relative fitness. We call this direction $\boldsymbol{\gamma_{min}}$. It is the direction given by the eigenvector with the minimum value of $|\lambda_i|$, which can be thought of as a *selective line of least resistance*. In all the plots shown in Figure 4.4, the selective lines of least resistance are inclined at about a 45 degree angle in trait space.

In the three-trait case, the variety of ISSs that can be approximated with quadratic surfaces is also very large (Phillips and Arnold 1989). Seven varieties are shown in Figure 4.5, where the canonical axes are denoted y_1, y_2, and y_3, but are not ordered by the size of their eigenvalues. Each of the nested surfaces in each figure represents equal values for relative fitness. Perhaps the easiest three-trait case to visualize is stabilizing selection on all three axes, shown in Figure 4.5b. Here the fitness optimum, $\boldsymbol{\theta}$, a stationary point, is situated at the intersection of the axes. Fitness falls off as concentric spheres about this

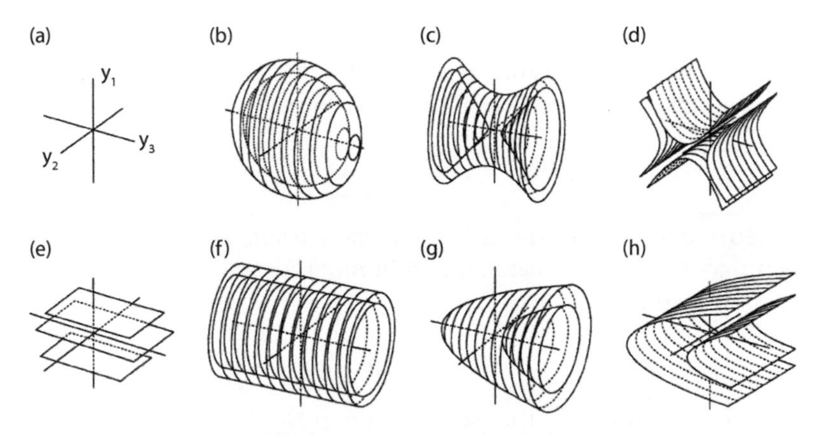

Figure 4.5 Quadratic selection on three traits. The three trait canonical axes are shown in a. Fitness is a fourth axis. All points on each surface have the same fitness. Two same-fitness surfaces are shown in each case. In b and c, maximum fitness is at the origin. In d and f maximum fitness is a line and in e it is a plane. (a) Axis labels. (b) Stabilizing selection on all three traits $(\lambda_1, \lambda_2, \lambda_3) = (-,-,-)$. (c) Stabilizing selection on y_1 and y_2, disruptive selection on y_3 $(-,-,+)$. (d) Stabilizing selection on y_1, no selection on y_2, disruptive selection on y_3 $(-,0,+)$. (e) Stabilizing selection on y_1, no selection on y_2 or y_3 $(-,0,0)$. (f) Stabilizing selection on y_1 and y_2, no selection on y_3 $(-,-,0)$. (g) Stabilizing selection on y_1 and y_2, strong directional selection on y_3 ($\theta_3 > 0$; centre of system is undefined, the surface is a rising ridge) $(-,-,0)$. (h) Stabilizing selection on y_1, no selection on y_2, strong directional selection on y_3 $(-,0,0)$.

From Phillips and Arnold (1989).

point. In Figure 4.5f, the optimum is a line corresponding to γ_3, with stabilizing selection on γ_1 and γ_2, but no selection along the γ_3-axis.

4.4 The Multivariate Adaptive Landscape

The adaptive landscape relates population mean fitness to average trait values in the multivariate case, just as it does in the univariate case. The slope and curvature of the AL, evaluated at the trait mean, $\bar{\mathbf{z}}$, are related to the selection gradients. In general and in the two-trait case,

$$\boldsymbol{\beta} = \frac{\partial \bar{W}}{\bar{W}\partial \bar{\mathbf{z}}} = \frac{\partial\, ln\bar{W}}{\partial \bar{\mathbf{z}}} = \begin{bmatrix} \frac{\partial\, ln\bar{W}}{\partial z_1} \\ \frac{\partial\, ln\bar{W}}{\partial z_2} \end{bmatrix} = \begin{bmatrix} \beta_1 \\ \beta_2 \end{bmatrix} \tag{4.7}$$

and

$$\boldsymbol{\gamma} - \boldsymbol{\beta}\boldsymbol{\beta}^{\mathrm{T}} = \frac{\partial^2 \bar{W}}{\bar{W}\partial \bar{\mathbf{z}}^2} = \frac{\partial^2 ln\bar{W}}{\partial \bar{\mathbf{z}}^2} = \begin{bmatrix} \frac{\partial^2\, ln\bar{W}}{\partial \bar{z}_1^2} & \frac{\partial^2\, ln\bar{W}}{\partial \bar{z}_1 \partial \bar{z}_2} \\ \frac{\partial^2\, ln\bar{W}}{\partial \bar{z}_1 \partial \bar{z}_2} & \frac{\partial^2\, ln\bar{W}}{\partial \bar{z}_2^2} \end{bmatrix} = \begin{bmatrix} \gamma_{11} - \beta_1^2 & \gamma_{12} - \beta_1\beta_2 \\ \gamma_{12} - \beta_1\beta_2 & \gamma_{22} - \beta_2^2 \end{bmatrix}. \tag{4.8}$$

The directional selection gradient, $\boldsymbol{\beta}$, gives the direction of steepest uphill slope (first derivatives) from the multivariate mean, $\bar{\mathbf{z}}$ (Lande 1979). The matrix $\boldsymbol{\gamma} - \boldsymbol{\beta}\boldsymbol{\beta}^{\mathrm{T}}$ describes the curvature (second derivatives) of the AL at the multivariate mean (Lande 1979; Lande and Arnold 1983). Notice that the signs of the elements $-\beta_i^2$ and $-\beta_1\beta_2$ are always negative, so we can conclude that the curvature of the AL, $\boldsymbol{\gamma} - \boldsymbol{\beta}\boldsymbol{\beta}^{\mathrm{T}}$, is always less than the curvature of the ISS, $\boldsymbol{\gamma}$.

To visualize the AL as a surface, we need to average the ISS over translations (lateral shiftings) of the trait distribution, $p(\mathbf{z})$. Building on our discussion of the univariate case (Section 3.3), this averaging is easy if $p(\mathbf{z})$ is multivariate normal and if the ISS is approximated by a *multivariate Gaussian surface*,

$$w(\mathbf{z}) = exp\left\{-\frac{1}{2}(\mathbf{z} - \boldsymbol{\theta})^{\mathrm{T}} \boldsymbol{\omega}^{-1} (\mathbf{z} - \boldsymbol{\theta})\right\} \tag{4.9}$$

(Lande 1979, 1980). The $\boldsymbol{\omega}^{-1}$ term is the inverse of the ω-matrix, which in the two-trait case takes the form

$$\boldsymbol{\omega} = \begin{bmatrix} \omega_{11} & \omega_{12} \\ \omega_{12} & \omega_{22} \end{bmatrix}, \tag{4.10}$$

with ω_{11} and ω_{22} analogous to variances. When these two coefficients are positive, the ISS may be a bell-shaped hill (or any one of the other surfaces shown in Figure 4.4). The off-diagonal term, ω_{12}, is analogous to a covariance. When it is positive, the hill is tilted upward in the z_1 by z_2 dimension. When ω_{12} is negative the hill is tilted downward. We can produce a multivariate AL by averaging this Gaussian ISS over a multivariate normal trait distribution. The resulting adaptive landscape is also multivariate Gaussian,

$$ln\bar{W} = exp\left\{-\frac{1}{2}(\bar{\mathbf{z}} - \boldsymbol{\theta})^{\mathrm{T}} \boldsymbol{\Omega}^{-1} (\bar{\mathbf{z}} - \boldsymbol{\theta})\right\}, \tag{4.11a}$$

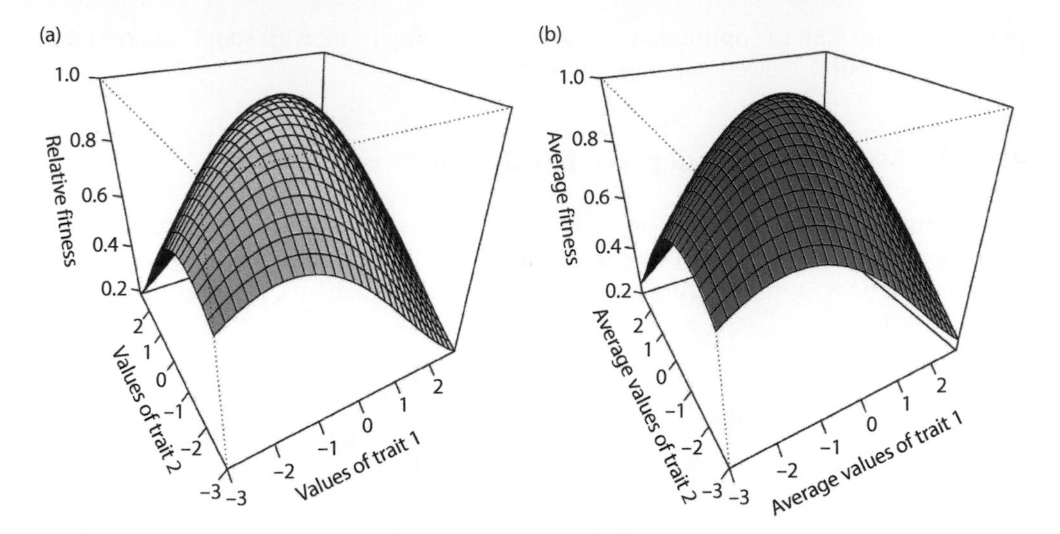

Figure 4.6 A Gaussian ISS (a) and corresponding Gaussian AL (b) are similar in configuration if stabilizing selection is weak, as in this example: Gaussian ISS with $\boldsymbol{\theta} = (0\ 0)^{\mathrm{T}}$ and $\boldsymbol{\omega} = (5\ \ 2.5,\ 2.5\ \ 5)$ and the corresponding Gaussian AL with $\boldsymbol{\Omega} = (6\ \ 2.5,\ 2.5\ \ 6)$, where $\boldsymbol{P} = (1\ \ 0,\ 0\ \ 1)$.

where $\boldsymbol{\Omega} = \boldsymbol{\omega} + \boldsymbol{P}$ is a shape and orientation matrix (Lande 1979). In the two-trait case,

$$\boldsymbol{\Omega} = \boldsymbol{\omega} + \boldsymbol{P} = \begin{bmatrix} \Omega_{11} & \Omega_{12} \\ \Omega_{12} & \Omega_{22} \end{bmatrix} = \begin{bmatrix} \omega_{11} + P_{11} & \omega_{12} + P_{12} \\ \omega_{12} + P_{12} & \omega_{22} + P_{22} \end{bmatrix}, \tag{4.11b}$$

where, as in $\boldsymbol{\omega}$, the diagonal terms are analogous to variances and the off-diagonal terms are analogous to a covariance, and both kinds of terms carry the same kinds of geometric interpretations. Because P_{11} and P_{22} are always positive, the AL will be somewhat flatter than the ISS, but its tilt may differ if ω_{12} and P_{12} differ in sign. If stabilizing selection is weak, however, so that $\boldsymbol{\omega} \gg \boldsymbol{P}$, the ISS and AL will be very similar in configuration (Figure 4.6).

A Gaussian form for the ISS is also useful in theoretical work because it enables us to solve for the multivariate trait distribution after selection. If the trait distribution, $p(\mathbf{z})$, is multivariate normal before selection and the ISS is Gaussian (4.11), the phenotypic trait distribution is multivariate normal after selection with means and variance-covariance matrix given by

$$\bar{\mathbf{z}}^* = \boldsymbol{\Omega}^{-1}(\boldsymbol{\omega}\bar{\mathbf{z}} + \boldsymbol{P}\boldsymbol{\theta}) \tag{4.12a}$$

and

$$\boldsymbol{P}^* = \boldsymbol{\Omega}^{-1}\boldsymbol{\omega}\boldsymbol{P}, \tag{4.12b}$$

(Lande 1981), and a directional selection gradient equal to

$$\boldsymbol{\beta} = \boldsymbol{\Omega}^{-1}(\boldsymbol{\theta} - \bar{\mathbf{z}}), \tag{4.12c}$$

the multivariate version of (3.10a). In these forms, we see that the AL influences $p^*(\mathbf{z})$ via its shape matrix, $\boldsymbol{\Omega} = \boldsymbol{\omega} + \boldsymbol{P}$.

To translate between γ and ω, we use the multivariate version of (3.10c),

$$\gamma = \beta\beta^{\mathrm{T}} - \Omega^{-1}. \tag{4.13a}$$

When directional and multivariate stabilizing selections are weak, we can approximate ω from the negative inverse of the γ-matrix,

$$\omega \approx -\gamma^{-1}, \tag{4.13b}$$

From this approximation of ω, we can easily approximate $\Omega = \omega + \boldsymbol{P}$. Ordering eigenvalues by their absolute values (Section 4.3), the ω- and γ-matrices have the same eigenvectors but in reverse order, so ω_{max} corresponds to γ_{min}, the direction in trait space in which the ISS has the weakest curvature. In other words, ω_{max} is the eigenvector of ω with the largest eigenvalue, giving the direction in trait space that is most forgiving with respect to selection, a *selective line of least resistance*. Similarly, Ω_{max} is the eigenvector of Ω with the largest eigenvalue, which is likely to be similar to ω_{max}, if multivariate stabilizing selection is weak (Figure 4.7).

In conclusion, we note while the ISS is useful as a local description of selection, the AL is useful because it is the surface on which $\bar{\mathbf{z}}$ evolves. Because of this distinction, we will need a vision of the AL in our later discussion of evolution. We adopt a Gaussian framework because it gives us that needed vision.

4.5 Examples of Quadratic Approximations of the ISS

We can appreciate the relationship of selection gradients to the ISS by returning to the example of how crawling speed in newborn garter snakes is affected by body and tail vertebral counts (Arnold 1988; Arnold and Bennett 1988). In Chapter 2 we showed how the gradients in this example could be estimated by comparing samples before and after selection (2.3, 2.10), but here we show estimation by linear and quadratic regression. The directional selection gradients were estimated by linear regression (4.3) and the nonlinear selection gradients were estimated by quadratic regression (4.4) (Table 4.1). Bootstrap estimates of 95% confidence limits suggest that the point estimates of β_1, β_2, γ_{11}, and γ_{22} are not different from zero. Point estimates of standard errors provided by regression analysis without bootstrapping provide a similar picture (Table 4.1). In contrast, the correlational selection gradient, γ_{12}, is positive and substantially different from zero.

To visualize the ISS that corresponds to these selection gradients, we plot the corresponding surface,

$$w(\mathbf{z}) = 1.0 + 0.031z_1 + 0.002z_2 + \frac{1}{2}(-0.011)\,z_1^2 + \frac{1}{2}(-0.006)\,z_2^2 + 0.079z_1z_2. \tag{4.14}$$

Note the selection gradient estimates in Chapter 2 were made assuming that the trait distribution before selection was multivariate normal, while the estimates in Table 4.1 did not make this assumption.

A contour plot of (4.14) reveals that the surface is saddle-shaped (Figure 4.8). The surface curves slightly upward from the bottom left-hand corner to the upper right-hand corner, and slightly downward from the upper left-hand corner to the bottom right-hand corner. A canonical analysis of the γ-matrix gives us a quantitative vision of the selection surface.

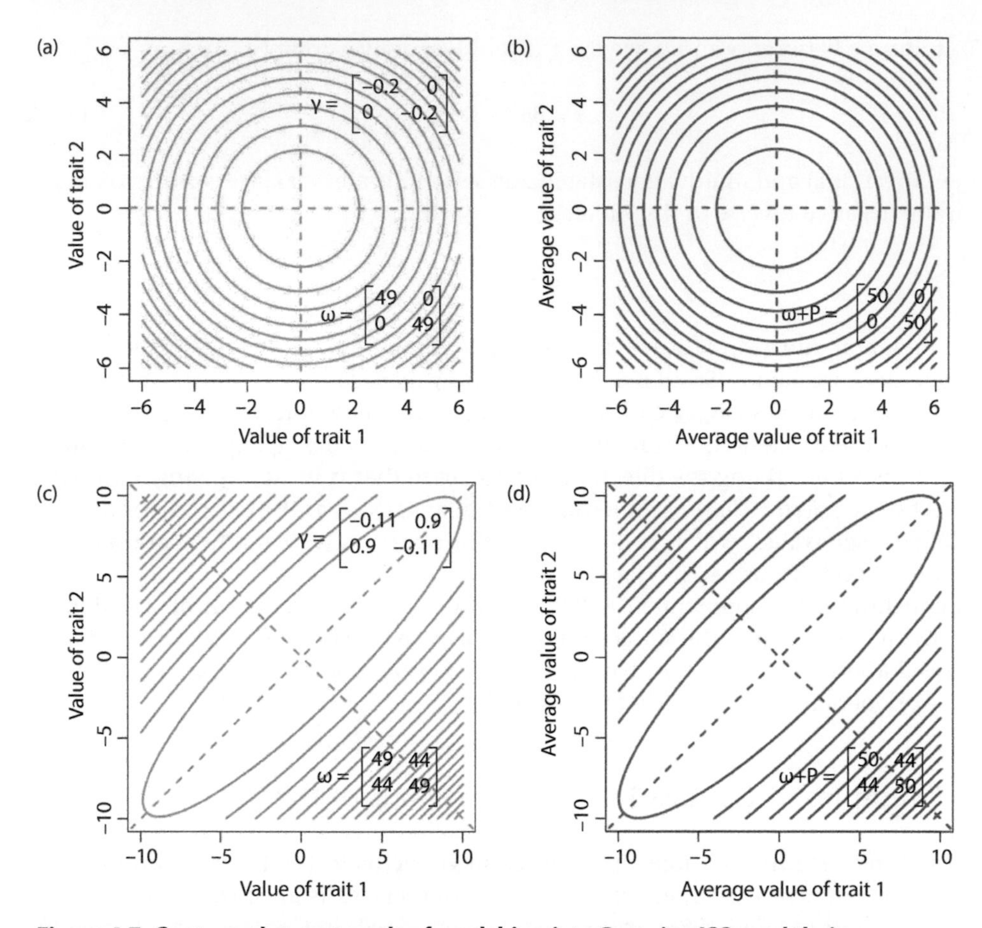

Figure 4.7 Contour plot portrayals of weak bivariate Gaussian ISSs and their corresponding adaptive landscapes. In each figure the optimum is at $z_1 = 0$ and $z_2 = 0$, and $\boldsymbol{P} = (1\ 0,\ 0\ 1)$. Eigenvectors are shown as dashed lines. Matrix representations are superimposed on each surface. (a) An ISS with equally strong stabilizing selection on each trait with no correlational selection. Contours show equal values of relative fitness, $w(\boldsymbol{z})$. (b) The AL corresponding to Figure 4.7a. Contours show equal values of average absolute fitness, $\bar{W}$. (c) An ISS with equally strong stabilizing selection on each trait and strong correlational selection, $r_\omega = 0.9$. (d) The AL corresponding to Fig. 4.7c

The new axes described by the eigenvectors of $\boldsymbol{\gamma}$ are biologically informative. Eigenvectors are linear combinations of the original axes. In the case of selection surfaces, each eigenvector consists of a series of weights (loadings) each of which describes how a particular trait contributes to the new axis. When all the weights are positive the weighting produces a combination that is the weighted average (sum) of the traits. When the weights have a mixture of signs, the weighting produces a combination that is a difference or contrast in traits. Thus, the leading eigenvector of the surface in Figure 4.8, with a slope close to one, represents the sum of the two vertebral counts, whereas the second eigenvector, with a slope close to minus one, represent the difference in the two vertebral counts. Curvature is about the same in both the sum and difference directions ($|0.071| \approx |-0.087|$). The

Table 4.1 Quadratic approximation of the crawling speed selection surface in newborn garter snakes, *T. radix*. The surface portrays crawling speed as a function of numbers of body and tail vertebrae. Data from Arnold and Bennett (1988).

Type of selection	Trait	Trait (symbol)	Coefficient	Value (±s.e.)
Directional	Body vertebrae	z_1	β_1	0.031 (0.032)
Directional	Tail vertebrae	z_2	β_2	0.002 (0.032)
Nonlinear	Body vertebrae	z_1^2	γ_{11}	−0.011 (0.040)
Nonlinear	Tail vertebrae	z_2^2	γ_{22}	−0.006 (0.046)
Correlational	Body and tail vertebrae	$z_1 z_2$	γ_{12}	0.079 (0.031)

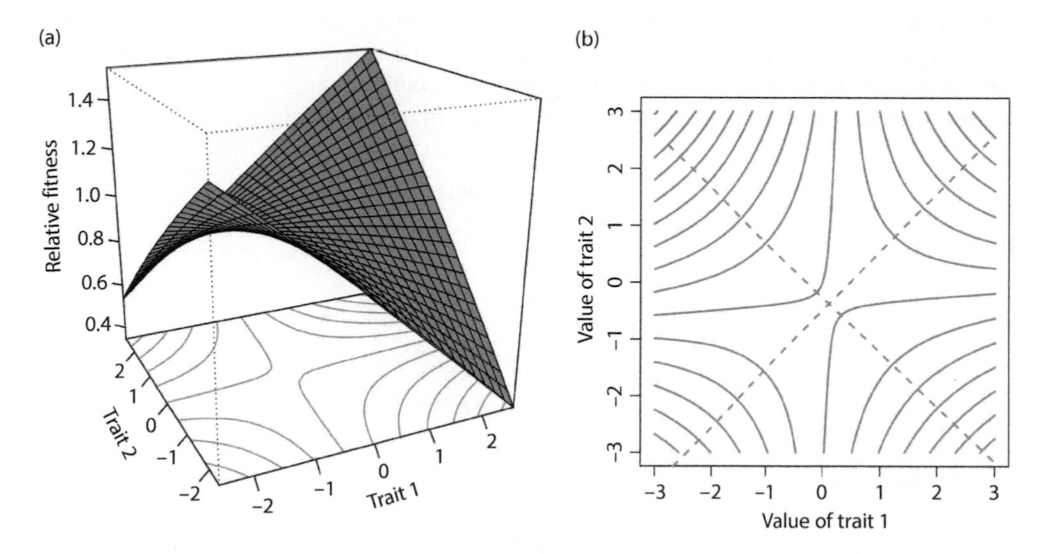

Figure 4.8 Quadratic approximation of a crawling speed performance surface in the garter snake *T. radix*. Body and tail vertebral counts are shown, respectively on the x- and y-axes. The three-dimensional surface (a) and the contour surface (b) were drawn using selection gradients estimated by linear and quadratic regression. Eigenvectors are shown as dashed lines. The stationary point, calculated using (4.5e), is at the intersection of the eigenvectrors, near the center of each bivariate plot. The slope of the leading eigenvector is approximately +1 while the slope of the trailing eigenvector is approximately −1.

values of β_1 and β_2 tell us little about the surface. Instead, they describe the slope of surface at the location of the bivariate mean, which in the present case is very close to the stationary point.

The illuminating properties of the eigenvectors and eigenvalues suggest that they are the best platform to explore the sampling properties of the ISS, however it is estimated. In particular, bootstrap sampling of the eigenvalues of γ (not shown) can tell us how often the ISS takes the form of a saddle (versus other shapes). In the present case, the ISS most often takes a saddle shape (λ_1 positive, λ_2 negative), but a bowl-shaped surface is relatively common (λ_1 positive, λ_2 positive). The distribution of slopes for the leading eigenvector is bimodal, and straddles zero with a slight preponderance of negative values. Bootstrap sampling is revealing because it tells us that our visualization of the ISS is fragile. Our vision of the ISS changes shape from boot to boot. This case study emphasizes the point that when curvature of the ISS is slight, our confidence in the shape of the ISS and in its major axis is compromised.

Finally, we use our regression estimate of the γ-matrix to illustrate approximations of $\boldsymbol{\omega}$ and $\boldsymbol{\Omega}$. Using (4.13b), we find that

$$\omega = -\gamma^{-1} = -\begin{bmatrix} -0.011 & 0.079 \\ 0.079 & -0.006 \end{bmatrix}^{-1} = \begin{bmatrix} -0.957 & -12.784 \\ -12.784 & -1.711 \end{bmatrix}. \tag{4.15a}$$

The negative inverse operation converts a matrix of second derivatives for a quadratic surface, $\boldsymbol{\gamma}$, into a matrix of variance- and covariance-like elements that describes a Gaussian surface, $\boldsymbol{\omega}$. When the variance-like diagonal elements of $\boldsymbol{\omega}$ are positive the Gaussian surface is convex, like a familiar bell curve, but when the diagonal elements are negative, the Gaussian surface is concave, like an upside-down bell curve.

The canonical form of $\boldsymbol{\gamma}$, with eigenvalues on the main diagonal, is

$$\Lambda = \begin{bmatrix} \lambda_1 & 0 \\ 0 & \lambda_2 \end{bmatrix} = \begin{bmatrix} -0.088 & 0 \\ 0 & 0.071 \end{bmatrix}, \tag{4.15b}$$

with the eigenvectors of $\boldsymbol{\gamma}$ given by the columns of

$$\mathbf{M} = \begin{bmatrix} -0.718 & -0.696 \\ 0.696 & -0.718 \end{bmatrix}. \tag{4.15c}$$

The canonical form of $\boldsymbol{\omega}$ is

$$\Lambda_\omega = \begin{bmatrix} -14.123 & 0 \\ 0 & 11.455 \end{bmatrix}, \tag{4.15d}$$

with eigenvectors given by the columns of

$$\mathbf{M}_\omega = \begin{bmatrix} 0.697 & -0.717 \\ 0.717 & 0.697 \end{bmatrix}. \tag{4.15e}$$

As expected, the eigenvectors of $\boldsymbol{\gamma}$ and $\boldsymbol{\omega}$ are the same but in reverse order. The selective line of least resistance is given by $\boldsymbol{\gamma}_{min} = \begin{bmatrix} -0.696 & -0.718 \end{bmatrix}^T$, or equivalently $\boldsymbol{\omega}_{max} = \begin{bmatrix} 0.697 & 0.717 \end{bmatrix}^T$.

Because our vertebral count distributions are approximately normal before selection, and because the trait variances have been standardized to one before selection, we can readily visualize the curvature of the Gaussian ω-surface in relation to our trait variances. The contrasting signs of the eigenvalues (4.15d) indicate a saddle-shaped surface. A first eigenvalue of –14.123 tells us that the Gaussian surface is concave in the direction of the first eigenvector with a 'variance' about 14 times larger than our trait variance. Similarly, a second eigenvalue of 11.455 tells us that the surface is convex in an orthogonal direction with a 'variance' about 11 times larger than our trait variances.

Our corresponding estimate of $\boldsymbol{\Omega}$, using the values in $\boldsymbol{P}$ from Table 4.1, is

$$\boldsymbol{\Omega} = \boldsymbol{\omega} + \boldsymbol{P} = \begin{bmatrix} -0.957 & -12.784 \\ -12.784 & -1.711 \end{bmatrix} + \begin{bmatrix} 1.000 & 0.073 \\ 0.073 & 1.000 \end{bmatrix} = \begin{bmatrix} 0.043 & -12.711 \\ -12.711 & -0.711 \end{bmatrix}, \tag{4.16a}$$

with a canonical form that indicates a saddle-shaped surface,

$$\boldsymbol{\Lambda}_{\Omega} = \begin{bmatrix} -13.051 & 0 \\ 0 & 12.383 \end{bmatrix}, \tag{4.16b}$$

and eigenvectors given by the columns of

$$\mathbf{M}_{\Omega} = \begin{bmatrix} \boldsymbol{m}_{\Omega 1} & \boldsymbol{m}_{\Omega 2} \end{bmatrix} = \begin{bmatrix} 0.697 & -0.718 \\ 0.718 & 0.697 \end{bmatrix}. \tag{4.16c}$$

As expected, the canonical forms of $\boldsymbol{\omega}$ and $\boldsymbol{\Omega}$ are very similar. The selective line of least resistance, determined from the canonical form of $\boldsymbol{\Omega}$, is $\boldsymbol{m}_{\Omega 1} = \begin{bmatrix} 0.697 & 0.718 \end{bmatrix}^{\mathrm{T}}$.

Brodie's (1992) study of how viability in garter snakes is affected by coloration and behaviour provides another illuminating example of selection surface analysis. This study was motivated by the observation that coloration pattern and antipredator behaviour coevolve in snakes, so that higher taxa fall out along a bivariate continuum (Jackson et al. 1976). At one end of the continuum are slow-moving snakes with blotched colour patterns that rely on crypsis to evade predators. At the other end are fast-moving snakes with stripes or no pattern that rely on speed and optical illusion to escape predators. Remarkably, this same coloration-behaviour continuum occurs within individual populations of the garter snake *Thamnophis ordinoides*. The population includes blotched snakes than tend to reverse directions and striped snakes that crawl in a straight line, but off-continuum snakes are also well represented (striped, reversing snakes and blotched, straight-crawling snakes). Despite this bivariate smear before selection, the connection to the interspecific continuum becomes clear when we visualize the viability selection surface. That visualization reveals that snakes that fall out along the continuum are favoured by selection, but selection acts against snakes with the maladaptive combinations of striped-reversal and blotched-straight crawl (Figure 4.9).

Turning to the selection coefficient that exerts the most influence on the portrayal of the surface, this example bears many similarities to the *T. radix* crawling speed-vertebral count example that we just discussed. As in the *T. radix* example, it is a correlational selection differential and corresponding gradient that are sufficiently large to attain statistical significance. Bivariate selection on coloration pattern and behaviour affects the correlation between those traits. In a later section we will discuss the significance of the fact that this bivariate selection has also resulted in a genetic correlation between these two traits (Brodie 1993).

4.6 Surveys of Quadratic Approximations to the ISS

In Chapter 2, we reviewed surveys of β and γ estimates, many of which were from multivariate analyses. The new dimension of parameter estimation introduced in the current

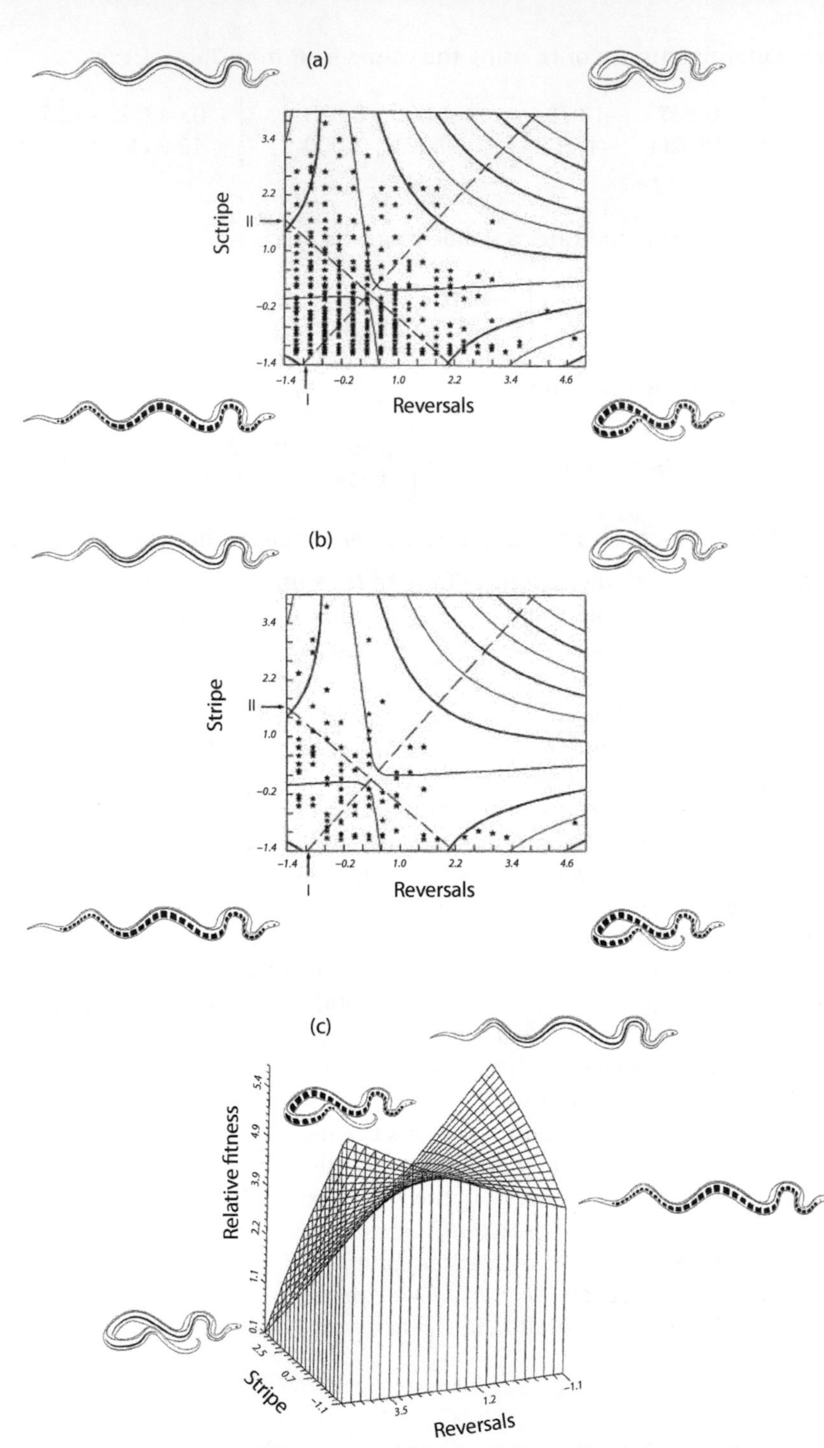

Figure 4.9 Viability selection as a function of colouration pattern and antipredator behaviour in the garter snake *T. ordinoides*. Survival in the field is shown as a function of reversals (the tendency to reverse directions during simulated predation exposure in the laboratory) and stripe (overall stripedness

(Continued)

chapter is correlational selection, γ_{12}, but unfortunately so few studies have estimated this parameter that no useful summary is available.

Surveys of multivariate estimates of mean-standardized $\boldsymbol{\beta}$ provide an important perspective on the ISS in nature. In Chapter 3 we reviewed a survey of univariate estimates of $\boldsymbol{\beta}$ by Hereford et al. (2004), here we consider their review of multivariate estimates. Hereford et al. (2004) surveyed 340 multivariate studies and computed both variance- and mean-standardized $\boldsymbol{\beta}$s and presented their results on both arithmetic and log 10 scales (Figure 4.10). The results for mean-standardized $\boldsymbol{\beta}$ on the arithmetic scale (Figure 4.10a) resemble the histogram of the larger sample assembled by Kingsolver et al. (2001). The distribution has a median value of $\boldsymbol{\beta}'$ close to zero (0.18 or a 18% change in relative fitness when the trait value is changed by one phenotypic standard deviation), a modal category even closer to zero and tail extending to larger values (>1.0). Viewing the same distribution on a log scale (Figure 4.10b), we see more clearly that only the larger values attain statistical significance. The distribution of mean-standardized gradients, β_μ, also has a mode close to zero on the arithmetic scale, with a median of 0.54 and a long tail extending to larger values. A correction for bias changes the median estimate to 0.31. In other words, the median value of β_μ represents a trait change of one-third of the value of the trait mean. On the log scale we see that those larger values form a distinct cluster, as they do in the case of $\boldsymbol{\beta}$ on the log scale, and only the largest values tend to be statistically significant.

We would like to know what multivariate form of selection is most prevalent in nature. From a theoretical standpoint we can expect quadratic approximations to take a variety of shapes (bowl, dome, saddle, rising ridge, etc.), but which of these shapes is most commonly encountered? Blows and Brooks (2003) took a major step towards answering this question by surveying studies that tackled the issue of shape in three or more dimensions (Table 4.2). The surprising answer was that in 17 of 19 cases the surface was saddle-shaped (one bowl and one dome accounted for the other two cases). This answer is surprising because a saddle is inherently unstable. For this very reason theoreticians often employ a multivariate dome to describe selection on continuous traits. Dome-shaped selection tends to move the trait mean towards a stable local optimum. Why is this multi-stabilizing form of selection so rare? One possible reason—among many—for the disconnect between theoretical expectation and empirical realization is that we nearly always measure components of fitness, not lifetime fitness in studies of selection surfaces. Arnold and Bennett (1988) and Brodie (1992), for example, assessed selection over the first few weeks or years of life, respectively, not over the entire lifespan. Another possibility is that when curvilinear selection is weak, a point estimate may produce convex curvature when the parametric curvature is actually concave, and vice versa. In other words, when curvilinear selection is routinely weak, empirically estimated curvature often flexes in the opposite direction from parametric curvature, yielding eigenvalues with a mixture of signs.

Figure 4.9 *(Continued)*
of the color pattern). Contours show relative fitness (survival) as a quadratic function of reversals and stripe. Eigenvectors are shown as dashed lines. Selection is stabilizing (concave) along the leading eigenvector (I) and disruptive (convex) along the second eigenvector (II). (a) The bivariate sample before selection (n = 646). (b) The bivariate sample after selection (n = 101). (c) Surface plot of the quadratic selection function.

From (Brodie 1992) with permission.

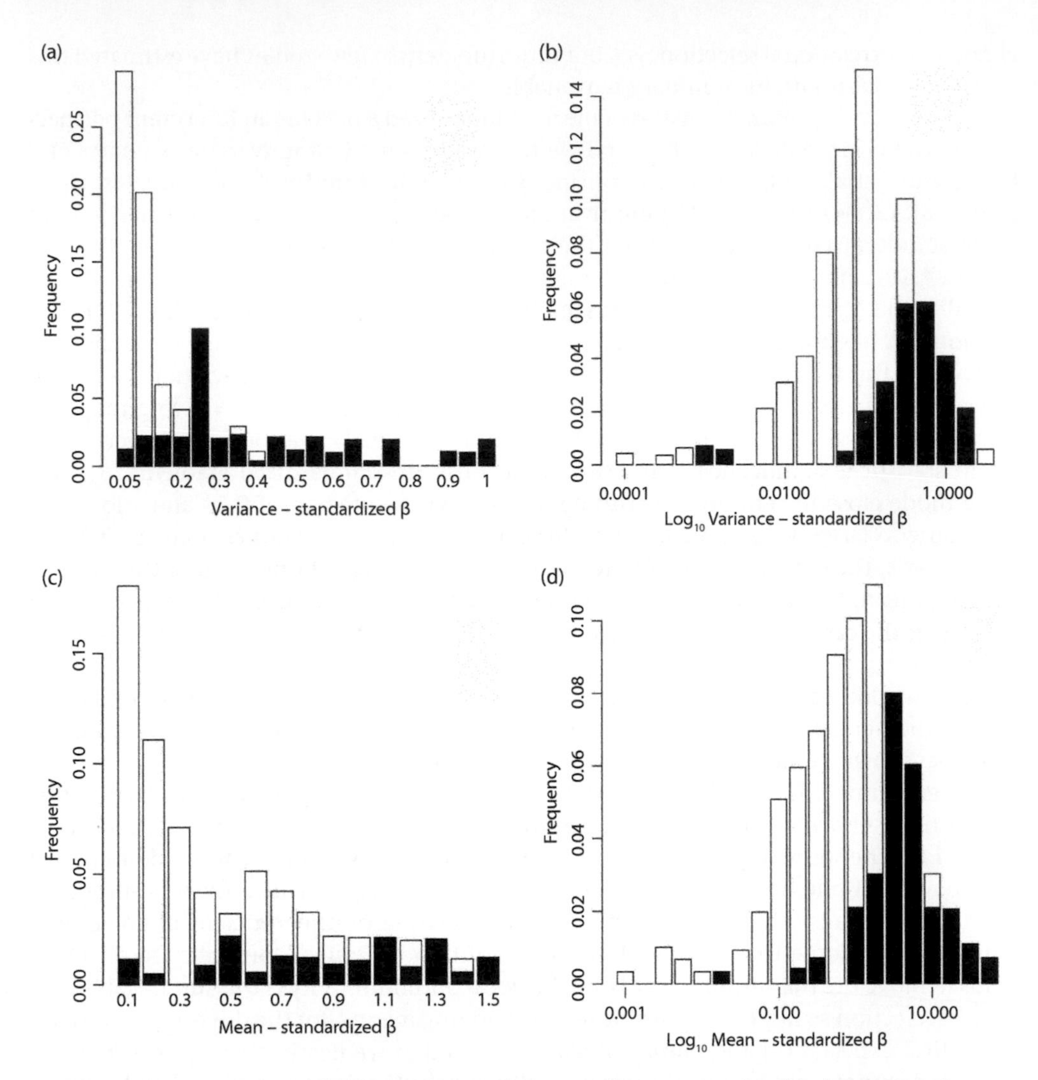

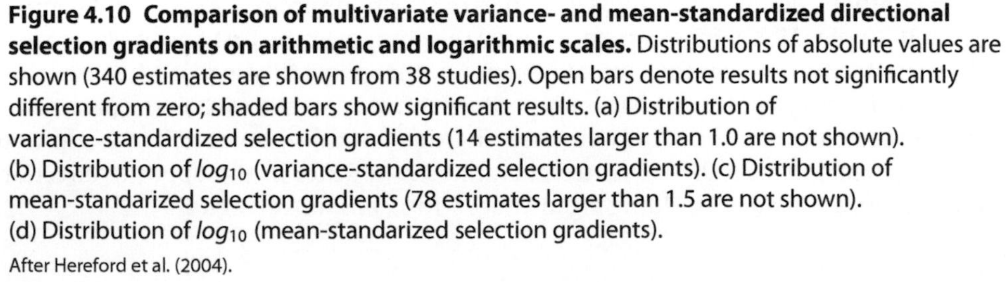

Figure 4.10 Comparison of multivariate variance- and mean-standardized directional selection gradients on arithmetic and logarithmic scales. Distributions of absolute values are shown (340 estimates are shown from 38 studies). Open bars denote results not significantly different from zero; shaded bars show significant results. (a) Distribution of variance-standardized selection gradients (14 estimates larger than 1.0 are not shown). (b) Distribution of log_{10} (variance-standardized selection gradients). (c) Distribution of mean-standarized selection gradients (78 estimates larger than 1.5 are not shown). (d) Distribution of log_{10} (mean-standarized selection gradients).

After Hereford et al. (2004).

4.7 Examples of Cubic Approximations of the ISS

The cubic spline solution to the problem of estimating the shape of the ISS (Section 3.3) can be generalized to multiple traits (Schluter and Nychka 1994). The essential idea in this approach is that certain trait dimensions tell us more about the selection surface than others. These most informative dimensions can be visualized as cross-sections of

Table 4.2 Survey of multivariate selection surfaces. Number of traits measured in each study = n. Largest eigenvalues of $\boldsymbol{\gamma} = \lambda_i$. Type of surface is based on eigenvalues (Phillips & Arnold 1989): peak (all eigenvalues negative), bowl (all eigenvalues positive), saddle (some eigenvalues negative, some positive). Type of selection (Kingsolver et al. 2001): S = survival, M = mating success, F = fertility/fecundity, O = others. From Blows and Brooks (2003) with permission.

n	Largest γ_{ii}	Largest λ_i	Type of surface	Type of selection	Reference
5	0.044	0.062	Saddle	O	Mitchell-Olds and Bergelson 1990
4	−0.457	−1.262	Saddle	F	Moore 1990
4	−0.550	0.714	Saddle	M	Moore 1990
4	−0.707	−1.093	Saddle	F	Moore 1990
4	0.498	−0.729	Saddle	M	Moore 1990
4	0.102	0.155	Saddle	M	Moore 1990
4	−0.538	−0.650	Saddle	F	Moore 1990
4	−0.122	−0.273	Saddle	S	Brodie 1992
3	−0.874	−0.875	Saddle	F	Nunez-Farfan and Dirzo 1994
3	0.370	0.552	Saddle	F	O'Connell and Johnston 1998
3	1.180	1.709	Bowl	F	O'Connell and Johnston 1998
3	0.770	1.124	Saddle	F	O'Connell and Johnston 1998
3	0.260	0.283	Saddle	F	O'Connell and Johnston 1998
3	0.200	0.305	Saddle	F	O'Connell and Johnston 1998
3	0.230	0.260	Saddle	F	O'Connell and Johnston 1998
5	0.994	0.999	Saddle	F	Simms 1990
3	−0.019	−0.021	Peak	S	Kelly 1992
4	0.016	0.027	Saddle	S	Kelly 1992
5	0.112	0.214	Saddle	F	Kelly 1992

the surface in directions of strongest selection, which can be discovered using *projection pursuit regression* (Friedman and Stuetzle 1981). Let the projection pursuit approximation of the true ISS be the function f,

$$f(\boldsymbol{z}) \approx f_1\left(\boldsymbol{a}_1^{\mathrm{T}}\boldsymbol{z}\right) + f_2\left(\boldsymbol{a}_2^{\mathrm{T}}\boldsymbol{z}\right) + \ldots + f_p\left(\boldsymbol{a}_p^{\mathrm{T}}\boldsymbol{z}\right) = f_1(x_1) + f_2(x_2) + \ldots + f_p(x_p), \qquad (4.17)$$

where the f_i are single-variable regressions (ridge functions), $\boldsymbol{z}$ is a column of vector of the original phenotypic traits (means set to zero), and the column vectors of constants $\boldsymbol{a}_i$ are the *projections* that identify the directions of each cross section. These directions are standardized so that $a_i^{\mathrm{T}} a_i = 1$. The process of finding the least squares directions (*pursuit*) and then estimating the f_i by cubic spline regression is described by Schluter and Nychka (1994). This analysis revealed a dome-shaped relationship between infant survival, infant mass, and gestation period (Figure 4.11). The subtleties of this surface would not have been apparent in a quadratic portrayal.

Bentsen et al. (2006) used the Schluter and Nychka (1994) approach to visualize female mate choice based on the calling characteristics of male crickets (*Teleogryllus commodus*). These authors began their analysis with linear and quadratic approximations of an ISS based on six call traits. To understand the resulting γ-matrix, they conducted a canonical analysis (Phillips and Arnold 1989) which identified six eigenvectors of the surface ($\boldsymbol{m}_1$, $\boldsymbol{m}_2$, ... $\boldsymbol{m}_6$). Statistically significant stabilizing selection acted along three of these

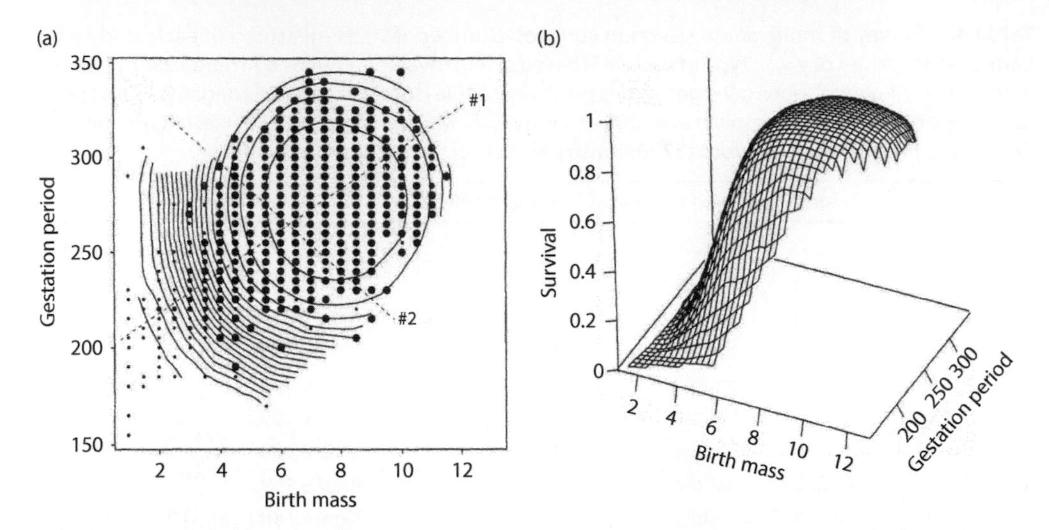

Figure 4.11 Survival probability in human infants as a function of birth mass (pounds) and gestation period (days). (a) Fitness contours (increments of 0.5) describe a dome rising steeply from the lower left to a broad plateau. The size of the symbols (filled circles) is proportional to probability of survival. The directions (a_1, a_2) used to approximate the surface are labelled #1 and #2. (b) A three-dimensional portrayal of the surface.

From Schluter and Nychka (1994) with permission.

eigenvectors ($\boldsymbol{m}_4$, $\boldsymbol{m}_5$, and $\boldsymbol{m}_6$). Three pair-wise visualizations of surface in these three dimensions are shown in Figure 4.12. This example demonstrates the point that parameter estimation using a quadratic approximation and visualization of the selection surface using other approaches provide complementary insights. They are not mutually exclusive.

4.8 Technical Issues in Estimating and Interpreting the ISS

Over the last few decades, the trend in selection studies has been to include portrayals of the ISS, as well as tables of $\boldsymbol{\beta}$ and $\boldsymbol{\gamma}$ estimates and their standard errors. In part this trend reflects increasing appreciation of the fact the ISS cannot be visualized from a table of $\boldsymbol{\beta}$ and $\boldsymbol{\gamma}$ estimates (Phillips and Arnold 1989). The other growing realization is that the ISS is an object of interest in its own right (Blows 2006). Accurate portrayals of the ISS can summarize complex modes of selection, provide a vision of the AL, and may point to important avenues of evolution during adaptive radiations (Chapter 18). Nevertheless, a disturbingly large fraction of selection studies fails to report coefficients of nonlinear selection (often estimates of γ_{ij} are missing, but sometimes estimates of γ_{ii} as well).

Using multivariate regression to capture a vision of the ISS immediately places us in a quandary of how many traits to include. The quandary arises because the vagaries of sampling force us to consider the overall shape of the ISS rather than just the values and significance of individual elements in $\boldsymbol{\beta}$ and $\boldsymbol{\gamma}$, because the actual targets of selection (the real subset of traits under the strongest directional or nonlinear selection) are usually unknown to us. To capture the real targets, we could include more traits in the selection analysis, but—because $\boldsymbol{\gamma}$ is a matrix—the number of coefficients to be estimated goes up as the square of the number of traits. Increasing the number of traits may increase our

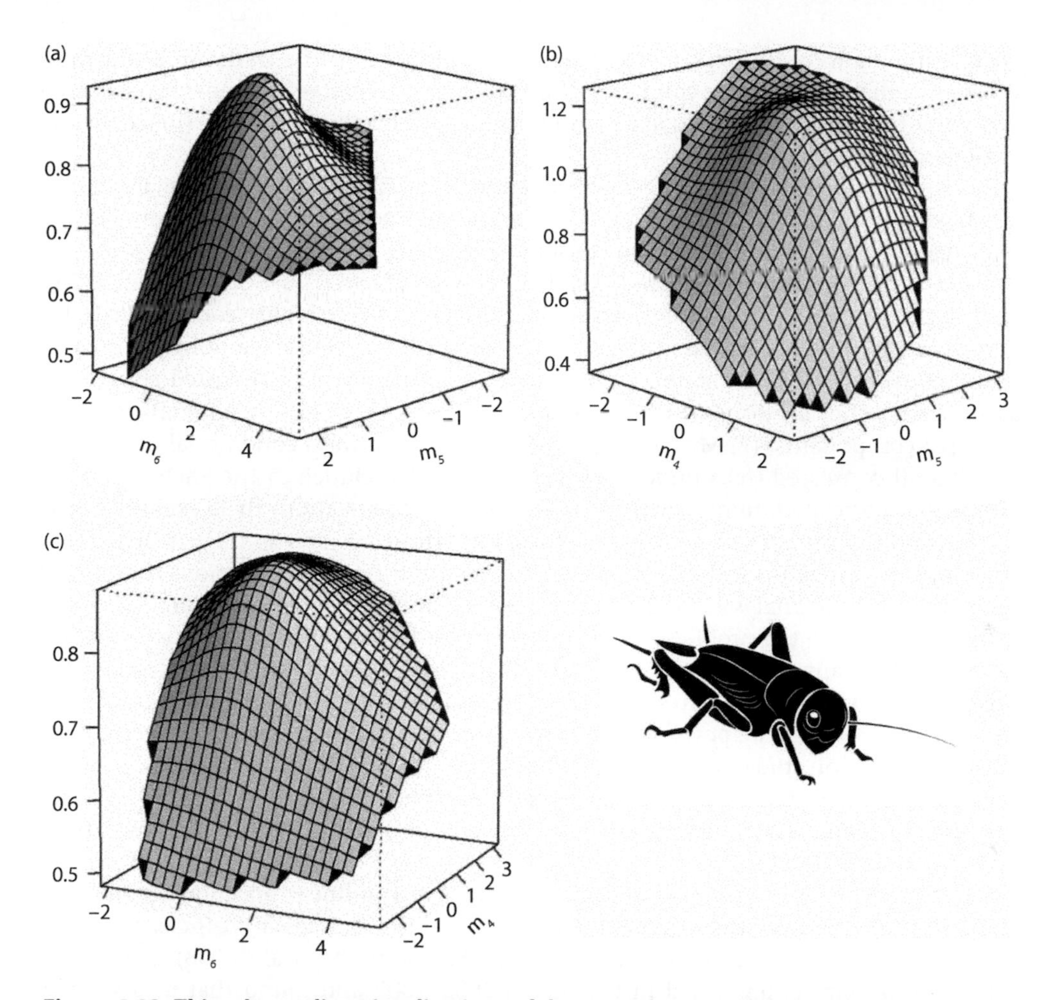

Figure 4.12 Thin-plate spline visualizations of the mate choice surface in crickets (*Teleogryllus commodus*), using call eigenvectors that demonstrated significant stabilizing selection as axes. (a) $\boldsymbol{m}_6$ and $\boldsymbol{m}_5$, (b) $\boldsymbol{m}_4$ and $\boldsymbol{m}_5$, and (c) $\boldsymbol{m}_6$ and $\boldsymbol{m}_4$. The vertical axes are proportional to numbers of females attracted using particular combinations of call parameters.
From Bentsen et al. (2006) with permission.

chances of including real targets, but as we attempt to estimate more coefficients our power to detect selection goes down (i.e., more of our estimates are nonsignificant).

Unfortunately, this multivariate quandary has no universal solution, although it can often be mitigated by two considerations. The first is that ecological or biomechanical arguments may enable us to narrow the field of possible traits. For example, in the case of Galapágos finches, a host of ecological observations and biomechanical analogies bolster a priori the choice of beak depth and length as probable targets of selection (Grant 1986; Grant and Grant 2011). On this basis, the investigators restricted their selection analysis to just two traits. The second consideration is that selection can sometimes be profitably viewed as acting on combinations of traits. Taking this approach, one can analyze selection on principal components or other linear combinations of traits and thereby reduce the trait number to one, two, or three. This tactic works best if the linear combinations are readily interpretable. For example, if the traits are linear measurements, the first principal component can often be interpreted as a measure of overall size, while components with small eigenvalues represent measures of shape (Jolicoeur and Mosimann 1960).

A variety of other issues are in general endemic to multivariate statistical analyses, not just to selection analysis. Although these issues are not peculiar to selection analysis by multiple regression, they should be considered in using this approach or in interpreting results (Lande and Arnold 1983; Mitchell-Olds and Shaw 1987; Brodie et al. 1995). (1) The problem of *multicollinearity* arises if a subset of traits is highly correlated. In this circumstance, the analysis can fail to fully account for correlations among traits, with resulting distortions of estimated coefficients. As in the quandary discussed above, two useful approaches are to reduce the number of traits (using outside criteria) or to use principal components. (2) Multiple regression results are always conditional on the *proviso* that all correlated traits under selection have been included in the analysis. To the extent that correlated, unmeasured traits are exposed to selection, their exclusion may have distorted estimates of selection coefficients. While this *proviso* may at first sound fatal, most investigators feel that it is better to account for some correlations than to fall back on univariate analyses. (3) Stepwise regression and related strategies are useful statistical solutions to the problem of sorting through a large field of possible explanatory variables. These approaches are problematic in selection analyses, however, because of the biases they cause in the estimates of $\boldsymbol{\beta}$ and $\boldsymbol{\gamma}$. (4) Variance in relative fitness (the *opportunity for selection*) can limit the power to detect selection and should be taken into account in designing selection studies (Hersch and Phillips 2004). (5) A dichotomous fitness measure (e.g., survivors and nonsurvivors) complicates the estimation of standard errors for β and γ coefficients. Janzen and Stern (1998) and Blanckenhorn et al. (1999) describe solutions that employ *logistic regression.*

A technical problem plagues published estimates of nonlinear selection coefficients. Recall that in the quadratic regression equation (4.3) the nonlinear coefficient for z_i^2 is $\frac{1}{2}\gamma_{ii}$ so that γ_{ii} is a second derivative of the ISS. Stinchcombe et al. (2008) queried the authors of 32 papers published in the period 2002–2007 and found that in a sizeable fraction of those papers (78%), the authors failed to take the factor of $\frac{1}{2}$ into account. The consequence is that the published estimates labelled γ_{ii} are actually $\frac{1}{2}\gamma_{ii}$. Because the diagonal elements of γ-matrix are under-estimated by a factor of 2, while the off-diagonal elements are not, portrayals of the ISS can be affected. Unless authors specifically state that the factor of $\frac{1}{2}$ was taken into account or reproduce (4.3), a cautious reader should assume that published values of γ_{ii} are likely to be in error.

Under certain circumstances one must assume that nonlinear selection is weak ($\boldsymbol{\omega} \gg \boldsymbol{P}$ or $\boldsymbol{P} \gg |\gamma|$). One such circumstance arises when one wishes to claim that a quadratic or Gaussian approximation provides a good representation of the ISS. If the actual ISS is asymmetric or if stabilizing or disruptive selection is strong, neither of these approximations is likely to be satisfactory.

4.9 Why the Multivariate View of Selection Is Important

Multivariate selection has properties that are not revealed by a collection of univariate analyses. Correlational selection is the fundamental property that demands that selection on multiple traits be treated with an actual multivariate analysis. At the same time, the large collection of selection coefficients that emerges in multivariate analysis can impede comprehension of the selection surface. Canonical analysis and plotting the surface can cut through the complexity of the multivariate approach.

4.10 Conclusions

- A bivariate stabilizing selection surface can be portrayed as a three-dimensional hill or as a contour plot of equal-fitness contour lines. The eigenvectors of the surface are similar to principal components and have useful properties.
- A bivariate trait distribution before selection can be portrayed as an ellipse with principal component axes. Selection can change the location, size, shape, and angle of the ellipse.
- Visualization of the ISS as a surface that relates the fitness of individuals to their trait values provides a summary and integration of many aspects of selection.
- The ISS is mathematically related to changes in means, variances, and covariances that are induced by selection.
- The slope and curvature of the individual selection surface (ISS) are key properties. The selection gradients (β and γ) can be expressed in terms of the slope and curvature of the ISS.
- Key properties of the ISS can be deduced by approximating this surface with simple functions.
- The selection gradients (β and γ) can be estimated using linear and quadratic approximations of the ISS.
- The shape of the ISS can be visualized using a quadratic function but more complicated functions (e.g., cubic) are useful when the ISS is complicated.
- The ISS is related to the adaptive landscape (AL), a surface in which individual fitness and trait values are averaged over the phenotypic distribution of the trait.
- The average slope and curvature of the AL are functions of the linear and nonlinear selection gradients.
- The AL is important because this surface helps us predict the long-term evolution of the trait mean.

Literature Cited

Arnold, S. J. 1988. Quantitative genetics and selection in natural populations: microevolution of vertebral numbers in the garter snake *Thamnophis elegans*. pp. 619–636 *in* B. S. Weir, E. J. Eisen, M. M. Goodman, and G. Namkoong, eds. Proceedings of the Second International Conference on Quantitative Genetics. Sinauer Associates, Inc., Sunderland.

Arnold, S. J., and A. F. Bennett. 1988. Behavioural variation in natural populations.V. Morphological correlates of locomotion in the garter snake (*Thamnophis radix*). Biol. J. Linn. Soc. 34:175–190.

Bentsen, C. L., J. Hunt, M. D. Jennions, and R. Brooks. 2006. Complex multivariate sexual selection on male acoustic signaling in a wild population of *Teleogryllus commodus*. Am. Nat. 167:E102–E116.

Blanckenhorn, W. U., M. Reuter, P. I. Ward, and A. D. Barbour. 1999. Correcting for sampling bias in quantitative measures of selection when fitness is discrete. Evolution 53:286.

Blows, M. W. 2006. A tale of two matrices: Multivariate approaches in evolutionary biology. J. Evol. Biol. 20:1–8.

Blows, M. W., and R. Brooks. 2003. Measuring nonlinear selection. Am. Nat. 162:815–820.

Brodie, E. D. III. 1992. Correlational selection for colour pattern and antipredator behaviour in the garter snake *Thamnophis ordinoides*. Evolution 46:1284–1298.

Brodie, E. D. III. 1993. Homogeneity of the genetic variance-covariance matrix for antipredator traits in two natural populations of the garter snake *Thamnophis ordinoides*. Evolution 47:844–854.

Brodie, E. D. III., A. J. Moore, and F. J. Janzen. 1995. Visualizing and quantifying natural selection. Trends Ecol. Evol. 10:313–318.

Friedman, J. A., and W. Stuetzle. 1981. Projection pursuit regression. J. Am. Stat. Assoc. 76: 817–823.

Grant, P. R. 1986. Ecology and Evolution of Darwin's Finches. Princeton University Press, Princeton.

Grant, P. R., and B. R. Grant. 2011. Causes of lifetime fitness of Darwin's finches in a fluctuating environment. Proc. Natl. Acad. Sci. USA 108:674–679.

Hazel, L. N. 1943. The genetic basis for constructing selection indices. Genetics 28:476–490.

Hereford, J., T. F. Hansen, and D. Houle. 2004. Comparing strengths of directional selection: How strong is strong? Evolution 58:2133–2143.

Hersch, E. I., and P. C. Phillips. 2004. Power and potential bias in field studies of natural selection. Evolution 58:479–485.

Jackson, J. F., W. I. Ingram, and H. W. Campbell. 1976. The dorsal pigmentation pattern of snakes as an antipredator strategy: A multivariate approach. Am. Nat. 110:1029–1053.

Janzen, F. J., and H. S. Stern. 1998. Logistic Regression for Empirical Studies of Multivariate Selection. Evolution 52:1564–1571.

Jolicoeur, P., and Mosimann. 1960. Size and shape variation in the painted turtle. A principal component analysis. Growth 24:339–354.

Kelly, C. A. 1992. Spatial and temporal variation in selection on correlated life history traits and plant size in *Chamaecrista fasciculata*. Evolution 46:1658–1673.

Kendall, M. G., and A. Stuart. 1973. The Advanced Theory of Statistics. Vol. 2. Inference and Relationship, 3rd ed. MacMillan, New York.

Kingsolver, J. G., H. E. Hoekstra, J. M. Hoekstra, D. Berrigan, S. N. Vignieri, C. E. Hill, A. Hoang, P. Gibert, and P. Beerli. 2001. The strength of phenotypic selection in natural populations. Am. Nat. 157:245–261.

Lande, R. 1979. Quantitative genetic analysis of multivariate evolution, applied to brain: bodysize allometry. Evolution 33:402–416.

Lande, R. 1980. Sexual dimorphism, sexual selection, and adaptation in polygenic characters. Evolution 34:292–305.

Lande, R. 1981. Models of speciation by sexual selection on polygenic traits. Proc. Natl. Acad. Sci. USA 78:3721–3725.

Lande, R., and S. J. Arnold. 1983. The measurement of selection on correlated characters. Evolution 37:1210–1226.

Mitchell-Olds, T., and J. Bergelson. 1990. Statistical genetics of an annual plant, *Impatiens capensis*. II. Natural selection. Genetics 124:417–421.

Mitchell-Olds, T., and R. G. Shaw. 1987. Regression analysis of natural selection: Statistical inference and biological interpretation. Evolution 41:1149–1161.

Moore, A. J. 1990. The evolution of sexual dimorphism by sexual selection: The separate effects of intrasexual selection and intersexual selection. Evolution 44:315–331.

Nunez-Farfan, J., and R. Dirzo. 1994. Evolutionary ecology of *Datura stramonium* L. in central Mexico: Natural selection for resistance to herbivorous insects. Evolution 48:423–436.

O'Connell, L. M., and M. O. Johnston. 1998. Male and female pollination success in a deceptive orchid, a selection study. Ecology 79:1246–1260.

Phillips, P. C., and S. J. Arnold. 1989. Visualizing multivariate selection. Evolution 43:1209–1222.

Schluter, D., and D. Nychka. 1994. Exploring fitness surfaces. Am. Nat. 143:597–616.

Simms, E. L. 1990. Examining selection on the multivariate phenotype: plant resistance to herbivores. Evolution 44:1177–1188.

Stinchcombe, J. R., A. F. Agrawal, P. A. Hohenlohe, S. J. Arnold, and M. W. Blows. 2008. Estimating nonlinear selection gradients using quadratic regression coefficients: double or nothing? Evolution 62:2435–2440.

CHAPTER 5

Inheritance of a Single Trait

Overview—When a trait is affected by many genes, a statistical concept of inheritance can be used to predict how the effects of selection will be transmitted to the next generation. This statistical concept (additive genetic variance) can also be used to model the effects of selection and drift over evolutionary time. Analyses of the best-studied traits show that dozens, if not hundreds of loci contribute to genetic variance in natural populations. The standing crop of genetic variation in a population can be visualized as a balance between the opposing effects of selection and mutation. Genetic variation is nibbled away each generation by selection, but variation is restored by mutation and other processes (recombination and migration). Studies of the mutation process reveal a measurable per-generation input to genetic variation in many traits. Because selection is often weak, mutation in conjunction with recombination and migration can compensate for selective losses. Consequently, the standing crop of genetic variation is substantial in most populations for many kinds of traits. The standardized genetic variance (heritability) of traits is typically in the range 0.2–0.6 (on a scale that ranges from 0–1).

With the addition of inheritance, the life cycle is as follows. (1) Diploid zygotes inherit genetic effects from their parents in the preceding generation, which together with new mutations affect the expression of zygotic phenotypic traits. (2) The zygotic cohort experience selection based on their phenotypic traits. (3) The cohort produces a new cohort of zygotes by sexual reproduction.

5.1 Phenotypic Resemblance Between Parents and Offspring Reveals Heritable Variation

Before the rediscovery of Mendelism in 1905, Francis Galton (1889) quantified inheritance with parent-offspring plots. Galton focused on a variety of human traits, but especially on easily measured attributes such as stature. Compiling records on 205 families, Galton plotted the average height of both parents (midparent) as a function of the average height of offspring in a family (midoffspring). His data are shown in Figure 5.1, which follows the contemporary convention of plotting offspring as a function of parents. Noting that the least-squares fit of such data always has a lesser slope than perfect inheritance, Galton referred to the fitted line as a *regression*. Focusing on the regression line, we see that the offspring of tall parents regress towards the average, and so do the offspring of short parents. The slope of Galton's regression line was later known as heritability. As we shall see, heritability is of key importance in summarizing the inheritance of a trait affected by many genes and in transmitting the effects of selection from one generation to the next.

Evolutionary Quantitative Genetics. Stevan J. Arnold, Oxford University Press. © Stevan Arnold (2023).
DOI: 10.1093/oso/9780192859389.003.0006

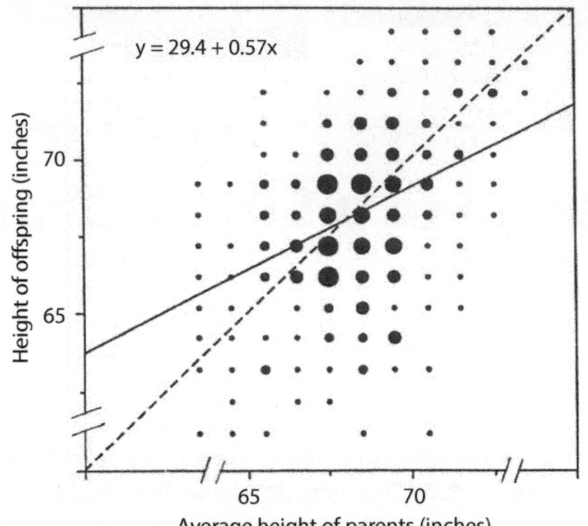

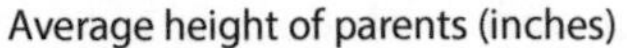

Figure 5.1 Galton's data shown as a human offspring versus parent plot of height in a British population. Each point represents a combination of average offspring (n = 928) and average parental values (n = 205 sets). The size of each point is proportional to the sample size of that trait combination. The dashed line shows perfect inheritance. The solid line is a least-squares regression.

From Arnold (1994).

5.2 A Model for Phenotypic Value

The statistical underpinnings for Galton's observation of parent-offspring resemblance were independently discovered by Weinberg (1908), Fisher (1918), and Wright (1921). Working in the period following the rediscovery of Mendelism and independently because of WWI, Weinberg, Fisher, and Wright took the same perspective on the kinds of continuously distributed traits studied and plotted by Galton. These investigators showed that Galton's regression was a natural consequence of multiple-factor (*polygenic*) inheritance. In this section we will sketch the inheritance models they used to reach that conclusion.

We now turn to a particular version of that polygenic model for inheritance with the aim of understanding the Mendelian basis for Galton's regression slope, which is called h^2 (Wright 1921). This section follows Lande (1976) parameterization of the additive model of inheritance. In particular, for the moment we will ignore dominance (within-locus interactions) and epistasis (between-locus interactions). Putting those genetic complications aside, the model takes a simple form. Let us assume for simplicity that the effects of alleles at those many polygenic loci are added together to constitute an individual's *additive genetic* or *breeding value*, x. We shall consider the genetic values of all individuals in a population to be a random variable with standard statistical properties (e.g., x has a distribution that can be characterized by a mean, variance, etc). We will also assume that the trait in question is affected by a host of environmental factors. We will call the sum of all those environmental factors, e, the individual's *environmental value*. Turning to the population, we assume as before that e is a random variable that can be characterized in the standard way. Finally, let us assume that an individual's phenotypic value, z, is simply the sum of its genetic and environmental values,

$$z = x + e. \tag{5.1a}$$

Although x and e are invisible to us, we can measure an individual's *phenotypic value*, z. This elementary fact, the observability of z, sets the stage for all that follows. Because z is a random variable that is the sum of two other random variables, x and e, we can infer the statistical properties of x and e from those of z. To proceed we shall make one more key assumption, that x and e are independent (uncorrelated) variables. From these assumptions we can show that the means and variances of our three random variables are related in simple, additive ways,

$$\bar{z} = \bar{x} + \bar{e}, \tag{5.1b}$$

$$P = G + E. \tag{5.2}$$

The three variances are *phenotypic* (P), *additive genetic* (G), and *environmental* (E). So far we have made no assumptions about the distributions of z, x, and e. It is natural to assume, however, that x and e are normally distributed, because we know from the *Central Limit Theorem* that the sum of many, identically-distributed variables is normally distributed, no matter what the distribution of those many random variables. Consequently because both x and e are likely to be the sum of many random variables, we expect their distributions to be normal and, because of (5.1a), z will be normally distributed as well (Figure 5.2).

We are now in a position to understand Galton's regression slope. For simplicity, consider first the covariance between the phenotypic values of one parent and midoffspring.

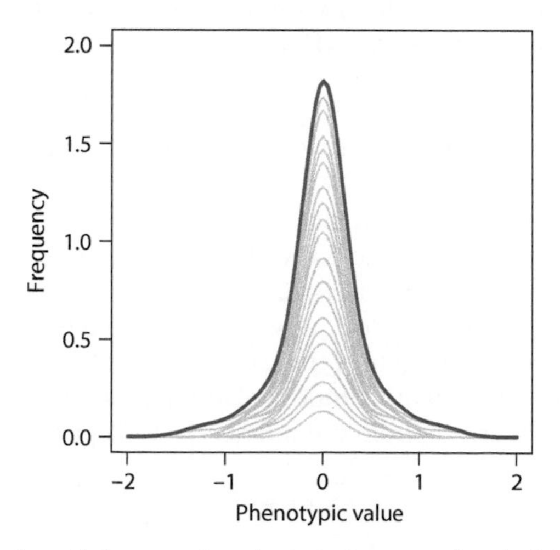

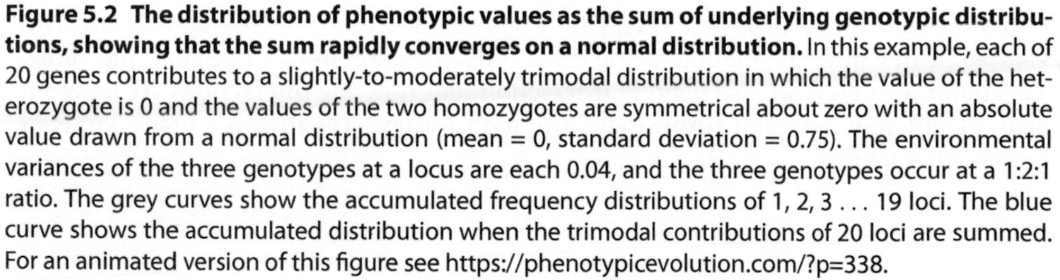
Figure 5.2 The distribution of phenotypic values as the sum of underlying genotypic distributions, showing that the sum rapidly converges on a normal distribution. In this example, each of 20 genes contributes to a slightly-to-moderately trimodal distribution in which the value of the heterozygote is 0 and the values of the two homozygotes are symmetrical about zero with an absolute value drawn from a normal distribution (mean = 0, standard deviation = 0.75). The environmental variances of the three genotypes at a locus are each 0.04, and the three genotypes occur at a 1:2:1 ratio. The grey curves show the accumulated frequency distributions of 1, 2, 3 . . . 19 loci. The blue curve shows the accumulated distribution when the trimodal contributions of 20 loci are summed. For an animated version of this figure see https://phenotypicevolution.com/?p=338.

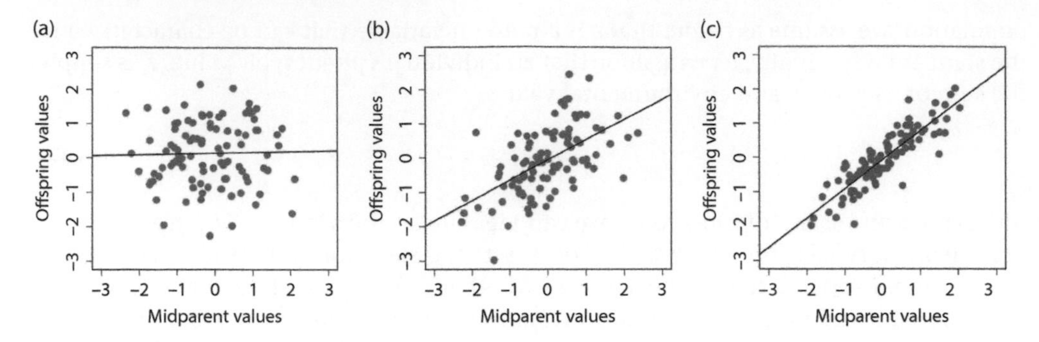

Figure 5.3 Hypothetical plots of offspring versus midparent values. Each plot portrays a sample of 100 offspring-parent values sampled from a bivariate normal distribution. (a) In the parametric distribution, $h^2 = 0$. In the sample, $h^2 = 0.03 \pm 0.09$ s.e. (b) In the parametric distribution, $h^2 = 0.5$. In the sample, $h^2 = 0.59 \pm 0.09$ s.e. (c) In the parametric distribution, $h^2 = 0.9$. In the sample, $h^2 = 0.85 \pm 0.05$ s.e.

If we substitute $x+e$ for both kinds of phenotypic values into our standard expression for a covariance (1.8), we produce four covariance terms, but all but one are zero. The one term that remains is one-half the covariance between the breeding values of parents and offspring which is equivalent to half the genetic variance. The algebra is only slightly more complicated if we use midparent values instead of single parent values, for then we obtain

$$h^2 = Cov\,(z_o, z_p)\,P^{-1} = Cov\,(x_o, x_p)\,P^{-1} = GP^{-1} = G/P, \tag{5.3}$$

where h^2 is the *heritability* of the trait, and the o and p subscripts denote midoffspring and midparents, respectively. In other words, the regression of midoffspring on midparental phenotypes estimates G/P, which we call the heritability of the trait, h^2 (Figure 5.3). From (5.2 and 5.3), we can see that h^2 varies from zero to one. From Galton's plot we estimate the heritability of human stature as 0.57 (Figure 5.1).

5.3 A General Expression for Resemblance Among Relatives

To obtain a general expression for covariance among relatives, we start with a more general model. This more general model includes the effects of dominance and epistasis, as well as additive genetic effects. In a later section, we will discuss how each of these effects represents contributions from many loci, as well as from the different alleles at a locus. For now we are concerned only with how these three kinds of aggregate effects contribute to phenotypic value and resemblance among relatives. In a model for phenotypic value that includes genetic parts arising from dominance, y, and epistasis, i,

$$z = x + y + i + e. \tag{5.4}$$

Dominance refers to nonadditive interaction between the alleles at the same locus. For, example, if one allele is dominant, the other allele has no effect on phenotypic value. In a similar way, *epistasis* also denotes interaction, but now the interactions are between alleles at different loci. Not surprisingly, the term for epistasis includes many kinds of possible interactions,

$$i = i_{AA} + i_{AD} + i_{DD} + i_{AAA} + i_{AAD} + i_{ADD} + i_{DDD} + \cdots, \tag{5.5}$$

in which AA denotes interactions between additive effects of different loci taken two at a time, AAA denotes three-way interactions between additive effects at three different loci, etc. Mixed subscripts (e.g., AD, AAD) denote interactions between the additive and dominance effects of different loci. This more complicated model for phenotypic value can be substituted into the formula for covariance (1.8). Algebraic manipulation leads to a general expression for the composition of phenotypic covariance that includes dominance and epistatic effects. In other words, this general theoretical result partitions phenotypic covariance between particular kinds of relatives (X and Y) (e.g., X = parents and Y = offspring, or X = uncles and Y = nephews) into genetic parts (*components of variance*) attributable to additive effects (A), dominance (D) and epistasis (e.g., AA, ADD):

$$Cov(X, Y) = rG + uG_D + r^2G_{AA} + ruG_{AD} + u^2G_{DD} + r^3G_{AAA} + r^2uG_{AAD} + ru^2G_{ADD} + u^3G_{DDD} + \cdots \tag{5.6}$$

Cockerham (1959). In this expression, r is the *coefficient of relationship* (the probability that relatives X and Y have the same allele through *identity by descent*), and u is *coefficient of consanguinity* (the probability that two alleles at a locus drawn at random, one each from X and Y, will be identical by descent). The first coefficient describes the inheritance on a single allele and hence additive effects, while the second describes the inheritance of pairs of alleles and hence dominance deviations (Crow and Kimura 1970; Falconer and Mackay 1996). For example, rG is the product of r and additive genetic variance, uG_D is the product of u and dominance variance, r^2G_{AA} is the product of r^2 and the covariance of additive effects at pairs of loci, etc. The two coefficients, r and u, can be determined for various kinds of relatives (e.g., full-sibs) from a pedigree are presented in Table 5.1.

The importance of G to parent-offspring resemblance can be understood by realizing that for these relatives $u = 0$, and consequently expression (5.6) is greatly simplified and involves only G, G_{AA}, G_{AAA}, etc. In other words, dominance makes no contribution to the phenotypic resemblance between parents and offspring. The collective contribution of epistasis to parent-offspring resemblance (i.e., G_{AA}, G_{AAA}, etc.) is likely to be small for

Table 5.1 Coefficients of relationship (r), consanguinity (u), and the contribution of maternal effects for various kinds of relatives. An asterisk (*) denotes a relatively minor contribution of maternal effect. From Willham (1963) with permission.

Relatives	*r*	*u*	Maternal effect contribution
Mother-offspring	½	0	yes
Father-offspring	½	0	no*
Full-sib	½	¼	yes
Paternal half-sib	¼	0	no
Maternal half-sib	¼	0	yes
Double first cousins (fathers full-sibs, mothers full-sibs)	¼	⅛	yes
Double first cousins (both opposite sexes are full-sibs)	¼	⅛	yes
Single first cousins (fathers full-sibs)	⅛	0	no
Single first cousins (mothers full-sibs)	⅛	0	yes

the following reason. The coefficient of relationship for parents and offspring is ½, consequently when it is raised to progressively higher powers as the coefficient for higher order terms describing additive-by-additive epistasis (*AA*, *AAA*, etc.), the net effect is a progressive lowering of contributions. In other words, in practical terms we can ignore the contributions of epistasis and view resemblance between parents and offspring, as in Galton's plots, as providing a virtually clean estimate of additive genetic variance. In a similar way, paternal half-sibs are useful for estimating G, because $r = 1/4$ and $u = 0$. In contrast, certain other relatives provide problematic estimates of G because dominance does contribute to phenotypic resemblance. Full-sibs are a case in point. Because full-sibs have two parents in common, $u = 1/4$ and $r = 1/2$. Ignoring epistasis, we see from (5.6) that the resemblance among full-sibs springs from two terms, $\frac{1}{2}G + \frac{1}{4}G_D$, not just from additive genetic variance as in the parent-offspring case. To make matters worse, full-sibs also share a maternal environment and that sharing can also contribute to phenotypic resemblance.

In general, maternal effects can contribute to resemblance between mothers and their offspring and to other kinds of relatives. The kinds of relatives that are affected in the *Willham model of maternal effects* are shown in Table 5.1 (Willham 1963, 1972). In the Willham model, one kind of trait expressed in mothers affects the same trait expressed in offspring. For example, milk yield in a female mammal affects milk yield in her female offspring. The model allows the trait to be heritable (with dominance effects, as well as additive genetic and environmental parts), so multiple maternal pathways can contribute to mother-daughter resemblance in, say, milk yield. Table 5.1 tell us that if we wish to avoid the complications that this kind of maternal effect exerts on an estimate of additive genetic variance, then three kinds of relatives are particularly useful (father-offspring, paternal half-sibs, and single first cousins in which fathers rather than mothers are full-sibs). The *Kirkpatrick & Lande model of maternal effects* allows one trait in mothers to affect the expression of another trait or the same trait in offspring (Kirkpatrick and Lande 1989, 1992). This more general model makes the same qualitative predictions about the usefulness of relatives as does Willham's model (Table 5.1)

Phenotypic resemblance arising from shared environments, maternal or otherwise, is a complication that must be considered in the estimation of additive genetic variance and similar genetic parameters. In terms of our model (5.6), the complication means that although we assumed that the covariance between the environments of two relatives was zero, when environments are shared that may not be a safe assumption. The genetic aspect of resemblance between relatives is rooted in Mendelism and so submits to a general treatment (5.6). In contrast, the environmental aspect of resemblance depends on the ecological peculiarities of individual species, making generalization difficult.

A standard approach to the issue of environmental resemblance is to experimentally manipulate rearing environments so that contributions to phenotypic resemblance can be isolated and assessed. For example, Smith and Dhondt (1980) used *cross-fostering* to study phenotypic resemblance in a song bird on a small island that allowed access to all nests. By swapping eggs between nests, they were able to compare actual parent-offspring resemblance with foster parent-foster offspring resemblance (Figure 5.4). In this case, the shared environment is the territory that shelters and nurtures successive generations. Apparently, however, the shared territory of foster parents and their foster offspring makes no contribution to phenotypic resemblance in body size. These and other kinds of experimental manipulations can be used to test for and evaluate the magnitude of the contribution from shared environments.

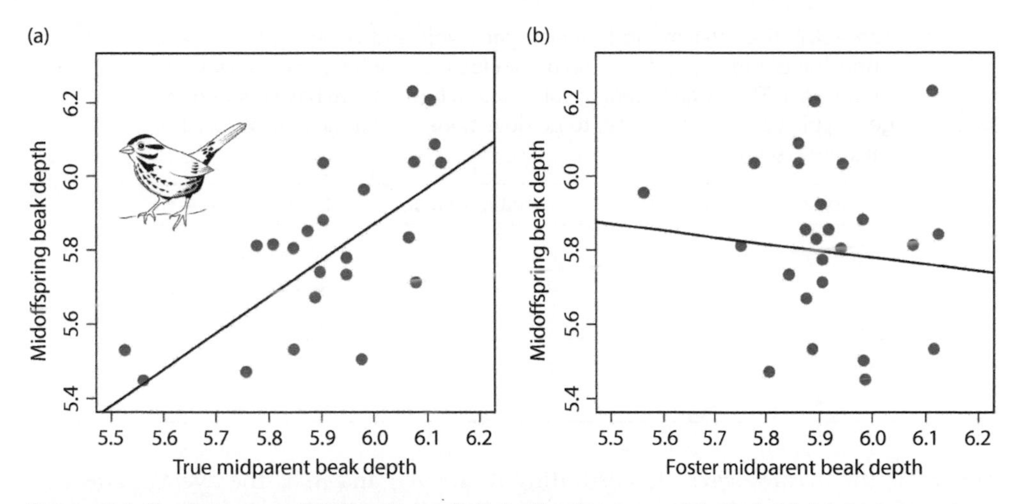

Figure 5.4 Song sparrow (*Melospiza melodia*) offspring resemble their biological parents but not their foster parents. (a) Beak depth of midoffspring as a function of the beak depths of their true (biological) parents ($y = -0.01 + 0.98\,x$, $r^2 = 0.496$). (b) Beak depth of midoffspring as a function of the beak depths of their foster parents ($y = 6.86 - 0.18\,x$, $r^2 = 0.011$).

After Smith and Dhondt (1980).

5.4 Additive Genetic and Dominance Effects at Individual Loci

In presenting a model for phenotypic values in Section 5.3, we cavalierly used variables that represented the sum of effects over all loci in the genome. It will help to make those concepts more concrete, if we consider those same effects at the level of individual loci, before the summation takes place. The following account is based on Fisher's (1958) somewhat opaque exposition and Falconer and Mackay's (1996) portrayal of that exposition.

For simplicity we will consider a locus, A, with just two alleles, A_1 and A_2. We define the *genotypic values* of each genotype as the average trait value of a large sample of individuals with that genotype. It is convenient to scale the trait values of the genotypes, so that trait value midway between the two homozygotes is zero. On this scale we will refer the genotypic values of the genotype with the lowest score, A_1A_1, as $-a$ and the value of the genotype with the highest score, A_2A_2, as $+a$. We shall denote the genotypic value of the heterozygote, A_1A_2, as d. Thus, in the absence of dominance, $d = 0$. These conventions for the contributions of a particular genotype to an individual's phenotypic value are straightforward, but a moment's reflection tells us that if we want to relate these genotypic values to average genetic value in the population, we will need to know the frequencies of the three genotypes. A simple way to proceed, is to let p denote the frequency of the A_1 allele, so that the frequency of the A_2 allele is $q = 1 - p$, and then assume Hardy-Weinberg proportions for the genotypes. Skipping over the details of the algebra, which are sketched in Falconer and Mackay (1996), we can show that the average effects on the trait mean of substituting one allele for another are

$$\alpha_1 = q\,[a + d\,(q - p)] \tag{5.7a}$$

$$\alpha_2 = -p\,[a + d\,(q - p)]\,, \tag{5.7b}$$

Table 5.2 The relationship between genotypic and breeding values at a single locus and the contribution of that locus to additive genetic variance for the trait. The mean breeding value is zero because we have defined the genotypic values, $-a$ and $+a$, to be deviations from a mean of zero. After Falconer (1960).

Genotype	Frequency	Genotypic value	Breeding value, x
A_1A_1	p^2	$+a$	$2\alpha_1 = 2q\alpha$
A_1A_2	$2pq$	d	$\alpha_1 + \alpha_2 = (q - p)\,\alpha$
A_2A_2	q^2	$-a$	$2\alpha_2 = -2p\alpha$
mean		$a\,(p - q) + 2dpq$	0
variance			G

where α_1 is the *average effect* of substituting A_1 for A_2, and α_2 is the average effect of substituting A_2 for A_1. Averaging the effects of these two kinds of substitutions across genotypes, we find that the average effect, α, of an allelic substitution at this locus is

$$\alpha = a + d\,(q - p) = \alpha_1 - \alpha_2. \tag{5.7c}$$

The point of these definitions and this algebra is that we are now in a position to see the relationship between the genotypic value and breeding value, x (Table 5.2). The genotypic value of a genotype is a simple quantity that does not change with gene frequency, for example, the genotypic value of A_1A_1 is $+\alpha$. In contrast, the *breeding value* of a genotype does change with gene frequency, taking account of the fact that the consequence of substituting one allele for another will depend on the frequencies of the two alleles, p and q. The allelic property that incorporates this frequency dependence is average effect. The definitions of the average effects of alleles at a locus may seem awkward, but they allow us to achieve the simple result that the breeding value of a genotype is simply the sum of average effects of its alleles. Furthermore, the contribution of allelic variety at a locus to additive genetic variance is simply the variance in breeding values at that locus. These same relationships hold if we allow multiple alleles at a locus, and if we allow multiple loci to affect the trait. Effects are summed over all alleles at a locus and across all loci to yield the total breeding value of an individual and additive genetic variance is the variance of those multi-locus breeding values.

Part of the complexity in distinguishing between genotypic value and breeding value arises from dominance. In the absence of dominance, $d = 0$, and $a = \alpha$, but of course breeding values are still a function of gene frequency. Ignoring epistasis, the difference between genotypic value and breeding value is known as the *dominance deviation*, y, of that genotype. Like breeding value, x, dominance deviation, y, represents the sum of effects over loci. Dominance deviation (y) is a function of gene frequency, unlike d, which is simple property of a heterozygous genotype, irrespective of gene frequency. *Dominance variance*, G_D, is variance in dominance deviations.

5.5 Estimating Additive Genetic Variance Using Covariance of Relatives

In the covariance approach, replicated sets of relatives (e.g., parents and offspring) are assembled either by direct sampling from nature or by conducting breeding designs.

Additive genetic variance, G, can then be estimated with varying degrees of fidelity, depending on the kinds of relatives that are used and the sample size that is employed. Estimation of G is a statistical procedure conducted by regression or analysis of variance (ANOVA). In this section we will give an example of the regression approach that is commonly used to analyze parent-offspring data. In the ANOVA approach, the focus is on estimating components of variances (e.g., the among-sire component of variance, covariance among paternal half-sibs) that bear a simple mathematical relationship to additive genetic variance. So, for example, four times the among-sire component of variance gives an estimate of additive genetic variance. Falconer and Mackay (1996) and Lynch and Walsh (1998) provide details for analyzing several important breeding designs.

We want an interval as well as a point estimate of G. In other words, we want to know how much confidence to place on the single value that we present as our estimate of G. In most implementations of the covariance approach, family size will vary, often appreciably. Unfortunately, analytical solutions for the standard error of G assume equal family sizes and solutions do not exist for the unbalanced case. This is one problem that spurred the development of the animal model, which is reviewed in Section 5.7. Another solution is to embed the ANOVA implementation in a resampling scheme such as bootstrapping that can estimate standard errors (e.g., Phillips and Arnold 1999).

Garter snake vertebral numbers provide an example of G estimation using the covariance approach (Arnold and Phillips 1999; Phillips and Arnold 1999). A set of 151 mothers and their 712 female offspring was assembled by capturing pregnant females at a single locality in nature and holding them in the laboratory until their litters were born. The number of body and tail vertebrae can be conveniently assessed by counting scales on the ventral surface, with the transition between the two regions demarcated by a distinctive vent scale. The counts do not change during the postnatal life of the individual, except that the tips of the tail can sometimes be lost during predation attempts or skin-shedding. In that case, the conical scale at the tip is missing and the count can be scored as missing. Vertebral counts in fishes are known to be affected by temperature during development, but garter snake females assiduously thermoregulate during pregnancy, with only modest differences among females. Furthermore, holding females at different constant temperatures during pregnancy reveals flat reaction norms (offspring vertebral counts as a function of temperature) (Arnold and Peterson 2002). These and other considerations suggest that the two counts are not subject to maternal effects. Consequently, we can reasonably estimate additive genetic variance for vertebral counts by regressing average daughter counts on the counts of the mothers (Figure 5.5). The resulting estimates of heritability for body and tail counts are 0.54 ± 0.10 s.e. and 0.48 ± 0.10 s.e., respectively. Standard errors were estimated by bootstrapping over families.

5.6 The Utility of Pedigrees

One needs replicated sets of relatives to estimate genetic variance using the covariance approach. With many kinds of organisms such sets are difficult to assemble or many different kinds of relatives are available. In either of these situations it may be practical to estimate the pairwise relationships of individuals using genetic markers and assemble that information into a pedigree that straddles multiple generations. Having phenotypic scores for a large sample of individuals of known pedigree opens the door to G estimation, as well as to studies of inbreeding depression and sociality (Pemberton 2008).

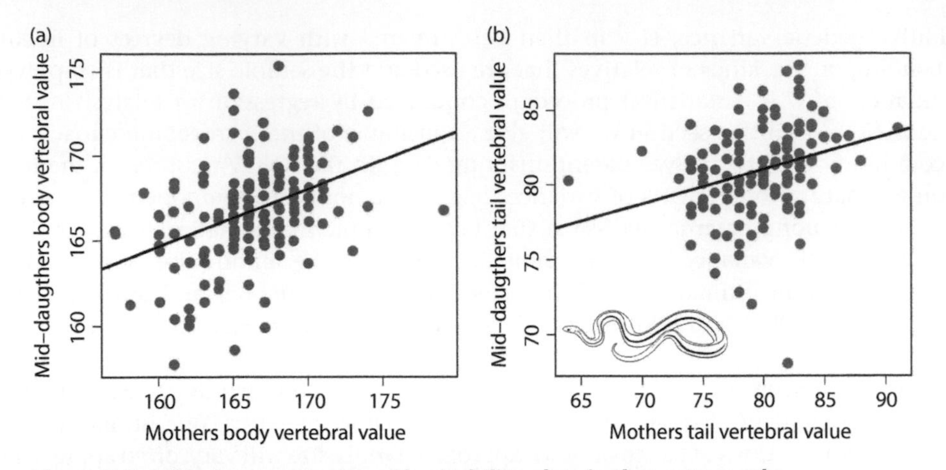

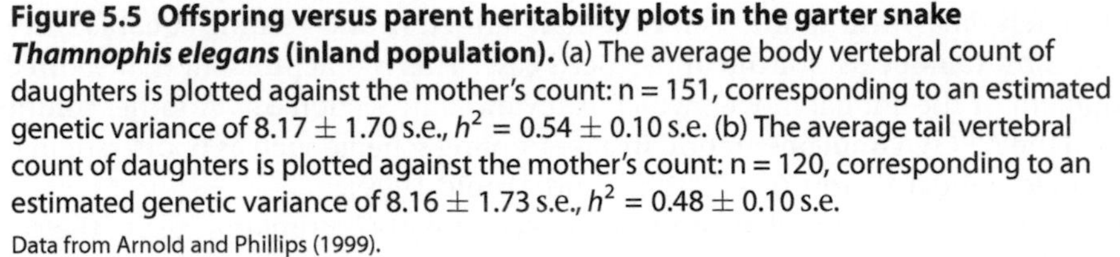
Figure 5.5 Offspring versus parent heritability plots in the garter snake *Thamnophis elegans* (inland population). (a) The average body vertebral count of daughters is plotted against the mother's count: n = 151, corresponding to an estimated genetic variance of 8.17 ± 1.70 s.e., h^2 = 0.54 ± 0.10 s.e. (b) The average tail vertebral count of daughters is plotted against the mother's count: n = 120, corresponding to an estimated genetic variance of 8.16 ± 1.73 s.e., h^2 = 0.48 ± 0.10 s.e.

Data from Arnold and Phillips (1999).

5.7 Estimating Genetic Variance Using Pedigrees: Introduction to the Animal Model

Pedigree analysis is a powerful alternative to using covariance among relatives to estimate genetic variance. The advantage of pedigree analysis is flexibility that allows it to simultaneously incorporate information from different kinds of relatives while accounting for a variety of fixed or random effects, such as sexual dimorphism, maternal effects, and geographic-specific environmental effects (Pemberton 2008; Tong and Thompson 2008). The analysis that accomplishes all of this is based on the *animal model*, which was pioneered especially by Henderson (1976, 1984). The animal model is a special case of a linear mixed model that includes both random and fixed effects. The animal model was inspired by the common situation in animal breeding that rearing often happens in different environments (herds) and in different times (years). Furthermore, family sizes are often different so that data are unbalanced (e.g., unequal litter sizes), creating problems in estimating variance components and attaching standard errors to variance components (Henderson 1984; Searle 1987). The following account of the animal model closely follows Henderson's (1984) description. Some readers may wish to bypass this detailed account by considering the case of estimating parameters with just a few individuals, because in that case the several vectors and matrices take a surprisingly simple form; see, for example, Hadfield (2010).

The animal model has two parts: one part that specifies the composition of the phenotypic values of a single trait, z_i, of each individual in a sample of n individuals and another part that specifies the distributions of the random variables that make up the column vector of phenotypic values of all individuals in the sample, $\boldsymbol{z}$. In a sample of n individuals, let the phenotypic trait value of the *ith* individual be

$$z_i = \mu_j + u_k + e_i \,, \tag{5.8a}$$

where μ_j is a fixed effect, u_k is a random effect, and e_i is a residual effect, not accounted for by the fixed and random effects in the model. The expressions for all the individuals in the sample can be written in matrix form as,

$$\mathbf{z} = \boldsymbol{X}\boldsymbol{b} + \boldsymbol{Z}\boldsymbol{u} + \boldsymbol{e}, \tag{5.8b}$$

where:

$\boldsymbol{b}$ is a fixed, usually unknown column vector of length p, where p is the number of fixed effects, giving the contributions of the fixed effect to $\mathbf{z}$;

$\boldsymbol{u}$ is a random, unknown column vector of additive values of length q, where q is the number of individuals in the pedigree, with zero means, giving the contribution of the random effect to $\mathbf{z}$;

$\boldsymbol{e}$ is a random, unknown column vector of length n, where n is the number of individuals with phenotypic measures, with zero means, giving the residual contribution to $\mathbf{z}$;

$\boldsymbol{X}$ is a known $n \times p$ design matrix (with zero and one entries) which assigns the fixed effects to individuals; and

$\boldsymbol{Z}$ is a known $n \times q$ design matrix (with zero and one entries) which assigns the random effects to individuals.

Focusing just on the column vectors and expressing them as their transposes, they are

$$\mathbf{z}^{\mathrm{T}} = \begin{pmatrix} z_1 & z_2 & z_3 \cdots z_n \end{pmatrix}, \tag{5.9a}$$

$$\boldsymbol{b}^{\mathrm{T}} = \begin{pmatrix} b_1 & b_2 & b_3 \cdots b_p \end{pmatrix}, \tag{5.9b}$$

$$\boldsymbol{u}^{\mathrm{T}} = \begin{pmatrix} u_1 & u_2 & u_3 \cdots u_q \end{pmatrix}, \tag{5.9c}$$

$$\boldsymbol{e}^{\mathrm{T}} = \begin{pmatrix} e_1 & e_2 & e_3 \cdots e_n \end{pmatrix}. \tag{5.9d}$$

Similarly, the matrices are

$$\boldsymbol{X} = \begin{bmatrix} X_{1,1} & X_{1,2} & \cdots & X_{1,p} \\ X_{2,1} & X_{2,2} & \cdots & X_{2,p} \\ \vdots & \vdots & \ddots & X_{n-1,p} \\ X_{n,1} & X_{n,2} & X_{n,p-1} & X_{n,p} \end{bmatrix} \tag{5.10a}$$

$$\boldsymbol{Z} = \begin{bmatrix} Z_{1,1} & Z_{1,2} & \cdots & Z_{1,q} \\ Z_{2,1} & Z_{2,2} & \cdots & Z_{2,q} \\ \vdots & \vdots & \ddots & Z_{n-1,p} \\ Z_{n,1} & Z_{n,2} & Z_{n,p-1} & Z_{n,q} \end{bmatrix}. \tag{5.10b}$$

Note that in (5.9a) the vector $\boldsymbol{z}$ has as its elements the phenotypic values of the n individuals in the sample, rather than the phenotypic values of multiple traits in a single individual, as in Chapter 2.

The second part of the model has to do with the distributions of the unknown random variables, $\boldsymbol{u}$ (which will usually be closely related to breeding values, $\boldsymbol{x}$) and $\boldsymbol{e}$, which has the same meaning it did in (5.4). We will denote the variance-covariance matrix for a vector of random variables as $Var()$, which describes the theoretical sampling variance that would arise if we made repeated draws from our statistical population. $Var(\boldsymbol{u}) = \boldsymbol{F}$ is a $q \times q$ symmetric matrix that is usually nonsingular (has a matrix inverse). We wish to estimate $\boldsymbol{F}$, because from it and $\boldsymbol{Z}$ we can obtain an estimate of G, the genetic variance of the trait. Commonly, the values of $\boldsymbol{u}$ will not be independent (e.g., when individuals are related) and we can take that non-independence into account in specifying the structure of the unknown matrix $\boldsymbol{F}$. $Var(\boldsymbol{e}) = \boldsymbol{R}$ is an $n \times n$ diagonal matrix with each individual's error on the diagonal and zeros elsewhere. The vectors $\boldsymbol{u}$ and $\boldsymbol{e}$ are also commonly assumed to follow multivariate normal distributions, and likelihood expressions are based on that assumption. The covariance between $\boldsymbol{u}$ and $\boldsymbol{e}$ is assumed to be zero.

Although $\boldsymbol{F}$ and $\boldsymbol{R}$ are unknown matrices, in general we will know or can assume something about each of their structures. For example, the random variables $\boldsymbol{u}$ of two related individuals are not independent, but if we know the pedigree of those individuals, we could use their relatedness to compute an appropriate value for a particular element of $\boldsymbol{F}$. $\boldsymbol{R}$ will often be a diagonal matrix (with zeros for off-diagonal elements), but errors that are correlated between individuals could also be accounted for.

Turning to the item of key interest, how can we use the animal model to estimate the genetic variance of the trait? Consider the simple case in which the random effects, $\boldsymbol{u}$, are parts of breeding values, which each individual inherits from a different, unrelated sire. Then

$$\boldsymbol{F} = Var(\boldsymbol{u}) = \boldsymbol{I}\sigma_S^2 = \begin{bmatrix} 1 & 0 & \cdots & 0 \\ 0 & 1 & \cdots & 0 \\ \vdots & \vdots & \ddots & 0 \\ 0 & & 00 & 1 \end{bmatrix} \sigma_S^2 = \begin{bmatrix} \sigma_S^2 & 0 & \cdots & 0 \\ 0 & \sigma_S^2 & \cdots & 0 \\ \vdots & \vdots & \ddots & 0 \\ 0 & & 00 & \sigma_S^2 \end{bmatrix}, \tag{5.11}$$

where σ_S^2 is the component of variance among-sire effects, which needs to be estimated. Each individual inherits half of its breeding value from its sire, $\boldsymbol{u} = \frac{1}{2}\boldsymbol{x}$, so the variance of those halves is ¼ G. In other words, $2\boldsymbol{u}$ estimates $\boldsymbol{x}$, the vector of breeding values, and $4\sigma_S^2$ estimates G, the genetic variance of the trait. Similar principles apply in more complicated cases. For example, if sires are related, a sire relationship matrix (with relationship coefficients for pairs of sires as elements) replaces $\boldsymbol{I}$ in the expression above, but σ_S^2 is still the key quantity that leads to an estimate of G.

Given the model just specified, we also have expressions for the trait mean and its sampling variance. The expected value, or mean, and variance of the trait are

$$E(\boldsymbol{z}) = \boldsymbol{Xb} \tag{5.12a}$$

$$Var(\boldsymbol{z}) = \boldsymbol{ZFZ}^{\mathrm{T}} + \boldsymbol{R}. \tag{5.12b}$$

We are not surprised to see that the trait mean is a function of only the fixed effects, because we have standardized the random effects means so that they are zero. The sampling variance-covariance of the mean is a function of the sampling variance-covariance of random effects, $\boldsymbol{ZFZ}^{\mathrm{T}}$, and residual contributions, $\boldsymbol{R}$.

The point of the animal model, with all of its matrix and vector bookkeeping, is that it can be used to estimate the unknown breeding values, $\boldsymbol{x}$, and their variance, G, together with their sampling properties, while accounting for one or many kinds of fixed effects. At the same time, the model estimates the values of fixed effects. All of these outputs can be very much worth the effort of setting up the precursor tables to $\boldsymbol{Z}$ and $\boldsymbol{X}$ that are required by estimation programs. Details of implementing the animal model are discussed by Meyer (2007) and Hadfield (2010). Furthermore, the approach can be extended to multiple traits, as we shall see in the next chapter.

Pedigree methods for estimating G and $\boldsymbol{x}$ have become increasingly flexible but have not escaped some key assumptions of normality. The original specifications of the animal model assumed that the distribution of the phenotypic trait, $\mathbf{z}$, was multivariate normal, as well as the breeding values, $\boldsymbol{x}$, and residual effects, $\boldsymbol{e}$. These distributional assumptions allowed likelihood functions to be specified in Gaussian terms, and the maxima of those functions could be found by analytical manipulations (Meyer 2007; Tong and Thompson 2008). More recently, *Markov chain Monte Carlo* (MCMC) methods allow one to maximize complex likelihood functions and so escape from the assumption that $\mathbf{z}$ is normally distributed (Sorensen and Gianola 2002; Hadfield 2010). These methods employ a link function (Nelder and Wedderburn 1972), so that the animal model is specified in terms of a derived variable, rather than $\mathbf{z}$. The derived variable can take any of a variety of specified distributional forms (Gaussian, Poisson, binomial, etc). Random effects can include interactions, and fixed effects may be continuous as well as categorical. Random, fixed, and residual effects are assumed to be multivariate normal with specified covariance structures. See Hadfield (2010) for an illuminating worked example.

5.8 Estimation of Genetic Variance Using a GWAS

A GWAS (*genome-wide association study*) uses population-level linkage disequilibrium to find statistical associations between *single-nucleotide polymorphisms* (SNPs) and sites affecting phenotypic traits (Hindorff et al. 2009; Yang et al. 2010; Visscher et al. 2012). In this approach, hundreds of thousands of SNPs are used to map close genomic associations with sites that may be causally related to a phenotypic score. A linear model is used to estimate G as a function of the additive effects of the trait-associated SNPs. The approach employs a relationship matrix that is estimated from the SNP data (Yang et al. 2010, 2011a,b). The triumph of the GWAS approach is that while simply summing the univariate contributions of individual SNPs accounts for only a small fraction of G (as estimated by covariance or standard pedigree methods), simultaneous (joint) estimation accounts for a much larger fraction. Applying this approach to human height revealed that hundreds of genes contribute to additive genetic variation in particular quantitative traits.

Yang et al. (2011b) used a sample of 11,576 human genomes scored for 565,040 autosomal SNPs to estimate the proportion of additive genetic variance explained by the entire set of SNPs, yielding a heritability estimate of 0.448 ± 0.029 *s.e.* This estimate

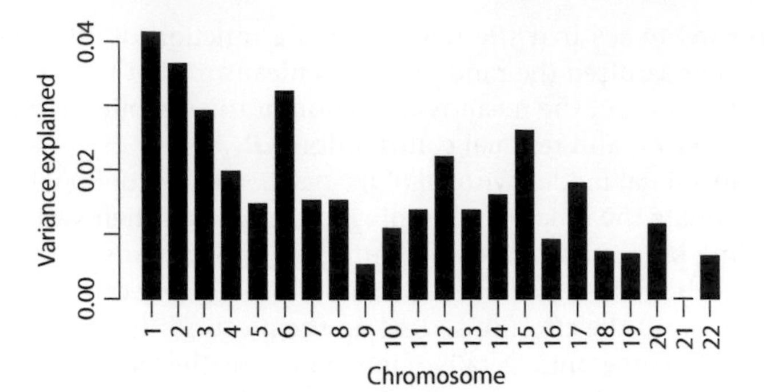

Figure 5.6 Estimates of the proportion of additive genetic variance in human height explained by genic regions of each chromosome using a joint analysis. The boundaries of protein-coding genes (genic regions) are defined as ± 50 kb of the 3′ and 5′ UTRs.
After Yang et al. 2011b.

is comparable to estimates based on parent-offspring covariance (e.g., Figure 5.1) and contrasts with GWAS heritability estimates of about 0.10 based on a simple sum of contributions. As in the case of human height, genomic and covariance estimates of heritability in cattle and Soay sheep are in close agreement (Bérénos et al. 2014).

Height-associated SNPs were distributed throughout the human genome on 21 out of 22 autosomal chromosomes. Although SNPs fell within intergenic as well as genic regions of the genome, genic regions explained more genetic variance in height (Figure 5.6). Furthermore, the amount of genetic variance explained by any given autosomal chromosomal segment is approximately proportional to the length of genes in that segment. A small additional contribution was detected on the X chromosome. In other words, hundreds (thousands?) of genes contribute to genetic variation in human height. This result validates the 40+ year old opinion of quantitative geneticists that the number of genes found to contribute to variation in a quantitative trait is directly proportional to the effort expended by the investigator (J. L. Jinks in a verbal rebuttal to Gottlieb 1984; also see Coyne and Lande 1985).

5.9 Technical Issues in Estimating Additive Genetic Variance

Complications arise in estimation because of the special nature of some traits. For example, some traits change as a function of age. One solution is to define age-specific traits and take measurements at specified intervals of age. Alternatively, age can be treated as a fixed or random effect using the animal model. If inheritance of the age-trajectory is the issue, the trajectory itself can be considered a trait in this approach (Kirkpatrick and Heckman 1989; Meyer and Hill 1997; Gomulkiewicz et al. 2018). The theory for *function-valued traits* also applies to reaction norms in which phenotype expression varies as a function of an environmental variable. In other words, the key to tackling both ontogeny and reactions norms is to take a multiple trait approach to the problem, using tools that are introduced in the next chapter.

The expression of some traits fluctuates through time. Courtship displays, heart rate, and body temperature are examples. If the trait value for an individual for such a trait is captured by a single measurement, the trait is bound to have a small heritability. From an ecological or evolutionary perspective, this small heritability may be misleading if the trait is expressed thousands or even millions of times during the individual's lifetime or during relevant selection episodes. One solution is to define the trait as the sum or average of some fixed number of measurements. One can readily show that both the *repeatability* and the heritability of the trait is an increasing function of the number of measurements (Arnold 1994).

5.10 Survey of Heritability Estimates

Surveys of heritability estimates for morphological characters reveal a broad distribution centred in the range 0.4–0.5 (Mousseau and Roff 1987; Postma 2014) (Figure 5.7). One might be tempted to conclude from such surveys, as some authors have, that in the absence of data, everything is heritable. That conclusion is unjustified on several grounds. In the first place, Figure 5.7 reports point estimates, while ignoring sampling properties. Secondly, publication bias undoubtedly inflates the right-hand side of the distribution, since statistically nonsignificant estimates may not be published as frequently as significant estimates. Finally, when we approach the problem from a multivariate perspective in the next chapter, we will find that heritable variation is so sparse as to be completely lacking in some regions of phenotypic space.

Houle (1992) outlined several problems with using h^2 as a comparative measure of genetic variability and recommended using the additive genetic coefficient of variation instead. Houle (1992) compiled and analyzed a large database representing many kinds of outbred populations to illustrate his arguments. The mean value of h^2 for morphological traits (Figure 5.8a) was 0.416 ± 0.008 s.e. (n = 540), which is similar to the Mousseau & Roff (1987) and Postma (2014) results. The focus on h^2 in this book, however, is on the use of h^2 in quantifying univariate response to selection. Problems in using h^2 in that context can be bypassed by using G and $\beta = P^{-1}s$ (which generalizes to the multivariate case) rather than h^2 and s (10.1 and 11.1). Furthermore, for the trait category of main concern in this book (morphological traits), h^2 does not show a dependency on the trait mean (Figure 5.8a). Houle (1992) makes strong argument for using the *additive genetic coefficient of variation*, $CV_G = 100\sqrt{G}/\bar{z}$, as a measure of genetic variation in comparative studies. Note, however, that this measure may be inversely related to the magnitude of the trait mean, as in Houle's (1992) sample (Figure 5.8b). For general arguments about the utility of coefficients of variation see Wright (1968), Lande (1977a), and Bryant (1986).

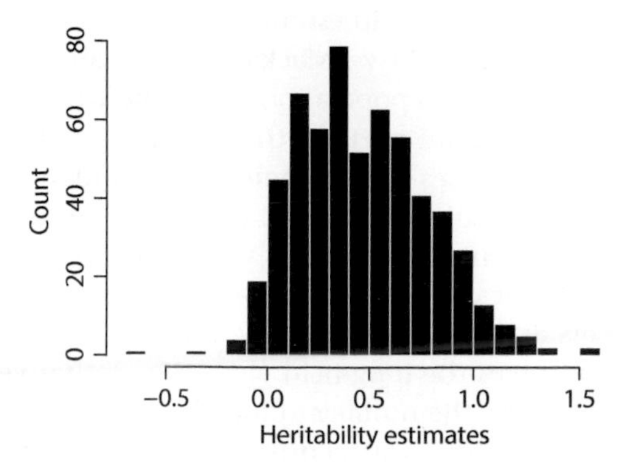

Figure 5.7 Histogram of a large sample (n = 580) of heritability estimates for morphological characters in vertebrate and invertebrate animals. Note that estimates outside the parameter range ($0 \leq h^2 \leq 1$) are possible with some estimation procedures. Mean = 0.47, median = 0.44, variance = 0.10. Based on data in Mousseau & Roff (1987), courtesy of D. Roff.

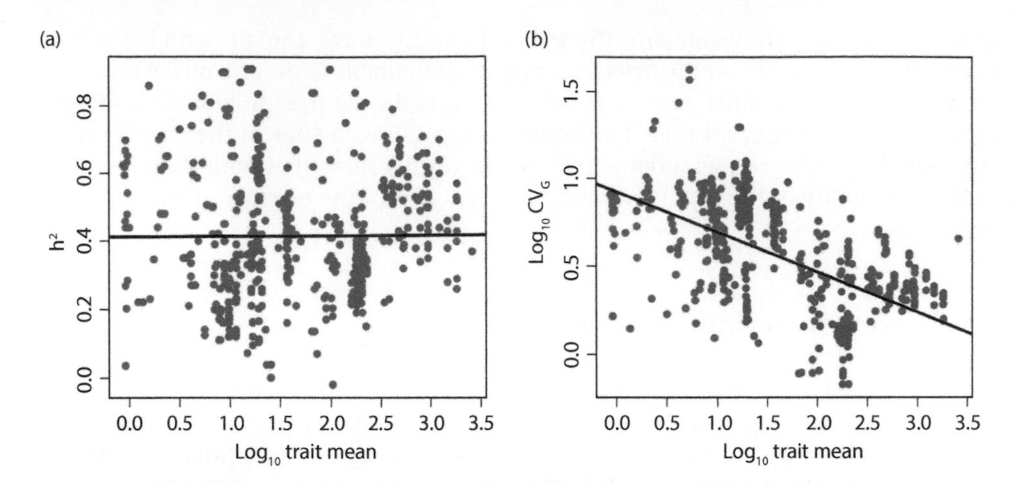

Figure 5.8 Heritability and additive genetic coefficient of variation as functions of $\log_{10}$ trait mean ($\log_{10}\bar{z}$) for morphological traits (meristic, size-related, and other morphological traits). Least squares regression lines are shown in black. (a) Heritability (h^2) as a function of log10 trait mean ($\log_{10}\bar{z}$) (n = 540). (b) Log_{10} additive genetic coefficient of variation ($CV_G = 100\sqrt{G}/\bar{z}$) as a function of $\log_{10}$ trait mean ($\log_{10}\bar{z}$) (n = 537). Based on data in Houle (1992), courtesy of D. Houle.

5.11 Detailed Analysis of Genetic Variation Within Populations

Because the genetic basis of natural variation in *Drosophila* bristle counts has been studied in depth from a number of perspectives, the lessons from this model system are especially informative (Mackay and Lyman 2005). Two overarching conclusions spring from the long tradition of work on this study system. First, certain results are consistent from study to study, even as increasingly sophisticated techniques are brought to bear. Second, the nature of quantitative variation appears to be more complex the more we study it.

Additive genetic variation is consistently abundant for counts of both abdominal and sternopleural bristles. Estimates of heritability are consistently about 0.50 for both traits and are largely attributable to additive genetic variance (Mackay and Lyman 2005). For example, Clayton and Robertson (1957) estimated the heritability of the abdominal count using three different covariances (offspring-parent, paternal half-sib, full-sib) with similar results (0.51 ± 0.07s.e., 0.48 ± 0.11s.e., 0.53 ± 0.07s.e.). The similarity of the offspring-parent and full-sib estimates to the half-sib estimate indicates that most of the genetic variance is additive. Mackay and Lyman (2005) analyzed samples of single chromosomes from a natural population and perpetuated in chromosome substitution lines. The resulting histogram portrays the contribution to the standing crop of variation by particular chromosomes (e.g., chromosomes 2 and 3). Notice that the variance among chromosomes of a particular type (i.e., 2 vs 3) is substantial (Figure 5.9).

A picture of increasingly complicated genetic architecture has emerged as ever more sophisticated techniques have used to analyze genetic variation in *Drosophila* bristle numbers and other quantitative traits. In the first place, the *Drosophila* experience strongly reinforces the long-held view of quantitative geneticists that the more a trait is studied, the larger the number of loci that are discovered. Noting the steady climb in the number of loci found to affect bristle numbers as the resolution of QTL mapping studies improved,

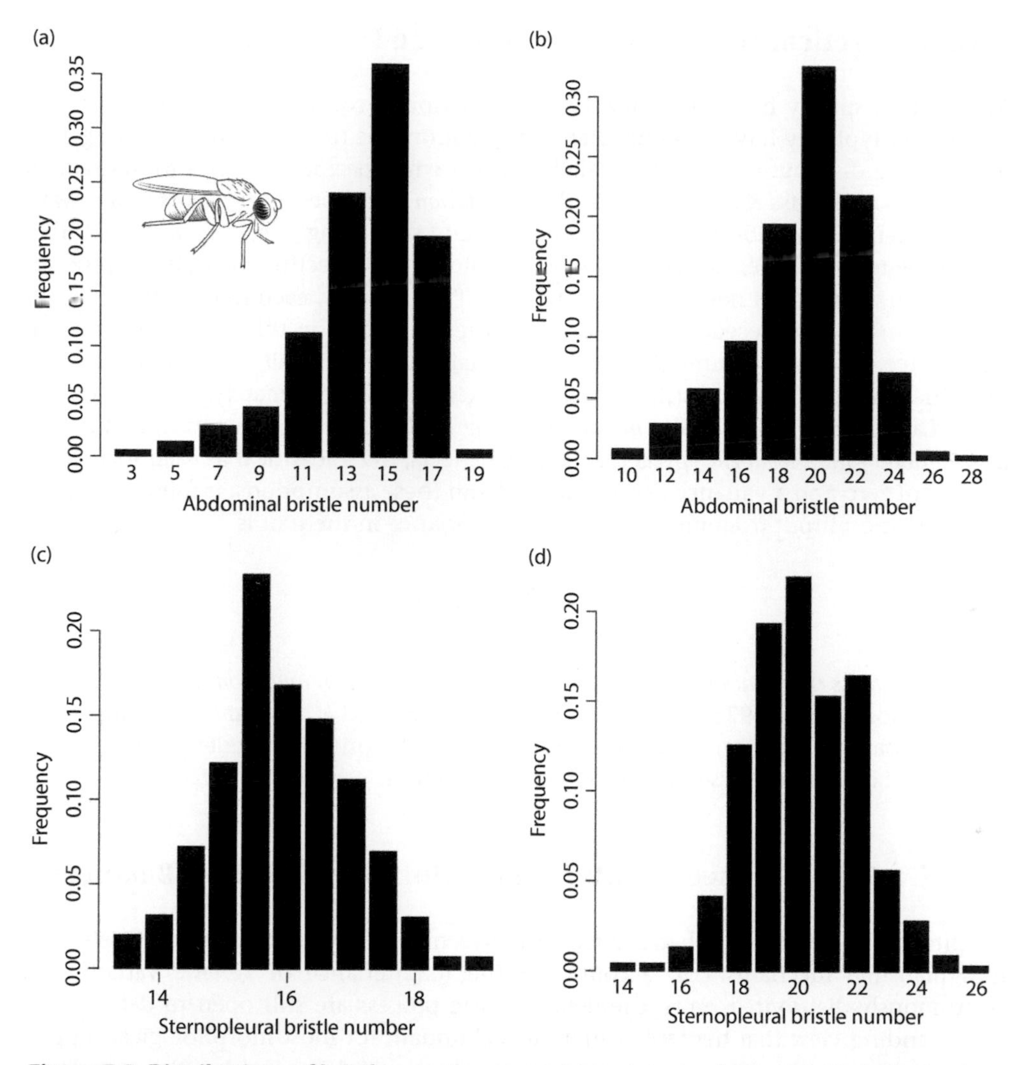

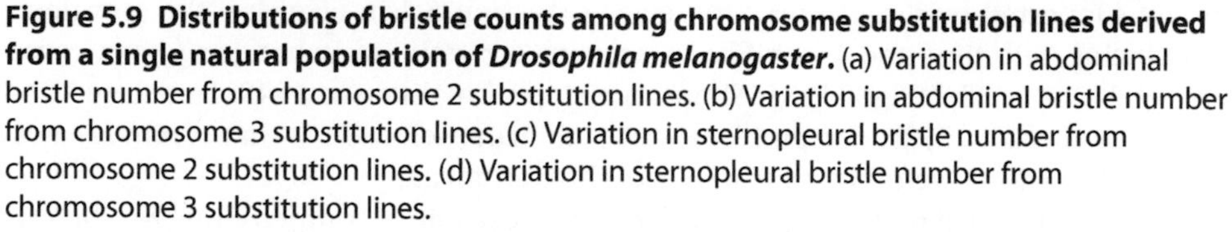

Figure 5.9 Distributions of bristle counts among chromosome substitution lines derived from a single natural population of *Drosophila melanogaster*. (a) Variation in abdominal bristle number from chromosome 2 substitution lines. (b) Variation in abdominal bristle number from chromosome 3 substitution lines. (c) Variation in sternopleural bristle number from chromosome 2 substitution lines. (d) Variation in sternopleural bristle number from chromosome 3 substitution lines.

After Mackay and Lyman (2005).

Mackay and Lyman (2005) concluded that over 100 loci potentially affect these counts. The distribution of contributions from these many loci appears to be normally distributed about zero, as we shall see in Section 5.12. Aside from the multiplicity of contributions, complexity also arises from a number of other sources: sex-specific effects, environment-specific effects, and epistasis. All of these complications are abundantly documented in the inheritance of bristle counts and many other quantitative traits in *Drosophila* (Mackay and Lyman 2005; Mackay et al. 2009).

5.12 Theoretical Input of Genetic Variance From Mutation

Models that specify how much new genetic variation is contributed by mutation each generation typically have two ingredients: a mutation rate (per haploid locus per generation) and a distribution of mutant (allelic) effects with specified variance for individual loci. We shall discuss *Kimura's infinite-alleles mutation model* because it leads to a powerful treatment of equilibrium between mutation and stabilizing selection (Kimura 1965). According to this model, each of n freely-recombining loci affecting the trait can produce an infinite sequence of new alleles by mutation. That is to say, each new mutation is different from all that preceded it. The effects of the mutations on the trait are assumed to be additive. If we also assume that the mutational effects are small, Kimura's model predicts that the equilibrium distribution of allelic effects is approximately Gaussian (Latter 1970; Lande 1976). Let μ be the *mutation rate per haploid locus per generation*, and let the *distribution of mutant allelic effects* (i.e., average effects, α) be identical at each locus with a mean of zero and a variance denoted as α^2. From these assumptions, it follows that the per-generation input from mutation to genetic variance in the trait is

$$U = 2n\mu \sum \alpha^2 = \zeta M, \tag{5.13}$$

where $\zeta = 2n\mu$ is the *total mutation rate*, and M is the *variance in mutational effects* summed across all loci (Lande 1976, 1977b). U is sometimes denoted σ_m^2. In a model of mutation-selection balance, this per-generation input from mutation to the genetic variance of a trait, U, is balanced by losses due to stabilizing selection.

5.13 Genetic Variance Maintained by Mutation-Selection Balance

Despite broad agreement that a balance between mutation and stabilizing selection is a plausible mechanism to account for the maintenance of genetic variation in continuously-distributed traits, the details of the process are still open to debate. The long standing view that heritable variation is abundant for most morphological traits is strongly supported by recent literature reviews (Section 5.10). Likewise, the prevalence of stabilizing selection and its ability to account for trait stasis has long been recognized (Haldane 1932; Wright 1935; Simpson 1944; Schmalhausen 1949; Lewontin 1964; Charlesworth et al. 1982). The problem with stabilizing selection is that it nibbles away at genetic variation, even though it stabilizes the trait mean.

Lande (1976) argued that polygenic mutation was capable of compensating for the variation nibbled away by stabilizing selection and showed that a *mutation-selection balance* (MSB) was consistent with data on both selection and mutation (Figure 5.10). Lande's argument was based on particular assumptions about mutation and selection, so one can ask how sensitive the conclusions are to those assumptions. One can also ask whether other sets of assumptions might account for the maintenance of variation while also accounting for related phenomena. The upshot of these considerations is that while Lande's (1976) proposal does a reasonable job of accounting for the maintenance of genetic variation, other more complicated sets of assumptions do at least as good a job on the maintenance score, while simultaneously accounting for more of the known features of polygenic mutation. At the same time, decisions about which assumptions to prefer are complicated by the fact that empirical data are still sketchy on the issue of how

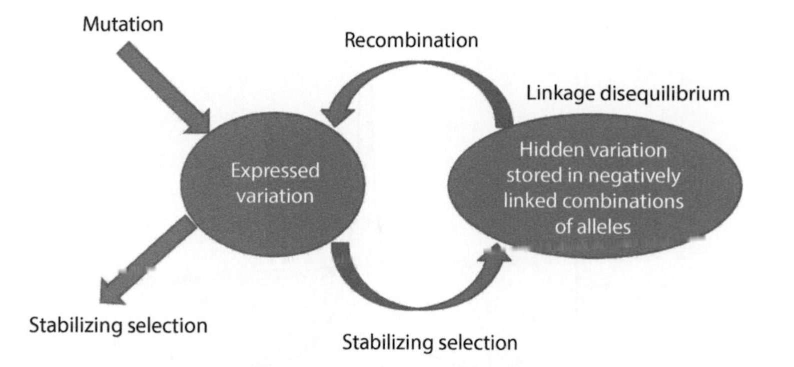

Figure 5.10 A flowchart view of mutation-selection balance. Recombination converts hidden variation into expressed variation by decreasing correlations between loci, while stabilizing selection has the reverse effect (centre). Mutation does not alter the hidden variation but contributes directly to the expressed genetic variation, while stabilizing selection depletes expressed variation (left). At equilibrium, the flow out of expressed variation into hidden variation equals the flow into expressed variation, regardless of the rate of recombination of the loci.
After Lande (1976).

mutational effects are distributed. In Lande (1976) and the MSB models discussed below, stabilizing selection is modelled as a Gaussian function, with width parameter ω (3.8).

The amount of genetic variation that can be maintained by mutation-selection balance is affected by the form of the distribution of mutational effects. In Section 5.12 we pointed out that Lande's (1976) model, based on Kimura's (1965) infinite-alleles model of mutation, implies that the equilibrium distribution of allelic effects at each locus is Gaussian. Other statistical models of mutation have been proposed (Crow and Kimura 1964; Kingman 1977, 1978; Zeng and Cockerham 1993) (Figure 5.11) and they can lead to quantitatively different predictions about how much genetic variation can be maintained.

Bürger's (2000) synthesis provides a useful summary of an extensive theoretical literature on mutation-selection balance. One simplifying result is that linkage relationships among loci have little effect on the equilibrium genetic variance in large populations. Consequently, single-locus results can be extrapolated to yield multiple locus conclusions. The Gaussian set of assumptions about mutation asserts that mutational effects are small relative to the single-locus genetic variances at equilibrium. The consequence is that most genetic variance arises from rare alleles of large effect. Under this view of mutation a variety of mathematical approaches predict that the equilibrium genetic variance is approximately $\sqrt{U\omega}$ (Lande 1976; Fleming 1979; Bürger 2000). Alternatively, if one assumes a *house of cards model of mutation*, which emphasizes the contribution of mutations of large effect (Kingman 1977, 1978), the equilibrium genetic variance is approximately $2U\omega$ (Turelli 1984; Bürger 2000).

A fundamentally different model of selection and mutation was proposed by (Robertson 1960). In this view, stabilizing selection arises from deleterious mutations that are disproportionally represented in genotypes that reside at the two ends of the trait distribution. According to this model, each mutation has two effects, one on the trait and the other on fitness itself. The trait can be viewed as selectively neutral or, in a generalization by Zhang and Hill (2002), as experiencing pure stabilizing selection, as well as the

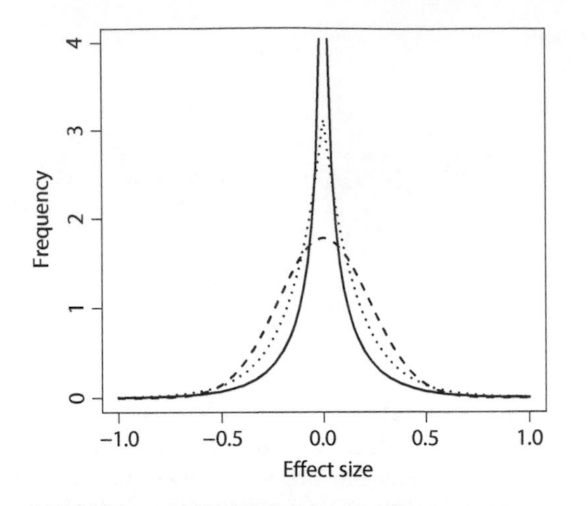

Figure 5.11 Three different candidate distributions for mutational effects. Each of these examples has a mean of zero and a variance of about 0.05: Gaussian distribution (dashed line), reflected exponential (Laplace) distribution (dotted line), and reflected gamma distribution (solid line).

After Bürger (2000).

apparent stabilizing selection induced by the pleiotropic effects of mutant alleles on fitness. Whatever the merit of Robertson's proposal, Zhang and Hill (2002) found that apparent stabilizing selection made a relatively small contribution to the genetic variance that was maintained at equilibrium.

Contemplating the concept of apparent stabilizing selection leads us to consider the effects of mutations on fitness and to include them in a model of mutation-selection balance. One well-established result is that most mutations are deleterious and tend to be recessive (Mukai et al. 1972; Simmons and Crow 1977; Mackay et al. 1992; Caballero and Keightley 1994; Keightley and Lynch 2003). Furthermore, partial dominance appears to be common in the inheritance of continuously-distributed traits. For both of these reasons, Zhang et al. (2004) included dominance in their investigation of mutation-selection balance, which used a joint-effect model that included both pure and apparent stabilizing selection, as in Zhang and Hill (2002). In line with some empirical results, Zhang et al. (2004) assumed that the distributions of allelic effects on the trait, $|a|$, and on fitness were more leptokurtic than a normal distribution and modelled them with Gamma distributions. Their general conclusion is that the inclusion of dominance in the model leads to the maintenance of a quantitatively higher level of additive genetic variance at equilibrium. In other words, a model without dominance leads to a conservative prediction about the maintenance of genetic variation.

Returning to our main theme of characterizing the opposing roles of stabilizing selection and mutation in maintaining genetic variation, we shall now focus on the results of the Gaussian model of allelic effects in a completely additive framework (no dominance or epistasis), not because this is the only possible model, but because its results are in broad agreement with more complicated—and perhaps more realistic—models.

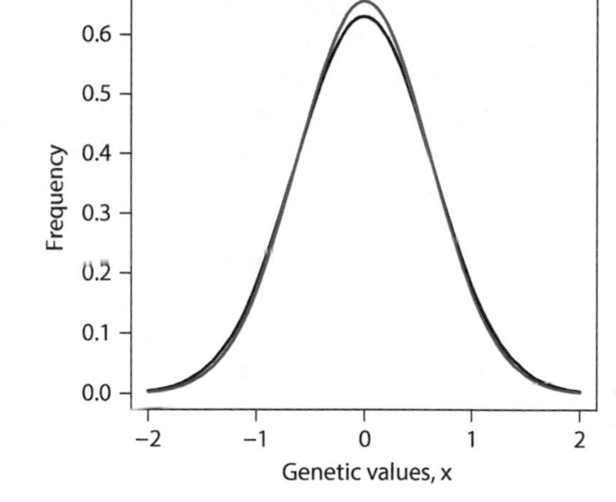

Figure 5.12 Curves showing the hypothetical distribution of breeding values before selection (blue; $G = 0.400$) and after selection ($G^* = 0.368$), when stabilizing selection is relatively strong, ($\gamma = -0.2$) and $\beta = 0$. In an equilibrium population, mutation would restore variation in the next generation.

Stabilizing selection reduces the standing crop of genetic variance within a generation by an amount given by

$$\Delta_S G = G^2 \left(\gamma - \beta^2\right) \tag{5.14}$$

(Lande 1976, 1980; Phillips and Arnold 1989). At equilibrium, the losses due to selection are balanced by the input from mutation and recombination, so that

$$U = \zeta M = -\Delta_S G \tag{5.15}$$

(Lande 1980). If both the phenotypic distribution, $p(z)$, and the ISS are Gaussian, and assuming weak selection, the equilibrium genetic variance, $\hat{G}$, is a function of the input from mutation and the curvature of the AL,

$$\hat{G} \cong \sqrt{U\tilde{\omega}}, \tag{5.16}$$

where $\tilde{\omega} \equiv \omega + E \cong \omega + P = \Omega$ (Lande 1976, 1977b, 1980). In particular, we see—as expected—more genetic variance at equilibrium when mutational input is large and stabilizing selection is weak (large ω). Substituting realistic values for G and γ, we see that the effect of stabilizing selection in diminishing genetic variation and corresponding restorative power of mutation are both relatively weak. Even when stabilizing selection is relatively strong, the amount of genetic variation lost each generation due to selection and restored by mutation is small (Figure 5.12).

5.14 Empirical Tests of Mutation-Selection Balance and Other Models for Maintenance of Genetic Variation

Although the simplicity of MSB makes it the obvious working hypothesis (Barton and Keightley 2002), it is not the only possible model. In an ideal world, we would have quantitative head-to-head tests of all the competing models with empirical data. The available evaluations of models fall far short of that ideal. The best tests of MSB and other models have focused on fitness and its components. Such life-history traits are perpetually under directional selection and so the problem of maintaining genetic variation is most severe. Charlesworth and Hughes (2000) reviewed 25 years of results on quantitative inheritance of life-history traits in *D. melanogaster* and used those results to evaluate three models: pure balancing selection at individual loci, directional selection (in opposing directions on different fitness components), and MSB. Although the *Drosophila* results suggest that the mutational contribution to variation is substantial, that contribution is probably not sufficient to explain observed levels of genetic variance and inbreeding depression. The apparent excess in maintained genetic variance could be a consequence of any of a variety of mechanisms (antagonistic pleiotropy, temporal or spatial variation in selection, or frequency-dependent selection). The possible roles of these competing mechanisms in the maintenance of genetic variance are discussed in detail by Mitchell-Olds et al. (2007) and Grieshop and Arnqvist (2018).

5.15 The Mutation Process as Revealed by Mutation Accumulation Experiments

It is a remarkable fact that mutation alone can supply enough genetic variation to sustain appreciable response to selection in polygenic traits. This result has been established by *mutation accumulation* (MA) *experiments* in which directional selection is imposed on highly inbred lines. Response to selection can then be attributed to new mutations. Mutational input to variation in the abdominal and sternopleural bristle counts of *Drosophila* has been analyzed with MA experiments. Starting from an inbred stock population, Mackay et al. (2005) established replicate lines and subjected each to truncation selection for high or low bristle counts for 206 generations. All of the selection lines showed appreciable responses in the anticipated directions (see Chapter 10). MA experiments can be used to estimate mutational variance from these responses to selection. Using the *infinitesimal model* (Bulmer 1971) and averaging over all 206 generations and both traits, the average estimate of mutational variance, M, is $0.00315 \pm 0.24 \times 10^{-3}$ s.e. Standardizing by the environmental variances, the average estimate of M/E is $0.00024 \pm 0.03 \times 10^{-3}$ s.e. The latter estimate is consistent with earlier estimates for these traits.

5.16 Estimates of Mutation Input to Trait Variance

Estimates of *standardized mutational variance* for morphological traits consistently fall in the range 0.002–0.02 M/E (Lynch 1988; Lynch and Walsh 1998; Halligan and Keightley 2009) (Table 5.3). These results imply that, beginning with a totally inbred population, approximately a thousand generations would be required to build genetic variance to a heritability level of 50%, so that $G \approx E$.

The statistical properties of the new mutants that arise in mutation accumulation experiments can be characterized, giving us a window on the mutational process. In a particularly revealing study, Dilda and Mackay (2002) selected on abdominal and sternopleural bristle numbers, starting with a highly inbred line which was initially free of P-elements. After producing high and low lines by truncation selection, Dilda and Mackay extracted single chromosomes from each selected line and perpetuated them in isogenic lines. Using interval mapping to determine genomic positions of bristle count QTLs, they characterized QTLs for sternopleural and abdominal bristle numbers. The combined distributions of additive and dominance effects are shown in Figure 5.13. The distribution

Table 5.3 Estimates of standardized mutational variance. The third column shows the ratio of mutational to environmental variance, multiplied by 1000. From Lynch and Walsh (1998) with permission.

Species	Trait	U/E ($\times 10^3$)	Reference
Drosophila melanogaster	Abdominal bristle number	3.5	See Lynch and Walsh 1998
	Sternopleural bristle number	4.3	See Lynch and Walsh 1998
	Ethanol resistance	0.9	Weber and Diggins 1990
	Body mass	4.7	Clark et al. 1995
	Wing dimensions	2.0	Santiago et al. 1992
	Viability	0.3	See Lynch and Walsh 1998
Tribolium castaneum	Pupal mass	9.1	Goodwill and Endfield 1971
Daphnia pulex	Life-history traits	1.7	Lynch 1985
Mouse	Lengths of limb bones	23.4	Bailey 1959
	Mandible measures	23.1	Festing 1973
	Skull measures	5.2	See Lynch and Walsh 1998
	Six-week mass	3.4	Caballero and Keightley 1995
Arabidopsis thaliana	Life-history traits	3.9	Schultz and Willis 1999
Maize	Plant size	11.2	Russell et al. 1963
	Reproductive traits	7.3	Russell et al. 1963
Rice	Plant size	3.0	Oka et al. 1958
	Reproductive traits	2.8	Sakai and Suzuki 1964
Barley	Life-history traits	0.2	Cox et al. 1987

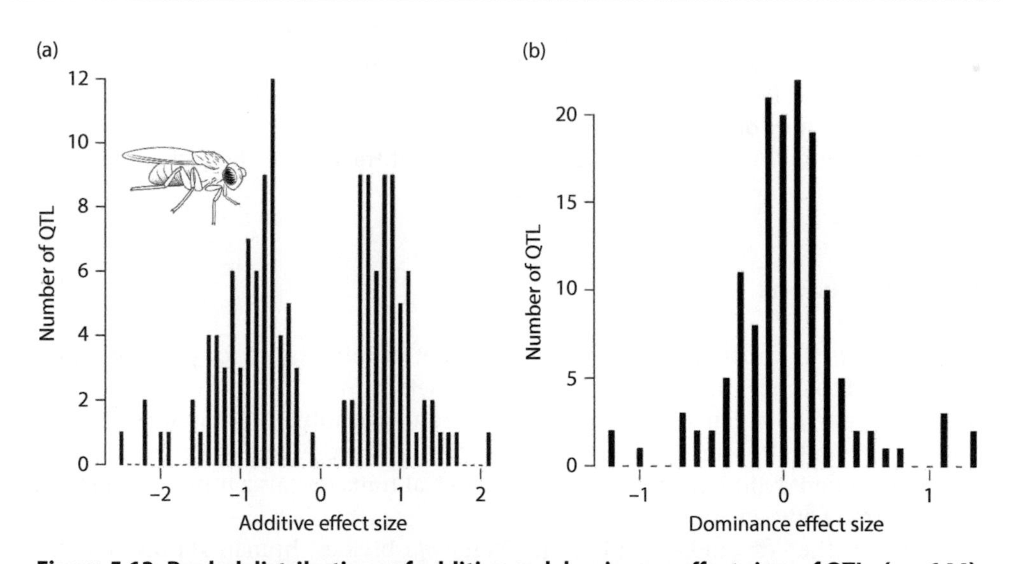

Figure 5.13 Pooled distributions of additive and dominance effect sizes of QTLs (n = 144) for abdominal and sternopleural bristle numbers in *Drosophila melanogaster*.
(a) Distribution of additive effect sizes shown as a ratio of estimated additive effect to additive genetic standard deviation, $a/\sqrt{G}$. (b) Distribution of dominance effect sizes, d/a.
After Dilda & Mackay (2002).

of additive effects (Figure 5.13a) shows a few large effects at both ends of the distribution, with an increasing frequency of smaller effects sizes toward the limits of detection near zero. The distribution of dominance effects (Figure 5.13b) is similarly centred about zero (but with no gap near zero) and with no indication of bias towards negative or positive values. Both distributions resemble the exponential distribution shown in Figure 5.9.

5.17 The Mutation Process as Revealed by Effects of P-Element Mutagenesis on Bristle Numbers in *Drosophila*

Special techniques to study the statistical properties of new mutations are available in *Drosophila melanogaster*. In particular, *P-element mutagenesis* takes advantage of the special properties of transposable elements to create stable lines that contain one or a few new mutations on particular chromosomes (Robertson et al. 1988). In specially created lines, the rate of P-element transposition occurs at a particularly high rate (two inserts per chromosome per generation), so that properties of even rare mutational events (i.e., rare alleles of large effect) can be characterized (Mackay et al. 1992). Furthermore, the statistical properties of P-element inserts mimic the natural mutation process. Indeed more than 50% of natural mutants in *D. melanogaster* represent P-element insertions (Finnegan 1992). Starting with a highly inbred host strain that is initially free of P-elements, replicate lines are produced that differ only in P-element insertions. The effects on these now stable insertions (QTLs) on particular phenotypes such as bristle counts can then be characterized.

Analyses based on P-element mutagenesis reveal that mutations at possibly hundreds of loci affect numbers of abdominal and sternopleural bristles. Two teams of investigators created nearly 2800 P-element insertion lines (Lyman et al. 1996; Spradling et al. 1999). Analysis and mapping of these two sets of lines revealed, respectively, 42 and 262 loci that affect bristle numbers (Norga et al. 2003). A substantial fraction of the loci identified in both screens are involved in neurogenic processes.

5.18 Conclusions

- Additive genetic variance is the statistical concept of inheritance for polygenic traits that is used to model the evolution by drift and selection.
- Additive genetic variance can be estimated from phenotypic covariance between relatives, from traits scored on a pedigree, or from trait scores and genomic data.
- A survey of heritability estimates for morphological traits reveals a unimodal distribution centred between about 0.4 and 0.5.
- Analyses of the best-studied traits (e.g., *Drosophila* bristles, human stature) indicate that dozens, if not hundreds of loci contribute to genetic variation in natural populations. These contributing loci are distributed on multiple chromosomes throughout the genome.
- The maintenance of genetic variance can be usefully modelled as a balance between gain from mutation and migration and loss from selection.
- Empirical studies of polygenic mutation reveal a small but measurable gain from mutation each generation. The input from mutation is typically in the range of one-fifth of a per cent to 2% of the environmental variance, E, of the trait in question.
- Mutation-selection balance in conjunction with migration can account for the standing crop of genetic variance observed in natural populations.

Literature Cited

Arnold, S. J. 1994. Multivariate inheritance and evolution: a review of concepts. pp. 17–48 *in* C. R. B. Boake, ed. Quantitative Genetic Studies of Behavioral Evolution. University of Chicago Press, Chicago.

Arnold, S. J., and C. R. Peterson. 2002. A model for optimal reaction norms: the case of the pregnant garter snake and her temperature-sensitive embryos. Am. Nat. 160:306–316.

Arnold, S. J., and P. C. Phillips. 1999. Hierarchical comparison of genetic variance-covariance matrices. II. Coastal-inland divergence in the garter snake, *Thamnophis elegans*. Evolution 53:1516–1527.

Bailey, D. W 1959. Rates of subline divergence in highly inbred lines of mice. J Hered. 50:26–30.

Barton, N. H., and P. D. Keightley. 2002. Understanding quantitative genetic variation. Nat. Rev. Genet. 3:11–21.

Bérénos, C., P. A. Ellis, J. G. Pilkington, and J. M. Pemberton. 2014. Estimating quantitative genetic parameters in wild populations: a comparison of pedigree and genomic approaches. Mol. Ecol. 23:3434–3451.

Bryant, E. H. 1986. On the use of logarithms to accomodate scale. Syst. Zool. 35:552–559.

Bulmer, M. G. 1971. The effect of selection on genetic variability. Am. Nat. 105:201–211.

Bürger, R. 2000. The Mathematical Theory of Selection, Recombination, and Mutation. John Wiley and Sons, New York.

Caballero, A., and P. D. Keightley. 1994. Inferences on genome-wide deleterious mutation rates in inbred populations of *Drosophila* and mice. Genetica 102/103:229–239.

Caballero, A., and P. D. Keightley. 1995. Accumulation of mutations affecting body weight in inbred mouse lines. Genet. Res. 65:145–149.

Charlesworth, B., and K. A. Hughes. 2000. The maintenance of genetic variation in life history traits. pp. 369–391 *in* R. S. Singh and C. B. Krimbas, eds. Evolutionary. Genetics from Molecules to Morphology. Cambridge University Press, Cambridge.

Charlesworth, B., R. Lande, and M. Slatkin. 1982. A neo-Darwinian commentary on macroevolution. Evolution 36:474.

Clark, A. G., L. Wang, and T. Hulleberg. 1995. Spontaneous mutation rate of modifiers of metabolism in *Drosophila*. Genetics 139:767–779.

Clayton, G. A., and R. Robertson. 1957. An experimental check on quantitative genetical theory. II. The long-term effects of selection. J. Genet. 55:152–170.

Cockerham, C. C. 1959. Partitions of hereditary variance for various genetic models. Genetics 44:1141–1148.

Cox, T. S., D. J. Cox, and K. J. Frey. 1987. Mutations for polygenic traits in barley under nutrient stress. Euphytica 36:823–829.

Coyne, J. A., and R. Lande. 1985. The genetic basis of species differences in plants. Am. Nat. 126:141–145.

Crow, J. F., and Kimura. 1970. An Introduction to Population Genetics Theory. Harper and Row, New York.

Crow, J. F., and M. Kimura. 1964. The theory of genetic loads. Proc XI Intl Cong Genet 2:495–505.

Dilda, C. L., and T. F. C. Mackay. 2002. The genetic architecture of *Drosophila* sensory bristle number. Genetics 162:1655–1674.

Falconer, D. S. 1960. An Introduction to Quantitative Genetics. Ronald Press, New York.

Falconer, D. S., and T. F. C. Mackay. 1996. Introduction to Quantitative Genetics. 4th ed. Longman Scientific & Technical. J. Wiley & Son, Essex.

Festing, M. F. W. 1973. A multivariate analysis of subline divergence in the shape of the mandible in C57BL/Gr mice. Genet. Res. 21:121–132.

Finnegan, D. J. 1992. Transposable elements. Curr. Opin. Genet. Dev. 2:861–867.

Fisher, R. A. 1918. The correlation between relatives on the supposition of Mendelian inheritance. Trans Roy Soc Edinb 52:399–433.

Fisher, R. A. 1958. The Genetical Theory of Natural Selection. Dover Publications, New York.

Fleming, W. H. 1979. Equilibrium distributions of continuous polygenic traits. SIAM J. Appl. Math. 36:148–168.

Galton, F. 1889. Natural Inheritance. MacMillan, London.
Gomulkiewicz, R., J. G. Kingsolver, P. A. Carter, and N. Heckman. 2018. Variation and evolution of function-valued traits. Annu. Rev. Ecol. Evol. Syst. 49:139–164.
Goodwill, R., and F. D. Endfield. 1971. Heterozygosity in inbred lines of *Tribolium castaneum*. Theor. Appl. Genet. 41:5–12.
Gottlieb, L. D. 1984. Genetics and morphological evolution in plants. Am. Nat. 123:681–709.
Grieshop, K., and G. Arnqvist. 2018. Sex-specific dominance reversal of genetic variation for fitness. PLOS Biol. 16:e2006810.
Hadfield, J. D. 2010. MCMC Methods for multi-response generalized linear mixed models: The MCMCglmm *R* Package. J. Stat. Softw. 33.
Haldane, J. B. S. 1932. The Causes of Evolution. Harper, New York.
Halligan, D. L., and P. D. Keightley. 2009. Spontaneous mutation accumulation studies in evolutionary genetics. Annu. Rev. Ecol. Evol. Syst. 40:151–172.
Henderson, C. R. 1976. A simple method for computing the inverse of a numerator relationship matrix used in prediction of breeding values. Biometrics 32:69–83.
Henderson, C. R. 1984. Applications of Linear Models in Animal Breeding. University of Guelph, Guelph.
Hindorff, L. A., P. Sethupathy, H. A. Junkins, E. M. Ramos, J. P. Mehta, F. S. Collins, and T. A. Manolio. 2009. Potential etiologic and functional implications of genome-wide association loci for human diseases and traits. Proc. Natl. Acad. Sci. 106:9362–9367.
Houle, D. 1992. Comparing evolvability and variability of quantitative traits. Genetics 130:195–204.
Keightley, P. D., and M. Lynch. 2003. Toward a realistic model of mutations affecting fitness. Evolution 57:683–685.
Kimura, M. 1965. A stochastic model concerning the maintenance of genetic variability in quantitative characters. Proc. Natl. Acad. Sci. 54:731–736.
Kingman, J. F. C. 1977. On the properties of bilinear models for the balance between genetic mutation and selection. Math. Proc. Camb. Phil. Soc. 81:443–453.
Kingman, J. F. C. 1978. A simple model for the balance between selection and mutation. J Appl. Prob. 15:1–12.
Kirkpatrick, M., and N. Heckman. 1989. A quantitative genetic model for growth, shape, reaction norms, and other infinite-dimensional characters. J. Math. Biol. 27:429–450.
Kirkpatrick, M., and R. Lande. 1989. The evolution of maternal characters. Evolution 43:485–503.
Kirkpatrick, M., and R. Lande. 1992. The evolution of maternal characters: Errata. Evolution 46:284.
Lande, R. 1976. The maintenance of genetic variability by mutation in a polygenic character with linked loci. Genet. Res. 26:221–235.
Lande, R. 1977a. On comparing coefficients of variation. Syst. Zool. 26:214–217.
Lande, R. 1977b. The influence of the mating system on the maintenance of genetic variability in polygenic characters. Genetics 86:485–498.
Lande, R. 1980. The genetic covariance between characters maintained by pleiotropic mutations. Genetics 94:201–215.
Latter, B. D. H. 1970. Selection in finite populations with multiple alleles. II. Centripetal selection, mutation, and isoallelic variation. Genetics 66:165–186.
Lewontin, R. C. 1964. The interaction of selection and linkage. II. Optimum models. Genetics 50:757–782.
Lyman, R. F., F. Lawrence, S. V. Nuzhdin, and T. F. C. Mackay. 1996. Effects of single P-element insertions on bristle number and viability in *Drosophila melanogaster*. Genetics 143:277–292.
Lynch, M. 1985. Spontaneous mutations for life-history characters in an obligate parthenogen. Evolution 39:804–818.
Lynch, M. 1988. The rate of polygenic mutation. Genet. Res. 51:137–148.
Lynch, M., and B. Walsh. 1998. Genetics and Analysis of Quantitative Traits. Sinauer Associates, Inc., Sunderland, Massachusetts.
Mackay, T. F., R. F. Lyman, and M. S. Jackson. 1992. Effects of P element insertions on quantitative traits in *Drosophila melanogaster*. Genetics 130:315–332.

Mackay, T. F. C., and R. F. Lyman. 2005. *Drosophila* bristles and the nature of quantitative genetic variation. Philos. Trans. R. Soc. B Biol. Sci. 360:1513–1527.

Mackay, T. F. C., R. F. Lyman, and F. Lawrence. 2005. Polygenic mutation in *Drosophila melanogaster*. Genetics 170:1723–1735.

Mackay, T. F. C., E. A. Stone, and J. F. Ayroles. 2009. The genetics of quantitative traits: challenges and prospects. Nat. Rev. Genet. 10:565–577.

Meyer, K., and W. G. Hill. 1997. Estimation of genetic and phenotypic covariance functions for longitudinal or 'repeated' records by restricted maximum likelihood. Livest. Prod. Sci. 47: 185–200.

Meyer, K. G. 2007. WOMBAT—A tool for mixed model analyses in quantitative genetics by restricted maximum likelihood (REML). J. Zhejiang Univ. Sci. B 8:815–821.

Mitchell-Olds, T., J. H. Willis, and D. B. Golstein. 2007. Which evolutionary processes influence natural genetic variation for phenotypic traits? Nat. Rev. Genet. 8:845–856.

Mousseau, T. A., and D. A. Roff. 1987. Natural selection and the heritability of fitness components. Heredity 59:181–197.

Mukai, T., S. I. Chigusa, L. E. Mettler, and J. F. Crow. 1972. Mutation rate and dominance of genes affecting viability in *Drosophila melanogaster*. Genetics 72 335–355.

Nelder, J. A., and R. W. M. Wedderburn. 1972. Generalized linear models. J. R. Stat. Soc. Ser. A 135:370–384.

Norga, K. K., M. C. Gurganus, C. L. Dilda, A. Yamamoto, R. F. Lyman, P. H. Patel, M. Rubin, R. A. Hoskins, T. F. C. Mackay, and H. J. Bellen 2003. Quantitative analysis of bristle number in *Drosophila* mutants identifies genes involved in neural development. Curr. Biol. 13:1388–1396.

Oka, H. I., J. Hayashi, and I. Shiojiri. 1958. Induced mutations of polygenes for quantitative characters in rice. J Hered. 49:11–14.

Pemberton, J. M. 2008. Wild pedigrees: The way forward. Proc. R. Soc. B Biol. Sci. 275:613–621.

Phillips, P. C., and S. J. Arnold. 1989. Visualizing multivariate selection. Evolution 1209–1222.

Phillips, P. C., and S. J. Arnold. 1999. Hierarchical comparison of genetic variance-covariance matrices. I. Using the Flury hierarchy. Evolution 53:1506–1515.

Postma, E. 2014. Four decades of estimating heritabilities in wild vertebrate populations: improved methods, more data, better estimates? pp. 16–33 *in* A. Charmantier, D. Garant, L. Kruuk, and E. B. Loeske, eds. Quantitative Genetics in the Wild. Oxford University Press, New York.

Robertson, A. 1960. A theory of limits in artificial selection. Proc Roy Soc Lond. B 153:234–249.

Robertson, H. M., C. R. Preston, R. W. Phillis, D. M. Johnson-Schlitz, W. K. Benz, and W. R. Engels. 1988. A stable genomic source of P element transposace in *Drosophila melanogaster*. Genetics 118:461–470.

Russell, W. A., G. F. Sprague, and L. H. Penny. 1963. Mutations affecting quantitative characters in long-term inbred lines of maize. Crop Sci 3:175–178.

Sakai, K.-I., and A. Suzuki. 1964. Induced mutation and pleiotropy of genes responsible for quantitative characters in rice. Rad Bot. 4:141–151.

Santiago, E. J., A. Albomoz, M. A. T. Dominguez, and C. Lopez-Fanjul. 1992. The distribution of spontaneous mutations on quantitative traits and fitness in *Drosophila melanogaster*. Genetics 132:771–781.

Schmalhausen, I. I. 1949. Factors of Evolution, the Theory of Stabilizing Selection. Blakiston, Philadelphia.

Schultz, S. T., and J. H. Willis. 1999. Individual variation in inbreeding depression: The roles of inbreeding and mutation. Genetics 141:1209–1223.

Searle, S. R. 1987. Linear Models for Unbalanced Data. John Wiley and Sons, New York.

Simmons, M. J., and J. F. Crow. 1977. Mutations affecting fitness in Drosophila populations. Ann Rev Genet. 11:49–78.

Simpson, G. G. 1944. Tempo and Mode in Evolution. Columbia University Press, New York.

Smith, J. N. M., and A. A. Dhondt. 1980. Experimental confirmation of heritable morphological variation in a natural population of song sparrows. Evolution 34:1155–1158.

Sorensen, D., and D. Gianola. 2002. Likelihood, Bayesian, and MCMC Methods in Quantitative Genetics. Springer, New York.

Spradling, A. C., D. Stern, A. Beaton, E. J. Rhem, T. Laverty, N. Mozden, S. Misra, and G. M. Rubin. 1999. The Berkeley *Drosophila* genome project gene disruption project: Single P-element insertions mutating 25% of vital *Drosophila* genes. Genetics 153:135–177.

Tong, L., and E. Thompson. 2008. Multilocus Lod scores in large pedigrees: Combination of exact and approximate calculations. Hum. Hered. 65:142–153.

Turelli, M. 1984. Heritable genetic variation via mutation-selection balance: Lerch's zeta meets the abdominal bristle. Theor Pop Biol 25:138–193.

Visscher, P. M., M. A. Brown, M. I. McCarthy, and J. Yang. 2012. Five Years of GWAS Discovery. Am. J. Hum. Genet. 90:7–24.

Weber, K. E., and L. T. Diggins. 1990. Increased selection response in larger populations II. Selection for ethanol vapor resistance in *Drosophila melanogaster* at two population sizes. Genetics 125:585–597.

Weinberg, W. 1908. Ueber den Nachweis der Vererbung beim Menschen. Jh Ver. vaterl. Naturk. Württemb. 64:369–382.

Willham, R. L. 1963. The covariance between relatives for characters composed of components contributed by related individuals. Biometrics 19:18–27.

Willham, R. L. 1972. The role of maternal effects in animal breeding. III. Biometrical aspects of maternal effects in animals. J. Anim. Sci. 35:1288–1293.

Wright, S. 1921. Systems of mating. I. The biometric relations between parents and offspring . Genetics 6:111–123.

Wright, S. 1935. Evolution in populations in approximate equilibrium. J. Genet. 30:257–266.

Wright, S. 1968. Evolution and the Genetics of Populations: a Treatise. University of Chicago Press, Chicago.

Yang, J., B. Benyamin, B. P. McEvoy, S. Gordon, A. K. Henders, D. R. Nyholt, P. A. Madden, A. C. Heath, N. G. Martin, G. W. Montgomery, M. E. Goddard, and P. M. Visscher. 2010. Common SNPs explain a large proportion of the heritability for human height. Nat. Genet. 42:565–569.

Yang, J., S. H. Lee, M. E. Goddard, and P. M. Visscher. 2011a. GCTA: A tool for genome-wide complex trait analysis. Am. J. Hum. Genet. 88:76–82.

Yang, J., T. A. Manolio, L. R. Pasquale, E. Boerwinkle, N. Caporaso, J. M. Cunningham, M. de Andrade, B. Feenstra, E. Feingold, M. G. Hayes, W. G. Hill, M. T. Landi, A. Alonso, G. Lettre, P. Lin, H. Ling, W. Lowe, R. A. Mathias, M. Melbye, E. Pugh, M. C. Cornelis, B. S. Weir, M. E. Goddard, and P. M. Visscher. 2011b. Genome partitioning of genetic variation for complex traits using common SNPs. Nat. Genet. 43:519–525.

Zeng, Z.-B., and C. C. Cockerham. 1993. Mutation models and quantitative genetic variation. Genetics 133:729–736.

Zhang, X.-S., and W. G. Hill. 2002. Joint effects of pleiotropic selection and stabilizing selection on the maintenance of quantitative genetic variation at mutation-selection balance. Genetics 162:459–471.

Zhang, X.-S., J. Wang, and W. G. Hill. 2004. Influence of dominance, leptokurtosis and pleiotropy of deleterious mutations on quantitative genetic variation at mutation-selection balance. Genetics 166:597–610.

CHAPTER 6

Inheritance of Multiple Traits

Overview—The G-matrix provides a useful way to model inheritance when multiple traits are affected by many genes and genetically coupled. The G-matrix is an m trait × m trait table with genetic variances on its main diagonal and genetic covariances elsewhere. The genetic covariance between two traits summarizes genetic connections arising from linkage disequilibrium and/or pleiotropy. Studies of multiple trait inheritance often reveal appreciable genetic correlations between traits, especially when those traits are size-related. The M-matrix summarizes the patterns of mutational effects on multiple traits. Just as covariances among relatives are the foundation for estimation of the G-matrix, the accumulation of mutations in inbred lines is the foundation for estimation of the M-matrix. Studies of G- and M-matrices suggest that these matrices have almost as many dimensions of variation as the number of characters included in the study. Consequently, these matrices are unlikely to constrain the course of evolution on long evolutionary timescales. The entire G-matrix, as well as its constituent genetic variances and covariances, can be visualized as achieving a stable balance between opposing forces. This stable equilibrium represents a compromise between alternative structures imposed by the processes of selection, mutation, recombination, and migration.

6.1 The Genotype-Phenotype Map for Multiple Traits

The relationship between genes and traits in quantitative genetics, the *genotype-phenotype map*, is a compromise between reality and tractability. In reality the causal path from a gene to a trait involves complex interactions with other genes and the environment. The implicit map that we will use is based on the ideas that traits are affected by many genes (*polygeny*) of small effect, individual genes affect more than one trait (*pleiotropy*), and the environment exerts a residual, nonheritable influence on trait expression (Figure 6.1). These simple ideas can be used to construct a useful and tractable model of inheritance that can also be used as a foundation for modelling more complicated paths from gene to trait. In the rest of this book, we will employ the model shown in Figure 6.1e, which incorporates multiple traits, polygeny, and pleiotropy but no additional complications.

Evolutionary Quantitative Genetics. Stevan J. Arnold, Oxford University Press. © Stevan Arnold (2023).
DOI: 10.1093/oso/9780192859389.003.0007

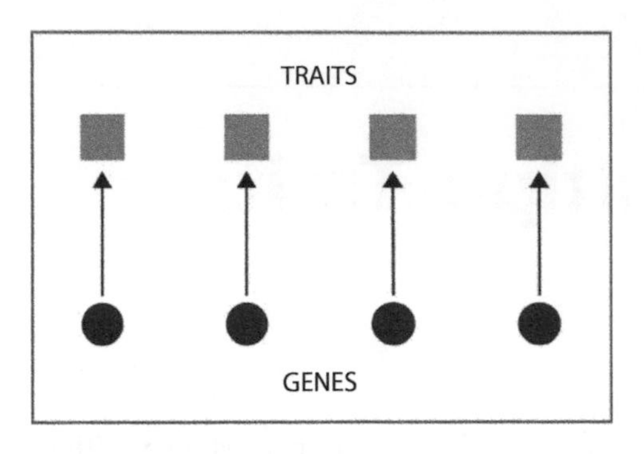

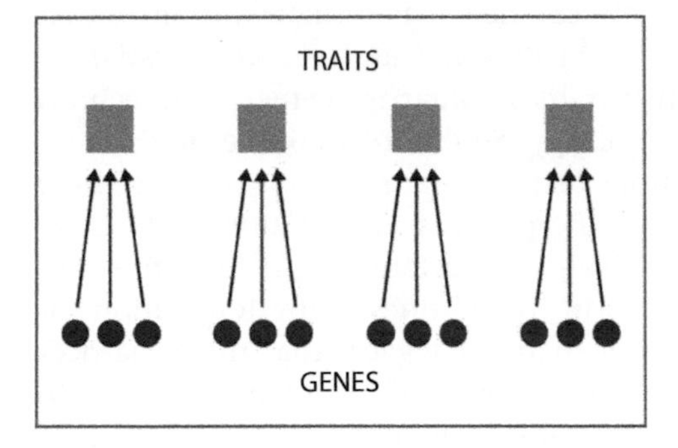

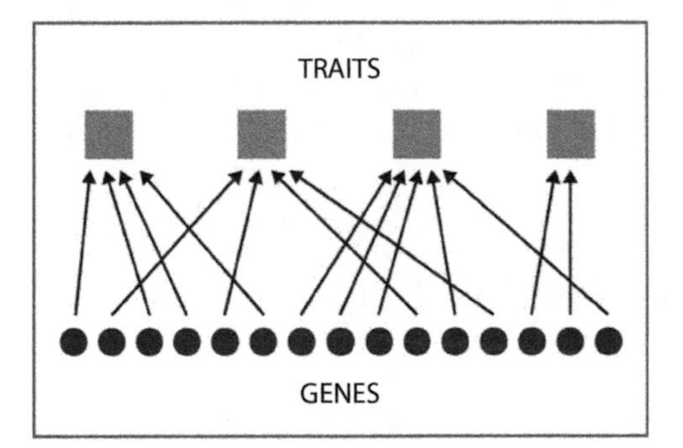

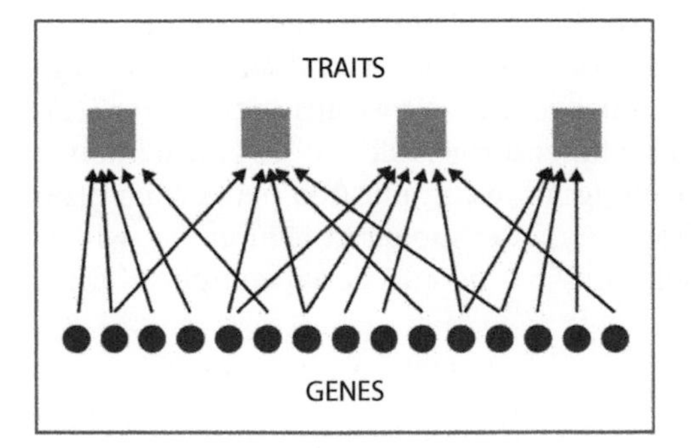

Figure 6.1 Alternative genotype-phenotype maps for multiple traits. (a) Each trait affected by a single, private gene. (b) Each trait affected by multiple private genes. (c) Each trait affected by multiple non-private genes, no pleiotropy. (d) Each trait affected by multiple non-private genes, no pleiotropic. (e) Each trait affected by multiple non-private genes, with pleiotropy.

6.2 Multivariate Resemblance Between Parents and Offspring, the *G*-matrix

The multivariate generalization of genetic variance is the *G*-matrix, the key player in models for the inheritance of multiple traits. The *G*-matrix summarizes multivariate information about parent-offspring resemblance, just as a single scatterplot captures heritability in the univariate case. Returning to our example of vertebral numbers in garter snakes (Figure 5.5) that we used to estimate heritabilities, Figure 6.2 shows cross-trait plots, one trait in mothers is plotted against another trait in daughters. These two plots can be used to estimate the genetic covariance between body and tail vertebral numbers.

Converting the information in Figs. 5.5 and 6.2 into three parameter estimates, we can display the information in the following table, a *G*-matrix,

$$\boldsymbol{G} = \begin{bmatrix} G_{11} & G_{12} \\ G_{12} & G_{22} \end{bmatrix} = \begin{bmatrix} 8.17 & 3.78 \\ 3.78 & 8.16 \end{bmatrix}, \tag{6.1}$$

in which $G_{11} = 8.17$ is the additive genetic variance for body vertebral number, $G_{22} = 8.16$ is the additive genetic variance for tail vertebral number, and $G_{12} = 3.78$ is the additive genetic covariance between body and tail vertebral numbers (Arnold and Phillips 1999; Phillips and Arnold 1999). The new element in this estimation exercise is genetic covariance, which describes how two traits run together in families. In the current example with only two traits, the *G*-matrix houses a single genetic covariance, but with m traits the *G*-matrix would be populated with $\left(m^2 - m\right)/2$ covariances. The *G*-matrix is a convenient repository for all this information. More importantly, in matrix form all of the

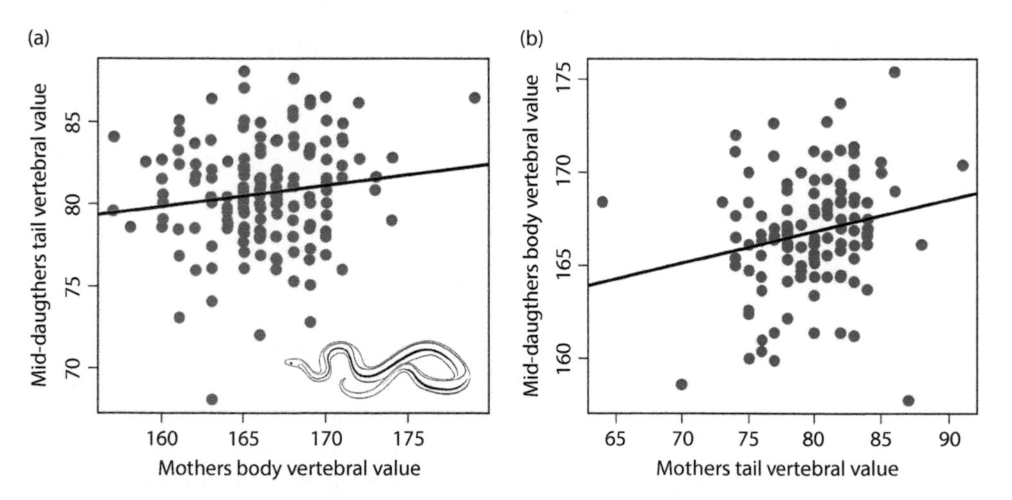

Figure 6.2 Examples of genetic covariance estimated from offspring-parent plots in the garter snake *T. elegans* (inland population). (a) The average tail vertebral count of daughters as a function of mother's body vertebral count, n = 154. (b) The average body vertebral count of daughters as a function of mother's tail vertebral count, n = 117. The estimate of genetic covariance based on both sets of data is 3.78 ± 1.67s.e.

Data from Phillips and Arnold (1999).

inheritance information can be used to make predictions about multivariate responses to selection and drift. To appreciate the important role that the *G*-matrix plays, we need to consider its conceptual underpinnings and its relationship to four other matrices, $\boldsymbol{P}$, $\boldsymbol{E}$, $\boldsymbol{M}$, and $\boldsymbol{U}$. None of these matrices is novel. We have seen each of them in univariate guise, as scalar variables in Chapter 5.

The multivariate analogue of our univariate expression for heritability (5.3) is

$$Cov(\boldsymbol{x}, \boldsymbol{z})\,\boldsymbol{P}^{-1} = \boldsymbol{G}\boldsymbol{P}^{-1}, \tag{6.2}$$

which is a matrix of partial regressions of breeding values on particular parental phenotypic values, holding other phenotypic values constant. We will return to this realization in Chapter 11, when we consider how the effects of selection in one generation are transmitted into the next generation.

6.3 The Additive Model for Multivariate Phenotypic Value

We assume the same polygenic model for inheritance as in Section 5.2, but we consider the case of multiple traits, each affected by multiple loci that may each effect multiple traits (Lande 1979). Assuming no dominance and no epistasis, the model takes a simple form. Under these assumptions, each of the m additive genetic values is a sum of allelic effects within and across loci. Phenotypic ($\boldsymbol{z}$), genetic ($\boldsymbol{x}$), and environmental values ($\boldsymbol{e}$) are each $m \times 1$ column vectors, so that in the two-trait case

$$\boldsymbol{z} = \boldsymbol{x} + \boldsymbol{e} = \begin{bmatrix} z_1 \\ z_1 \end{bmatrix} = \begin{bmatrix} x_1 \\ x_1 \end{bmatrix} + \begin{bmatrix} e_1 \\ e_1 \end{bmatrix}. \tag{6.3}$$

As before, we will assume no correlation or interaction between genetic and environmental values, so that $\boldsymbol{z}$, $\boldsymbol{x}$, and $\boldsymbol{e}$ are multivariate normal in distribution with means,

$$\bar{\boldsymbol{z}} = \bar{\boldsymbol{x}} + \bar{\boldsymbol{e}} \tag{6.4a}$$

and variance/covariance matrices

$$\boldsymbol{P} = \boldsymbol{G} + \boldsymbol{E}, \tag{6.4b}$$

where $\boldsymbol{G}$, $\boldsymbol{P}$, and $\boldsymbol{E}$ denote, respectively, additive genetic, phenotypic, and environmental variance-covariance matrices. The variance elements of $\boldsymbol{P}$, $\boldsymbol{G}$, and $\boldsymbol{E}$, which occur on the main diagonals, are familiar from Chapter 5. The off-diagonal elements in $\boldsymbol{P}$ are phenotypic covariances that arise from covariance in genetic and environmental values. We will consider *genetic covariance* in detail because of its importance in affecting evolutionary responses to selection and finite population size.

Is it reasonable to assume that the distributions of $\boldsymbol{x}$ and $\boldsymbol{e}$ are multivariate normal? The Central Limit Theorem tells us that if a random variable is the sum of random variables, the distribution of the sum variable will converge on the normal distribution as the number of underlying variables increases, no matter what the distribution of the underlying variables which are independent and identically distributed. Consequently, we expect $\boldsymbol{x}$

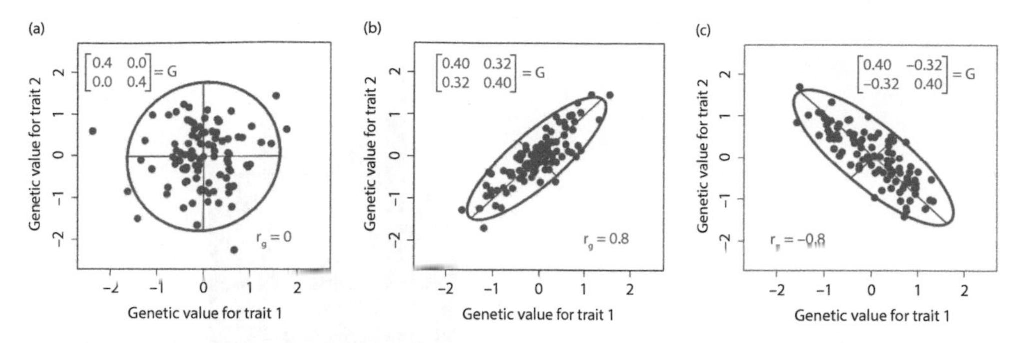

Figure 6.3 Hypothetical bivariate samples of genetic values (n = 100) from bivariate normal distributions. Genetic correlations, r_g, of (a) 0, (b) 0.8, and (c) −0.8. Sample 95% confidence ellipses for the bivariate mean are shown. Straight lines show the eigenvectors for the matrices shown in each figure, rather than for each sample.

and $\boldsymbol{e}$ to be normally distributed so long as each is the sum of many random variables. Furthermore, it is generally possible to find a transformation of $\boldsymbol{z}$ that renders it multivariate normally distributed.

In principle, we can determine the genetic value of an individual by assessing the phenotypic value of offspring produced by breeding that individual to a large, random sample of mating partners. In the multivariate case, that individual will have multiple genetic values, one for each of m traits. If we plot the additive genetic values (breeding values) for two traits for many individuals, we will obtain the kind of statistical clouds shown in Figure 6.3. Such scatterplots can reveal patterns in the covariance between pairs of traits. To quantify those patterns, it is often useful to calculate the *additive genetic correlation* between two traits,

$$r_g = G_{ij}/\sqrt{G_{ii}G_{jj}}\ , \qquad (6.5)$$

where G_{ij} is the additive genetic covariance for the two traits, and G_{ii} and G_{jj} are their additive genetic variances. For example, from the garter snake example above, we can determine that the additive genetic correlation between body and tail vertebral numbers is $r_g = 3.78/\sqrt{8.17 * 8.16} = 0.46$. As is usual for correlations, $-1 \leq r_g \leq 1$. If genetic values for the two traits are independent, $r_g = 0$, the cloud will be elliptical or circular (if the two traits have the same genetic variance) with no inclination (Figure 6.3a). If genetic values are positively correlated, $r_g > 0$, then the cloud has a positive inclination, as in Figure 6.3b. A negative correlation, $r_g < 0$, means that the cloud has a negative inclination (Figure 6.3c).

6.4 The Prevalence of Genetic Correlation

Morphological traits are often genetically correlated with other morphological traits (Figure 6.4). The striking tendency of these correlations to be skewed towards positive values reflects the likelihood that a large number of traits in a study are size-related (morphometric).

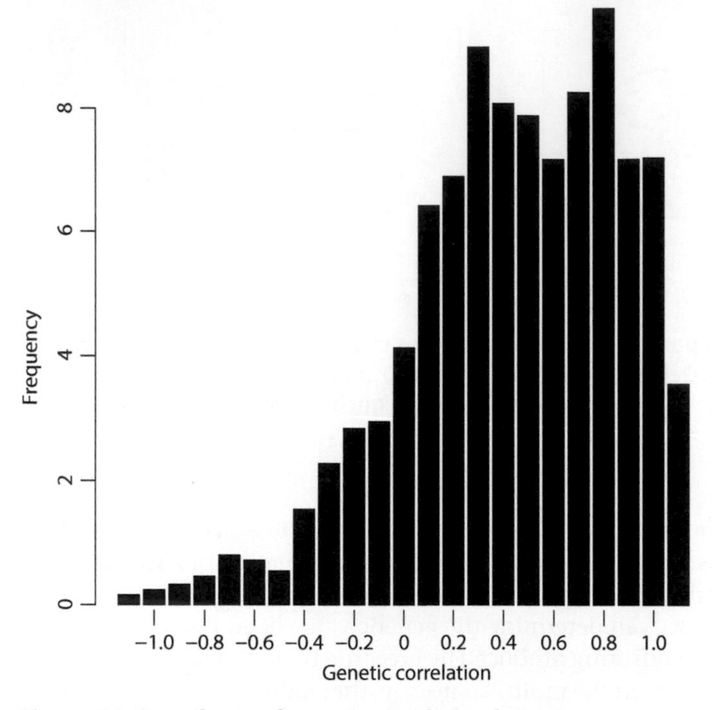

Figure 6.4 Prevalence of genetic correlation between morphological traits. The histogram summarizes 1210 estimates from 38 species. Frequency is shown as a percent of observations. Mean = 0.47 ± 0.009s.e.

After Roff (1997).

6.5 The Causes of Genetic Covariance

Genetic covariance has two distinct proximate causes, pleiotropy and linkage disequilibrium. *Pleiotropy* means that the alleles at a particular locus have effects on each of the traits in question. Pleiotropy is an almost universal feature of mutation (Caspari 1952) and will be considered in detail in the next section. Aggregating such pleiotropic effects within and across loci can produce positive or negative genetic covariance or even cancellation so that the aggregate gives little or no indication of pleiotropy. *Linkage disequilibrium*, the second cause of genetic correlations, refers to a statistical relationship between two loci (Section 5.8), but in the present case we are concerned with loci that affect two different traits. For example, linkage disequilibrium is positive if a positive effect on a trait by an allele at one locus is positively correlated with a positive effect on another trait by an allele at a second locus. Linkage disequilibrium can contribute to a negative genetic covariance if the alleles at the two loci have opposite effects on the two traits.

Linkage disequilibrium can arise from and be maintained by correlational selection which works in opposition to erosion by recombination. A hypothetical example of this kind of selection is shown in Figure 6.5.

Diagnosing the cause of genetic correlation is usually a matter of guesswork, because the requisite discriminating experiments are difficult to conduct. Pleiotropy is the likely (but unproven) cause of genetic covariance between body and tail vertebrae in the garter

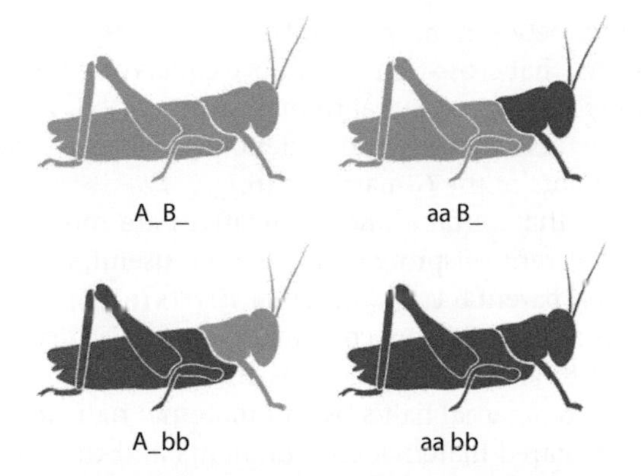

Figure 6.5 Correlational selection can create and maintain linkage disequilibrium. In this hypothetical example, the *A* locus affects the colour of the head of a grasshopper, so that *A_* (*AA* and *Aa*) grasshoppers have green heads and *aa* grasshoppers have brown heads. The *B* locus affects the colour of the body, so that *B_* (*BB* and *Bb*) grasshoppers have green bodies and *bb* grasshoppers have brown bodies. Correlational selection acting on this inheritance scheme can produce linkage disequilibrium. Suppose, for example, that colour-matched grasshoppers are favoured in the population because they are harder for predators to detect. In contrast, colour miss-matched grasshoppers stand out against both green and brown backgrounds and fall prey to predators on both backgrounds. This selection scheme creates linkage disequilibrium between the two loci such that the haplotypes *AB* and *ab* are disproportionally represented in the population. This disequilibrium is expressed as a genetic correlation between colours in the head and body.

snake example; we can easily imagine that some genes affect vertebral counts in both regions. Likewise, the positive genetic correlations that are frequently observed in the sizes of homologous structures probably reflect pleiotropic gene action (Lande 1979). On the other hand, a genetic correlation between traits that are unlikely to share developmental pathways may reflect linkage disequilibrium. The example of escape behaviour and colouration in garter snakes is a case in point (Brodie 1992, 1993). Pleiotropy seems an unlikely explanation for the genetic correlation observed between these different kinds of traits. Furthermore, linkage disequilibrium seems probable because the genetic correlation has the same sign as the correlational selection that has been documented in the same population (see Figure 4.9). We expect exactly this kind of correspondence if a type of correlational selection, multivariate stabilizing selection, shaped the pattern of genetic covariance arising from linkage disequilibrium, as in the hypothetical grasshopper example.

6.6 Estimating the *G*-matrix From Covariance Among Relatives

The theory and procedures for estimating the *G*-matrix are straightforward extensions of the univariate covariance approach (Mode and Robinson 1959). In particular, our general expression for resemblance between relatives (5.6) still applies. In the cross-trait approach, one trait is expressed in one kind of relative (e.g., offspring) and another trait is expressed in the other kind of relative (e.g., parents). Alternatively, a single kind of relative (e.g., paternal half-sibs) can be used to estimate genetic covariance. In either case, all of the variance terms on the right side of (5.6) become covariance terms. For example, consider the cross-trait resemblance between mothers and daughters displayed in Figure 6.3. Ignoring epistatic contributions, that cross-trait covariance equals one-half the additive genetic covariance between body and tail vertebral numbers, since $r = 1/2$ and $u = 0$. This mapping between phenotypic resemblance and genetic parameters, provided by (5.6), was used to estimate the elements of the G-matrix in (6.1).

We can now appreciate that certain kinds of relatives are more useful that others in estimating the G-matrix. Parent-offspring data are often useful, especially if one or both of the parents do not exert parental environmental effects (e.g., maternal environmental effects), because dominance effects do not contribute to resemblance and because epistatic contributions are limited to additive interactions and are in any case likely to be small. These same virtues apply to paternal half-sibs, but maternal half-sibs may yield problematic estimates because of shared maternal environments and the possibility of maternal effects. As a general rule, large family sizes lead to better estimates of additive genetic values and better estimates of $\boldsymbol{G}$ (i.e., smaller standard errors). The rule is not absolute, however. A cap on the total number of individuals that can be scored will mean that the investigator must choose whether to increase the number of families or the number of individuals sampled in each family. Formulas and guidelines are available to design an optimal allocation of effort (Robertson 1959a,b; Lynch and Walsh 1998).

The biological characteristics of each study system also play a large role in estimating and interpreting G-matrices. In garter snakes, for example, litter sizes average about 10, so optimizing an estimation design means scoring all the littermates that are available and maximizing the number of litters. Although mothers thermoregulate during pregnancy and pass nutrients to developing embryos, maternal effects are not inevitable. Experimental manipulations show that scale count means (e.g., body and tail vertebral counts) are remarkably immune to temperature effects (Arnold and Peterson 2002), and neonatal thermoregulation is unaffected by the mother's temperature during pregnancy (Arnold et al. 1995). These results mean that some of the best a priori candidates for maternal effects do not complicate the interpretation of estimates based on mother-offspring resemblance. However, the best estimation design and the interpretation of estimates must be decided on a case-by-case basis that takes into account biological circumstances and experimental evidence.

6.7 Estimating the *G*-matrix With the Animal Model

The advantages that the animal model offers over the covariance approach carries over to the multivariate case. In particular the animal model offers great flexibility in accounting for fixed effects (e.g., spatial and temporal differences in rearing condition, sex, age, etc.), as well as for multiple kinds of relationship.

The model for the value of a phenotypic trait looks much the same as it did in the univariate case (Chapter 5),

$$\boldsymbol{z} = \boldsymbol{X}\boldsymbol{b} + \boldsymbol{Z}\boldsymbol{u} + \boldsymbol{e}, \tag{6.7}$$

$$Var(\boldsymbol{z}) = \boldsymbol{Z}\boldsymbol{F}\boldsymbol{Z}^T + \boldsymbol{R}, \tag{6.8}$$

$$\boldsymbol{F} = Var(\boldsymbol{u}), \tag{6.9}$$

$$\boldsymbol{R} = Var(\boldsymbol{e}), \tag{6.10}$$

but now vectors representing each trait are stacked one on top of the other in the column vectors $\boldsymbol{z}$, $\boldsymbol{b}$, $\boldsymbol{u}$, and $\boldsymbol{e}$, and the design matrices $\boldsymbol{X}$ and $\boldsymbol{Z}$ are partitioned into submatrices, according to relations within and between traits. So, for example, in the simple case of five related individuals, each with observations on two traits, let the additive genetic and environmental variance-covariance matrices be

$$\boldsymbol{G} = \begin{bmatrix} G_{11} & G_{12} \\ G_{12} & G_{22} \end{bmatrix} \tag{6.11}$$

and

$$\boldsymbol{E} = \begin{bmatrix} E_{11} & E_{12} \\ E_{12} & E_{22} \end{bmatrix}. \tag{6.12}$$

The design matrix is

$$\boldsymbol{X}^{\mathrm{T}} = \begin{bmatrix} 1 & 1 & 1 & 1 & 1 & 0 & 0 & 0 & 0 & 0 \\ 0 & 0 & 0 & 0 & 0 & 1 & 1 & 1 & 1 & 1 \end{bmatrix}. \tag{6.13}$$

The contributions to phenotypic values by fixed effects are

$$\boldsymbol{b}^{\mathrm{T}} = \begin{bmatrix} \mu_1 & \mu_2 \end{bmatrix}. \tag{6.14}$$

In $\boldsymbol{u}$, the five breeding values for trait 1 (individuals 1 through 5) are stacked on top of the five breeding values for trait 2 (individuals 1 through 5), with analogous stacking in $\boldsymbol{z}$ and $\boldsymbol{e}$. The design matrix for $\boldsymbol{u}$ is equivalent to the identity matrix,

$$\boldsymbol{Z} = \boldsymbol{I}. \tag{6.15}$$

The variance-covariance matrix for $\boldsymbol{u}$ is $\boldsymbol{F}$, which in the two-trait case is

$$\boldsymbol{F} = \begin{bmatrix} \boldsymbol{A}G_{11} & \boldsymbol{A}G_{12} \\ \boldsymbol{A}G_{12} & \boldsymbol{A}G_{22} \end{bmatrix}, \tag{6.16}$$

where $\boldsymbol{A}$ is the symmetric 5×5 relationship matrix for the five individuals (Henderson 1984). Likewise the variance/covariance matrix for $\boldsymbol{e}$ is $\boldsymbol{R}$, which in the two-trait case is

$$\boldsymbol{R} = \begin{bmatrix} \boldsymbol{I}E_{11} & \boldsymbol{I}E_{12} \\ \boldsymbol{I}E_{12} & \boldsymbol{I}E_{22} \end{bmatrix}, \tag{6.17}$$

where $\boldsymbol{I}$ is an identity matrix of order 5 (Henderson 1984). With more traits, the same bookkeeping conventions are retained, the vectors are stacked higher, and the matrices have more partitions. Details for implementing the multivariate animal model, with estimation of the G-matrix, are described by Meyer (2007) and Hadfield (2010). Assumptions of the approach are briefly discussed in Chapter 5.

6.8 Dimensionality of the *G*-matrix

Analysis of the principal components (eigenvectors) of the *G*-matrix is useful for a number of reasons. First, if a large number of traits are being studied, *principal component analysis* (PCA) can be used to compress a large number of traits into a smaller number of eigenvectors without discarding statistical information. Second, the eigenvectors themselves may be of interest, because they may represent bundles (pleiades) of correlated traits with a shared genetic basis. Third, because the eigenvectors are ranked in order of their genetic variance, PCA tells us which weighted combinations of traits have the most and which have the least genetic variance. Finally, the genetic variance profile of the eigenvectors (eigenvectors as a function of eigenvector rank) tells us how many independent dimensions reside in by the original trait *G*-matrix, as we shall see in the next example.

To know how many independent dimensions are required to statistically explain the total variation in the genetic cloud represented by the *G*-matrix, we need an inheritance study in which many traits are included. Given a *G*-matrix for many traits, the issue of dimensionality is usually approached with a principal component analysis which identifies the eigenvectors of the matrix and estimates how much variation each explains. Mezey and Houle (2005) conducted an exemplary analysis of this kind using 12 landmarks on the wing of *Drosophila melanogaster* (Figure 6.6).

Additive genetic variances and covariances for the position of these landmarks were estimated from parent-offspring, as well as from full-sib and half-sib data using isofemale lines sampled from a single natural population. The position of each landmark was represented as x- and y-coordinates and consequently genetic variance and covariance was assessed for a total of 24 traits in each sex. A dimensionality portrayal of the *G*-matrix for males is shown in Figure 6.7. Note that Figure 6.7 shows the logarithm of eigenvalue size corresponding to the eigenvectors (ranked from the largest to the smallest eigenvalue). Eigenvalue size is proportional to the amount of trait genetic variance explained by a particular eigenvector (dimension of variation).

The big message from Figure 6.7 is that although contribution to total variation falls off exponentially as we proceed down the ranked list of eigenvectors, 20 orthogonal dimensions are required to account for the 24 landmark traits in males. In other words, nearly all the orthogonal dimensions of the *G*-matrix possess genetic variation and can respond to selection. Analysis of female data produced a similar result. Although many dimensions explain variation, a few dimensions make very large contributions.

A common result of PCA of trait measurements is that the first eigenvector (PC 1) accounts for a large proportion of total variation and represents overall size (with positive loadings by most if not all traits), while the remaining eigenvectors accounts for progressively smaller proportions of total variation and represent shape combinations

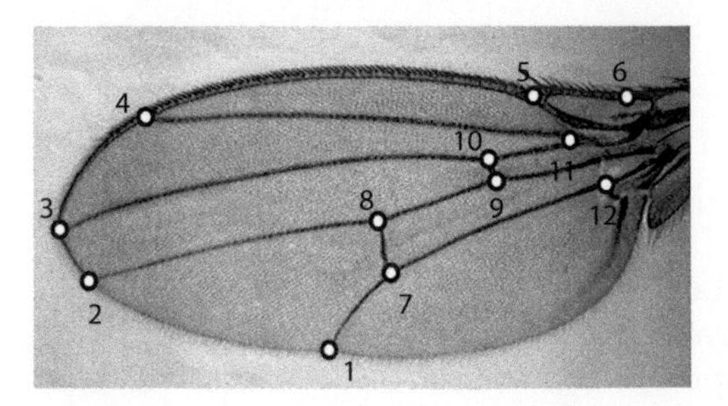

Figure 6.6 Landmarks on the wing of *Drosophila melanogaster*.

From Mezey and Houle (2005) with permission.

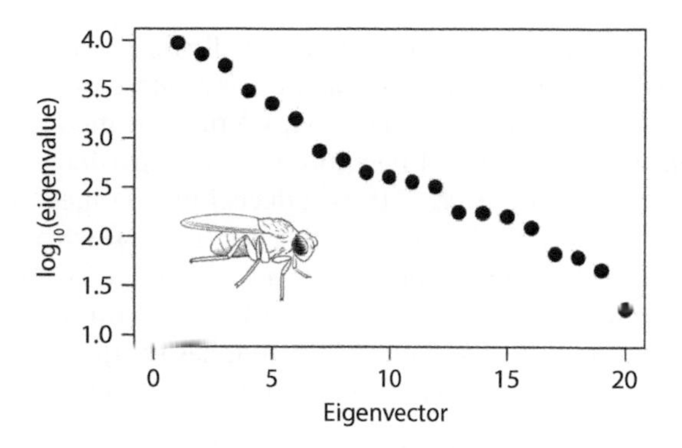

Figure 6.7 Dimensionality analysis of the *G*-matrix for 24 landmark traits on the wing of male *Drosophila melanogaster*. Solid circles show the eigenvalue estimates for each eigenvector.
After Mezey and Houle (2005).

(with mixtures of positive and negative loadings). Faced with this result, one might be tempted to confine attention and subsequent analysis to the first few eigenvectors (e.g., PC 1 and PC 2). The Mezey and Houle (2005) result (Figure 6.7) cautions us to avoid this trap and instead consider the possibility that a very large number of trait combinations might have appreciable genetic variance and so be capable of responding to selection.

6.9 Multivariate Maternal Effects

Consideration of multivariate maternal effects requires a more complicated model than the Willham model (1963, 1972) discussed in Section 5.3. Kirkpatrick and Lande (1989, 1992) proposed a model in which one kind of trait in mothers affects another kind of trait in offspring, namely

$$z_{oi} = x_{oi} + e_{oi} + \sum_{j=1}^{m} m_{ij} z_{ij}^{*}, \tag{6.18}$$

where the subscript *oi* denotes the *ith* trait in offspring, m_{ij} is the maternal effect of the *jth* maternal trait on the *ith* offspring trait, and z_{ij}^{*} is the *jth* maternal trait after selection. In this model, the covariance between the traits of mothers and the traits of offspring is no longer simply $\boldsymbol{G}$ but is instead a function of a matrix of covariances between additive genetic and phenotypic values, $\boldsymbol{C_{az}}$, and a matrix of coefficients, $\boldsymbol{m}$, that represent the strength of the effect of one maternal trait on the same or a different trait in offspring (see discussion in Arnold 1994).

6.10 Theoretical Input of Genetic Variance From Mutation, the *M*-matrix

Mutation is the well-spring for inheritance, the ultimate source for genetic variation described by the *G*-matrix. Our formulation for input to genetic variation from mutation will consist of two fundamental ingredients, just as it did in the univariate case: a total, genome-wide mutation rate and a multivariate distribution of allelic effects. As in the univariate case, we will adopt *Kimura's infinite-alleles model* for mutations at a locus (Kimura 1965). In the multivariate case, however, we need to incorporate *pleiotropy* into our mutation model. In other words, a new mutation at a particular locus may affect a number of phenotypic traits, and each of those effects must be specified. We will assume that the per-locus distribution of effects is multivariate normal with a mean of zero and a variance-covariance matrix given by *the mutational effects matrix*, $\boldsymbol{M}$. The diagonal elements in this $m \times m$ symmetric M-matrix are the variances in mutational effects for each of the m traits. The off-diagonal elements are covariances in mutational effects for pairs of traits. In the bivariate case, the M-matrix takes the following form

$$\boldsymbol{M} = \begin{bmatrix} M_{11} & M_{12} \\ M_{12} & M_{22} \end{bmatrix}, \tag{6.19}$$

where, for example, M_{11} is the mutational variance for trait 1, and M_{12} is the mutational covariance for traits 1 and 2, a reflection of pleiotropy. It will prove useful to express this covariance as a *mutational correlation*

$$r_\mu = M_{12}/\sqrt{M_{11}M_{22}}. \tag{6.20}$$

Summing up, we can think of mutation as producing effects on two traits, so that the resulting statistical cloud of effects forms a bivariate normal distribution which in turn is summarized by the M-matrix (Figure 6.8). In the two-trait case, with equal mutational variances, the shape of this bivariate cloud is determined by the mutational correlation, r_μ.

So far we have considered the mutational effects of a single, typical locus. A simple way to extend this model across the entire genome is to assume that although the mutations at each locus are independent, their effects are drawn from identical multivariate distributions. We make a companion assumption about mutation rate, so that the total mutation

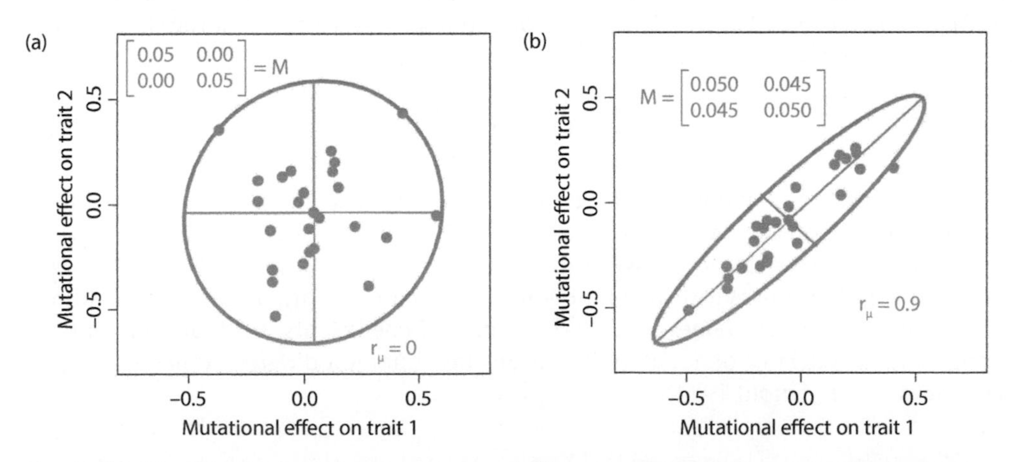

Figure 6.8 The distribution of new mutational effects on two traits from a particular locus can be represented as a cloud of values. Samples of 25 hypothetical alleles drawn from bivariate normal distributions are plotted along with the 95% confidence ellipses for the means. Axes inside each ellipse are eigenvectors (principal components). (a) A cloud of mutational effects with no correlation, $r_\mu = 0$. (b) A cloud of mutational effects with a strong positive correlation, $r_\mu = 0.9$.

After Arnold et al. (2008).

rate across entire diploid genome for our traits is a scalar variable (zeta), $\zeta = 2n\mu$ (5.13). The total input of genetic variation to our entire set of traits from mutation each generation is given by *the total mutational input matrix*, $\boldsymbol{U}$, which represents the product of the total mutation rate across the genome and the M-matrix,

$$\boldsymbol{U} = \zeta\boldsymbol{M} \tag{6.21}$$

(Lande 1980). The assumptions that lead to (6.21) are not as hazardous as they may seem. Even if mutational effects were produced by a set of distributions that varied from place to place in the genome with a spatially variable mutation rate, we could still aggregate those rates and effects into a single U-matrix. The assumptions of spatial homogeneity en route to (6.21) are problematic only when we are forced to decompose the U-matrix in particular applications.

6.11 Estimating the *M*-matrix

In Section 5.15, we described a mutation accumulation experiment in which a set of lines had been purged of genetic variation by inbreeding. Directional selection was then imposed, so that any response to selection could be attributed to new mutations and hence could be used to estimate $\boldsymbol{M}$. An alternative approach is to maintain the inbred lines at very small N_e and with continued brother-sister mating, so that mutational variation and covariation (i.e., the M-matrix) can be estimated from the variation and covariation of lines as they diverge by drift rather than selection (Houle and Nuzhdin 2004). This approach is particularly well suited for estimating an M-matrix of high rank, because it does not require multiple regimes of directional selection.

Our empirical knowledge of the M-matrix is limited because the requisite experiments have focused on only a few kinds of traits in a small sample of taxa (Houle et al. 1996; Halligan and Keightley 2009). Houle and Fierst (2013) estimated the M-matrix for the *Drosophila* wing landmark traits shown in Figure 6.6. The M-matrix has high dimensionality that parallels the dimensionality of the G-matrix (Figure 6.7).

One of the most comprehensive multivariate analyses is provided by the Houle et al. (2017) study of M- and G-matrices of landmark traits in the *Drosophila* wing. The estimates in this study refer to the two traits (x- and y-coordinates) associated with each landmark, not to the full 24×24 matrices studied by Mezey and Houle (2005) and Houle and Fierst (2013), which included covariation among landmarks. The two estimates of the M-matrix for each landmark are very similar (see Houle et al. 2017 for details). The estimates of 2×2 $\boldsymbol{M}$ and $\boldsymbol{G}$ for each trait are aligned and extremely similar in shape, suggesting the kind of selection-driven alignment process described by (Jones et al. 2014, Chapter 9).

6.12 Multivariate Genetic Variance Maintained by Mutation-Migration-Selection Balance

It is natural to assume that the entire G-matrix equilibrates in response to the opposing forces of mutation and stabilizing selection, just as we assumed for the equilibration of genetic variance in the univariate case. As before, our results pertain only to the case of infinite population size. Using the multi-trait generalization of Kimura's (1965) infinite-alleles mutation model that we have just described, we can consider a population experiencing multivariate stabilizing selection and ask about the state of *G*-matrix at equilibrium (Lande 1980). Not surprisingly, the expression for the equilibrium *G*-matrix is a function of the total mutational input and the curvature of the adaptive landscape, but the expression is considerably more complicated than in the univariate case (5.16). If we assume that the trait distribution and the AL are multivariate Gaussian, the equilibrium contribution of the *ith* locus to the *G*-matrix is

$$\hat{\boldsymbol{G}}_{ii} \cong \Omega^{1/2}\sqrt{\Omega^{-1/2}\boldsymbol{U}_i\Omega^{-1/2}}\,\Omega^{1/2}, \tag{6.22}$$

where ω is a matrix of Gaussian stabilizing selection parameters (see Chapter 4) and $\tilde{\omega} \equiv \omega + \boldsymbol{E} \cong \omega + \boldsymbol{P} = \Omega$ (Lande 1980). Despite the intimidating complexity of this expression, its message meshes with our intuition. The size of the equilibrium $\boldsymbol{G}$ is enhanced by large mutation input, $\boldsymbol{U}$, and trimmed by strong stabilizing selection (small Ω). The shape of the equilibrium *G*-matrix is a compromise between the shapes of the *M*-matrix (via $\boldsymbol{U}$) and the AL (via Ω).

We can derive expressions for $\boldsymbol{G}$ as it approaches its equilibrium, but those expressions depend on the details of per locus distributions of mutational effects and do not lend themselves to broad generalization (Lande 1980; Barton 1989). The change in $\boldsymbol{G}$ induced by selection within a generation, however, is approximately

$$\Delta_S\boldsymbol{G} = \boldsymbol{G}\left(\gamma - \beta\beta^{\mathrm{T}}\right)\boldsymbol{G} = -\boldsymbol{G}\Omega^{-1}\boldsymbol{G}, \tag{6.23}$$

where γ is a matrix of nonlinear selection parameters (see Chapter 4), and the last expression for the change in $\boldsymbol{G}$ assumes a Gaussian ISS (Lande 1980; Phillips and Arnold 1989). At equilibrium, the losses due to selection are balanced by the input from mutation and recombination, so that

$$\boldsymbol{U} = -\Delta_S\boldsymbol{G} \tag{6.24}$$

(Lande 1980). However, in this expression we assume that genetic variation is lost only by selection and not from drift. In Chapter 9 we will consider the combined losses from selection and drift. An implication of (6.23), which we will explore in later chapters, is that multivariate selection will tend to shape the *G*-matrix in its own image.

6.13 Conclusions

- The *G*-matrix provides a useful model for the inheritance of polygenic traits that can be used to model multivariate evolution in response to the drift and selection.
- The off-diagonal elements of the *G*-matrix summarize the genetic connections between traits that arise from linkage disequilibrium and pleiotropy.
- The *G*-matrix can be viewed as a balance between gains from mutation and migration and losses from recombination and selection.
- Empirical results favour a genotype-phenotype map with prevalent pleiotropy.
- Genetic correlations are common and are particularly expected among elements in functional complexes.
- The dimensionality of the *G*-matrix maybe nearly as large as the number of traits in the study. This result suggests that even traits with low levels of genetic variance will respond to selection on long evolutionary timescales.

- The *M*-matrix summarizes the per generation mutational input to the genetic variation of multiple traits. Like the *G*-matrix, its dimensionality may be very high.
- Mutation-migration-selection balance can account for the standing crop of multivariate genetic variance and covariance in natural populations.

Literature Cited

Arnold, S. J. 1994. Multivariate inheritance and evolution: a review of concepts. pp. 17–48 *in* C. R. B. Boake, ed. Quantitative Genetic Studies of Behavioral Evolution. University of Chicago Press, Chicago.

Arnold, S. J., R. Bürger, P. A. Hohenlohe, B. C. Ajie, and A. G. Jones. 2008. Understanding the evolution and stability of the G-matrix. Evolution 62:2451–2461.

Arnold, S. J., and C. R. Peterson. 2002. A model for optimal reaction norms: the case of the pregnant garter snake and her temperature-sensitive embryos. Am. Nat. 160:306–316.

Arnold, S. J., C. R. Peterson, and J. Gladstone. 1995. Behavioural variation in natural populations. VII. Maternal body temperature does not affect juvenile thermoregulation in a garter snake. Anim. Behav. 50:623–633.

Arnold, S. J., and P. C. Phillips. 1999. Hierarchical comparison of genetic variance-covariance matrices. II. Coastal-inland divergence in the garter snake, *Thamnophis elegans*. Evolution 53:1516–1527.

Barton, N. 1989. The divergence of a polygenic system subject to stabilizing selection, mutation and drift. Genet. Res. 54:59–78.

Brodie, E. D. III. 1992. Correlational selection for color pattern and antipredator behavior in the garter snake *Thamnophis ordinoides*. Evolution 46:1284.

Brodie, E. D. III. 1993. Homogeneity of the genetic variance-covariance matrix for antipredator traits in two natural populations of the garter snake *Thamnophis ordinoides*. Evolution 47:844.

Caspari, E. 1952. Pleiotropic gene action. Evolution 6:1–18.

Hadfield, J. D. 2010. MCMC Methods for multi-response generalized linear mixed models: The MCMCglmm *R* Package. J. Stat. Softw. 33.

Halligan, D. L., and P. D. Keightley. 2009. Spontaneous mutation accumulation studies in evolutionary genetics. Annu. Rev. Ecol. Evol. Syst. 40:151–172.

Henderson, C. R. 1984. Applications of Linear Models in Animal Breeding. University of Guelph.

Houle, D., G. H. Bolstad, K. van der Linde, and T. F. Hansen. 2017. Mutation predicts 40 million years of fly wing evolution. Nature 548:447–450.

Houle, D., and J. Fierst. 2013. Properties of spontaneous mutational variance and covariance for wing size and shape in *Drosophila melanogaster*: Mutational covariance in *Drosophila* wings. Evolution 67:1116–1130.

Houle, D., B. Morikawa, and M. Lynch. 1996. Comparing mutational variabilities. Genetics 143:1467–1483.

Houle, D. and S. Y. Nuzhdin. 2004. Mutation accumulation and the effect of *copia* insertions in *Drosophila melanogaster*. Genet. Res. Camb. 83: 7-18.

Jones, A. G., R. Bürger, and S. J. Arnold. 2014. Epistasis and natural selection shape the mutational architecture of complex traits. Nat. Commun. 5:3709.

Kimura, M. 1965. A stochastic model concerning the maintenance of genetic variability in quantitative characters. Proc. Natl. Acad. Sci. 54:731–736.

Kirkpatrick, M., and R. Lande. 1989. The evolution of maternal characters. Evolution 43:485–503.

Kirkpatrick, M., and R. Lande. 1992. The evolution of maternal characters: Errata. Evolution 46:284.

Lande, R. 1979. Quantitative genetic analysis of multivariate evolution, applied to brain: body size allometry. Evolution 33:402–416.

Lande, R. 1980. The genetic covariance between characters maintained by pleiotropic mutations. Genetics 94:201–215.

Lynch, M., and B. Walsh. 1998. Genetics and Analysis of Quantitative Traits. Sinauer Associates, Inc., Sunderland, Massachusetts.

Meyer, K. C. 2007. WOMBAT—A tool for mixed model analyses in quantitative genetics by restricted maximum likelihood (REML). J. Zhejiang Univ. Sci. B 8:815–821.

Mezey, J. G., and D. Houle. 2005. The dimensionality of genetic variation for wing shape in *Drosophila melanogaster*. Evolution 59:1027–1038.

Mode, C. J., and H. F. Robinson. 1959. Pleiotropism and the genetic variance and covariance. Biometrics 15:518–537.

Phillips, P. C., and S. J. Arnold. 1989. Visualizing multivariate selection. Evolution 43:1209–1222.

Phillips, P. C., and S. J. Arnold. 1999. Hierarchical comparison of genetic variance-covariance matrices. I. Using the Flury hierarchy. Evolution 53:1506–1515.

Robertson, A. 1959a. Experimental design in the evaluation of genetic parameters. Biometrics 15:219–226.

Robertson, A. 1959b. The sampling variance of the genetic correlation coefficient. Biometrics 15:469–485.

Roff, D. A. 1997. Evolutionary Quantitative Genetics. Springer, Netherlands.

Willham, R. L. 1963. The covariance between relatives for characters composed of components contributed by related individuals. Biometrics 19:18–27.

Willham, R. L. 1972. The role of maternal effects in animal breeding. III. Biometrical aspects of maternal effects in animals. J. Anim. Sci. 35:1288–1293.

CHAPTER 7

Modularity, Performance, and Functional Complexes

Overview—Phenotypic traits do not exist as isolated entities. Instead, traits are bundled together in ensembles (modules) of interacting traits that form functional complexes. The bundling of traits into complexes is accomplished by correlational selection, a kind of multivariate stabilizing selection. One can measure the correlational selection that acts on a functional complex by scoring the overall performance of the complex and the values of the component traits in a sample of individuals. In the ensuing selection analysis, one fits a statistical model in which individual values of performance are related to the individual values of traits in the complex with the goal of estimating the γ-matrix. The γ-matrix in turn plays a key role in the evolution of modularity and modules. For example, on a deep evolutionary timescale, trait ensembles that form functional complexes often show a distinctive pattern of multivariate stasis and conservatism. As we shall see in later chapters, this multivariate pattern of stasis cannot be accounted for by drift or unspecified forces of inertia. These considerations suggest that the patterns of multivariate selection that are responsible for the maintenance of functional complexes also tend to persist through deep time and shape long-term evolutionary patterns.

7.1 Traits as Parts of a Functional Complex

Students of morphological evolution have long argued for a holistic perspective in which individual morphological traits are seen as interacting and evolving as parts of the whole organism (Waddington 1957; Riedl 1978; Gould and Lewontin 1979; Cheverud 1982; Klingenberg 2008). Because of their interactions and correlations, individual traits do not evolve in isolation from other traits, but as parts of *functional complexes*, ensembles of interacting traits. From this perspective, functional interaction between characters is a ubiquitous feature of life, and our attention is especially drawn to sets of tightly interacting traits that show low correlations with other sets of traits. Such trait ensembles are called *modules* (Wagner et al. 2007).

In the next few sections we will review examples of morphological, molecular, physiological, and behavioural traits participating in functional complexes. Our intent is two-fold. First, we make the point that many functional complexes are composed of different kinds of interacting traits. Our second goal in reviewing these examples is that we wish to lay the groundwork for a discussion of evolutionary process at the end of this

Evolutionary Quantitative Genetics. Stevan J. Arnold, Oxford University Press. © Stevan Arnold (2023).
DOI: 10.1093/oso/9780192859389.003.0008

chapter and in later chapters. As we shall see, the trait composition of a complex and the functional interactions among those traits may induce complicated coevolutionary trajectories. Evolutionary models that include those functional interactions can account for both the capacity of functional complexes to diversify, as well as their fundamental conservatism.

The organization of interacting characters into functional complexes or modules is particularly obvious if we focus on complexes that are dedicated to well-defined ecological functions. For example: (1) Spiders in several families use specialized silk glands in their abdomens to construct webs of characteristic design (e.g., orb, sheet, funnel). In each family, the web is produced by a behavioural algorithm that is guided by tactile cues. The spider also uses tactile cues to detect and respond to prey that contact or become snared in the web (Gosline et al. 1999; Cardoso et al. 2011). (2) Viperid snakes envenomate their prey with large, erectile fangs (Figure 7.1). Venom is produced in specialized gland and delivered at high pressure through a duct leading to a hollow fang. The venom itself is a complex cocktail of proteins that immobilizes the prey and begins the digestive process. The strike that envenomates the prey is triggered by thermal, visual, and chemical cues (Klauber 1956; Jackson 2003; Fry et al. 2008; Cundall 2009). (3) Angiosperms use characteristic visual and chemical cues in their flowers to attract pollinators (i.e., arthropods and other animals). A variety of floral traits are used to transfer pollen to and from pollinators (Schemske and Ågren 1995; Wilson 1995 p. 199; O'Connell and Johnston 1998; Caruso 2000; Armbruster et al. 2005; Benitez-Vieyra et al. 2006). (4) Some bolitoglossine salamanders capture prey by tongue projection that involves several tightly interacting muscles and bony elements (Figure 7.2).

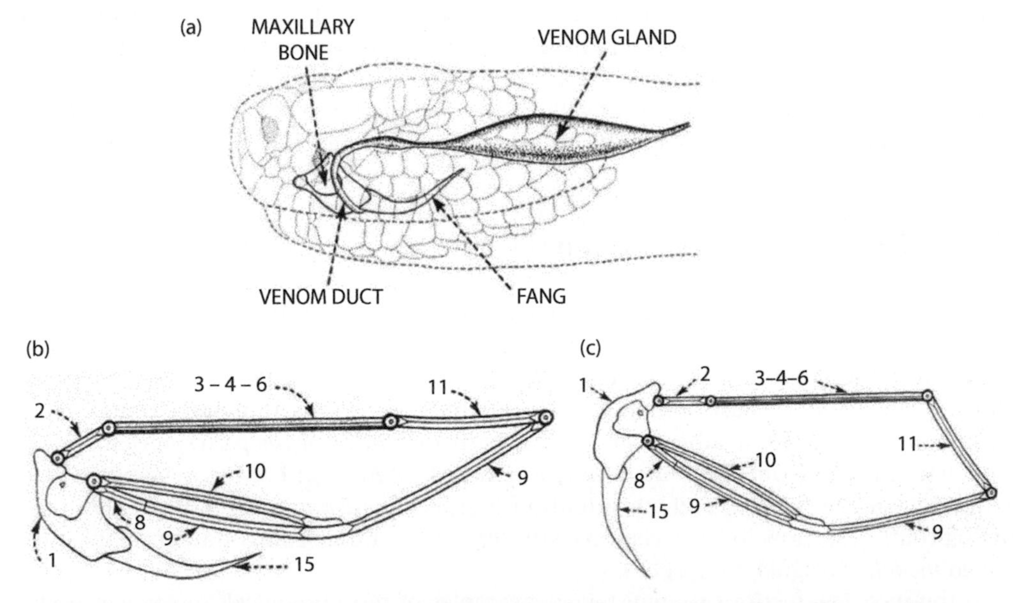

Figure 7.1 The venom production and delivery module of a rattlesnake. (a) The resting positions of the venom gland, venom duct, maxillary bone, and fang in the head of the viper. (b) A biomechanical model for fang erection showing the resting position. (c) Erect position of the fang (15), and the 11 linked bones responsible for erecting the fang: maxillary (1), prefrontal (2), frontal (3), parietal (4), squamosal (5), palatine (8), pterygoid (9), ectopterygoid (10), and quadrate (11).

From Klauber (1956) with permission.

Figure 7.2 The tongue projection mechanism of a plethodontid salamander. The tongue skeleton (black) is normally folded inside the body cavity but is projected forward during prey capture by contracted protractor muscles (dark red). Retractor muscles (light red), attached to the tongue skeleton, run the full length of the body and attach to the pelvis. The contraction of these muscles pulls the tongue and attached prey back into the body.
After Deban et al. (1997).

In each of these four examples, morphological, behavioural, and chemical traits in the functional complex of one organism (salamander, spider, viper, angiosperm) interact with traits in another species (prey or pollinator). In these examples, the case for interaction and modularity is especially clear, but of course not all functional complexes mediate interactions between species. Indeed, one could argue that all quantitative traits, regardless of whether they are involved in species interactions, are parts of functional complexes. In this chapter we argue that correlational selection (Sinervo and Svensson 2002, Svensson et al. 2021), especially within modules, is the key to understanding the evolution and operation of these functional complexes. In Chapters 18–19 we model the evolutionary processes engendered by correlational selection that acts within and between species. Our argument applies to behavioural and physiological traits as well as morphology.

Many fields implicitly if not explicitly recognize that interacting traits are organized in modules that are in turn organized into hierarchies. Ethology and physiology are two notable examples. Ethologists have long recognized modular organization in the complex courtship rituals of animals. (1) In stickleback courtship, as in many animal courtships, the sequence of behavioural acts by the male is triggered at each step by a specific female response (Tinbergen 1952). (2) In birds of paradise, the elaborate courtship displays of the male are organized into modules with complex temporal structure (Scholes 2008). Each module typically includes characteristic repetitive choreography, vocalizations, and the display of particular features of plumage. Likewise, the courtship of salamanders shows modular structure (Houck and Arnold 2003). (3) In plethodontid salamanders, the module used by males to deliver courtship pheromones to the female has been especially well studied (Arnold et al. 2017; Wilburn et al. 2017). During a tail-straddling walk that aligns sexual partners prior to sperm transfer, the male slaps a specialized gland across the female's nares (Figure 7.3), delivering his pheromone to chemosensors in her nasal cavity (i.e., to her vomeronasal organ). The pheromone, a mixture of three protein families,

Figure 7.3 Courtship pheromone delivery during the tail-straddling walk of the salamander *Plethodon shermani*. The male (left) holds his mental gland over the female's head before slapping it across her nares and delivering a pheromone cocktail.

increases the sexual receptivity of the female. In other words, the male module consists of a gland that produces a complex chemical message and a behavioural algorithm for delivering the message to the female. The female module consists of a behavioural algorithm that interacts with the male's algorithm, as well as neural processors and circuitry that produce a response to the male's message. In each of these three examples, the functional complexes of sexual partners interact and affect the sexual success of both males and females. In Chapter 13 we will explore why trait interactions within and between sexual partners can result in long-term stasis or rapid diversification of traits.

Functional complexes are equally recognizable when a modules of characters participates in well-understood physiological functions. Furthermore, as in the case of functional complexes that interact with mates and prey, the performance of physiological character complexes can often be summarized by one or a few metrics that capture the operation of the entire complex of traits: (1) In vertebrates, the heart pumps blood to oxygenation sites in the lungs or gills, as well as to sites of metabolism in the peripheral issues, through a vascular system of vessels and capillaries. Oxygenation is affected by transport molecules (haemoglobin). Metabolism is accomplished by a host of pathways, some of which involve the mitochondria. Despite its complexity, the operation of the entire cardio-vascular system can be captured my measuring the maximum rate of oxygen consumption during sustained exercise ($\dot{V}O_2$ max). $\dot{V}O_2$ max is an important measure of cardio-vascular performance because it sets limits on sustainable activity (Bennett 1980). (2) In vascular plants, transpiration is the process of water moving upward from the roots, through the plant, and evaporating at leaves, stems, and flowers. Water flows through a system of vascular tubules (xylem). The performance of the entire vascular system under particular environmental conditions (e.g., temperature, wind, humidity) can be captured by measuring transpiration rate (Sinha 2004).

In these and many other physiological trait complexes, the physiological traits of individuals can be related to whole organismal performance. As we argued in Chapters 3–4, both the interactions between traits and trait effects on performance can be visualized as surface, which in turn can be used to model adaptive radiations. Before we turn to that endeavour, we will consider the effects that trait interactions might have on phenotypic and genetic correlations among traits. Historically, this focus on the connection between trait interaction and correlation has been used to validate and diagnose the process of trait integration.

7.2 Correlations Between Interacting Traits as an Expression of Integration

The evolutionary process that bundles traits into a functionally interacting complex was called *integration* by (Olson and Miller 1958). Perhaps the most tractable prediction arising from the hypothesis of integration is that interacting traits should be correlated (Terentjev 1931; Olson and Miller 1958; Berg 1960). Indeed, the earliest studies successfully established a correspondence between interaction and correlation by comparing correlations between interacting and non-interacting traits. Olson and Miller (1958) illustrated their approach by analyzing several morphological systems. For example, consider their analysis of 50 osteological measurements in a sample of 20 frogs (*Rana pipiens*) from a single population. The traits in their sample were distributed among five functional complexes: jaw movement, head orientation, axial movement and orientation, forelimb locomotion, and hindlimb locomotion. Three traits straddled to or more of these functional groups. Olson and Miller's study was conducted at a time before multivariate statistics were routinely used by zoologists. For example, principal component analysis, which is now one of the primary tools for studying patterns of integration (Klingenberg 2008), was first used in a zoological study by Jolicoeur and Mosimann (1960). Consequently, the Olson and Miller (1958) analysis may strike the modern reader as ad hoc. They first calculated 1225 product moment correlations among their traits and then searched for clusters of highly, moderately, and lowly correlated traits. The anticipated correspondence between functional groups and correlation was only dimly visible in the *R. pipiens* example. Nevertheless, across the their entire set of empirical examples, Olson and Miller (1958) were able to make a case that the functional organization of traits was reflected in phenotypic correlations, a result reproduced in many subsequent studies. For more on the contribution of Olson and Miller (1958) to contemporary perspectives, see Chernoff and Magwene (1999).

From a quantitative genetic perspective, one can ask if the correspondence between functional interaction and phenotypic correlation also holds for genetic and environmental correlation (Bailey 1956; Leamy 1977; Arnold 1981; Atchley et al. 1981; Cheverud 1982). In other words, are the *P*-, *G*-, and *E*-matrices comparably integrated into a pattern that reflects functional complexes? In an early study of this kind, Cheverud (1982) used functional and developmental studies of the human cranium to recognize five functional complexes in the primate cranium and then measured 56 traits distributed in those complexes in a sample of over 400 rhesus macaques of known pedigree. The expected pattern of higher correlation within functional complexes was obvious for $\boldsymbol{P}$, apparent for $\boldsymbol{E}$, and barely present in $\boldsymbol{G}$. Likewise, principal component analysis of $\boldsymbol{P}$ revealed groups that mapped onto functional complexes, with poorer mapping by $\boldsymbol{E}$, and worse mapping by $\boldsymbol{G}$. Cheverud's success in detecting functional groups in $\boldsymbol{P}$ has been reproduced in other studies (Berg 1960; Cheverud 1995, 1996; Armbruster et al. 1999; Marroig and Cheverud 2001; Klingenberg et al. 2004; Young and Hallgrímsson 2005; Porto et al. 2009).

Equating correlation with functional interaction is, however, complicated by development. Traits that are connected by a common developmental basis might be correlated because of shared developmental processes regardless of whether they interact functionally (Hall 1999; Hallgrímsson et al. 2007, 2009; Klingenberg 2008). Consequently, a few studies have attempted to separate the contribution of function and development to trait correlation (Simon and Marroig 2017). The basic idea is to classify traits into developmental modules based on shared developmental pathways and into functional modules based on biomechanical interactions. These two trait classifications can be used to test the hypotheses that the trait *P*-matrix resembles functional rather than developmental modularity on the argument that selection on function will eventually overwhelm developmental connections. Using this approach in a comparative study of anuran skulls that included size variation, Simon and Marroig (2017) found that functional modularity was more apparent in the *P*-matrix than was developmental modularity, perhaps because metamorphosis helped erase developmental signals. Despite this success, the complication of development means that phenotypic covariance patterns alone are unlikely to be a trustworthy guide to the causes of integration (Armbruster et al. 1999, Hallgrímsson et al. 2009). Riedl (1977) and Cheverud (1984) have argued that functional interactions will be built into the developmental system by selection on pleiotropy and so will frustrate attempts to separate the roles of selection and development.

Performance is the missing ingredient in the perspective and examples that we have just reviewed. The usual approach to morphological integration focuses on trait correlation as an outcome without dissecting the causes of correlation. To study the operation and evolution of functional complexes, however, we need to measure the successful operation of the complex. That measure of operation is called performance.

7.3 Morphology, Performance, and Fitness

Performance is a measure of the successful operation of a functional complex. For example, the proportion of successful attempts at prey capture in a salamander using tongue projections could be used as a measure of performance. Likewise, maximum rate of oxygen consumptions ($\dot{V}O_2$ max) is routinely used as a measure of cardio-vascular performance in vertebrates (Bennett 1980). In general, how do we concoct a relevant and useful measure of performance? A first step is to study the operation of the complex by dissection, observation, and experimentation (e.g., Gans 1974). In some cases, we can also take the next step and express the trait-performance relationship as a biomechanical or physiological model (Figure 7.1b). Finally, we need to express the performance measure using the elements of an actual or hypothetical model. In other words, we need a mathematical expression that relates performance of individuals to the values of particular traits in the functional complex.

To see the utility of a model for performance, consider the case of swallowing performance in snakes (Arnold 1983). Because snakes ingest their prey whole, ability to engulf prey is a simple function of the lengths of the bones and ligaments that encircle the prey during swallowing. In the particular case of egg-eating snakes, Gans (1952) made a detailed observational and anatomical study of the egg-eating process. The upshot of the Gans study is summarized with a simple set of five traits that can be used to model swallowing performance (Figure 7.4). In other words, our model for performance, f, is a simple additive function of five measurements,

$$f = \beta_{fz_1} z_1 + 2\beta_{fz_2} z_2 + 2\beta_{fz_3} z_3 + 2\beta_{fz_4} z_4 + \beta_{fz_5} z_5 + \varepsilon_f, \tag{7.1}$$

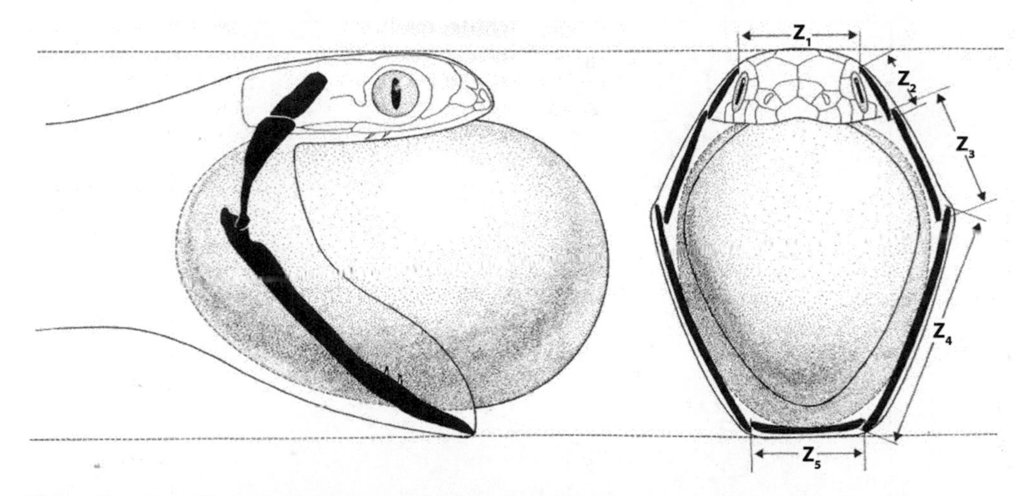

Figure 7.4 An African egg-eating snake (*Dasypeltis*) in the process of swallowing its prey. The lengths of five structural elements determine the maximal cross-sectional area of prey than can be ingested: width of the brain case (z_1), and the lengths of the supratemporal (z_2), quadrate (z_3), mandible (z_4), and mandibular symphysis (z_5).

Based on Gans (1952), from Arnold (1983).

where the subscripted β_s represent directional coefficients describing the effect of each of the five traits on performance, and the factors of 2 account for traits that participate twice in effects on performance. These β-coefficients are not linear selection gradients, which would represent effects of traits on fitness, we will call them *linear performance gradients*.

The relationship between a performance gradient and an ordinary selection gradient becomes clear if we construct a path diagram (Wright 1934; Li 1975). In Figure 7.5, we have made our example slightly more general by including two performance measures, f_1 and f_2, and by increasing the number of traits from five to k. From a statistical standpoint, Figure 7.5 portrays a nested regression model. The diagram takes our standard expression for multivariate directional selection,

$$w(\mathbf{z}) = \beta_1 z_1 + \beta_2 z_2 \cdots + \beta_k z_k + \varepsilon_w, \tag{7.2}$$

and converts it into a model in which all of the k traits exert their effects on fitness via two performance measures. In other words,

$$w(\mathbf{z}) = \left(\beta_{f_1 z_1}\beta_{wf_1} z_1 + \beta_{f_2 z_1}\beta_{wf_2} z_1\right) + \left(\beta_{f_1 z_2}\beta_{wf_1} z_2 + \beta_{f_2 z_2}\beta_{wf_2} z_2\right) \cdots + \left(\beta_{f_1 z_k}\beta_{wf_1} z_k + \beta_{f_2 z_k}\beta_{wf_2} z_k\right) + \varepsilon_w, \tag{7.3}$$

where the terms such as β_{wf_1} represent directional effects of performance on fitness, which we shall call a *fitness gradient* and terms such as $\beta_{f_1 z_1}$ represent effects of a trait on performance, which we will call a *performance gradient*. In the derivation of (7.3) we used a *product rule*: each directional selection gradient is the sum of products of a performance gradient and a fitness gradient. For example, selection on trait 1 via effects on performance f_1 is the product of a performance gradient $\beta_{f_1 z_1}$ and a fitness gradient β_{wf_1},

$$\beta_1 = \beta_{f_1 z_1}\beta_{wf_1}. \tag{7.4}$$

The simplicity of this formulation is apparent in a multivariate path diagram (Figure 7.5).

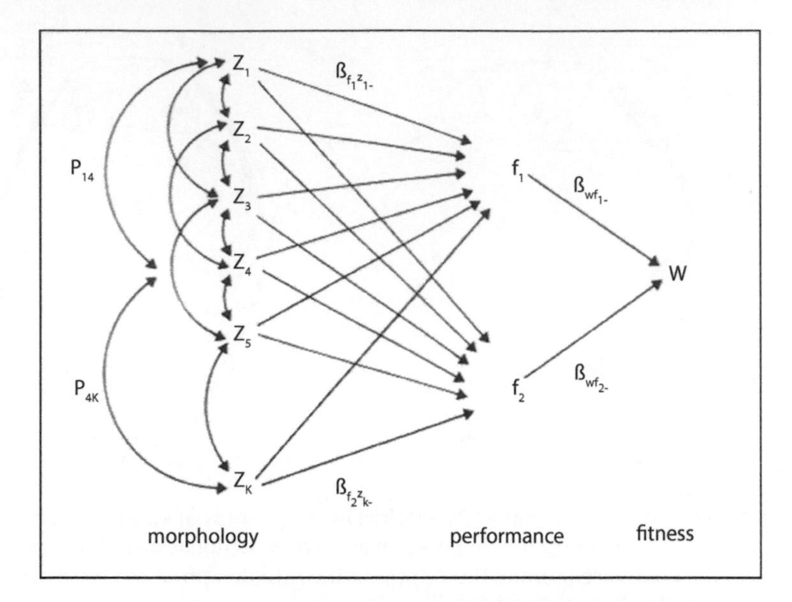

Figure 7.5 Path diagram showing causal relationships (paths with single arrowhead) between *k* morphological traits (z_1, z_2, $\cdots$ z_k), two measures of performance (f_1 and f_2), and relative fitness (w). Double-headed arrows represent phenotypic covariances between traits, before selection. The covariances between the first and fourth traits and between the first and kth traits are labelled, respectively, P_{14} and P_{1k}.

From Arnold (1983).

The significance of this product rule (Wright 1921, 1934, 1968) is that even when it is not feasible to estimate selection gradients or fitness gradients, it may often be possible to estimate performance gradients. Performance gradients are more routinely tractable because they can be estimated in the laboratory by performance testing, whereas fitness gradients require assessment of fitness or its components under field conditions (Jayne and Bennett 1990). Under ideal circumstances, both kinds of gradients can be estimated in a combined laboratory and field study (Garland and Losos 1994).

Morphological and other kinds of traits can have curvilinear as well as directional effects on performance, so we need to consider a more general model that includes those effects. The algebra of such a model is detailed by Arnold (2003). Without recounting those details here, we can note that in general the linear model that we have just considered can be represented by a multivariate path diagram in which the traits and performance are vectors (Figure 7.6). If we add curvilinear effects of traits on performance, that extended model can be represented by the path diagram shown in (Figure 7.7). Although this model is tractable and estimable, it is obviously a considerable simplification. We have assumed that fitness is a planar rather than a curvilinear function of performance. In other words, we have ruled out the possibility of an intermediate optimum. Although that assumption is justified and appropriate if our goal is to estimate directional fitness gradients, we need empirical inquires to establish whether the surface relating performance to fitness can indeed be approximated by a plane.

In general, we want to consider a general model in which performance is a curvilinear function of trait values (Figure 7.7) and fitness is a curvilinear model of performance

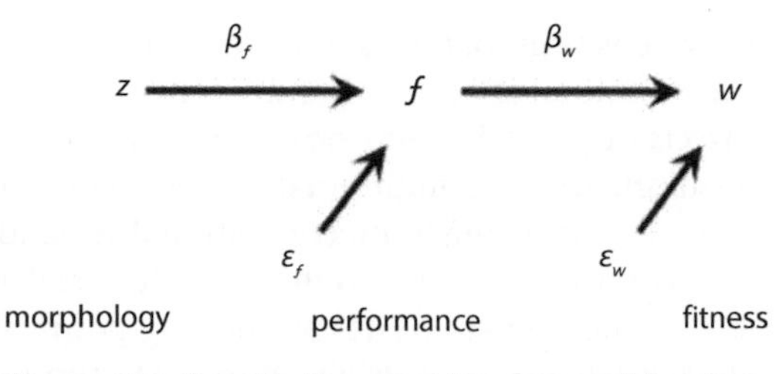

Figure 7.6 A path diagram showing multivariate linear performance and fitness gradients, β_f and β_w, that relate trait values to performance and performance values to fitness. Column vectors of trait and performance values are denoted $\boldsymbol{z}$ and $\boldsymbol{f}$. The residual value of performance, unaccounted for by trait values, is denoted ε_f. Fitness and the residual values of fitness, unaccounted for by trait values, are scalars and are denoted respectively as $w = w(\boldsymbol{z})$ and ε_w.

From Arnold (2003).

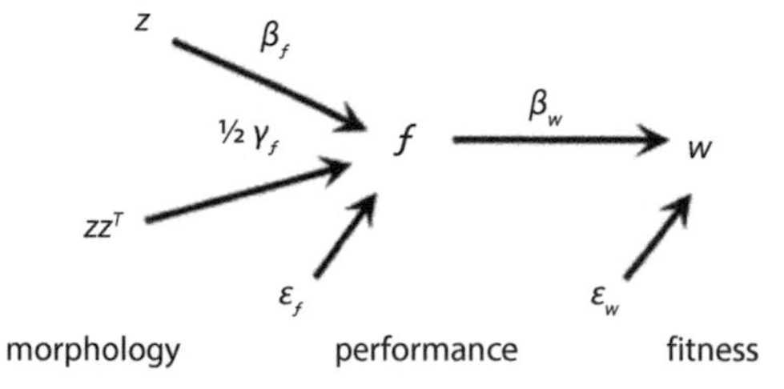

Figure 7.7 A path diagram showing the multivariate linear (β_w, a vector) and nonlinear (γ_w, a matrix) performance gradients that relate trait values to performance. Other conventions as in Figure 7.6.

From Arnold (2003).

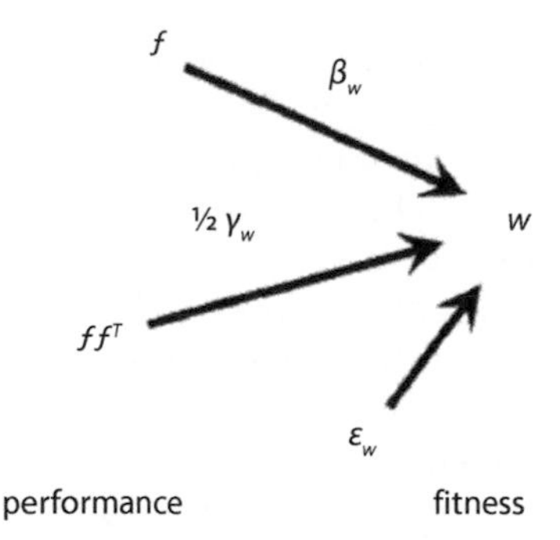

Figure 7.8 Path diagram showing the multivariate linear (β_w, a vector) and nonlinear (γ_w, a matrix) fitness gradients that relate performance to fitness. Other conventions as in Figure 7.6.

From Arnold (2003).

(Figure 7.8). The first of these two functions captures the key prediction that selection mediated through performance will build correlations between interacting traits, viz.,

$$f = \boldsymbol{\beta_f}^{\mathrm{T}}\boldsymbol{z} + \frac{1}{2}\boldsymbol{z}^{\mathrm{T}}\boldsymbol{\gamma_f}\,\boldsymbol{z} + \epsilon_f \tag{7.5a}$$

or, in the two trait case

$$f = z_1\beta_{fz_1} + z_2\beta_{fz_2} + \frac{1}{2}z_1^2\gamma_{fz_1^2} + \frac{1}{2}z_2^2\gamma_{fz_2^2} + z_1z_2\gamma_{fz_1z_2} + \epsilon_f. \tag{7.5b}$$

In general, *correlational selection*—symbolized with a coefficient of the form $\gamma_{fz_iz_j}$—acts through performance, *f*, and builds correlation between two interacting traits, z_i and z_j.

Building correlation by selection happens on two timescales. In the short term, correlational selection creates linkage disequilibrium between allelic effects at different loci. That is, between effects on z_i and effects on z_j. That creation is balanced by losses due to recombination, resulting in an equilibrium contribution to the genetic correlation between z_i and z_j (Chapter 6). In the long term, correlational selection changes the distribution of pleiotropic effects on the two traits, so that genetic correlation is established on a more permanent basis (Chapters 15–16). The overarching point is that a focus on performance and performance gradients gives us a way to study the process that integrates traits and makes them correlated.

7.4 Characterizing Trait Interactions in a Functional Complex Using the γ–matrix

We can view trait correlations as the product of selection that occurs when traits work together to perform a common function (Olson and Miller 1958, Hallgrímsson et al. 2009). Functional interactions of this kind can help to build trait correlations (covariances) (Lande 1980; Cheverud 1982; Wagner et al. 2007). Consequently, it follows that we should be able to detect functional interactions by assessing the effects of selection on trait covariances. The tools for that assessment are the off-diagonal elements of the C-matrix and especially of the γ-matrix. Although the causal basis of integration should reside in the off-diagonal elements of the γ-matrix, which measure correlational selection, the sheer size of the γ-matrix can frustrate our attempt to visualize multivariate correlational selection. The use of canonical analysis of $\boldsymbol{\gamma}$ to assess functional interaction of five traits is illustrated in the next example.

A study of the morphological determinants of bite force in the lizard *Tropidurus hispidus* provides an example of how trait interactions can be assessed using the γ-matrix (Simon et al. 2019). The authors assessed bite force by allowing live lizards to bite a force-measuring plate. They also measured five bones in the lower jaw and constructed biomechanical models of bite force based on functional considerations (Figure 7.9). In each of these models, muscles attach to an in-lever on one side of the fulcrum or pivot point of the lower jaw. When these muscles contract, they either close the jaw, exerting bite force, or open the jaw. In mechanical terminology, the muscles act on an *in-lever* and exert force at the end of an *out-lever*. Bite force can be expressed as *mechanical advantage*, which is the ratio of in-lever to out-lever (i.e., $IN_1 = a/d$ and $IN_2 = c/d$). Likewise, the mechanical advantage of jaw opening can be expressed as $IN_3 = p/d$. These relationships lead to the specific hypotheses that correlational selection would act within the in-levers (changing the lengths of c, a, and p in the same direction) or within the out-lever (changing the lengths of d and s in the same direction, but opposite to the in-levers) to increase bite force. Using their observations of bite force, the authors estimated linear and nonlinear performance gradients using the quadratic approximation described in Section 4.2. Their canonical analysis of the resulting γ-matrix (Table 7.1) helps us evaluate the evidence for correlational selection.

In a *canonical analysis* of this kind (Chapter 4), the γ-matrix is transformed into a diagonal matrix, $\boldsymbol{\Lambda}$, with the five eigenvalues of $\boldsymbol{\gamma}$ on its main diagonal (Table 7.1). These eigenvalues measure stabilizing selection (when the sign is negative) or disruptive selection (when the sign is positive) in trait directions corresponding to the five eigenvectors of the γ-matrix. When the loadings on a particular eigenvector are substantial, their signs

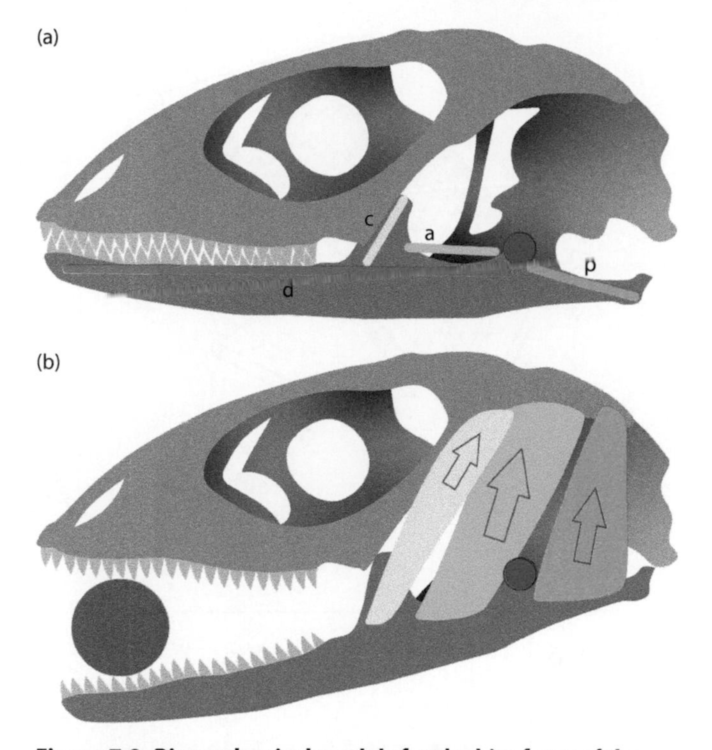

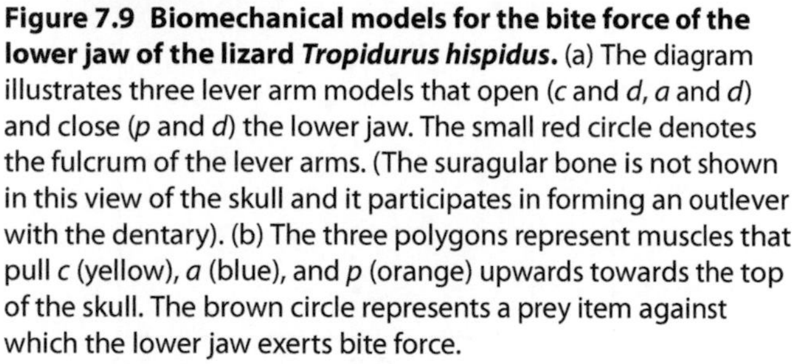

Figure 7.9 Biomechanical models for the bite force of the lower jaw of the lizard *Tropidurus hispidus*. (a) The diagram illustrates three lever arm models that open (*c* and *d*, *a* and *d*) and close (*p* and *d*) the lower jaw. The small red circle denotes the fulcrum of the lever arms. (The suragular bone is not shown in this view of the skull and it participates in forming an outlever with the dentary). (b) The three polygons represent muscles that pull *c* (yellow), *a* (blue), and *p* (orange) upwards towards the top of the skull. The brown circle represents a prey item against which the lower jaw exerts bite force.

Based on diagrams in Gröning et al. (2013) and Simon et al. (2019).

indicate two kinds of correlational selection. For example, the signs for $\boldsymbol{m}_1$ indicate negative correlational selection between coronoid and surangular, whereas the signs for $\boldsymbol{m}_2$ indicate positive correlational selection between coronoid and surangular and negative correlational selection between these bones and the angular. However, the only direction for which the eigenvalue gives strong evidence for nonlinear selection is $\boldsymbol{m}_5$, which corresponds to the jaw opening, in-level/out-lever contrast between prearticular and dentary (IN_3). Two other directions correspond to in-lever mapping ratios for jaw closing, and hence to the production of bite force, namely $\boldsymbol{m}_1$ and $\boldsymbol{m}_2$.

Visualization of the bite force performance surface corresponding to the coefficients in Table 7.1 for $\boldsymbol{m}_3$ and $\boldsymbol{m}_5$ helps us understand the overall picture provided by the canonical analysis of the γ-matrix (Figure 7.10). The performance surface is dome-shaped with an intermediate optimum near extreme values for $\boldsymbol{m}_3$ (circa 2.5), but close to the mean value for $\boldsymbol{m}_5$ (circa -0.05). In this visualization both $\boldsymbol{m}_3$ and $\boldsymbol{m}_5$ are under stabilizing selection towards an intermediate optimum that is relatively close to the bivariate mean.

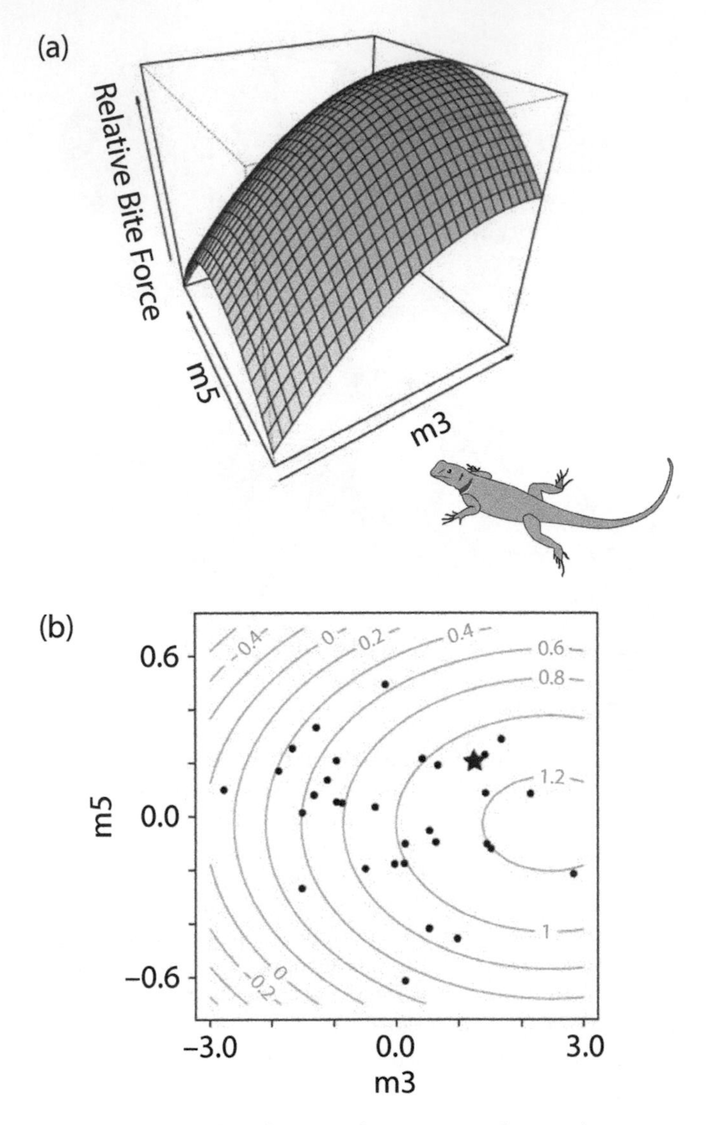

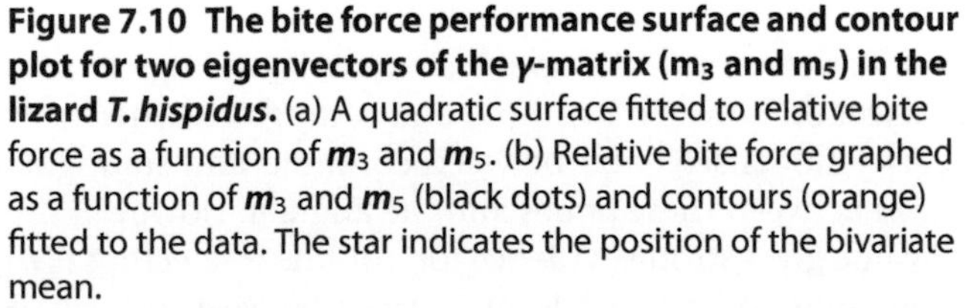

Figure 7.10 The bite force performance surface and contour plot for two eigenvectors of the γ-matrix (m_3 and m_5) in the lizard *T. hispidus*. (a) A quadratic surface fitted to relative bite force as a function of $\boldsymbol{m}_3$ and $\boldsymbol{m}_5$. (b) Relative bite force graphed as a function of $\boldsymbol{m}_3$ and $\boldsymbol{m}_5$ (black dots) and contours (orange) fitted to the data. The star indicates the position of the bivariate mean.

From Simon et al. (2019).

In summary, this study supports the optimism that the γ-matrix, especially its canonical form and visualization, can bridge between a biomechanical model for a functional complex and the selective forces that maintain the complex and shape its evolution. We now turn the issue of how modularity and individual modules evolve and how those evolutionary processes are in turn related to the γ-matrix.

Table 7.1 Eigenvectors and eigenvalues of a γ-matrix based on bite-force performance in the lizard *T. hispidus* and their biomechanical mappings (Simon et al. 2019). Boldface values indicate the traits that make the largest contributions to each eigenvector. Statistical significance of eigenvalues was evaluated with permutation tests.

	Eigenvectors of γ				
Traits, abbreviations	m_1	m_2	m_3	m_4	m_5
coronoid, c	**0.851**	**−0.463**	0.113	−0.103	−0.194
angular, a	0.066	**0.403**	0.179	**−0.890**	−0.095
dentary, c	0.035	0.149	**0.935**	**0.235**	**0.219**
surangular, s	**−0.504**	**−0.759**	**0.242**	**−0.323**	−0.088
prearticular, p	−0.126	0.159	0.152	0.195	**−0.947**
Eigenvalue (λ_i)	5.997 ns	2.468 ns	−0.162 ns	−3.263 ns	−6.104*
In-lever mapping	IN_1	IN_2	no	no	IN_3

∗ denotes P = 0.07, *ns* denotes P > 0.17. Highlighted cells indicate the expected participation of bony elements in in-levers.

7.5 Evolution of Modularity and Functional Complexes

A useful perspective on the modular design of traits is to view a module as a network of functional interactions that is relatively autonomous (Wagner et al. 2007). This concept of modularity in phenotypic traits is aligned with modularity concepts in many other contexts (e.g., communication, sociality, ecology). Taking an evolutionary perspective on phenotypes, we want to know how modularity is established and how modules evolve (Wagner et al. 2007, Hallgrimson et al. 2009). In the rest of this section, we will briefly consider questions that will be considered in greater detail in later chapters. The point in reviewing them now is to raise the prospect of connections between γ-matrices based on performance and specific evolutionary patterns.

What accounts for the evolutionary conservatism of many functional complexes? Fundamental conservatism in basic design is a defining characteristic of many functional complexes. Olson and Miller (1958), for example, give examples from the fossil record of Devonian amphibians in which the basic design of the skull is retained despite extensive shape transformations. Likewise, the bones and their characteristic arrangements in the skulls of reptiles remained fundamentally unchanged during radiations lasting tens to hundreds of millions of years (Romer 1956). These observations support the argument that understanding the selective forces that maintain patterns of integration (i.e., the γ-matrix) is fundamental to understanding the causes and mechanism of change in those patterns. The problem of long-lasting conservatism in functional complexes is discussed in Chapter 22. Part of that discussion is a comparative study of functional complexes in the courtship of salamanders and birds of paradise. Among the outstanding issues that we wish to address are the following:

1. How can some components of a complex diversify while other components remain stable? Returning to the example of pheromone delivery in plethodontid salamanders, we find that evolutionary pattern of the mental gland that produces the pheromone is one of long-term stability. In contrast, if we consider the pheromone itself on an evolutionary timescale, we see a perpetually, roiling cocktail of proteins that evolves in response to positive selection (Watts et al. 2004; Wilburn et al. 2017). How can

two compartments of a single functional complex with such different evolutionary personalities be described with the γ-matrix? The implication of the split in personality is that two compartments are largely independent. Does that independence spring directly from the γ-matrix?

2. How can we characterize the process by which functional complexes change? We can visualize fundamental change with diagrams showing, for example, new independence between the forelimbs and hind limbs (Young and Hallgrímsson 2005; Bell et al. 2011), but how exactly has selection changed and provoked this transition? Following the argument of this chapter, we need a model for dynamic change in selection that is cast in terms of the γ-matrix. We expect this model of dynamic change in the γ-matrix to reflect a change in modularity. For example, a set of traits can become a module if the γ-matrix changes so that it promotes positive within-module correlational selection that did not occur in the ancestral population.
3. How is the evolution of the compartments affected when a functional complex is at the crux of interactions between sexual partners or between species? Rapidity often characterizes the evolution of molecular and phenotypic traits involved in interactions between sexual partners (Arnold and Houck 2016; Wilburn and Swanson 2016). Similarly, rapid evolutionary arms races can occur in traits that mediate the interactions of different species (Thompson 2005). In these instances, how does modularity, encapsulated by the γ-matrix, affect rapid evolution and vice versa? We return to this question in Chapter 22.

7.6 Using the *P*-matrix Instead of the *G*-matrix

Recognizing that the *G*-matrix is both central to equations for response to selection and drift but often unavailable or difficult to estimate, Cheverud (1988) argued that $\boldsymbol{P}$ might be used in place of $\boldsymbol{G}$ in a variety of circumstances. Cheverud supported his argument with a meta-analysis that compared 41 pairs of *P*- and *G*-matrices. His analysis showed that the two kinds of matrices are often similar, especially if $\boldsymbol{G}$ is well estimated. The result is not surprising if we recall the standard result using the additive model that $\boldsymbol{P} = \boldsymbol{G} + \boldsymbol{E}$, which reminds us that $\boldsymbol{G}$ is contained in $\boldsymbol{P}$. On the other hand, one can argue that $\boldsymbol{E}$ might make a contribution to $\boldsymbol{P}$ that obscures the resemblance to $\boldsymbol{G}$ (Willis et al. 1988). Cheverud (1988) argued that $\boldsymbol{G}$ and $\boldsymbol{E}$ might have similar structure, so $\boldsymbol{P}$ versus $\boldsymbol{G}$ resemblance would not be obscured. Alternatively, $\boldsymbol{E}$ might be a virtually diagonal matrix with very small off-diagonal elements (as in Arnold and Phillips 1999), a situation that will also promote $\boldsymbol{P}$ versus $\boldsymbol{G}$ resemblance. Regardless, the case for substitution rests on the empirical evidence for resemblance, which appears to grow as data accumulate (Roff 1995; Koots and Gibson 1996; Arnold and Phillips 1999; Marroig and Cheverud 2001; Sodini et al. 2018).

7.7 Performance Is Not a Surrogate for Fitness

Measures of performance have sometimes been called 'surrogates for fitness' (Endler 1986). This characterization is unfortunate because it suggests that we are pretending that performance is fitness. No such deception is being practiced. Performance mediates the effects of traits on fitness. It is an aspect of fitness. By recognizing this relationship, we can characterize the different parts of selection that acts on functional complexes.

7.8 Conclusions

- Phenotypic traits are not isolated entities but are instead bundled into ensembles (modules) of interacting traits. These ensembles are called functional complexes.
- The most proximate effect on fitness of a functional complex is mediated through a component of fitness known as performance.
- The bundling (integration) of traits into a functional complex is accomplished by multivariate stabilizing selection, especially correlational selection.
- The evolutionary fates of traits within a complex are bound together by multivariate stabilizing selection and genetic correlation.
- Trait ensembles often retain their identities for hundreds of millions of years. This persistence of identity suggests that the correlational selection responsible for bundling of the complex likewise persists through deep evolutionary time.

Literature Cited

Armbruster, W. S., L. Antonsen, and C. Pelabon. 2005. Phenotypic selection on *Dalechampia* blossoms: honest signaling affects pollination success. Ecology 86:3323–3333.

Armbruster, W. S., V. S. Di Stilio, J. D. Tuxill, T. C. Flores, and J. L. Velasquez Runk. 1999. Covariance and decoupling of floral and vegetative traits in nine neotropical plants: a re-evaluation of Berg's correlation-pleiades concept. Am. J. Bot. 86:39–55.

Arnold, S. J. 1981. Behavioral variation in natural populations. I. Phenotypic, genetic and environmental correlations between chemoreceptive responses to prey in the garter snake, *Thamnophis elegans*. Evolution 489–509.

Arnold, S. J. 1983. Morphology, performance and fitness. Am. Zool. 23:347–361.

Arnold, S. J. 2003. Performance surfaces and adaptive landscapes. Integr. Comp. Biol. 43:367–375.

Arnold, S. J., and L. D. Houck. 2016. Can the Fisher-Lande process account for birds of paradise and other sexual radiations? Am. Nat. 187:717–735.

Arnold, S. J., K. M. Kiemnec-Tyburczy, and L. D. Houck. 2017. The evolution of courtship behavior in plethodontid salamanders, contrasting patterns of stasis and diversification. Herpetologica 73:190.

Arnold, S. J., and P. C. Phillips. 1999. Hierarchical comparison of genetic variance-covariance matrices. II. Coastal-inland divergence in the garter snake, *Thamnophis elegans*. Evolution 53:1516–1527.

Atchley, W. R., J. J. Rutledge, and D. E. Cowley. 1981. Genetic components of size and shape. II. Multivariate covariance patterns in the rat and mouse skull. Evolution 35:1037–1055.

Bailey, D. W. 1956. A comparison of genetic and environmental principal components of morphogenesis in mice. Growth 20:63–74.

Bell, E., B. Andres, and A. Goswami. 2011. Integration and dissociation of limb elements in flying vertebrates: a comparison of pterosaurs, birds and bats: Integration and dissociation of limb elements in flying vertebrates. J. Evol. Biol. 24:2586–2599.

Benitez-Vieyra, S., A. M. Medina, E. Glinos, and A. A. Cocucci. 2006. Pollinator-mediated selection on floral traits and size of floral display in *Cyclopogon elatus*, a sweat bee-pollinated orchid. Funct. Ecol. 20:948–957.

Bennett, A. F. 1980. The metabolic foundations of vertebrate behavior. BioScience 30:452–456.

Berg, R. L. 1960. The ecological significance of correlation pleiades. Evolution 14:171–180.

Cardoso, P., S. Pekár, R. Jocqué, and J. A. Coddington. 2011. Global patterns of guild composition and functional diversity of spiders. PLoS ONE 6:e21710.

Caruso, C. M. 2000. Competition for pollination influences selection on floral traits of *Ipomopsis aggregate*. Evolution 54:1546–1557.

Chernoff, B., and P. M. Magwene. 1999. Morpological integration: forty years later. Pp. 316–360 *in* E. C. Olson and R. L. Miller, Morphological Integration. University of Chicago Press, Chicago.

Cheverud, J. M. 1982. Phenotypic, genetic, and environmental morphological integration in the cranium. Evolution 36:499.

Cheverud, J. M. 1984. Quantitative genetics and developmental constraints on evolution by selection. J. Theor. Biol. 110:155–171.

Cheverud, J. M. 1988. A comparison of genetic and phenotypic correlations. Evolution 42: 958–968.

Cheverud, J. M. 1995. Morphological integration in the saddle-back tamarin (*Saguinus fuscicollis*) cranium. Am. Nat. 145:63–89.

Cheverud, J. M. 1996. Developmental integration and the evolution of pleiotropy. Am. Zool. 36:44–50.

Cundall, D. 2009. Viper fangs: Functional limitations of extreme teeth. Physiol. Biochem. Zool. 82:63–79.

Deban, S. M., D. B. Wake, and G. Roth. 1997. Salamander with a ballistic tongue. Nature 389: 27–28.

Endler, J. A. 1986. Natural Selection in the Wild. Princeton University Press, Princeton.

Fry, B. G., H. Scheib, L. van der Weerd, B. Young, J. McNaughtan, S. F. R. Ramjan, N. Vidal, R. E. Poelmann, and J. A. Norman. 2008. Evolution of an arsenal. Mol. Cell. Proteomics 7:215–246.

Gans, C. 1952. The functional morpology of the egg-eating adaptations in the snake genus *Dasypeltis*. Zoologica 37:209–244.

Gans, C. 1974. Biomechanics: an Approach to Vertebrate Biology. Lippincott, Philadelphia.

Garland Jr, T., and J. B. Losos. 1994. Ecological morphology of locomotor performance in squamate reptiles. Pp. 240–302 *in* P. C. Wainwright and S. M. Reilly, eds., Ecological Morphology: Integrative Organismal Biology. University of Chicago Press, Chicago.

Gosline, J. M., P. A. Guerette, C. S. Ortleipp, and K. N. Savage. 1999. The mechanical design of spider silks. J. Exp. Biol. 202:3295–3303.

Gould, S. J., and R. C. Lewontin. 1979. The spandrels of San Marco and the panglossian paradigm: A critique of the adaptationist programme. Proc. R. Soc. Lond. B Biol. Sci. 205:581–598.

Gröning, F., M. E. H. Jones, N. Curtis, A. Herrel, P. O'Higgins, S. E. Evans, and M. J. Fagan. 2013. The importance of accurate muscle modelling for biomechanical analyses: A case study with a lizard skull. J. R. Soc. Interface 10:20130216.

Hall, B. K. 1999. The Neural Crest in Development and Evolution. Springer Science and Business Media, New York.

Hallgrímsson, B., H. Jamniczky, N. M. Young, C. Rolian, T. E. Parsons, J. C. Boughner, and R. S. Marcucio. 2009. Deciphering the palimpsest: Studying the relationship between morphological integration and phenotypic covariation. Evol. Biol. 36:355–376.

Hallgrímsson, B., D. E. Lieberman, W. Liu, A. F. Ford-Hutchinson, and F. R. Jirik. 2007. Epigenetic interactions and the structure of phenotypic variation in the cranium: Epigenetic interactions in the cranium. Evol. Dev. 9:76–91.

Houck, L. D., and S. Arnold. 2003. Courtship and mating behavior. Pp. 383–424 in Reproductive biology and phylogeny of Urodela. M/s Science Publications, Endfield.

Jackson, K. 2003. The evolution of venom-delivery systems in snakes. Zool. J. Linn. Soc. 137: 337–354.

Jayne, B. C., and A. F. Bennett. 1990. Selection on locomotor performance capacity in a natural population of garter snakes. Evolution 44:1204–1229.

Jolicoeur, P., and Mosimann. 1960. Size and shape variation in the painted turtle. A principal component analysis. Growth 24:339–354.

Klauber, L. M. 1956. Rattlesnakes, Their habits, Life Histories, and Influence on Mankind. University of California Press, Berkeley.

Klingenberg, C. P. 2008. Morphological integration and developmental modularity. Annu. Rev. Ecol. Evol. Syst. 39:115–132.

Klingenberg, C. P., L. J. Leamy, and J. M. Cheverud. 2004. Integration and modularity of quantitative trait locus effects on geometric shape in the mouse mandible. Genetics 166: 1909–1921

Koots, K. R., and J. P. Gibson. 1996. Realized sampling variances of estimates of genetic parameters and the difference between genetic and phenotypic correlations. Genetics 143:1409–1416.

Lande, R. 1980. The genetic covariance between characters maintained by pleiotropic mutations. Genetics 94:201–215.

Leamy, L. 1977. Genetic and environmental correlations of morphometric traits in random bred house mice. Evolution 31:357–369.

Li, C. C. 1975. Path Analysis: a Primer. Boxwood Press, Pacific Grove.

Marroig, G., and J. M. Cheverud. 2001. A comparison of phenotypic variation and covariation patterns and the role of phylogeny, ecology, and ontogeny during cranial evolution in New World monkeys. Evolution 55:2576–2600.

O'Connell, L. M., and M. O. Johnston. 1998. Male and female pollination success in a deceptive orchid, a selection study. Ecology 79:1246–1260.

Olson, E. C., and R. L. Miller. 1958. Morphological Integraton. University of Chicago Press, Chicago.

Porto, A., F. B. de Oliveira, L. T. Shirai, V. De Conto, and G. Marroig. 2009. The evolution of modularity in the mammalian skull I: Morphological integration patterns and magnitudes. Evol. Biol. 36:118–135.

Riedl, R. 1977. A systems-analytical approach to macro-evolutionary phenomena. Quarterley Rev. Biol. 52:351–370.

Riedl, R. 1978. Order in Living Systems: Systems Analysis of Evolution. John Wiley and Sons, New York.

Roff, D. A. 1995. The estimation of genetic correlations from phenotypic correlations: A test of Cheverud's conjecture. Heredity 74:481–490.

Romer, A. S. 1956. The Vertebrate Body. W. B.Saunders, Philadelphia.

Schemske, D. W., and J. Ågren. 1995. Deceit pollination and selection on female flower size in *Begonia involucrata*: An experimental approach. Evolution 49:207–214.

Scholes III, E. 2008. Evolution of the courtship phenotype in the bird of paradise genus *Parotia* (Aves: Paradisaeidae): homology, phylogeny, and modularity. Biol. J. Linn. Soc. 94:491–504.

Simon, M. N., Brandt, T. Kohlsddorf, and S. J. Arnold. 2019. Bite performance surfaces of three ecologically divergent Iguanidae lizards: relationships with lower jaw bones. Biol. J. Linn. Soc. 127:810–825.

Simon, M. N., and G. Marroig. 2017. Evolution of a complex phenotype with biphasic ontogeny: contribution of development versus function and climatic variation to skull modularity in toads. Ecol. Evol. 7:10752–10769.

Sinervo, B. and E. Svensson. 2002. Correlational selection and the evolution of genomic architecture. Heredity 89:329–338.

Sinha, R. K. 2004. Modern Plant Physiology. Alpha Science, Pangbourne.

Sodini, S. M., K. E. Kemper, N. R. Wray, and M. Trzaskowski. 2018. Comparison of genotypic and phenotypic correlations: Cheverud's conjecture in humans. Genetics 209: 941–948.

Svensson, E. I., S. J. Arnold, R. Burger, K. Csilléry, J. Draghi, J. M. Henshaw, A. G. Jones, S. De Lisle, D. A. Marques, K. McGuigan, M. N. Simon, and Anna Runemark. 2021. Correlational selection in the age of genomics. Nature Ecology & Evolution 5:562–573.

Terentjev, P. V. 1931. Biometrische Untersuchungen über die morphologischen Merkmale von *Rana ribibunda* Pall. (Amphibia, Salientia). Biometrika 23:23–51.

Thompson, J. N. 2005. The Geographic Mosaic of Coevolution. University of Chicago Press, Chicago.

Tinbergen, N. 1952. The curious behavior of the stickleback. Sci. Am. 187:22–27.

Waddington, C. H. 1957. The strategy of the Genes: A Discussion of Some Aspects of Theoretical Biology. Routledge, Abingdon.

Wagner, G. P., M. Pavlicev, and J. M. Cheverud. 2007. The road to modularity. Nat. Rev. Genet. 8:921–931.

Watts, R. A., C. A. Palmer, R. C. Feldhoff, P. W. Feldhoff, L. D. Houck, A. G. Jones, M. E. Pfrender, S. M. Rollmann, and S. J. Arnold. 2004. Stabilizing selection on behavior and morphology masks positive selection on the signal in a salamander pheromone signaling complex. Mol. Biol. Evol. 21:1032–1041.

Wilburn, D. B., S. J. Arnold, L. D. Houck, P. W. Feldhoff, and R. C. Feldhoff. 2017. Gene duplication, co-option, structural evolution, and phenotypic tango in the courtship pheromones of plethodontid salamanders. Herpetologica 73: 206–219.

Wilburn, D. B., and W. J. Swanson. 2016. From molecules to mating: rapid evolution and biochemical studies of reproductive proteins. J. Proteomics 135:12–25.

Willis, J. H., J. A. Coyne, and M. Kirkpatrick. 1991. Can one predict the evolution of quantitative characters without genetics? Evolution 45: 441–444.

Wilson, P. 1995. Selection for pollination success and the mechanical fit of *Impatiens* flowers around bumblebee bodies. Biol. J. Linn. Soc. 55:355–383.

Wright, S. 1921. Systems of mating. I. the biometric relations between parents and offspring. Genetics 6:111–123.

Wright, S. 1934. The method of path coefficients. Ann Math Stat 5:161–215.

Wright, S. 1968. Evolution and the Genetics of Populations: a Treatise. University of Chicago Press, Chicago.

Young, N. M., and B. Hallgrímsson. 2005. Serial homology and the evolution of mammalian limb covariation structure. 59:2691–2704.

CHAPTER 8

Drift of a Single, Neutral Trait

Overview—Random sampling of parents can change gene frequencies and cause the trait mean to fluctuate. Such fluctuations are most severe in small populations and in the absence of selection. The tendency of the trait mean to fluctuate by drift is captured by N_e, the effective size of a population. Long-term estimates of N_e based on nucleotide diversity range up to tens of millions but are much lower in some species. These numbers represent the effective size of a set of evolving, interbreeding lineages that may compose an entire species. Theory for drift in the trait mean is based on a model in which many replicate populations are derived from a single ancestral population and thereafter evolve independently. Such stochastic models allow us to predict the variation among replicate lineages at any generation in the future, as a function of genetic variance and effective population size. In nature, populations diverge more rapidly in the short term than is predicted by drift. In the long term, populations and species diverge less than is predicted by drift. These considerations indicate that drift is not a sufficient explanation for trait differentiation on nearly all timescales. Nevertheless, it is important to build a model of drift so that we can assess its contribution to diversification in relation to selection and other factors.

With the addition of a constraint on population size (N_e) and migration from other populations, the life cycle is as follows. (1) Diploid zygotes inherit genetic effects from their parents in the preceding generation, which together with new mutations affect the expression of zygotic phenotypic traits. (2) The zygotic cohort experience selection based on their phenotypic traits. (3) Random choice of a fixed number of adults from the survivors of selection. (4) Migrants from other populations join the pool of reproducing adults. (5) The pool of adults produces a new cohort of zygotes by sexual reproduction.

8.1 The Concept of Effective Population Size, N_e

Population size has a fundamental effect on random variation in the trait mean from generation to generation. Such random or stochastic effects can be conveniently modelled using a single number called *effective population size*, N_e (Wright 1931; Lande and Barrowclough 1987; Charlesworth 2009). In the *Fisher-Wright model*, with no selection and an approximate Poisson distribution of family sizes, this number simply represents the number of parents. In more complicated models, effective population size can incorporate the sex ratio of breeding parents and other distributions of family size (Crow and Kimura 1970). In all cases, however, the basic idea is that a single number, N_e, can account for stochastic variation in genetic properties that arises each generation from finite sampling of parents. If we model the effects of temporal fluctuation in population size, it turns out that *harmonic mean* population size provides a good overall approximation of

Evolutionary Quantitative Genetics. Stevan J. Arnold, Oxford University Press. © Stevan Arnold (2023).
DOI: 10.1093/oso/9780192859389.003.0009

sampling effects (Wright 1969). Bottlenecks have a disproportionately large effect that is not captured by the arithmetic mean but is captured by the harmonic mean. Consequently, a single number, call it N_e, can be used to represent population size even when size fluctuates.

Two of the most popular methods of estimating N_e are based on samples of genetic markers that are thought to be selectively neutral (Charlesworth 2009). In the *temporal genetic method*, genetic markers are sampled from a single local population at two or more points in time. The effective size of the population is inferred from observed change in the frequencies genetic markers. In the *nucleotide diversity method*, silent site diversity is estimated from samples that may be taken over the entire geographic range of a species. Effective population size is estimated using the argument that product of mutation rate per nucleotide site and N_e has reached an equilibrium.

The two methods produce results that have very different properties and hence different domains of application. The temporal genetic method produces a snapshot of N_e over a brief interval of time in a single, local population. In contrast, the nucleotide diversity method can yield an estimate of N_e that pertains to the entire phylogenetic history of a set of populations, arrayed over a large geographic area. The use of each kind of estimate depends on the kind of question being asked.

8.2 Short- and Long-term Estimates of N_e

Not surprisingly, short term estimates of N_e from a single local population are commonly smaller than N_e estimates based on the *nucleotide diversity method*, because they typically focus on populations of limited spatial extent and short timescales. For example, the Palstra and Ruzzante (2008) survey of estimates using the *temporal genetic method* shows a strong mode in the range 100–300 but with a tail extending to >10,000 (Figure 8.1). Populations targeted for conservation attention are over-represented in the vicinity of the mode, suggesting that the survey's results are biased downwards.

In contrast, Charlesworth's (2009) survey of results using the nucleotide diversity methods yielded many estimates of N_e that exceed 10,000 (Table 8.1). Indeed, estimates for fruit flies and nematodes ranged as high as 10^6, and N_e was estimated to be 10^7 for *E. coli*. We can reconcile results from the two kinds of estimates using a metapopulation model

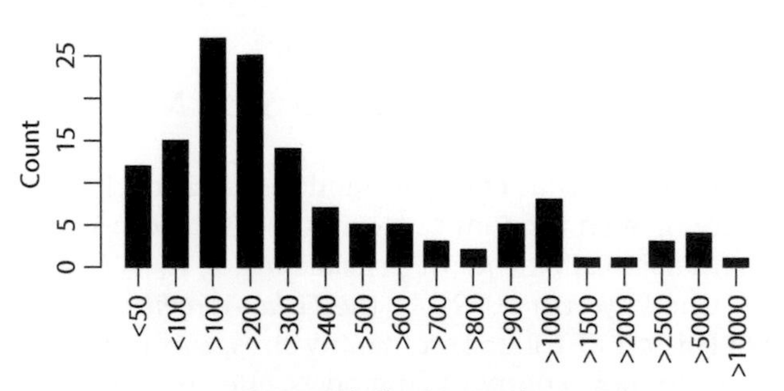

Figure 8.1 Histogram of N_e estimates using the temporal genetic method (n = 144). Most estimates are for vertebrate and invertebrate animals.

After Palstra and Ruzzante (2008).

Table 8.1 Estimates of effective population size from DNA sequence diversity. From Charlesworth (2009) with permission.

Species	N_e	Data	References
Humans	10,400	50 nuclear sequences	Yu et al. 2004
Bonobo	12,300	50 nuclear sequences	Yu et al. 2004
Chimpanzee	21,300	50 nuclear sequences	Yu et al. 2004
Gorilla	25,200	50 nuclear sequences	Yu et al. 2004
Gray whale	34,410	9 nuclear gene introns	Alter et al. 2007
Caenorhabditis elegans	80,000	6 nuclear genes	Cutter 2005
Plasmodium falciparum	255,000	204 nuclear genes	Mu et al. 2007
Drosophila melanogaster	1,150,000	252 nuclear geness	Shapiro et al. 2007
Caenorhabditis remanei	1,600,000	6 nuclear genes	Cutter et al. 2006
Escherichia coli	25,000,000	410 genes	Charlesworth and Eyre-Walker 2006

that combines or extrapolates results from individual populations into an estimate that pertains to an entire evolving lineage.

Another important genomic perspective is that many spatial and temporal patterns in gene frequencies may arise not from direct selection but from selection on linked sites in conjunction with drift. See Charlesworth (2009) for a discussion of background selection and hitchhiking in finite populations.

8.3 N_e Estimates for Metapopulations, Lineages, and Species

An evolving lineage is more than a single panmictic population with a particular N_e. In particular, we want to consider the case of a population subdivided into a set of demes of various effective sizes and exchanging migrants. As we follow this *metapopulation* through evolutionary time, some demes go extinct and their territory is colonized by other, more successful demes. Our intuition tells us that the total N_e for such a lineage composed of 10 demes, each with an N_e of 100, must be more than 100, perhaps more than 1000. For *Wright's island model* of migration between demes, this intuition is known to be correct (Wright 1943).

In the island model, migration occurs at the same rate between all pairs of island populations, which are of equal size. Wright (1939) showed for the island model that the effective size of the total metapopulation is

$$N_e = \frac{n\bar{N}}{(1 - F_{ST})}, \tag{8.1}$$

where n is the number of demes, $\bar{N}$ is their average effective size, and F_{ST} and $(1 - F_{ST})$ are, respectively, the among- and within-deme components of variance in gene frequency. In our thought experiment, N_e could be as small as 1000, if all of the variation is within demes ($F_{ST} = 0$), or large as 100,000, if nearly all of the genetic variation is among demes ($F_{ST} = 0.99$). When F_{ST} approaches 1, migration is slight, and the demes behave as nearly independent entities.

Under more general assumptions, Whitlock and Barton (1997) showed that a metapopulation has effective size

$$N_e = \frac{1}{\{1 + Var(\vartheta)\} \langle \frac{(1-F_{STi})}{N_i n} \rangle}, \quad (8.2)$$

where ϑ_i is the eventual contribution of the *ith* deme to the whole population, $Var(\vartheta)$ is the variance in those eventual contributions (which are scaled so that $\sum_i \vartheta_i = n$), and $\langle\rangle$ denotes an average over demes weighted by ϑ_i^2, N_i is the effective size of the *ith* deme, and F_{STi} is the average of pair-wise F_{ST}s for the ith deme (Whitlock and Barton 1997). The values of ϑ_i are given by the leading eigenvector of the migration matrix $\boldsymbol{m}$, which has elements m_{ij} representing the per-generation migration rate from deme j to deme i (the proportion of individuals in deme i newly arrived from deme j). Each element on the main diagonal of $\boldsymbol{m}$ is one. In expression (8.1) $\bar{N}$ is an average weighted by ϑ_i^2. We see that (8.2) reduces to (8.1), if the eventual contributions of all demes are equal, $Var(\vartheta) = 0$, but in general $Var(\vartheta)$ decreases the total N_e below the simple expectation of $n\bar{N}$.

A study using genetic markers in multiple demes of the garter snake (*Thamnophis elegans*) provides an opportunity to use expressions (8.1 and 8.2) to estimate N_e for a metapopulation. Manier and Arnold (2005) estimated N_e, F_{ST}, and migration rates among 20 demes of this garter snake using microsatellites and coalescent methods (Beerli and Felsenstein 2001). Effective population size averaged 327.625 in these 20 demes (variance = 21,065), with a range of 150–765. Overall F_{ST} was 0.024. Using these values in (8.1), N_e for the entire array of demes is 6714. This estimate is only slightly larger than 6552, the result we get by assuming $F_{ST} = 0$. In other words, microsatellite differentiation among these demes is so weak that we get a reasonably good estimate of N_e for the entire metapopulation by multiplying the number of demes by the average N_e. Taking migration into account to estimate the size of the metapopulation using (8.2), we find that the variance among demes in eventual contributions to effective size is 0.004176. The weighted average in (8.2) is 0.000141165, yielding a value for the metapopulation N_e of 7054.

Different effective sizes are appropriate for modelling neutral evolution at different temporal and spatial scales. In our garter snake example, the N_e estimate of 328 would be appropriate for modelling the neutral evolution in a set of replicate local demes. If we wish to consider the neutral evolution of an entire metapopulation (i.e., replicate sets of about 20 demes), then N_e in the range 6500–7000 would be appropriate. In the Manier and Arnold (2005) study, 20 demes occupied about 50 km^2 (an entire drainage basin) in one corner of Lassen Co., CA. The entire range of *T. elegans* occupies an area of about 2,551,127km^2 in western North America (Rossman et al. 1996). The five major lineages that make up that species each occupy areas averaging about 510,225km^2 (Uyeda 2012). To model the neutral evolution of sets of lineages, one might use an N_e of about 69 million, (510,225/50)∗6750, and an N_e of about five times that value, 345 million, to model the evolution of replicate sets of garter snake species. Of course these extrapolations from the 50km^2 study area assume that the population structure there is representative of all the major lineages in the species. In fact, the study area was chosen because *T. elegans* demes were both abundant and well-stocked, which might bias the extrapolations upwards. But even if the lineage and species-wide estimates of N_e are inflated by an order of magnitude, a major lineage N_e of 7 million and a species-wide N_e of 35 million are still vastly greater than the individual deme estimate of 328 or the metapopulation estimate of 6750.

8.4 Estimates of the Census Size of Populations

Although an estimate of the census size of a population does not account for variation in reproductive success and bottlenecks, census size has been estimated for many species in a diverse range of taxa. Buffalo (2021) compiled estimates of census size for 172 species representing arthropods, chordates, echinoderms, and molluscs. Census size ranged over 11 orders of magnitude, from 10^4 to 10^{15}, with a mode at 10^{10}. Census size varied among taxa, with chordates having a mode at about 10^8 and arthropods having a mode at about 10^{12}. In other words, these estimates agree with the preceding section in suggesting that species-wide population sizes are commonly in the range of tens to hundreds of millions of individuals.

8.5 Sampling From a Normal Distribution of Breeding Values

If we take repeated samples of size N from a normal distribution with variance V, the variance among the means of those samples will be V/N. This result reinforces our intuition that differences due to sampling will be appreciable with small samples, but inconsequential if the samples are large. We can apply this result to model the consequences of drift for replicate populations that obey the same rules of inheritance and sampling. We imagine a base population in which trait breeding values are normally distributed with mean $\bar{x} = \bar{z} = 0$ and variance G. Suppose we establish replicate populations, each of size N_e, from this base population. The variance among those replicates in mean breeding value at generation 0 will be

$$Var(\bar{z}_0) = \frac{1}{N_e}G \tag{8.3}$$

and the overall trait mean will be zero (Lande 1976).

8.6 Projecting the Distribution of Breeding Values Into the Future

Now allow each of the replicate populations to constitute a lineage. In each lineage, G remains the same, and the same process of sampling occurs each generation. If we follow a single lineage through time, its phenotypic mean each generation, $\bar{z}_{t+1}$, deviates from the mean in the preceding generation, $\bar{z}_t$, by an amount that can be specified by taking a random draw from normal distribution of breeding values with zero mean and variance G/N_e. Following the mean through time, we see a random walk, a kind of Brownian motion (Figure 8.2a). Each replicate lineage undergoes an independent random walk that obeys the same rules. We now wish to consider the statistical properties of the entire ensemble of replicate lineages at some generation, t, in the future. We began in generation 0 with an among-replicate distribution of means that was normally distributed with a mean of 0 and a variance of G/N_e (8.3). The distribution of means in generation 2 is the product of that distribution and normal distribution with the same properties, a so-called *convolution of normal distributions*, which is itself a normal distribution. By extending this logic, we can conclude that the *among-replicate distribution of means* at any future generation t, is a normal distribution of lineage trait means,

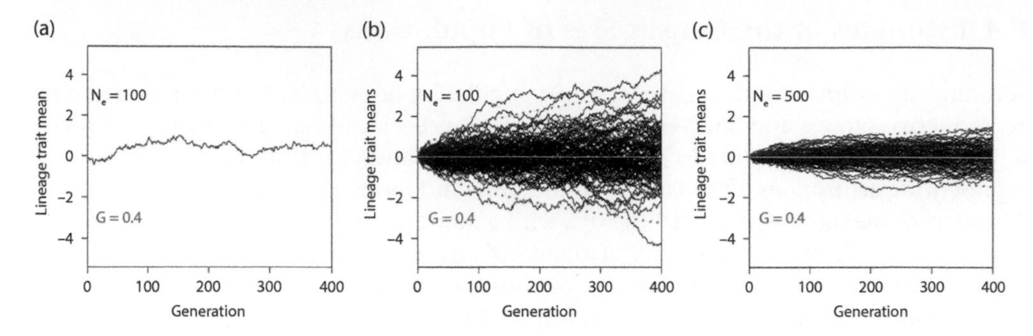

Figure 8.2 Genetic drift causes the phenotypic mean of a lineage to undergo a random walk. In these simulations of 400 generations of genetic drift, the lineage trait mean, $\bar{z}$, is shown in units of within-population phenotypic standard deviation, $\sqrt{P}$. (a) The random walk of a single lineage of small effective size. The random walks of 100 replicates lineages are shown for small (b) and moderate sized (c) populations. The theoretical 99% confidence limits for the distribution of lineage means, based on (8.5), are shown as dotted blue curves. The deterministic trait mean is shown as a white line.

$$\Phi\left(\bar{z}_t\right) = \frac{exp\left\{-\frac{1}{2}(\bar{z}_t - \bar{z}_0)^2/D\left(t\right)\right\}}{\sqrt{2\pi D\left(t\right)}}, \tag{8.4}$$

with a mean of zero, $\bar{z}_0 = 0$, and a variance given by

$$D\left(t\right) = \frac{1}{N_e} Gt, \tag{8.5}$$

where $D\left(t\right)$ stands for the dispersion (variance) of lineage trait means at generation t (Lande 1976). In other words, the variance among replicates increases linearly with time. This linear property means that the 99% confidence limits for the mean of means are ever-expanding, so that their graph forms a conic section at each temporal sampling point (Figure 8.2b, c). The distribution of replicate means is normal with a variance that increases with time (Figure 8.3). The results shown in Figures 8.2 and 8.3 are obviously simplifications since we have ignored the effects of sampling on G, a problem to which we will return in Chapter 9. We note, however, that (8.4) might be a reasonable approximation if N_e is large enough so that the sampling effects on G are relatively small. We have also used a single number to represent population size throughout time in all replicate lineages, another obvious simplification.

8.7 Tests for Neutral Evolution With Large Data Sets

It is instructive to test for the adequacy of a neutral model of trait evolution using large data sets. Using a data set assembled by Gingerich (2001), Estes and Arnold (2007) plotted divergence in phenotypic trait mean as a function of the length of the time interval over which the divergence was measured (Figure 8.4). The plots show bounded evolution in which, regardless of the length of the time interval, most divergence is less than $6\sqrt{P}$ of the mean at the start of the interval (Figure 8.4a), where $\sqrt{P}$ is a within-population trait standard deviation. Fitting models of drift to these plots reveals two kinds of model failures. Very small N_e (≤ 10) are required to account for the divergence that is observed at

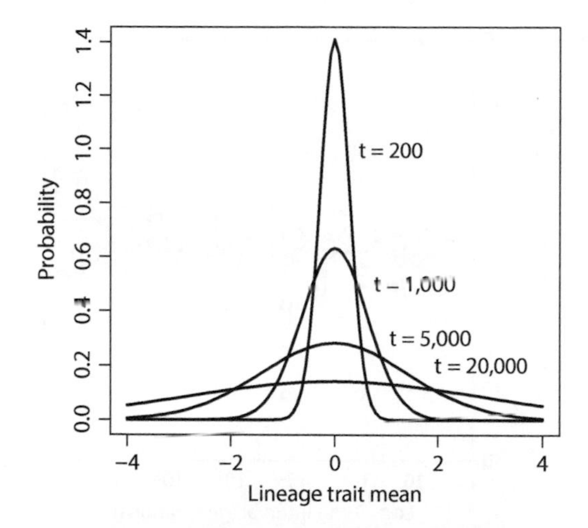

Figure 8.3 Distributions of lineage trait means under neutrality. Probability, $\Phi(\bar{z}_t)$, is plotted as a function of trait mean, $\bar{z}_t$. Each curve represents the distribution of lineage means after some number of elapsed generations, *t*. Additive genetic variance = 0.4 and N_e = 1000.

short timescales (1 – 100 generations), but such small size causes far too much divergence after just 100 generations. On the other hand, to account for the modest divergence that is observed at long timescales ($10^6 - 10^7$ generations), one needs to invoke extremely large values for N_e ($10^5 - 10^6$), but such very large populations show only a small fraction of the divergence that is observed on shorter timescales ($10^0 - 10^4$ generations). In other words, models of genetic drift cannot account for the broad, time-independent band of divergence that is observed in the Gingerich data (Estes and Arnold 2007).

The problem that genetic drift faces as a model of evolution can be visualized if we graph the process on long timescales. For example in Figure 8.4 we examine the consequences of drift on a moderately long time scale, using the same parameters that we employed in Figure 8.2c on a short time scale. Although the process appeared restrained on a time scale of hundreds of generations, when we expand the time scale to thousands of generations, a substantial proportion of lineages diverge beyond the limits observed in large data sets (Figure 8.5). On a time scale of millions of generations, this problem is exacerbated. Even populations with relatively large population sizes ($N_e = 10^6$) and a modal value for genetic variances ($G = 0.4P$) can drift beyond observed limits after 10^5 generations (Estes and Arnold 2007).

8.8 Drift on a Phylogeny

Another simplification in (8.5) is that we imagined that the replicate lineages diverged from a single common ancestor at generation 0. In other words, the replicates evolved on a so-called star phylogeny in which all the branch lengths are *t* generations. In applying neutral theory to evolving populations we will often want to allow for a more complex phylogeny.

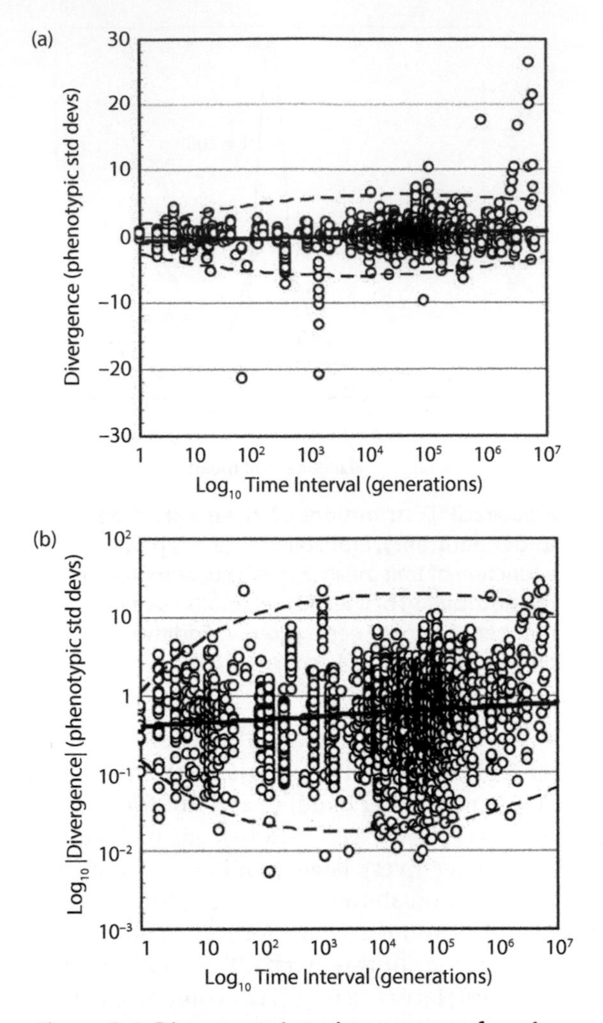

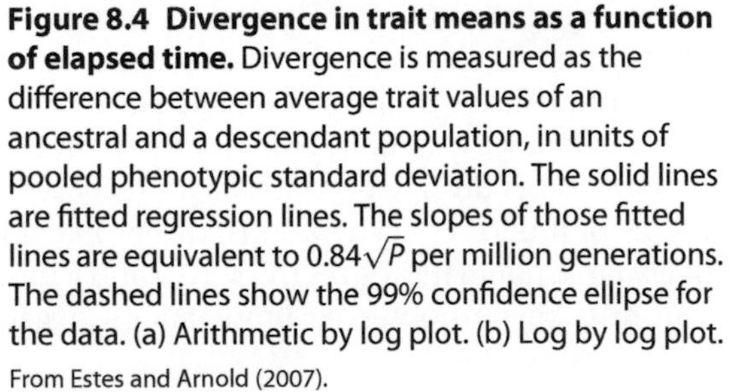

Figure 8.4 Divergence in trait means as a function of elapsed time. Divergence is measured as the difference between average trait values of an ancestral and a descendant population, in units of pooled phenotypic standard deviation. The solid lines are fitted regression lines. The slopes of those fitted lines are equivalent to $0.84\sqrt{P}$ per million generations. The dashed lines show the 99% confidence ellipse for the data. (a) Arithmetic by log plot. (b) Log by log plot.

From Estes and Arnold (2007).

Independent contrasts represent one way to take account of phylogeny (Felsenstein 1985). To apply this method, one converts data at the tips into contrasts (differences between means) that can be assumed to be independent if the stochastic process of trait evolution on the tree is *Brownian motion* (e.g., drift, as characterized above). For the tree shown in Figure 8.6, the independent contrasts would be $(\bar{z}_b - \bar{z}_c)$ and $(\bar{z}_a - \bar{z}_{bc})$, where $\bar{z}_{bc}$ is the trait mean of the common ancestor of b and c, which we will call bc. In data analysis, under the assumption of Brownian motion (BM), these two contrasts are treated

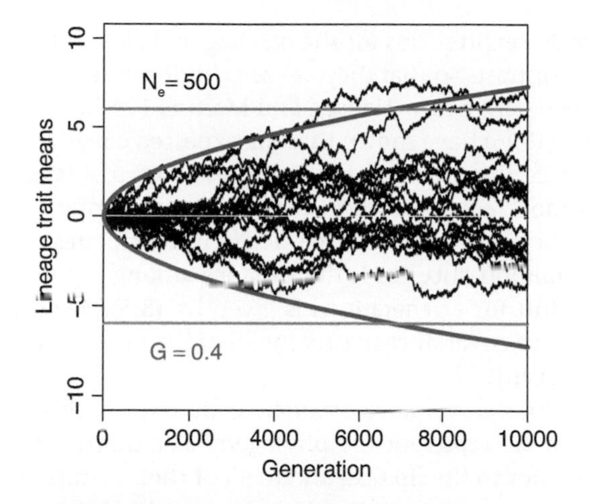

Figure 8.5 A simulation of drift in 20 replicate populations for 10,000 generations. The blue lines show the theoretical 99% confidence limits for lineage trait means. The white line shows the expected average of lineage means as a function of time. The violet lines show the limits of phenotypic divergence that are observed on timescales less than 1 million generations (Estes and Arnold 2007; Uyeda et al. 2011), namely, less than ± 6 within-population phenotypic standard deviations, $6\sqrt{P}$.

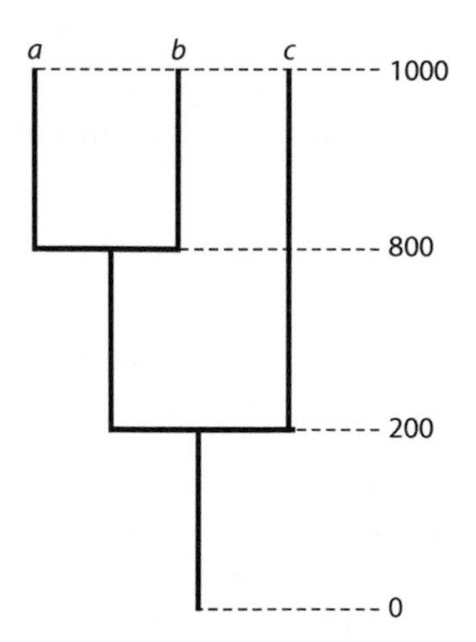

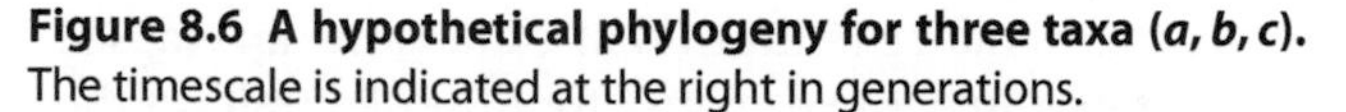
Figure 8.6 A hypothetical phylogeny for three taxa (*a*, *b*, *c*). The timescale is indicated at the right in generations.

as independent observations of evolutionary outcome and scaled so that they have the same expected trait variance (Felsenstein 1985). See Walsh and Lynch (2018) for a detailed discussion of Brownian motion processes.

An alternative approach avoids the assumption of independent contrasts as well as the need to reconstruct the means of common ancestors. In this alternative approach, we reconstruct the trait variance-covariance matrix for taxa evolving on the tree. This matrix could be based on Brownian motion or any other well-characterized stochastic process.

In other words, we derive expressions for the off-diagonal elements of this matrix rather than transforming to contrasts so that they are zero (independent). A general solution to those elements has been derived by (Hansen and Martins 1996).

Hansen and Martins (1996) have shown that the expected covariance of the trait means of two taxa evolving according to a stochastic process on a tree is equal to the expected trait variance of their most recent common ancestor. Suppose that an ancestor is t generations removed from the ancestor of the tree. Then, if the stochastic process is drift, the trait variance of the common ancestor is the variance among replicates of that common ancestor evolving by drift for t generations is given by (8.5). We can use this result to calculate the trait variance-covariance matrix for all of the taxa represented on the tips of the branches of a phylogeny.

To get an expression for that variance-covariance matrix, which we will call $\boldsymbol{A}$, we proceed in two steps. First, we represent the phylogeny as a matrix that gives the elapsed time from the tree ancestor to the tip taxa and to all of their common ancestors. Suppose there are r tip taxa. We can represent the phylogeny of those taxa with an $r \times r$ matrix $\boldsymbol{T}$, in which the off-diagonal *ijth* element is the time (in generations) from the root of the tree to the most recent common ancestor of taxon i and taxon j, and the diagonal elements are the time from the root to extant taxon i. $\boldsymbol{T}$ is a matrix of shared ancestry in the sense that the elements represent the number of elapsed generations in which pairs of taxa experienced shared ancestry (Hansen and Martins 1996; Martins 1996).

For example, the phylogeny of three taxa (*a*, *b*, *ca*, *b*, *c*) in Figure 8.6 can be represented by

$$\boldsymbol{T} = \begin{bmatrix} T_{aa} & T_{ab} & T_{ac} \\ T_{ba} & T_{bb} & T_{bc} \\ T_{ca} & T_{cb} & T_{cc} \end{bmatrix} = \begin{bmatrix} 1000 & 800 & 200 \\ 800 & 1000 & 200 \\ 200 & 200 & 1000 \end{bmatrix}. \tag{8.6}$$

In the second step, we describe the outcome of drift on a phylogeny from generation 0 to present time, at the tips of the branches, by producing $\boldsymbol{A}$, an $r \times r$ matrix analogue of (8.5), viz.,

$$\boldsymbol{A} = \frac{1}{N_e} G\boldsymbol{T}, \tag{8.7}$$

where G represents the average additive genetic variance for the trait across time and across populations, N_e is a similar average for effective population size, and $\boldsymbol{T}$ is the $r \times r$ matrix of shared ancestry (Hansen and Martins 1996; Hohenlohe and Arnold 2008). The off-diagonal, *ij*th element of $\boldsymbol{A}$ is the expected covariance in trait mean values between two taxa, i and j, with a shared coancestry equal to the *ij*th element of $\boldsymbol{T}$. Similarly, the diagonal, *ii*th element of $\boldsymbol{A}$ is the expected variance in mean trait values after an elapsed time given by the *ii*th element of $\boldsymbol{T}$. For example, the A-matrix for drift of a trait on the three taxon phylogeny shown in Figure 8.6, with $N_e = 100$, is

$$\boldsymbol{A} = \begin{bmatrix} \left(\frac{1000}{100}\right)G & \left(\frac{800}{100}\right)G & \left(\frac{200}{100}\right)G \\ \left(\frac{800}{100}\right)G & \left(\frac{1000}{100}\right)G & \left(\frac{200}{100}\right)G \\ \left(\frac{200}{100}\right)G & \left(\frac{200}{100}\right)G & \left(\frac{1000}{100}\right)G \end{bmatrix} = \begin{bmatrix} 10G & 8G & 2G \\ 8G & 10G & 2G \\ 2G & 2G & 10G \end{bmatrix}. \tag{8.8}$$

In other words, if a series of replicate populations of taxon a evolved by drift on the phylogeny, the expected variance among their trait means would be ten times the genetic variance of the trait, $10G$. If a series of replicate pairs of taxa a and b evolved by drift on the phylogeny, the expected covariance of their trait means would be $2G$.

With replicate taxa that have evolved to the tips of a phylogeny given by $\boldsymbol{T}$, the distribution of traits means is multivariate normal,

$$\Phi(\bar{\mathbf{z}}) = \frac{exp\left\{-\frac{1}{2}(\bar{\mathbf{z}} - \bar{\mathbf{z}}_0)^{\mathrm{T}}\boldsymbol{A}^{-1}(\bar{\mathbf{z}} - \bar{\mathbf{z}}_0)\right\}}{\sqrt{(2\pi)^r|\boldsymbol{A}|}}, \tag{8.9}$$

where $\bar{\mathbf{z}}$ is a vector of trait means for a single trait evolving by Brownian motion along the branches of a tree specified by $\boldsymbol{A}$ and observed at the r tips of the tree, and $\bar{\mathbf{z}}_0$ is a vector of identical ancestral values of the mean at the base of the tree.

8.9 Tests of Neutral Evolution on Trees

Harmon et al. (2010) assessed the ability of three models to account for diversification in size- and shape-related traits in 46 clades of squamates, birds, fish, mammals, amphibians, and arthropods. They found that Brownian motion (BM) generally gave a better fit to the data than two other models (Ornstein-Uhlenbeck and Early Burst). Harmon et al. (2010) estimated the BM *rate diversification parameter,* σ^2, by maximum likelihood. This parameter can be equated with N_e/G in a model of genetic drift for a quantitative trait (8.5). The authors standardized their trait values, so that $P = 1$. Consequently, using their maximum likelihood values of σ^2 and a modal value for G (0.4), we can solve for the values of N_e implied by their results. Focusing the results for body size across 42 clades, the average value of σ^2, expressed as a per-million year rate, was 0.74 with a range of 0.00032–10.9. If we assume 10 years per generation, then the average rate per generation is $\sigma^2 = 7.4 \times 10^4$ within-population standard deviations per generation, giving an N_e estimate of 3.1×10^4. In other words, the average estimated rate of diversification is consistent with a genetic drift version of Brownian motion (BM), in the sense that it implies an average effective population size of about ten thousand. Despite this consistency, we cannot rule out the possibility that some other version of BM outcompeted the two other models. For example, BM might mean Brownian motion of an intermediate optimum, a model that we will consider in a Chapter 14.

8.10 Maintenance of Genetic Variance by Mutation-Migration-Drift Balance in a Metapopulation

We now consider a metapopulation subdivided into a series of populations that exchange migrants each generation and assume that the input of variation from mutation and the tendency to homogenize population differences by migration have reached a steady state. Once the metapopulation has equilibrated, how will the total genetic variation in a selectively neutral trait in this metapopulation be apportioned among its constituent populations? We can get a useful answer to this question by making a series of simplifying assumptions. Let us assume that there are n populations each of effective size N and that the rate of migration between populations (the fraction of individuals that move from one population to another population each generation), m, is the same in both directions between all pairs of populations. We will also assume our standard additive genetic model for inheritance and that mutation contributes a constant amount, U, to the genetic variance in each population each generation. Under these conditions, (Lande 1992) showed that the genetic variances within-populations (G) and among-populations (G_a) equilibrate so that

$$\hat{G} = 2nNU \tag{8.10a}$$

and

$$\hat{G}_a = \frac{n-1}{m}U. \tag{8.10b}$$

Twice the genetic variance within populations is converted into genetic variance among populations by the process of random drift (Wright 1951, 1969). Consequently, the total genetic variance in the metapopulation is $G_a + 2G$ rather than $G_a + G$.

Several important conclusions can be drawn from these equilibrium results. First, the migration rate, m, does not affect the level of genetic variance within populations. This result contradicts our intuition that migration should enhance variation within populations, but we have assumed that equal numbers of migrants enter and leave each population. Second, the level of genetic variation maintained among populations is enhanced by increasing the number of populations, eroded by increasing migration rate, but is not affected by population size, N. The rates of approach to these equilibria (not shown) are negative exponential functions, so that the approach is rapid at first, followed by a slow, asymptotic period. With a large number of populations, that final approach takes a few times $n/2m$ or $2nN$ generations, whichever is larger (Lande 1992). Using our *T. elegans* example (Manier and Arnold 2005; Manier et al. 2007), for 20 populations of effective size 328, exchanging $mN = 0.4$ migrants per generation ($m = 0.0012$), more than 30 thousand generations would be required for the system to equilibrate in mutation-migration-drift balance.

It is useful to express the among-population genetic variance as a fraction of the total,

$$Q_{ST} = G_a / (G_a + 2G). \tag{8.11}$$

This fraction is comparable to Wright's F_{ST}, a measure of population differentiation in gene frequencies (Wright 1969; Lande 1992; Spitze 1993; Whitlock 1999). For more detail on Q_{ST}-F_{ST} methodology see (Whitlock and Guillaume 2009; Whitlock and Gilbert 2012).

8.11 Tests for Neutral Evolution in the *Thamnophis elegans* Metapopulation Using F_{ST}-Q_{ST} Comparisons

A study by Manier et al. (2007) illustrates the insights that an F_{ST}-Q_{ST} study can provide. In this study, F_{ST} estimates for nine microsatellite loci provide a neutral benchmark for evaluating Q_{ST} estimates for six scalation traits and 13 coloration traits in a metapopulation of nine *Thamnophis elegans* demes in the Eagle Lake basin of Northern California. The study was undertaken because these demes showed conspicuous spatial variation life history and coloration (Bronikowski and Arnold 1999). In particular, the colours of meadow and lake-shore snakes matched the background colours of their respective habitats. Scalation traits were also assessed because *T. elegans* shows pronounced geographic variation in scalation (Arnold and Phillips 1999), and the authors wished to determine whether scalation

Table 8.2 F_{ST} - Q_{ST} comparisons in a garter snake (*Thamnophis elegans*) metapopulation in which nine demes were sampled (Manier et al. 2007). Standard errors are given in parentheses. The Q_{ST} columns shows the average of male and female values. Separate P values for males and females are shown for scalation and coloration traits. Significance levels indicated as: ns, > 0.05; *, < 0.05; **, < 0.01; ***, < 0.001; ****, < 0.0001.

Microsat. locus	F_{ST}	P	Scalation traits	Q_{ST}	P	Coloration traits	Q_{ST}	P
1	0.02 (0.01)	**	VENT	0.31	****,****	DORRED	0.13	*,ns
2	0.06 (0.02)	****	SUB	0.35	****,****	DORGRN	0.16	*,ns
3	0.04 (0.01)	****	MID	0.21	****,****	DORBLU	0.39	**,**
4	0.01 (0.01)	ns	ILAB	0.44	****,****	LATRED	0.39	**,ns
5	0.00 (0.01)	ns	SLAB	0.09	*,**	LATGRN	0.15	ns, ns
6	0.06 (0.03)	****	POST	0.28	*,*	LATBLU	0.13	*,ns
7	0.05 (0.02)	****	Average	0.28 (0.05)		DORHUE	0.02	ns, ns
8	0.05 (0.02)	****				DORSAT	0.23	*,*
9	0.03 (0.03)	**				DORLT	0.26	**,*
Average	0.04 (0.01)					LATHUE	0.21	ns, ns
						LATSAT	0.34	**,ns
						LATLT	0.06	ns, ns
						BKGRD	0.71	***,***
						Average	0.25 (0.05)	

also varied on a local, interdemic scale. Both kinds of phenotypic traits were generally heritable within demes. The phenotypic traits were analyzed separately in males and females, because of sexual dimorphism.

Interdemic differentiation in phenotypic traits was about six times more pronounced than in molecular markers (Table 8.2). While on the average interdemic differences accounted for about 26 – 28% of total variation in phenotypic traits, they accounted for only 4% of total variation in microsatellites. In other words, on the short evolutionary time scale represented by local demes, phenotypic differentiation is more extensive than one would expect from a neutral model.

8.12 Surveys of Tests for Neutral Evolution in Metapopulations Using F_{ST}-Q_{ST} Comparisons

The result that phenotypic differentiation is more extensive than differentiation in neutral genetic markers, as in the Manier et al. (2007) case study, is typical. In one survey of 29 studies and another of 143 studies, Q_{ST} tends to be larger than F_{ST} (Figures 8.7 and 8.8). Taking the F_{ST} results as a surrogate for the neutral model, we see that this model generally predicts too little phenotypic differentiation on the spatial and temporal scales represented by these within-species studies. Generally speaking, that temporal scale is less than 10^6 generations, often much less. In other words, phenotypic differentiation happens faster and on a finer geographic scale than we would expect by genetic drift. At the same time, we see pronounced statistical spread in Q_{ST} in Figures 8.7 and 8.8, perhaps reflecting both heterogeneity in selection and the long times required for equilibration.

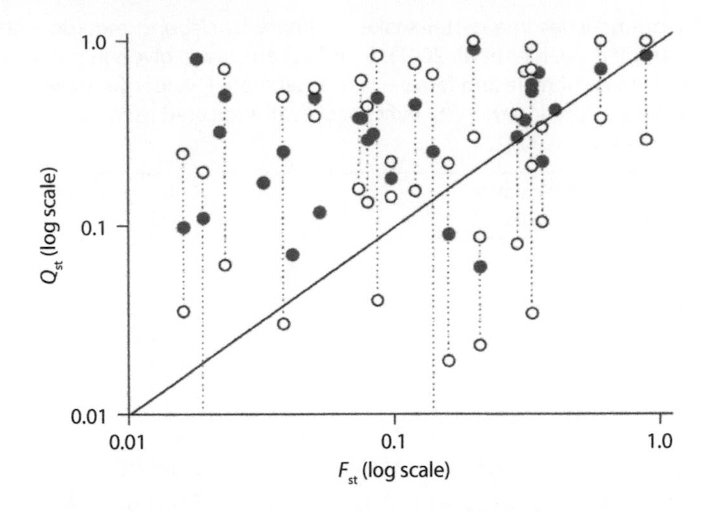

Figure 8.7 Q_{ST} as a function of F_{ST} in a sample of 29 studies. Each study estimated the two variation measures for multiple traits in a single species. Average values across traits are indicated with solid circles. The range of Q_{ST} values across multiple traits sampled in each species is shown with open circles. The solid line shows $Q_{ST} = F_{ST}$.

From McKay and Latta (2002) with permission.

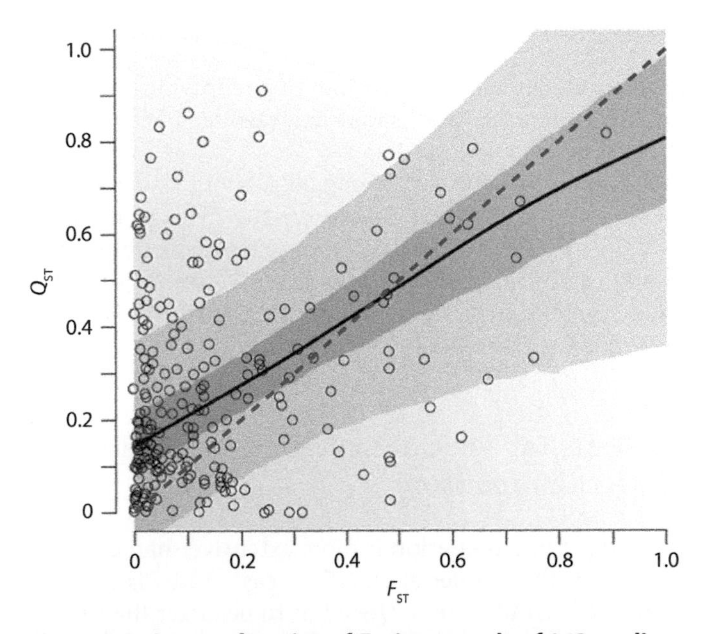

Figure 8.8 Q_{ST} as a function of F_{ST} in a sample of 143 studies. Points representing different traits in the same study are included. The dashed red line shows $Q_{ST} = F_{ST}$. A fitted relationship between Q_{ST} and F_{ST} is shown as a solid black line. Dark and light gray areas show 50% and 95% posterior density intervals.

From Leinonen et al. (2013) with permission.

8.13 Conclusions

- Random sampling of parents can cause the trait mean to fluctuate from generation to generation.
- Effective population size, N_e, is the fundamental parameter that captures the effects of that random sampling on population genetic attributes, including the trait mean.
- Long term estimates of N_e based on nucleotide diversity are typically in the range 10,000 to 25 million. These numbers represent the effective size of a set of evolving, interbreeding lineages.
- Estimates of species-wide population size based on metapopulation models and surveys of census sizes suggest size estimates in the tens to hundreds of millions of individuals.
- Genetic drift of the trait mean can be described as a stochastic process in which replicate lineages descend from a common ancestor at generation 0. The mean of each lineage undergoes an independent random walk, a kind of Brownian motion. The distribution of the lineage means at any given time is normally distributed with a variance equal to the additive genetic variance of the trait divided by N_e and multiplied by the number of elapsed generations. This expression allows us to predict how much divergence to expect by drift as a function of time.
- Genetic drift alone fails to account for trait divergence observed in nature.
- On short timescales, drift predicts too little divergence among lineages, while on long timescales, it predicts too much divergence.
- Nevertheless, in conjunction with selection, drift may contribute to trait differentiation.

Literature Cited

Alter, S. E., E. Rynes, and S. R. Palumbi. 2007. DNA evidence for historic population size and past ecosystem impacts of gray whales. Proc. Natl. Acad. Sci. USA 104:15162–15167.

Arnold, S. J., and P. C. Phillips. 1999. Hierarchical comparison of genetic variance-covariance matrices. II. Coastal-inland divergence in the garter snake, *Thamnophis elegans*. Evolution 53: 1516–1527.

Beerli, P., and J. Felsenstein. 2001. Maximum likelihood estimation of a migration matrix and effective population sizes in *n* subpopulations by using a coalescent approach. Proc. Natl. Acad. Sci. 98:4563–4568.

Bronikowski, A. M., and S. J. Arnold. 1999. The evolutionary ecology of life history variation in the garter snake *Thamnophis elegans*. Ecology 80:2314–2325.

Buffalo, V. 2021. Quantifying the relationship between genetic diversity and population size suggests natural selection cannot explain Lewontin's Paradox. eLife 10:e67509.

Charlesworth, B. 2009. Effective population size and patterns of molecular evolution and variation. Nat. Rev. Genet. 10:195–205.

Charlesworth, J. and A. Eyre-Walker. 2006. The rate of adaptive evolution in enteric bacteria. Mol. Biol. Evol. 23: 1348–1356.

Crow, J. F., and Kimura. 1970. An Introduction to Population Genetics Theory. Harper and Row, New York.

Cutter, A. D. 2005. Nucleotide polymorphism and linkage disequilibrium in wild populations of the partial selfer *Caenorhabditis elegans*. Genetics 172: 171–184.

Estes, S., and S. J. Arnold. 2007. Resolving the paradox of stasis: models with stabilizing selection explain evolutionary divergence on all timescales. Am. Nat. 169:227–244.

Felsenstein, J. 1985. Phylogenies and the comparative method. Am. Nat. 125:1–15.

Gingerich, P. D. 2001. Rates of evolution on the time scale of the evolutionary process. pp. 127–144 *in* A. P. Hendry and M. T. Kinnison, eds. Microevolution Rate, Pattern, Process. Springer, Dordrecht.

Hansen, T. F., and E. P. Martins. 1996. Translating between microevolutionary process and macroevolutionary patterns: The correlation structure of interspecific data. Evolution 50:1404.

Harmon, L. J., J. B. Losos, T. Jonathan Davies, R. G. Gillespie, J. L. Gittleman, W. Bryan Jennings, K. H. Kozak, M. A. McPeek, F. Moreno-Roark, T. J. Near, A. Purvis, R. E. Ricklefs, D. Schluter, J. A. Schulte II, O. Seehausen, B. L. Sidlauskas, O. Torres-Carvajal, J. T. Weir, and A. Ø. Mooers. 2010. Early bursts of body size and shape evolution are rare in comparative data. Evolution 64:2385–2396.

Hohenlohe, P. A., and S. J. Arnold. 2008. MIPoD: a hypothesis-testing framework for microevolutionary inference from patterns of divergence. Am. Nat. 171:366.

Lande, R. 1976. Natural selection and random genetic drift in phenotypic evolution. Evolution 30:314–334.

Lande, R. 1992. Neutral theory of quantitative genetic variance in an island model with local extinction and colonization. Evolution 46:381–389.

Lande, R., and G. F. Barrowclough. 1987. Effective population size, genetic variation, and their use in population management. Pp. 87–124 *in* M. E. Soulé, ed. Viable Populations for Conservation. Cambridge University Press, Cambridge.

Leinonen, T., R. J. S. McCairns, R. B. O'Hara, and J. Merilä. 2013. QST–FST comparisons: Evolutionary and ecological insights from genomic heterogeneity. Nat. Rev. Genet. 14:179–190.

Manier, M., C. Seyler, and S. Arnold. 2007. Adaptive divergence within and between ecotypes of the terrestrial garter snake, *Thamnophis elegans*, assessed with FST-QST comparisons. J. Evol. Biol. 20:1705–1719.

Manier, M. K., and S. J. Arnold. 2005. Population genetic analysis identifies source-sink dynamics for two sympatric garter snake species (*Thamnophis elegans* and *Thamnophis sirtalis*). Mol. Ecol. 14:3965–3976.

Martins, E. P. 1996. Phylogenies, spatial autoregression, and the comparative method: A computer simulation test. Evolution 50:1750–1765.

McKay, J. K., and R. G. Latta. 2002. Adaptive population divergence: markers, QTL and traits. Trends Ecol. Evol. 17:285–291.

Mu, J. et al. 2002. Chromosome-wide SNPs reveal an ancient origin for *Plasmodium falciparum*. Nature 418: 323–326

Palstra, F. P., and D. E. Ruzzante. 2008. Genetic estimates of contemporary effective population size: what can they tell us about the importance of genetic stochasticity for wild population persistence? Mol. Ecol. 17:3428–3447.

Rossman, D. A., N. B. Ford, and R. A. Seigel. 1996. The Garter Snakes, Evolution and Ecology. University of Oklahoma Press, Norman.

Shapiro, J. A. et al. 2007. Adaptive genic evolution in the *Drosophila* genome. Proc. Natl. Acad. Sci. USA 104: 2271–2276.

Spitze, K. 1993. Population structure in *Daphnia obtusa*: Quantitative genetic and allozymic variation. Genetics 135:367–374.

Uyeda, J. C. 2012. Connecting microevolutionary processes to macroevolutionary patterns across space and time. PhD thesis, Oregon State University, Corvallis, Oregon.

Uyeda, J. C., T. F. Hansen, S. J. Arnold, and J. Pienaar. 2011. The million-year wait for macroevolutionary bursts. Proc. Natl. Acad. Sci. 108:15908–15913.

Walsh, B., and M. Lynch. 2018. Evolution and Selection of Quantitative Traits. Oxford University Press, New York.

Whitlock, M. C. 1999. Neutral additive genetic variance in a metapopulation. Genet. Res. 74: 215–221.

Whitlock, M. C., and N. H. Barton. 1997. The effective size of a subdivided population. Genetics 146:427–441.

Whitlock, M. C., and K. J. Gilbert. 2012. Q_{ST} in a hierarchically structured population. Mol. Ecol. Resour. 12:481–483.

Whitlock, M. C., and F. Guillaume. 2009. Testing for spatially divergent selection: Comparing QST to FST. Genetics 183:1055–1063.

Wright, S. 1931. Evolution in Mendelian populations. Genetics 16:97–159.

Wright, S. 1939. Statistical genetics in relation to evolution. pp. 1–69 *in* Exposés de biométrie et de statistique biologique, 13.; Actualités scientifiques et industrielles, 802. Hermann et Cie, Paris.

Wright, S. 1943. Isolation by distance. Genetics 28:114–138.

Wright, S. 1951. The genetical structure of populations. Ann Eugen. 15:323–354.

Wright, S. 1969. Evolution and the Genetics of Populations. Volume 2. The Theory of Gene Frequencies. University of California Press, Chicago.

Yu, N., M. I. Jensen-Seaman, L. Chemnick, O. Ryder, and W. H. Li. 2004. Nucleotide diversity in gorillas. Genetics 166: 1375-1383.

CHAPTER 9

Drift of Multiple, Neutral Traits

Overview—Drift of the multivariate mean provides an important baseline against which we can gauge the effects of multivariate selection. Multivariate drift is a function of elapsed time, effective population size, and the G-matrix. In the two-trait case at any given time, we can visualize the trait means of replicate populations as a cloud of points in bivariate trait space. Although each mean takes an erratic path through trait space as time unfolds, the statistical behaviour of the cloud is highly predictable. The dispersion cloud retains its shape and orientation, as it gradually expands in size from generation to generation. Moreover, the cloud shares its characteristic shape and orientation with the G-matrix. Tests of these predictions in natural populations reject drift as an explanation for trait radiation. In particular, the observed dispersion cloud is smaller, sometimes much smaller than expected by neutral drift of the trait means. The need for models that include selection as well as drift is also suggested by the empirical result that the major axis of the dispersion cloud can be aligned with the major axes of inheritance and/or selection.

9.1 Sampling From a Multivariate Distribution of Breeding Values

The only new concept needed when we consider multiple traits evolving by drift is the covariance among replicate lineages in the means of two different traits. If we take repeated samples of size N from a normal bivariate distribution with variance-covariance matrix $\boldsymbol{V}$, the trait means of the samples will be normally distributed with a variance-covariance of $\boldsymbol{V}/N$. The same idea extends to m multiple traits whose within-replicate distribution of breeding values is multivariate normal with variance-covariance matrix $\boldsymbol{G}$. If we establish replicates of size N_e from an ancestral population with that distribution, the trait means of those replicates will have a multivariate normal distribution with a $m \times m$ variance-covariance matrix given by

$$Var(\bar{\mathbf{z}}_1) = \frac{1}{N_e}\boldsymbol{G} \tag{9.1}$$

at generation t, where the 1 indicates one generation beyond the ancestral generation at $t = 0$. The diagonal elements in this matrix give the variance among replicate means

Evolutionary Quantitative Genetics. Stevan J. Arnold, Oxford University Press. © Stevan Arnold (2023).
DOI: 10.1093/oso/9780192859389.003.0010

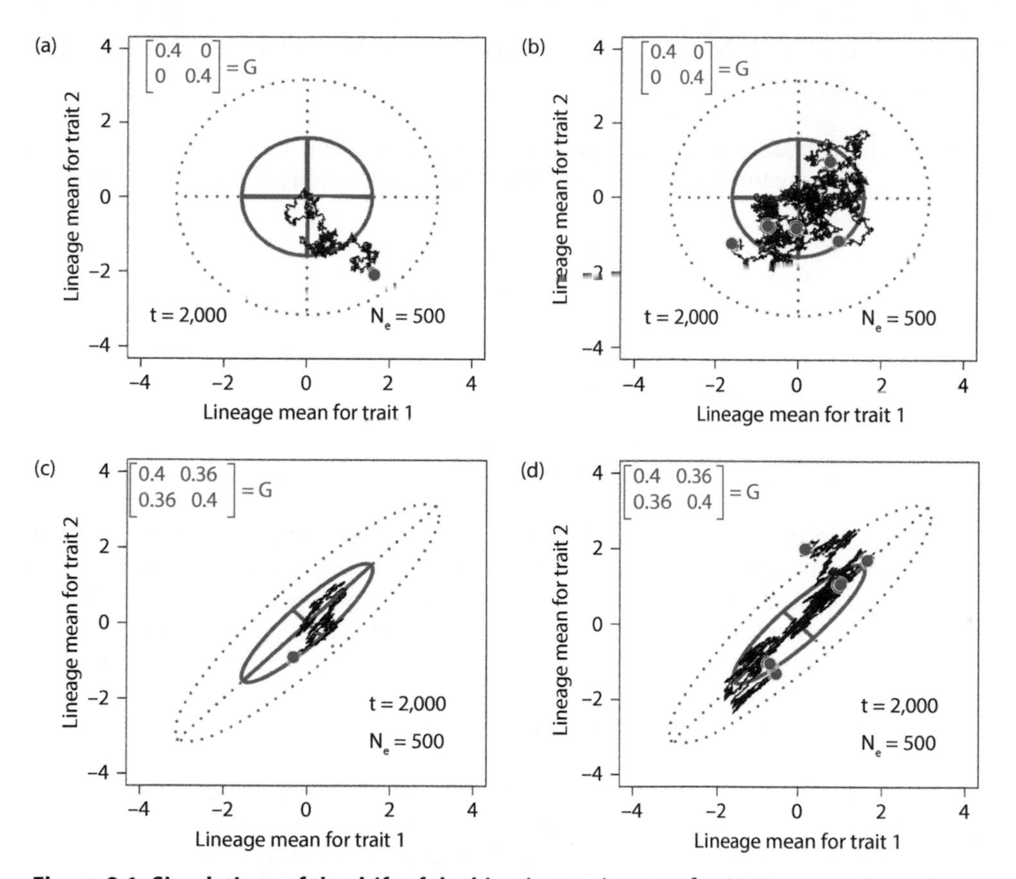

Figure 9.1 Simulations of the drift of the bivariate trait mean for 2000 generations. The effective population size is 500. The 95% confidence ellipses for the *G*-matrix is shown in solid blue. The 95% confidence ellipses for the dispersion of lineage means after 2000 generations are shown in dotted blue. Evolutionary paths of the bivariate trait mean are shown as thin black lines. The red dots show the position of the bivariate trait mean at generation 2000. (a) Drift of a single lineage with no genetic correlation. (b) Drift of five replicate lineages with no genetic correlation. (c) Drift of one lineage with a genetic correlation of 0.9. (d) Drift of five replicate lineages with a genetic correlation of 0.9.

and the off-diagonal elements give the covariances among the replicate means for different traits. The distribution of replicate means has the same mean as the ancestral population.

If we follow a lineage through time as its bivariate mean evolves by drift, we see a chaotic path in two dimensions as the process draws increments of change each generation according to (9.1) (Figure 9.1). The effect of the *G*-matrix is to constrain the path of drift, so that while lineages with no genetic trait correlation drift away from the starting point in all directions (Figure 9.1a, b), lineages with genetic trait correlation tend to drift along the main axis (leading eigenvector) of the *G*-matrix (Figure 9.1c, d). Although the individual paths appear chaotic, their ensemble behaviour follows statistical rules that we shall now consider.

9.2 Projecting the Multivariate Distribution of Breeding Values Into the Future

If we continue the same program of finite sampling, generation after generation, the distribution of replicate lineage means for m traits at any generation t is multivariate normal,

$$\Phi(\bar{\mathbf{z}}_t) = \frac{exp\left\{-\frac{1}{2}(\bar{\mathbf{z}}_t - \bar{\mathbf{z}}_0)^{\mathrm{T}}\boldsymbol{D}(t)^{-1}(\bar{\mathbf{z}}_t - \bar{\mathbf{z}}_0)\right\}}{\sqrt{(2\pi)^m\,|\boldsymbol{D}(t)|}}, \tag{9.2}$$

with a mean the same as the ancestral population, $\bar{\mathbf{z}}_0$, where T denotes transpose, and the variance-covariance matrix or *dispersion matrix*

$$\boldsymbol{D}(t) = \frac{1}{N_e}\boldsymbol{G}t \tag{9.3}$$

(Lande 1979; Hohenlohe and Arnold 2008). These results follow, just as they do in the univariate case, from the properties of a convolution of identical normal distributions. In the present case, multivariate normal distributions are convoluted. We see from (9.3) the important result that the dispersion matrix for a set of replicate populations drifting on a star phylogeny is a scalar multiple of $\boldsymbol{G}$ and hence proportional to the G-matrix. Furthermore, Lande (1979) points out that (9.3) holds for a population evolving on a stochastic phylogeny in which the lineage branching and extinction rates are independent of the mean phenotype, $\bar{\mathbf{z}}_t$.

We can visualize the dispersion matrix, $\boldsymbol{D}(t)$, as a 95% confidence ellipse. At generation t, 95% of the bivariate means of replicate lineages are expected within this ellipse. The confidence ellipses are proportional to $\boldsymbol{G}$, so if we draw those ellipses at, say, generation 1000, 2000, and 5000, they form a nested series of ellipses, all with same shape as the ellipse for $\boldsymbol{G}$ (Figure 9.2). A useful rule is to note from (9.3) that when $t = N_e$, $\boldsymbol{D}(t) = \boldsymbol{G}$, and their confidence ellipses are identical as well. For example, suppose that $N_e = 500$, as

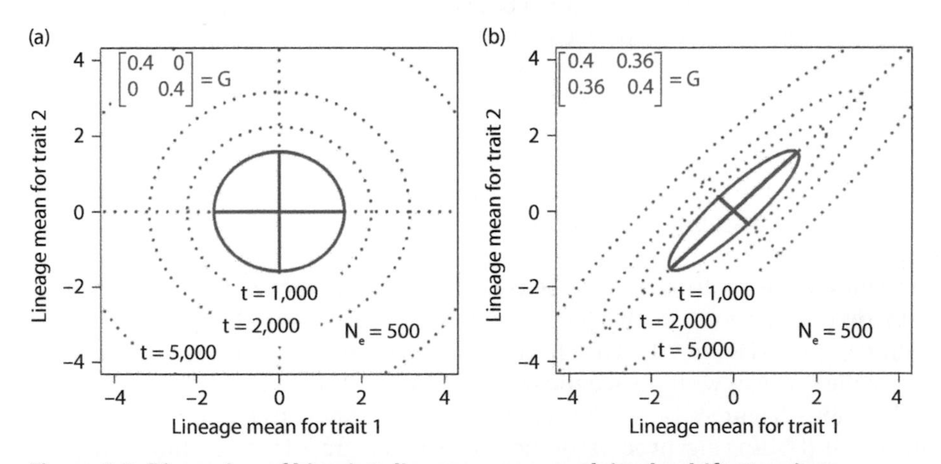

Figure 9.2 Dispersion of bivariate lineage means evolving by drift at various elapsed intervals from time zero. The 95% confidence ellipse for the G-matrix is shown as a solid ellipse. The 95% confidence ellipses for the dispersion of lineage means are shown in dotted blue. (a) Dispersion of bivariate means for replicate lineages with no genetic correlation. (b) Dispersion of bivariate means for replicate lineages with a genetic correlation of 0.9.

in Figure 9.2. After 500 generations of drift, 95% of lineage means are expected to reside inside the corresponding confidence ellipse for $\boldsymbol{G}$. But, on a geological time scale, the dispersion of lineage means will be located inside an ellipse that is gigantic compared with $\boldsymbol{G}$.

9.3 Multivariate Drift on a Phylogeny

A phylogeny more complicated than a star introduces complexities that we have already treated in Section 8.8. In data analysis we need to consider populations evolving by multivariate drift on a particular phylogeny. In a phylogenetic model of drift, time is now played out on a particular phylogenetic tree that can be represented by the $r \times r$ matrix $\boldsymbol{T}$ described in Section 8.8. Typically, the variances and covariances of trait means need to be predicted and evaluated at a single point in time, at the tips of the branches. We will represent the trait means of those r tip populations as a stacked column vector, $\bar{\bar{\mathbf{z}}}$, in which the m means of the first population are stacked above the m means of the second population and so on to form a vector with mr elements. This vector of means, $\bar{\bar{\mathbf{z}}}$, has a multivariate normal distribution,

$$\Phi\left(\bar{\bar{\mathbf{z}}}\right), = \frac{exp\left\{-\frac{1}{2}(\bar{\bar{\mathbf{z}}}-\bar{\bar{\mathbf{z}}}_0)^{\mathrm{T}}\boldsymbol{A}^{-1}\left(\bar{\bar{\mathbf{z}}}-\bar{\bar{\mathbf{z}}}_0\right)\right\}}{\sqrt{(2\pi)^{mr}\left|\boldsymbol{A}\right|}}, \tag{9.4}$$

with a mr-element mean vector, $\bar{\bar{\mathbf{z}}}_0$, with elements the same as the ancestral population, and a $mr \times mr$ variance-covariance matrix

$$\boldsymbol{A} = \frac{1}{N_e}\boldsymbol{G} \otimes \boldsymbol{T}, \tag{9.5}$$

where $\otimes$ represents the multiplication of each element in $\boldsymbol{T}$ by $\boldsymbol{G}$, this multiplication is a *Kronecker product*, not the usual matrix product (Hohenlohe and Arnold 2008). For example, if the G-matrix for two traits is

$$\boldsymbol{G} = \begin{bmatrix} 4 & 2 \\ 2 & 6 \end{bmatrix}, \tag{9.6}$$

then the A-matrix for the three taxon phylogeny shown in Figure 8.6 is 6×6 or

$$\boldsymbol{A} = \begin{bmatrix} 10\begin{bmatrix} 4 & 2 \\ 2 & 6 \end{bmatrix} & 8\begin{bmatrix} 4 & 2 \\ 2 & 6 \end{bmatrix} & 2\begin{bmatrix} 4 & 2 \\ 2 & 6 \end{bmatrix} \\ 8\begin{bmatrix} 4 & 2 \\ 2 & 6 \end{bmatrix} & 10\begin{bmatrix} 4 & 2 \\ 2 & 6 \end{bmatrix} & 2\begin{bmatrix} 4 & 2 \\ 2 & 6 \end{bmatrix} \\ 2\begin{bmatrix} 4 & 2 \\ 2 & 6 \end{bmatrix} & 2\begin{bmatrix} 4 & 2 \\ 2 & 6 \end{bmatrix} & 10\begin{bmatrix} 4 & 2 \\ 2 & 6 \end{bmatrix} \end{bmatrix} = \begin{bmatrix} 40 & 20 & 32 & 16 & 8 & 4 \\ 20 & 60 & 16 & 48 & 4 & 12 \\ 32 & 16 & 40 & 20 & 8 & 4 \\ 16 & 48 & 20 & 60 & 4 & 12 \\ 8 & 4 & 8 & 4 & 40 & 20 \\ 4 & 12 & 4 & 12 & 20 & 60 \end{bmatrix} \tag{9.7}$$

We can use (9.4) to construct a maximum likelihood framework for assessing the various kinds of departures from neutrality that might be observed in distribution of trait means at the tips of a phylogenetic tree (Hohenlohe and Arnold 2008).

9.4 Tests of the Multivariate Neutral Model

The first tests of the multivariate neutral model used the prediction from (9.3) that the dispersion matrix, $\boldsymbol{D}$, should be proportional to the *G*-matrix (Lande 1979). In other words, all the elements of $\boldsymbol{D}$ are predicted to be a scalar multiple of the elements of $\boldsymbol{G}$. In one of the earliest tests of this prediction, Lofsvold (1988) studied within- and among-population phenotypic variation in 15 morphological traits in deer mice (*Peromyscus*). He assessed similarity in the two matrices by computing the vector correlation between their first five eigenvectors. In test cases with three different species the correlation was low (< 0.27), suggesting little structural similarity between the two matrices. In a similar analysis, Hunt (2007) compared within-locality variation and among-lineage divergence in landmark-based morphological traits in the fossil ostracode *Poseidonamicus*. In particular, he compared the leading eigenvector of the *within-locality phenotypic matrix* ($\boldsymbol{p}_{max}$) with the angle of evolutionary trait divergence and was able to reject the hypothesis on the grounds of non-concordance. Both of these analyses seize on similarity of matrix eigenvectors and ignore other aspects of matrix similarity implied by (9.3). The next generation of tests for the proportionality result (9.3) compared both the eigenvalues and the eigenvectors of the two matrices using a perspective developed by Flury (1988).

To compare two variance-covariance matrices, Flury (1988) proposed a hierarchy that involves comparing both the eigenvalues and the eigenvectors of the matrices. We will discuss this hierarchy in detail in Chapter 16 as a vehicle for comparing *G*-matrices. For the moment we will concentrate on the implementation of this hierarchy as a way to tease apart the different aspects of proportionality between $\boldsymbol{D}$ and $\boldsymbol{G}$ (Hohenlohe and Arnold 2008).

If we have estimates of all the elements in our proportionality prediction ($\boldsymbol{D}$, $\boldsymbol{G}$, N_e, and t), as well as an estimate of the ω-matrix (see Chapter 4 for an introduction to this matrix), how can we combine all this information in a single analysis? The answer is to use a maximum likelihood framework. The key in using that framework is the realization that we expect the distribution of species means to be multivariate normal (9.3) with a variance-covariance matrix given by $\boldsymbol{A}$ (9.5), which is the tree-based analogue of $\boldsymbol{D}$ (9.3). The essence of the maximum likelihood approach is to take the species means data as given and use the ML technique to estimate the values of parameters that make up the neutral model. In particular, we can determine which of the parameters is responsible for a poor fit of the data to the likelihood of the parameter set. We expect to reject the idea that $\boldsymbol{A}$ is proportional to $\boldsymbol{G}$, but what is it about $\boldsymbol{G}$ that leads to this rejection? We will use the eigenvalues and eigenvectors of $\boldsymbol{G}$ to characterize the rejection.

A variance-covariance matrix can be characterized by its eigenvalues and eigenvectors (Chapter 2). A 2×2 *G*-matrix, for example, has two eigenvectors and each has an associated eigenvalue. In Figure 9.2b, we see those eigenvectors and eigenvalues portrayed as an ellipse. For the purposes of the analysis we wish to conduct, it will be useful to express those eigenvectors and eigenvalues as three new parameters: *size* (the sum of the eigenvalues), *shape* (the ratio of the largest eigenvalue to the sum), and *orientation* (angle between the leading eigenvector, $\boldsymbol{g}_{max}$, and the axis of the first trait). We can express $\boldsymbol{G}$ as a function of these three parameters, $\boldsymbol{G} = f$(size, shape, orientation), and consequently the expected variance-covariance matrix $\boldsymbol{A}$ can be written as a function of these three parameters. We can now ask which of the $\boldsymbol{G}$ parameters (size, shape, or orientation) contributes to our rejection of the neutral model.

In a test case, we will determine whether the neutral model can account for the evolution of body and tail vertebral counts in garter snakes and water snakes (Figure 9.3).

We have the following data: vertebral counts for 39 species, an mtDNA-based phylogeny (Figure 9.3a) and hence an estimate of time, two estimates of the relevant G-matrix (Arnold and Phillips 1999), an estimate of N_e based on data analysis by Manier and Arnold (2005), and two estimates of $\boldsymbol{\omega_{max}}$ based on selection analyses by Arnold (1988) and Arnold and Bennett (1988). A plot of mean tail vertebral counts as a function of body vertebral counts is shown in Figure 9.3b, which suggests correlated evolution of the two traits. From the bivariate dispersion of points in this plot, we can compute its leading eigenvector (first principal component), which we will call $\boldsymbol{d_{max}}$. Using our phylogeny we can also compute a phylogeny-corrected analogue, $\boldsymbol{d^*_{max}}$ (Figure 9.3b). A central focus of our analysis is to determine whether $\boldsymbol{d^*_{max}}$ aligns with $\boldsymbol{\omega_{max}}$, thus revealing evolution along a selective line of least resistance (SLLR) (Chapter 18). Or whether $\boldsymbol{d^*_{max}}$ aligns with $\boldsymbol{g_{max}}$, as predicted by (Schluter 1996).

Turning to the test results, at the first step (a size test, Figure 9.4a), we find that a much smaller G-matrix does a better job ($P < 0.0001$) of explaining the divergence data than the direct estimate of $\boldsymbol{G}$. In other words, neutrality predicts too much divergence, suggesting that stabilizing selection or some other impeding force has restrained divergence.

At the second step (a shape test, Figure 9.4b), we find that a more elongate G-matrix (blue) does a better job ($P = 0.012$) fitting the divergence data than the estimate in Figure 9.4a. In other words, neutrality favours about the same degree of divergence in the both eigenvector directions, but some other force biases divergence along the long axis of both ellipses.

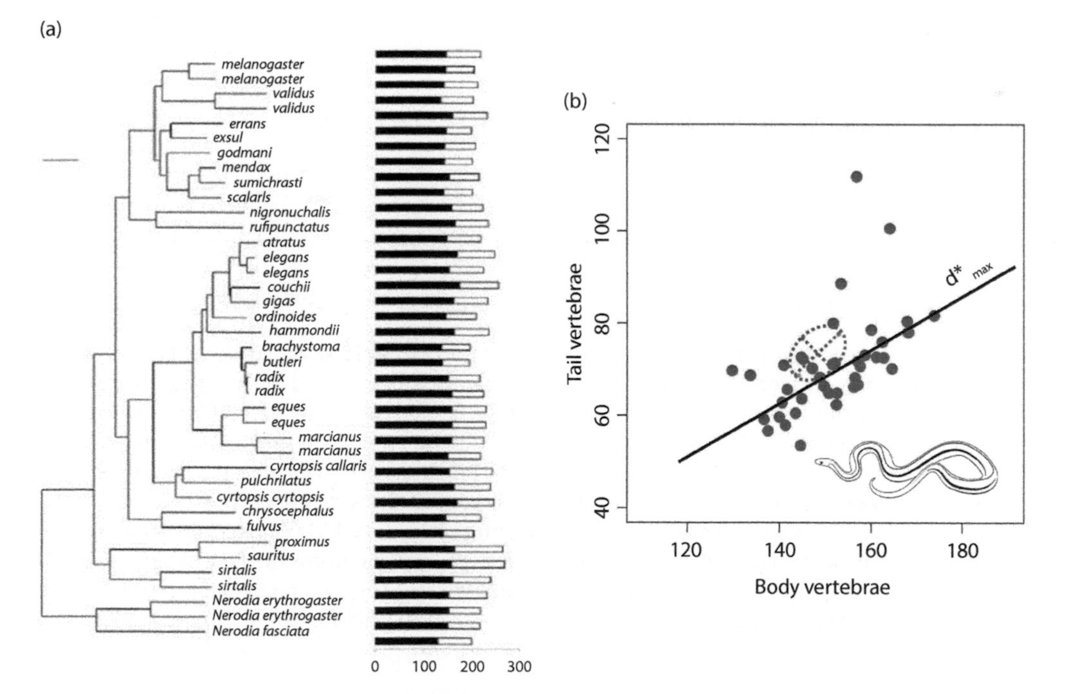

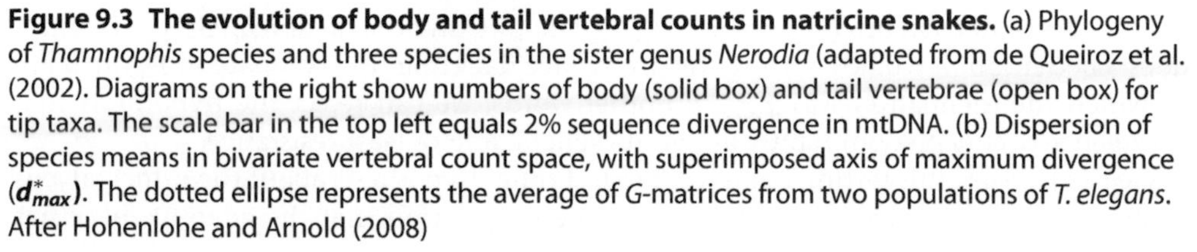

Figure 9.3 The evolution of body and tail vertebral counts in natricine snakes. (a) Phylogeny of *Thamnophis* species and three species in the sister genus *Nerodia* (adapted from de Queiroz et al. (2002). Diagrams on the right show numbers of body (solid box) and tail vertebrae (open box) for tip taxa. The scale bar in the top left equals 2% sequence divergence in mtDNA. (b) Dispersion of species means in bivariate vertebral count space, with superimposed axis of maximum divergence ($\boldsymbol{d^*_{max}}$). The dotted ellipse represents the average of G-matrices from two populations of *T. elegans*. After Hohenlohe and Arnold (2008)

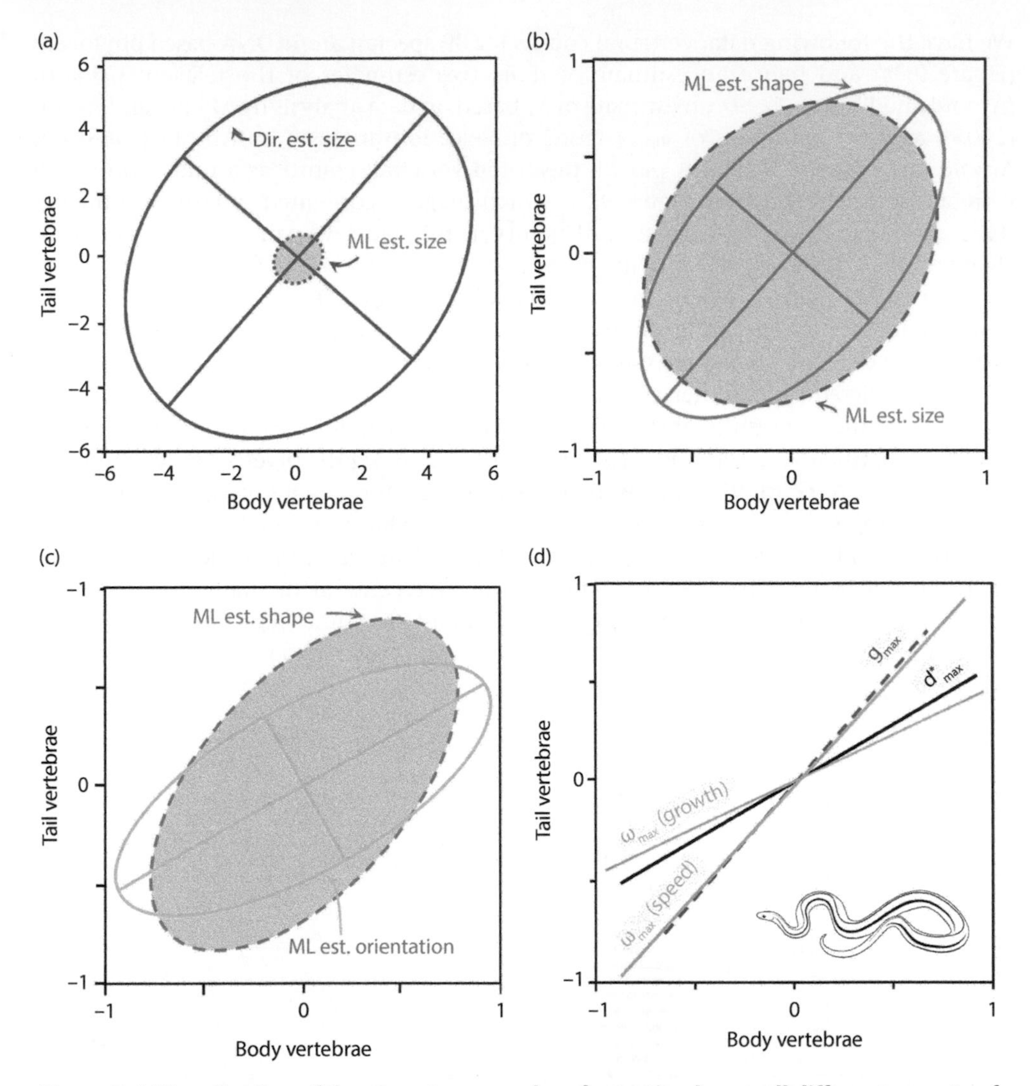

Figure 9.4 Visualization of the stepwise procedure for testing how well different aspects of the G-matrix account for trait divergence. (a) Comparison of direct and ML estimates of matrix size. The direct estimate of the *G*-matrix is shown as the large blue ellipse, based on the average *G*-matrix for *T. elegans*. The ML estimate for size (Σ) is shown as the much smaller blue, gray-filled ellipse. (b) Comparison of the ML estimate for size (dashed blue, gray-filled ellipse) with the ML estimate of shape (ε, solid blue ellipse). (c) Comparison of the ML estimate for shape (dashed blue, gray-filled ellipse) with the ML estimate of orientation (φ, solid blue ellipse). (d) Estimates of orientation of $\boldsymbol{\omega}_{max}$ from growth rate in *T. elegans* (orange line) and from crawling speed in *T. radix* (orange line) plotted along with the direct estimate of $\boldsymbol{g}_{max}$ (dashed dark blue line) and $\boldsymbol{d}^*_{max}$ (solid black line, the tree-corrected version of $\boldsymbol{d}_{max}$). After Hohenlohe and Arnold (2008)

At the third step (an orientation test, Figure 9.4c), we find that a more inclined *G*-matrix does a better job (P = 0.001) fitting the divergence data than the estimate in Figure 9.4b, in which the inclination was $\boldsymbol{g}_{max}$. In other words, we can reject the hypothesis that divergence occurs predominantly along the genetic line of least resistance.

We can also use this testing framework to test other hypothesis about the principal axis of divergence. In particular we wish to test the proposition that the adaptive peak has moved along a selective line of least resistance (Chapter 18). We have two estimates of such selective lines, one from a study of crawling speed as a function of vertebral numbers and one from a study of growth rate as a function of vertebral numbers (Figure 9.4d). In both cases, the selective line of least resistance represents the leading eigenvector of the selection surface. The estimate of ω_{max} for growth rate is closely aligned with the major axis of divergence (d^*_{max}), and we cannot show that the two estimates are different ($P = 0.264$). In contrast, we can show that ω_{max} for crawling speed differs from d^*_{max} ($P = 0.0003$), although this ω_{max} is closely aligned with the direct estimate of g_{max} (Figure 9.3D). We will return to such instances of alignment in Chapter 18.

McGlothlin et al. (2018) found a strong association between the principal axes of divergence and the principal axes of the *G*-matrix in a radiation of morphometric traits in the lizard genus *Anolis* that spanned 20–40 million years. On its face, the result supports the supposition of divergence along genetic lines of least resistance, but as the authors point out, the result could arise from divergence along conserved selective lines of least resistance, with the observed alignment of ***D*** and ***G*** arising as a secondary effect (see Chapter 18).

Houle et al. (2017) made a particularly strong test of the hypothesis that the multivariate mean evolves by drift, using data on 12 landmark traits on the wings of 112 species of drosophilid flies. The landmark data were summarized as wing size and 20 orthogonal wing shape traits. Using these transformed trait-data, the authors estimated *M*-, *G*-, and *D*-matrices. Several conclusions emerged from the analysis of these data. First, all 21 transformed traits diverged across the species sample. Second, divergence in species means was remarkably constrained. The amount of divergence that was observed was much smaller than would be expected under drift alone. Indeed, the observed rate of divergence is ten thousand times smaller than the amount expected by drift. Third, the patterns of mutational, additive genetic, and among-species variation are very similar, despite some conspicuous exceptions (Figure 9.5).

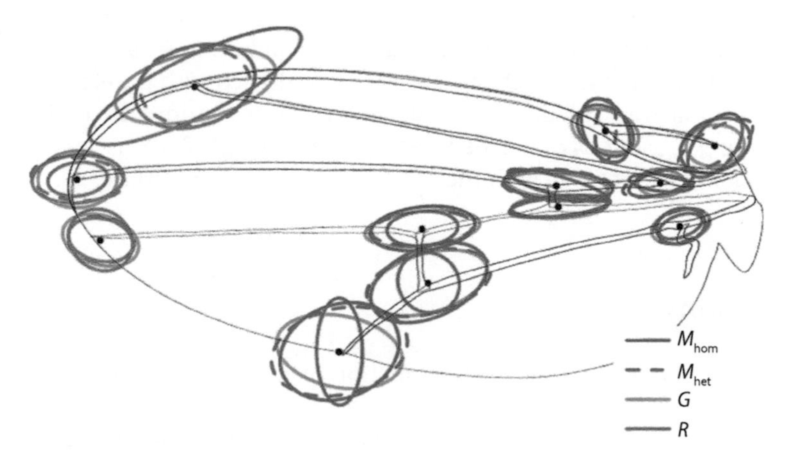

Figure 9.5 Variation ellipses contrasting the patterns of mutational variation (M_{hom}, M_{het}), standing genetic variation (*G*), and variation in divergence (*R*) around each landmark. The *R*-matrix was estimated from the among-species variance in means and is equivalent to the *D*-matrix, discussed above. Matrices are scaled to the same size, so that areas of ellipses are relative to the total variance within each matrix.

From Houle et al. (2017) with permission.

The second result, gross underestimation of overall divergence, is enough to rule out simple models of mutational and genetic constraint, such as drift, as the explanation for divergence in drosophilid wing venation (Houle et al. 2017). Furthermore, the remarkable similarity of the patterns of variation in mutation, inheritance, and divergence (Figure 9.5) implicates selection as a powerful constraining influence on venation evolution. For the moment, we will leave the observed similarity and alignment of the *M*-, *G*- and *R*-matrices as a puzzle, one that we will examine in detail in Chapter 16.

9.5 Maintenance of Multivariate Genetic Variance by Mutation-Drift Balance in a Metapopulation

Using the same model and assumptions as in Section 8.10, we can determine the equilibrium values for within- and among-population variation for a set of selectively neutral traits in a metapopulation. By analogy with the univariate case, we expect the equilibrium within- population ($\widehat{\boldsymbol{G}}$) and among-population ($\widehat{\boldsymbol{G}}_{\boldsymbol{a}}$) genetic variance-covariance matrices to be scalar multiples of the *U*-matrix and to be proportional to each other (8.10). A useful multivariate measure of population differentiation in our neutral quantitative traits, comparable to F_{ST}, is the $m \times m$ trait matrix

$$\boldsymbol{Q_{ST}} = [\boldsymbol{G_a} + 2\boldsymbol{G}]^{-1/2} \ \boldsymbol{G_a} \ [\boldsymbol{G_a} + 2\boldsymbol{G}]^{-1/2}, \tag{9.8}$$

which contains the among-population components of genetic variance and covariance corrected for genetic covariance among traits (Kremer et al. 1997; Chenoweth and Blows 2008). Using data on phentypic trait values within and among populations, one employs a two-level multivariate analysis of variance to estimate the elements of the G_a-matrix as the among-population component of genetic variance and covariance. $\boldsymbol{G}$ might be estimated from a single population in the usual way or, better, by pooling such estimates from two or more populations. The eigenvectors of $\boldsymbol{Q_{ST}}$ can be used to determine trait dimensions with the greatest proportional differentiation among populations (Chenoweth and Blows 2008). One issue is whether the eigenvectors of $\boldsymbol{Q_{ST}}$ are aligned with the eigenvectors of $\boldsymbol{G}$, but if selection surfaces are not included in the analysis, one cannot differentiate between evolution along $\boldsymbol{g_{max}}$ and $\boldsymbol{\omega_{max}}$ (or between $\boldsymbol{g_{max}}$ and $\boldsymbol{\gamma_{min}}$). For this reason, comparisons of the eigenvectors of $\boldsymbol{Q_{ST}}$ with the major axes of selection surfaces are particularly needed.

9.6 Conclusions

- Multivariate drift of the trait mean is a function of elapsed time, effective population size and the *G*-matrix.
- In the two-trait case we can visualize multivariate drift as a random walk (Brownian motion) in bivariate space.
- If we follow a set of independent replicate populations descended from a common ancestor through time, the distribution of their trait means will be multivariate normal at time t with a dispersion (variance-covariance) matrix that is proportional to $\boldsymbol{G}$.
- A similar result characterizes the distribution of the multivariate trait mean as it drifts on a phylogeny with non-independent lineages.

- Analyses of multi-trait divergence in natural populations typically reject the neutral model (drift of the trait mean) expectation of proportionality between divergence and the *G*-matrix.
- Case studies of multiple traits in drosophilid flies and natricine snakes show that the cloud of diverged trait means is much smaller than expected under neutral drift of the multivariate mean, suggesting that a restraining force, such as stabilizing selection, is in play.
- If the cloud of diverged means is aligned with the major axes of inheritance ($\boldsymbol{G}$) and selection ($\boldsymbol{\omega}$), these alignment patterns suggest that selection has played an active role in producing alignment.

Literature Cited

Arnold, S. J. 1988. Quantitative genetics and selection in natural populations: microevolution of vertebral numbers in the garter snake *Thamnophis elegans*. pp. 619–636 *in* B. S. Weir, E. J. Eisen, M. M. Goodman, and G. Namkoong, eds. Proceedings of the Second International Conference on Quantitative Genetics. Sinauer Associates, Inc., Sunderland.

Arnold, S. J., and A. F. Bennett. 1988. Behavioural variation in natural populations. V. Morphological correlates of locomotion in the garter snake (*Thamnophis radix*). Biol. J. Linn. Soc. 34:175–190.

Arnold, S. J., and P. C. Phillips. 1999. Hierarchical comparison of genetic variance-covariance matrices. II. Coastal-inland divergence in the garter snake, *Thamnophis elegans*. Evolution 1516–1527.

Chenoweth, S. F., and Mark. W. Blows. 2008. Q_{ST} meets the G matrix: the dimensionality of adaptive divergence in multiple correlated quantitative traits. Evolution 62:1437–1449.

de Queiroz, A., R. Lawson, and J. A. Lemos-Espinal. 2002. Phylogenetic relationships of North American garter snakes (*Thamnophis*) based on four mitochondrial genes: How much DNA sequence is enough? Mol. Phylogenet. Evol. 22:315–329.

Flury, B. D. 1988. Common Principal Components and Related Multivariate Models. John Wiley and Sons, New York.

Hohenlohe, P. A., and S. J. Arnold. 2008. MIPoD: a hypothesis-testing framework for microevolutionary inference from patterns of divergence. Am. Nat. 171:366.

Houle, D., G. H. Bolstad, K. van der Linde, and T. F. Hansen. 2017. Mutation predicts 40 million years of fly wing evolution. Nature 548:447–450.

Hunt, G. 2007. Evolutionary divergence in directions of high phenotypic variance in the ostracode genus *Poseidonamicus*. Evolution 61:1560–1576.

Kremer, A., A. Zanetto, and A. Ducousso. 1997. Multilocus and multitrait measures of differentiation for gene markers and phenotypic traits. Genetics 145:1229–1241.

Lande, R. 1979. Quantitative genetic analysis of multivariate evolution, applied to brain:body size allometry. Evolution 33:402–416.

Lofsvold, D. 1988. Quantitative genetics of morphological differentiation in *Peromyscus*. II. Analysis of selection and drift. Evolution 42:54–67.

Manier, M. K., and S. J. Arnold. 2005. Population genetic analysis identifies source–sink dynamics for two sympatric garter snake species (*Thamnophis elegans* and *Thamnophis sirtalis*). Mol. Ecol. 14:3965–3976.

McGlothlin, J. W., M. E. Kobiela, H. V. Wright, D. L. Mahler, J. J. Kolbe, J. B. Losos, and E. D. Brodie. 2018. Adaptive radiation along a deeply conserved genetic line of least resistance in *Anolis* lizards. Evol. Lett. 2:310–322.

Schluter, D. 1996. Adaptive radiation along genetic lines of least resistance. Evolution 50:1766–1774.

CHAPTER 10

Response of a Single Trait to Selection

Overview—The evolution of the trait mean in response to directional selection is an important ingredient of adaptive radiation and in the improvement of domestic stocks. Selection response is a function of genetic variance and the magnitude of directional selection. Experimenters have imposed deliberate selection for many generations to improve domesticated plants and animals and to gauge response in experimental lines. Steady change in the mean has often been observed for 10 – 120 generations, although sometimes the response is asymptotic. These observations reinforce the view that abundant genetic variation is usually available for most kinds of traits, even in small natural populations exposed to intense selection. The genetic underpinnings of selection response have been assessed in studies of quantitative trait loci (QTL and GWAS). A common finding is that traits are affected by dozens to hundreds of genes with locations sprinkled about the genome. Studies of geographic variation in quantitative traits suggest the trait mean has tracked spatial variation in the adaptive peak. Analysis of trait divergence on the scale of millions of generations indicates that long-continued directional selection of minute magnitude could account for the observed divergence. These diverse lines of evidence point to selection as the driver of adaptive radiations.

10.1 Response to Selection as a Regression Problem

In Chapter 1 we considered the effects of selection within a generation. We now wish to consider how those effects are transmitted from one generation to the next. Consider the plot of offspring averages as a function of parental averages in Figure 10.1a. The regression of offspring means on parental means is equivalent to a regression of breeding values on phenotypic values or $GP^{-1} = h^2$, the slope of the line in Figure 10.1a. That figure also shows the implementation of a truncation selection scheme in which we suppose that only midparents with a value ≥ 0 are allowed to breed and become the actual parents of the next generation. A result due to Pearson (1903) tells us that if we select on one trait, call it x, that selection may change the variance of x, the variance of a correlated trait y, and the covariance of x and y, but it will not change the regression of y on x. Consequently, the regression of breeding values on phenotypic values for the set of selected parents in Figure 10.1a is h^2, the same as it is in the entire set of potential parents. The upshot of this result is that we can with justification draw the regression plot, as in Figure 10.1b, so that the difference in average breeding values caused by the selection differential s, is

Evolutionary Quantitative Genetics. Stevan J. Arnold, Oxford University Press. © Stevan Arnold (2023).
DOI: 10.1093/oso/9780192859389.003.0011

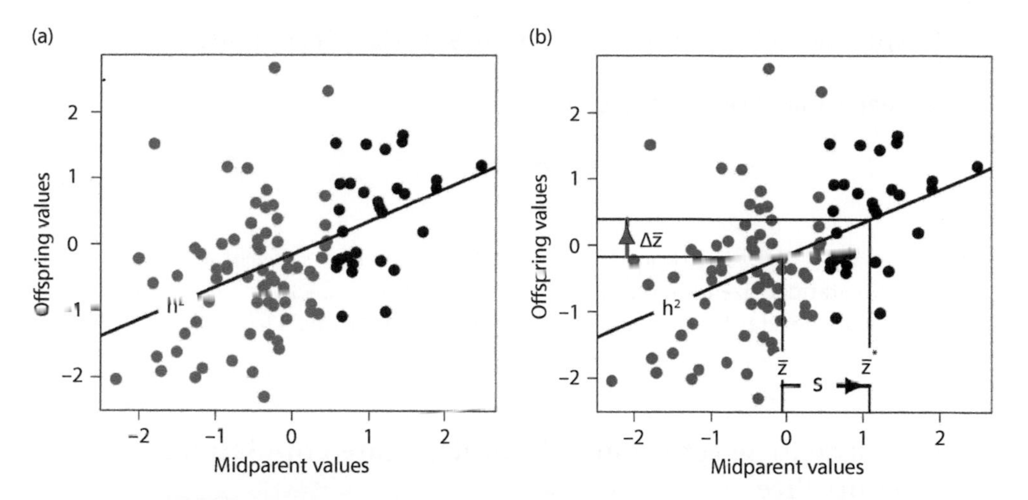

Figure 10.1 Response to selection as a regression problem. (a) A plot of hypothetical offspring trait values as a function of hypothetical mid-parental trait values, showing h^2, the slope of the regression line (heavy line). (b) Truncation selection acts so that one set of parents (points shown in black) becomes the actual parents of the next generation. Their trait mean is $\bar{z}^*$, whereas the mean of all potential parents (blue and black points) is $\bar{z}$. The vertical lines project the trait means before and after selection up to the regression, which yields the two expected offspring means (horizontal lines), and hence the expected response to selection.

a regression. Furthermore, we can equate the difference in breeding values of actual and potential parents with an induced difference in phenotypic values in the next generation. In other words, the change in phenotypic trait mean in the next generation caused by directional selection, the *response to selection*, is

$$\Delta\bar{z} = GP^{-1}s = h^2 s. \tag{10.1}$$

This equation describes the response to truncation selection, but it also describes the response to whatever form of selection is associated with a directional selection differential of magnitude s.

If $GP^{-1} = h^2$ remains constant through time, and the same selection differential, s, is applied generation after generation, after t generations we expect the *accumulated response to selection* to be

$$\Delta\bar{z}_t = tGP^{-1}s = th^2 s. \tag{10.2}$$

Even minuscule directional selection, if long-continued, can produce huge effects. Suppose that the same slight selection differential, $s = 0.01$, is applied generation after generation for 6000 generations, the approximate elapsed time since modern humans evolved (assuming $P = 1$). A selection differential this small would not be detectable in field studies. Nevertheless, with a typical heritability of 0.4, the trait mean would increase by 24 phenotypic standard deviations over that period of time. This simple example illustrates the point that although the measurement of evolutionary rates is an important endeavour (Hendry and Kinnison 1999), selection in one direction is probably not sustained for periods of dozens much less hundreds of generations. Nevertheless, on short timescales, the power of directional selection has been established by plant and animal breeders (Darwin 1868), as we shall see below.

10.2 Response to Selection on a Mean-Standardized Trait

The mean-standardized version of (10.1) is

$$\frac{\Delta\bar{z}}{\bar{z}} = \frac{h^2 s}{\bar{z}} = I_A\,\beta_\mu\,, \tag{10.3}$$

where $I_A = CV_A^2 = G/\bar{z}^2$ is the square of the *additive genetic coefficient of variation*, and β_μ is the mean-standardized selection gradient introduced in Section 3.6 (Hereford et al. 2004). The mean-standardized form gives the response to selection in units equivalent to the trait mean.

10.3 Response to Selection in a Set of Replicate Populations of Finite Size

The response to selection in (10.1) is the so-called *deterministic response to selection*, the change we expect in a population of infinite size or the average response we expect to see across a series of replicate populations of finite size. Both selection and drift can happen each generation within a lineage and their effects can be combined into a *total per generation change in the trait mean,* $\bar{z}$. For example, each generation we can model the response to selection with (10.1) and then draw an random sample of parents from a normal distribution with mean zero and variance G/N_e to yield a change in the mean after selection due to drift. Adding that *stochastic change in the mean* to the deterministic change due to selection gives us, each generation, the combined effects of selection and drift. If we start a set of finite replicate populations at generation 0 with the same zero mean, their trajectories through time form an expanding cone around the deterministic response, th^2s (Figure 10.2). The smaller the effective size, the wider the cone at any generation t. In other words, response to selection is not a simple 'march of the frequency distribution'. When populations are of finite size, as they must be in the real world, stochastic dispersion of means is superimposed on the simple deterministic march of the means.

10.4 Response to Deliberate Selection in Laboratory and Agricultural Settings

Response to deliberate directional selection has been systematically studied in laboratory and agricultural populations for nearly 100 years. These quantitative studies are motivated by three practical questions: (1) Will a particular trait respond to selection? (2) How much change can be accomplished per generation? and (3) How long can the response be sustained? On a more theoretical level, studies of response to deliberate selection are designed to: (1) estimate the heritability or genetic variance of the selected trait, (2) estimate the mutational variance for the trait, (3) understand the basis of asymmetry and plateau in selection response, and (4) produce divergent lines that can be used to analyze polygeny in the selected trait. Furthermore, all of these motivations and issues have multivariate counterparts that will be discussed in Chapter 11.

Our discussion of response to selection in the preceding section provides a guide to what we might expect in a selection experiment. The selection response should be a function of trait heritability and selection intensity (10.1). But, since our selected population will

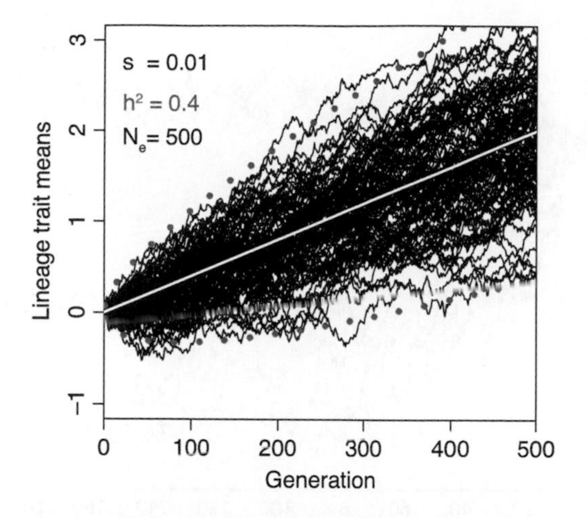

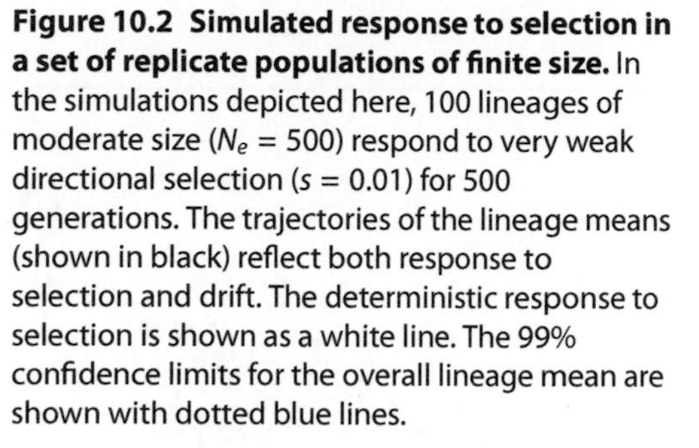

Figure 10.2 Simulated response to selection in a set of replicate populations of finite size. In the simulations depicted here, 100 lineages of moderate size (N_e = 500) respond to very weak directional selection (s = 0.01) for 500 generations. The trajectories of the lineage means (shown in black) reflect both response to selection and drift. The deterministic response to selection is shown as a white line. The 99% confidence limits for the overall lineage mean are shown with dotted blue lines.

necessarily be finite in size, we need to take account of drift which will produce stochastic dispersion among replicate selected lines (Hill 1982). The two examples that follow will give a feel for the issues of design and interpretation that are confronted in selection experiments.

Mackay et al. (2005) selected on bristle numbers in *D. melanogaster* to produce divergent lines that could be used to identify quantitative trait loci (QTL). Their experiment started with a highly inbred stock so that all of the selection response could be attributed to new mutations at loci that affect abdominal and sternopleural bristle counts. Because they selected for high and low counts in separate lines, they can evaluate the symmetry of response. They also maintained replicate selected lines in each direction, so that they could detect trends (i.e., slope and curvature of response in trait means) and distinguish them from drift (which causes increasing dispersion about a trend with zero slope and no curvature). Using values of N_e = 14 and i = 0.915 appropriate to their experimental design, Mackay et al. 2005 also used their results to estimate G and mutational input to G, ($V_m = U$), for each trait.

Impressive change in the means of the two traits was accomplished by 206 generations of directional selection (Figure 10.3). The abdominal bristle count diverged by 13.1 bristles ($5.7\sqrt{P}$) and the sternopleural count diverged by 15.2 bristles ($8.8\sqrt{P}$). Although average rate of response per generation for the two traits were about $0.03\sqrt{P}$ and $0.04\sqrt{P}$, some initial rates were faster than these. By generations 120 – 160, some rates were fluctuating around zero. These pronounced trends in rate of response, as well as the asymmetries between up and down responses, which are apparent but opposite in the two traits, are not predicted by (10.2).

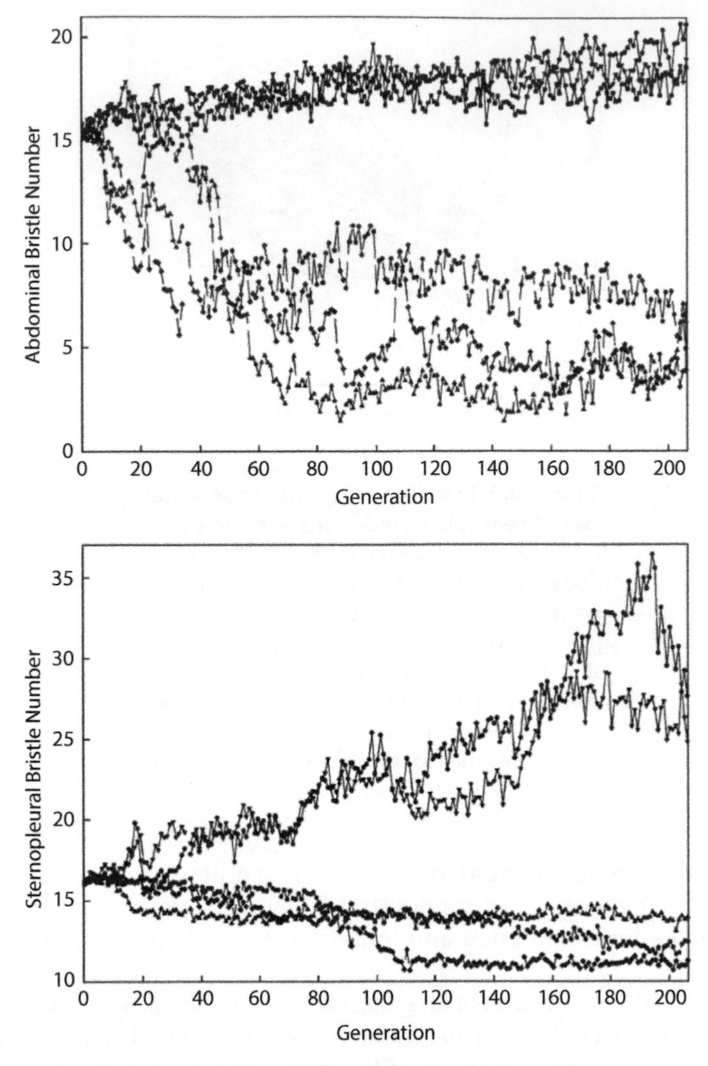

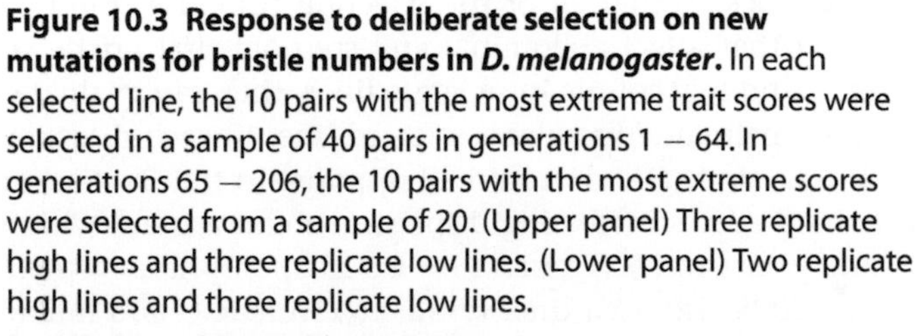

Figure 10.3 Response to deliberate selection on new mutations for bristle numbers in *D. melanogaster*. In each selected line, the 10 pairs with the most extreme trait scores were selected in a sample of 40 pairs in generations 1 – 64. In generations 65 – 206, the 10 pairs with the most extreme scores were selected from a sample of 20. (Upper panel) Three replicate high lines and three replicate low lines. (Lower panel) Two replicate high lines and three replicate low lines.

From Mackay et al. (2005) with permission.

Natural selection is a leading candidate explanation for both the declining rate of response to selection and the response asymmetries observed by Mackay et al 2005. Other explanations can be sought by invoking special explanations in terms of gene action (Mackay et al. 2005), but contravening stabilizing selection offers the simplest explanation. As deliberate selection pulls the trait mean further and further away from its optimum, a restoring force of natural selection slows and eventually halts further

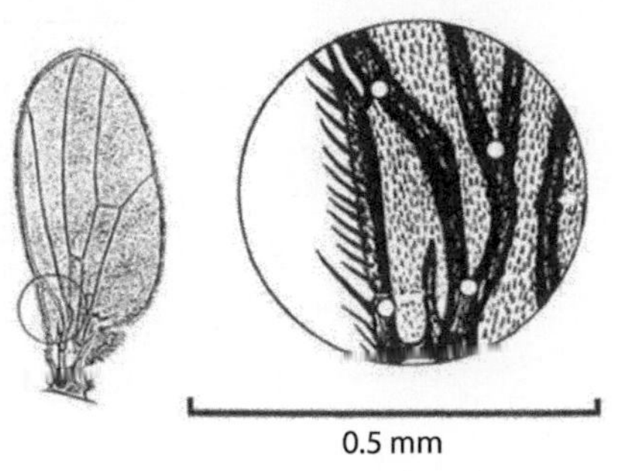

Figure 10.4 Landmark traits that were used to construct targets of deliberate selection in *D. melanogaster*. (Left) Full wing, showing the trait region. (Right) Detail of the trait region. D_1 is the distance between two upper points. D_2 is the distance between the two lower points.

From Weber (1992) with permission.

departure. Response asymmetry implies that stabilizing natural selection is weaker in one direction away from the optimum. Other examples of natural selection opposing the action of deliberate selection are discussed by Falconer and Mackay (1996) and Walsh and Lynch (2018).

Weber (1990, 1992) devised automated devices to impose deliberate selection. These innovations allowed Weber to achieve huge sample sizes before and after selection, virtually eliminating both sampling error and drift in his results. For example, Weber (1992) used deliberate selection to determine whether a lower limit exists for small-scaled, localized change in the parts of an organism. To motivate his experiment, Weber (1992) contrasts the perspectives of developmental biologists—who argue for course genetic control of epigenetic mechanisms—and functional morphologists who make a case for fine-scaled, autotomous optimization of the parts of complex structures. Only the functional perspective predicts that deliberate selection will produce relatively autotomous responses at a very localized spatial scale. With this prediction in mind, Weber (1992) selected in antagonistic directions in a small (< 0.6 mm^2) region of the *Drosophila* wing (Figure 10.4).

Weber (1992) achieved antagonistic responses to selection on a small spatial scale. Deliberate selection was relatively strong each generation on an initially outbred population. In each selection line, the most extreme 20% of 100 – 150 males were selected in each of 10 generations. This value for truncation selection corresponds to a selection intensity $i = 1.46$ (see Section 1.3). Over the course of 11 generations, the average offset in the up-selected line diverged from the control average by $4.54\sqrt{P}$, while in the down-selected line, the comparable divergence was $2.98\sqrt{P}$ (Figure 10.5). The total divergence averaged over sexes was $6.7\sqrt{P}$. The responses were linear over 11 generations with no sign of response plateaus. Despite the dramatic change of the angular trait that amounted to nearly $7\sqrt{P}$, linear dimensions in the vicinity of the target landmarks changed only slightly (average of 11%). Weber (1992) argues that these results argue for genetic variance on a small spatial scale that can respond to selection with minimum spatial reverberation. The results also support the view that wing venation phenotype is highly polygenic on a small spatial scale (see also Bailey 1985, 1986).

10.5 Response to Selection in Natural Populations

10.5.1 *Response to selection on ecological timescales*

In recent decades, many rapid and dramatic examples of response to selection in natural populations have been catalogued on ecological timescales (<100 years) (Thompson 1998;

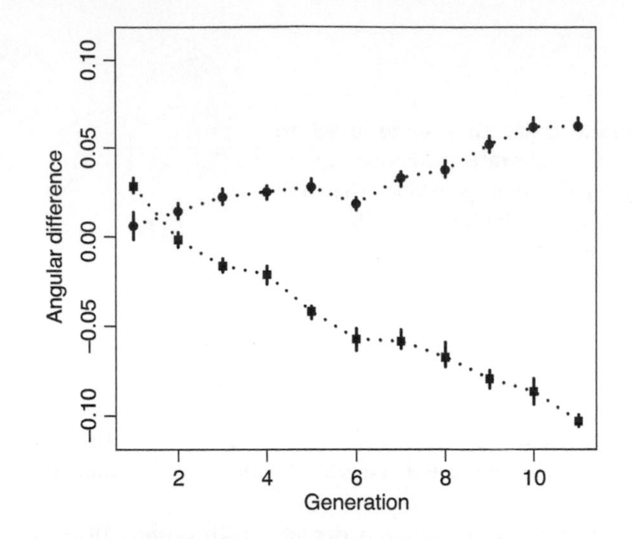

Figure 10.5 Response to deliberate selection on angular traits in *D. melanogaster*. Response to selection for large D_1/small D_2 (circles) and to small D_1/large D_2 (squares), ± s.e. (vertical lines). Responses are shown as the angular difference from control lines. Realized heritabilities estimated from these selection responses were 0.12 ± 0.02 s.e. (up line) and 0.24 ± 0.02 s.e. (down line). After Weber (1992).

Hendry and Kinnison 1999; Bone and Farres 2001; Palumbi 2002; Hendry 2016). These results indicate that many natural populations rapidly respond to selection resulting from temporal change and spatial gradients in ecological conditions.

Reznick and Ghalambor (2001) surveyed 47 studies of rapid responses to selection in nature on timescales ranging from a few to as many as a 100 generations in a search for the ecological causes of rapid selection responses. The organisms in this survey included bacteria, algae, fungi, angiosperms, arthropods, annelids, molluscs, fish, birds, and mammals. The studies fell into two ecological settings: colonization of new environments or in situ adaptation to local environments. Rapid evolution after colonization (or introduction) was associated with a novel host or food resource, a new biophysical environment, a new predator(s), or a new competitor. Reznick and Ghalambor (2001) point out that the common denominator in these cases of colonization events (e.g., the introduction of rabbits into Australia) or in situ adaptation (e.g., the evolution of industrial melanism in moths) is that rapid evolution was triggered by directional selection combined with an opportunity for population growth. In the terminology of the moving optimum model, the common denominator is that the adaptive peak is displaced from the trait mean (or vice versa) and the population responds to the directional selection associated with the displacement by rapidly evolving towards a new peak in trait space. We will return to the interesting idea that population growth is associated with the response to peak displacement in Chapter 21. In the following sections, we consider spatial differences in the trait mean that have evolved on longer timescales, typically much greater than 100 years.

10.5.2 *Geographic variation in phenotypic trait means*

In contrast with Section 8.12, which used tests with neutral theory expectations to infer selection, and with the descriptive documentation in Section 10.5.1, in this section we explicitly model the evolutionary response of the trait mean to an intermediate optimum that varies in space.

Theory for the response of the trait mean to spatial variation in its optimum incorporates the role of dispersal between populations. Intuitively we can imagine that the geographic variation in the trait mean might reflect geographic variation in the trait optimum, but dispersal among populations might obscure that reflection. We need a model to evaluate this intuition. In this section we rely on models developed by Nagylaki (1975), Slatkin (1978), and Lande (1982).

Following Lande (1982), we make the following simplifying assumptions: (1) Relevant environmental variables (i.e., selective agents) vary in just one spatial dimension. (2) Populations are uniformly distributed along that spatial dimension and are large enough that drift of the trait mean can be ignored. (3) Gaussian stabilizing selection occurs at each spatial position, x, towards a local optimum $\theta(x)$. Dispersal occurs freely along the environment dimension such that *individual dispersal distance* each generation has a mean of zero and a variance of l^2 at each spatial position. (4) The magnitude of nonheritable variation for the trait is the same in each local population, and $\bar{z}$ varies only a small amount over distances less than l. Under these conditions, G achieves the same mutation-selection equilibrium at every spatial position (Slatkin 1978). Furthermore, if directional selection is relatively weak at any given location (i.e., the trait mean is relatively close to its optimum in trait space), the evolutionary response of the trait mean to selection and migration at generation t can be modelled with a diffusion equation of the following form (Nagylaki 1975),

$$\frac{\partial \bar{z}}{\partial t} = G\beta(x) + \frac{l^2}{2}\frac{\partial^2 \bar{z}}{\partial x^2}, \tag{10.4a}$$

where $\beta(x)$ is the directional selection gradient at spatial position x, which under the assumption of a Gaussian ISS is

$$\beta(x) = \Omega^{-1}\left[\theta(x) - \bar{z}\right] \tag{10.4b}$$

(3.10a) and l^2 is variance in dispersal distance. The final fraction in (10.4a) is the curvature of the geographic variation surface that relates the trait mean to spatial position, i.e., the *curvature of geographic variation*. When this surface is strongly curved, dispersal can make a large contribution to the evolutionary response. Analysis of this equation shows that the equilibrium geographic variation in the trait mean is the spatial average of geographic variation in the optima, $\theta(x)$, weighted by an exponentially decreasing function with a *characteristic length* given by

$$L = l\sqrt{\omega/G}, \tag{10.5}$$

where l is standard deviation in dispersal distance (Slatkin 1978, Lande 1982). Characteristic length is the minimum length over which the geographic pattern in the trait mean will respond to geographic differences in selection (i.e., positions of the optimum) and not be swamped by dispersal between populations (Slatkin 1978). For example, if l = 5km, ω = 10, and G = 0.4 (with P standardized to 1), then L = 25km. In this example, we would not expect $\bar{z}$ to track differences in $\theta(x)$ on a scale much smaller than 25km. Conversely, we would expect the geographic differences in $\bar{z}$ to reflect differences in $\theta(x)$ for locations that are 25, 100, or 1000km apart. After a brief review of empirical studies of geographic variation, we will use (10.5) to interpret one well-documented example of geographic variation.

Geographic trait variation is a common if not ubiquitous feature of wide-ranging species (Mayr 1963; Gould and Johnston 1972; Endler 1977). Geographic variation has been documented in a wide variety of phenotypic traits encompassing morphology, physiology, behaviour, and life history (Mayr 1963; Foster and Endler 1999). Geographic variation is interesting in its own right, but it can also be viewed as a stepping stone to differentiation among species and higher taxa.

Detailed studies of geographic variation on a continental scale in vertebrate species suggest underlying geographic differences trait optima. In a particularly revealing study of this kind, James (1970) found that six of 12 species of birds showed concordant geographic size variation. In each of these six passerine species, wing length decreased from north to south in the Appalachian Mountains, and increased from south to north in the Mississippi River valley (Figure 10.6). Correlations between environmental variables and wing size in each species suggest the geographic variation in size optima is related to heat loss and energy budgets.

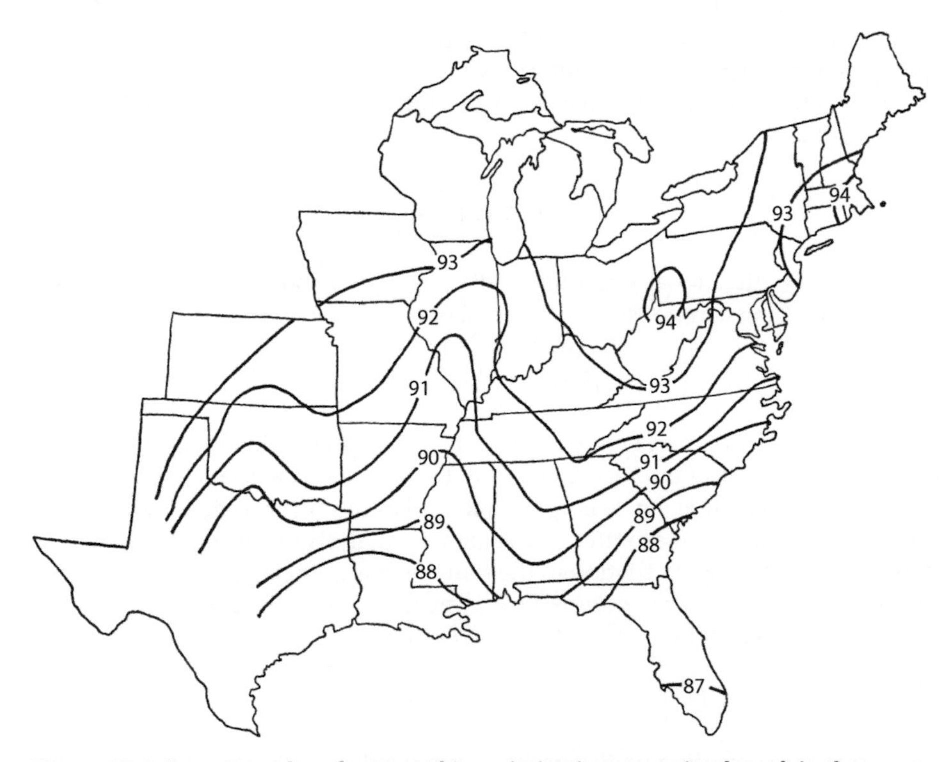

Figure 10.6 A contour plot of geographic variation in mean wing length in the downy woodpecker (*Dryobates pubescens*) in eastern USA. Contours were fitted with an iterative methodology. Contour interval is 1 mm, n = 1038 specimens, $P \approx 4$.

From James (1970) with permission.

Is it plausible that geographic size variation in the Downy Woodpecker and five other species has tracked geographic variation in size optima? We can use (10.5) to help answer this question. Studies of natal dispersal in birds (Greenwood and Harvey 1982; Paradis et al. 1998) suggest that $l = 2$km is a reasonable value for small passerines. Using values of $\omega = 10$, and $G \approx 0.4 * 4.0 = 1.6$ with (10.5), the expected characteristic length is $L = 5$km. In Figure 10.6, the distance between NE Illinois (93mm) and NE Louisiana (88mm) is about 1200km with five contour intervals in between, spaced an average of 240km apart, which is about 24 times the characteristic length of 10km. Even if we let stabilizing selection be much weaker ($\omega = 50$) and dispersal distance appreciably longer ($l = 20$km), $L = 112 < 240$km; we would not expect migration to swamp tracking of geographic variation in size optima by the trait mean. In other words, it is completely plausible that the size mean would be able to track size optima with a spacing of optima corresponding to the contours in Figure 10.6.

Two features of the James (1970) study reinforce the inference that geographic trait variation is an evolutionary response to geographic variation in trait optima. First, concordance in geographic trait variation across species argues against the view that the geographic pattern is random, arising—say—from drift of the trait mean. We don't expect concordance in outcome from independent processes of random drift.

Turning to a second example of concordance, (Klauber 1941) found parallel geographic variation in body vertebral numbers of 12 out of 13 species of snakes sampled across an inland to coast transect in Southern California. Studies of snake locomotion by Jayne (1988, 2020) and Kelley et al. (2003) suggest that the associations between more vertebrae in inland sites (with sparse vegetation) and fewer vertebrae in coastal sites (with dense vegetation) may have a biomechanical explanation. Second, statistical association between environmental variables that are plausibly related to selection on the trait in question supports the spatial variation-in-optimum hypothesis. Both the James (1970) and Klauber (1941) studies have this element to support a spatial variation in selection interpretation of their results.

10.5.3 *Common garden and transplant experiments*

A geographic difference in traits means can have genetic and environmental parts, and consequently the nature of geographic variation in phenotypes is inherently ambiguous. Common garden and transplant experiments can be used to test the hypothesis that the differences between population phenoypes in nature have a genetic basis (Turesson 1922; Clausen et al. 1940, 1948; Antonovics and Primack 1982; Niewiarowski and Roosenburg 1992; Johnson et al. 2021). Geographic variation is very commonly found to have a genetic basis when it is assayed with common garden and transplant experiments (Stebbins 1950).

In a *common garden experiment*, genotypic differences are revealed by growing phenotypes under the same environmental conditions. Let the trait mean of population i grown in its native environment i be $\bar{z}_{ii}$ with genotypic part $\bar{g}_1$ and environmental part $\bar{e}_1$, then in the two environment case we have

$$\bar{z}_{11} = \bar{g}_1 + \bar{e}_1 \tag{10.6a}$$

$$\bar{z}_{22} = \bar{g}_2 + \bar{e}_2 \,, \tag{10.6b}$$

where $\bar{g}_i = \bar{x}_i + \bar{y}_i + \bar{i}_i$ (5.4). If we observe only $\bar{z}_{11}$ and $\bar{z}_{22}$, we can't tell whether the difference between them is genotypic or environmental. But if we rear zygotes in a common environment, then the trait means of individuals from the two populations are

$$\bar{z}_{1cg} = \bar{g}_1 + \bar{e}_{cg} \tag{10.6c}$$

$$\bar{z}_{2cg} = \bar{g}_2 + \bar{e}_{cg}\,, \tag{10.6d}$$

where the subscript *cg* denotes common garden, and the difference between the two population means is a difference in genotypes,

$$\bar{z}_{1cg} - \bar{z}_{2cg} = (\bar{g}_1 + \bar{e}_{cg}) - (\bar{g}_2 + \bar{e}_{cg}) = \bar{g}_1 - \bar{g}_2. \tag{10.6e}$$

A limitation arises if the common garden is not a natural, native environment for then $\bar{e}_{cg}$ is not representative of any environment effect in nature and the $\bar{g}_i$ may not represent the expression of genotypes in nature. These problems can be circumvented in a reciprocal transplant experiment.

In a *reciprocal transplant experiment*, individuals are reared in other natural environments as well as in their native environment. In the case of two natural environments, the control trait means are

$$\bar{z}_{11} = \bar{g}_1 + \bar{e}_1 \tag{10.7a}$$

$$\bar{z}_{22} = \bar{g}_2 + \bar{e}_2\,, \tag{10.7b}$$

while the experimental trait means are

$$\bar{z}_{12} = \bar{g}_1 + \bar{e}_2 \tag{10.7c}$$

$$\bar{z}_{21} = \bar{g}_2 + \bar{e}_1\,. \tag{10.7d}$$

We have four equations with four unknowns, and consequently we can solve for the two genotypic means and the two environmental means. We also have some revealing contrasts that we did not have in the common garden experiment. For example, the difference between the trait mean of a population when individuals are reared in its own environment and when they are reared in a foreign environment is equal to the difference in the environmental effects of the two environments,

$$\bar{z}_{11} - \bar{z}_{12} = (\bar{g}_1 + \bar{e}_1) - (\bar{g}_1 + \bar{e}_2) = \bar{e}_1 - \bar{e}_2 \tag{10.7e}$$

In a statistical model with *genotype x environment interaction*, the effect of a genotype depends on the environment in which it is reared. The model for the trait mean of a genotype from population *i* reared in environment *j* is now

$$\bar{z}_{ij} = \bar{g}_i + \bar{e}_j + \delta_{ij}\,, \tag{10.8}$$

where $\bar{g}_i$ is the main effect of the *ith* genotype, $\bar{e}_j$ is the main effect of the *jth* environment, and δ_{ij} is the interaction effect resulting from rearing genotypes from population *i* in environment j. If we conduct a replicated, fully factorial experiment in which zygotes from each of p environments are reared in each of p environments, we can use (10.8) to test for the average effects of genotype, environment, and their interaction and evaluate the size of their contributions to the trait mean. By combining a reciprocal transplant experiment with a Mendelian cross (Section 10.8), and assaying fitness in nature, one can test a series of hypotheses about the height of adaptive peaks and other issues dealing with adaptation (Schluter 2000; Rundle and Whitlock 2001).

A graphical portral of the additive model for two environment transplant experiment (10.7) is shown in Figure 10.7. The model is additive in the sense that trait means of transplants are simply combinations of the genotypic and environmental parts of natives. The absence of an interaction term results in parallel transplant reaction norms (dotted black lines in the second vertical panel).

Berven (1982) reciprocally transplanted juvenile wood frogs (*Rana sylvatica*) between mountain and lowland ponds in Virginia and analyzed environmental and genotypic effects on body size at first reproduction. A factorial analysis of variance revealed highly significant main effects of site of origin and site of rearing ($P < 0.001$) and a much less significant interaction terms ($P < 0.45$) (Figure 10.8). An analysis of male body size at first reproduction had similar results, except that the interaction term was not significant. In other words, the geographic differences in female and male body size at first reproduction

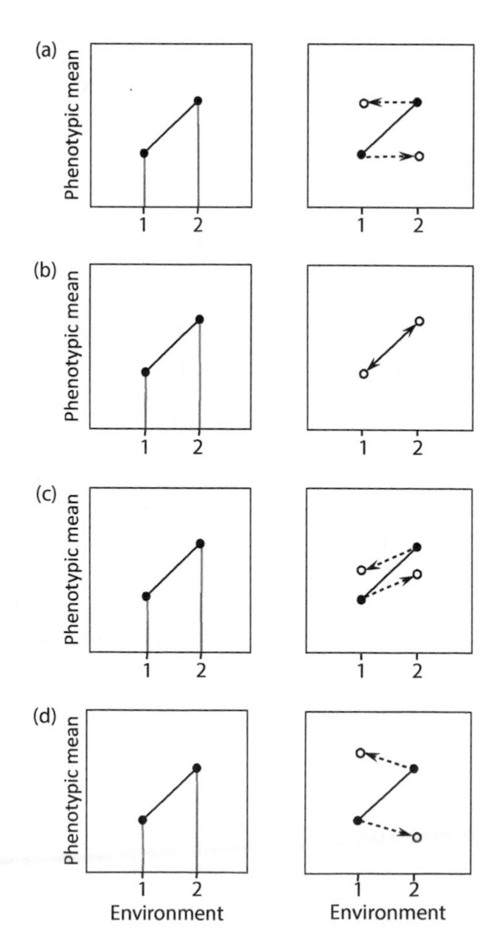

Figure 10.7 Graphical illustrations of the additive theory of reciprocal transplants between two environments (geographic sites). The phenotypic trait mean in each environment is the sum of a genotypic and an environmental part. Four cases are shown in the panels on the left side with phenotypic trait means shown as solid black dots. Genotypic parts are shown in blue and environmental parts are shown in orange. The geographic difference in trait means is shown as a solid black line ($\bar{z}_2 > \bar{z}_1$ in all cases). The results of transplants (dotted black lines) are shown on the right side with trait means in native environments shown as solid circles and trait means in foreign environments shown as open circles. (a) The geographic difference in the trait mean is wholly genotypic: $\bar{g}_2 > \bar{g}_1$ and $\bar{e}_2 = \bar{e}_1$. (b) The geographic difference in the trait mean is wholly environmental: $\bar{g}_2 = \bar{g}_1$ and $\bar{e}_2 > \bar{e}_1$. On the right, the dotted transplant lines are hidden by the solid geographic line. (c) Cogradient basis for the geographic difference in the trait mean: $\bar{g}_2 > \bar{g}_1$ and $\bar{e}_2 > \bar{e}_1$. (d) Countergradient basis for the geographic difference in the trait mean: $\bar{g}_2 > \bar{g}_1$ and $\bar{e}_2 < \bar{e}_1$. See Berven et al. (1979) for an alternative graphical analysis.

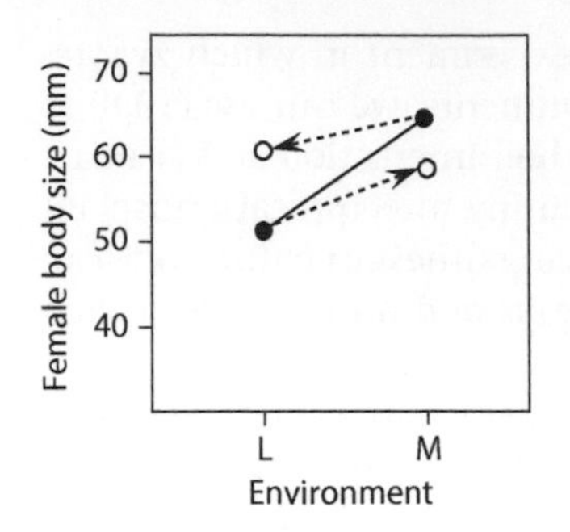

Figure 10.8 Graphical summary of body size at first reproduction of female wood frogs (*Rana sylvatica*) reciprocally transplanted between lowland (L) and mountain (M) geographic sites. Graphical conventions as in Figure 10.7. After Berven (1982).

were based on both genotypic and environmental differences with little or no genotype x environment interaction.

Notice the similarity between the female body size results (Figure 10.8) and Figure 10.7c in which $\bar{g}_2 > \bar{g}_1$ and $\bar{e}_2 > \bar{e}_1$. In a case such as this, geographic gradients for genotypic and environmental parts are in the same direction, a situation that was termed *cogradient selection* by Berven et al. (1979). In contrast, if the geographic gradients of the two parts are in opposite directions (e.g., $\bar{g}_2 > \bar{g}_1$ and $\bar{e}_2 < \bar{e}_1$, as in Figure 10.7d), the result is termed *countergradient selection* (Berven et al. 1979).

James (1983) conducted reciprocal transplant experiments with red-wing blackbirds (*Agelaius phoeniceus*) as a follow-up to the James (1970) study of geographic variation in avian body size. Morphometric shape variables that maximally discriminated between birds from the source localities were used in the analyses. Eggs were reciprocally transplanted between Tallahassee and the Everglades in Florida and from Colorado to Minnesota, but not in the opposite direction. In all three cases, average shape is shifted in the direction of the population of the foster parents, suggesting that environmental effects had a major effect on the expression of shape traits.

10.5.4 *The bearing of plasticity experiments on the interpretation of geographic variation in trait means*

In a *plasticity experiment*, individuals are reared in a series of environments that differ in single environmental factor (e.g., temperature). Usually the rearing environments are in the laboratory or greenhouse, and the variable environmental factor has a plausible, causal connection to geographic differences in the trait mean. A plasticity experiment assesses the population's capacity to alter phenotype in response to differences in environmental factors such as temperature, salinity, or food supply. The results of such experiments are often portrayed as *reaction norms* that graph the trait mean as a function of a particular environmental factor. Relevance to geographic variation in trait means depends on the plausibility of a causal connection to the environmental factor and whether the extent of plasticity is capable of accounting for geographic differences in the trait mean.

Coyne and Beecham (1987) analyzed the effect of rearing temperature on wing length and abdominal bristle number in multiple populations of *Drosophila melanogaster*. Ten populations were sampled along a north to south transect in the USA (Maine to Florida) and reared in the laboratory at three different temperatures (18, 24, 30 deg C). Wing length and bristle number in samples of wild-caught males (n = 96 – 200 per population) showed similar north to south clines, with larger wings and more bristles in northern populations. Population reaction norms were linear for wing length with similar negative slopes (i.e., larger wings at lower temperatures). The main difference among populations was in the intercepts of reaction norms. Northern populations tended to have higher interceps than southern populations (i.e., northern populations had larger wings at all temperatures). In contrast, reaction norms for bristle number were Λ–shaped, with higher bristle counts at the intermediate temperature. Northern populations tended to have reaction norms with higher intercepts (i.e., more bristles at all temperatures). The authors concluded that genotypic differences made a major contribution to geographic variation and that the clines in both traits were responses to selection. Similar clines have been reported on different continents in other species of *Drosophila* (Coyne & Beecham 1987). Kingsolver and Huey (2008) present a conceptual model for understanding responses to selection in temperature-sensitive traits.

In a second example, we return to Klauber's (1941) study of parallel evolution of body vertebral numbers across an ecological gradient in Southern California by multiple species of snakes (Section 10.5.2). Systematic herpetologists shied away from a genetic interpretation of such geographic variation in vertebral counts in snakes, following claims that temperature treatments imposed during pregnancy could dramatically influence vertebral counts in a live-bearing snake (Fox 1948; Fox et al. 1961). Later experiments, with better temperature control and statistical analysis of results, showed that mean vertebral counts were actually developmentally buffered against temperature effects. Because of this buffering, geographic variation in these traits in snakes is typically too extensive to be accounted for by geographic differences in temperature (Arnold and Peterson 2002).

10.6 Minimum Selective Mortality Required to Account for Trait Divergence

It is illuminating to consider the minimum amount of selective mortality required to account for the magnitude of known divergence events (Lande 1976). Suppose we know or can infer the means of a normally distributed trait at two points in time, separated by t generations. How strong must truncation selection be each generation to account for observed magnitude of divergence (difference in means) on that timescale? Let the mean of the trait distribution before selection each generation be $\bar{z}$ with *point of truncation* located at $\bar{z} = -b\sqrt{P}$. Using expression (10.1), the expected change in the mean each generation resulting from this mode of selection is

$$\Delta\bar{z} = \pm\frac{G}{\sqrt{2\pi}}exp\left(-b^2/2\right) \tag{10.9}$$

Lande (1976). If we accumulate this response to selection for t generations and equate it with the absolute value of the observed divergence in mean, we have

$$\frac{|\bar{z}_t - \bar{z}_0|}{\sqrt{P}} = \frac{tG}{P\sqrt{2\pi}}exp\left\{-b^2/2\right\}, \tag{10.10a}$$

where $\bar{z}_0$ and $\bar{z}_t$ are, respectively, the trait means at generations 0 and t. Solving for the truncation point, we obtain

$$b = \pm\sqrt{-2\ln\left(\sqrt{2\pi P}\,\frac{|\bar{z}_t - \bar{z}_0|}{tG}\right)} \tag{10.10b}$$

Lande (1976). In other words, if we set the truncation point at distance b from the mean and suppose that all individuals beyond that point die, and repeat this procedure for t generations, we can account for the observed divergence.

Surprisingly weak truncation selection can account for particular cases of divergence that are observed in the fossil record. In the following example, we assume a standard value for G (0.5), with $P = 1$. Lande (1976) analyzed 21 cases of divergence in mammalian tooth dimensions from the fossil record. Divergence ranged from 0.8 to 44 phenotypic standard deviations over time intervals ranging from 0.3 – 1.7 million generations. In all cases the observed divergence could be accounted for by truncation selection of magnitude: 10^{-6} to 10^{-7} per generation. In other words, only one in ten million to one in a million individuals would have to die a selective death per generation to account for these instances of divergence. Of course, the magnitude of selection is small in part because the deaths are spread out over the entire extent of the elapsed time interval. But even if we imagine that the selective deaths occurred during a fraction of the total time available in each interval, the magnitude of necessary truncation selection is still very small.

10.7 The Distribution of Genetic Effects Accumulated During a Response to Selection

What kinds of genetic (allelic) effects contribute to the responses to selection that we have just reviewed? More precisely, what is the distribution of effects of those genes that are selected to fixation during a bout of adaptation? Could those effects simply be a random sample of mutations so that the distribution we seek resembles the distribution of mutational effects? This answer seems unlikely because upon reflection we realize that an allele of large favourable effect might be rapidly swept to fixation, so that such genes would be disproportionally represented. Clearly we need a model; see Walsh and Lynch 2018, Chapter 27 for a review of such models.

A simple model of selection during a bout of adaptation is one in which an intermediate optimum has suddenly moved to a new position some distance from the phenotypic mean. In a later chapter we will refer to this situation as the *displaced optimum model* and consider its consequences in some detail. For now, we simply need to specify that the selection regime that provokes adaptation is an instantaneous displacement of the optimum of a Gaussian ISS. Given that this event has occurred, what kinds of alleles will be swept to fixation during the time it takes for the phenotypic mean to evolve to the new position of the optimum?

Orr (1998, 1999) modelled and simulated the process just specified to determine the distribution of genetic effects that become fixed during the approach to the new optimum. This problem has an interesting conditional effect. Whether a particular allele is favourable or not depends on its effect size and the distance from the phenotypic mean to the optimum. Early in the process of adaptation, alleles with large effect may be favoured, but when the mean is closer to the optimum, alleles of smaller effect will be favoured, because they will not cause the population to overshoot the optimum (Fisher 1930; Kimura 1983). Orr (1998) showed that the proportional reduction in the distance to the optimum by each expected mutational substitution is nearly a constant. Secondly, the distribution of effect sizes for the mutations that are fixed during approach to the optimum is approximately negative exponential (Figure 10.9). Furthermore, this distribution holds regardless of how many trait dimensions are involved in the displacement

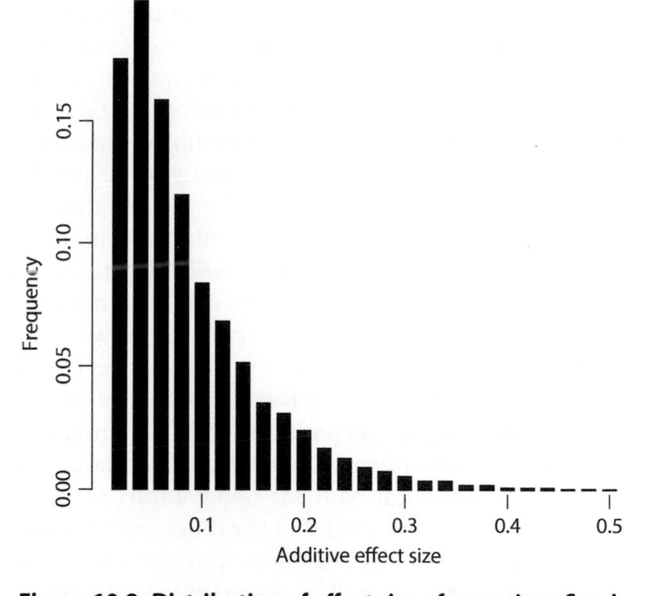

Figure 10.9 Distribution of effect size of mutations fixed during a simulation study of evolution towards a new trait optimum. The proportion of observations is graphed as a function of effect size. After Orr (1999).

of the optimum. Finally and most importantly, the distribution is negative exponential regardless how mutational effects are distributed.

Orr's (1998, 1999) exponential result has important implications for our understanding of the genetic differences that accumulate during adaptation. Our first expectation is that the process should be polygenic. On a probabilistic basis, we would not expect the first mutation to take the phenotypic mean all the way and exactly to the optimum. Instead, we expect mutant alleles of various effect sizes to be swept to fixation. Secondly, while a few genes may be of large effect, we expect that most will be of intermediate and small effects. In the next section we will consider QTL analysis, an empirical approach that has the potential to test these expectations.

10.8 The Analysis of Divergent Lines Produced by Deliberate Selection

In the decade following the rediscovery of Mendelism in 1905, geneticists pursued the possibility that single-factor inheritance might be widespread. Experimental crosses were made using many kinds of traits in a variety of organisms. Although many examples of single-factor inheritance of discrete polymorphism were discovered, a much larger number of cases failed to yield to a simple Mendelian explanation (East 1910; Wright 1968). Often these traits were normally distributed in each of two parental populations with means that had been pushed far apart by deliberate or natural selection.

Deviations in the positions of the mean in F_1 and second generation hybrids from additive expectations can provide evidence for dominance, maternal effects, and sex-linkage, depending on the nature of the deviations (Wright 1968, 1977). For example, deviation of the trait means of both sexes in the F_1 toward one of the parental means indicates *directional dominance*. That is to say, one or more underlying factors show dominance in effects on the trait, causing an overall shift in the mean away from intermediacy.

Steady improvement in the analysis of second generation hybrids has been made since 1910. Early tests for polygeny were based on comparison of variance between first and second generation hybrids and were plagued with many assumptions, some of which can be relaxed (Castle 1921; Wright 1968; Lande 1981). Later analyses used chromosome markers to associate phenotypic differences with particular linkage groups (Sax 1923). The use of genome-wide markers for association mapping will be discussed in the next section.

Consider the case of characid fishes of the *Astyanax* complex that exist in both surface- and cave-dwelling populations in Mexico. The inheritance of eye size has been studied by crossing these two ecotypes (Wilkins 1971). The F_1 mean deviates towards the eyed parental mean (Figure 10.10), indicating directional dominance for larger eye size. The failure of the F_2 sample to encompass the positions of the parental means suggests that several genetic factors may be responsible for ecotypic differentiation in eye size.

To better understand the theoretical basis for observations such as those shown in Figure 10.10, consider a simple additive model for the differences between the trait means of two parental populations (Lande 1981). Denote the difference in average effects at locus i in two parental populations as $\delta_i = mean\,(P_{i2}) - mean\,(P_{i1})$. Assuming that this difference has the same sign across all loci, the difference in the two trait means is

$$mean\,(P_2) - mean\,(P_1) = 2\sum_{i=1}^{n}\delta_i = 2n\bar{\delta}, \tag{10.11}$$

where $\bar{\delta}$ is the mean value of δ_i average across all relevant loci. Consequently, the extra genetic variance appearing in the F_2 progeny beyond that in the F_1 progeny, the *segregational variance*, is

$$\sigma_S^2 = \frac{1}{2}\sum_{i=1}^{n}\delta_i^2 = \frac{1}{2}n\left[\sigma_\delta^2 + \left(\bar{\delta}\right)^2\right] = Var\,(F_2) - Var\,(F_1)\,. \tag{10.12}$$

Solving for n, using (10.11) and (10.12), the actual number of factors (loci) contributing to the mean trait difference between samples of the parental populations raised in a common environment is

$$n = \frac{[mean\,(P_2) - mean\,(P_1)]^2}{8\sigma_S^2}\left[1 + \left(\sigma_\delta/\bar{\delta}\right)^2\right]. \tag{10.13a}$$

The last term in parentheses is generally unknown but must be positive, therefore an estimate of the minimum number of genetic factors is

$$n_E = [mean\,(P_2) - mean\,(P_1)]^2/\left(8\sigma_S^2\right) \le n\,. \tag{10.13b}$$

This estimator is known as the *effective number of factors* with equal magnitudes of effect and producing the same pattern of segregation as observed in the cross (Lande 1981). Using standard methods to estimate the sampling variance in a large sample, the sampling variances of n_E and σ_S^2 are

$$Var\,(n_E) \cong {n_E}^2\left\{\frac{4\,[Var\,(P_1)\,/N_{P_1} + Var\,(P_2)\,/N_{P_2}]}{[mean\,(P_2) - mean\,(P_1)]^2} + \frac{Var\left(\sigma_S^2\right)}{\sigma_S^4}\right\} \tag{10.13c}$$

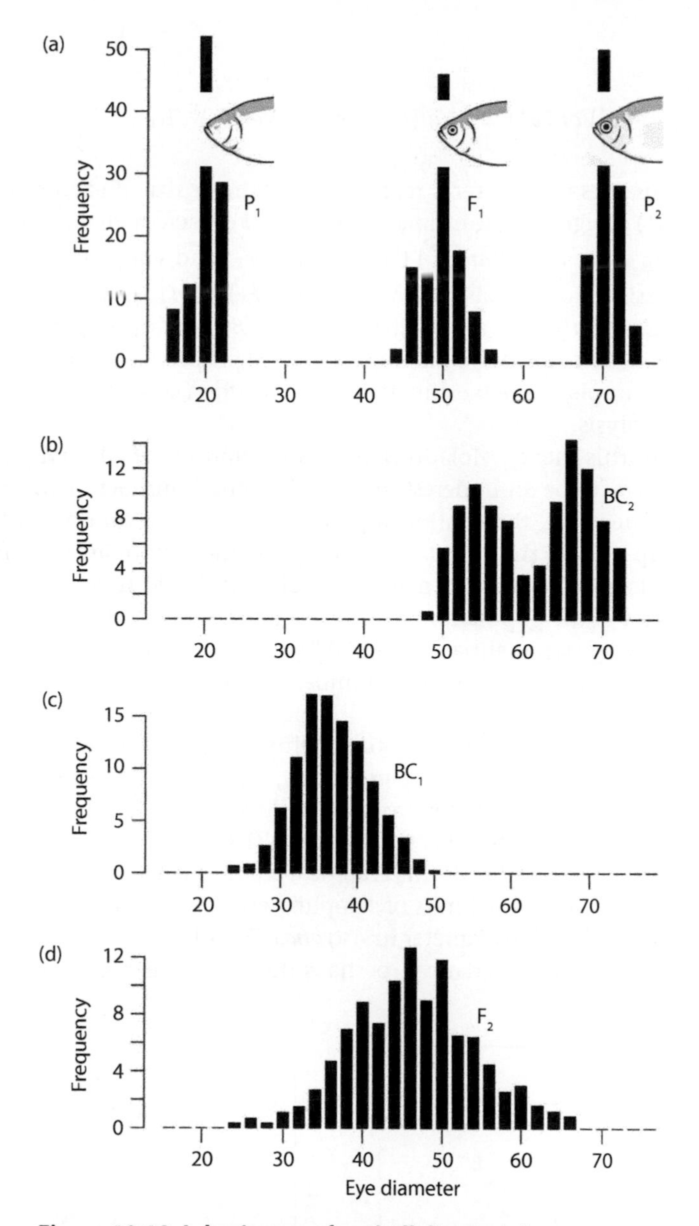

Figure 10.10 Inheritance of eyeball diameter in a cross between surface- and cave-dwelling populations of fish (*Astyanax*). Frequency is shown as a percent of observations. (a) Distributions of eye size in cave-dwelling (P_1, n = 30) and surface-welling (P_2, n = 30) populations and their hybrids (F_1, n = 30). (b) Eye size distribution in BC_2 (n = 142), a cross between P_2 and F_1. (c) Eye size distribution in BC_1 (n = 450), a cross between P_1 and F_1. (d) Eye size distribution in the F_2 (n = 702). After Wilkins (1971).

and

$$Var\left(\sigma_S^2\right) \cong 2Var(F_2)^2/N_{F_2} + 2Var(F_1)^2/N_{F_1}, \tag{10.13d}$$

where the denominators in each term represent the sample size of each progeny (Lande 1981). Lande (1981) designates the estimator of number of factors given by (10.13a) as n_1, with corresponding estimate of $Var\left(\sigma_S^2\right)$ based on $Var(F_2)$ and $Var(F_1)$ (10.13d). Likewise, he also gives expressions for: n_2 [based on $Var(F_2)$, $Var(F_1)$, $Var(P_1)$, and $Var(P_2)$], n_3 [based on $Var(F_2)$, $Var(BC_1)$, $Var(BC_2)$], and n_4 [based on $Var(BC_1)$, $Var(BC_2)$, $Var(F_1)$, $Var(P_1)$), and $Var(P_2)$)]. These estimators have different statistical properties and uses (Wright 1968; Lande 1981). For example, some n-estimates are useful when some kinds of progenies are not available for analysis.

In applications of this theory, violations of the assumptions of additivity and equality of effects will cause n_E to be an underestimate of the actual number of contributing loci (Wright 1968). Furthermore, the relationship of n_E to the actual number of loci is much like the relationship of N_e to the actual size of the population. Both are effective sizes that tell us how many loci or individuals making equal contributions would be required to mimic the properties of the population. In the case of n_E, we expect that the distribution of effect sizes is negative exponential (Figure 10.9) rather than uniform. Consequently, if our estimate of n_E is, say 10, then the actual number of loci is probably orders of magnitude larger.

A simple visual test for the adequacy of the additive model of inheritance can be conducted by plotting the variances of the parental, F_1, and second generation hybrids as a function of their means. The prediction from the additive model is that on a suitable scale the resulting graph will form a triangle (Lande 1981) (Figure 10.11).

A sample of six traits analyzed with this graphical technique shows reasonably good fits to the additive model, within the limits of sampling error, but there are some exceptions (Figure 10.12, Lande 1981). Eye diameter in *Astyanax* (Figure 10.12b), for example, shows discrepancies in the backcross variances, perhaps due to the segregation of a gene with

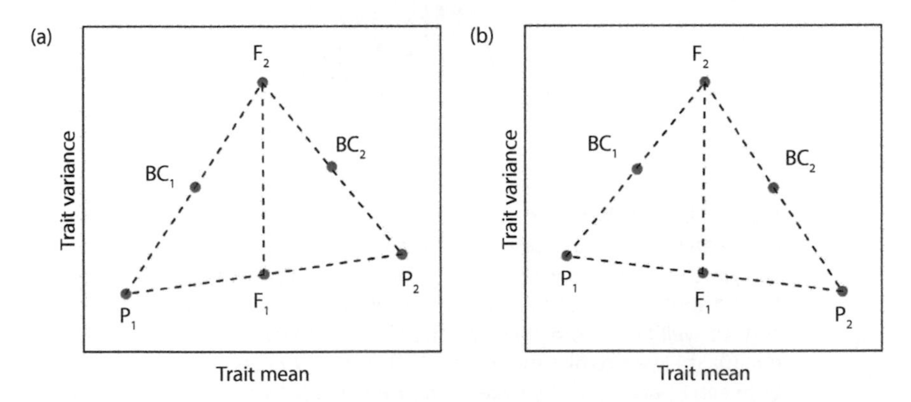

Figure 10.11 Theoretical means and variances of a quantitative trait in a cross between two parental populations according to the additive model of inheritance. Trait variance is graphed as a function of trait mean in parental populations and their derived hybrids. (a) An example case in which the variance in P_2 is greater than in P_1. (b) An example case in which the variance in P_1 is greater than in P_2. After Lande (1981).

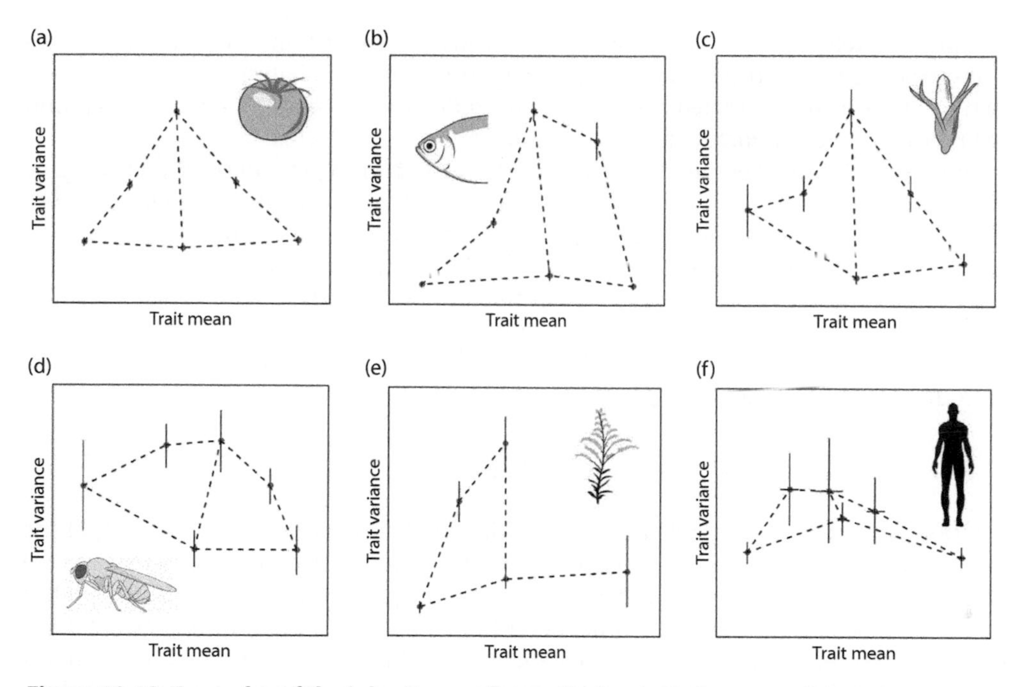

Figure 10.12 Examples of the inheritance of quantitative traits in crosses between divergent parental populations. Graphical conventions as in Figure 10.11. Thin vertical and horizontal lines show ± 1 standard error. Standard errors for the trait means for the first five cases were small and are not shown. Trait measurement is denoted as *x*. Standard errors (s.e.) of n_2 and n_4 are the square roots of corresponding estimates of $Var\,(n_E)$. (a) Tomato, fruit weight (gm), $\log\,(x - 0.153)$, data from Powers (1942), scale from Wright (1968), $n_2 = 10.0 \pm 0.7$ *s.e.* (b) *Astyanax*, eye diameter, *x* is an arbitrary unit of measurement, data from Wilkins (1971), $n_2 = 5.9 \pm 0.4$ *s.e.* (c) Maize, percent oil in kernels, $\log\,(x + 1.87)$, data from Sprague and Brimhall (1949), scale from Wright (1968)), $n_2 = 19.4 \pm 3.1$ *s.e.* (d) Hawaiian *Drosophila*, female head shape, arctan (head width/head length), data from Val (1977), scale from Templeton (1977), $n_2 = 9.4 \pm 3.6$ *s.e.* (e) Goldenrod (*Solidago*), date of anthesis, date relative to August, from Goodwin (1944), $n_2 = 6.4 \pm 1.4$ *s.e.* (f) Human, skin colour, antilog (reluctance at 685 $m\mu$), data from Harrison and Owen (1964), $n_4 = 4.6 \pm 3.3$ *s.e.* Analysis and graphics after Lande (1981).

major effect in one of the backcrosses. In the case of human skin colour (Figure 10.12f), the fit is so poor (primarily because of small F_2 sample) that the causes are hard to diagnose.

The estimates of effective number of factors for the six cases shown in Figure 10.12 (see caption) range from about 5 to 20. Likewise, in a partially overlapping sample of 23 cases, Wright (1968) obtained a range of 5 to 19. In other words, a median of about 10 equal-contribution loci underly divergence in trait means. This typical number is a minimum estimate, and the actual number of loci making small as well as large contributions to divergence is undoubtedly much larger.

10.9 QTL Analysis of Divergent Lines Produced By Deliberate Selection

Analyses of second generation hybrids (backcrosses and F_2) have advanced to finer and finer scale mapping and the use of maximum likelihood (ML) for statistical inference.

Maximum likelihood is an approach that will figure regularly in later chapters, so we will introduce the method here by recounting its use in QTL analysis. Genomic regions that are statistically associated with phenotypic differences in second (or later) generation hybrids are called *quantitative trait loci* (QTLs).

The ML method makes use of correlations between phenotypic trait score and genomic markers in backcross, F_2 or other derived progenies. These correlations are in turn produced by the linkage disequilibrium generated by mixing the P_1 and P_2 genomes. The trick is to use ML to estimate the position of the QTLs. Using the *Lander and Botstein (1989) likelihood model,* we wish to evaluate the evidence for a QTL at a particular marker position in the genome. We will use a backcross progeny obtained by crossing the F_1 with one of the parental populations, P_1. We use a linear model for the phenotypic score of the *ith* individual in that backcross progeny

$$z_i = a + bg_i + \varepsilon, \tag{10.14}$$

where g_i is an indicator variable that denotes the number of alleles from P_2 (0 or 1) at the marker locus, ε is a normally distributed error contribution with a mean of 0 and variance σ^2, and b is an unknown parameter that represents the phenotypic effect of a single allele substitution at a putative QTL. To use the ML approach, we need an expression for a probability in terms of the unknown parameters. We have already made an assumption about the distribution of residuals, ε. To proceed we need to rearrange (10.14) so that we have an expression for ε in terms of z_i, a, b, and σ^2. We now wish to estimate a, b, and σ^2 by maximum likelihood. In other words, we wish to find the values of these parameters that maximize the probability $L(a, b, \sigma^2)$ that the observed data would have occurred. Given our model (10.14) for a regression with normally distributed residuals, the probability of the observed data is

$$L\left(a,\ b, \sigma^2\right) = \prod_i p\left[z_i - (a + bg_i)\,, \sigma^2\right], \tag{10.15}$$

where $p[x_i, \sigma^2]$ is the probability density for the normal distribution of ε with mean 0 and variance σ^2. Notice that to do this calculation for a set of individuals in the backcross progeny, we need to know only each individual's phenotype (z_i) and the number of P_2 alleles it carries at the marker locus (i.e., g_i). We call the likelihood that maximizes this function $L(\hat{a}, \hat{b}, \hat{\sigma}^2)$, where $(\hat{a}, \hat{b}, \hat{\sigma}^2)$ are the values of the parameters that give this maximum value (i.e., the *unconstrained maximum likelihood estimates*).

We can use a likelihood ratio to access the evidence for a QTL. The *constrained maximum likelihood estimates* under the assumption that $b = 0$ are $(\bar{z}_1, 0, \sigma^2_{B1})$, where $\bar{z}_1$ is the mean phenotype in P_1 and σ^2_{B1} is the variance of the backcross population (i.e., the cross between P_1 and F_1). This set of parameters is used to compute a value for the likelihood, $L(\bar{z}_1, 0, \sigma^2_{B1})$, that is constrained in the sense that the markers have no effect on the phenotype z_1. The evidence for a QTL is then given by the so-called *LOD score,* the log of the ratio of the unconstrained and constrained likelihoods,

$$LOD = log_{10}[L\left(\hat{a}, \hat{b}, \hat{\sigma}^2\right) / L\left(\bar{z}_1, 0, \sigma^2_{B1}\right)], \tag{10.16}$$

which indicates how much more probable the data are when one assumes that a QTL is present versus assuming that one is absent. *LOD* (times a constant) is χ^2-distributed with 1 degree of freedom. Figure 10.13 shows an example of *LOD* as a function of genomic

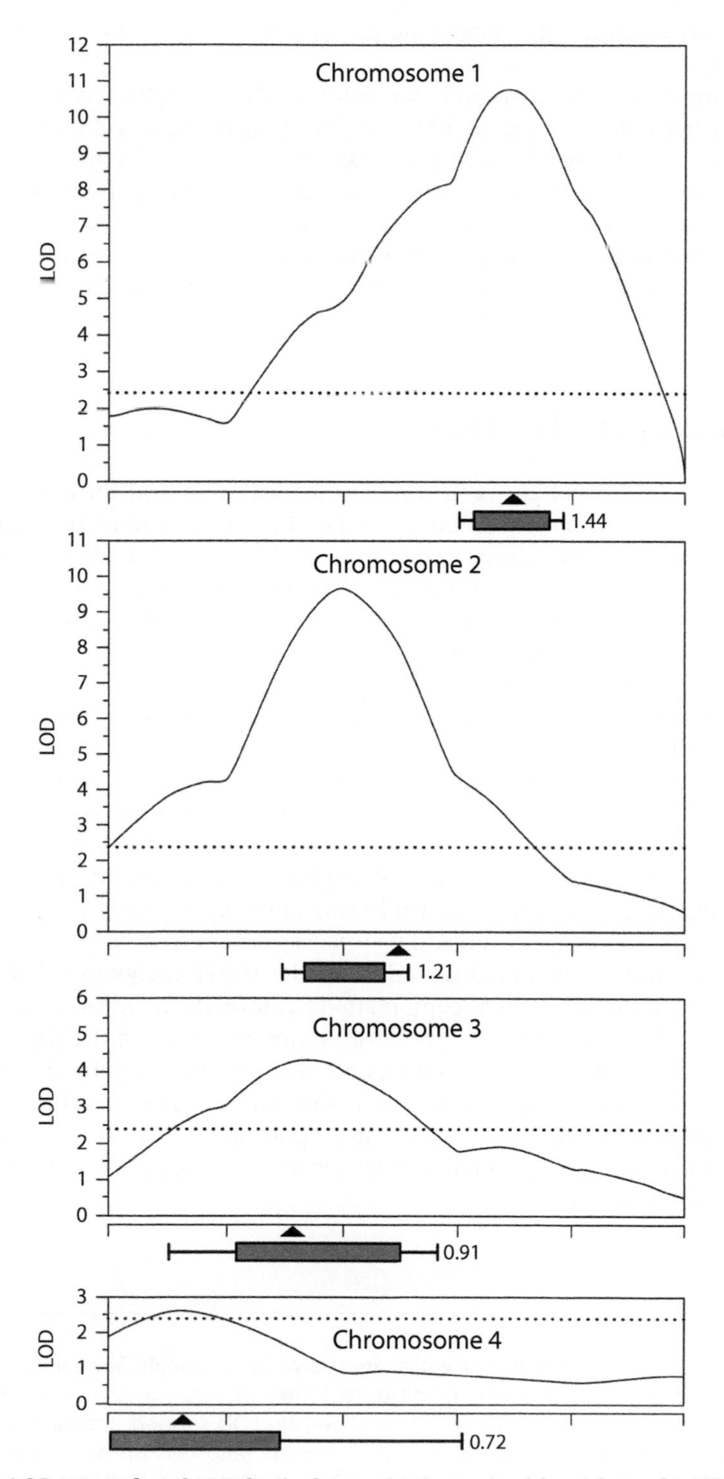

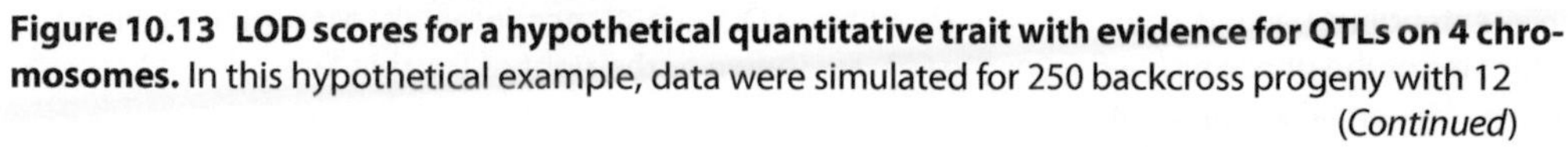

Figure 10.13 LOD scores for a hypothetical quantitative trait with evidence for QTLs on 4 chromosomes. In this hypothetical example, data were simulated for 250 backcross progeny with 12

(*Continued*)

position. In this example, the data show strong evidence for a QTL on each of four chromosomes.

The blooming of genetic variation in second-generation hybrids has made them popular with geneticists for more than 100 years, but that popularity does not mean that genetic variation in F_1 and backcross generations is typical of standing variation in natural populations. Usually the parental lines have been driven apart by natural or deliberate selection, so that the genes we detect in F_2 and backcrosses are ones that have responded to directional selection and have been pushed to or near fixation. The loci involved in divergence may or may not be responsible for genetic variation within natural populations.

10.10 Summary of QTL Results

The results of QTL studies have been reviewed several times and those reviews should be consulted for details and discussion of the fact that these studies are biased towards detecting QTLs of large effect (Kearsey and Farquhar 1998; Lynch and Walsh 1998; Goring et al. 2001; Mackay 2001a, b, 2004; Barton and Keightley 2002; Doerge 2002; Abiola 2003). Because of this bias, QTL analysis does not directly estimate the distribution of allelic effects that are fixed during bouts of adaptation. Consequently this section focuses on overarching generalizations that bear on adopting a moving optimum view of evolution rather than on a distribution of effect sizes that could be compared with Orr's theoretical distribution (Figure 10.9).

A summary of the number of QTL detected in multiple studies of *Drosophila* is shown in Table 10.1. Most of these studies represent analysis of variation within outbred populations of *Drosophila melanogaster*. The exception is shown in the last row of the table which is a QTL analysis of morphological divergence between *D. simulans* and *D. mauritana*. The large, ambitious studies shown in this summary typically detected dozens of loci contributing to variation within and between species.

Kearsey and Farquhar (1998) summarized the results of QTL analyses in 47 plant studies, representing 176 trait-environment combinations. Although the mean number of QTLs reported is about 4 and the mode is 2, the distribution is skewed to the right, with a few studies reporting more than 12. The authors point out that the results are undoubtedly biased towards loci of large effect and that at the time of the summary it was technically difficult to detect more than 12 QTL. The proportion of phenotypic variation in the analyzed (hyper-segregating) population averaged about 46%, but the proportion varied hugely among studies.

Figure 10.13 *(Continued)*
chromosomes, each 100cM long, four of which are shown in the panels. See Lander and Botstein (1989) for the details of the simulation. The dotted lines at $LOD = 2.4$ shows the required significance level to detect a QTL. The grey bars show one LOD support interval for the position of the QTLS (solid triangle). Outside this region, the odds ratio falls off by a factor of 10. The lines extending beyond the grey bars show two LOD confidence intervals. Maximum likelihood estimates of the phenotypic effect of the QTL are shown to the right of the thin lines.

From Lander and Botstein (1989) with permission.

Table 10.1 Variation in quantitative traits is due to multiple loci. References: 1 = Shrimpton and Robertson (1988); 2 = Gurganus (1998), (Nuzhdin et al. (1999); 3 = Nuzhdin et al. (1997), Vieira et al. (2000), Leips and Mackay (2000); 4 = Weber et al. (1999); 5 = Fry et al. (1998); 6 = Wayne et al. (2001); 7 = Zeng et al. (2000). From Mackay (2001a) with permission.

Trait	Chromosomes	Number of QTLs	References
Sternopleural bristle number	3	17	1
Sternopleural bristle number	1,2,3	22	2
Abdominal bristle number	1,2,3	26	2
Longevity	3	19	3
Wing shape	3	11	4
Competitive fitness	1,2,3	6	5
Reproductive success	1,2,3	2	6
Male genital arch	1,2,3	19	7

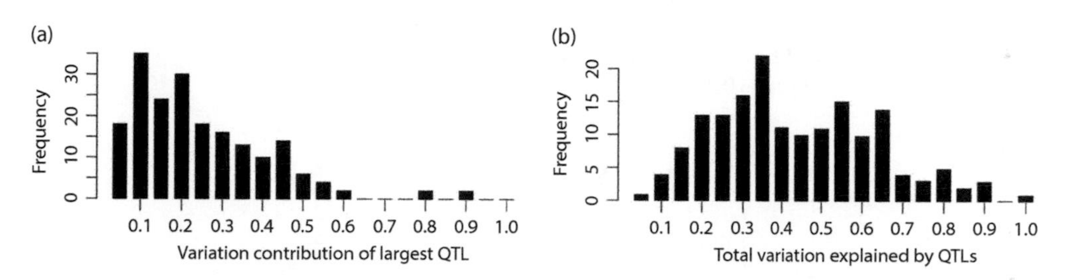

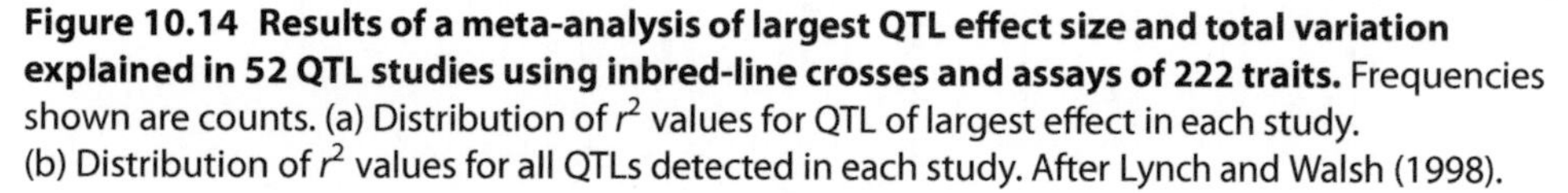

Figure 10.14 Results of a meta-analysis of largest QTL effect size and total variation explained in 52 QTL studies using inbred-line crosses and assays of 222 traits. Frequencies shown are counts. (a) Distribution of r^2 values for QTL of largest effect in each study. (b) Distribution of r^2 values for all QTLs detected in each study. After Lynch and Walsh (1998).

In a meta-analysis of multiple studies, Lynch and Walsh (1998) found that the QTL of largest effect typically explained less that 25% of the phenotypic variance in a trait (Figure 10.14a), and taken together all of the QTLs in each study typically accounted for less than half of the phenotypic variance (Figure 10.14b). Although the per cent of variation explained ranged from nearly 0 to nearly 100% and the number of detected QTLs ranged from 1 to 19, there was no statistical relationship between the per cent of variation explained by QTLs and the number of QTLs that had been detected (Figure 10.15). These results imply that a large number of QTLs of small effect typically underlie backcross and F_2 variation in the crop traits that were analyzed.

QTLs for particular traits are usually scattered throughout the genome rather than clumped in one region. This scattered pattern is conspicuous for QTLs that affect growth in house mice (*Mus musculus*) (Figure 10.16).

In summary, QTL analysis of deliberately selected, phenotypically divergent lines suggest that fixation or near fixation of alleles of small effect at many loci scattered throughout the genome typically underlie divergence. Sometimes dominant alleles are fixed and epistatic interactions are common.

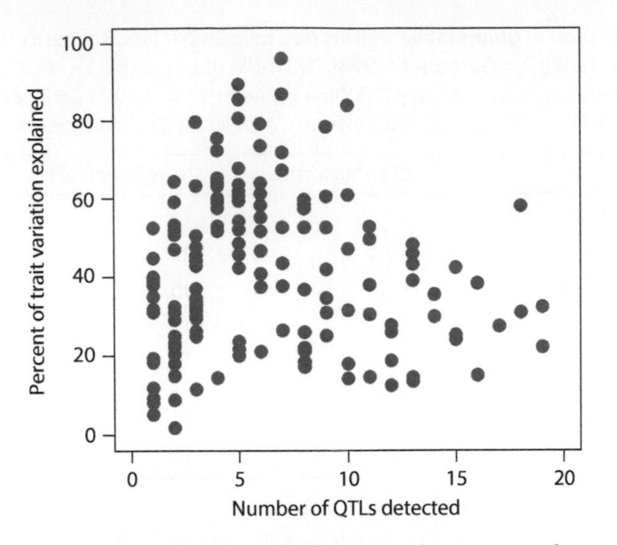

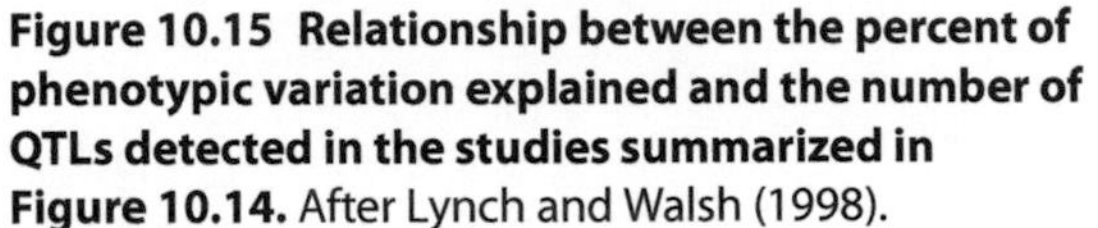

Figure 10.15 Relationship between the percent of phenotypic variation explained and the number of QTLs detected in the studies summarized in Figure 10.14. After Lynch and Walsh (1998).

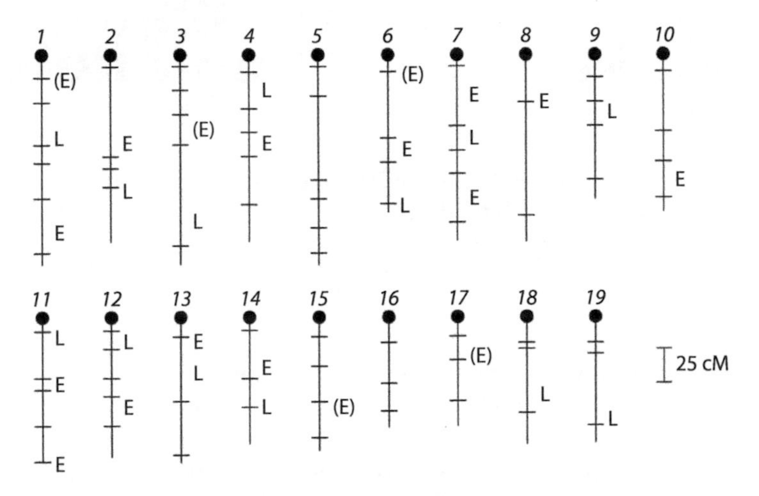

Figure 10.16 Genomic positions of 75 QTLs that affect growth in mice on each of 19 linkage groups. E and L denote effects on early and late growth respectively. Entries in parentheses affect early or late weight but not growth itself.

From Cheverud et al. (1996) with permission.

10.11 Genomic Analysis of Response to Selection in Nature

One aim of genomic analysis is to identify genomic sites that have responded to selection during particular episodes of adaptation. As a first example, consider how genomic analysis has informed our understanding of the reduction in armour that accompanies adaptation to freshwater habitats by sticklebacks (*Gasterosteus aculeatus*). In our conceptual model for this example, an ancestral population seasonally moves between two environments with adaptive peaks that are far apart in trait space. Derived populations take up year-round residence in one of these environments and thereafter evolve in response to just one of the two adaptive peaks.

10.11.1 *Armour loss in fish populations*

Sticklebacks are named for spines that they erect to impede attack and ingestion by predators (Hoogland 1951; Hoogland et al. 1957). These spines are part of a functional complex that governs interactions between lateral plates, the spines, and spine supports that allow the spines to be locked in vertical positions (Figure 10.17). In their erect, locked positions, the spines are buttressed by lateral plates and other body elements. Lateral plates are bony elements that have a one-to-one relationship with underlying muscle segments

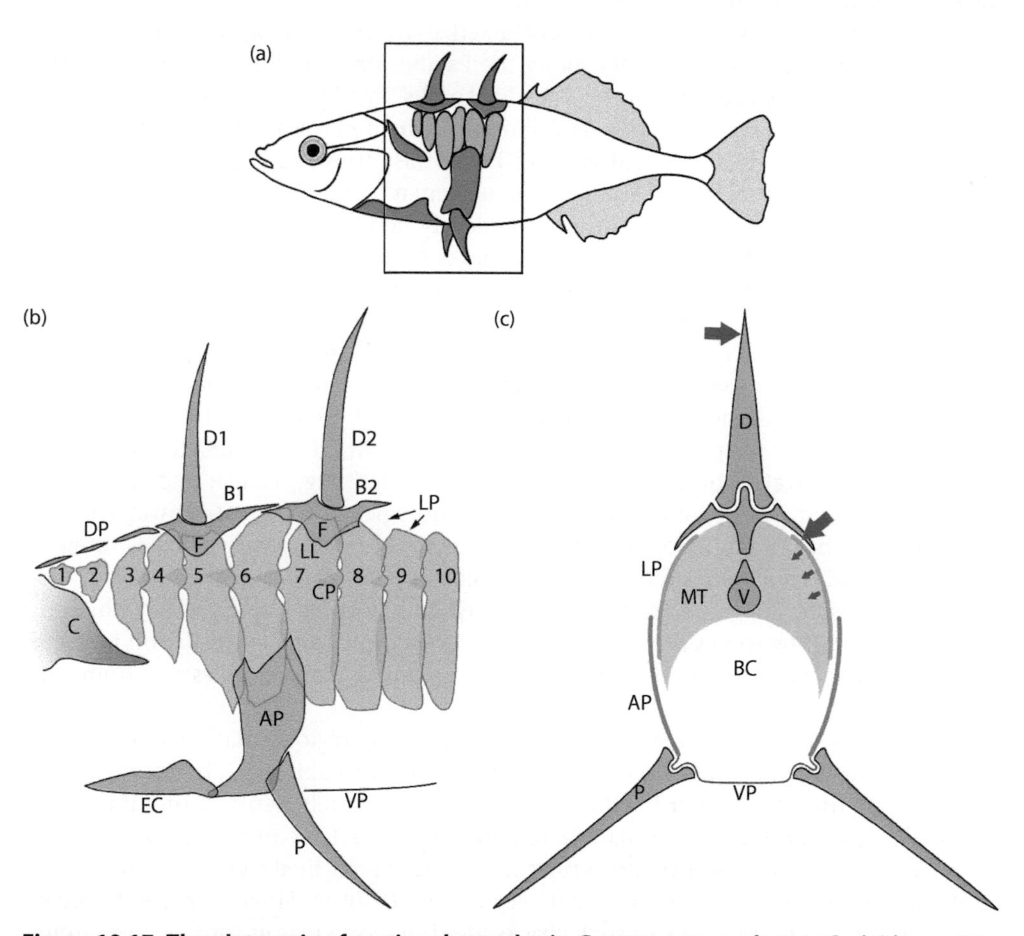

Figure 10.17 The plate-spine functional complex in *Gasterosteus aculeatus*. C, cleithrum; D1, first dorsal spine; D2, second dorsal spine; DP, dorsal plates; B1, first basal plate; B2, second basal plate; Lateral plates, 1–10; AP, ascending process of the pelvis; VP, ventral plates; EC, ectocoracoid; P, pelvic spine; V, vertebral column; BC, body cavity: MT, musculature. (a) Location of the complex. (b) Lateral view of the complex. (c) Cross-section through the complex. Red arrows show transmission of lateral deflection of dorsal spine (large, right-pointing arrow) to lateral plates (large, left and down-pointing arrow) and to musculature (small arrows). After Reimchen (1983, 1994).

(myomers). The spines are attached to underlying bony elements with a peg and socket articulation (Figure 10.17b) that allows the spines to be folded or held erect while resisting lateral movement. These bony elements (basal plates for dorsal spines, ascending process of the pelvis for the pelvic spines) are in turn attached to adjacent lateral plates (P4-P8). In addition lateral plates (P3-P8) articulate to the flanges of adjacent plates. In summary, a total of 21 body elements, plus associated muscles and tendons, participate in this functional complex (Reimchen 1983, 1994). Reimchen (1983) showed that lateral force to a dorsal spine is transmitted through the complex (in the sequence D1 $\rightarrow$ B1 $\rightarrow$ LP) and diffused in the thoracic musculature (Fig 10.17c). A variety of observations and experiments shows that the functional complex protects sticklebacks from fish, bird, and mammalian predation (Hoogland et al. 1957; Hagen and Gilbertson 1973a; Reimchen 1983, 1992, 1994; Reimchen and Nosil 2002), although the spines may be a liability in withstanding attack by odonate nymphs (Reimchen 1980).

A repeated pattern of parallel evolution characterizes the invasion of freshwater habitats by oceanic populations of stickleback along the Pacific coast of North America and the south western Atlantic coast of Europe. All sticklebacks breed in freshwater habitats, but andadromous ('oceanic') populations return to the ocean after freshwater breeding season that lasts a few months in spring and summer. In contrast, 'freshwater' populations reside in freshwater habitats year round. Studies with molecular markers indicate that freshwater populations have been repeatedly and independently derived from oceanic populations along the Pacific coast of North America, from Alaska to California (Bell 2001). Freshwater and oceanic populations also differ in the armour that protects them from predators.

While oceanic populations have well-developed spines and lateral plates, both of these aspects of the functional complex can be much reduced in freshwater populations (Bell 1976, Figure 10.18). In extreme freshwater cases, the functional complex literally falls apart, so that all the elements are small and disarticulated (Reimchen 1983). Although ecology of predation on oceanic populations is unknown, several detailed studies of predation on freshwater sticklebacks have been conducted (e.g., Hagen and Gilbertson 1973b). In addition, comparative studies of morphologically diverse freshwater populations suggest that the functional complex remains robust where predation is intense, but reduced and even nonfunctional where predation is less severe (Reimchen 1983, 1994). These results suggest that a well-developed functional complex with numerous lateral plates and long spines is maintained in oceanic populations by predation pressure and that some freshwater populations reduce their armour when they escape from this pressure.

Crosses and QTL analyses have helped identify genomic regions that have responded to selection during the stickleback invasion of freshwater habitats. Laboratory crosses of sticklebacks from freshwater habitats show that particular elements of the functional complex (e.g., number of lateral plates) are heritable and affected by genes with major effects (Hagen and Gilbertson 1973b). Other studies have identified a gene of major effect on lateral plate number (*Eda*) on LG IV (Colosimo et al. 2005). Finally, crosses between freshwater Alaskan populations of sticklebacks identified one region (on Linkage Group XVII) with major effects on plate number and a second independently segregating region that affects pelvic armour (on LG VII). In the same study, QTL analysis identified three additional regions affecting the functional complex: two affecting lateral plates (on LG XII and XVI) and one affecting pelvic spines (on LG VIII) (Cresko et al. 2004). In other words, change at five or more genomic regions underlies the varieties of reduction in the functional complex that are apparent among freshwater populations.

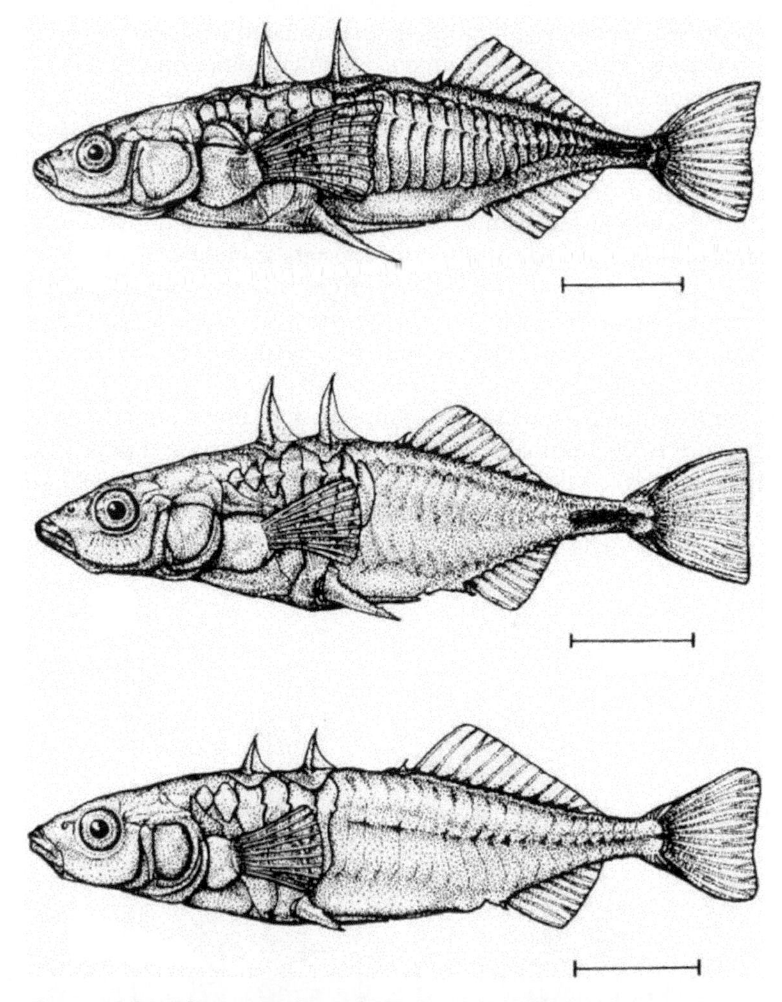

Figure 10.18 Alternative lateral plate and spine morphology in sticklebacks (*Gasterosteus aculeatus*). Reference bars are 1 cm long. (Top) Completely-plated morph, characteristic of oceanic populations. (Middle) Partially plated morph, rare in oceanic and freshwater populations. (Bottom) Low-plated morph with reduced dorsal and pectoral spines, common in freshwater populations.

From Bell (1976) with permission.

A genome-wide scan has helped define the full extent of responses to selection during establishment of freshwater residence by sticklebacks (*Gasterosteus aculeatus*) in Alaska. These responses include osmoregulatory adaptation to freshwater, as well as armour reduction. The Alaskan freshwater populations were probably established within the last 10,000 years and are now geographically isolated (Orti et al. 1994; Cresko 2000; Cresko et al. 2004). Hohenlohe et al. (2010) analyzed 100 individuals from five populations (two oceanic and three freshwater) using Illumina sequencing of *Restriction-site Associated DNA* (RAD tags). The approach yielded haplotype information on several thousand SNP markers. These sites were evenly distributed throughout the genome. Each tag was sequenced five to 10 times in every individual.

Results supported the standing hypothesis that isolated freshwater populations were independently derived from a large panmictic oceanic population. Even though the two sampled oceanic populations were over 1000km apart, they were only slightly differentiated (F_{ST} = 0.0076). In contrast, the pairwise F_{ST}s among the freshwater populations and between freshwater and oceanic populations were substantially larger (0.05–0.15), although modest. These results set the stage for identification of sites that responded to selection during the establishment of freshwater resident populations.

Nine genomic peaks associated with freshwater establishment were identified by testing for statistical elevation of F_{ST} as a function of genomic position (Figure 10.19, middle). In contrast to comparisons between the two oceanic populations (Figure 10.19, top shown), F_{ST} was elevated at just a few genomic sites in the among-freshwater population comparisons (Figure 10.19, bottom).

The details of freshwater population responses to selection in one linkage group (IV) are shown in Figure 10.20. Comparisons among the three freshwater populations indicate differences in responses to selection (e.g., the second peak in Figure 10.20, top). In contrast, the third peak, which is the most substantial peak in the overall oceanic-freshwater comparison (F_{ST} = 0.515), is similar in all three freshwater populations. In other words,

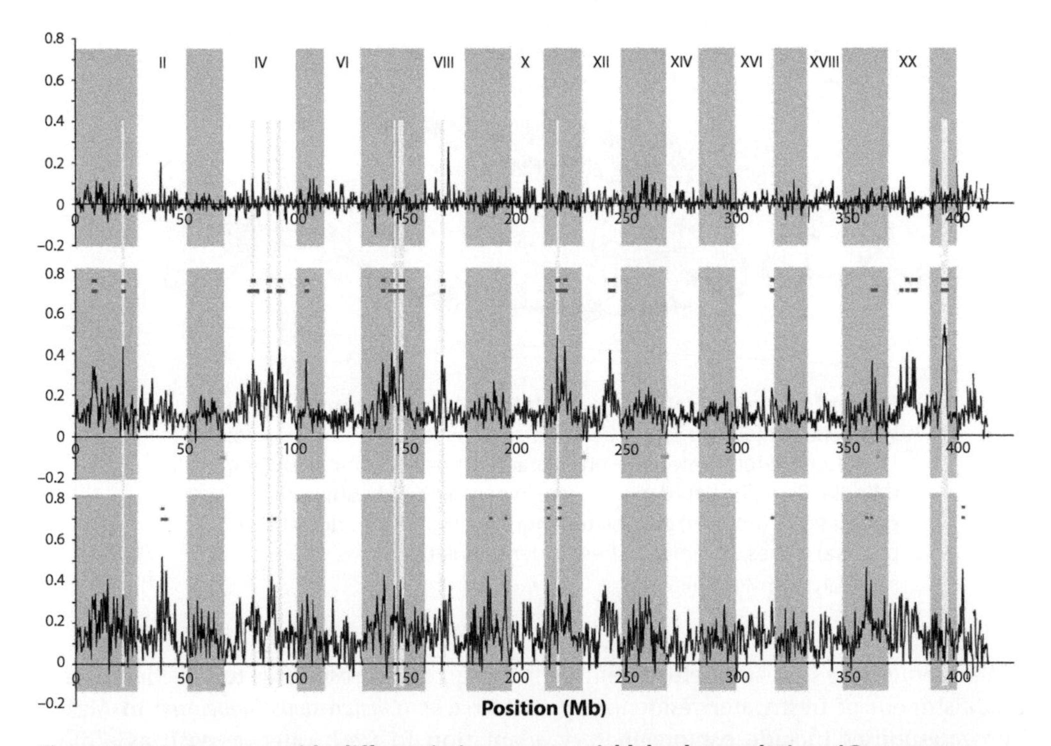

Figure 10.19 Genome-wide differentiation among stickleback populations (*Gasterosteus aculeatus*) assessed with F_{ST}. Linkage group labels are shown at the top of the figure. Coloured dots indicated significantly elevated ($P \leq 10^{-5}$, blue; $P \leq 10^{-7}$, red) and reduced ($P \leq 10^{-5}$, green) values of F_{ST}. Vertical grey shading indicates boundaries of linkage groups (I–XXI) and unassembled shaffolds. Yellow shading indicates nine peaks of substantial population differentiation. (Top) F_{ST} between the two oceanic populations. (Middle) Overall F_{ST} between the oceanic and freshwater populations. (Bottom) F_{ST} among the three freshwater populations.

From Hohenlohe et al. (2010) with permission.

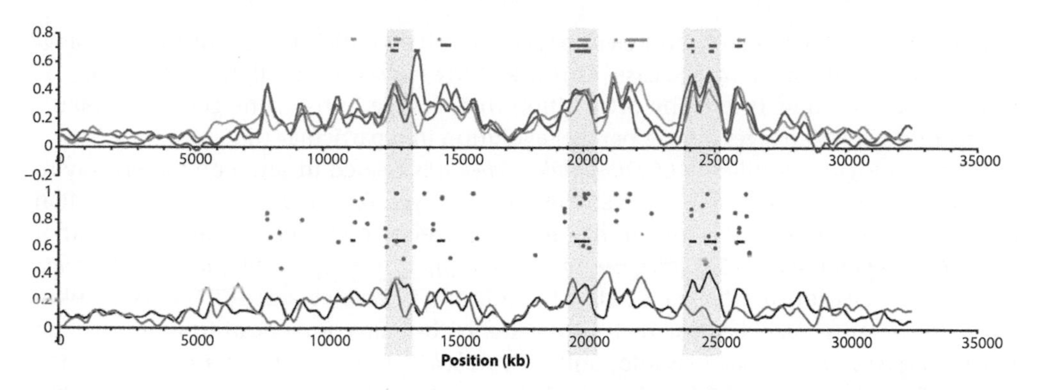

Figure 10.20 Details of differentiation (F_{ST}) among freshwater populations and between oceanic and freshwater populations of stickleback (*Gasterosteus aculeatus*) in Linkage Group IV. (Top) F_{ST} for each of three freshwater populations compared with the oceanic populations. Coloured bars show regions of bootstrap significant ($P \leq 10^{-5}$) for each population (indicated in red, blue, and green). (Bottom) F_{ST} among the three freshwater populations (orange) and F_{ST} for the overall oceanic–freshwater comparison (black). Black bars show corresponding regions of significance ($P \leq 10^{-5}$).

From Hohenlohe et al. (2010) with permission.

some responses to selection during freshwater establishment are unique to individual populations, but other responses are common denominators.

Hohenlohe et al. (2010) also used annotations of the stickleback genome to identify candidate genes related to morphology and osmoregulation that responded to selection during freshwater establishment. Nearly 600 genes (open reading frames) are located in the nine genomic regions identified in Figure 10.19. Of these, 24 (including *Eda*) are related to osteoblast differentiation or bone density and mineralization, and hence may play a role in armour reduction during freshwater establishment.

In summary, while we can be confident that at least nine genomic regions are involved in freshwater establishment, we have only a rough understanding of how the genomic signatures of selection during establishment relate to the details of functional complex that defends stickleback from their predators (Figure 10.18). Morphological and genomic analyses remain disarticulated in this example and in many others. A goal of future research will be to combine them in morphological and genomic analyses.

One important *proviso* was emphasized by Hohenlohe et al. (2010). Their genomic analysis pertains to the loss of morphology rather than the *de novo* assembly of a complex functional complex. The point here is that the genomic basis of morphological loss might be relatively simple, involving for example, blocking of developmental pathways. In contrast, the evolutionary assembly and maintenance of an intricate complex of traits is likely to involve correspondingly complex, step-wise genomic change. In other words, while the loss of stickleback armour appears to involve relatively few genomic regions, the evolutionary origin of armour in its functional context may have involved dozens if not hundred of genes.

10.11.2 *Loss of larval bristles in* Drosophila

In a second example we turn to a particularly detailed analysis of the polygenic changes that underlie the loss of bristles in *Drosophila*. One might anticipate that a single mutation in a regulatory gene would be sufficient to accomplish evolutionary loss of a morphological structure, but this is not the case. This case study is also of special interest because, as in the previous study, it tests the proposition that change in the same genetic elements occurs in separate lineages during parallel evolution in a particular trait.

A morphological peculiarity of *Drosophila sechellia* is related in some unknown way to its peculiar ecology. *D. sechellia* is restricted to the Seychelles Archipelago in the Indian Ocean (Legrand et al. 2011) and unlike its closest relative *D. simulans* and most other related *Drosophila*, its larvae feed on a single host plant, the ripe fruit (noni) of *Morinda citrifolia* (Rubiaceae). *M. citrifolia* is native to India and adjacent countries, but its distribution is now pantropical because its fruit is eaten by humans. Ripe noni fruit are toxic to *D. melanogaster* and *D. simulans* and, unlike *D. sechellia*, *D. melanogaster* are not attracted to noni fruit (Legal et al. 1992). The peculiar morphological feature of *D. sechellia* is that its larvae lack the numerous small bristles (trichomes) that cover the larval cuticle in most *Drosophila*. The loss of trichosome is probably is one of many adaptations associated with dietary specialization on noni, but larval ecology has not been studied in enough detail to say why the loss is adaptive. Trichosomes on the ventral surfaces are used for locomotion (Chanut-Delalande et al. 2006), so loss of those trichosomes hints at a derived mode of larval movement in ripe noni fruit.

Developmental studies identified Shavenbaby (Svb), a transcription factor as playing a major role in the morphological patterning of the *Drosophila* larval cuticle. Furthermore, mutants of *svb* have an essentially naked cuticle (Payre et al. 1999; Chanut-Delalande et al. 2006). Frankel et al 2011 assayed the genetic changes responsible for the larval trichome difference between *D. sechellia* and *D. melanogaster.* Those assays revealed changes in five cis-regulatory regions of the ca. 110 kb *svb* gene that encode a transcription factor which orchestrates trichome morphogenesis (Figure 10.21). Analysis of the changes at one of these ca. 5 kb transcriptional enhancers revealed 14 single-nucleotide changes, each of which had a small effect in reducing trichome number (Frankel et al. 2011). Taken together these substitutions accounted for the *melanogaster-sechellia* difference in trichome number, but the changes were not simply additive and involved substantial instances of epistasis. Rate tests based on DNA sequence analysis confirm our suspicions that the molecular differentiation was driven by positive selection (Frankel et al. 2011),

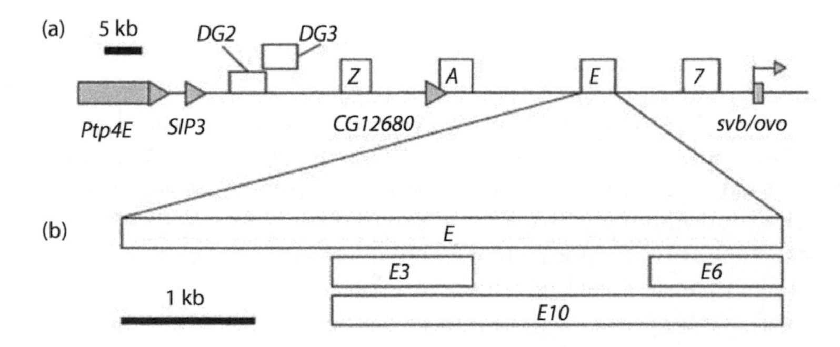

Figure 10.21 Divergence in the pattern of trichosomes between two *Drosophila* species reflects change in the *E* enhancers of the *svb* gene. (a) Locations of the six enhancers of *svb* (open boxes). (b) Details of the *E* enhancer in *D. melanogaster*.

From Frankel et al (2011) with permission.

which at the morphological level presumably coincided with some change in larval ecology. Changes in these same regulatory elements are responsible of the independent loss of trichomes in the *Drosophila virilis* group (Sucena et al. 2003). This case study, by virtue of its detailed dissections, provides a powerful example of the rule that morphological evolution proceeds by many small steps at the genomic level, rather than by a single jump.

10.11.3 *Lessons from morphological loss in sticklebacks and* Drosophila

The overarching lesson from these two examples is that ecology, functional morphology and genomics offer complementary perspectives on adaptation. When one of these perspectives is missing or sketchy, we have an incomplete understanding of the adaptive process. While both case studies are exceptionally complete, both are plagued by gaps in one or more perspective.

In the stickleback case study, a detailed ecological picture included the insights that predators differ in how they interact with the stickleback's spine-plate functional complex (Hoogland et al 1957) and that the degree to which the functional complex is retained in resident freshwater populations is related to the particular combination of predators faced by the stickleback population (Reimchen 1994). Unfortunately, this detailed understanding of how the function complex operates in its ecological setting is not matched by a genomic vision of the genetic architecture of the complex. Genomic analysis hints at a polygenic basis with a broad genomic distribution, but the genotype-to-phenotype map is sketchy. In the *Drosophila* case study, we have a detailed understanding of how one developmental pathway regulates the loss of larval bristles. We know that many small genetic steps underpin the complete loss of bristles in *D. sechellia*. Unfortunately, we don't know how bristle loss relates to the peculiar larval ecology of *D. sechellia*.

In both case studies, a moving optimum model (see Chapter 14) helps to integrate the perspectives of ecology, functional morphology, and genomics. In both cases, an ancestral population confronts an adaptive peak located a considerable distance away from the ancestral peak in trait space. In the stickleback case, the ancestral (oceanic) population encountered two peaks at different times of year: a marine peak during most of the year that favoured heavy armour and a freshwater peak during the relatively short breeding season that favoured light or absent armour. Derived (freshwater resident) populations escaped from the marine peak and confronted the freshwater peak year round during an interval no longer than 10,000 years. Hunt et al. (2008) used the moving optimum model to interpret the loss of armour in a fossil freshwater lineage of sticklebacks (*Gasterosteus doryssus*), related to *G. aculeatus*, that apparently invaded a predator-free environment (Bell et al. 2006). Loss and reduction of dorsal spines and plates occurred rapidly over the first 4000 years after the invasion and then more slowly for the next 4000 years. Loss and reduction in the pelvic parts of the complex showed the same pattern but its onset was delayed by 4000 years (see Section 14.5). In this case, complete remodelling of the functional complex required 12,000 years.

The *D. sechellia* case, also involves shifting between adaptive peaks but the dynamics are different and poorly understood. In this case, the ancestral larval dietary peak involves exploitation of multiple host plants (as in *D. simulans*). The derived population shifted to an adaptive peak that involved exploitation of a single, novel host plant on an isolated oceanic island. We know that this shift occurred sometime during an interval that lasted hundreds of thousands of years (Kliman et al. 2000), but we don't know whether the shift occurred near the beginning or the end of this long interval. Incidentally, it would be a mistake to view *D. sechellia* as an evolutionary dead end because of its geographic isolation. Its larval host plant (nonia fruit) is widely distributed in the tropic and ripe for colonization by *D. sechellia*.

10.12 Conclusions

- Deterministic response to selection on a single rate is a simple function of heritability and directional selection.
- Finite population size adds an element of stochasticity to selection response, which needs to be taken into account in designing selection experiments and interpreting response to selection in nature.
- Selection experiments and the experience of plant and animal breeders reinforce the view that genetic variation is generally available in many—if not all—directions of phenotypic space. It is often possible to move the trait mean by 5 – 20 phenotypic standard deviations in 10 – 100 generations of deliberate selection.
- Observation of natural populations on an ecological timescale sometimes reveals rapid divergence, often in response to anthropomorphic disturbance. This result suggests that populations can rapidly respond to novel, substantial peak movement.
- Geographic variation in trait means is commonly observed within species on regional and continental spatial scales. Analysis of individual cases suggests that this geographic variation mirrors spatial variation in adaptive peaks with limited smoothing due to gene flow.
- Studies of the genomic underpinnings of selection driven-divergence suggest that at least 10 and possibly hundreds of loci at diverse locations in the genome contribute to the selection response.
- On a geological timescale, long-continued directional selection of minute magnitude can account for divergence in the fossil record.
- These observations suggest that selection rather than particular mutations will usually be the limiting ingredient in adaptive radiation.

Literature Cited

Abiola, O. et al. 2003. The nature and identification of quantitative trait loci: a community's view. Nat. Rev. Genet. 4:911–916.

Antonovics, J., and R. B. Primack. 1982. Experimental ecological genetics in *Plantago*. VI. The demography of seedling transplants of *P. lanceolata*. J. Ecol. 70: 55–75.

Arnold, S. J., and C. R. Peterson. 2002. A model for optimal reaction norms: the case of the pregnant garter snake and her temperature-sensitive embryos. Am. Nat. 160:306–316.

Bailey, D. W. 1985. Genes that affect the shape of the murine mandible. J Hered. 76:107–114.

Bailey, D. W. 1986. Genes that affect morphogenesis of the murine mandible. J Hered. 77:17–25.

Barton, N. H., and P. D. Keightley. 2002. Understanding quantitative genetic variation. Nat. Rev. Genet. 3:11–21.

Bell, M. A. 1976. Evolution of phenotypic diversity in *Gasterosteus aculeatus* superspecies on the Pacific coast of North America. Syst. Zool. 25:211–227.

Bell, M. A. 2001. Lateral plate evolution in the threespine stickleback: getting nowhere fast. pp. 445–461 *in* A. P. Hendry and M. T. Kinnison, eds. Microevolution: Rate, Pattern, Process. Springer, Dordrecht.

Bell, M. A., M. P. Travis, and D. M. Blouw. 2006. Inferring natural selection in a fossil threespine stickleback. Paleobiology 32:562–577.

Berven, K. A. 1982. The genetic basis of altitudinal variation in the Wood Frog *Rana sylvatica*. I. An experimental analysis of life history traits. Evolution 36:962–983.

Berven, K. A., D. E. Gill, and S. J. Smith-Gill. 1979. Countergradient selection in the Green Frog, *Rana clamitans*. Evolution 33:609–623.

Bone, E., and A. Farres. 2001. Trends and rates of microevolution in plants. Genetica 112:165–182.

Castle, W. E. 1921. An improved method of estimating the number of genetic factors concerned in cases of blending inheritance. Science(Washington) 54:223–223.

Chanut-Delalande, H., I. Fernandes, F. Roch, F. Payre, and S. Plaza. 2006. Shavenbaby couples patterning to epidermal cell shape control. PLoS Biol. 4:e290.

Cheverud, J. M., E. J. Routman, F. A. M. Duarte, B. van Swinderen, K. Cothran, and C. Perel. 1996. Quantitative trait loci for murine growth. Genetics 142:1305–1319.

Clausen, J., D. D. Keck, and W. H. Hiesey. 1940. Experimental studies on the nature of species. I. Effect of varied environments on western North American plants. Carnegie Instit Wash. Publ. 520:452.

Clausen, J., D. D. Keck, and W. H. Hiesey. 1948. Experimental studies on the nature of species. III. Environmental responses of climatic races of *Achillea*. Carnegie Instit Wash. Publ. 581:129.

Colosimo, P. F., K. E. Hosemann, S. Balabhadra, G. Villarreal, M. Dickson, J. Grimwood, J. Schmutz, R. M. Myers, D. Schluter, and D. M. Kingsley. 2005. Widespread parallel evolution in sticklebacks by repeated fixation of ectodysplasin alleles. Science 307:1928–1933.

Coyne, J. A., and E. Beecham. 1987. Heritability of two morphological characters within and among natural populations of *Drosophila melanogaster*. Genetics 117:727–737.

Cresko, W. A. 2000. The ecology and geography of speciation: A case study using an adaptive radiation of threespine stickleback in Alaska. Ph.D. thesis, Clark University, Worcester, MA.

Cresko, W. A., A. Amores, C. Wilson, J. Murphy, M. Currey, P. Phillips, M. A. Bell, C. B. Kimmel, and J. H. Postlethwait. 2004. Parallel genetic basis for repeated evolution of armor loss in Alaskan threespine stickleback populations. Proc. Natl. Acad. Sci. 101:6050–6055.

Darwin, C. 1868. The Variation of Animals and Plants under Domestication. 2 Volumes. John Murray, London.

Doerge, R. W. 2002. Mapping and analysis of quantitative trait loci in experimental populations. Nat. Rev. Genet. 3:43–52.

East, E. M. 1910. A Mendelian interpretation of variation that is apparently continuous. Am. Nat. 44:65–82.

Endler, J. A. 1977. Geographic Variation, Speciaiton and Clines. Princeton University Press, Princeton.

Falconer, D. S., and T. F. C. Mackay. 1996. Introduction to Quantitative Genetics. 4th ed. Longman Scientific & Technical, J. Wiley & son, Essex.

Fisher, R. A. 1930. The Genetical Theory of Natural Selection. Oxford University Press, Oxford.

Foster, S. A., and J. A. Endler. 1999. Geographic Variation in Behavior: Perspectives on Evolutionary Mechanisms. Oxford University Press, Oxford.

Fox, W. 1948. Effect of temperature on development of scutellation in the garter snake, *Thamnophis elegans atratus*. Coepia 1948:252–262.

Fox, W., C. Gordon, and M. H. Fox. 1961. Morphological effects of low temperature during embryonic development of the garter snake, *Thamnophis elegans*. Zoologica 46:57–71.

Frankel, N., D. F. Erezyilmaz, A. P. McGregor, S. Wang, F. Payre, and D. L. Stern. 2011. Morphological evolution caused by many subtle-effect substitutions in regulatory DNA. Nature 474:598–603.

Fry, J. D., S. V. Nuzhdin, E. G. Pasyukova, and T. F. C. Mackay. 1998. QTL mapping of genotype-environment interaction for fitness in *Drosophila melanogaster*. Genet. Res. 71:133–141.

Goodwin, R. H. 1944. The inheritance of flowering time in a short-day species, *Solidago sempervirens* L. Genetics 503–519.

Goring, H. H., J. D. Terwilliger, and J. Blangero. 2001. Large upward bias in estimation of locus-specific effects from genomewide scans. Am. J. Hum. Genet. 69:1357–1369.

Gould, S. J., and R. F. Johnston. 1972. Geographic variation. Annu. Rev. Ecol. Evol. Syst. 3: 457–498.

Greenwood, P. J., and P. H. Harvey. 1982. The natal and breeding dispersal of birds. Annu. Rev. Ecol. Syst. 13:1–21.

Gurganus, M. C. 1998. Genotype-environment interaction at quantitative trait loci affecting sensory bristle number in *Drosophila melanogaster*. Genetics 149:1883–1898.

Hagen, D. W., and L. G. Gilbertson. 1973a. Selective predation and the intensity of selection acting upon the lateral plates of threespine sticklebacks. Heredity 30:273–287.

Hagen, D. W., and L. G. Gilbertson. 1973b. The genetics of plate morphs in freshwater threespine sticklebacks. Heredity 31:75–84.

Harrison, G. A., and J. J. T. Owen. 1964. Studies on the inheritance of human skin color. Ann Hum. Genet 28:27–37.

Hendry, A. P. 2016. Eco-evolutionary Dynamics. Princeton University Press, Princeton.

Hendry, A. P., and M. T. Kinnison. 1999. Perspective: The pace of modern life: measuring rates of contemporary microevolution. Evolution 53:1637–1653.

Hereford, J., T. F. Hansen, and D. Houle. 2004. Comparing strengths of directional selection: How strong is strong? Evolution 58:2133–2143.

Hill, W. G. 1982. Predictions of response to artificial selection from new mutations. Genet. Res. 40:255–278.

Hohenlohe, P. A., S. Bassham, P. D. Etter, N. Stiffler, E. A. Johnson, and W. A. Cresko. 2010. Population genomics of parallel adaptation in threespine stickleback using sequenced RAD tags. PLoS Genet. 6:e1000862.

Hoogland, R. D. 1951. On the fixing-mechanism in the spines of *Gasterosteus aculeatus* L. K. Ned. Akad Wet. Amst. Proc 54:171–180.

Hoogland, R. D., D. Morris, and N. Tinbergen. 1957. The spines of sticklebacks (*Gasterosteus* and *Pygosteus*) as a means of defence against predators (*Perca* and *Esox*). Behaviour 10:205–236.

Hunt, G., M. A. Bell, and M. P. Travis. 2008. Evolution toward a new adaptive optimum: Phenotypic evolution in a fossil stickleback lineage. Evolution 62:700–710.

James, F. C. 1970. Geographic size variation in birds and its relationship to climate. Ecology 51:365–390.

James, F. C. 1983. Environmental component of morphological differentiation in birds. Science 221:184–186.

Jayne, B. C. 1988. Muscular mechanisms of snake locomotion: An electromyograpic study of lateral undulations of the Florida banded water snake (*Nerodia fasciata*) and the yellow rat snake (*Elaphe obsoleta*). J. Morphol. 197:1159–1181.

Jayne, B. C. 2020. What defines different modes of snake locomotion? Integr. Comp. Biol. 60: 156–170.

Johnson, L. C., M. B. Galliart, J. D. Alsdurf, B. R. Maricle, S. G. Baer, N. M. Bello, D. J. Gibson, and A. B. Smith. 2021. Reciprocal transplant gardens as gold standard to detect local adaptation in grassland species: New opportunities moving into the 21st century. J. Ecol. 1365–2745.13695.

Kearsey, M. J., and A. G. L. Farquhar. 1998. QTL analysis in plants; where are we now? Heredity 137–142.

Kelley, K., S. Arnold, and J. Gladstone. 2003. The effects of substrate and vertebral number on locomotion in the garter snake *Thamnophis elegans*. Funct. Ecol. 11:189–198.

Kimura, M. 1983. The Neutral Theory of Molecular Evolution. Cambridge University Press, Cambridge.

Kingsolver, J. G., and R. B. Huey. 2008. Size, temperature, and fitness: three rules. Evol. Ecol. Res. 10:251–268.

Klauber, L. M. 1941. Four papers on the applications of statistical methods to herpetological problems. Zoological Society of San Diego Bulletin No. 17, San Diego.

Kliman, R. M., P. Andolfatto, J. A. Coyne, F. Depaulis, M. Kreitman, A. J. Berry, J. McCarter, J. Wakeley, and J. Hey. 2000. The population genetics of the origin and divergence of the *Drosophila simulans* complex species. Genetics 156:1913–1931.

Lande, R. 1976. Natural selection and random genetic drift in phenotypic evolution. Evolution 30:314–334.

Lande, R. 1981. The minimum number of genes contributing to quantitative variation between and within populations. Genetics 99:541–553.

Lande, R. 1982. Rapid origin of sexual isolation and character divergence in a cline. Evolution 36:213–223.

Lander, E. S., and D. Botstein. 1989. Mapping mendelian factors underlying quantitative traits using RFLP linkage maps. Genetics 121:185–199.

Legal, L., J. R. David, and J. M. Jallon. 1992. Toxicity and attraction effects produced by *Morinda citrifolia* fruits on the *Drosophila melanogaster* complex of species. Chemoecology 3:125–129.

Legrand, D., D. Vautrin, D. Lachaise, and M.-L. Cariou. 2011. Microsatellite variation suggests a recent fine-scale population structure of *Drosophila sechellia*, a species endemic of the Seychelles archipelago. Genetica 139:909–919.

Leips, J., and T. F. C. Mackay. 2000. Quantitative trait loci for life span in *Drosophila melanogaster*: Interactions with genetic background and larval density. Genetics 155:1773–1788.

Lynch, M., and B. Walsh. 1998. Genetics and Analysis of Quantitative Traits. Sinauer Associates, Inc., Sunderland, Massachusetts.

Mackay, T. F. C. 2001a. Quantitative trait loci in *Drosophila*. Nat. Rev. Genet. 2:11–20.

Mackay, T. F. C. 2001b. The genetic architecture of quantitative traits. Ann Rev Genet. 35:303–339.

Mackay, T. F. C. 2004. The genetic architecture of quantitative traits: Lessons from *Drosophila*. Curr. Opin. Genet. Dev. 14:253–257.

Mackay, T. F. C., R. F. Lyman, and F. Lawrence. 2005. Polygenic mutation in *Drosophila melanogaster*. Genetics 170:1723–1735.

Mayr, E. 1963. Animal Species and Evolution. Belknap Press, Harvard University Press, Cambridge.

Nagylaki, T. 1975. Conditions for the existence of clines. Genetics 80:595–615.

Niewiarowski, P. H., and W. Roosenburg. 1992. Reciprocal transplant reveals sources of variation in growth rates of the lizard *Sceloporus undulatus*. Ecology 74:1992–2002.

Nuzhdin, S. V., C. L. Dilda, and T. F. C. Mackay. 1999. The genetic architecture of selection response: Inferences from fine-scale mapping of bristle number quantitative trait loci in *Drosophila melanogaster*. Genetics 153:1317–1331.

Nuzhdin, S. V., E. G. Pasyukova, C. L. Dilda, Z.-B. Zeng, and T. F. C. Mackay. 1997. Sex-specific quantitative trait loci affecting longevity in *Drosophila melanogaster*. Proc Natl Acad Sci USA 94:9734–9739.

Orr, H. A. 1998. The population genetics of adaptation: The distribution of factors fixed during adaptive evolution. Evolution 52:935.

Orr, H. A. 1999. The evolutionary genetics of adaptation: a simulation study. Genet. Res. 74: 207–214.

Orti, G., M. A. Bell, T. E. Reimchen, and A. Meyer. 1994. Global survey of mitochondrial DNA sequences in the threespine stickleback: Evidence for recent migrations. Evolution 48:608–662.

Palumbi, S. R. 2002. The Evolution Explosion: How Humans Cause Rapid Evolutionary Change. Norton, New York.

Paradis, E., S. R. Baillie, W. J. Sutherland, and R. D. Gregory. 1998. Patterns of natal and breeding dispersal in birds. J. Anim. Ecol. 67:518–536.

Payre, F., A. Vincent, and S. Carreno. 1999. ovo/svb integrates Wingless and DER pathways to control epidermis differentiation. Nature 400:271–275.

Pearson, K. 1903. Mathematical contributions to the theory of evolution. XI. On the influence of natural selection on the variability and correlation of organs. Philos. Trans. R. Soc. Lond. A 200:1–66.

Powers, L. 1942. The nature of the series of environmental variances and the estimation of the genetic variances and geometric means in crosses involving species of *Lycopersicon*. Genetics 27:561–575.

Reimchen, T. E. 1980. Spine deficiency and polymorphism in a population of *Gasterosteus aculeatus*: An adaptation to predators? Can J Zool 58:1232–1244.

Reimchen, T. E. 1983. Structural relationships between spines and lateral plates in threespine stickleback (*Gasterosteus aculeatus*). Evolution 37:931–946.

Reimchen, T. E. 1992. Reimchen, T. E., 1992. Injuries on stickleback from attacks by a toothed predator (*Oncorhynchus*) and implications for the evolution of lateral plates. Evolution 46: 1224–1230.

Reimchen, T. E. 1994. Predators and morphological evolution in threespine stickleback. pp. 240–276 *in* M. A. Bell and S. A. Foster, The Evolutionary Biology of the Threespine Stickleback. Oxford University Press, Oxford.

Reimchen, T. E., and P. Nosil. 2002. Temporal variation in divergent selection on spine number in threespine stickleback. Evolution 56:2472–2483.

Reznick, D. N., and C. K. Ghalambor. 2001. The population ecology of contemporary adaptations: What empirical studies reveal about the conditions that promote adaptive evolution. pp. 183–198 *in* A. P. Hendry and M. T. Kinnison, eds. Microevolution: Rate, Pattern, Process. Springer, Dordrecht, Netherlands.

Rundle, H. D., and M. C. Whitlock. 2001. A genetic interpretation of ecologically dependent isolation. Evolution 55:198–201.

Sax, K. 1923. The association of size differences with seed-coat pattern and pigmentation in *Phaseolus vulgaris*. Genetics 8:552.

Schluter, D. 2000. The Ecology of Adaptive Radiation. Oxford University Press, Oxford.

Shrimpton, A. E., and A. Robertson. 1988. The isolation of polygenic factors controlling bristle score in *Drosophila melanogaster*. II. Distribution of third chromosome bristle effects within chromosome sections. Genetics 118:445–459.

Slatkin, M. 1978. Spatial patterns in the distributions of polygenic characters. J. Theor. Biol. 70:213–228.

Sprague, G. F., and B. Brimhall. 1949. Quantitative inheritance of oil in the corn kernel. Agron. J. 30–33.

Stebbins, G. L. Jr. 1950. Variation and Evolution in Plants. Columbia University Press, New York.

Sucena, E., I. Delon, I. Jones, F. Payre, and D. L. Stern. 2003. Regulatory evolution of shavenbaby/ovo underlies multiple cases of morphological parallelism. Nature 424:935–938.

Templeton, A. R. 1977. Analysis of head shape differences between two interfertile species of Hawaiian *Drosophila*. Evolution 13:330–341.

Thompson, J. N. 1998. Rapid evolution as an ecological process. Trends Ecol. Evol. 13:329–332.

Turesson, G. 1922. The genotypical response of the plant species to the habitat. Hereditas 3: 211–350.

Val, F. C. 1977. Genetic analysis of the morphological differences between two interfertile species of Hawaiian *Drosophila*. Evolution 31:611–629.

Vieira, C., E. G. Pasyukova, Z.-B. Zeng, J. B. Hackett, R. F. Lyman, and T. F. C. Mackay. 2000. Genotype-environment interaction for quantitative trait loci affecting life span in *Drosophila melanogaster*. Genetics 154:213–227.

Walsh, B. M. and M. Lynch. 2018. Evolution and Selection of Quantitative Traits. Oxford University Press, New York.

Wayne, M. L., J. B. Hackett, C. L. Dilda, S. V. Nuzhdin, E. G. Pasyukova, and T. F. C. Mackay. 2001. Quantitative trait locus mapping of fitness-related traits in *Drosophila melanogaster*. Genet. Res. 77:107–116.

Weber, K., R. Eisman, L. Morey, A. Patty, J. Sparks, M. Tausek, and Z.-B. Zeng. 1999. An analysis of polygenes affecting wing shape on chromosome 3 in *Drosophila melanogaster*. Genetics 153: 773–786.

Weber, K. E. 1990. Selection on wing allometry in *Drosophila melanogaster*. Genetics 126:975–989.

Weber, K. E. 1992. How small are the smallest selectable domains of form? Genetics 130:345–353.

Wilkins, H. 1971. Genetic interpretation of regressive evolutionary processes: Studies on hybrid eyes of two *Astyanax* cave populations (Characidae, Pices). Evolution 25:530–544.

Wright, S. 1968. Evolution and the Genetics of Populations: a Treatise. University of Chicago Press, Chicago.

Wright, S. 1977. Evolution and the Genetics of Populations. Vol 3. Experimental Results and Evolutionary Deductions. University of Chicago Press, Chicago.

Zeng, Z.-B., J. Liu, L. F. Stam, C.-H. Kao, J. M. Mercer, and C. C. Laurie. 2000. Genetic architecture of a morphological shape difference between two *Drosophila* species. Genetics 154:299–310.

CHAPTER 11

Response of Multiple Traits to Selection

Overview—In the short term, the response of multiple traits to directional selection is a function of the G-matrix and the vector of selection gradients, $\boldsymbol{\beta}$. Directional selection on a particular trait is expected to change the mean of that trait, but selection may also induce responses in genetically correlated traits. This phenomenon is familiar to plant and animal breeders, who counteract the undesirable correlated responses to selection by selecting on a special linear combination of traits. In the common case in which natural selection alone acts on the multivariate phenotype, the initial response may be biased towards directions with abundant genetic variation. On longer timescales, evolution may be temporarily constrained for certain kinds of traits, especially those that represent trait combinations with low mutation rates and little standing genetic variation. For most traits, however, evolution in the long term is not constrained by genetic variation. Calculations of minimum selective mortality indicate that only a small amount of directional selection each generation is required to account for even appreciable trait change that is observed in the fossil record.

11.1 Multivariate Response to Directional Selection

The response of multiple traits to selection is a function of the G-matrix and the vector of selection gradients, $\boldsymbol{\beta}$,

$$\Delta\bar{\boldsymbol{z}} = \boldsymbol{G}\boldsymbol{P}^{-1}\boldsymbol{s} = \boldsymbol{G}\boldsymbol{\beta}\,. \tag{11.1}$$

This multivariate expression for response to selection from (Lande 1979), cast in terms of the directional selection gradient ($\boldsymbol{\beta} = \boldsymbol{P}^{-1}\boldsymbol{s}$), is often and erroneously called the breeder's equation. The breeder's equation, however, refers to the regression characterization of response to selection illustrated in Figure 10.1, with no specified connections to $\boldsymbol{\beta}$ and the adaptive landscape (Lynch and Walsh 1998). Focusing on the two-trait case, we see that the response to selection of each trait is composed of two terms,

$$\begin{bmatrix} \Delta\bar{z}_1 \\ \Delta\bar{z}_2 \end{bmatrix} = \begin{bmatrix} G_{11} & G_{12} \\ G_{12} & G_{22} \end{bmatrix} \begin{bmatrix} \beta_1 \\ \beta_2 \end{bmatrix} = \begin{bmatrix} G_{11}\beta_1 + G_{12}\beta_2 \\ G_{12}\beta_1 + G_{22}\beta_2 \end{bmatrix}, \tag{11.2}$$

a *direct response* to selection on the trait in question (e.g., $G_{11}\beta_1$) and a *correlated response* due to selection on the other trait ($G_{12}\beta_2$). The direct response is what we might expect

Evolutionary Quantitative Genetics. Stevan J. Arnold, Oxford University Press. © Stevan Arnold (2023).
DOI: 10.1093/oso/9780192859389.003.0012

Figure 11.1 A correlated response to selection when the genetic correlation is negative. In these hypothetical data, truncation selection acts only on trait 1. The sample of actual parents, after selection, is shown in black, other conventions as in Figure 10.1. Even though selection favours higher values of trait 1, trait 2 evolves in the opposite direction because of a negative genetic correlation. Recall that the regression slope is a function of the genetic covariance between the two traits.

from the univariate case, $G_{ii}\beta_i$. The correlated response, on the other hand, is mediated through the genetic covariance between the two traits, G_{12}. As a consequence multivariate selection can produce surprises in the response to selection. Consider the case in which β_1 is positive. The direct response is then necessarily positive if $G_{11} > 0$; it can't be negative. The correlated response could be positive, reinforcing the direct response. Or the correlated response could be negative, if either G_{12} or β_2 are negative (but not both), contradicting the direct response (Figure 11.1). In extreme cases, the correlated response could overwhelm the direct response, so that the trait evolves in the opposite direction to its selection gradient. The possibility of non-obvious response is especially obvious when we consider the response of one trait, z_1, to selection on an entire set of m traits,

$$\Delta\bar{z}_1 = G_{11}\beta_1 + G_{12}\beta_2 + \cdots + G_{1m}\beta_m = G_{11}\beta_1 + \sum\nolimits_{i=2}^{m} G_{1i}\beta_i. \quad (11.3)$$

Because the summation term represents the entire set of correlated responses of z_1 to selection on the other $m - 1$ traits, it might easily amplify or overwhelm the direct response, depending on the signs and magnitudes of the elements in both the first row of the G-matrix and in $\boldsymbol{\beta}$.

The game of pool provides a useful analogy for developing intuition about how two correlated traits respond to selection. The response of the ball to the cue stick represents $\Delta\bar{\mathbf{z}}$ (Figure 11.2). The angle of the cue and its force in hitting the ball represent, respectively, the angle and length of the vector $\boldsymbol{\beta}$. If the two traits are not genetically correlated and have the same genetic variances, the response of the ball to the cue is exactly what we would expect from a conventional pool table. If we hit the ball at a 25 deg angle, it moves away at a 25 deg angle, a hit at 75 deg yields a 75 deg response, etc. (Figure 11.2a). Genetic correlation changes the behaviour of the ball, but in predictable ways (Figure 11.2b and c). Even with genetic correlation, the ball responds in the normal way if our cue is aligned with the major or the minor eigenvectors of the G-matrix (Figure 11.2d). We get an especially strong response when the cue is aligned with the major eigenvector, but a much reduced response when it is aligned with the minor eigenvector. If we hit the ball at any other angle, the ball moves at an angle biased towards the direction of the major eigenvector. In other words, at any other angle, correlated responses to selection come into play so that $\Delta\bar{\mathbf{z}}$ is not proportional to $\boldsymbol{\beta}$.

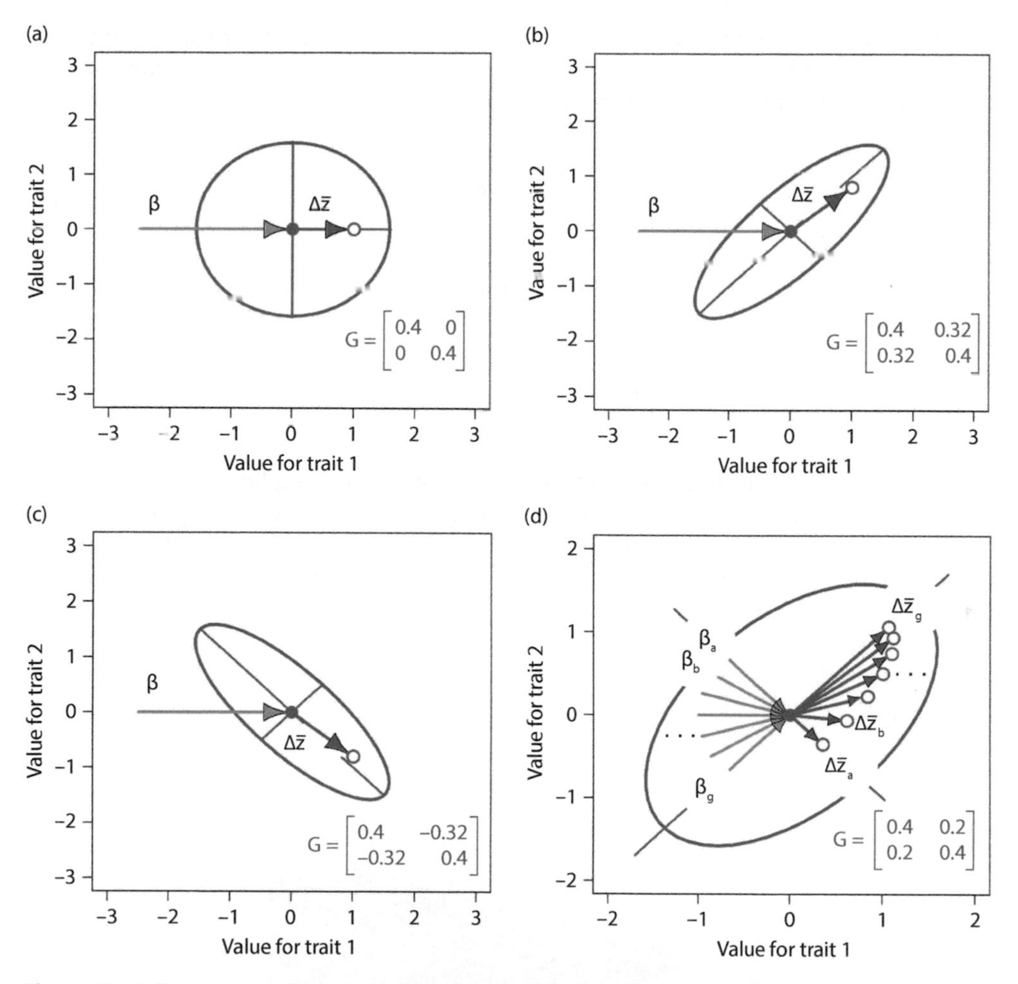

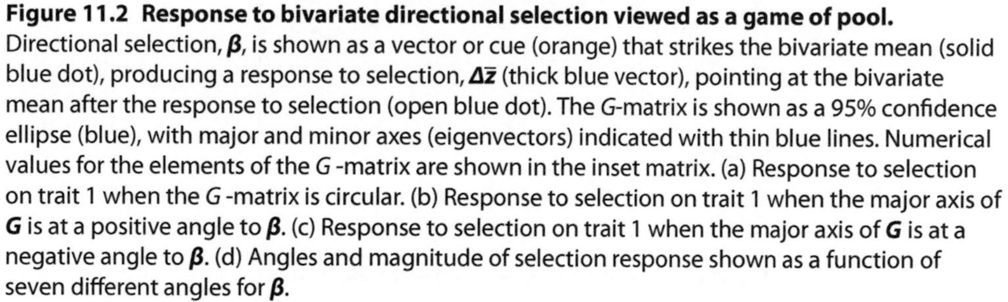

Figure 11.2 Response to bivariate directional selection viewed as a game of pool. Directional selection, $\boldsymbol{\beta}$, is shown as a vector or cue (orange) that strikes the bivariate mean (solid blue dot), producing a response to selection, $\boldsymbol{\Delta\bar{z}}$ (thick blue vector), pointing at the bivariate mean after the response to selection (open blue dot). The *G*-matrix is shown as a 95% confidence ellipse (blue), with major and minor axes (eigenvectors) indicated with thin blue lines. Numerical values for the elements of the *G* -matrix are shown in the inset matrix. (a) Response to selection on trait 1 when the *G* -matrix is circular. (b) Response to selection on trait 1 when the major axis of **G** is at a positive angle to $\boldsymbol{\beta}$. (c) Response to selection on trait 1 when the major axis of **G** is at a negative angle to $\boldsymbol{\beta}$. (d) Angles and magnitude of selection response shown as a function of seven different angles for $\boldsymbol{\beta}$.

11.2 Multivariate Response to Directional Selection in a Set of Replicate Populations of Finite Size

In the single-trait case, we saw that drift produced a normal distribution of replicate means about the deterministic response to selection (Figure 10.2). In the multiple trait case, after t generations with a per-generation vector of selection, $\boldsymbol{\beta}$, and a constant *G*-matrix, $\mathbf{G}$, the total deterministic response is

$$\sum\nolimits_{o}^{t} \boldsymbol{\Delta\bar{z}} = t\mathbf{G}\boldsymbol{\beta}. \tag{11.4}$$

This multivariate response is composed of direct and correlated responses to selection, as we saw in the two-trait case, over a single generation (11.2). If selection acts only on trait 1, we obtain a direct response in that trait and a correlated response in the second trait

$$\begin{pmatrix} \Delta\bar{z}_1 \\ \Delta\bar{z}_2 \end{pmatrix} = \begin{pmatrix} G_{11}\beta_1 \\ G_{12}\beta_1 \end{pmatrix}. \tag{11.5}$$

Substituting into (11.4), we obtain the total direct and correlated responses to selection over t generations

$$t\begin{pmatrix} \Delta\bar{z}_1 \\ \Delta\bar{z}_2 \end{pmatrix} = t\begin{pmatrix} G_{11}\beta_1 \\ G_{12}\beta_1 \end{pmatrix}. \tag{11.6}$$

If we plot just the correlated response to selection in a set of replicate populations, we obtain Figure 11.3, the companion to Figure 10.2 (which shows just the direct response to selection). In this simulation the genetic covariance between the selected trait and the trait plotted here is $G_{12} = 0.032$; the genetic correlation is 0.08. With such a modest genetic correlation, it would be difficult to distinguish the correlated response from pure drift, in which case the white line in Figure 11.3 would have no slope. At any given generation, t, the expected, deterministic value of the total correlated response is $tG_{12}\beta_1$ and the variance among replicate populations is tG_{12}/N_e.

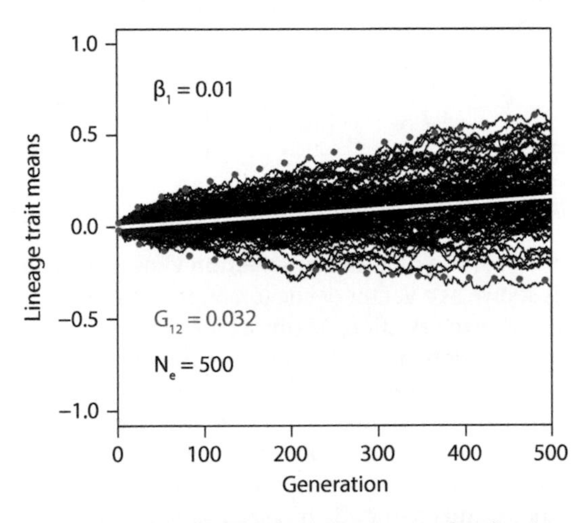

Figure 11.3 Correlated response to selection in a set of replicate populations of finite size. In the simulation depicted here, 100 lineages of moderate size (N_e = 500) respond to a very weak directional selection (β_1 = 0.01) on another weakly correlated trait (G_{12} = 0.032). The trajectories of the correlated lineage responses (shown in black) reflect both response to selection and drift. The 99% confidence limits for the overall lineage response are shown with dotted blue lines. The deterministic correlated response to selection is shown as a white line.

In general, at any given generation the distribution of replicate correlated responses is multivariate normal. As time unfolds, the multivariate responses steadily increase about this deterministic response, i.e., the stochastic cloud of responses by replicate populations forms a perpetually expanding cone. Against the simplicity of these theoretical expectations, we now consider the empirical experience of multivariate response to selection, first in the laboratory and then in agricultural and field settings.

11.3 Genetic Slippage in Response to Selection for Multiple Objectives

The remarkable multivariate response that can be achieved by deliberate selection was familiar to (Darwin 1868), who especially remarked on diversification of domestic breeds of pigeons, which involved osteology and behaviour, as well as plumage. Darwin's own experience with pigeon breeding, as well as interviews with other animal and plant breeders, convinced him that modification of domestic stocks had been accomplished by deliberate selection. But despite the triumphs of domestication and improvement from Darwin's time to the present, plant and animal breeders have always contended with correlated responses that frustrate progress towards goals in stock improvement. Figure 11.4 provides a particular vivid example because the correlated loss in male body weight (Figure 11.4b) was completely unexpected.

Dickerson's (1955) memorable phrase about slippage (the heading for this section) refers to the stalling of response to selection that can occur when multivariate selection is in conflict with one or more genetic covariance. To account for this and other troublesome aspects of correlated response in a program of deliberate selection, we can select

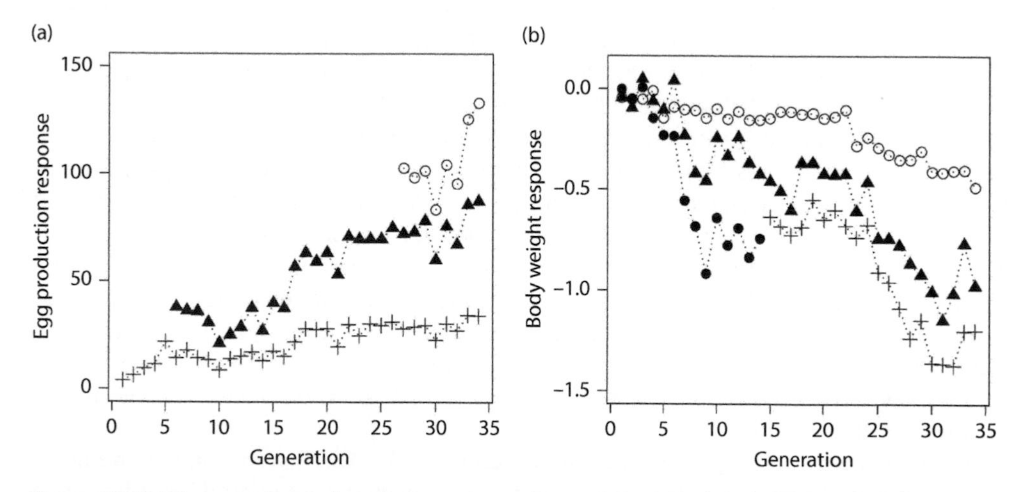

Figure 11.4 Direct and correlated responses to long-term selection for increased egg production in turkeys. (a) Direct response to selection for increased egg production (eggs/hen, expressed as deviations from a control line). Egg production is shown at three hen ages: 84 days (crosses), 180 days (solid triangles), and 260 days (open circles). (b) Unexpected correlated response of male body weight to selection for increased egg production (kg, expressed as deviations from a control line). Weights are shown at four male ages: 8 weeks (open circles), 18 weeks (solid triangles), 24 weeks (solid circles), and 20 weeks (crosses). After Nestor et al. (1996).

on a *desired gains index*, I, that gives the desired vector of responses while accounting for genetic covariances (Dickerson et al. 1954; Walsh and Lynch 2018, Chapter 37),

$$I = \boldsymbol{\beta}^{\mathrm{T}}\mathbf{z} = \Delta\bar{\mathbf{z}}^{\mathrm{T}}\boldsymbol{G}^{-1}\mathbf{z}. \tag{11.7}$$

(Lande 1979). In this expression, we see that if we wish to produce a response to selection in a particular desired direction, $\Delta\bar{\mathbf{z}}$, and we know the inverse of $\boldsymbol{G}$, we can solve for a set of selection gradients, $\boldsymbol{\beta}$, and associated index, I, that will produce that response.

11.4 Correlated Response to Deliberate Selection in the Laboratory

In the 1940s and 1950s, practitioners of quantitative genetics took an experimental approach to multivariate response to selection. The motivation for these experiments was the practical one of determining whether theoretical expectations, such as the ones we just reviewed, would be a trustworthy guide for plant and animal improvement by deliberate selection. In Chapter 10, we reviewed the phenomena of selection asymmetries and plateaus that characterize the responses of single traits to selection. Here we focus on correlated responses to selection, the new phenomenon introduced by multivariate selection.

In these multivariate experiments, selection was imposed on a single trait, but responses were monitored in multiple traits, as well as in the selected trait. A good introduction to the revelations of the 1950s can be gained by focusing on the results of Clayton et al. (1956, 1957).

Clayton and his colleagues (1956, 1957) selected on abdominal bristle count in *D. melanogaster* and monitored the response in that count, as well as the correlated response in sternopleural bristle count (Figure 11.5). Over the course of seven generations, abdominal bristles showed a dramatic direct response to upward selection that averaged 14.4 genetic standard deviations at generation 7 (not shown). Likewise, downward selection produced a direct response of 10.6 genetic standard deviations (not shown). The correlated responses in sternopleural bristle counts showed a striking asymmetry (Figure 11.5a). The correlated response to upward selection was modest, averaging 1.9 genetic standard deviations (although one line failed to respond), but downward selection produced no correlated response, averaging less than 0.2 genetic standard deviations. The authors conclude that the modest correlated response and other evidence are consistent with a very low genetic correlation between the two bristle counts (≤ 0.1). They attribute much of the variation among selected lines to drift arising especially from the family (full-sib) selection scheme but are at a loss to explain absence of a correlated response in the downward selected lines.

The main conclusions apparent to Clayton et al. (1956, 1957) were that: (1) replicate lines are essential to evaluate the effects of drift, (2) selected lines should be of large size and inbreeding should be avoided, and (3) 50 generations of selection may be required to detect correlated responses to selection when genetic correlation is weak (e.g., $r_g = 0.20$). This advice and conclusions are mirrored in subsequent quantitative studies of correlated response to selection (e.g., Sheldon and Milton 1972; Travisano et al. 1995) and amplified by (Hill 1980, 1984).

Allometry and other pairwise relationships among traits are often treated as constraints on evolutionary change (Gould 1966), but that perspective flies in the face of experimental

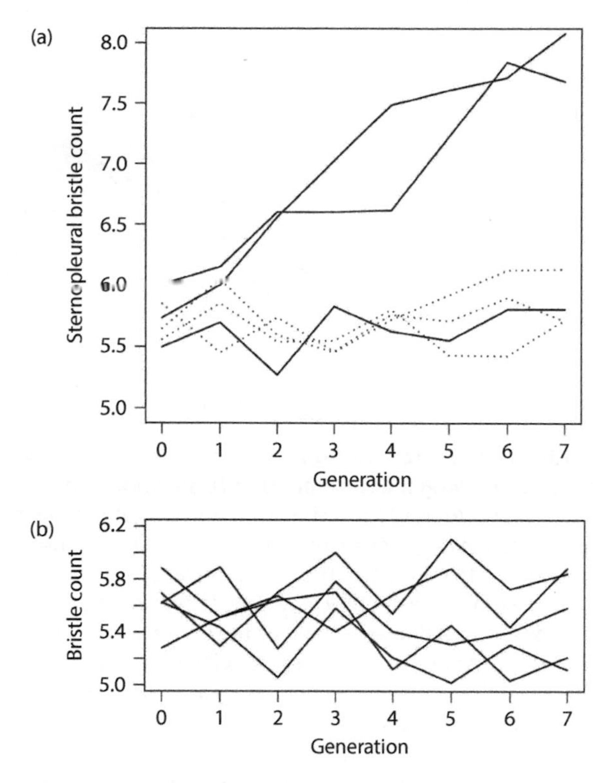

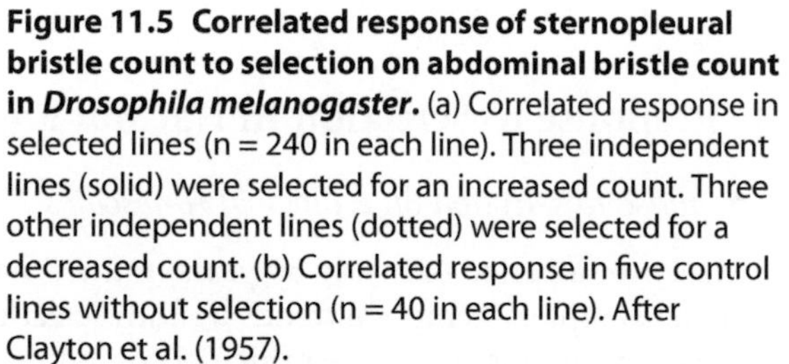

Figure 11.5 Correlated response of sternopleural bristle count to selection on abdominal bristle count in *Drosophila melanogaster*. (a) Correlated response in selected lines (n = 240 in each line). Three independent lines (solid) were selected for an increased count. Three other independent lines (dotted) were selected for a decreased count. (b) Correlated response in five control lines without selection (n = 40 in each line). After Clayton et al. (1957).

results. Using selection experiments, we can ask whether allometry itself responds to selection. In a pioneering example of this approach, (Weber 1990) selected on bivariate relationships between wing dimensions in *Drosophila melanogaster* and succeeded in changing the relationships in predictable ways (one relationship is shown in Figure 11.6). Weber documented a baseline that accurately described bivariate relationships in environmental treatments, wild-caught flies from seven localities, and their laboratory-reared offspring. Phenotypes from all these sources were tightly clustered about each baseline. In each of five bivariate relationships, Weber performed truncation selection (most extreme 20% in samples of 100 males and 100 females) in two directions away from the baseline. After 16 generations of selection, the mean distance (n = 7) between the offsets of the two selected lines was 14.7 phenotypic standard deviations. These impressive responses to selection, as well as measurements of drift among long-maintained isofemale lines, indicate that considerable additive genetic variation in bivariate relationships exists in natural population. Allometry is indeed malleable under the force of directional selection.

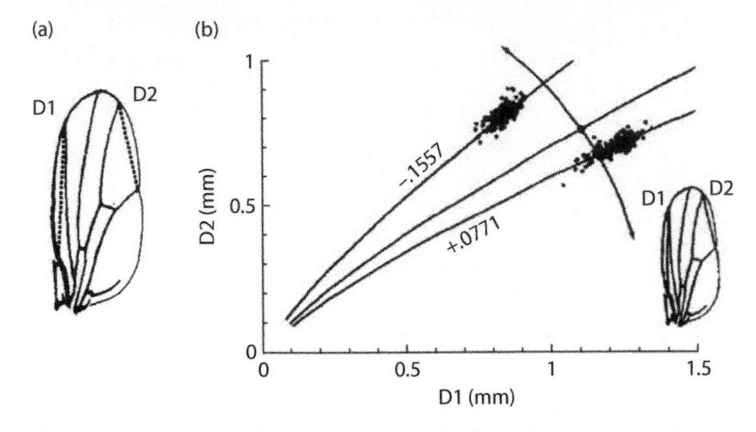

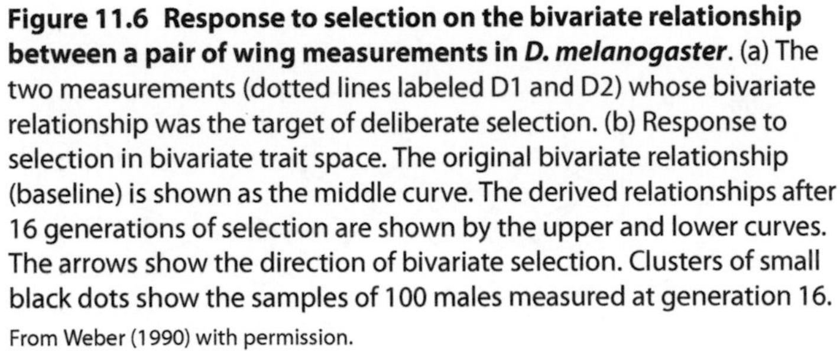

Figure 11.6 Response to selection on the bivariate relationship between a pair of wing measurements in *D. melanogaster*. (a) The two measurements (dotted lines labeled D1 and D2) whose bivariate relationship was the target of deliberate selection. (b) Response to selection in bivariate trait space. The original bivariate relationship (baseline) is shown as the middle curve. The derived relationships after 16 generations of selection are shown by the upper and lower curves. The arrows show the direction of bivariate selection. Clusters of small black dots show the samples of 100 males measured at generation 16.
From Weber (1990) with permission.

11.5 Multivariate Response to Selection in Natural Populations

11.5.1 *Multivariate response to selection on ecological timescales*

Studies of *Geospiza* finches over a period of nearly 30 years in the Galápagos Islands provide a particularly detailed example of temporal trends in both selection and response to selection in multiple traits. During this period the principal components of body size and beak measurements fluctuated in a complex pattern (Figure 11.7). All of the traits composing these principal components have been shown to be highly heritable (Grant and Grant 1995; Keller et al. 2001) and under multivariate directional selection (Figure 11.8). Consequently, the major trends in Figure 11.7 probably represent responses to selection, although a contribution from drift cannot be excluded.

The temporal pattern in directional selection differentials (Figure 11.8) reflects in part periodic episodes of drought that affected food supply and hence selection on body size and beak morphology. During some of these drought episodes, both selection gradient analysis on beak measurements (Price et al. 1984) and the selection differential analysis on beak principal components (Figure 10.5) revealed directional selection for larger and more pointed beaks in both species. Ecological observations indicated that handling drought-available *Tribulus* fruits was enhanced by large, pointed beaks. Episodes of directional selection often lasted a few generations and then reversed direction (Gibbs and Grant 1987; Grant and Grant 1995, 2002). The authors embraced a model in which ecological conditions favour different morphology optima in different years (Grant et al. 1976). As the optimum changes, selection fluctuates in both intensity and direction (Gibbs and Grant 1987).

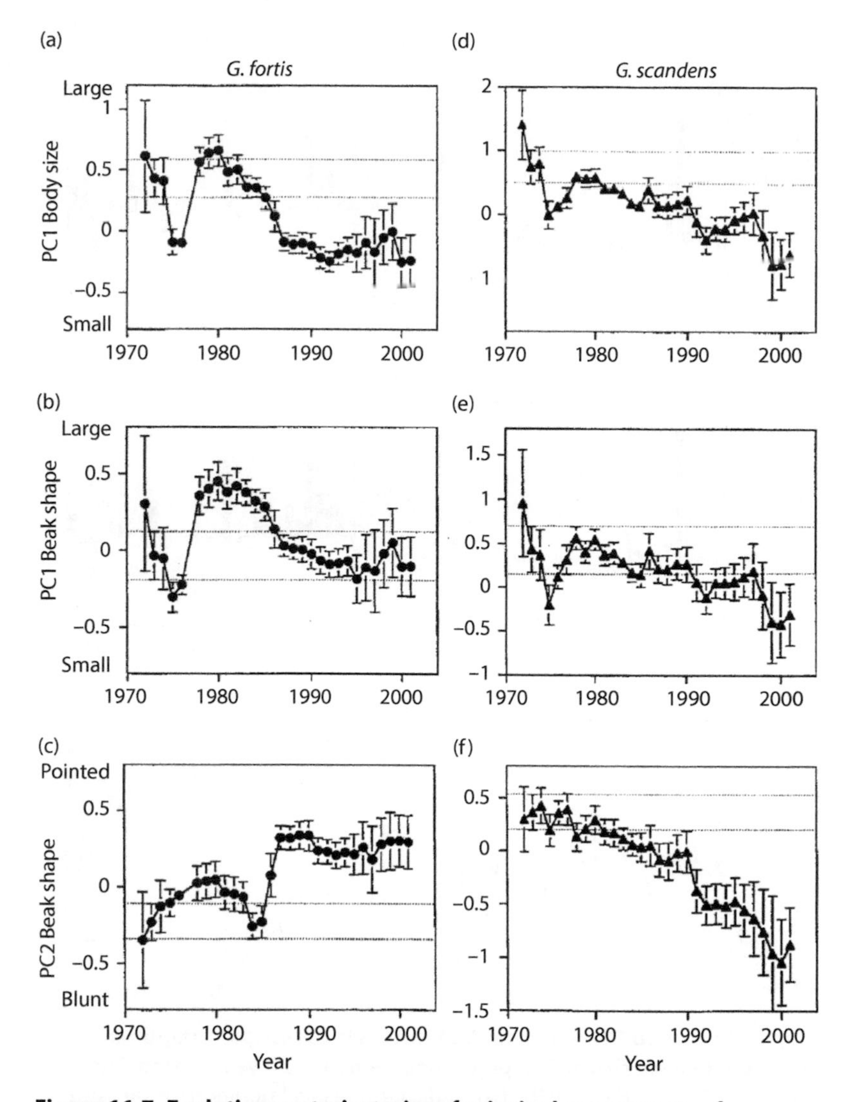

Figure 11.7 Evolutionary trajectories of principal components of morphology in two species of Galápagos finches over a 30 year period. Mean values for principal component scores are shown with solid circles (*G. fortis*) and triangles (*G. scandens*). Confidence intervals for means (95%) are shown as brackets.

From Grant and Grant (2002) with permission.

11.5.2 *Geographic variation in multiple phenotypic traits*

Many of the conceptual points and empirical generalization that we developed in an earlier discussion of geographic variation in single traits (Section 10.5b) also apply in the case of multiple traits. A key observation that emerges in studies of geographic variation in multiple traits is that the geographic pattern can differ from trait to trait (Wilson and Brown 1953). A vivid case in point is provided by a study of geographic variation in the coloration pattern of milk snakes (*Lampropeltis triangulum*).

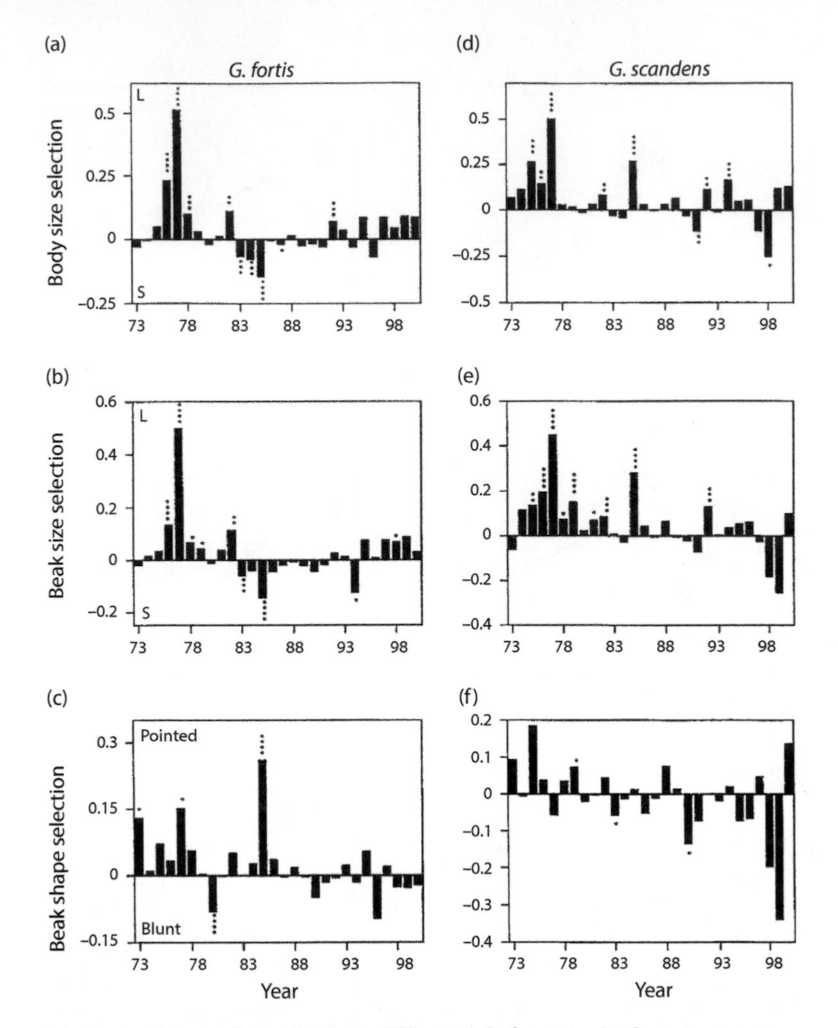

Figure 11.8 Viability selection differentials for principal component scores in two species of Galápagos finches over a 30 year period. The asterisks above and below the bars indicate statistical significance, ranging from * $P < 0.05$ to **** $P < 0.001$.

From Grant and Grant (2002) with permission.

The milk snake shows striking geographic variation in colour pattern over its extensive geographic range, which extends from southern Ontario, Canada to northern Peru. Milk snakes have dark brown blotches on a lighter brown background in the Great Lakes region and New England (Figure 11.9). This coloration pattern is shared with other sympatric eastern North American snakes (e.g., *Pituophis melanoleucus, Pantherophis vulpinus, Lampropeltis callisgaster*) and is generally interpreted as cryptic on the forest floor. Over most of its range, however, milk snakes have an aposematic coloration pattern in which white bands are bounded by black bands and separated by red bands (Figure 11.9). In southeastern USA and from Mexico south to Peru milk snakes are sympatric with venomous coral snakes (*Micrurus* sp.), which are presumed to be the models for the mimetic

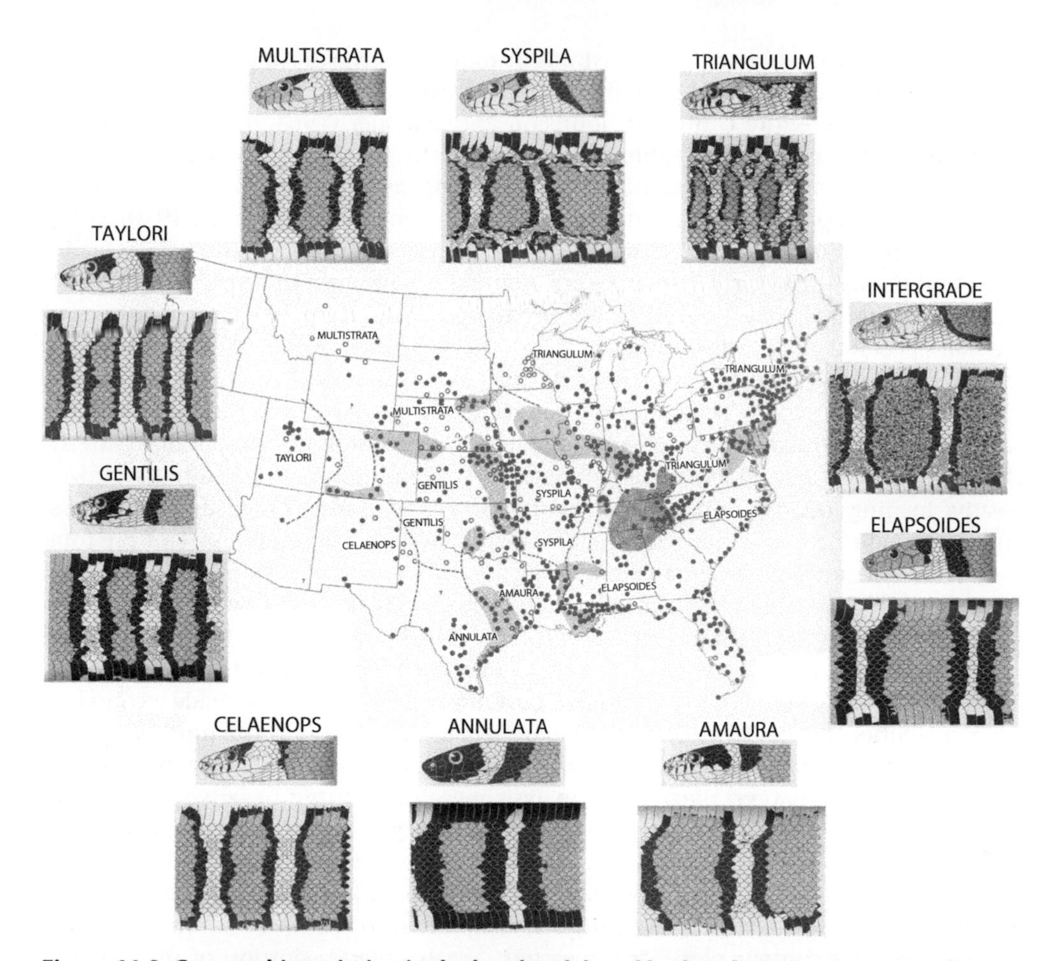

Figure 11.9 Geographic variation in the head and dorsal body colouration patterns of the milk snake (*Lampropeltis triangulum*) in the United States. Dotted lines show the boundaries between the ranges of adjacent subspecies. Areas of intergradation are indicated with stippling. Solid circles indicate locations of specimens examined by the author, open circles indicate locations of literature records or museum specimens. Insets show coloration patterns on the head/neck and the dorsum of the body. The form found on the coastal plain of North Carolina, South Carolina, Georgia, and Alabama and throughout Florida is generally recognized as a separate species (*L. elapsoides*).

From (Williams 1978).

milk snakes (Greene and McDiarmid 1981). Indeed, at many localities in Middle and South America, milk snakes are precise mimics of particular sympatric species of *Micrurus*. The problem for a general application of this coral snake mimicry hypothesis derives from the fact that throughout much of the Midwest and western USA milk snakes are not sympatric with coral snakes. Several explanations have been offered for this discrepancy (Smith 1969). Despite this wrinkle in the coral snake mimicry story, *L. riangulum* certainly shows complex geographic variation that involves multiple aspects of its coloration pattern. A standard approach to the analysis of such multifaceted differentiation is to use multivariate statistics that account for correlations among traits, as in the next example.

The house sparrow (*Passer domesticus*) was introduced in North America from England in the period 1852–1960 and over the next 100 years (50–115 generations) expanded its range over the entire continent. Johnston and Selander (1964, 1971) undertook a comprehensive study of geographic variation in North America to determine whether it mirrored the geographic differentiation of *P. domesticus* in its European homeland, as well as the geographic variation of native North American passerines (James 1970). Using an approach similar to that of James (1970), Johnston & Selander (1971) assembled a collection of 1824 whole sparrow skeletons from nearly 100 localities throughout North America. Using this collection, they measured 16 morphometric traits on the bones of the skull, pectoral girdle, wings, and legs. All 16 traits showed statistically significant geographic variation in both sexes in univariate analyses of variance that did not account for correlations among traits. Separate principal component analyses of males and females based on unstandardized measurements of individual specimens revealed similar patterns of loading on the first principal component (PC 1). All but one of the 16 traits showed positive loadings on PC 1 in both sexes. In other words, PC 1 can be considered an assessment of overall size (Jolicoeur & Mosimann 1960). PC 1 accounted for 54% of the total variance in 16 morphometric traits in females.

Geographic trends in PC 1 of females are plotted in Figure 11.10, which shows contours fitted to principal component scores for individual localities based on standardized locality means for all 16 traits: i.e., the trait mean at each locality was standardized using its within-population phenotypic standard deviation so that each trait made a comparable contribution to the PC 1 score. The largest females are found in the northern plains and the smallest females are found in coastal localities (Figure 11.10). Males showed a similar geographic pattern. Regression analysis revealed a strong negative relationship between female PC 1 and winter temperature (January maximum wet-bulb temperature). In other words, the localities with the coldest winter temperatures had the largest females, as might be expected from heat loss considerations. Thus, geographic variation in PC 1 appears to be, at least partly, an adaptation to spatial variation in climate. Furthermore, the contour plots for the eastern North America portion of the range of *P. domesticus* roughly resemble those for multiple species of passerines studied by James (1970).

To determine whether the plot in Figure 11.10 can reasonably be considered a surface of optimal values for PC1 in light of dispersal between localities, we perform an analysis similar to one we conducted for geographic variation of wing length in the downy woodpecker (Section 10.5b). Using values for dispersal distance, stabilizing selection, and genetic variance (l = 20, ω = 50, and $G \approx 0.4$) with expression (10.5), the expected characteristic length is L = 224km. The distance from coastal California to the central plains spans 10 contour intervals in Figure 11.10 or about $2.3\sqrt{P}$ over a distance of about 2000km, which is about nine times the characteristic length of 224km. In other words, even allowing for a substantial individual dispersal distance each generation (20km) and very weak stabilizing selection on PC 1, populations should be able to track differences in the PC 1 optimum from the centre of the continent to its coasts, and even when localities are more than 224km apart.

Johnston and Selander (1971) found that even though geographic variation in body size was substantial across North America, it was less than in the native European range of *P. domesticus*. The authors attribute this difference to the shorter interval of time available for response to selection in North America.

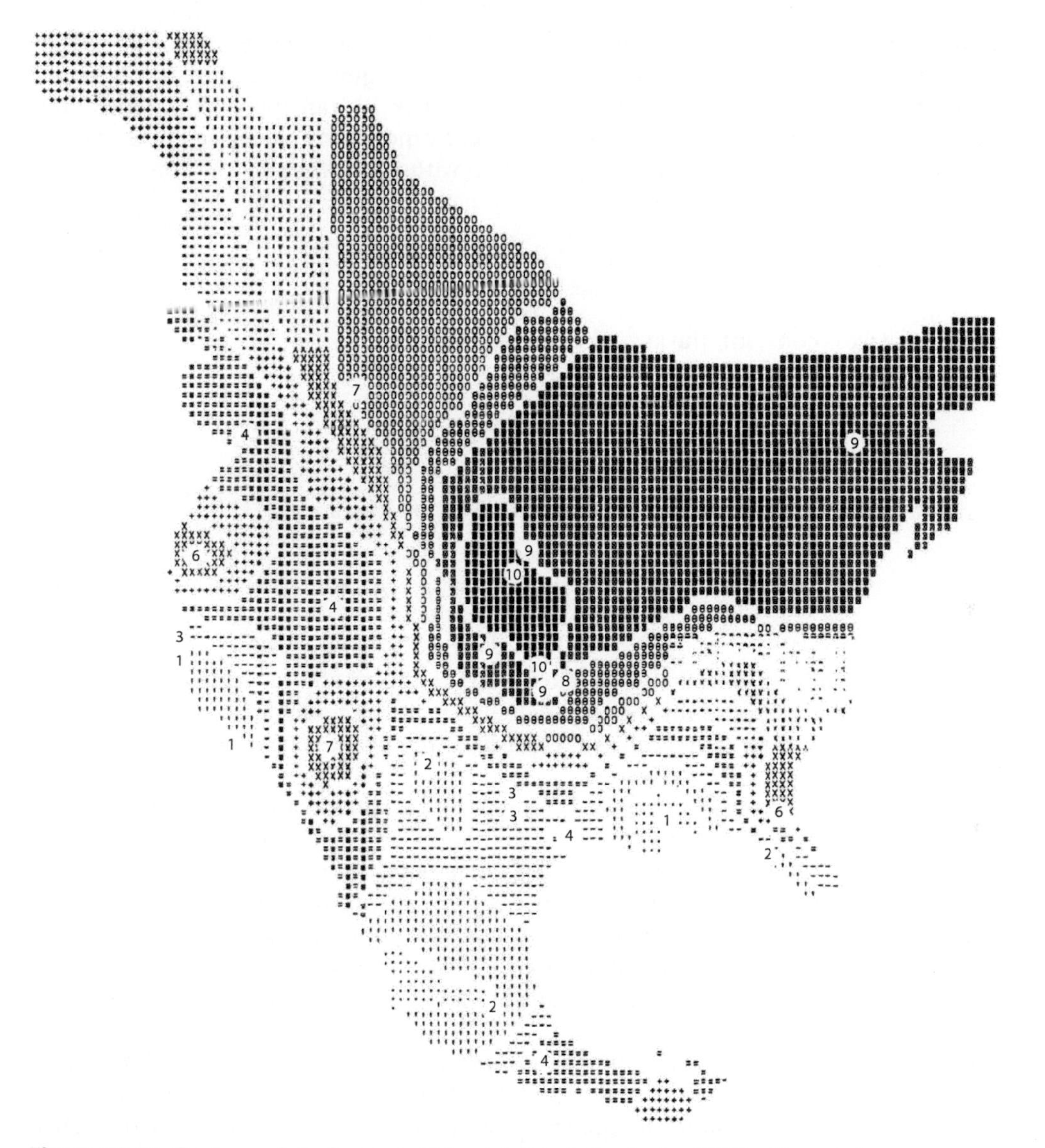

Figure 11.10 Contour plot of geographic variation in body size (PC I) of female house sparrows (*Passer domesticus*). A generalized contour plot of the geographic surface of locality mean values for PC I is shown for females (n=594 individuals from 85 localities). The largest specimens are found in areas with the darkest contours. The contour interval is 0.23 $\sqrt{P}$, where P is the within-population variance in PC I.

From Johnston and Selander (1971) with permission.

11.6 The Net Selection Gradient

Net selection analysis provides a way to characterize multivariate directional selection over an interval of evolutionary time, even when the number of elapsed generations is unknown. If we know how much the trait mean has changed in an evolving lineage, we can reconstruct a selection vector that could account for that change. To do the reconstruction we need to know the G-matrix for the traits in question, and we will need to make some assumptions about the stability of that matrix. We can apply (11.1) generation by generation to predict the total change in $\bar{\mathbf{z}}$ after t generations, denoting the G-matrix, β-vector, and trait mean response each generation with i subscripts,

$$\sum_{i=0}^{t-1} \mathbf{G}_i \boldsymbol{\beta}_i = \sum_{i=0}^{t-1} \Delta \bar{\mathbf{z}}_i = [\bar{\mathbf{z}}_t - \bar{\mathbf{z}}_0] . \tag{11.8}$$

If the G-matrix is constant, the algebra is simplified so that

$$\mathbf{G} \sum_{i=0}^{t-1} \boldsymbol{\beta}_i = [\bar{\mathbf{z}}_t - \bar{\mathbf{z}}_0] . \tag{11.9}$$

We rearrange this expression and define the *net selection gradient*

$$\boldsymbol{net\beta} \equiv \sum_{i=0}^{t-1} \boldsymbol{\beta}_i = \mathbf{G}^{-1} [\bar{\mathbf{z}}_t - \bar{\mathbf{z}}_0] , \tag{11.10}$$

(Lande 1979). This net selection gradient represents the minimum force of directional selection needed to account for the total net change in the multivariate trait mean. We can also think of this last expression as a *genetic linear discriminant function* (Lofsvold 1988; Cheverud 1996).

If the evolutionary trajectory of $\bar{\mathbf{z}}$ is complicated, rather than linear, responses to selection in opposite directions will tend to cancel and will not be represented in the reconstructed gradient. Just as $[\bar{\mathbf{z}}_t - \bar{\mathbf{z}}_0]$ represents the net change in the mean, $\mathbf{G}^{-1} [\bar{\mathbf{z}}_t - \bar{\mathbf{z}}_0]$ estimates the net selection needed to produce that change. Returning to the pool analogy of Figure 11.2, we can reconstruct the pool shot that would accomplish the entire change in multivariate mean (***netβ***), but in actuality, selection will often be composed of a long sequence of pool shots at various angles and generally of smaller magnitude than ***netβ***.

Objections to Lande's (1979) formulation of ***netβ*** (11.10) have been offered by critics but refuted by Jones et al. (2004, 2012). Turelli (1988) argued that (11.10) lacks a covariance term that might accumulate due to stochastic fluctuations in the G-matrix. This objection was echoed and elaborated by Shaw et al. (1995). However, in their extensive simulations of stochastic variation in the *G-matrix,* Jones et al. (2004, 2012) computed the size of this covariance term and showed that it was small and inconsequential in a variety of scenarios (see Chapter 17).

Several authors have used calculations of ***netβ*** to test the proposition that directional selection has shaped particular features of morphological divergence (Lofsvold 1988; Cheverud 1996; Marroig and Cheverud 2005; Marroig et al. 2012). Cheverud (1996), for example, used ***netβ*** to analyse divergence in the skulls of two species of neotropical monkeys (tamarins). The analysis suggested that selection for food processing using the incisors was a major cause of skull divergence between these two species.

11.7 Minimum Selective Mortality Required to Account for Trait Divergence

The selection index, I, that has proved so useful in plant and animal breeding (Section 11.3), can also be used to compute the *minimum selective mortality* that would be required to accomplish a given amount of evolutionary change in multivariate trait space. Let $\Delta\bar{\mathbf{z}}$ be the amount of change in the trait mean over an interval of t generations. If $\boldsymbol{G}$ is the average G-matrix over that same interval, we can solve for the t-generation selection gradient required to accomplish the change using

$$\boldsymbol{\beta} = \boldsymbol{G}^{-1}\Delta\bar{\mathbf{z}} \tag{11.11}$$

Then, using this β-vector, we construct a *selection index* of the following form

$$I = \boldsymbol{\beta}^{\mathrm{T}}\mathbf{z} = \beta_1 z_1 + \beta_2 z_2 + \cdots + \beta_q z_q = \Delta\bar{\mathbf{z}}^{\mathrm{T}}\boldsymbol{G}^{-1}\mathbf{z} \tag{11.12}$$

(Lande 1979). This index converts the problem of selection on the vector $\mathbf{z}$ into the univariate problem of selection on I. Consequently, the formulas in Section 10.6 can be used to calculate the minimum selection mortality needed to produce a particular magnitude of change in I by truncation selection. Using this approach on individual traits, Lande (1976) found that weak truncation selection can easily account for appreciable trait change observed in the fossil record (Section 10.6). The approach could be extended to multivariate selection, but—of course—one would need to justify a particular choice for average $\boldsymbol{G}$.

11.8 Is Long-term Multivariate Evolution Constrained by a Lack of Genetic Variance?

In the long term, the answer to this question is no, but in the short term multivariate responses to selection are shaped by the G-matrix. The lesson from the pool game analogy (Figure 11.2) is that the G-matrix biases the direction of the selection response. However, it would be a mistake to extrapolate this simple, short-term bias into deep evolutionary time. In the next few chapters we will use the adaptive landscape to explore the roles of selection and $\boldsymbol{G}$ deep into time. As we shall see, the deeper we go, the weaker the role of $\boldsymbol{G}$.

11.9 Conclusions

- The experience of animal and plant breeders is that genetic variation is available in a surprisingly diverse array of phenotypic directions. Likewise, bewildering phenotypic change and variety has been produced rapidly in many domestic radiations (e.g., maize, pigeons, domestic fowl, dogs), as noted by Darwin (1868).
- In the short term (a few generations), genetic covariances can frustrate responses to deliberate selection imposed by breeders and experimenters.

- However, selection on a linear combination of traits (i.e., a selection index) can compensate for the effects of genetic covariance. Furthermore, bivariate directional selection can change the allometric relationships between traits.
- Longitudinal studies of multivariate selection over dozens of years in nature reveal both apparent trends and fluctuations in directional selection.
- Appreciable trait change observed in the fossil record can be accounted for by directional selection that is weak on a per-generation basis.
- As we argued in the univariate case (Chapter 10), all of these observations suggest that multivariate selection rather than inheritance is the creative force in adaptive radiations.
- Evolution in deep evolutionary time is probably not constrained by genetic variation. Indeed, as we will see in Chapter 13, the influence of the G-matrix on the shape of an adaptive radiation decays exponentially with time.

Literature Cited

Cheverud, J. M. 1996. Developmental integration and the evolution of pleiotropy. Am. Zool. 36:44–50.

Clayton, G. A., G. R. Knight, J. A. Morris, and A. Robertson. 1957. An experimental check on quantitative genetical theory. III. Correlated responses. J. Genet. 55:171–180.

Clayton, G. A., J. A. Morris, and A. Robertson. 1956. An experimental check on quantitative genetical theory. I. The short-term response to selection. J. Genet. 55:131–151.

Darwin, C. 1868. The Variation of Animals and Plants under Domestication. John Murray, London.

Dickerson, G. E. 1955. Genetic slippage in response to selection for multiple objectives. Cold Spring Harb. Symp. Quant. Biol. 20:213–224.

Dickerson, G. E., C. T. Blunn, A. B. Chapman, R. M. Kottman, J. L. Krider, E. J. Warwick, and J. A. Whatley Jr. 1954. Evaluation of selection in developing inbred lines of swine. Coll. Agric. Res. Bull. 551, University of Missouri. Columbia.

Gibbs, H. L., and P. R. Grant. 1987. Oscillating selection on Darwin's finches. Nature 327:511–513.

Gould, S. J. 1966. Allometry and size in ontogeny and phylogeny. Biol. Rev. 41:587–638.

Grant, P. R., and B. R. Grant. 1995. Predicting microevolutionary responses to directional selection on heritable variation. Evolution 49:241–251.

Grant, P. R., and B. R. Grant. 2002. Unpredictable evolution in a 30-year study of Darwin's finches. Science 296:707–711.

Grant, P. R., B. R. Grant, J. N. Smith, I. J. Abbott, and L. K. Abbott. 1976. Darwin's finches: population variation and natural selection. Proc. Natl. Acad. Sci. USA 73:257–261.

Greene, H. W., and R. W. McDiarmid. 1981. Coral snake mimicry: does it occur? Science 213: 1207–1212.

Hill, W. G. 1980. Design of quantitative genetic selection experiments. pp. 1–13 *in* A. Robertson, ed. Selection experiments in laboratory and domestic animals: proceedings of a symposium held at Harrogate, UK, on 21st–22nd July 1979. Commonwealth Agricultural Bureau, Slough, England.

Hill, W. G. 1984. Quantitative Genetics, Part 2: The Selection. Van Nostrand Reinhold, NY.

James, F. C. 1970. Geographic size variation in birds and its relationship to climate. Ecology 51:365–390.

Johnston, R. F., and R. K. Selander. 1964. House sparrows: Rapid evolution of races in North America. Science 3618:548–550.

Johnston, R. F., and R. K. Selander. 1971. Evolution in the house sparrow. II. Adaptive differentiation in North American populations. Evolution 25:1–28.

Jolicoeur, P. and J. E. Mosimann. 1960. Size and shape variation in the Painted Turtle. a principal component analysis. Growth 24: 339–354.

Jones, A. G., S. J. Arnold, and R. Bürger. 2004. Evolution and stability of the G-matrix on a landscape with a moving optimum. Evolution 58:1639–1654.

Jones, A. G., R. Bürger, S. J. Arnold, P. A. Hohenlohe, and J. C. Uyeda. 2012. The effects of stochastic and episodic movement of the optimum on the evolution of the G-matrix and the response of the trait mean to selection. J. Evol. Biol. 25:2210–2231.

Keller, L. F., P. R. Grant, B. R. Grant, and K. Petren. 2001. Heritability of morphological traits in Darwin's finches: Misidentified paternity and maternal effects. Heredity 87:325–336.

Lande, R. 1976. Natural selection and random genetic drift in phenotypic evolution. Evolution 30:314–334.

Lande, R. 1979. Quantitative genetic analysis of multivariate evolution, applied to brain:body size allometry. Evolution 33:402–416.

Lofsvold, D. 1988. Quantitative genetics of morphological differentiation in *Peromyscus*. II. Analysis of selection and drift. Evolution 42:54–67.

Lynch, M., and B. Walsh. 1998. Genetics and Analysis of Quantitative Traits. Sinauer Associates, Inc., Sunderland, Massachusetts.

Marroig, G., and J. M. Cheverud. 2005. Size as a line of least evolutionary resistance: Diet and adaptive morphological radiation in New World monkeys. Evolution 59:1128–1142.

Marroig, G., D. A. R. Melo, and G. Garcia. 2012. Modularity, noise, and natural selection. Evolution 66:1506–1524.

Nestor, K. E., D. O. Noble, J. Zhu, and Y. Moritsu. 1996. Direct and correlated responses to long-term selection for increased body weight and egg production in turkeys. Poult. Sci. 75: 1180–1191.

Price, T., Grant, P. R., Gibbs, H. L., and Boag, P. T. 1984. Recurrent patterns of natural selection in a population of Darwin's finches. Nature 309:787–789.

Shaw, F. H., R. G. Shaw, G. S. Wilkinson, and M. Turelli. 1995. Changes in genetic variances and covariances: G WHIZ! Evolution 49:1260–1267.

Sheldon, B. L., and M. K. Milton. 1972. Studies on the scutellar bristles of *Drosophila melanogaster*. II. Long-term selection for high bristle number in the Oregon RC strain and correlated responses in abdominal chaetae. Genetics 71:567–595.

Smith, N. G. 1969. Avian predation of coral snakes. Coepia 1969:402–404.

Travisano, M., J. A. Mongold, A. F. Bennett, and R. E. Lenski. 1995. Experimental tests of the roles of adaptation, chance, and history in evolution. Science 267:87–90.

Turelli, M. 1988. Phenotypic evolution, constant covariances, and the maintenance of additive variance. Evolution 42:1342–1347.

Walsh, B., and M. Lynch. 2018. Evolution and Selection Of Quantitative Traits. Oxford University Press, New York.

Weber, K. E. 1990. Selection on wing allometry in *Drosophila melanogaster*. Genetics 126:975–989.

Williams, K. L. 1978. Systematics and natural history of the American Milk Snake. Milwaukee Public Mus. Publications in Biology and Geology, No. 2. Milwaukee.

Wilson, E. O., and W. L. Brown. 1953. The subspecies concept and its taxonomic application. Syst. Zool. 2:97–111.

CHAPTER 12

Evolution of a Single Trait on a Stationary Adaptive Landscape

Overview—The adaptive landscape is a powerful tool for conceptualizing and modelling long-term responses to selection. Here we focus on a single trait evolving on an adaptive landscape with a single, stationary peak. To visualize the stochastic models that we will use, imagine a set of replicate populations of identical size that descend all at once from a common ancestor and thereafter evolve on identical adaptive landscapes. A ∩-shaped adaptive landscape results in a stable equilibrium distribution of the replicate trait means. Stabilizing selection tends to pull the mean towards a peak on the landscape, but that tendency is balanced by drift of the trait mean away from the peak. As a consequence of this push-pull dynamic, once all the replicates have achieved an equilibrium, variation among their trait means is inversely proportional to effective population size. The smaller the size of each replicate, the more variable the response to selection. Tests of model predictions with large data sets indicate that this Ornstein-Uhlenbeck (OU) model is undoubtedly an element in the general explanation for stasis, but other factors are also in play. In particular, the OU model with realistic parameter produces only a fraction of the differentiation that is observed in nature. Parameter fits are even worse for peak-shift models (Estes and Arnold 2007). These results reinforce arguments made by Hansen and Houle (2004 in their discussion of the paradox of stasis. As we shall see, stabilizing selection helps explain stasis, but we also need a model for the movement of the peak. These are the topics of Chapter 13.

Because stabilizing selection and stasis figure so prominently in this chapter, we will begin with a historical perspective. The idea that stasis is a common mode of evolution in the fossil record is older than the idea of punctuated equilibrium popularized by Gould and Eldredge (1972) (Pennell et al. 2014). For example, Simpson (1944) recognized several different modes of evolution and viewed them as expressions of a process constrained by adaptive zones (Figure 12.1). Simpson wished to explain the constrained phenotypic evolution of families and genera. Weasels (mustelids) and civet cats (viverids), for example, are easily recognizable despite their extensive radiations. Evolution in each of these families is bounded and constrained—more so in some traits than others—so that traces of ancestry are carried across entire family and generic level radiations. Similar conceptualizations are presented by Futuyma (1987, 2010) and in a diagram by Eldredge et al. (2005) but on a shorter timescale.

Two ideas are common to the diagrams of Simpson (1944) and Eldredge et al. (2005), presented over 50 years apart. First, stasis is not literal in these diagrams. Trait means are

Evolutionary Quantitative Genetics. Stevan J. Arnold, Oxford University Press. © Stevan Arnold (2023).
DOI: 10.1093/oso/9780192859389.003.0013

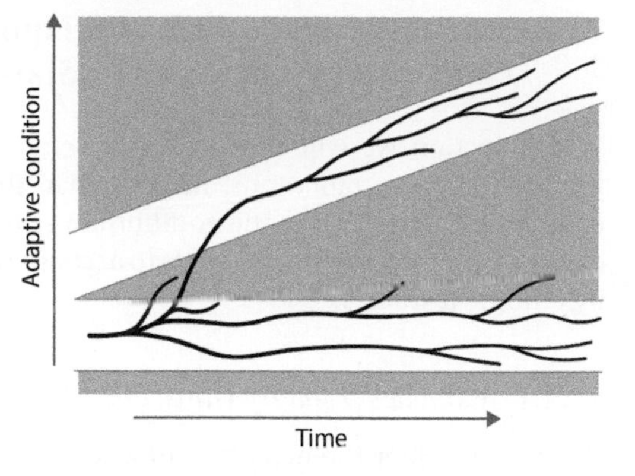

Figure 12.1 Simpson's concept of trait evolution within and between adaptive zones. Evolution of a trait mean is indicated with lines that show branchings (speciation) and terminations (extinctions) inside and outside of adaptive zones. The lower band represents a stable zone within which lineages proliferate but trait evolution is bounded. The upper, slanting band represents a steadily changing range of environmental conditions or opportunity. Within this adaptive zone evolving trait means show a pronounced trend that occurs in parallel in related lineages. After Simpson (1944).

constantly changing within and among related lineages, but usually within bounds or limits. Second, pronounced change in trait means is both rare and rapid.

Hansen and Houle (2004) discuss the historical and contemporary arguments that link stasis to stabilizing selection. While stabilizing selection emerges as the most plausible basis for stasis (see, for example, Charlesworth et al. 1982; Lande 1986), we need more than a knee-jerk invocation of this particular mode of selection, because we are confronted with a troubling paradox (Hansen and Houle 2004). The very traits that are prone to long term stasis are often heritable in nature. As our focus in explaining stasis is drawn away from inheritance to models of stabilizing selection, we need to understand the processes responsible for constraining the position of trait optima for long intervals of evolutionary time. In this chapter we shall build up to the puzzle of long-constrained optima by first considering the necessity and the limitations of accounting for stasis with models that include stabilizing selection.

12.1 Tendency to Evolve Uphill on the Adaptive Landscape

Consider the single trait version of Lande's (1979) argument that trait evolution will occur in an uphill direction on the adaptive landscape. Recall that the change in the trait mean after a single generation of selection and inheritance is

$$\Delta\bar{z} = G\beta = G\frac{\partial ln\bar{W}}{\partial\bar{z}}. \tag{12.1}$$

The squared phenotypic distance between the trait means at two points in time, a and b, is

$$\Delta\bar{z}^2 = (\bar{z}_a - \bar{z}_b)^2, \tag{12.2}$$

which is always positive. Therefore, rearranging (12.1), the change in the log mean fitness is

$$\Delta ln\bar{W} \cong \Delta\bar{z}^2 G^{-1} \geq 0. \tag{12.3}$$

In other words, the movement of the trait mean on the adaptive landscape under the force of directional selection is always uphill (Lande 1979).

12.2 Stochastic Dynamics and Equilibrium of Trait Mean on a Landscape With a Single, Stationary Adaptive Peak

In this section we will review the capacity of models with a single stationary peak to produce adaptive radiations. In particular, the analytical properties of these models at equilibrium help specify the equilibrium size of radiations. Then in the next section we explore the ability of these models to account for the actual size of radiations as established by empirical studies.

12.2.1 *A single stationary Gaussian adaptive peak*

We now consider the evolution of a single normally distributed trait in a population of finite size on a Gaussian adaptive landscape. We introduced the properties of normally distributed trait on a Gaussian AL in Section 3.4, and we will now use them to construct a model of drift-selection balance about a single adaptive peak, which is a special case of the *Ornstein-Uhlenbeck (OU) process* (Uhlenbeck and Ornstein 1930), following the model developed by Lande (1976).

Assume that the population is finite in size, so that, according to the results in Chapter 8, the amount that the trait mean drifts away from the optimum each generation will be equivalent to a draw from a normal distribution with mean of zero and a variance of G/N_e. Each generation the trait mean is pulled back towards the optimum by a force of directional selection that is proportional to the distance from the optimum, θ, to the trait mean,

$$\beta = \Omega^{-1}(\theta - \bar{z}), \tag{12.4}$$

where $\Omega = \omega + P$ is the width of the Gaussian adaptive landscape. In other words, specifying the opposing effects of directional selection towards the optimum (the adaptive peak) and drift, the value of the mean at generation $t+1$ will be

$$\bar{z}_{t+1} = \bar{z}_t + \left[\frac{\theta - \bar{z}_t}{\Omega}\right] G + N(0, G/N_e). \tag{12.5}$$

For simplicity, let the position of the optimum be zero so that after t elapsed generations, the expected value of the mean of lineage means is zero with lineage means normally distributed about that expectation with a variance of

$$Var(\bar{z}_t) = \frac{\Omega}{2N_e}\left[1 - exp\left\{-2\left(\frac{G}{\Omega}\right)t\right\}\right] \tag{12.6}$$

(Lande 1976). An example of a lineage drifting about its optimum is shown in Figure 12.2a. Each time the trait mean drifts away from the optimum, it is pulled back, and so, as a long term average, the mean resides at the optimum. An ensemble of 10 replicate populations, buzzing about identical optima, are shown in Figure 12.2b. The outer limits of their paths are accurately described by the 99% confidence limits shown in blue.

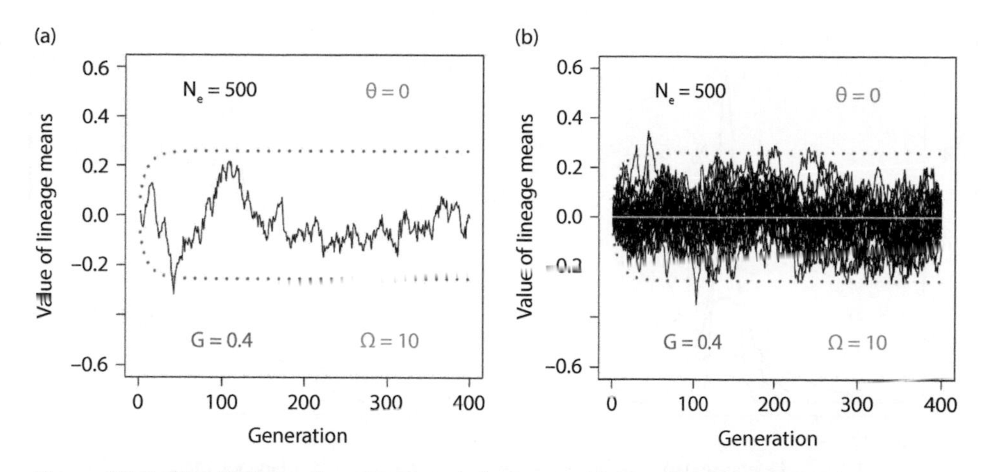

Figure 12.2 Simulations of replicate populations evolving about a stationary intermediate optimum. 99% confidence limits, calculated using (12.6) are shown as dotted blue curves. (a) A single lineage mean in a population of finite size evolving according to an OU process. (b) The lineage means of 25 replicate populations evolving according to the same process.

At equilibrium, trait means are normally distributed with a mean of means at the optimum, θ, and a variance of lineage means given by

$$Var(\bar{z}_\infty) = \frac{\Omega}{2N_e}, \tag{12.7}$$

so that the equilibrium distribution of means is given by

$$\Phi(\bar{z}_\infty) = \frac{1}{\sqrt{2\pi Var(\bar{z}_\infty)}} exp\left\{\frac{-(\bar{z}-\theta)^2}{2Var(\bar{z}_\infty)}\right\} \tag{12.8}$$

Lande (1976). In other words, variation in lineage means is greater if stabilizing selection is weak (large Ω) and population effective size is small. Smaller populations can drift further from the optimum, especially if the restraining force of selection is weak.

We can use (12.7) to assess the model's capacity to produce an adaptive radiation. For example, let $P = 1$, $\Omega = 100$, and $N_e = 5000$. The expected variance among the means of replicate lineages is 0.01, which is only 1% as large than the variance within a single lineage. The small scale of the radiation under these conditions is symptomatic of the problem that this model faces in producing an adaptive radiation. To produce a radiation in which the variance among lineage means is, say, 10–100 times larger than the variance within a lineage, one must invoke unrealistically weak stabilizing selection (large Ω), unrealistically small N_e, or both (Figure 12.3). (See Chapters 3 and 8, respectively, for a discussion of empirically realistic values of Ω and N_e.) Even without applying the model to real data, we can conclude that a single, stationary adaptive peak is an inadequate model for adaptive radiation. Nevertheless, this model could explain stasis of a trait in a lineage or a set of lineages.

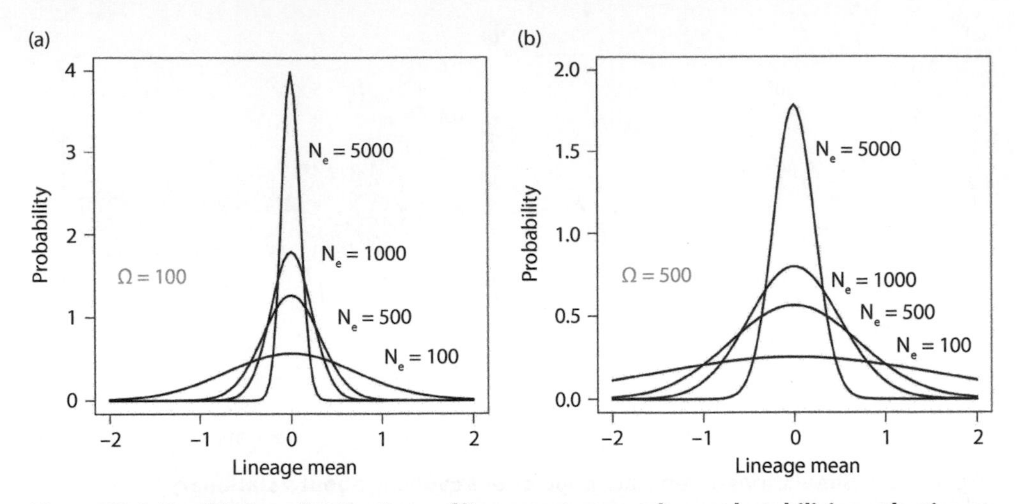

Figure 12.3 Equilibrium distributions of lineage means under weak stabilizing selection as a function of population size. Distributions of lineage means, $\Phi\,(\bar{z}_\infty)$, as a function of effective population size, N_e, calculated using (12.8). (a) Weak stabilizing selection, $\Omega = 100$. (b) Very weak stabilizing selection, $\Omega = 500$.

12.2.2 *Early-burst model*

In the *early-burst model* (EB) introduced by (Blomberg et al. 2003), also known as the *ACDC model*, variance among lineages increases through time but quickly reaches an asymptotic value. The early-burst process model for the trait mean is

$$\bar{z}_{t+1} = \bar{z}_t + N\,(0,\ Var), \tag{12.9}$$

where $Var = (G/N_e)\ exp\,\{rt\}$, with $r < 0$ describing decreasing fluctuation within-lineage variance (Blomberg et al. 2003; Harmon et al. 2010). The expected value of the lineage means is zero with normally distributed variation about that value with the variance in trait means at generation t given by

$$Var\,(\bar{z}_t) = G/N_e \left[\frac{\exp\{rt\} - 1}{r}\right] \tag{12.10a}$$

(Blomberg et al. 2003, Harmon et al. 2010). When $r < 0$, after many generation this variance converges on

$$Var\,(\bar{z}_\infty) = G/(|r|\,N_e). \tag{12.10b}$$

In comparison, this among-lineage variance continues to increase under BM of the trait mean, so that at generation t it is Gt/N_e (8.4), while under OU the limiting variance is $\Omega/2N_e$ (12.6). Because we have set $P = 1$, $0 > G < 1$, and under weak selection $\Omega \gg 1$, we can conclude that the EB model is likely to produce less diversification than OU (12.6), when $r < 0$.

Using the same parameters for population size, selection, and inheritance as in the last example, Figure 12.4 portrays an example of results from the early-burst model.

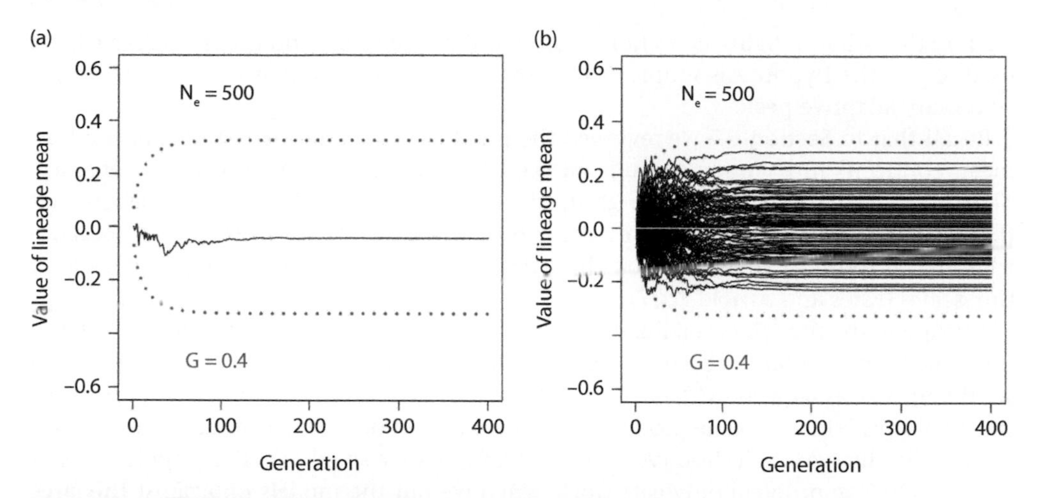

Figure 12.4 Simulations of lineages evolving according to the early-burst model of Blomberg et al. (2003). 99% confidence limits calculated using (12.10a), other conventions as in Figure 12.2. The parameter describing diminution of drift variance is set to $r = -0.05$. (a) A simulation of one evolving lineage. (b) A simulation with 100 replicate lineages.

This simulation quickly produced a radiation, but the among-lineage variation reached its asymptote after about 100 generations. The asymptotic extent of the radiation is extremely modest, amounting to about $0.4\sqrt{P}$, about the same as the OU radiation shown in Figure 12.3. An expected but curious feature of the model is that within-lineage fluctuation in trait means disappears after about 120 generations and thereafter each lineage mean has a fixed value. The r parameter has suppressed variation arising from drift, but the biology of this suppression remains unexplained by the model.

12.2.3 *Squashed stabilizing selection*

Haller and Hendry (2014) outlined a complicated model of stabilizing selection that incorporates negative frequency-dependent selection so that the fitness function is squashed. In other words, selection vacillates between stabilizing and disruptive, between decreasing and increasing trait variance. Although the consequences of this model for adaptive radiation have not been explored (but see Chapter 21), the virtue of this model is that it offers an explanation for the fact that disruptive and stabilizing selection are almost equally prevalent in surveys of phenotypic selection (Chapter 1–4).

12.3 Attempts to Account for Divergence Data With Models That Employ Drift and/or a Single Stationary Peak

In Section 3.5 we discussed estimating parameters of the single Gaussian peak model on a relatively short timescale of 30 years (Chevin et al. 2015). In that example of egg-laying date in a passerine bird, the best fitting model was a Gaussian peak with minor fluctuations in position, but with no long-term trend in position. A stationary peak on such a short timescale may not be surprising, so we wish to examine this hypothesis of stationarity over longer stretches of time. Gingerich (2001) and Uyeda et al. (2011) assembled divergence data on timescales ranging up to hundreds of millions of years, and those data have been used to test the hypothesis simple drift in the trait mean, as well as the idea of a single stationary adaptive peak.

Recall that in Section 8.6 we reviewed the problems that a model for drift of the trait mean confronts in trying to account for the Gingerich (2001) data on trait divergence. Those data show trait divergence grouped in a band, regardless of timescale, so that the absolute divergence is less than about $6\sqrt{P}$. Drift failed as an explanation because it produced too little divergence on short timescales and too much divergence on long timescales (Estes and Arnold 2007).

Turning to the models of stabilizing selection reviewed in the preceding section, we find the same kind of failure in both models. The Gingerich (2001) data tell us that we should see divergence as large as $6\sqrt{P}$ on all timescales, but models of stabilizing selection with a stationary peak have trouble producing even a $\sqrt{P}$ level of divergence, unless we invoke extraordinarily weak selection (very large Ω) and/or very small effective population size ($N_e < 200$). The problem only gets worse when we put the models up against the large data set assembled by Uyeda et al. (2011).

The Uyeda et al. (2011) data are similar to the Gingerich (2001) data, but they are more extensive and span longer time intervals. Nevertheless, they too show a distinct band of divergence that lies $\pm 6\sqrt{P}$ on either side of the zero value of the trait mean (Figure 12.5). The band is apparent at the left end of the timescale, but divergence values pile up inside the band at all timescales (Arnold 2014). This band-like feature, evidence for constrained

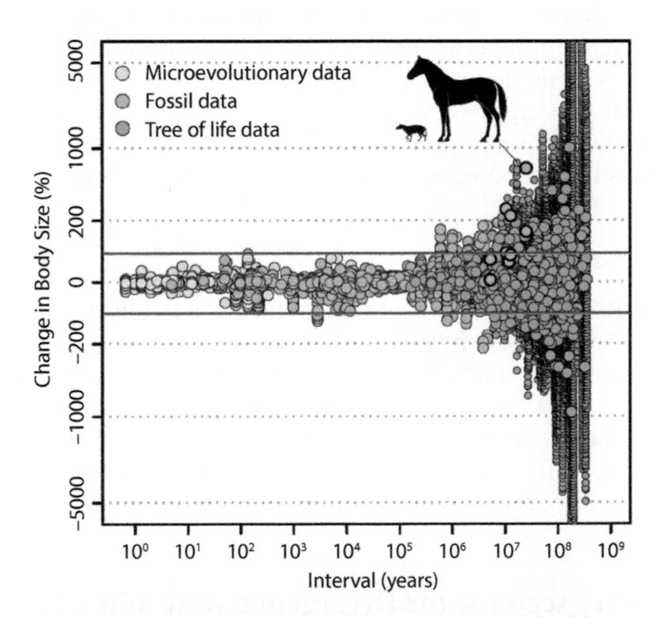

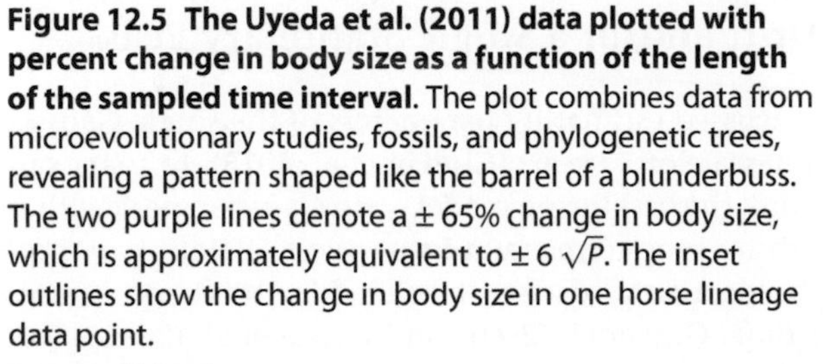

Figure 12.5 The Uyeda et al. (2011) data plotted with percent change in body size as a function of the length of the sampled time interval. The plot combines data from microevolutionary studies, fossils, and phylogenetic trees, revealing a pattern shaped like the barrel of a blunderbuss. The two purple lines denote a ± 65% change in body size, which is approximately equivalent to $\pm 6\sqrt{P}$. The inset outlines show the change in body size in one horse lineage data point.

From Arnold (2014).

evolution, is one of the strong arguments for including stabilizing selection in models for macroevolution. The new feature in the plot, however, is the appearance of very large divergence values (> 200%) for time intervals greater than about one million years, producing a blunderbuss pattern in the divergence values. This flaring of the blunderbuss barrel is fatal for the single, stationary peak models. In the next chapter we will attempt to account for this pattern with models of peak movement.

12.4 Shifting of the Trait Mean Between Two Adaptive Peaks

Up to this point we have considered only a stationary AL with a single peak. Now let us suppose that the AL has more than one peak, and for simplicity we consider the simplest beginning point, two stationary peaks. If the two peaks are stationary, we need to invoke some process to move the trait mean away from one peak, against the force of selection. The obvious choice for the needed process is drift of the trait mean (Lande 1985, 1986). Under the simplifying conditions depicted in Figure 12.6, Lande (1985) showed that the expected time for the trait mean, initially situated between points a and b, to evolve to either of the peaks is approximately

$$T \cong \frac{2\pi\Omega}{G}\left(\frac{\bar{W}_a}{\bar{W}_v}\right)^{2N_e}. \tag{12.11}$$

In other words, the expected number of generations required for the trait mean to evolve from the first peak (a) to the second (b) is proportional to the ratio of the height of the first peak to the height of the valley (v) raised to the power $2N_e$. Thus, a shallow valley can present a formidable barrier even if the effective population size is as small as 200.

Estes and Arnold (2007) used Lande's (1985) results to derive an approximate expression for the expected variance among replicate trait means at time t, so that the results of this model could be compared to the ones we reviewed above. Treating the peak shifting as a

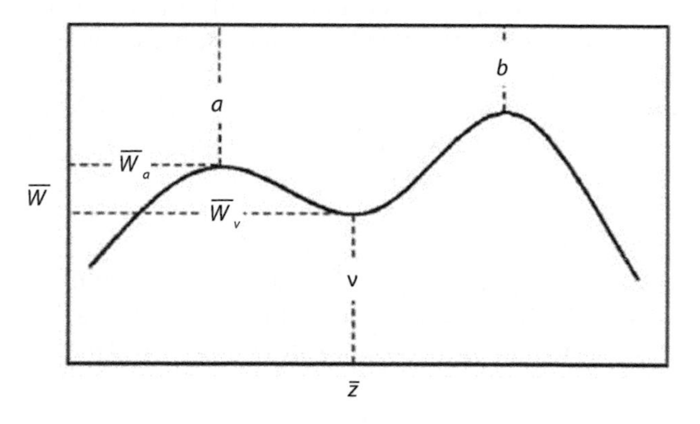

Figure 12.6 Lande's adaptive landscape model with two adaptive peaks. A critical parameter in the model is the ratio of the height of the first adaptive peak at a, ($\bar{W}_a$), to the height of the valley between the peaks at v, ($\bar{W}_v$).
From Estes and Arnold (2007).

Poisson process and supposing that the peaks are separated by a distance $d = b - a$, with the value of the mean at position a set to zero, the expected value of the trait mean at time t is

$$E(\bar{z}_t) = d(1 - p), \tag{12.12}$$

where $p = exp\{-t/T\}$ is the proportion of replicates residing at position a, while a proportion $1 - p$ reside at position b. The variance among replicates at generation t is approximately

$$Var(\bar{z}_t) = p\bar{z}_t + (1 - p)(d - \bar{z}_t)^2 \ (\Omega/2N_e) \tag{12.13}$$

(Estes and Arnold 2007).

Comparing the model's predictions to the Gingerich (2001) data revealed several model limitations (Estes and Arnold 2007). Initial comparisons focused on empirically central values for heritability and the strength of stabilizing selection ($h^2 = 0.4$, $\Omega = 4.2$). Under those conditions, and over a wide range of values for d, the model required relatively small effective population sizes ($N_e = 250 - 750$) and even then it underestimated divergence on short timescales. Weaker strengths of stabilizing selection made the fits worse. The general problem with this model is that it requires a delicate balancing of parameters to produce occupancy of both adaptive peaks.

12.5 Conclusions

- The simple OU (drift-stabilizing selection balance) model with one stationary peak may be a satisfactory model for local adaptation, but not for most kinds of diversification among species or for geographic variation within species.
- A basic failing of the simple OU model is that its production of diversification is too constrained. To account for actual radiations with this model one needs to invoke unrealistically small population size, so that diversification can be driven by drift. Alternatively, one needs to invoke stabilizing selection that is much weaker than selection routinely observed in nature.
- The early-burst model provides an especially poor fit to different kinds of data. The fit is especially poor because the model predicts the exact opposite of the pattern seen in large data sets, namely diminishment rather than exuberance of differentiation on long timescales.
- A peak-shift model with two peaks also shows a poor fit to actual data. A delicate balance of parameters is required to produce substantial residence of the trait mean near both adaptive peaks. In the absence of that hard-to-produce balanced residence, the model—like the simple OU model—produces too little differentiation.

Literature Cited

Arnold, S. J. 2014. Phenotypic evolution: the ongoing synthesis. Am. Nat. 183:729–746.

Blomberg, S. P., T. Garland, and A. R. Ives. 2003. Testing for phylogenetic signal in comparative data: Behavioral traits are more labile. Evolution 57:717–745.

Charlesworth, B., R. Lande, and M. Slatkin. 1982. A neo-Darwinian commentary on macroevolution. Evolution 36:474.

Chevin, L.-M., M. E. Visser, and J. Tufto. 2015. Estimating the variation, autocorrelation, and environmental sensitivity of phenotypic selection. Evolution 69:2319–2332.

Eldredge, N., J. N. Thompson, P. M. Brakefield, S. Gavrilets, D. Jablonski, J. B. C. Jackson, R. E. Lenski, B. S. Lieberman, M. A. McPeek, and W. Miller. 2005. The dynamics of evolutionary stasis. Paleobiology 31:133–145.

Estes, S., and S. J. Arnold. 2007. Resolving the paradox of stasis: Models with stabilizing selection explain evolutionary divergence on all timescales. Am. Nat. 169:227–244.

Futuyma, D. J. 1987. On the role of species in anagenesis. Am. Nat. 130:465–473.

Futuyma, D. J. 2010. Evolutionary constraint and ecological consequences. Evolution 64: 1865–1884.

Gingerich, P. D. 2001. Rates of evolution on the time scale of the evolutionary process. pp. 127–144 *in* A. P. Hendry and M. T. Kinnison, eds. Microevolution: Rate, Pattern, Process. Springer, Dordrecht

Gould, S. J., and N. Eldredge. 1972. Punctuated equilibria: An alternative to phyletic gradualism. pp. 82–115 *in* T. J. M. Schopf, ed. Models in Paleobiology. Freeman and Cooper, San Francisco.

Haller, B. C., and A. P. Hendry. 2014. Solving the paradox of stasis: Squashed stabilizing selection and the limits of detection: selection and the limits of detection. Evolution 68:483–500.

Hansen, T. F., and D. Houle. 2004. Evolvability, stabilizing selection and the problem of stasis. pp. 129–135 *in* M. Pigliucci and K. Preston (eds.), Phenotypic Integration: Studying the Ecology and Evolution of Complex Phenotypes. Oxford University Press, Oxford.

Harmon, L. J., J. B. Losos, T. Jonathan Davies, R. G. Gillespie, J. L. Gittleman, W. Bryan Jennings, K. H. Kozak, M. A. McPeek, F. Moreno-Roark, T. J. Near, A. Purvis, R. E. Ricklefs, D. Schluter, J. A. Schulte II, O. Seehausen, B. L. Sidlauskas, O. Torres-Carvajal, J. T. Weir, and A. Ø. Mooers. 2010. Early bursts of body size and shape evolution are rare in comparative data. Evolution 64:2385–2396.

Lande, R. 1976. Natural selection and random genetic drift in phenotypic evolution. Evolution 30:314–334.

Lande, R. 1979. Quantitative genetic analysis of multivariate evolution, applied to brain: bodysize allometry. Evolution 33:402–416.

Lande, R. 1985. Expected time for random genetic drift of a population between stable phenotypic states. Proc. Natl. Acad. Sci. USA 82:7641–7645.

Lande, R. 1986. The dynamics of peak shifts and the pattern of morphological evolution. Paleobiology 12:343–354.

Pennell, M. W., L. J. Harmon, and J. C. Uyeda. 2014. Is there room for punctuated equilibrium in macroevolution? Trends Ecol. Evol. 29:23–32.

Simpson, G. G. 1944. Tempo and Mode in Evolution. Columbia University Press, New York.

Uhlenbeck, G. E., and L. S. Ornstein. 1930. On the theory of the Brownian motion. Phys. Rev. 36:823–841.

Uyeda, J. C., T. F. Hansen, S. J. Arnold, and J. Pienaar. 2011. The million-year wait for macroevolutionary bursts. Proc. Natl. Acad. Sci. USA 108:15908–15913.

CHAPTER 13

Evolution of Multiple Traits on a Stationary Adaptive Landscape

Overview—The theoretical evolution of the multivariate trait mean has been most explored in the case of a hill-shaped adaptive landscape that has a consistent position, shape, and orientation. Under particular but general conditions, the mean tends to evolve uphill on this stationary landscape, towards the adaptive peak. When two or more traits are genetically correlated, the evolutionary path towards the peak is generally curved rather than straight. The big challenge is to develop generalizations about how multivariate evolution will proceed on different kinds of adaptive peaks. The effect of peak configuration on multivariate trait radiations has been explored by assuming that the peak is multivariate Gaussian. Theory is especially well developed for systems in which a single male trait evolves in response to sexual selection exerted by a second, female trait (sexual preference). Under these conditions, the equilibrium of the bivariate mean may be a stable point, a line or an elliptical cycle. Unstable dynamics are also possible, but probably unlikely. Finite population size adds an element of uncertainty to the outcome, changing stable points into clouds, for example. Predictions from these models are consistent with important features of actual sexual radiations, but only a few discriminating, quantitative tests have been conducted.

Although it is early days in testing the adequacy of Gaussian models of stationary adaptive radiations, some conclusions seem inescapable even at this early juncture. In particular, the potential scope of adaptive radiation appears to be modest unless we assume weak stabilizing selection and/or small effective population size. Tests with actual data also bear out this expectation. As we shall see at the end of this chapter, one escape from this quandary is to abandon the notion of a stationary peak and instead allow the peak to move. But, before we get ahead of our story, we need proceed stepwise through the details of this overview, beginning with the fundamental point of uphill evolution on a stationary adaptive landscape.

13.1 Tendency to Evolve Uphill on the Multivariate Adaptive Landscape

To understand the multivariate version of Lande's (1979) argument that the trait mean will evolve in an uphill direction on the AL, we need to consider his deterministic equation for the response of the multivariate mean to multivariate directional selection (11.1),

Evolutionary Quantitative Genetics. Stevan J. Arnold, Oxford University Press. © Stevan Arnold (2023).
DOI: 10.1093/oso/9780192859389.003.0014

$$\Delta\bar{\mathbf{z}} = \boldsymbol{G}\beta = \frac{\partial ln\bar{W}}{\partial\bar{\mathbf{z}}}. \tag{13.1}$$

In multivariate trait space, the squared distance between the trait means of two populations, a and b, is represented by a *generalized phenotypic distance*, a version of Mahalanobis distance,

$$\Delta\bar{\mathbf{z}}^2 = \left[(\bar{\mathbf{z}}_a - \bar{\mathbf{z}}_b)^{\mathrm{T}}\boldsymbol{P}^{-1}(\bar{\mathbf{z}}_a - \bar{\mathbf{z}}_b)\right]^{\frac{1}{2}}, \tag{13.2}$$

which accounts for covariances among traits and differences in variance. To obtain *generalized genetic distance*, we replace $\boldsymbol{P}$ with $\boldsymbol{G}$ in this expression. Therefore, rearranging (13.1), we see that the change in log mean fitness is approximately equal to generalized genetic distance, which is always positive (i.e., equal to or greater than zero),

$$\Delta ln\bar{W} \cong \Delta\bar{\mathbf{z}}^{\mathrm{T}}\boldsymbol{G}^{-1}\Delta\bar{\mathbf{z}} \geq 0. \tag{13.3}$$

Consequently, we can conclude that the trait mean evolves uphill on the adaptive landscape under the force of directional selection (Lande 1979).

13.2 Deterministic Evolution of the Multiple Trait Mean, the Spaceship Model

The case of two traits evolving on a stationary adaptive landscape illustrates many properties of the general multivariate case, so we will use it as a starting point. In the case of a single trait, the process under discussion consists of the 'boring march of the frequency distribution' (Luria et al. 1981), but in the two-trait case, the process is much more interesting. In particular, we shall consider the process when the bivariate mean lies some considerable distance from the optimum. Under this circumstance we will see that the process consists of two distinct phases: an initial phase during which the mean evolves rapidly towards the optimum, and a second phase, often with a distinctly different trajectory, during which the mean slowly approaches the optimum.

The per-generation deterministic change in the trait mean when the AL is Gaussian is a special case of the general expression, $\Delta\bar{\mathbf{z}} = \boldsymbol{G}\beta$,

$$\Delta\bar{\mathbf{z}} = \boldsymbol{G}\beta = \boldsymbol{G}\Omega^{-1}(\theta - \bar{\mathbf{z}}), \tag{13.4}$$

which follows from (4.12c). We examined the two-trait version of this expression in Chapter 11 (11.2).

It will be useful to remember the two-trait version of this last equation as the *spaceship model*. Think of a bivariate mean located some considerable distance from a single, stationary adaptive peak. The mean is surrounded by a cloud of genetic values that can be visualized as a blue ellipse, our spaceship. We keep the overall configuration of the cloud constant as the bivariate mean responds to selection exerted by the distant peak and moves towards it. The particular trajectory that the spaceship takes on its journey is a function of its starting position, as well as the configurations of the genetic cloud and the AL (Figure 13.1).

Figure 13.1 illustrates three important properties of the lineage mean's response to a distant adaptive peak (Lande 1980). First, responses are greater when the mean is further

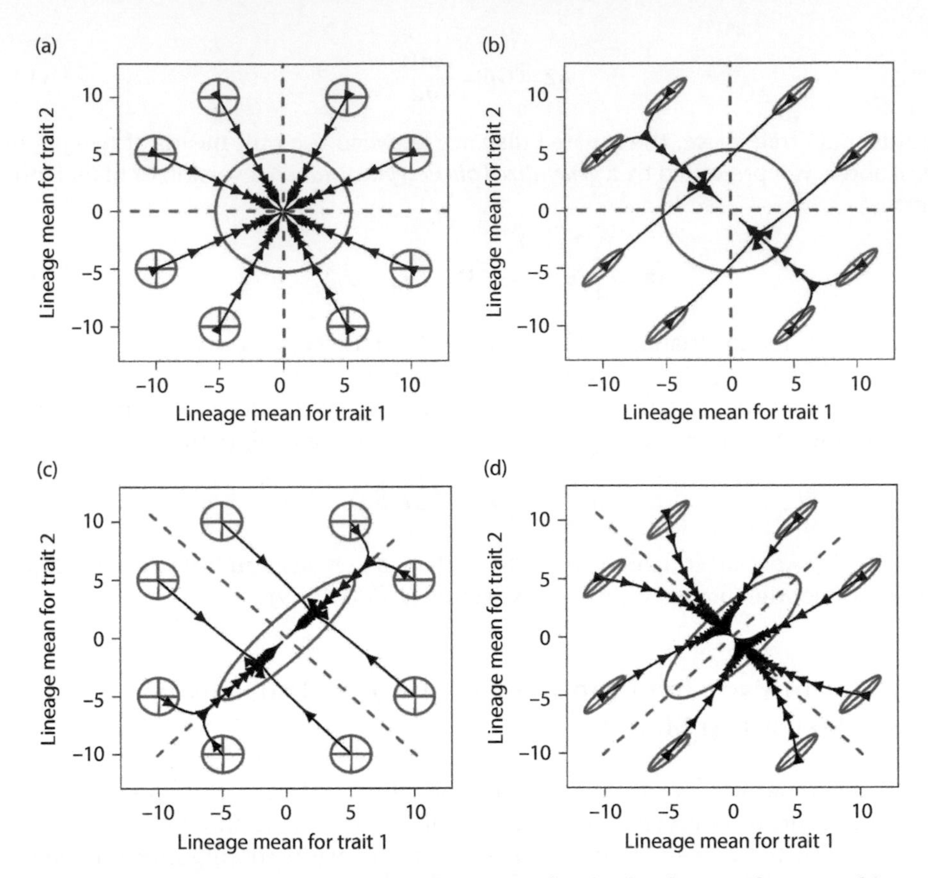

Figure 13.1 Bivariate evolution on a Gaussian adaptive landscape, the spaceship model. The adaptive landscape is shown in red: the ellipse represents the equivalent 50% confidence region, with eigenvectors shown as dashed red lines. Stabilizing selection of equal magnitude acts on each trait: $\Omega_{11} = \Omega_{22} = 50$. *G*-matrices (represented as 95% confidence ellipses) are shown in blue. In each case, the genetic variance of both traits is 0.4. Evolutionary trajectories of the bivariate mean are shown in black, with arrows at regular intervals of elapsed generations. (a) $r_g = r_\Omega = 0$. Each evolutionary trajectory is 500 generations in duration, with arrowheads every 50 generations. (b) $r_g = 0.9$, $r_\Omega = 0$. Each evolutionary trajectory is 1500 generations in duration, with arrowheads every 300 generations. (c) $r_g = 0$, $r_\Omega = 0.9$. Each evolutionary trajectory is 500 generation in duration, with arrowheads every 50 generations. (d) $r_g = 0.9$, $r_\Omega = 0.8$. Each evolutionary trajectory is 700 generations in duration, with arrowheads every 50 generations. Animations of the panels in this figure are available at https://phenotypicevolution.com/?p=102.

from the adaptive peak. This greater response reflects the fact that directional selection is stronger the further away the mean is from the optimum. Second, genetic covariance and correlational selection can each cause evolutionary trajectories to be curved. Third, the evolving trait mean can over- or under-shoot the optimum so that the population experiences long periods of maladaptation (Figure 13.1b, c). This phenomenon of maladaptation is likewise a consequence of genetic covariance and correlational selection. We will now examine these aspects of the spaceship's trajectory and journey in greater detail.

We see in Figure 13.1 that the shapes of the *G*-matrix and the AL can dramatically affect the evolutionary trajectory of the mean during both the initial and secondary phases of approach to the optimum. In the absence of both genetic correlation and correlational selection, both the cloud of genetic values and the AL are circular in cross-section. Under this circumstance (Figure 13.1a), the mean evolves directly towards the adaptive peak. Evolution is rapid at first and then gradually decelerates, until the mean ceases its evolution at the peak. If the AL is circular but the *G*-matrix is elongate, the rapid initial phase is in the direction of $\boldsymbol{g}_{max}$, the long axis of the *G*-matrix (Figure 13.1b). However, when the trait mean is located away from the optimum along the vector $\boldsymbol{g}_{min}$, the short axis of the *G*-matrix, the trajectory turns in that direction and evolution decelerates dramatically as the mean completes its slow approach to the optimum. The difference in initial and secondary phases makes intuitive sense: a rapid initial phase fuelled by abundant genetic variance, then a slow secondary phase in the direction of least genetic variance. This intuition is expanded when we consider a third case, in which the *G*-matrix and the AL have reverse configurations (Figure 13.2c). Now the genetic cloud is circular, so the elongate AL governs the trajectories of approach to the optimum. The rapid initial phase is along $\boldsymbol{\Omega}_{min}$, the short axis of the AL, and the slow, second phase is along $\boldsymbol{\Omega}_{max}$, the long axis of the AL. When genetic variance is the same in all directions, the shape of the AL dictates the difference in trajectories during the initial and secondary phases of evolution. Finally, in a revealing but unillustrated case, consider the outcome when the *G*-matrix and the AL are perfectly aligned ($\boldsymbol{g}_{max} = \boldsymbol{\Omega}_{max}$) and proportional (i.e., the corresponding eigenvalues of the two matrices differ by the same constant of proportionality). Under this circumstance of alignment and proportionality, the effects of the two processes cancel and the trajectories of the mean resemble the star-like pattern seen in Figure 13.1a. In retrospect, we could have anticipated this result from (13.4); because $c\boldsymbol{G} = \boldsymbol{\Omega}$, where c is a scalar constant, $c\boldsymbol{G}\boldsymbol{\Omega}^{-1} = \boldsymbol{I}$, where $\boldsymbol{I}$ is the identity matrix. Because this special case requires a precise combination of parameters, a more general case is illustrated in Figure 13.1d. Here, $\boldsymbol{G}$ and the AL are aligned but not exactly the same shape, with the result that both inheritance and selection affect the trajectories of the mean.

An arbitrary feature of these illustrations of the spaceship model is that we began each trajectory with the mean situated far from the optimum. How commonly does the trait mean reside 5–10 phenotypic standard deviations away from the optimum? Under the Gaussian assumptions used in Figure 13.1, the magnitude of the selection gradient, $\boldsymbol{\beta}$, is directly proportional to the distance to the optimum and inversely proportional to the

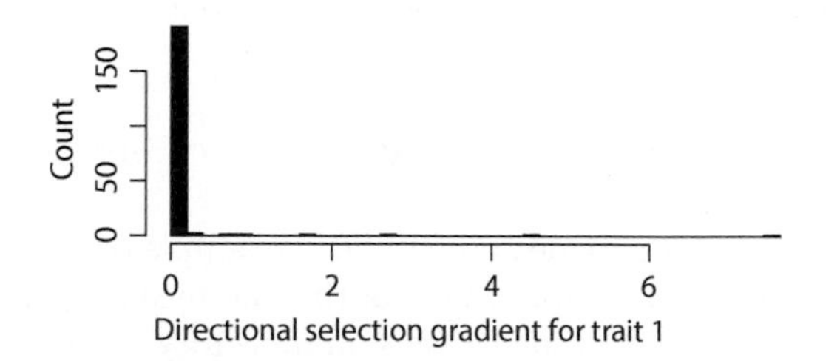

Figure 13.2 Values of the directional selection gradient during a simulated approach of the trait mean to a stationary adaptive peak. The values for directional selection on trait 1 (n=199) are plotted for the trajectory beginning at ($\bar{z}_1 = -10, \bar{z}_2 = 5$) in Figure 13.1c.

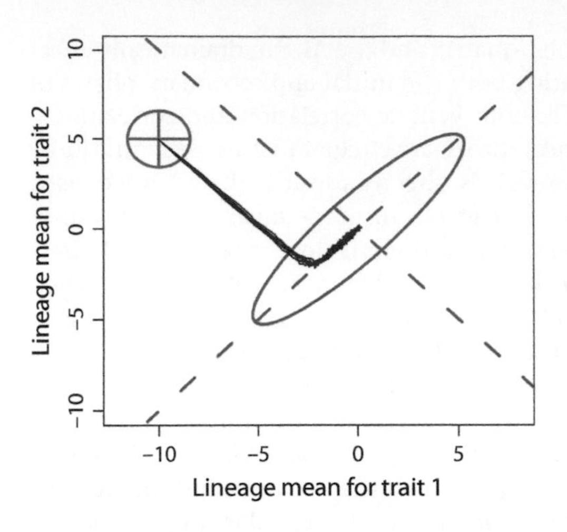

Figure 13.3 Stochastic response of the bivariate mean to a Gaussian adaptive landscape for 500 generations in eight replicate lineages. No genetic correlation, r_s = 0.9 and N_e = 500. Other conventions as in Figure 13.1c, except that elapsed intervals are not indicated with arrowheads.

curvature of the AL, $\mathbf{\Omega}$. Consequently, we can produce a histogram of β-values for any of the trajectories shown in Figure 13.1 and compare that histogram to the histogram of β-values estimated in natural populations (Figure 2.5c). A representative histogram from a trajectory in Figure 13.1 is shown in 13.2. That trajectory produced a few values of $\boldsymbol{\beta}$ that are much larger than those typically seen in natural populations and of course these values were produced during the first few generations of the first phase, when the mean was far from the optimum. The histogram also shows an overabundance of values very close to the optimum. In other words, the process of deterministic approach to an optimum from far away, as in Figure 13.2, is not consistent with the picture of evolution implied by direct estimates of directional selection (Figure 3.12). This result highlights a problem with the supposition of *evolution along genetic lines of least resistance* as an explanation for adaptive radiations (Schluter 1996). While trajectories may roughly align with $\boldsymbol{g}_{max}$ when the trait mean is far from a peak, when the mean is close to the peak—where it probably resides most of the time—the shape of the AL is the dominant controlling influence.

So far we have considered the deterministic evolution of the bivariate mean. Finite population size will cause stochasticity in the sampling of parents each generation and produce a distribution of trajectories. One such case is shown in Figure 13.3. In general, as in this case, stochasticity is tempered by directional and stabilizing selection, so that the variation among lineage means is considerably less than in the neutral case.

13.3 Sexual Coevolution and the Evolution of Sexual Dimorphism

Models of sexual coevolution provide an instructive example of how a persistent genetic correlation can affect evolutionary trajectories as the trait mean evolves towards a stationary peak. Two relevant examples of such effects on trajectory were shown in Figure 13.1a, b, but here we will explore an interesting empirical application of those examples. In particular, these model applications also illuminate the evolution of *sexual dimorphism* (a difference between the sexes in the means of homologous traits).

Following Lande's (1980) treatment of evolving sexual dimorphism, we will assume that multiple autosomal loci affect expression of the trait differently in the two sexes. Let z and y be normally distributed traits in males and females, respectively, of a dioecious species with within-sex genetic variances G_m and G_f, and a between-sex genetic covariance of B. The per-generation evolution of the means, $\bar{z}$ and $\bar{y}$, is a special case of (12.1), namely

$$\begin{bmatrix} \Delta\bar{z} \\ \Delta\bar{y} \end{bmatrix} = \frac{1}{2}\begin{bmatrix} G_m & B \\ B & G_f \end{bmatrix}\begin{bmatrix} \beta_m \\ \beta_f \end{bmatrix} = \frac{1}{2}\begin{bmatrix} (G_m\beta_m + B\beta_f) \\ (B\beta_m + G_f\beta_f) \end{bmatrix}, \tag{13.5}$$

where the factor of $\frac{1}{2}$ accounts for sex-limited expression; β_m and β_f are the selection gradients acting on the two sexes. We assume that the two traits are under Gaussian viability selection towards intermediate optima (θ_m and θ_f). Each generation viability selection is followed by a process of assortative mating and frequency-dependent sexual selection. The total selection gradients in (13.5) are composed of two terms, representing the directional forces of viability and sexual selection.

$$\begin{bmatrix} \beta_m \\ \beta_f \end{bmatrix} = \begin{bmatrix} \dfrac{(\theta_m - \bar{z})}{\Omega_m} + c_m \\ \dfrac{(\theta_f - \bar{y})}{\Omega_f} + c_f \end{bmatrix}, \tag{13.6}$$

where $\Omega_m = \omega_m + P_m$ and $\Omega_f = \omega_f + P_f$. If sexual selection does not affect mean fitness and only affects the distribution of offspring among the individuals of each sex, the adaptive landscape for the two sexes is defined solely by natural selection. Furthermore, if no individual expresses both z and y, the phenotypic covariance between the two traits is undefined, and there is correlational selection between the two traits, although stabilizing selection may be stronger in one sex, so that $\omega_m \neq \omega_f$. As a consequence, the adaptive landscape is circular, as in Figure 13.2a, b, or elliptical with no tilt (Lande 1980). For simplicity, we will assume that the forces of sexual selection acting on the two sexes, c_m and c_f, are constant from generation to generation.

Setting the two terms in expression (13.5) equal to zero, we find that evolution ceases at an equilibrium given by

$$\bar{z} = \theta_m + \Omega_m c_m \tag{13.7a}$$

$$\bar{y} = \theta_f + \Omega_f c_f, \tag{13.7b}$$

where Ω_m and Ω_f are the 'variances' of the Gaussian adaptive landscape in the $\bar{z}$ and $\bar{y}$ dimensions, and the second terms are the distances that the equilibrium means are pulled off the natural selection optima by sexual selection. We see from (13.7) that the two traits will equilibrate on their adaptive peaks if there is no sexual selection ($c_m = 0$ and $c_f = 0$) (Figure 13.2). On the other hand, if sexual selection acts only on males ($c_f = 0$), males will be pulled off their adaptive peak at equilibrium, but females will not be pulled off their peak.

To analyse the evolutionary approach to the equilibrium, it will be useful to consider a special case. For simplicity, let us focus on the stage early in the process of evolving sexual dimorphism, when the variances the two sexes are equal, $P_m = P_f = P$ and $G_m = G_f = G$, and the two sexes experience similar strengths of stabilizing selection, $\omega_m = \omega_f = \omega$. It will also be useful to define the *sexual average* as $a = \frac{1}{2}(\bar{z} + \bar{y})$ and *sexual dimorphism* as $d = (\bar{z} - \bar{y})$. These new traits, a and d, constitute a 90 deg rotation of the axes in Figure 13.1.

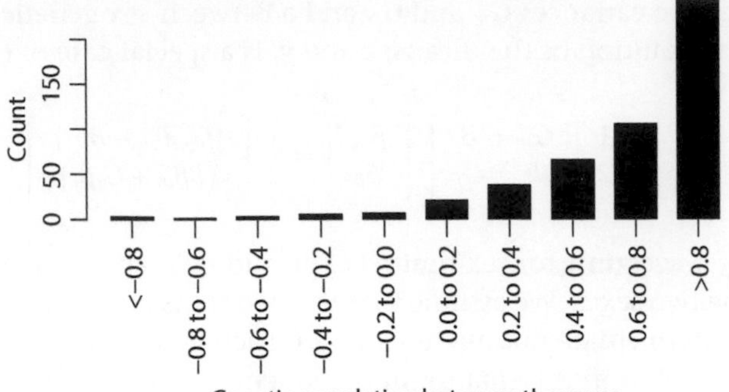

Figure 13.4 Distribution of estimates of genetic correlations between the sexes for homologous traits. The histogram summarizes 488 estimates from 114 sources, including both plants and animals. After Poissant et al. (2010).

Substituting these definitions of a and d into (13.5), we obtain equations for the evolution of the sexual average and sexual dimorphism

$$\Delta a = \frac{1}{2}(G+B)\left\{\frac{\frac{1}{2}(\theta_m+\theta_f)-a}{\Omega}+\frac{1}{2}(c_m+c_f)\right\} \tag{13.8a}$$

$$\Delta d = \frac{1}{2}(G-B)\left\{\frac{(\theta_m-\theta_f)-d}{\Omega}+(c_m-c_f)\right\}. \tag{13.8b}$$

We see from (13.8) and especially from a comparison of Figure 13.1a and b, that the genetic covariance between the sexes, B, amplifies evolution of the sexual average but impedes evolution of sexual dimorphism.

Empirical studies reveal a striking pattern of high genetic correlations between the sexes for homologous traits (Figure 13.4). In half of the estimates, the genetic correlation was greater than 0.8. These results, in conjunction with Lande's (1980) model, suggest that a strong genetic correlation between the sexes will often affect the sexual coevolution. The effect of such a strong genetic correlation, which we saw in the case of two homologous traits (Figure 13.1b), will also carry over to the case of multiple homologous traits.

13.4 Maintenance of Genetic Variance by a Balance Between Drift and Stabilizing Selection on a Gaussian Landscape With a Single, Stationary Peak

Up to his point we have considered deterministic evolution on an adaptive landscape. Now we will consider the effect of finite population size in opposing the effects of multivariate stabilizing selection. The following model is the multivariate version of the drift-selection balance model (single-peak OU) that we considered in Chapter 12. In this more general version, we see that the fact of multiple traits complicates both the adaptive landscape and the contribution of drift. The account given here expands on Lande's (1976) univariate model and on Hansen and Martin's (1996) multivariate formula.

Each generation, the overall change in the vector of lineage trait means is the sum of two contributions: drift, which tends to move the trait mean away from the optimum, and stabilizing selection, which tends to pull the trait mean back towards the optimum,

$$\bar{\mathbf{z}}_{t+1} = \bar{\mathbf{z}}_t + \boldsymbol{G}\Omega^{-1}\left[\boldsymbol{\theta} - \bar{\mathbf{z}}_t\right] + N\left(\mathbf{0}, \boldsymbol{G}/N_e\right). \tag{13.9}$$

The last term gives the contribution of drift, which is a random draw from a multivariate normal distribution with mean vector of zeros and a variance-covariance matrix, $\boldsymbol{G}/N_e$. We recognize the second term on the right as a response to directional selection, $\boldsymbol{G\beta}$. At generation t, the lineage trait means are multivariate normally distributed about a mean vector of zeros with a variance-covariance matrix given by

$$Var\left(\bar{\mathbf{z}}_{t+1}\right) = \frac{\Omega}{2N_e}\left[1 - exp\left\{-2t\boldsymbol{G}\Omega^{-1}\right\}\right]. \tag{13.10}$$

The first term involving N_e represents the balance between drift and stabilizing selection, while the second represents responses to directional selection. In the last term, we see exponential decay in the influence of the G-matrix, so that the dispersion matrix eventually achieves a limiting value given by

$$Var\left(\bar{\mathbf{z}}_{\infty}\right) = \frac{\Omega}{2N_e}. \tag{13.11}$$

Consequently, on long timescales, the probability distribution of lineage means is multivariate normal and converges on

$$\Phi\left(\bar{\mathbf{z}}_{\infty}\right) = \sqrt{(2\pi)^{-m}\left|Var(\bar{\mathbf{z}}_{\infty})\right|}^{\,-1} exp\left\{-\frac{1}{2}(\bar{\mathbf{z}} - \boldsymbol{\theta})^{\mathrm{T}} Var(\bar{\mathbf{z}}_{\infty})^{-1}(\bar{\mathbf{z}} - \boldsymbol{\theta})\right\}. \tag{13.12}$$

Notice that while the AL appears in the last two equations, the G-matrix does not, indicating that the long term pattern of lineage means is primarily shaped by the AL (Figure 13.5).

The adaptive landscape exerts a strong influence on the outcome of drift-selection balance. Figure 13.5 illustrates relevant cases of bivariate stabilizing selection, showing that even very weak stabilizing selection strongly constrains the extent to which lineage means depart from the immediate vicinity of the adaptive peak. To appreciate the remarkable constraint exerted by the AL, compare Figure 13.5 with Figure 9.1, which shows the 95% confidence ellipse for lineage means evolving by drift alone. Those ellipses are considerably larger than the comparable ellipse for the G-matrix. In contrast, even very weak stabilizing selection constrains evolving lineage means so that they occupy a bivariate area that is a fraction of the size of the ellipse that represents genetic values dispersed around the bivariate mean (Figure 13.5a, b). The AL also strongly constrains the shape of the lineage dispersion. Those dispersion patterns (shown as dotted blue ellipses), always mirror the shape of the AL and are almost completely independent of the shape of the G-matrix, as shown by (13.5d). Notice in Figure 13.5d that although the long-term expectation for dispersion (blue dotted ellipse) mirrors the AL and not $\boldsymbol{G}$, the sample path appears constrained by the narrow width of the G -ellipse. This appearance of constraint

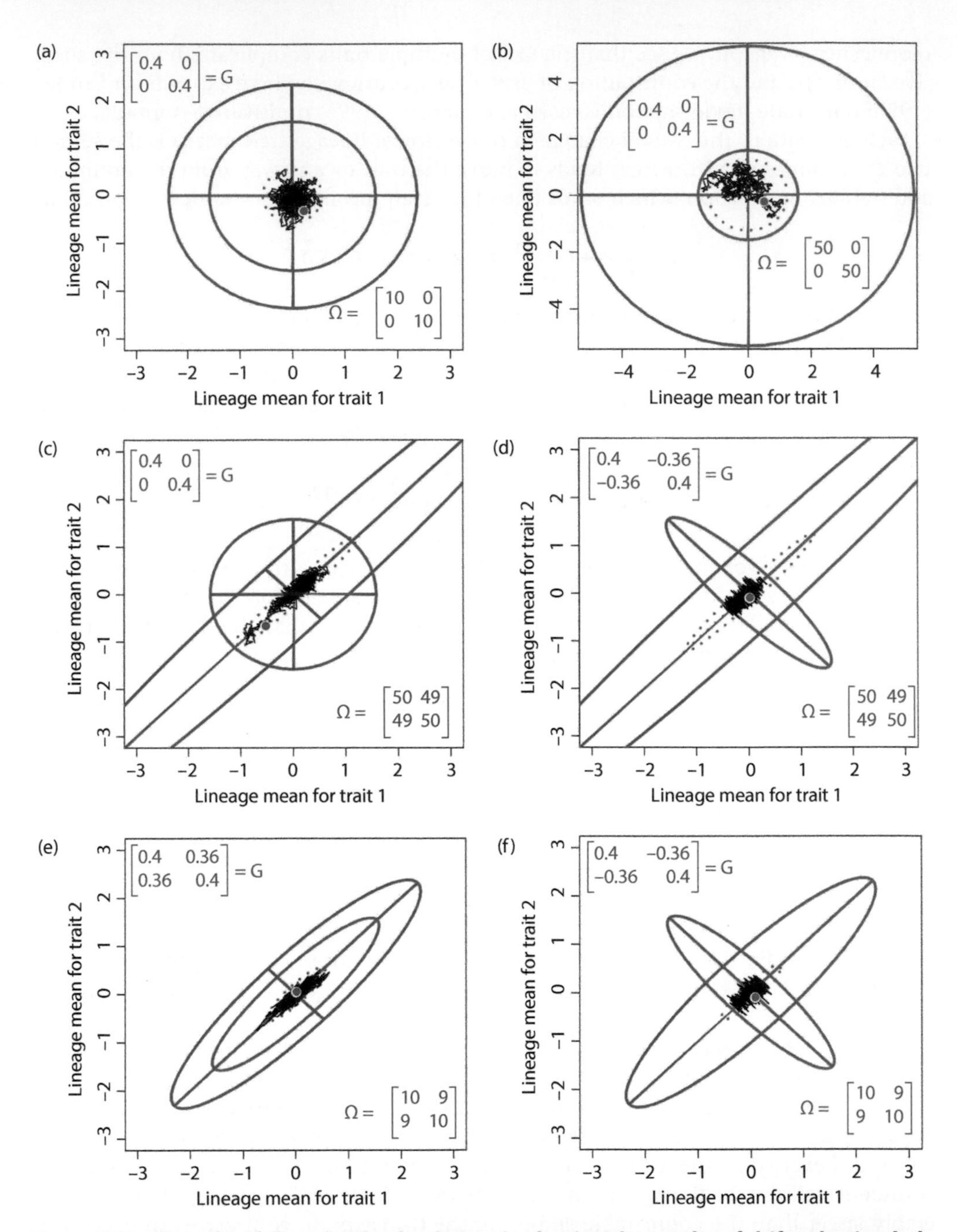

Figure 13.5 The role of the adaptive landscape in shaping the results of drift-selection balance. In each panel the black line shows the results of 1000 generations of simulated evolution of the lineage mean, $\bar{\mathbf{z}}_t$, in a single lineage under the conditions specified in each panel. Each trajectory begins at the origin and ends at a point denoted by a solid red dot. Effective population size is 100. The 95% confidence ellipse for genetic values is shown in solid blue and represents the *G*-matrix. The adaptive landscape is bivariate stabilizing in all cases and is represented by its 25% confidence ellipse in red. Such a low level ellipse must be employed, because the limits of the 95% ellipse would be far outside the scale limits of the figure. The limiting pattern for dispersion of lineage means,

(Continued)

will vanish on a longer timescale (i.e., $\gg$ 1000 generations). Finally, notice that particularly favourable cases for lineage dispersion are presented in Figure 13.5 by using a very small effective population size (N_e = 100). For more realistic sizes, say N_e in the range 500 to 5000, the lineage dispersion ellipses would be minuscule and the paths of simulated lineage paths could not be visually resolved. It is no exaggeration to say that, in the absence of peak movement, even very weak stabilizing selection eliminates the possibility of appreciable radiation among lineages.

Does the process of drift-selection balance produce the kind of distribution of directional selection gradients that we see in nature? A representative sample of directional selection gradients for a bivariate trait mean at quasi-equilibrium for 1000 generations is shown in Figure 13.6. Compared with the survey of β-values illustrated in Figure 3.12a, the distribution does show a shortage of values very close to zero, corresponding to trait means very close to the optimum, but despite this discrepancy, the overall distribution is much closer to empirical observations than one produced by approach to the optimum from afar (Figure 13.2).

13.5 Hill Climbing With Two Adaptive Peaks

Evolution on a landscape with two peaks in a two-trait space presents some complications that are not easy to anticipate from the univariate trait case. Complication arises especially from the probably common situation in which both $\boldsymbol{G}$ and the AL are elongate ellipses. As we see in Figure 13.1f, this situation causes curved trajectories, especially far from the optimum. If the landscape has two peaks, this curved feature could very much affect the long-term outcome of evolution.

13.6 The Coevolution of Sexual Preference and Sexually Selected Traits

Darwin (1859, 1871) proposed a special process, *sexual selection*, to account for the evolution of traits that aid males in obtaining mates by sexual persuasion of females or by physical offence and defence in contests with sexual rivals. In Darwin's view, ordinary viability selection could not account for the elaboration of such traits, and indeed that selective agency would often act in the opposite direction, towards diminution. While sexual preference of females for elaborate male traits seemed inevitable to Darwin, it

Figure 13.5 (*Continued*)
Var ($\bar{\mathbf{z}}_\infty$), is represented by its 95% confidence ellipse in dotted blue, calculated using (13.11). (a) Weak stabilizing selection with no selectional correlation and no genetic correlation. (b) Very weak stabilizing selection with no selectional correlation and no genetic correlation. Notice the change in scale. (c) Very weak stabilizing selection with strong positive selectional correlation and no genetic correlation. (d) Very weak stabilizing selection with strong positive selectional correlation and a strong negative genetic correlation. (e) Weak stabilizing selection with strong positive selectional correlation and a strong positive genetic correlation. (f) Weak stabilizing selection with strong positive selectional correlation and a strong negative genetic correlation. Animations of these figures are available at https://phenotypicevolution.com/?page_id=33.

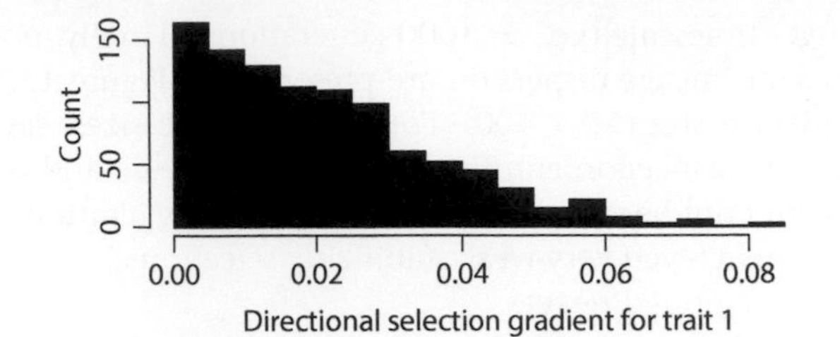

Figure 13.6 Absolute values of the directional selection gradient during a simulation of drift-selection balance. The values for directional selection on trait 1 (n=1000) are for the trajectory beginning at $\bar{z}_1 = 0$, $\bar{z}_2 = 0$ in Figure 13.5c and lasting for 1000 generations.

Figure 13.7 Predators and mates exert opposing forces of selection on a male ornament. In this hypothetical example, selection by the peahen on the peacock's tail favours a larger, more complex tail, while selection by a predator (the tiger) favours small tails that promote more rapid escape. Drawing by Dafila Scott. From Stearns and Hoekstra (2000) with permission.

remained controversial for decades into the twentieth century. Fisher (1915, 1930), however, sketched a model for Darwin's process of mate choice that became famous for its prediction of *runaway evolution* of female preference and male trait. In this section, we summarize Lande's (1981) model of the Darwin-Fisher proposal, with some simplifications in the interest of getting quickly to the main conclusions.

Lande (1981) modelled the joint evolution of two traits with sex-limited expression: a male trait, that we will call the *ornament* (e.g., tail size in peacocks), and a female trait, *sexual preference* for the male ornament (Figure 13.7). This model has the interesting feature that evolution results in a line of equilibria, rather than a single unique outcome. The line of equilibria is a consequence of specifying two types of selection on the male ornament: viability selection represented by the tiger in Figure 13.7, and mating preferences, represented by the peahen in the same figure.

Let z be the phenotypic value of a normally distributed male ornament. Before selection the mean of z is $\bar{z}$ and its variance is P_m. The ornament is affected by many autosomal genes of small effect so that its breeding value is normally distributed with genetic variance G. Each generation Gaussian viability selection acts on the ornament in a stabilizing mode with an intermediate optimum θ_m. This viability selection changes the distribution of z so that after selection its mean and variance are

$$\bar{z}^* = (\bar{z}\omega + \theta P_m) / \Omega_m \tag{13.13}$$

$$P_m^* = \omega P_m / \Omega_m \tag{13.14}$$

where ω is the 'variance' of the viability ISS function, and the selection differential due to viability selection is

$$\bar{z}^* - \bar{z} = (\theta - \bar{z}) / \Omega_m. \tag{13.15}$$

Each generation, following viability selection, the population of surviving males is subjected to sexual selection by females that choose their mates on the basis of the male ornament. Female preference, y, is a normally distributed trait with phenotypic mean $\bar{y}$ and variance τ^2. We suppose that female preference is affected by many genes, so that breeding values are normally distributed with variance H. Female choice of males occurs in the following way. The tendency of the female with preference value y to mate with a male with ornament value z is

$$\psi(z|y) \propto exp\left\{-(z-y)^2/2v^2\right\}. \tag{13.16}$$

In other words, a female is most likely to mate with an encountered male if his ornament matches her preference, so that $z = y$, and her tendency to mate falls off as a Gaussian function with width v as his ornament deviates from that optimum, y. Averaging this preference function over the distribution of female preferences, we obtain a Gaussian function that gives us the overall probability of females mating as a function of male ornament value, z,

$$\psi(z) = b\,exp\left\{-\left(z-\bar{y}\right)^2/2\left(\tau^2+v^2\right)\right\} \tag{13.17}$$

where b is a constant (Arnold et al. 1996). After sexual selection by a subset of female with preference y, the ornament mean $\bar{z}^*$ will be shifted by an amount

$$(y - \bar{z}^*)\, P_m^* / \left(v^2 + P_m^*\right). \tag{13.18}$$

Averaging these shifts over the entire female population yields the selection differential due to sexual preference, $\bar{z}^{**} - \bar{z}^*$. The total selection differential is the sum of viability and sexual differentials, giving a total selection gradient for the ornament

$$P_m^{-1}s = P_m^{-1}\,[(\bar{z}^{**} - \bar{z}^*) + (\bar{z}^* - \bar{z})] = P_m^{-1}\,(\bar{z}^{**} - \bar{z}) \tag{13.19a}$$

$$P_m^{-1}s \approx \frac{\bar{y}/\alpha - \left(1 + \frac{1}{\alpha}\right)\bar{z} + \theta}{\omega} \tag{13.19b}$$

where $\alpha \cong v^2/\omega$

Turning to females, let us assume that a female's fitness (progeny count) is unaffected by her mate choice, with the consequence that no selection acts on female preference, y. As a result, average preference, $\bar{y}$, will evolve only as a correlated response to selection on the ornament. We realize from earlier results (13.5), that such a correlated response will depend on a genetic covariance between the two traits, B, a genetic covariance between the sexes. This genetic covariance is unlikely to be a consequence of pleiotropy, because in general it is difficult to imagine a contributing locus that affects a male ornament will also have a pleiotropic effect on female preference, or vice versa. On the other hand, B could reflect linkage disequilibrium. Mate choice in the model leads to assortative mating between the z and y phenotypes as well as selection that favours particular combinations of z and y. Lande (1981) reports a submodel showing that linkage disequilibrium does result from this combination of assortative mating and sexual selection and accumulates across loci to constitute a genetic covariance between the sexes, B. Furthermore, the primary determinate of the magnitude of B is the ratio of mutation rates affecting female preferences versus the male ornament.

Putting these results together with an expectation that G and H will equilibrate in mutation-selection balance, Lande (1981) concludes that the G-matrix for our two characters,

$$\begin{bmatrix} G & B \\ B & H \end{bmatrix}, \tag{13.20}$$

is likely to be stable if stabilizing selection on the preference is weak ($\omega \gg P_m$).

Using our standard expression for the response to two traits to selection (12.1) and recalling that we have not allowed direct selection on y, we have

$$\Delta\bar{z} = \frac{1}{2} G P_m^{-1} s \tag{13.21a}$$

$$\Delta\bar{y} = \frac{1}{2} B P_m^{-1} s \, . \tag{13.21b}$$

In other words, the ornament mean evolves in response to viability and sexual selection on the ornament, whereas the preference evolves as a correlated response to that selection.

To find the equilibrium, we set the total selection gradient equal to zero and find that the equilibrium is a line

$$\bar{y} = (\alpha + 1)\,\bar{z} - \alpha\theta \, . \tag{13.22}$$

In other words, at any position along the equilibrium line, the force of viability selection pulling $\bar{z}$ towards the optimum θ is exactly balanced by a force of sexual selection in the opposite direction (Figure 13.8).

To analyse the stability properties of this equilibrium, it is useful to shift the coordinate system by a constant in the $\bar{z}$-dimension by defining

$$\tilde{z} = \bar{z} - \theta / \left(1 + \frac{1}{\alpha}\right) \tag{13.23a}$$

$$\tilde{y} = \bar{y} \, . \tag{13.23b}$$

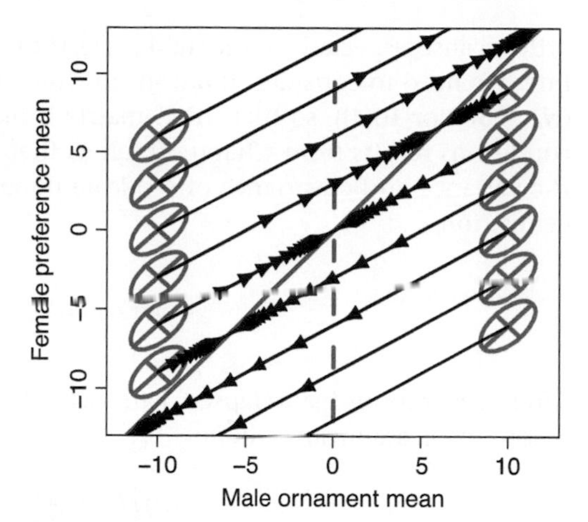

Figure 13.8 Lande's (1981) model for the deterministic evolution of a male ornament and female mating preference based on that ornament. The stable case is illustrated. The solid red line is the line of equilibria. The dashed red line shows the position of the viability selection optimum, θ, for the male ornament. Solid black lines show evolutionary trajectories for 12 starting positions, with arrows, denoting the response to selection, spaced every generation. The blue ellipses at the starting points are 95% confidence ellipses for the G-matrices. Parameter values are: $G = H = 0.4; B = 0.24; \omega = 4, \alpha = 0.1, P = 1, \theta = 0$. The similarity between g_{max} and the line of equilibria in this figure is a coincidence.

In this new coordinate system, the selection gradient is

$$\beta = P_m^{-1} s = \frac{\tilde{y}/\alpha - \tilde{z}\left(1 + \frac{1}{\alpha}\right)}{\omega}, \tag{13.24}$$

and consequently the line of equilibria is

$$\tilde{y} = (\alpha + 1)\,\tilde{z}. \tag{13.25}$$

Substituting (13.23) into (13.21), we obtain a simple expression for evolution in our new coordinate frame,

$$\begin{bmatrix} \Delta\tilde{z} \\ \Delta\tilde{y} \end{bmatrix} = \frac{1}{2\alpha\omega} \begin{bmatrix} -(\alpha+1)\,G & G \\ -(\alpha+1)\,B & B \end{bmatrix} \begin{bmatrix} \tilde{z} \\ \tilde{y} \end{bmatrix} = \frac{1}{2\alpha\omega} \begin{bmatrix} -(\alpha+1)\,G\tilde{z} + G\tilde{y} \\ -(\alpha+1)\,B\tilde{z} + B\tilde{y} \end{bmatrix}. \tag{13.26}$$

From (13.21) we see that $\Delta\bar{y}/\Delta\bar{z} = B/G$, indicating that evolution will occur along straight line trajectories with a slope given by the genetic regression B/G. From (13.26), we obtain the same result. Taking the eigenvectors of the matrix in (13.26), we find that one eigenvector corresponds to B/G, and its eigenvalue is

$$\lambda = [B - (\alpha + 1)\,G] \,/\, (2\alpha\omega). \tag{13.27}$$

Consequently, we can conclude that after t elapsed generations, the vector of trait means evolving along a B/G trajectory will have changed by an amount $(1 + \lambda)^t$. If $B/G < (\alpha + 1)$, then λ is negative, and evolving populations approach the line of equilibria with ever decreasing speed; the equilibrium is stable (the *walk-towards* shown in Figure 13.8). But if $B/G > (\alpha + 1)$, λ is positive, and populations evolve away from the line of equilibrium at ever increasing speed (the *runaway* proposed by Fisher 1958); the equilibrium is unstable. The other eigenvector is the line of equilibrium itself with an associated eigenvalue equal to zero, indicating responses to selection cease on that line. The simplest change in ecology that could trigger the unstable, runaway case is a relaxation in stabilizing selection (an increase in ω) so that $B/G > \left(v^2/\omega\right) + 1$.

So far, we have considered only deterministic evolution of preference and ornament, ignoring the possibility of drift. The existence of a line of equilibria invites us to consider drift because, especially in the stable case, the bivariate mean might drift up and down that line. We need to consider situations in which effective population size is relatively large ($N_e \geq 500$ or 1000), so that the G-matrix is likely to be relatively constant. Under these conditions results from Chapter 9 tell us that random sampling of parents will increase the variance and covariance of replicate population means by a constant amount each generation,

$$Cov(\bar{z}, \bar{y}) = \frac{1}{N_e} \begin{bmatrix} G & B \\ B & H \end{bmatrix}. \tag{13.28}$$

After t generations have elapsed, Lande (1981) shows that in the stable case the dispersion among replicate means will be

$$\boldsymbol{D}(t) \approx \boldsymbol{K} + \frac{H\left(1 - r_g^2\right) t}{N_e(\alpha + 1 - B/G)^2} \begin{bmatrix} 1 & \alpha + 1 \\ \alpha + 1 & (\alpha + 1)^2 \end{bmatrix}, \tag{13.29}$$

where $\boldsymbol{K}$ is a constant matrix and $r_g = B/\sqrt{GH}$. Thus, genetic variance for preference, H, which played no role in deterministic responses to selection, plays a major role in the drift of both preference and ornament. This result makes intuitive sense. The preference distribution is not under direct selection and is free to drift along the line of equilibria. As it drifts, that preference distribution exerts directional selection on the male ornament distribution, so that the drift of $\bar{z}$ is paced by the inheritance of y, i.e., by H.

Simulations by Uyeda et al. (2009) confirm Lande's (1981) interpretation that drift along the line of equilibria can produce rapid evolutionary diversification, even with N_e as large as $5000 - 10,000$. Those same simulations reveal that $\bar{z}$ and $\bar{y}$ are extremely close together as a lineage drifts along the equilibrium line, departures from the line are minuscule, and that the two means often switch positions, so that sometimes $\bar{y} > \bar{z}$ and sometimes $\bar{z} > \bar{y}$.

Iwasa and Pomiankowski (1995) described a process that yields perpetual evolution of a male ornament and female mating preference. Their model differs from Lande's (1981) model in a couple of important respects. First, viability selection on the male trait is a fourth-power function, $exp\left\{c\bar{z}^4\right\}$ where c is a constant, so that the viability function is more flat topped than a Gaussian or quadratic function. The optimum trait value with respect to viability selection is arbitrarily set to 0. Female mating preference for a particular male is an exponential function of his trait value, $\{a\,(z - \bar{z})\,\bar{y}\}$ $\exp\{a(z - \bar{z})\bar{y}\}$, where a is a constant. Secondly, selection acts on female mating preferences, so that female fitness with respect to her preference value y is $exp\left\{-by^2\right\}$. Ordinarily, the effect of selection on female preferences is to collapse a line of equilibria to a point, as in Figure 13.8, so that in the stable case, the joint evolution of ornament and preference has a single unique outcome (Iwasa et al. 1991; Pomiankowski et al. 1991). In the present model, however, when the assumption of selection on preferences is combined with a flat-topped viability selection function for the male trait, the combined effect is to produce an equilibrium curve that is a cubic function. This cubic function has the remarkable property of producing a stable limit cycle in a population of infinite size (Iwasa and Pomiankowski 1995). This regular cyclical behaviour of the model disappears in populations of finite size even if they are very large (Figure 13.9), but nevertheless evolution appears to consist of perpetual swings between alternative stable zones on either side of the viability optimum, $\bar{z} = 0$. For example, in Figure 13.9d, two lineages make the transit between alternative stable zones, but on a longer timescale, such transits are relatively common. Alternatively,

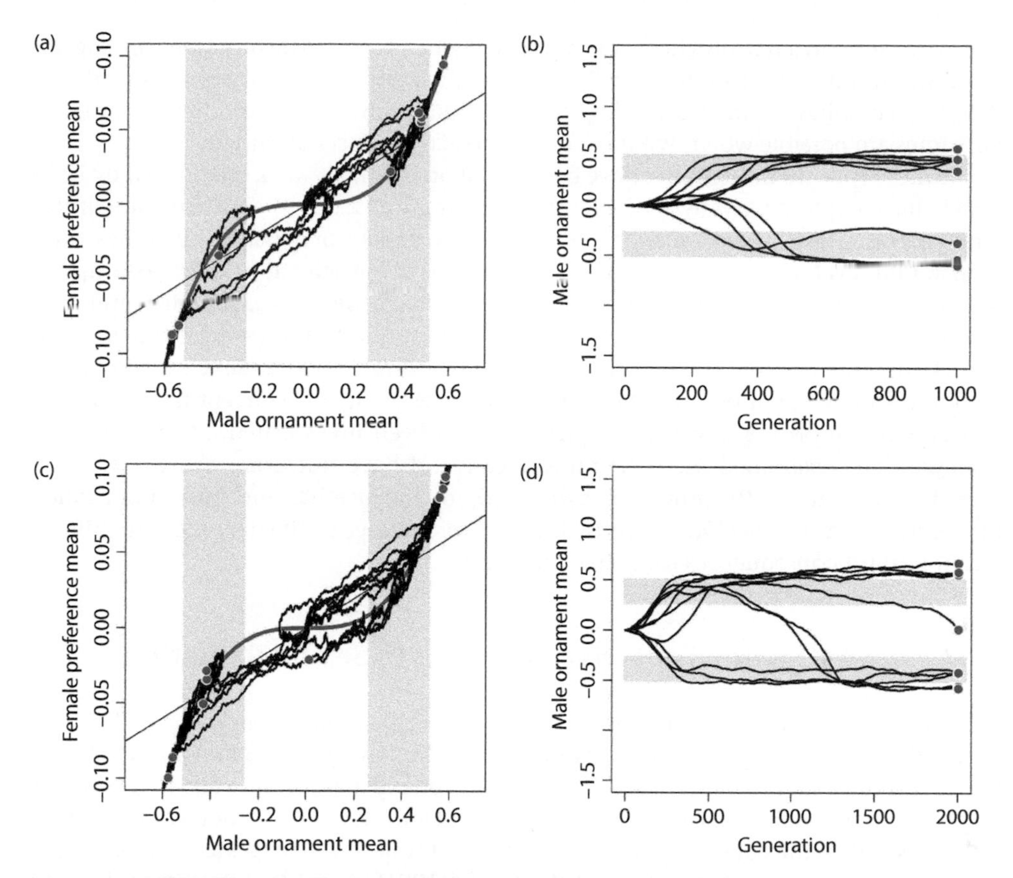

Figure 13.9 Simulations of Iwasa and Pomiankowski's (1995) model for perpetual evolution of a male ornament and a female mating preference. The evolutionary trajectories of two sets of 10 replicate lineages are shown with origination at $\bar{z} = \bar{y} = 0$, with simulation durations of 1000 (above) and 2000 generations (below). Red dots show the positions of each bivariate lineage mean at the end of the simulation. In this model the line of equilibria, the red curves in the panels on the left, is a cubic function. This line of equilibria consists of stable and unstable zones. The transition to flanking stable zones is shown as grey bands. The central unstable zone is located between the grey bands. Populations off the line of equilibria evolve deterministically along trajectories with a slope of B/G; one such line (thin black) is shown through the origin. (a and b) On a 1000 generation timescale, the tendency of lineages to stall within the zones of stability (at or beyond the grey bands) is apparent. (c and d) The 2000 generation timescale in this simulation provides more opportunities for transit between the two stable zones. Indeed, two lineages have evolved across the unstable zone, from one stable zone to the other, and another lineage has entered the unstable, central zone. Parameter values are: $N_e = 10^6$, $G = H = 0.5$, $B = 0.05$, $a = 0.4$, $b = 0.001$, $c = 0.05$, $u = 0$ (unbiased mutation).

with N_e smaller but still large (e.g., 10,000), transits are common on a timescale ≥ 1000 generations.

In several respects the Iwasa and Pomiankowski (1995) model should be considered a work in progress. For one thing, because of the model's parameterization, we do not know whether its assumptions about forms and intensity of selection on the two traits match empirical reality. Secondly, we suspect, but do not know, if its signature feature of limit cycles depends on a narrow choice of parameter values. Finally, the function that specifies the distribution of trait means as a function of time and N_e is unknown. Nevertheless, the model emphasizes the essential point, discussed below, that a variety of evolutionary outcomes are possible when two traits have interactive effects on fitness.

The more than 30 models that have been based on the framework established by Lande (1981) have explored a variety of alternative assumptions and established some important generalizations (Mead and Arnold 2004). The variety of possible outcomes (lines, points, limit cycles) depends mainly on assumptions about selection. Other assumptions have been used to capture the essential features of 'good genes', sexual conflict, and costly female choice. Nevertheless, despite the variety assumptions and possible outcomes, all models reveal that the outcome can be stable or unstable, depending on the balance between genetic variation and selection. In the face of all this conceptual exploration, it is perhaps surprising that some topics have not been investigated. Most models, for example, have examined the deterministic case and have not examined the effects of drift. Uyeda et al (2009) showed that finite population size has enormous implications for speciation, as Lande (1981) surmised, and as we shall see in the next section, drift also has important consequences for sexual radiations.

13.7 Accounting for Trait Diversification in Sexual Radiations

Despite the proliferation and discussion of models of the *Fisher-Lande process* (FLP) based on Lande's (1981) model, actual tests of model predictions have remained rare. This rarity reflects the dual problems of specifying predictions that are empirically accessible and matching those predictions with empirical data that are relevant. In other words, the models must speak to the data and vice versa. Arnold and Houck (2016) tackled both problems. First, they used a stochastic version of the Lande (1981) model that predicted the scope of sexual radiation on specified timescales. Second, they compared those predictions with actual data on sexually selected traits evolving on a time-calibrated phylogeny. The test case uses the iconic example of evolution by sexual selection, plumage diversification in birds of paradise.

The version of Lande's (1981) model used by Arnold and Houck (2016) was designed to match empirical realities in actual sexual radiation by incorporating three features: finite population size, multiple ornament traits, and selection on female preferences. They termed this version the *Phenotypic Tango Model* (PTM) for reasons that will become apparent later. (1) Making N_e a parameter in the model, allowed Arnold and Houck to develop their model in the now familiar framework of stochastic trait diversification (Chapter 12). (2) Using a model with two male ornament traits and two corresponding female mating preferences takes a step towards the emerging empirical finding that courtship and mating displays typically involve multiple traits and multiple channels of sexual communication (Andersson 1994). (3) Allowing natural selection on female preference traits reflects a growing consensus that although such selection is probably inevitable, even if weak (Pomiankowski 1987; Kirkpatrick and Barton 1997). However, these incorporations came at a cost. The PTM is sufficiently complicated that no solution is available for its predictions. That is to say, the expression for expected trait variance among lineage means at time t is unknown. Nevertheless, we can simulate the PTM under specified parameter values and evaluate its predictions.

The stochastic response to selection equation for the PTM, i.e., for the per-generation joint evolution of ornament traits and preferences, is

$$\begin{pmatrix} \Delta\bar{\mathbf{z}} \\ \Delta\bar{\mathbf{y}} \end{pmatrix} = \begin{pmatrix} \Delta\bar{z}_1 \\ \Delta\bar{z}_2 \\ \Delta\bar{y}_1 \\ \Delta\bar{y}_2 \end{pmatrix} = \frac{1}{2}\begin{pmatrix} \boldsymbol{G} & \boldsymbol{B} \\ \boldsymbol{B} & \boldsymbol{H} \end{pmatrix}\begin{pmatrix} \boldsymbol{\beta}_{\mathbf{z}} \\ \boldsymbol{\beta}_{\mathbf{y}} \end{pmatrix} + N\left[\mathbf{0}, \frac{1}{2N_e}\begin{pmatrix} \boldsymbol{G} & \boldsymbol{B} \\ \boldsymbol{B} & \boldsymbol{H} \end{pmatrix}\right], \quad (13.30)$$

where $\bar{z}_1$ and $\bar{z}_2$ are the means of the two male ornament traits, $\bar{y}_1$ and $\bar{y}_2$ are the means of the two female preference traits, with corresponding inheritance matrices ($\boldsymbol{G}$ and $\boldsymbol{H}$), $\boldsymbol{B}$ is the between-sex genetic covariance matrix, $\boldsymbol{\beta}_{\mathbf{z}}$ is the two-element vector total selection gradients for the two male ornament traits (each element is composed of a natural and a sexual selection gradient), and $\boldsymbol{\beta}_{\mathbf{y}}$ is the two-element vector of natural selection gradients for the two female preference traits (Arnold and Houck 2016). The first part of the expression gives the deterministic response to selection by the four traits, which is the four-variable version of (13.21), but with selection on all the elements of both $\mathbf{z}$ and $\mathbf{y}$. The second part of the expression represents the contributions of drift each generation, which are drawn from a four-variable normal distribution with mean vector of zeros and a genetic variance-covariance matrix shown in partitioned format.

Although we don't have a solution for the time course of this stochastic process, we know that in males the process should equilibrate at some distance from the natural selection optima, the deviation from that optimum arising from sexual selection (Lande 1980, 1981). We proceed by ignoring sexual selection and the deviation it produces, so that we can characterize the stochastic process arising from natural selection alone and use that characterization as a baseline. Accordingly, we set the trait optimum of a stationary natural selection AL at zero. After t generations in the absence of sexual selection, the means of the ornament traits of the replicate lineages will be multivariate normally distributed with a mean of zero and variance-covariance matrix given by

$$Var\left[\bar{\mathbf{z}}(t)\right] = \frac{\boldsymbol{\Omega}_{\mathbf{z}}}{2N_e}\left[1 - exp\left\{-2\left(\boldsymbol{\Omega}_{\mathbf{z}}{}^{-1}\boldsymbol{G}\right)t\right\}\right] \quad (13.31)$$

(Hansen et al. 2008).

If the natural selection optimum is not stationary and instead moves by Brownian motion, we can solve expressions that summarize the dynamics of the process in the absence of sexual selection, but the variance expression is more complicated than in the stationary case. In particular, in the case of genetically and phenotypically uncorrelated ornament traits, the distribution of lineage means for each trait at generation t is normal with a mean given by the common position of the optimum at generation 0 and with a variance given by the following univariate expression for dispersion of ornament means,

$$Var\left[\bar{z}_t\right] = \frac{\sigma_\theta^2 + \frac{G}{N_e}}{2a}\left[1 - exp\left\{-2at\right\}\right] + \sigma_\theta^2 t\left[1 - 2\frac{(1 - exp\left\{at\right\})}{at}\right], \quad (13.32)$$

where $a = (\omega_z + P)^{-1}G$ and σ_θ^2 is the variance in the position of the ornament optimum, θ (Hansen et al. 2008). This expression rapidly converges to

$$Var\left[\bar{z}_t\right] = t\sigma_\theta^2 . \quad (13.33)$$

Similar arguments apply to the stochastic evolution of the female preference means in response to natural selection (Arnold and Houck 2016, Appendix).

Animations of (13.30), e.g., those at https://phenotypicevolution.com/?page_id=33, reveal important characteristics of the full PTM with both natural and sexual selection in

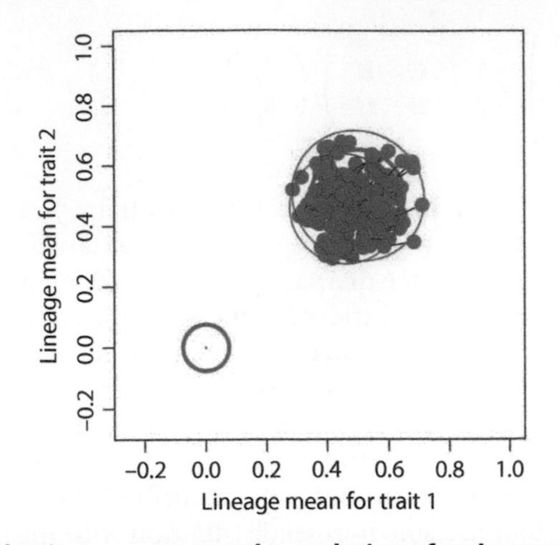

Figure 13.10 Sexual selection exaggerates the evolution of male ornaments and female preferences, producing a phenotypic tango. The evolution of the two male ornament means (blue dots) and the two female preference means (red dots) are superimposed in two-dimensional trait space. Trait means for a particular lineage are connected by thin black line. Means of 100 lineages are shown after 500 generations of evolution in response to natural and sexual selection. The 95% confidence limits for this sample of male ornament and female preference means are shown respectively as light blue and light red ellipses. The expected 95% confidence limits for male ornament and female preference means evolving in response to natural selection alone are shown as heavy ellipses (optima are shown as dots at the center of these ellipses). Crucial parameters are: $N_e = 5000$; $P_{ii} = Q_{ii} = 1$; $G_{ii} = H_{ii} = 0.4$; $B_{ii} = 0.24$; $\boldsymbol{\omega_z} = (4, 0, 0, 4)^{\mathrm{T}}, \boldsymbol{\omega_y} = (19, 0, 0, 19)^{\mathrm{T}}$. $\boldsymbol{Q}$ is the female analogue of the male phenotypic matrix $\boldsymbol{P}$. For other parameters and details see Arnold and Houck (2016), Figure 4B.

play. For sexual selection to exaggerate the male trait means beyond the bounds expected under natural selection, for example, stabilizing natural selection on preferences must be weaker than corresponding selection on ornament traits, as in Figure 13.10. In this figure we also see that the ornament means (blue dots) equilibrate in a cloud near the preference optima, which in this example is in a different location than the ornament optimum. This cloud is substantially larger than the ellipse (heavy blue) expected under natural selection alone. This enlarged dispersion is the signature of ornament exaggeration by sexual selection. In contrast, the preference means (red dots) equilibrate in a cloud centred about the natural selection optimum (at the centre of the heavy red ellipse) and only slightly larger than we expect without sexual selection on ornament traits (heavy red ellipse). Finally, the male and female means in a particular lineage remain close together throughout their joint evolution and appear to dance about in phenotypic tango. This motion is apparent in animations, https://phenotypicevolution.com/?page_id=33.

The analytical version of the PTM (and corresponding animations) establish that the Fisher-Lande process can produce evolutionary exaggeration of male ornament traits in multivariate space. The question remains, however, whether exaggeration happens as fast and to the same extent as in real sexual radiations. To answer this question, Arnold and Houck (2016) used simulations to determine whether the PTM could produce the quantitative characteristics of the sexual radiation of plumage in birds of paradise. Plumage (but not preference) measurements were available for one genus (*Paradisaea*). Those measurements (LeCroy 1981) indicate that the among-population and species radiation in two

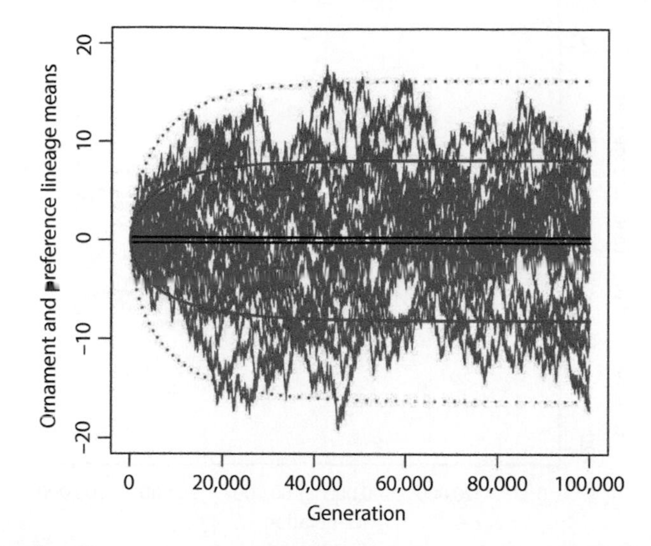

Figure 13.11 Simulated ornament and preference means evolving for 100,000 generations showing that the PTM can match observed divergence in birds of paradise ornament traits. Ornament (blue) and preference (red) means are shown in units of within-population standard deviation ($\sqrt{P_{ii}}$ and $\sqrt{Q_{ii}}$). Black curves show the expected 99% confidence limits for ornament evolution in response to natural selection alone. Bold red lines show the comparable limits for preference evolution. Dotted red lines show two times the 99% confidence limits for preferences. Simulation of 25 lineages with $N_e = 500$; $P_{ii} = Q_{ii} = 1$; $G_{ii} = H_{ii} = 0.4$; $B_{ii} = 0.24$; $\boldsymbol{\omega_z} = (9, 0, 0, 9)^{\mathrm{T}}$, $\boldsymbol{\omega_y} = (9999, 0, 0, 9999)^{\mathrm{T}}$. See Arnold and Houck (2016) for more details and parameter values.

male plumage traits spans about $10 - 12\sqrt{P}$ on either side of the bivariate mean. Using the time-calibrated phylogeny of Irestedt et al. (2009), this span in trait divergence happened over a 9 million year time period (1.8 million generations). In other words, if we run simulations of the PTM and produce plots of trait evolution in replicate lineage, the 99% confidence limits for traits values need to span $\pm 10 - 12\sqrt{P}$ on a timescale less than or equal to about 2 million generations. The challenge, of course, is to find a set of realistic parameter values that produce the requisite amount of divergence in the allotted time. Two examples of such success are shown below.

In Figure 13.11, we see several results that were typical over a broad range of parameter values. First, the ornament and preference means of a lineage are tightly coupled during evolution. Notice in particular that the each evolving ornament mean (blue) mirrors the evolution of its lineage preference mean (red). This close correspondence mean that the observed 99% confidence limits of the ornament mean accurately approximates the about two times the 99% confidence limits for evolving preference (which are unobserved in our bird of paradise example). Second, sexual selection exaggerates the evolving ornament mean, carrying it far beyond natural selection expectations for both ornament (bold black lines) and preferences (bold red lines). Third, that ornament exaggeration can match the divergence target of $\pm 10 - 12\sqrt{P}$ in as little as 20,000 generations. Fourth, with a stationary optimum, we must invoke extraordinarily very weak natural selection on preferences to produce the requisite amount of ornament exaggeration (i.e., $\omega_{yii} \approx 10,000$). In other words, to account for the observed ornament divergence with the PTM using stationary natural selection preference optimum, we must invoke such weak stabilizing selection on preference (and/or such a small N_e) that the model enters the domain of implausibility.

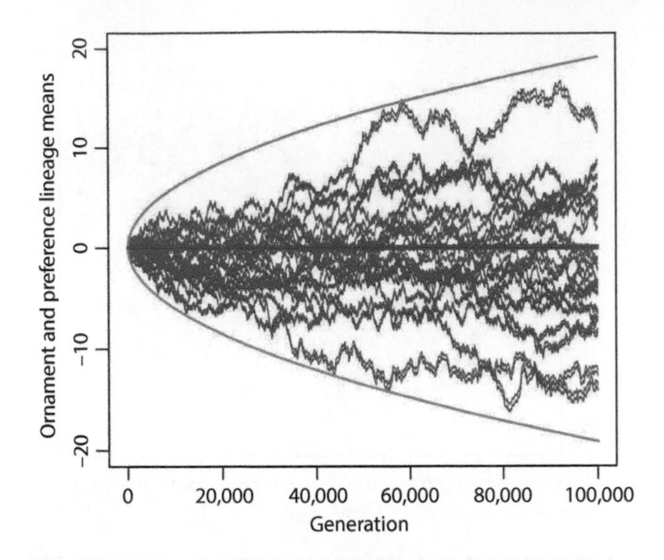

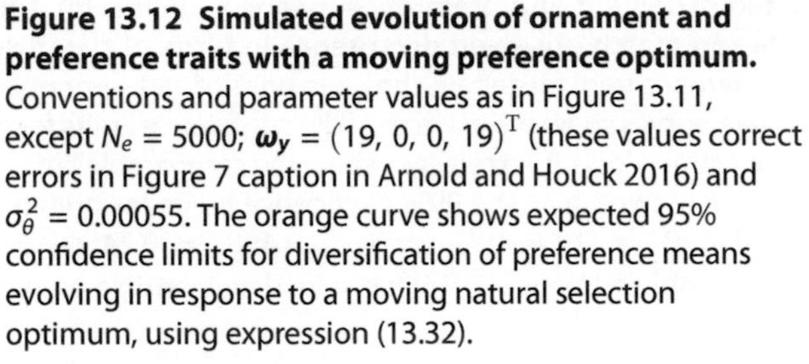

Figure 13.12 Simulated evolution of ornament and preference traits with a moving preference optimum. Conventions and parameter values as in Figure 13.11, except $N_e = 5000$; $\boldsymbol{\omega_y} = (19, 0, 0, 19)^{\mathrm{T}}$ (these values correct errors in Figure 7 caption in Arnold and Houck 2016) and $\sigma_\theta^2 = 0.00055$. The orange curve shows expected 95% confidence limits for diversification of preference means evolving in response to a moving natural selection optimum, using expression (13.32).

From Arnold and Houck (2016).

In the second example of a successful simulation (Figure 13.12), we escape from the implausibility that plagued the first example. Here we produce the requisite amount of divergence in the ornament trait without invoking small population size or implausibly weak stabilizing selection on the preference. In this second example, population size is 5000, stabilizing selection on preferences is relatively strong, and the natural selection optimum moves a very small amount each generation by Brownian motion. Nevertheless, male ornament divergence achieves the bird of paradise level in only 20,000 generations. Clearly, the Fisher-Lande process, in the form of the PTM, can account for pace and magnitude of sexual radiations such as the birds of paradise with benign assumptions about parameters.

Arnold and Houck (2016) used the OUwie hypothesis-testing framework devised by Beaulieu et al. (2012) to test alternative models of Brownian motion (BM) and Ornstein-Uhlenbeck stabilizing selection (OU) as explanations for the plumage radiation in *Paradisaea* (Figure 13.13). The advantage of this framework, as opposed to the simulation approach that we just reviewed, is that we can incorporate the topology of the bird of paradise phylogeny, and we can use maximum likelihood to compare fits to alternative models of process. In addition, we can estimate process parameters ($N_e, \omega_z, \sigma_\theta^2$) and use those to evaluate the empirical reality of each model fit. Two versions each of BM (BMS, BM1) and OU (OUM, OU1) were evaluated: one in which the parameters were constant across the entire tree, and one in which parameters could change from one a priori defined tree region to another (i.e., from Regime 1 to Regime 2). The resulting estimates of process parameters are shown in Table 13.1; see Arnold and Houck (2016) for estimates of

Figure 13.13 Two male birds of paradise (*Paradisaea apoda*) displaying to a female, showing the role of the wings and tail in forming a static shape. The evolution of the lengths of the male wings across the *Paradisaea* radiation are analyzed in Table 13.1.
From Arnold and Houck (2016).

Table 13.1 Comparison of Brownian motion and Ornstein-Uhlenbeck model fits to data on wing length evolution in the *Paradisaea* radiation using OWwie. OU1 and BM1 estimate one set of parameters across the entire tree. BMS and OUM models estimate separate sets of parameters for Regime 1 and Regime 2. From Arnold and Houck (2016).

Model	AICc	Regime 1	Regime 2
BMS	0.00	$N_e = 25,157$; $\sigma_\theta^2 = 0.000016$	$N_e = 722$; $\sigma_\theta^2 = 0.00055$
OUM	1.11	$\omega_z = 6,404$	$\omega_z = 6,404$
OU1	3.80	$\omega_z = 25,477$	
BM1	6.08	$N_e = 1,105$; $\sigma_\theta^2 = 0.00036$	

the generic parameters in the OUwie models (i.e., σ^2, stochastic diversification rate; α, restoring force in OU models; and θ, position of the OU optimum).

The two models that allowed regime-specific parameter values (BMS, OUM) gave appreciably better fits than the two models that fit one set of values across the entire tree (Table 13.1). In particular, OUM fit two values for the trait optimum (not shown in Table 13.1) and consequently estimated a somewhat more realistic value for stabilizing selection, ω_z, than OU1; estimates of ω_z were extracted using the relationship $\alpha = tG/(\omega_z + P)$. Two interpretations of Brownian motion are possible. Under the genetic drift of the trait mean interpretation (in which case $\sigma^2 = tG/N_e$), BMS estimates an implausible change in N_e between tree regimes. In contrast, under the Brownian motion of the optimum interpretation (in which case, $\sigma^2 = t\sigma_\theta^2$), BMS estimates modest and hence plausible values for σ_θ^2 that are consistent with the value used in the successful simulation depicted in Figure 13.13. Pulling all of these results together, the OUwie results and the simulation work suggest that Brownian motion of the natural selection preference optimum, together with exaggeration by sexual selection, is the most plausible process underlying plumage diversification in birds of paradise.

13.8 Accounting for Other Trait Radiations With Multivariate Models

Houle et al. (2017) assessed the ability of several classes of models to account for multivariate diversification of wing venation in 112 drosophilid fly species and five outgroup taxa. Trait variation was assessed in multiple orthologous dimensions (8 or 18, depending on the estimate of the *M*-matrix). A remarkable aspect of the study is that mutational (***M***), standing genetic variation (***G***), and divergence matrices were all estimated. Skipping to the primary findings, divergence was remarkably constrained and so soundly rejected the expectation of the neutral model. The results were also compared with predictions of various moving optimum models, revealing that slow movement by Brownian motion provided the best overall fit.

13.9 Conclusions

- The analytical result that the multivariate mean will tend to evolve uphill on an adaptive landscape is fundamental to our understanding of phenotypic evolution in both the short and long term. Because of this hill-climbing feature, the multivariate mean will evolve towards a single stationary peak and remain in its vicinity. The hill-climbing tendency also means that the mean will chase a moving peak, an outcome we will consider in the next chapter.
- The deterministic evolution of multivariate trait mean in response to a stationary Gaussian peak is profoundly affected by the *G*-matrix and the *Ω-matrix*. Nonzero, off-diagonal elements in both of these matrices cause the mean to evolve along curved evolutionary trajectories on its approach to the optimum. In the case of a non-diagonal *G*-matrix, the initial stage of approach is rapid and biased towards the leading eigenvector of ***G*** ($\boldsymbol{g_{max}}$), but the final phase is slow and biased towards $\boldsymbol{g_{min}}$. In contrast, when the *Ω-matrix* is non-diagonal, the initial, rapid phase is biased towards the trailing eigenvector and the final, slow phase is in the direction of the leading eigenvector.
- Sexual dimorphism provides an instructive example of how genetic covariance can shape evolutionary trajectories. In particular, because sexually homologous traits often have large genetic correlations, the final approach of the mean to the joint optimum can be both slow and nonadaptive.
- Theoretical work suggests that models with a stationary peak will not account for the magnitude of actual adaptive radiations. Just as a univariate OU model fails to account for actual radiation, multivariate OU models fail for the same reason. One must invoke unrealistically small population size (or unrealistically weak stabilizing selection) to enable the multivariate mean to drift away from its stationary peak and produce a radiation of appreciable size.
- A similar reliance on unrealistically powerful drift carries over to OU models with two adaptive peaks. In particular, in the two-peak model, a population must be unrealistically small to escape from the attractive force of one peak, so that it can transit to the other peak.
- Stochastic models show that the ultimate extent and shape of adaptive radiations is primarily determined by selection and drift and to a minor extent by inheritance. In these models, influence of the *G*-matrix decays exponentially with time, so that the

equilibrium configuration of the radiation is proportional to the Ω-matrix, and the size of the radiation is inversely proportional to N_e.

- Many aspects of the Fisher-Lande process have been explored with Gaussian models and some surprising generalities have emerged. For example, regardless of basic assumptions, the outcome of the FLP can be stable (walk-towards) or unstable (runaway). Despite this ubiquitous bimodality, the stable outcome appears to be more probable in the sense that it requires less extreme assumptions. Another surprising result is that selection on female preferences exerts a huge effect on the dynamics of the FLP. Selection on preferences can change the deterministic equilibrium from a line into a point. In stochastic models with selection on preferences and finite population size, the mean equilibrates in an elongated cloud around the equilibrium point.
- The analytical expectation that multivariate OU models with a stationary peak will fail to account to magnitude of adaptive radiations is verified by empirical test cases. In the case of birds of paradise plumage, a stationary OU peak model fails to account for the extraordinary magnitude of the radiation (Arnold and Houck 2016). In the case of wing venation in drosophilid flies, the trait radiation is remarkably constrained, but nevertheless a moving optimum model is required to account for its magnitude (Houle et al. 2017).

Literature Cited

Andersson, M. 1994. Sexual Selection. Princeton University Press, Princeton.

Arnold, S. J., and L. D. Houck. 2016. Can the Fisher-Lande process account for birds of paradise and other sexual radiations? Am. Nat. 187:717–735.

Arnold, S. J., P. A. Verrell, and S. G. Tilley. 1996. The evolution of asymmetry in sexual isolation: A model and a test case. Evolution 50:1024–1033.

Beaulieu, J. M., D.-C. Jhwueng, C. Boettiger, and B. C. O'Meara. 2012. Modeling stabilizing selection: Expanding the Ornstein-Uhlenbeck model of adaptive evolution. Evolution 66:2369–2383.

Darwin, C. 1859. The Origin of Species by Means of Natural Selection or the Preservation of Favoured Races in the Struggle for Life. John Murray, London.

Darwin, C. 1871. The Descent of Man and Selection in Relation to Sex. John Murray, London.

Fisher, R. A. 1915. The evolution of sexual preference. Eugen. Rev. 7:184–192.

Fisher, R. A. 1930. The Genetical Theory of Natural Selection. Oxford University Press, Oxford.

Fisher, R. A. 1958. The Genetical Theory of Natural Selection. Dover Publications, New York.

Hansen, T. F., and E. P. Martins. 1996. Translating between microevolutionary process and macroevolutionary patterns: The correlation structure of interspecific data. Evolution 50:1404–1417.

Hansen, T. F., J. Pienaar, and S. H. Orzack. 2008. A comparative method for studying adaptation to a randomly evolving environment. Evolution 62:1965–1977.

Houle, D., G. H. Bolstad, K. van der Linde, and T. F. Hansen. 2017. Mutation predicts 40 million years of fly wing evolution. Nature 548:447–450.

Irestedt, M., K. A. Jønsson, J. Fjeldså, L. Christidis, and P. G. P. Ericson. 2009. An unexpectedly long history of sexual selection in birds-of-paradise. BMC Evol. Biol. 9: 235–246.

Iwasa, Y., and A. Pomiankowski. 1995. Continual change in mate preferences. Nature 377: 420–422.

Iwasa, Y., A. Pomiankowski, and S. Nee. 1991. The evolution of costly mate preferences. II. The 'Handicap' Principle. Evolution 45:1431.

Kirkpatrick, M., and N. H. Barton. 1997. The strength of indirect selection on female mating preferences. Proc. Natl. Acad. Sci. USA 94:1282–1286.

Lande, R. 1976. Natural selection and random genetic drift in phenotypic evolution. Evolution 30: 314–334.

Lande, R. 1979. Quantitative genetic analysis of multivariate evolution, applied to brain:body size allometry. Evolution 33:402–416.

Lande, R. 1980. Sexual dimorphism, sexual selection, and adaptation in polygenic characters. Evolution 34:292–305.

Lande, R. 1981. Models of speciation by sexual selection on polygenic traits. Proc. Natl. Acad. Sci. USA 78:3721–3725.

LeCroy, M. 1981. The genus *Paradisaea*: display and evolution. American Museum Novitates 2714:1–52

Luria, S. E., S. J. Gould, and S. Singer. 1981. A View of Life. Benjamin/Cummings, Menlo Park.

Mead, L. S., and S. J. Arnold. 2004. Quantitative genetic models of sexual selection. Trends Ecol. Evol. 19:264–271.

Poissant, J., A. J. Wilson, and D. W. Coltman. 2010. Sex-specific genetic variance and the evolution of sexual dimorphism: A systematic review of cross-sex genetic correlations. Evolution 64: 97–107.

Pomiankowski, A. 1987. The costs of choice in sexual selection. J. Theor. Biol. 128:195–218.

Pomiankowski, A., Y. Iwasa, and S. Nee. 1991. The evolution of costly mate preferences I. Fisher and biased mutation. Evolution 45:1422.

Schluter, D. 1996. Adaptive radiation along genetic lines of least resistance. Evolution 50:1766–1774.

Stearns, S. C., and R. F. Hoekstra. 2000. Evolution, An Introduction. Oxford University Press, Oxford.

Uyeda, J. C., S. J. Arnold, P. A. Hohenlohe, and L. S. Mead. 2009. Drift promotes speciation by sexual selection. Evolution 63:583–594.

CHAPTER 14

Trait Evolution on Dynamic Adaptive Landscapes

Overview—Although a variety of models have been proposed to explain adaptive radiations, a common feature of relatively successful models is peak movement. Nevertheless, some of these models fail to explain temporal patterns in large data sets assembled by Philip Gingerich and others. For example, models in which the optimum steadily moves in the same direction (with or without superimposed white noise fluctuation) fail to account for longterm data. The models that come closest to accounting for the data are ones in which the optimum moves rarely, but is then capable of substantial excursions (0–6 phenotypic standard deviations). Likewise, Brownian movement of the peak does a reasonable job accounting for actual adaptive radiations. These results direct our attention to the problem of formulating and testing alternative models for the *peak controller process* (PCP), providing a research agenda for the future.

14.1 White Noise Motion of the Adaptive Peak

In this model, an OU process tracks a moving optimum. The simplest model of peak movement is white noise fluctuation of the optimum about a stationary position. In this tracking model (Hansen et al. 2008), the position of the optimum is $\theta_{t+1} = \varepsilon_\theta$, a random variable drawn from a normal distribution with a mean of zero and variance of σ_θ^2.

$$\theta_{t+1} = N\left(0, \sigma_\theta^2\right). \tag{14.1a}$$

Notice that the optimum carries none of its past history, as it does in the case of Brownian motion (next section). Instead, the optimum undergoes more erratic motion. Each generation the change in trait mean of a lineage is composed of two terms, response to recent movement of the optimum and genetic drift,

$$\Delta\bar{z}_{t+1} = \left(\frac{\theta_t - \bar{z}_t}{\Omega}\right) G + N(0, G/N_e). \tag{14.1b}$$

The trait means of replicate lineages are normally distributed with an expected value that corresponds to the expected value for the optimum, which we have assumed is zero.

Evolutionary Quantitative Genetics. Stevan J. Arnold, Oxford University Press. © Stevan Arnold (2023).
DOI: 10.1093/oso/9780192859389.003.0015

The variance among the trait means of replicate lineages at any particular generation t is

$$Var(\bar{z}_t) = \left[\frac{G\sigma_\theta^2}{2\Omega} + \frac{\Omega}{2N_e}\right]\left[1 - exp\left\{-2\frac{G}{\Omega}t\right\}\right], \tag{14.2a}$$

which as t goes to infinity, converges on

$$Var(\bar{z}_\infty) = \frac{G\sigma_\theta^2}{2\Omega} + \frac{\Omega}{2N_e}. \tag{14.2b}$$

(Hansen et al. 2008). Notice that the variance equilibrates at a constant value (14.2b), in contrast to the Brownian motion case (next section).

A particular example of white noise motion of the optimum and the evolution it evokes is shown in Figure 14.1, which employs intermediate values for genetic variance and population size and imposes moderately weak stabilizing. Notice in Figure 14.1a that the evolving mean fails to track the rapid, erratic motion of the optimum. As a consequence, the ensemble properties of many replicate populations are a much dampened version of their moving optima (Figure 14.1b). Another revealing contrast with Brownian motion is the general result that variance among replicate means rapidly achieves its asymptotic value instead of expanding endlessly. The consequence of this stationary variance is bounded evolution that reflects the principal feature observed in large data sets (e.g., Uyeda et al. 2011).

A variety of sources suggest that movement of the adaptive peak is conservative enough to be responsible for the stasis that is observed for many lineages in the fossil record. At the same time the adaptive peak must not be literally stationary, because small scale movements can be detected in ecological studies. At the finest temporal scale, we can estimate the speed of movement from studies that record trait means from generation to generation. To equate the movement of the mean with movement of the optimum, we must suppose that the mean closely tracks the optimum or that lag has stabilized so that the position of the mean mirrors the position of the optimum. In a sample of four such studies (Table 14.1), trait means moved an average of 0.09 phenotypic standard deviations per generation over an average of 11 generations. The average per-generation variance in movement of the optimum, σ_θ^2, estimated from the same studies, was 0.10.

We see from the range of values for variance in the position of the optimum in Table 14.1, using expression (14.2b) for equilibrium variance in the trait mean, that white noise fluctuation in the optimum cannot account for an adaptive radiation in which the trait means of species diverge by even a few, much less several, within-population standard deviations. For example, if we take the largest value for σ_θ^2 in Table 14.1, namely 0.582, and let $P = 1$, $G = 0.4$, $N_e = 500$ and $\Omega = 10$, then $Var(\bar{z}_\infty) = (0.233/20) + (20/1000) = 0.032$. The 99% confidence region for this radiation would be $\pm 2.57\sqrt{0.032} = \pm 0.457$. In such a constrained radiation, it would be unusual for a trait mean to stray more than half of a within-population phenotypic standard deviation from the ancestral mean. Furthermore, because we have equated movement of the optimum with movement of the trait mean in Table 14.1, and ignored error in estimated the trait mean, we have probably over-estimated σ_θ^2. In other words, white noise motion of the optimum will produce an unrealistically limited adaptive radiation.

The challenge that faces us is to find a model for peak movement capable of accounting for the ecological results that we have just reviewed, but that can also account for the evidence of stasis and evolutionary bursts that springs from data on a paleontological

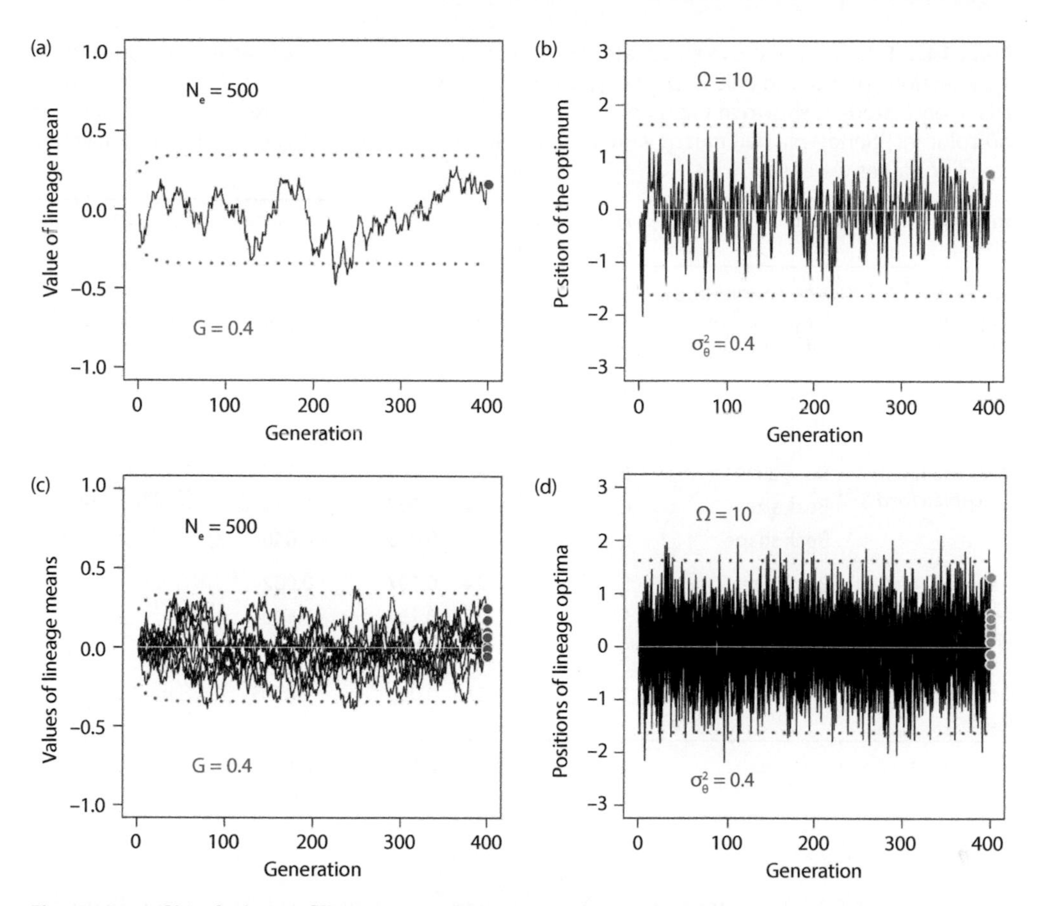

Figure 14.1 Simulations of lineages evolving in response to white noise motion of the optima of the AL. In these simulations the long-term average position of the optimum and the lineage means is zero, indicated with thin white lines. (a) The trait mean of a single lineage (above) evolving in response the movement of the optimum of its AL, shown in the next panel. The red dot shows the position of the trait mean at the end of the simulation. The blue dotted lines show the upper and lower 99% confidence limits of the average trait mean, calculated using (14.2b). (b) The position of the optimum as a function of time. The orange dot shows the position of the optimum at the end of the simulation. The red dotted lines show the upper and lower 99% confidence limits of the optimum. (c) The trait means of 10 replicate lineages evolving in response the movement of their optima, shown in the next panel. The red dots show the position of the trait means at the end of the simulation. (d) The positions of 10 replicate optima as a function of time. The orange dots show the position of the optima at the end of the simulation.

timescale. In the sections that follow, we review a set progressively more complicated models that pursue this dual goal.

14.2 Brownian Motion of the Adaptive Peak

In Chapter 12 we showed that the drift of populations of finite size on an adaptive landscape with a single stationary adaptive peak does not explain the diversification commonly seen in evolving clades. The basic problem is that a stationary peak does not permit

Table 14.1 Estimates of the average rate ($\Delta\theta$) and sample variance (σ_θ^2) of per-generation shifts in the selection optimum, in units of phenotypic standard deviation. The number of change-time intervals in each study is shown in the n column. Absolute values of average rates are presented in the $\Delta\theta$ column. Phenotypic trait means were used for all estimates except the great tit data set, where breeding values were used.

Species	Trait	Number of generations	n	Average rate ($\Delta\theta$)	Variance (σ_θ^2)	Reference
Grayling (*Thymallus thymallus*)	Age-specific body length at five different ages	13	6	0.184	0.582	Haugen and Vøllestad (2001) Figure 3
		13	7	0.092	0.118	
		13	6	0.010	0.036	
		13	6	0.021	0.060	
		13	6	0.045	0.079	
Darwin's finch (*Geospiza fortis*)	Body size	6	24	0.068	0.067	Grant and Grant (2002), Figure 1
	Beak size	6	24	0.000	0.061	
	Beak shape	6	25	0.078	0.040	
Darwin's finch (*Geospiza scandens*)	Body size	5	24	0.197	0.062	
	Beak size	5	25	0.137	0.100	
	Beak shape	5	24	0.290	0.025	
Great tit (*Parus major*)	Fledgling mass	19	34	0.003	0.0003	Garant et al. (2004), Figure 3
Mean (s.e.)		10.8		0.094 (0.026)	0.103 (0.045)	

enough variation among lineage means to provide good fits to empirical data. We can imagine solving this problem by allowing the peak to move in some pattern. In this section we explore Brownian motion of the peak and evaluate the diversification produced by that motion. The following account is a summary of results by Hansen et al. (2008). Bartoszek et al. (2012) present multiple trait versions of the Hansen et al. (2008) model in which optima move by Brownian motion or according to an OU process. See also Bartoszek et al. (2017) and Mitov et al. (2019).

By Brownian motion of the optimum we mean that the current position of the optimum deviates from its position in the preceding time interval and by an amount that is a normally distributed random variable: $\theta_{t+1} = \theta_t + \varepsilon_\theta$, where the last term is a random variable drawn from a normal distribution with a mean of zero and variance of σ_θ^2. We want to determine how a set of replicate populations will evolve in response to this mode of peak movement. If we focus on a single evolving lineage, the change in the trait mean in a particular generation $(t+1)$,

$$\Delta\bar{z}_{t+1} = \left(\frac{\theta_t - \bar{z}_t}{\Omega}\right) G + N(0, G/N_e), \tag{14.3}$$

has two components, a deterministic response (the first term on the right) that is proportional to the distance to the optimum and a stochastic, drift response that is exacerbated by small population size. The deviation of the optimum from its position in the previous generation is a draw from a normal distribution with a mean of zero and a variance of σ_θ^2,

$$\theta_{t+1} = \theta_t + N\left(0, \sigma_\theta^2\right). \tag{14.4}$$

The distribution of the trait means of replicate lineages is normally distributed about the long-term average position of the optimum with an ever-increasing variance,

$$Var\left[\bar{z}\left(t\right)\right] = \frac{\sigma_\theta^2 + \frac{G}{N_e}}{2a}\left[1 - exp\left\{-2at\right\} + \sigma_\theta^2 t\left[1 - 2\frac{\left(1 - exp\left\{-at\right\}\right)}{at}\right]\right], \tag{14.5}$$

where $a = G/\Omega$ (Hansen et al. 2008). Hansen et al. (2008) find that the variance among replicate lineages in the position of the optimum at generation t is

$$Var\left[\theta_t\right] = \sigma_\theta^2 t. \tag{14.6}$$

We see that the expression for the variance among replicate means rapidly converges on $\sigma_\theta^2 t$, which is the variance in the position of the optimum at generation t. We briefly discussed an example simulation using this model in Section 13.7 (Figure 13.12).

An example of lineage responses to Brownian motion of adaptive peaks is provided in Figure 14.2. Using reasonable values for selection and inheritance, we see that the dispersion of lineage means is largely a reflection of variation in the position of optima. In this particular example, the trait mean closely tracks the moving optimum (see the two panels on the left in Figure 14.2).

Several studies have shown that a model of Brownian motion of an intermediate optimum does a good job of accounting for trait divergence on a variety of timescales (Uyeda et al. 2011a; Hunt et al. 2015; Arnold and Houck 2016). The analysis of Arnold and Houck (2016) suggests that best interpretation is Brownian motion of an intermediate optimum rather than Brownian motion arising purely from genetic drift of the trait mean. We will explore this and other model comparisons later in this chapter.

Although Estes and Arnold (2007) rejected the BM peak movement model as an explanation for the Gingerich data, (Arnold 2014) pointed out that they did so on erroneous grounds.

14.3 Steady, Linear Movement of the Adaptive Peak

In this model, in contrast to the models discussed so far, the trait mean shows a long-term linear trend. Our model of the position of the optimum is $\theta_t = kt + \varepsilon_\theta$, where k is a constant denoting the rate of deterministic change in the optimum and ε_θ is a normally distributed random variable with a mean of zero and variance of σ_θ^2. The expected value for the trait mean at generation t is

$$E\left(\bar{z}_t\right) = kt - k\left(\frac{\Omega}{G}\right)\left[1 - exp\left\{-\frac{G}{\Omega}t\right\}\right], \tag{14.7}$$

with a variance of

$$Var\left(\bar{z}_t\right) = \left(\frac{G\sigma_\theta^2}{2\Omega} + \frac{\Omega}{2N_e}\right)\left[1 - exp\left\{-2\frac{G}{\Omega}t\right\}\right], \tag{14.8}$$

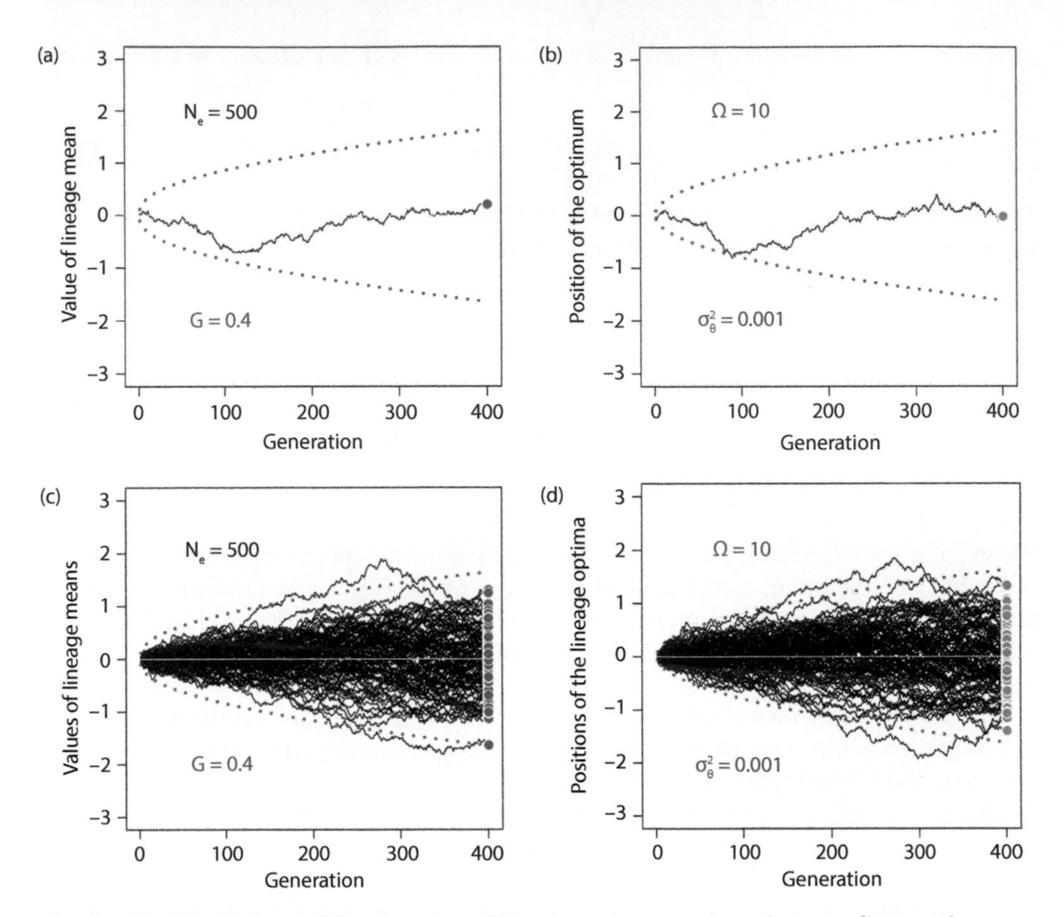

Figure 14.2 Simulations of the response of lineage trait means to an intermediate optimum that undergoes Brownian motion. Graphic conventions as in Figure 14.1. (a) The trait mean of a single lineage mean closely tracks the moving optimum of its AL, shown in the next panel. (b) The movement of the trait optimum. (c) The trait means of 100 replicate lineage means closely track the moving optima of their ALs, shown in next panel. (d) The movement of 100 trait optima.

(Lynch and Lande 1993). The first term on the right represents the variance contribution from response to directional selection elicited by peak movement with the denominator representing the negative contribution of stabilizing selection. The second term on the right represents the contribution of drift-selection balance. The final, exponential term represents a temporary variance contribution that decays on the approach to equilibrium. As t goes to infinity, the distribution of trait means equilibrates at the following values for overall mean and variance,

$$E(\bar{z}_\infty) = kt - k\left(\frac{\Omega}{G}\right) \tag{14.9}$$

$$Var(\bar{z}_\infty) = \frac{G\sigma_\theta^2}{2\Omega} + \frac{\Omega}{2N_e}. \tag{14.10}$$

In other words, the trait mean is always a linear function of time, while the variance equilibrates at a constant value.

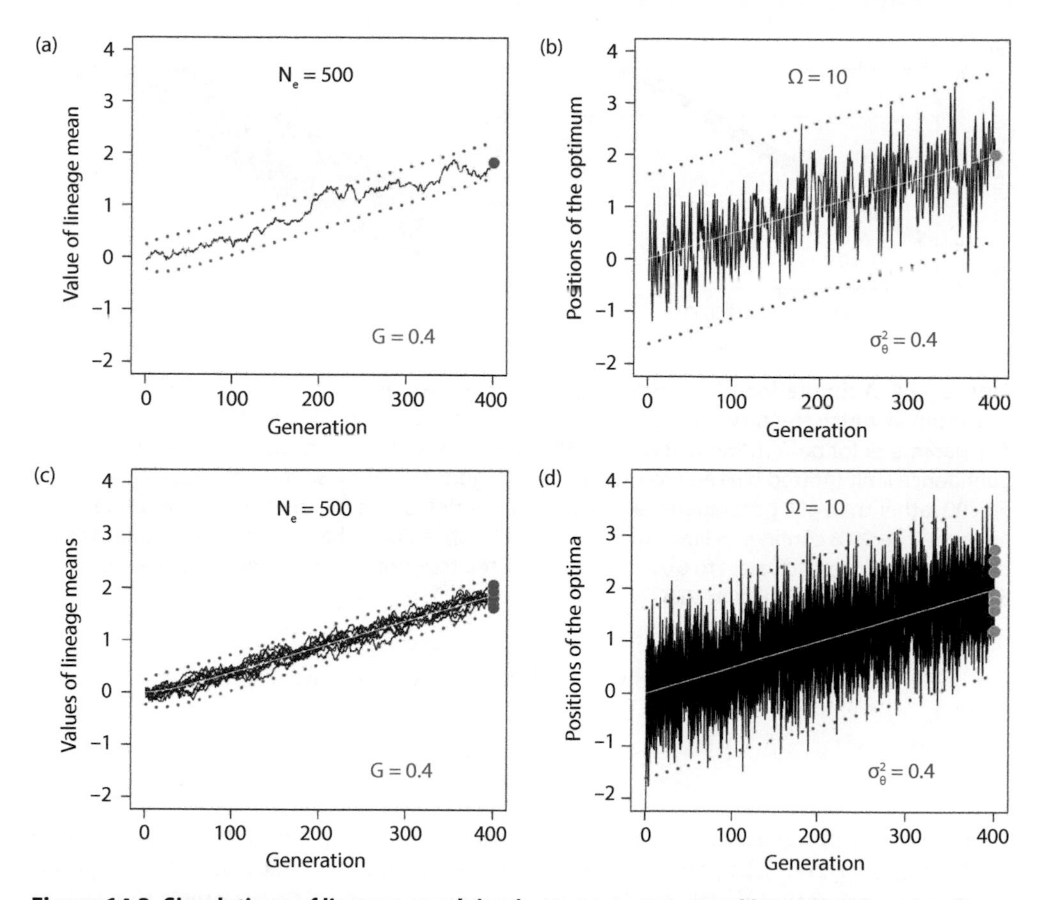

Figure 14.3 Simulations of lineages evolving in response to a steadily moving intermediate optimum with superimposed white noise fluctuations in position. Conventions as in Figure 14.1. The position of the optimum increases in value at a rate of 0.005 within-population standard deviation per generation ($k = 0.005$). (a) The trait mean of a single lineage evolving in response to a moving optimum, shown in the next panel. (b) Movement of the optimum of the AL. (c) The trait means of 10 replicate lineages evolving in response to the movement of their optima, shown in the next panel. (d) Movement of the optima of 10 replicate moving optima.

An example of lineages evolving in response to steady linear movement of the adaptive peak is provided in Figure 14.3. Notice that, as before, substantial white noise movement of the peak evokes only moderate excursions in the evolving trait mean.

The Achilles heel of the steadily moving optimum model is that, in the fullness of time, the trait mean is bound to evolve outside the observed empirical boundaries for divergence. For example, even if the optimum moves at a rate that would be undetectable in microevolutionary analyses ($k = 0.0002$), the average lineage will have diverged about six standard deviations after 30,000 generations (Figure 14.4). Although this excursion past the $6\sqrt{P}$ boundary may not seem egregious, a steadily moving optima leads to impossible levels of divergence on a timescales that exceed 100,000 generations, under all combinations of realistic parameter values (Estes and Arnold 2007).

Displacement of the optimum will cause a *lag*, a temporary exaggeration of distance between mean and the adaptive peak. If the optimum moves at a steady rate, this lag

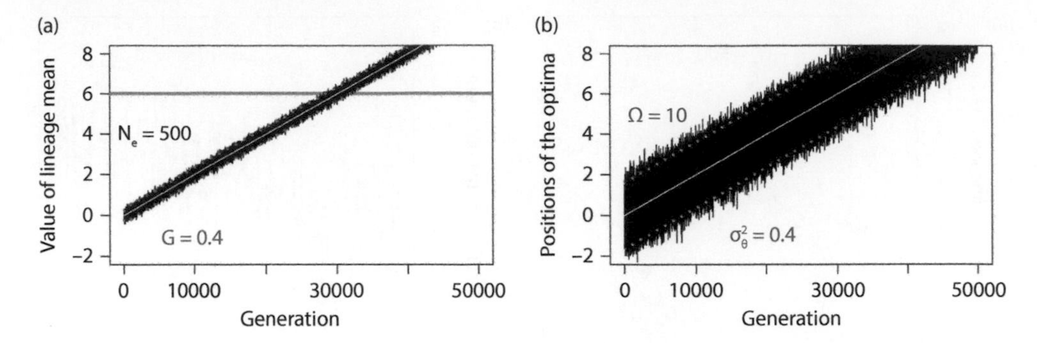

Figure 14.4 A simulation of a trait mean evolving in response to its steadily moving optimum in which the rate of motion of the optimum is 25 times slower than in Figure 14.3. The parameter for peak movement is $k = 0.0002$. The means of more lineages exceed the 99% confidence limit (dotted blue and red lines) than in Figure 14.3 because the simulation lasts 50,000 rather than 400 generations. (a) Simulated evolution of the trait means of 10 replicate lineages. The 99% confidence limits of the mean (thin white line) is barely visible. The violet line shows divergence equivalent to $6\sqrt{P}$. (b) The simulated trait optima of 100 replicate lineages. Same conventions as in Figure 14.3.

equilibrates at a value that is larger when stabilizing selection is weak and genetic variance is small,

$$L \equiv \theta_t - \bar{z}_t = \Omega G^{-1} \Delta\theta, \tag{14.11}$$

where $\Delta\theta$ is the per-generation change in the position of the optimum (Lynch and Lande 1993; Jones et al. 2004). Lag can have deleterious demographic consequences for the population. The further the mean is displaced from the adaptive peak, the greater is the decline in the population's mean fitness. The resulting impairment in population growth rate can lead to population extinction. Gomulkiewicz and Houle (2009) show how the resulting risk to the population can be expressed in critical values of genetic parameters.

14.4 Adaptive Peak Governed by an OU Process

Let us suppose that the trait optimum itself is controlled by an OU process rather than a BM process. In their general characterization of such a process, Reitan et al. (2012) describe the time course of a partially measured variable in visible layer 1 (X_1) and the time course of an unmeasured variable in hidden layer 2 (X_2), which at any given time represents the optimum of X_1. This two-layered double OU process is described by coupled stochastic differential equations

$$dX_1(t) = -\alpha_1 (X_1(t) - X_2(t))\, dt + \sigma_1 dW_1(t) \tag{14.12}$$

$$dX_2(t) = -\alpha_2 (X_2(t) - \mu_0)\, dt + \sigma_2 dW_2(t), \tag{14.13}$$

where α_1 is the strength of pull towards the optimum in layer 1, W_1 is a white noise variable with standard deviation σ_1, α_2 is the strength of pull towards the optimum in layer 2, μ_0 is the optimum in layer 2, and W_2 is a white noise variable with standard deviation σ_2.

In other words, at any given time, t, X_1 tends to move towards its optimum by an amount equal to the product of α_1 and the deviation of X_1 from its optimum plus a contribution from a white noise distribution. At the same time, X_2 tends to move towards its optimum by an amount equal to the product of α_2 and the deviation of X_1 from its optimum plus a contribution from another white noise distribution. Focusing on X_1, we can characterize the covariance between its values at time zero and time t as

$$Cov\left[X_1(0), X_1(t)\right] = \frac{\sigma_1^2}{2\alpha_1} exp\left\{-\alpha_1 t\right\} + \frac{\alpha_2^2 \sigma_1^2}{\alpha_1^2 - \alpha_2^2}\left[\frac{1}{2\alpha_2} exp\left\{-\alpha_2 t\right\} - \frac{1}{2\alpha_1} exp\left\{-\alpha_1 t\right\}\right] \quad (14.14)$$

(Reitan et al. 2012, expression 6). To obtain values for the first two moments of the limiting distribution, we set $t = 0$ and, for convenience, let $\alpha_1 = 1/t_1$ and $\alpha_2 = 1/t_2$, where t_1 and t_2 are *characteristic times*,

$$E(X_1) = \mu_0 \quad (14.15)$$

$$Var(X_1) = \sigma_1^2 t_1 + \frac{\sigma_2^2 t_2^2}{(t_1 + t_2)} \quad (14.16)$$

(Reitan, pers. comm.).

We are interested in a particular application of this model, viz., one in which the trait mean is pulled towards an optimum, $\theta(t)$, which is itself a variable being pulled towards its optimum, Ψ (Figure 14.5).

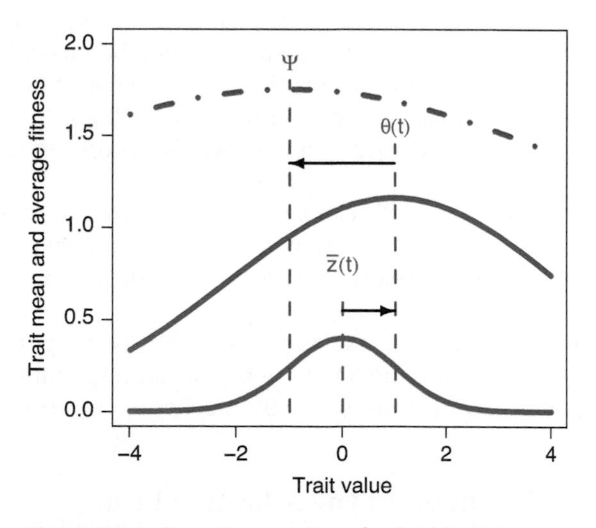

Figure 14.5 Gaussian version of a double OU process. The trait distribution (blue) has a mean $\bar{z}(t) = 0$ at time t and a variance of $P = 1$. The Gaussian adaptive landscape (red) has an optimum $\theta(t) = 1$ at time t and a width of $\Omega = 10$. The Gaussian peak controller function (purple) has an optimum $\Psi = -1$ and a width of $\Upsilon = 30$.

Writing the discrete time versions of our previous differential equations, we have

$$\Delta\bar{z}(t+1) = \frac{G}{\Omega}[\theta(t) - \bar{z}(t)] + N(0, G/N_e) \quad (14.17)$$

$$\Delta\theta(t+1) = \frac{1}{\Upsilon}\left[(\Psi - \theta(t)) + N\left(0, \sigma_\theta^2\right)\right], \quad (14.18)$$

where Ω is the variance-like width of the Gaussian AL function governing the evolution of $\bar{z}(t)$, and Υ is the variance-like width of a Gaussian function governing the change in the optimum $\theta(t)$ as it is continually pulled back towards its optimum, Ψ. The $N(mean, variance)$ terms denote draws from normal distributions, representing genetic drift in the case of $\Delta\bar{z}(t+1)$ and a stochastic contribution to the position of the optimum in the case of $\Delta\theta(t+1)$. Substituting into the results above, we obtain the limiting values for the expectation of the trait mean and its variance,

$$E[(\bar{z}(\infty)] = \Psi \quad (14.19)$$

$$Var[\bar{z}(\infty)] = \frac{1}{2}\left[\frac{\Omega}{N_e} + \frac{\sigma^2\Upsilon^2}{\Omega G^{-1} + \Upsilon}\right]. \quad (14.20)$$

In the version of the process illustrated in Figure 14.5, with $N_e = 1000$, $G = 0.4$, and $\sigma_\theta^2 = 0.00055$,

$$E[\bar{z}(\infty)] = \Psi = -1 \quad (14.21)$$

$$Var[\bar{z}(\infty)] = 0.095, \quad (14.22)$$

which describes is a very narrow radiation, with an among-lineage trait variance that is a small fraction of the within-population phenotypic variance. Because all other parameters are in the plausible range, this result suggests that the best chance of achieving a broad adaptive radiation will be by making Υ substantially larger than 30. In other words, a double OU process might permit a broad adaptive radiation if the restraining force of the *peak controller process* is very weak.

The elementary feature of double OU can be combined with other features to produce more complicated models of trait means chasing moving adaptive peaks. For example, Reitan et al. 2012 detail a model with three layers in which the optimum in one layer has a directional trend.

We now turn to models in which the adaptive occasionally jumps according to some other process. We will begin with a model of the trait mean responding to a single jump and then consider models in which we specify the stochastic process of jumping.

14.5 A Single Displacement of the Adaptive Peak

The evolutionary response of the trait mean to a single displacement of the adaptive peak is an important element in more complicated models of trait evolution. Lande (1976) provides useful results for this elementary process. Note that drift-stabilizing selection balance about a stationary peak is a special case of this model. Here we follow Lande's parameterization of the model, so that the displacement of the peak is to a position of $\bar{z} = 0$, whereas the initial position of the trait mean is $\bar{z}_0$, and $d = -\bar{z}_0$ is the displacement of the optimum. At time t, the expected value of the trait mean is

$$E(\bar{z}_t) = \bar{z}_0\, exp\left\{-\left(G\Omega^{-1}\right)t\right\} \tag{14.23}$$

and the variance among replicate trait means is

$$Var(\bar{z}_t) = \frac{\Omega}{2N_e}\left[1 - exp\left\{-2\left(G\Omega^{-1}\right)t\right\}\right] \tag{14.24}$$

which converges on

$$Var(\bar{z}_\infty) = \frac{\Omega}{2N_e} \tag{14.25}$$

(Lande 1976). A simulation of a displaced optimum and the evolutionary response of the trait mean is shown in Figure 14.6. The time course for a distribution of lineage means responding to a single displacement is illustrated by Lande (1976).

Bell et al. (2006) and Hunt et al. (2008) studied a fossil lineage of sticklebacks (*Gasterosteus doryssus*) that was sampled over a period of about 14,000 generations. Over this interval the fossil series shows progressive reduction and loss in three aspects of the spine-plate functional complex (Figure 10.17, Figure 14.7). The pattern of reduction is rapid at first and then gradually decelerates as the populations approach what appears to be a new optimum. This pattern shows a good qualitative match to the displaced optimum model. In this case the displacement may represent the loss of predators or the invasion of a predator-free environment.

The potential of the displaced optimum model to account for the bounded evolution that is observed in large, long-term datasets is illustrated in Figure 14.8. Here we relax

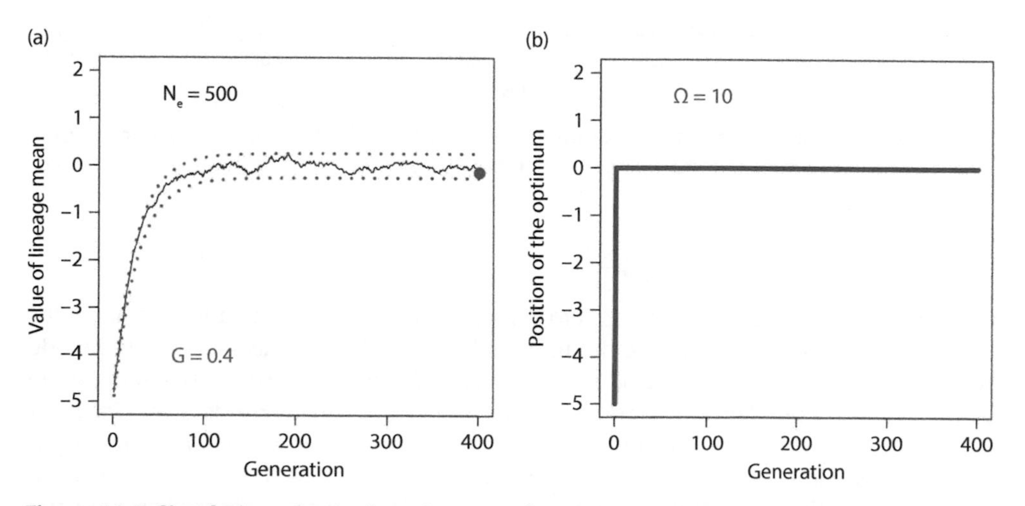

Figure 14.6 Simulation of a single trait mean of a single population evolving in response to a single displacement of its adaptive peak. (a) The evolving trait mean (thin black line) and its 99% confidence limits (dotted blue curve), calculated using (14.25). (b) The displacement of the adaptive peak by $5\sqrt{P}$ occurs instantaneously in generation 1. Conventions as in Figure 14.3.

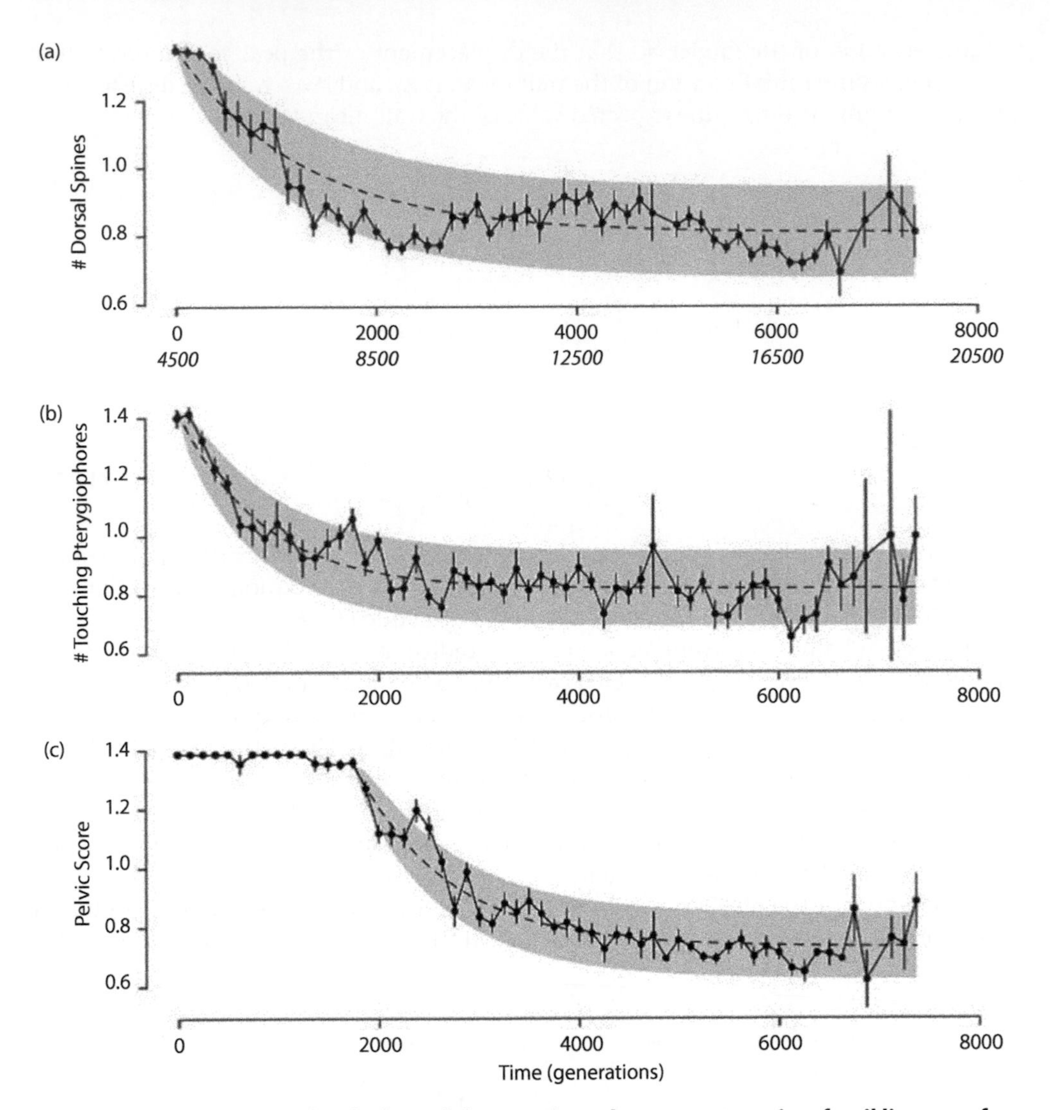

Figure 14.7 Time course for the loss of three anti-predator structures in a fossil lineage of stickleback. (a) Number of dorsal spines. (b) Number of touching pterygiophores. (c) Pelvic score.

From Hunt et al. (2008) with permission.

the assumption that the optimum is zero before displacement and in addition allow displacements of different magnitudes in different lineages. These adjustments to the model allow evolution outcomes that produce a band of evolving means that reside indefinitely within specified boundaries, matching a pattern seen in a long-term data set (Estes and Arnold 2007). Nevertheless, even with these adjustments, the model is unsatisfactory in a few respects. Allowing optima displacements only in generation 1 is obviously artificial. Secondly, we would like both the timing and magnitude of displacements to be generated by a stochastic process, rather than imposed by fiat. The next model grants both of those wishes.

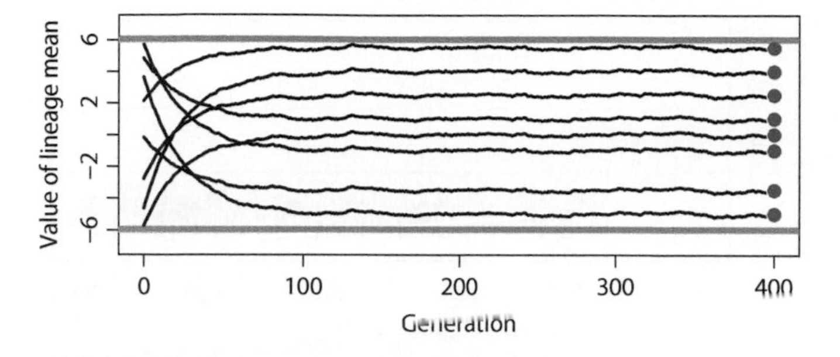

Figure 14.8 Simulations of 8 lineages evolving in response to different displacements of their intermediate optima. At generation 0, the position of the optima ranges from −6 to 6 and at generation 1 those optima are displaced by 3.5 to $9\sqrt{P}$. Red dots show the positions of lineage means at the end of the simulation.

14.6 Single-burst Model for Jumps by the Adaptive Peak

This model corrects the deficiencies of the displaced optimum model with the aim of producing the kind of evolutionary pattern observed by (Uyeda et al. 2011). For simplicity in this and the next model, we will equate the position of the trait mean with the position of its intermediate optimum. In other words, we will introduce additional realism into the movement of the optimum, but we will assume that the trait mean tracks those movements with no lag. We will consider the consequences of this modelling shortcut at the end of this section.

A key innovation of this model is that, instead of restricting occurrence of displacements at the beginning of time intervals, we will incorporate stochastic variation in the timing of displacements. The timing of the displacements is generated at random by a Poisson process, but as in the displaced optimum model, only a single displacement is allowed (Figure 14.8).

In a *Poisson process*, occurrences of a certain event, e.g., displacement, are both rare and independent of time. Thus, for a single lineage the probability that no displacement has occurred in elapsed time t is $exp\{-\lambda t\}$, where λ, the average rate of occurrence, is a constant. The probability that a first displacement occurs by time t is $1 - exp\{-\lambda t\}$. Once a displacement happens, its magnitude, d, is determined by a draw from a normal distribution with a mean of zero and a variance of σ_d^2. Even in the absence of a displacement, the position of the optimum varies from generation to generation, so that the steady state variance among trait means is normally distributed with a mean of zero and a variance of σ_θ^2, assuming instantaneous tracking of the optimum. The resulting distribution of trait means for a set of replicate lineages is the sum of two normal distributions, each weighted by a probability that a displacement does or does not occur, and is given by

$$\Phi(\bar{z}_t) = \frac{[1 - exp\{-\lambda t\}]}{\sqrt{2\pi(\sigma_\theta^2 + \sigma_d^2)}} exp\left\{\frac{-\bar{z}^2}{2(\sigma_\theta^2 + \sigma_d^2)}\right\} + \frac{exp\{-\lambda t\}}{\sqrt{2\pi\sigma_\theta^2}} exp\left\{\frac{-\bar{z}^2}{2\sigma_\theta^2}\right\} \tag{14.26}$$

(Uyeda et al. 2011). As $t \to \infty$, this distribution converges on a normal distribution with mean zero and a variance of $\sigma_\theta^2 + \sigma_d^2$ (Figure 14.9).

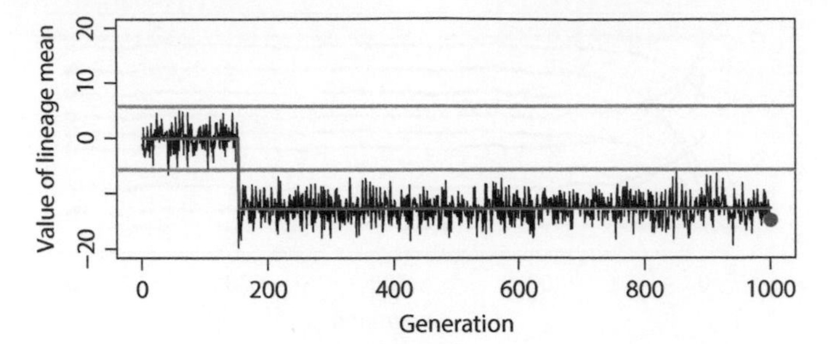

Figure 14.9 Simulation of the single-burst model against a background of white noise movement of the adaptive peak. The red line shows the position of the optimum. Parameter values are $\sigma_\theta^2 = 5$, $\sigma_d^2 = 37$, and $\lambda = 0.005$. The average waiting time for a displacement of the optimum is $1/\lambda = 200$ generations. The 99% confidence limits for a normal distribution with mean zero and variance $\sigma_\theta^2 = 5$ are shown in violet. In this run, a single displacement of the optimum occurs at about generation 150. Conventions as in Figure 14.8.

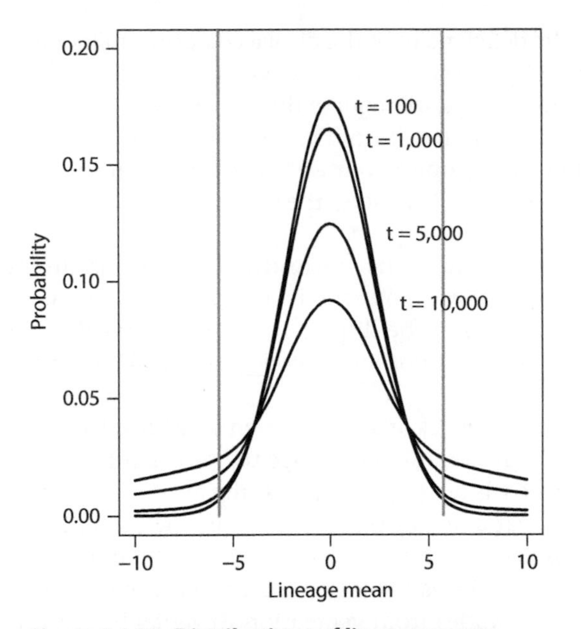

Figure 14.10 Distributions of lineage means after various elapsed times according to the single-burst model. Parameter values are $\sigma_\theta^2 = 5, \sigma_d^2 = 87.5$, and $\lambda = 0.0001$. The average waiting time for a displacement of the optimum is $1/\lambda = 10,000$ generations. The 99% confidence limits for a normal distribution with mean zero and variance $\sigma_\theta^2 = 5$ are shown in violet.

When the rate of occurrence, λ, is very small, lineages evolve within a bounded region for thousands of generations. The width of that bounded region is determined by the confidence limits of a normal distribution with a mean at zero and a variance of σ_θ^2 (Figure 14.10). On a longer timeframe, the bounded region expands so that its width corresponds to the confidence limits of a normal distribution with a mean at zero and a variance of $\sigma_\theta^2 + \sigma_d^2$.

In the preceding section we used the single-burst model to explain departures from stasis that persists on a long timescale (for hundreds of thousands to millions of generations), but the model can also be used to represent evolution on a much shorter time scale. For example, just as we used the displaced optimum model to explain divergence within but not outside the normal bounds of data (Figure 14.10), the single-burst model can also be employed to the same end. In Figure 14.11 we have chosen a small value for σ_θ^2 to represent small amplitude fluctuations in the position of the optimum, and a modest value for σ_d^2 so that single displacements of the optimum are unlikely to carry the trait mean more than about five standard deviations away from the clade's long-term average.

Although this model makes important and useful innovations in how the optimum moves, it takes some shortcuts in modelling the evolution of the trait mean. In the first place, we have simply equated the movement of the trait mean with the movement of the optimum. This economy in modelling is obviously a simplification, because we know from earlier models (e.g., Section 13.6) that the trait mean is likely to take dozens of generations to track a sudden, large displacement of the optimum. Secondly, we have modelled relatively large displacements of the optimum, but ignored the possibility of fluctuations in position on short timescales. Instead, we employed a white noise process to generate short-term stochastic variation in the trait mean via the variance parameter σ_θ^2. Again, we know that this is an unrealistic simplifying assumption, because we discovered in Sections 14.1 and 14.2 that even white noise movement of the optimum does not evoke white noise movement of the trait mean. In a white noise process, the deviation that is drawn from a normal distribution is from the long-term average, not just from the trait mean in the last generation (as in Brownian motion). The consequence is that under a white noise process, the mean would be rapidly whipped from one position to

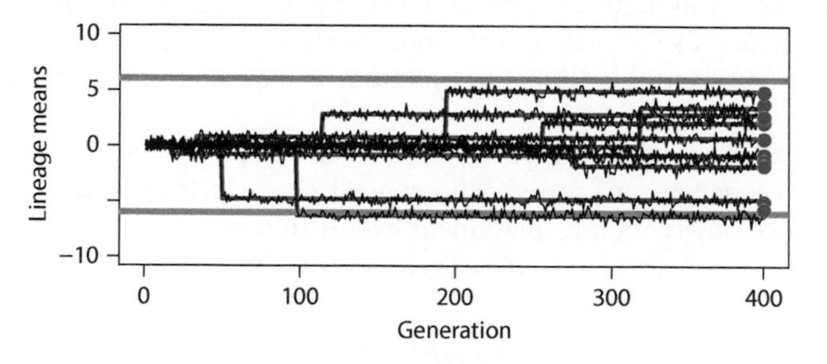

Figure 14.11 Simulation of 10 replicate lineage trait means evolving according to the single-burst model. The red lines show the position of the optima, which undergo a single displacement in each lineage. Parameter values are $\sigma_\theta^2 = 0.1$, are $\sigma_d^2 = 10$, and $\lambda = 0.01$. The average waiting time for a displacement of the optimum is $1/\lambda = 100$ generations. Typical bounds for actual data are shown in violet.

the next every generation. The normal rules of quantitative inheritance prevent this from happening in the real world. A real mean meanders through time, it does not whip back and forth. In other words, σ_θ^2 should be taken as a surrogate of variation produced by some process that needs to be more realistically modelled. White noise is an assumption of convenience that produces a desired result. The desired result is that variance produced by the process does not increase with time, enabling us to account for the long-term band of stasis that is observed in long-term data sets. The upshot is that while the single-burst model produces some key features that characterize data (long-term stasis or long delays in the appearance of substantial diversification), the model is still a work in process that awaits realistic modelling for short-term behaviour of the mean. The same realization applies to the model that we shall consider next.

14.7 Synapomorphy Resulting From a Rare Displacement of an Adaptive Peak

A trait that changes just once in the phylogenetic history of a clade constitutes an ideal indicator of phylogenetic relationships and is known as a *synapomorphy* (Hennig 1966). A plausible model of the process that underlies synapomorphy is the one discussed in Section 14.5, a single displacement of an adaptive peak. Under standard assumptions about process parameters (G, P, ω, d), the trait mean should rapidly evolve to the new position of the optimum. The most puzzling aspect of *synapomorphic peak movement* is why the displacement should occur just once. But whatever the causes of synapomorphic peak movement, the phenomenon is sufficiently common that systematists are routinely able to find characters that adhere to this mode of evolution.

14.8 Multiple-burst Model for Jumps by the Adaptive Peak

This model corrects another deficiency of the displaced optimum model that was not addressed by the single-burst model by allowing multiple displacements of the optimum. 'Multiple bursts' refers to this key feature of the model. Multiple displacements of an adaptive peak can occur at random during the lifespan of a lineage. Each of these displacements may evoke a burst of evolution, but these bursts are interspersed by potentially long intervals during which the optimum is stationary and evolution is static (Figure 14.12).

As in the single-burst model, the occurrence of the displacements is generated at random intervals by a Poisson process. The general characteristic of the process is that the probability of observing n displacement events in some time interval of length t is given by

$$p_n(t) = exp\{-\lambda t\} \frac{(\lambda t)^n}{n!}, \tag{14.27}$$

where λ is a constant, the long-term average rate of displacement (Bailey 1964). The expected number of displacements in an interval of length t is λt. Once a displacement (jump) occurs, its magnitude, d, is determined by a draw from a normal distribution with a mean of zero and a variance of σ_d^2. Let the trait mean be zero at the start of a time interval

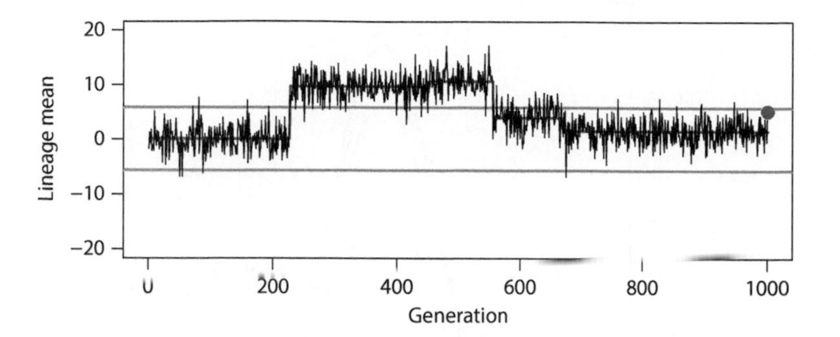

Figure 14.12 Simulation of the multiple-burst model for evolution of the trait mean in a single lineage. The red line shows the position of the optimum evolving by multiple bursts. Parameter values are $\sigma_\theta^2 = 5$, $\sigma_d^2 = 37$, and $\lambda = 0.005$. The average waiting time for a displacement of the optimum is $1/\lambda = 200$ generations. The 99% confidence limits for a normal distribution with mean zero and variance $\sigma_\theta^2 = 5$ are shown in violet. In this example, three displacements occur, at about 220, 550, and 675 generations.

of length t. At the end of that interval the probability distribution for the magnitude of divergence, $\bar{z}$, is a compound distribution given by

$$\Phi\left(\bar{z}_t\right) = \sum_{n=0}^{\infty} \frac{exp\left\{\frac{\bar{z}^2}{2n\sigma_d^2}\right\}}{\sqrt{2\pi\sigma_d^2}} exp\left\{-\lambda t\right\} \frac{(\lambda t)^n}{n!} \tag{14.28}$$

(Uyeda et al. 2011). As before, we add an element of white noise variation, σ_θ^2, to the trait mean that represents, for example, variance arising from the balance between drift and stabilizing selection, as well as contributions from other sources. Consequently, our probability distribution becomes

$$\Phi\left(\bar{z}_t\right) = \sum_{n=0}^{\infty} \frac{exp\left\{\frac{-\bar{z}^2}{2(\sigma_\theta^2+n\sigma_d^2)}\right\}}{\sqrt{2\pi(\sigma_\theta^2+n\sigma_d^2)}} exp\left\{-\lambda t\right\} \frac{(\lambda t)^n}{n!} \tag{14.29}$$

(Uyeda et al. 2011).

The behaviour of this model differs from the single-burst model in an important way. Instead of the distribution converging on a limiting variance, the variance expands perpetually. Nevertheless, this model can capture the essential features of the blunderbuss pattern, if displacements are extremely rare. Under that condition, lineage means can reside for tens of thousands of generations within boundaries approximated by the confidence limits of a normal distribution with variance σ_θ^2. On longer time scales, lineage means can diverge beyond those limits. Figure 14.13 shows the distributions of lineage means under the same parameters as in Figure 14.12, but here the average waiting time for displacement of the optimum is much longer, 100,000 generations, rather than 200. If displacements of the optimum are appreciably rarer than in Figure 14.11, an initial period of relative stasis can last hundreds of thousands or millions of generations, followed by a period of more substantial divergence that can closely approximate the pattern in Figure 14.12.

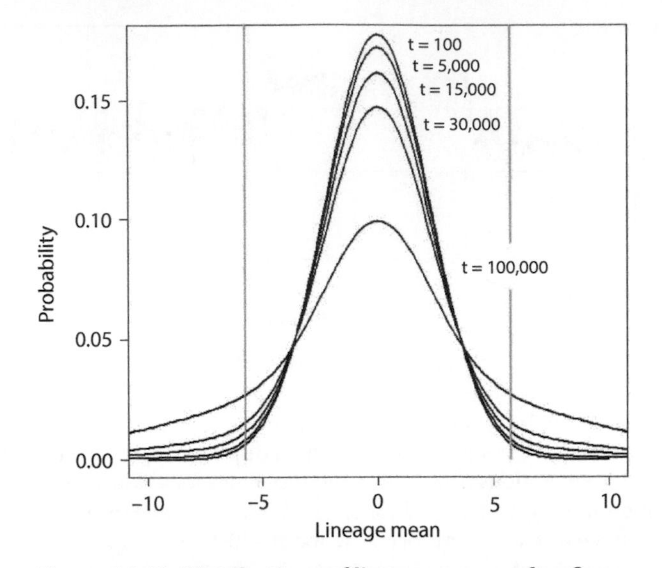

Figure 14.13 Distributions of lineage means after five different elapsed times according to the multiple-burst model. Parameter values are $\sigma_\theta^2 = 5, \sigma_d^2 = 37$, and $\lambda = 0.00001$. The average waiting time for a displacement of the optimum is $1/\lambda = 100,000$ generations. The 99% confidence limits for a normal distribution with mean zero and variance $\sigma_\theta^2 = 5$ are shown in violet.

14.9 Comparing the Fit of the Multiple-Burst Model to Other Models of Peak Movement

Uyeda et al (2011) compiled a large data set consisting of divergence values on timescales ranging from 1 to a few hundred million years. A surprising feature of the data was the *blunderbuss pattern* apparent in Figure 14.14a, in which divergence is strongly bounded on timescales ranging up to about 1 million years, but then rapidly expands on longer timescales.

Uyeda et al. (2011) used maximum likelihood to fit models, including the one just described, to the data shown in Figure 14.14a and estimated that $1/\hat{\lambda} = 10^{7.3976}$. In other words, the average waiting time to a displacement of the optimum was nearly 25 million years. The distribution of bursts size is centred at zero with a standard deviation of 27% change in body size. Thus, substantial displacements of the optimum appear to be very rare indeed. Divergence of the scale of a ± 200% change that was observed in time intervals longer than 10 million years (Figure 14.14), apparently arises as a consequence of repeated instances of rare movements of the optimum, each displacement being of modest magnitude. Thus, substantial macroevolutionary divergence arises from the accumulation of repeated, rare, modest bursts of evolution.

The multiple-burst model was the best performing of three models compared by Uyeda et al. 2011. Multiple-burst and single-burst models fit the data better than Brownian motion as judged by AIC. The better performance of multiple-burst is reflected in the failure of single-burst to account for the most extreme values of divergence at long time

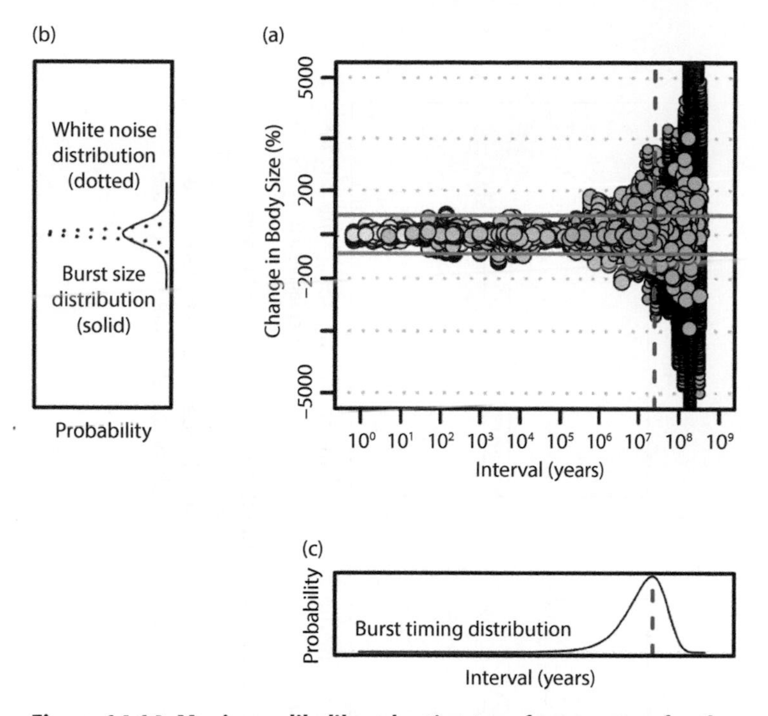

Figure 14.14 Maximum likelihood estimates of parameters for the best-fitting multiple-burst model. (a) Plot of the Uyeda et al. 2011 data. (b) Gaussian curves representing the white noise distribution (mean = 0, standard deviation = 0.096) and the burst size distribution (mean = 0, standard deviation = 0.272). (c) Burst timing distribution, showing the waiting times between bursts ($1/\lambda$) obtained by bootstrapping over studies (2000 replicates). The vertical red dashed lines show the position of the average waiting time on the actual and simulated data plots. The two violet lines show ± 65 percent change in size-related traits, equivalent to $\pm 6\sqrt{P}$ divergence.

From Arnold (2014).

intervals (Figure 14.15). Although Brownian motion appears to fit the data as well as multiple-burst, it predicts a normal distribution of data, whereas the data are clustered closer than normal to the zero point on the y-axis. Multiple-burst does a better job of fitting this leptokurtic distribution.

Before leaving this introduction to the multiple-burst model, we need to recall that this version shares certain simplifying features with its more constrained sibling, the single-burst model. Both of these models employ a white noise process because of the property that its contribution of among-lineage variance will be time invariant, even though its characterization of short-term behaviour of the mean is unrealistic. Secondly, as before, we have assumed that the response of the trait mean to the displaced optimum is instantaneous rather than asymptotic. Both of these features need to be replaced by more realistic assumptions, which will inevitably mean more complicated probability expressions to replace 14.28 and 14.29.

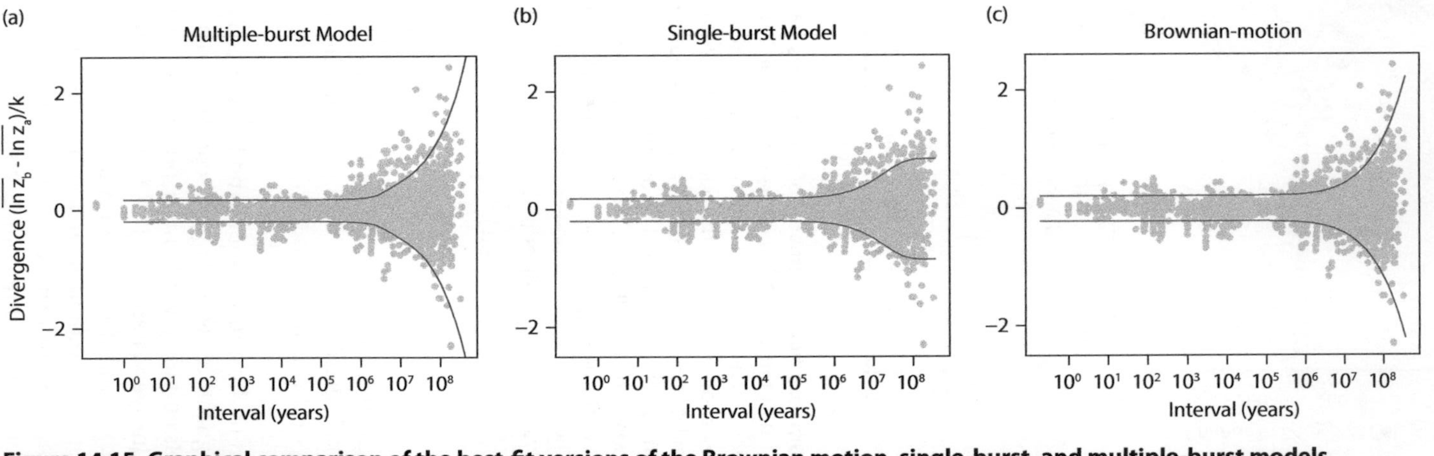

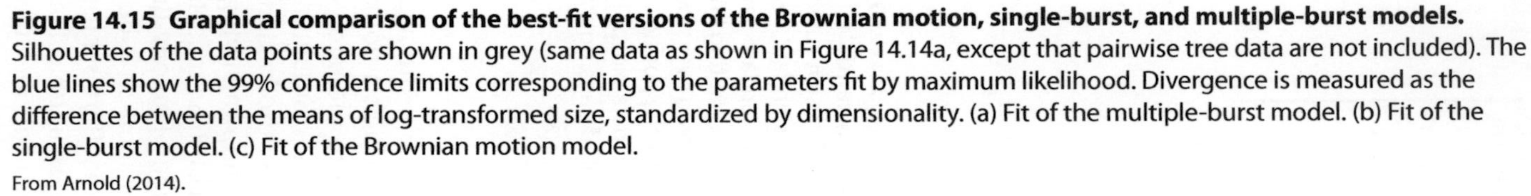

Figure 14.15 Graphical comparison of the best-fit versions of the Brownian motion, single-burst, and multiple-burst models. Silhouettes of the data points are shown in grey (same data as shown in Figure 14.14a, except that pairwise tree data are not included). The blue lines show the 99% confidence limits corresponding to the parameters fit by maximum likelihood. Divergence is measured as the difference between the means of log-transformed size, standardized by dimensionality. (a) Fit of the multiple-burst model. (b) Fit of the single-burst model. (c) Fit of the Brownian motion model.

From Arnold (2014).

14.10 Jumping Adaptive Peak Governed by a Lévy Process

Landis et al. (2013) and Landis and Schraiber (2017) used a Lévy process (Kallenberg 2010) to formally describe the multiple-burst model of Uyeda et al. 2011. They provide an analytical solution for their Lévy model, as well as a maximum likelihood testing framework for parameter estimation and analytical comparison with other models.

For the purposes of introduction, we will consider a *Lévy process* composed of just two component processes: Brownian motion of the peak (described by rate variance σ_θ^2) and stochastic jumps in the position of the peak governed by a Poisson process, with the size of the jumps described by a statistical distribution. Landis et al. 2013 call their basic model JN (jump normal). In the *JN model*, the peak jumps with Poisson rate λ and the size of those jumps is normally distributed with a mean of zero and a standard deviation of σ_d. The variance among trait means of replicate lineages undergoing BM and JN motion of the peak (BM+JN) at generation t, assuming instantaneous tracking of the peak by the trait mean, is

$$Var(\bar{z}(t)) = \left(\sigma_\theta^2 + \lambda\sigma_d^2\right) t. \tag{14.30}$$

The distribution of the trait means at generation t is complex and non-normal (Landis et al. 2013, expression 7) with excess kurtosis given by

$$K[\bar{z}(t)] = \frac{3\lambda\sigma_d^4}{\left(\sigma_\theta^2 + \lambda\sigma_d^2\right)^2 t}. \tag{14.31}$$

Landis and Schraiber (2017) used a comparative body size dataset to evaluate the performance of four classes of model: BM, OU with stationary peak, early burst, and four different versions of multiple bursts (JN, NIG, BM+JN, and BM+NIG). The model fits to 66 vertebrate clades show substantial variability among clades. In many clades (34%) no single model was decisively supported by the data. Multiple-burst evolution was the best fitting model in 32% of clades, with BM, early burst, and stationary peak OU performing as best fitting model in 18%, 14%, and 2% of clades, respectively.

Turning to the details of the parameter estimation using the best fitting multiple-burst models, the waiting time for a substantial jump in trait mean ($\geq$ 2 phenotypic standard deviations) ranged from 1–100 million years with a median waiting time of 25 million years. These values are very close to those observed in the Uyeda et al. (2011) analysis.

14.11 Conclusions

- Despite its simplicity, Brownian movement of the adaptive peak does a surprisingly good job of accounting for actual adaptive radiations.
- Rare bursts of peak movement, superimposed on Brownian motion, improve the fit.
- Rare displacement of an adaptive peak can result in synapomorphy and facilitate the recognition of sister taxa.
- The success of the Brownian peak motion model mirrors the success of the *neutral theory in community ecology* (Hubbell 2005). In both cases a much simplified model provides a good fit without fully illuminating underlying mechanisms.
- Situations in which success is improved by superimposing a jump process on Brownian motion of the peak promise insight into the factors that generate adaptive radiation.

In particular, the parameters of the peak jump process may match the parameters of candidate controller processes in the natural world.

- Model fitting results using processes with and without peak jumps differ from clade to clade. More work is needed to sort out the causes of this variation.

Literature Cited

Arnold, S. J. 2014. Phenotypic Evolution: The Ongoing Synthesis. Am. Nat. 183:729–746.

Arnold, S. J., and L. D. Houck. 2016. Can the Fisher-Lande process account for birds of paradise and other sexual radiations? Am. Nat. 187:717–735.

Bailey, N. T. J. 1964. The Elements of Stochastic Processes with Applications to the Natural Sciences. John Wiley, New York.

Bartoszek, K., S. Glémin, I. Kaj, and M. Lascoux. 2017. The Ornstein-Uhlenbeck process with migration: Evolution with interactions. J. Theor. Biol. 429:35–45.

Bartoszek, K., J. Pienaar, P. Mostad, S. Andersson, and T. F. Hansen. 2012. A phylogenetic comparative method for studying multivariate adaptation. J. Theor. Biol. 314:204–215.

Bell, M. A., M. P. Travis, and D. M. Blouw. 2006. Inferring natural selection in a fossil threespine stickleback. Paleobiology 32:562–577.

Estes, S., and S. J. Arnold. 2007. Resolving the paradox of stasis: Models with stabilizing selection explain evolutionary divergence on all timescales. Am. Nat. 169:227–244.

Garant, D., L. E. B. Kruuk, R. H. McCleery, and B. C. Sheldon. 2004. evolution in a changing environment: A case study with great tit fledging mass. Am. Nat. 164:E115–E129.

Gomulkiewicz, R., and D. Houle. 2009. Demographic and genetic constraints on evolution. Am. Nat. 174:E218–E229.

Grant, P. R., and B. R. Grant. 2002. Unpredictable evolution in a 30-year study of Darwin's Finches. Science 296:707–711.

Hansen, T. F., J. Pienaar, and S. H. Orzack. 2008. A comparative method for studying adaptation to a randomly evolving environment. Evolution 62:1965–1977.

Haugen, T. O., and L. A. Vøllestad. 2001. A century of life-history evolution in Grayling. Pp. 475–491 *in* A. P. Hendry and M. T. Kinnison, eds. Microevolution: Rate, Pattern, Process. Springer Netherlands, Dordrecht.

Hennig, W. 1966. Phylogenetic Systematics. University of Illinois Press, Urbana.

Hubbell, S. P. 2005 Neutral theory in community ecology and the hypothesis of functional equivalence. Functional Ecology 19:166–172.

Hunt, G., M. A. Bell, and M. P. Travis. 2008. Evolution toward a new adaptive optimum: Phenotypic evolution in a fossil stickleback lineage. Evolution 62:700–710.

Hunt, G., M. J. Hopkins, and S. Lidgard. 2015. Simple versus complex models of trait evolution and stasis as a response to environmental change. Proc. Natl. Acad. Sci. 112:4885–4890.

Jones, A. G., S. J. Arnold, and R. Burger. 2004. Evolution and stability of the G-matrix on a landscape with a moving optimum. Evolution 58:1639–1654.

Kallenberg, O. 2010. Foundations of Modern Probability. Springer, New York.

Lande, R. 1976. Natural selection and random genetic drift in phenotypic evolution. Evolution 30:314–334.

Landis, M. J., and J. G. Schraiber. 2017. Pulsed evolution shaped modern vertebrate body sizes. Proc. Natl. Acad. Sci. USA 114:13224–13229.

Landis, M. J., J. G. Schraiber, and M. Liang. 2013. Phylogenetic analysis using Lévy Processes: Finding jumps in the evolution of continuous traits. Syst. Biol. 62:193–204.

Lynch, M., and R. Lande. 1993. Evolution and extinction in response to environmental change. Pp. 234–250 *in* P. Kareiva, J. Kingsolver, and R. Huey, eds. Biotic Interactions and Global Change. Sinauer Associates, Inc., Sunderland, Massachusetts.

Mitov, V., K. Bartoszek, and T. Stadler. 2019. Automatic generation of evolutionary hypotheses using mixed Gaussian phylogenetic models. Proc. Natl. Acad. Sci. USA 116:16921–16926.

Reitan, T., T. Schweder, and J. Henderiks. 2012. Phenotypic evolution studied by layered stochastic differential equations. Ann. Appl. Stat. 6:1531–1551.

Uyeda, J. C., T. F. Hansen, S. J. Arnold, and J. Pienaar. 2011. The million-year wait for macroevolutionary bursts. Proc. Natl. Acad. Sci. USA 108:15908–15913.

CHAPTER 15

Evolution of Genetic Variance

Overview—We can imagine a conceptual world in which theory for the evolution of the trait mean is accompanied by theory for the evolution of the trait's genetic and phenotypic variances. Such a theory would provide linked equations for per-generation change in the trait's first two genetic moments. A tractable dual theory of this kind does not exist, so we must make do with either the simplistic idea that genetic variance is relatively constant or the more realistic perspective that genetic variance equilibrates under a specified set of opposing forces. Theory and simulation studies of equilibration of genetic variance suggest that relative constancy might be achieved under a wide variety of conditions that include large population size, weak selection, and migration among subpopulations. Empirical comparisons of genetic variance among related populations and species support the view that phenotypic and genetic variances are less variable than trait means. In other words, although equilibration of genetic variance can be viewed as a necessary convenience, it is also true that the equilibration perspective is supported by several lines of evidence.

15.1 The Quest for an Evolutionary Theory of Genetic Variance

We can view the equilibration of genetic variance under a prevailing selection regime as a useful adjunct to a theory for how the trait mean evolves in response to selection rather than as a contrivance to get around the absence of a dual theory. In the first place, stabilizing selection is a plausible selection regime both because it yields stasis of the trait mean—a dominant mode of evolution in the fossil record—and because it supports stability of both phenotypic and genetic variances. Theory for the immediate effect of selection on phenotypic variance (Chapter 1) forms the core for a theory of how selection affects additive genetic variance. The bigger problem in specifying the long-term effects of stabilizing selection on G is to account for the combined effects of selection and drift. As we shall see in the next section, analytical theory takes us only part of the way in solving this problem. However, we reach a number of useful conclusions using computer simulations.

15.2 Genetic Variance Under Mutation-Selection-Drift Balance

Prior to 1988, theory for the genetic variance under mutation and stabilizing selection was based largely on the case with no genetic drift (infinite population size). If population size is finite, we need to understand stochasticity in the distribution of breeding values that arises from sampling. However, the combination of drift and selection makes analytical

Evolutionary Quantitative Genetics. Stevan J. Arnold, Oxford University Press.
DOI: 10.1093/oso/9780192859389.003.0016

results difficult to achieve. See Bürger and Lande (1994) for a fuller account of the results that follow. But, before we tackle the combined case of stabilizing selection and finite population size, it will be useful to review the expected equilibrium value and variance of G under drift-mutation balance.

Lynch and Hill (1986) showed that the expected value of G at equilibrium, $\hat{G}$, under drift-mutation balance, without selection, is approximately

$$\hat{G}\,(N) \approx 2N_eU, \tag{15.1}$$

where the N in parentheses on the right denotes the Lynch and Hill neutral model, and N_e is effective population size. $U = 2n\mu\alpha^2$ is the expected per generation input of variation to the trait from mutation under the Kimura and Crow (1964) *continuum-of-alleles model*, recalling from Chapter 5 that n is the number of loci, μ is mutation rate, and α^2 is the variance in mutant additive effects (U is also known as V_m). Furthermore, the *expected equilibrium variance in* G among replicate populations is approximately

$$Var\left[\hat{G}\,(N)\right] \approx 2N_eU\alpha^2 \tag{15.2}$$

(Lynch and Hill 1986).

Now suppose that the ISS takes the form of a stabilizing selection function of Gaussian shape, with an optimum at zero and a width equal to $\sqrt{\omega}$ (3.7). Under a combined mutation-drift-selection process, what will be the mean and variance of G in a set of replicate populations at equilibrium? Lande (1976) derived results for the mean and variance of the trait mean among replicate populations (Chapter 12), but what about $\hat{G}$, the equilibrium value of G? Surprisingly, the same result for $\hat{G}$ was derived in 1988 and 1989 by four groups of authors (see Bürger and Lande 1994) and is known as the *stochastic house-of-cards* (SHC) approximation for the within-population equilibrium genetic variance,

$$\hat{G}\,(SHC) \approx \frac{4n\mu\alpha^2N_e}{1 + \left[\alpha^2N_e/\,(\omega + E)\right]} = \frac{2N_eU}{1 + \left[\alpha^2N_e/\,(\omega + E)\right]}. \tag{15.3}$$

In these expressions, the numerator is the contribution to equilibrium G from mutation-drift balance, while the denominator gives the proportional reduction in G from drift and selection. The variance in G among populations at equilibrium is a more difficult problem, but was solved by Barton (1989) under simplifying assumptions,

$$Var(G)_B \approx \frac{\hat{G}\,(SHC)\,\alpha^2}{1 + \left[\alpha^2N_e/\,(\omega + E)\right]}. \tag{15.4}$$

While these analytical results allow us to glimpse the equilibrium distribution of G under mutation-drift-selection balance, they are based on many simplifying assumptions, and they provide no results on the skewness and kurtosis of the distribution. With these limitations in mind, a few authors have used simulations to validate analytical results and to examine the details of the distribution (Keightley and Hill 1988, 1989; Bürger 1989; Bürger et al. 1989). In the next section, we will summarize only the more extensive results of Bürger and Lande (1994).

15.3 Simulation Studies of Equilibrium Genetic Variance

Bürger and Lande (1994) studied the evolution of genetic variance under mutation-drift-selection balance with stochastic simulations. In those simulations, the genotype of each individual was determined by 50 loci with no dominance or epistasis of alleles. Mutation was specified by the Crow and Kimura (1964) continuum-of-alleles model. To implement this mutation model, draws were made from a Gaussian distribution (Lande 1976; Turelli 1984) or a reflected gamma distribution (Keightley and Hill 1988) (Figure 5.11). To specify phenotypic value, an environmental value was drawn from a normal distribution with mean of zero and a variance of one and added to the genotypic value. The simulation implemented a three-stage life cycle consisting of: (1) random breeding of survivors from the preceding generation, (2) a production of offspring that included mutation and recombination, and (3) viability selection specified by (3.7) with the optimum set at zero. As a consequence of other details in the model, N_e took values between 11 and 683. The number of replicate runs per parameter combination varied from 10 to 100, inversely with the number of breeding adults (512 to 8).

The Bürger and Lande (1994) simulations validated several of the approximate expressions in Section 15.2. For example, the SHC expression (15.3) was generally accurate although it tended to slightly overestimate the true value of $\hat{G}$, while Barton's expression (15.4) consistently underestimated the variance of G by about 10%. The distributions of the trait mean ($\bar{z}$) and breeding values were usually closely approximated by Gaussian distributions.

Perhaps the most surprising result was the tendency of simulated G to show an erratic time course under some parameter combinations. Each generation, a simulation run produces a distribution of breeding values. The time course for the variance (G) and kurtosis of those breeding value distributions are shown for two sample runs in Figure 15.1, after

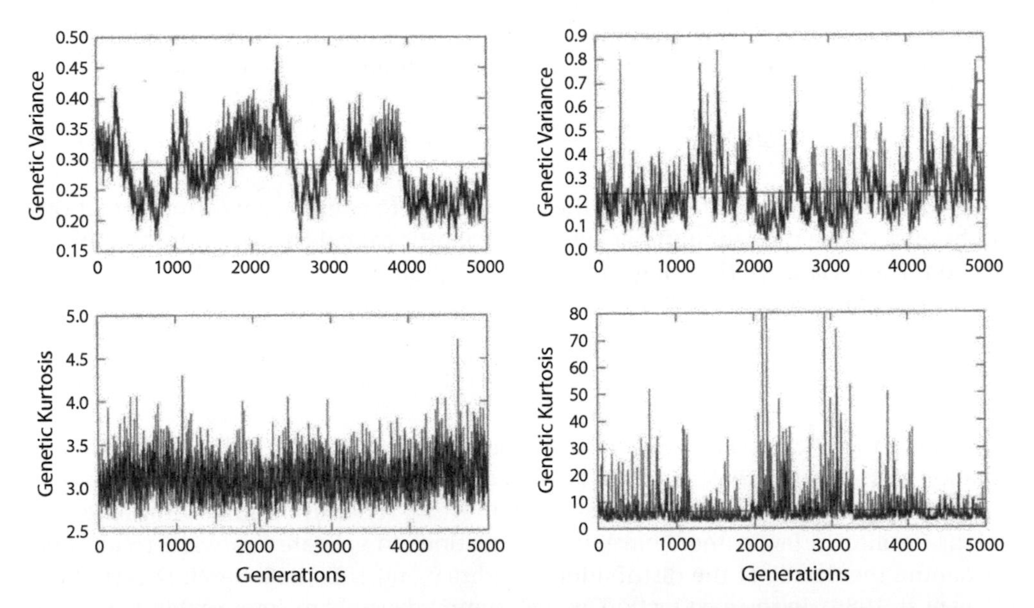

Figure 15.1 Simulated evolution of genetic variance (upper panels) and genetic kurtosis (lower panels) under mutation-drift-selection balance over 5000 generations. (Left side) The straight lines show the average values for genetic variance (0.285) and genetic kurtosis (3.12) for a single run with N_e = 683, $(\omega + 1)$ = 10, μ = 0.0002, and a normal distribution of mutant alleles with α^2 = 0.05. (Right side) The average values for genetic variance and genetic kurtosis are 0.233 and 6.27, respectively, for a single run with N_e = 171, $(\omega + 1)$ = 10, μ = 0.0002, and a reflected gamma distribution of mutant alleles with α^2 = 0.05.

From Bürger and Lande (1994) with permission.

mutation-drift-selection balance had been achieved. The time course for G in the upper left panel is especially chaotic and shows a complicated autocorrelation structure. In this run, the average value for kurtosis is almost exactly 3, which is what we expect for a normal distribution of breeding values. In contrast, the time course for G in the upper-right panel may be less complicated. This run used a reflected gamma distribution of mutant alleles. The average value of kurtosis in the run was 6.27, which is close to what we expect for a reflected gamma distribution.

Bürger and Lande (1994) measured the autocorrelation time for G in their simulation data and showed that it was in good agreement for their autocorrelation model, which was a function of N_e and μ. The requirements for a good fit were weak stabilizing selection and a relatively small number of mutant alleles per generation. They concluded that the autocorrelation observed in G was largely a function of stochasticity in the number of new mutants each generation, rather random departures from Hardy-Weinberg equilibrium or linkage disequilibrium.

15.4 Testing the Adequacy of Models for Genetic Variance Equilibration

How adequate is mutation-selection balance as an explanation for the amount of genetic variance in actual populations? Employing an interpretation made by Crow (1979), Houle et al. (1996) argued that G/U can be viewed as the persistence time of mutations under mutation-selection balance in a population of infinite size. Furthermore, according to that model, G/U should be $\ll 1000$, and the values of the ratio should be smaller for life history traits than for morphological traits. Compiling estimates from the literature, Houle et al. (1996) found that average G/U was 50 generations for life history traits and 100 generations for morphological traits. In other words, mutation-selection balance passed an adequacy test based on two criteria.

15.5 Comparative Studies of Phenotypic Variance

The idea that the trait variance is less variable than the trait mean is a generalization that emerges from systematic studies. This result is a practical one; the field of systematics focuses on traits with small variation within populations that highlights differences among populations (Simpson et al. 1960). Indeed, homogeneity of the trait variance across populations and species is such a routine expectation that heterogeneity of variance is often treated as a problem of scale. Wright (1968), for example, argued that the original scale might be abandoned for any of a number of reasons, including pursuit of homogeneity in variance. On standard measurement scales, the variance of a trait commonly increases with the mean, but this dependence can be removed by regressing the standard deviation on the mean or with a log transformation of the original scale. This pragmatic view of scale is a first cousin to the idea that similar variances might result from equilibration under similar regimes of stabilizing selection.

15.6 Comparative Studies of Genetic Variance

See Chapter 5 for a discussion of surveys of heritability estimates.

15.7 Conclusions

- Computer simulations have bypassed the difficulty of finding analytical solutions for a model of genetic variance evolving under mutation-drift-selection balance.
- Those simulations verify analytical results for the equilibrium values of genetic variance. Simulations also reveal apparently complicated patterns of autocorrelation in the evolution of *G* under some parameters of population size and selection. Despite their apparent complexity these patterns are consistent with a simple model of autocorrelation that involves only stochasticity in the number of mutant alleles entering the population each generation.
- Empirical studies lend support to the idea that genetic variance equilibrates under the influence of opposing forces.

Literature Cited

Barton, N. 1989. The divergence of a polygenic system subject to stabilizing selection, mutation and drift. Genet. Res. 54:59–78.

Bürger, R. 1989. Mutation-selection models in population genetics and evolutionary game theory. pp. 75–89 in A. B. Kurzhanski and K. Sigmund, Evolution and Control of Biological Systems. Springer, Dordrecht

Bürger, R., and R. Lande. 1994. On the distribution of the mean and variance of a quantitative trait under mutation-selection-drift balance. Genetics 138:901–912.

Bürger, R., G. P. Wagner, and F. Stettinger. 1989. How much heritable variation can be maintained in finite populations by mutation-selection balance? Evolution 43:1748–1766.

Crow, J. F. 1979. Minor viability mutants in *Drosophila*. Genetics 92:s165–z172.

Houle, D., B. Morikawa, and M. Lynch. 1996. Comparing mutational variabilities. Genetics 143:1467–1483.

Keightley, P. D., and W. G. Hill. 1988. Quantitative genetic variability maintained by mutation-stabilizing selection balance in finite populations. Genet. Res. 52:33–43.

Keightley, P. D., and W. G. Hill. 1989. Quantitative genetic variability maintained by mutation-stabilizing selection balance: sampling variation and response to subsequent directional selection. Genet. Res. 54:45–57.

Kimura and Crow. 1964. The number of alleles that can be maintained in finite population. Genetics 49: 725–738

Lande, R. 1976. The maintenance of genetic variability by mutation in a polygenic character with linked loci. Genet. Res. 26:221–235.

Lynch, M., and W. G. Hill. 1986. Phenotypic evolution by neutral mutation. Evolution 40: 915–935.

Simpson, G. G., A. Roe, and R. C. Lewontin. 1960. Quantitative Zoology. Harcourt, Brace and Co.

Turelli, M. 1984. Heritable genetic variation via mutation-selection balance: Lerch's zeta meets the abdominal bristle. Theor Pop Biol 25:138–193.

Wright, S. 1968. Evolution and the Genetics of Populations: a Treatise. University of Chicago Press, Chicago.

CHAPTER 16

Evolution of the *G*-matrix on a Stationary Adaptive Landscape

Overview—Existing analytical theory reveals little about the dynamics of $\boldsymbol{G}$, because the problem is regarded to be too complex to be mathematically tractable. Computer simulations and empirical comparisons of *G*-matrices, however, provide useful guidelines. Simulations show that the *G*-matrix tends to evolve towards alignment with the *M*-matrix and the adaptive landscape. Simulations also show that the orientation of the *G*-matrix is stabilized by large population size and between-trait correlations in mutation and stabilizing selection. An evolving *M*-matrix promotes the evolution of a triple alignment between the matrices that characterize mutation, inheritance, and selection. Empirical comparisons of *G*-matrices often reveal common principal components, suggesting that *G*-matrices have been shaped by a shared orientation of adaptive landscapes.

16.1 The Effect of Inbreeding on the *G*-matrix

As we might have anticipated from the univariate case, finite population size is expected to reduce the size of the *G*-matrix by a factor of $1/N_e$ each generation, or equivalently, by a factor of $1-F$, where F is the inbreeding coefficient of the population (Wright 1951; Lande 1980; Phillips et al. 2001). As in earlier accounts (Chapters 8 and 9), we now consider just the effects of random sampling, in the absence of selection and mutation. In other words, if we establish a series of replicate populations, each of size N_e, the expected *G*-matrix for all of those replicates will shrink in size by a constant each generation. This simple rule governs just the average $\boldsymbol{G}$. The *G*-matrices of individual replicates might expand or shrink. Random sampling will also affect among-replicate variation in $\boldsymbol{G}$. Although we lack analytical expressions for among-replicate variation, we can see a vivid portrayal of that variation in an experiment conducted by Phillips et al. (2001).

Phillips et al. (2001) produced 52 inbred lines from an outbred base population of *Drosophila melanogaster* by one generation of brother-sister mating and analysed about 90 families per inbred line. Ten wing landmark traits were scored on eight daughters in each control and inbred line family. Additive genetic variances and covariances were estimated by midoffspring on midparent regression. Standard errors were estimated by bootstrapping over families.

The average *G*-matrix of inbred replicates is smaller in size but shows no change in shape compared to outbred controls, as expected (Figure 16.1). However, variation in $\boldsymbol{G}$ among

Evolutionary Quantitative Genetics. Stevan J. Arnold, Oxford University Press. © Stevan Arnold (2023).
DOI: 10.1093/oso/9780192859389.003.0017

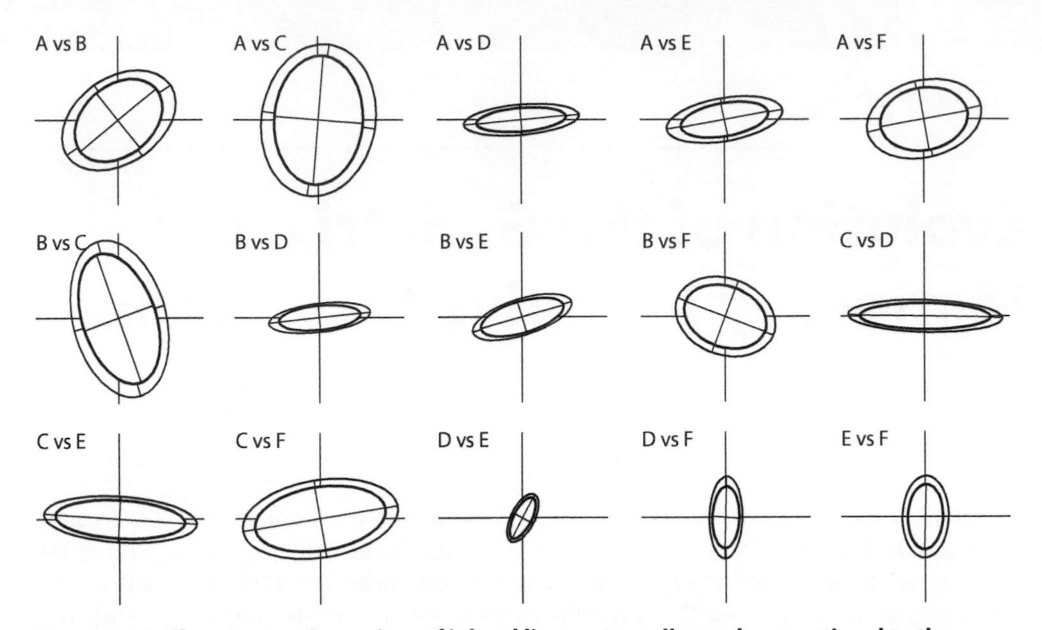

Figure 16.1 The average *G*-matrices of inbred lines are smaller and proportional to the *G*-matrices of outbred control lines. Each diagram shows the *G*-matrices (95% confidence ellipses) for a pair of traits (see labels for principal axes above each diagram). Each outer ellipse represents the *G*-matrix for outbred controls, while each inner ellipse the average *G*-matrix for inbred lines.

From Phillips et al. (2001) with permission.

inbred replicates can be substantial, as illustrated in Figure 16.2. This second result cautions against rejecting the hypothesis of drift when a single pair of populations has non-proportional *G*-matrices.

16.2 The *G*-matrix in Mutation-Drift Balance

Although the distribution of the *G*-matrix among-replicate populations in mutation-drift balance is an unsolved theoretical problem, we do have an approximate expression for the expected *G*-matrix at equilibrium,

$$\widehat{\boldsymbol{G}} \approx 2N_e\boldsymbol{U} = 2N_e\zeta\boldsymbol{M}, \tag{16.1}$$

where $\boldsymbol{U}$ is the expected per generation input of genetic variation to multiple traits from mutation, and ζ is the total mutation rate across the genome (Lande 1979; Lynch and Hill 1986). Rearranging this result, we obtain

$$\boldsymbol{U} \approx \widehat{\boldsymbol{G}}/2N_e. \tag{16.2}$$

Notice that under mutation-drift balance we expect $\boldsymbol{U}$ to be a positive small scalar fraction of $\widehat{\boldsymbol{G}}$.

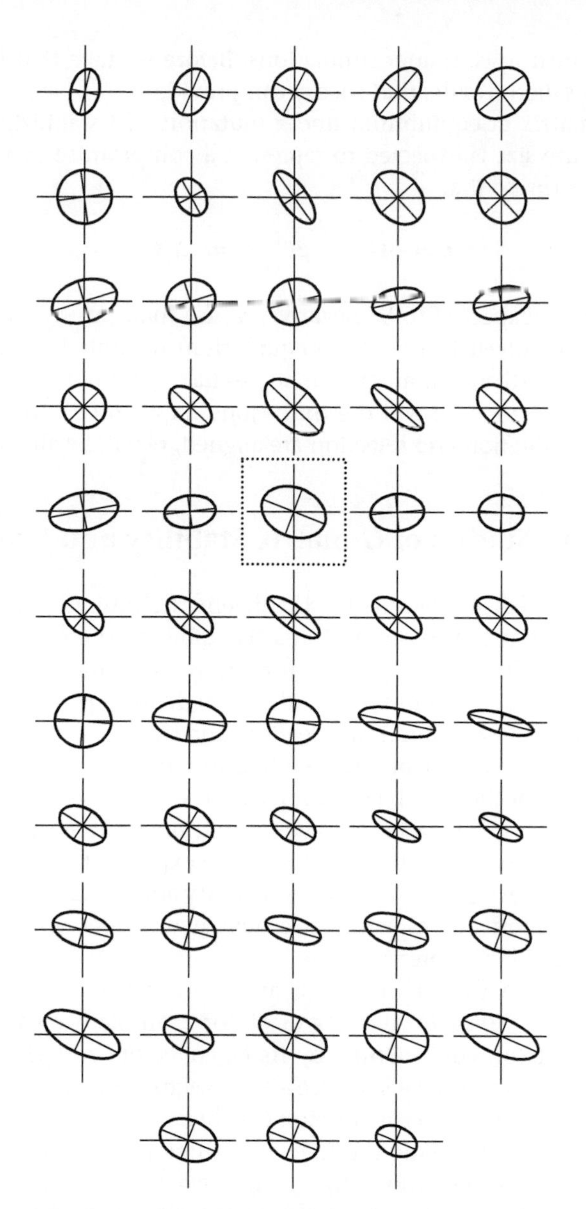

Figure 16.2 Variation in *G*-matrices for a single pair of traits (B vs. F) among inbred lines (n = 52). The ellipse for the outbred control is shown at the centre of the panel. Conventions as in Figure 16.1.
From Phillips et al. (2001) with permission.

16.3 The *G*-matrix in Mutation-Stabilizing Selection Balance

We now consider the time course of the *G*-matrix in a population experiencing mutation, selection, and random sampling of parents. Furthermore, we wish to understand the time course of a set of replicate populations under those specified conditions, not just the behaviour of the among-replicate average ***G***. These demands exceed what current theory can deliver, so we turn to computer simulations. Before we take that step, however, we should recall a few salient results that theory can provide.

The average *G*-matrix at equilibrium under mutation and stabilizing selection in an population of infinite size is expected to represent a compromise between the forces of mutation and selection, so that

$$\boldsymbol{U} = -\widehat{\boldsymbol{G}}\left(\boldsymbol{\gamma} - \boldsymbol{\beta}\boldsymbol{\beta}^{\mathrm{T}}\right)\widehat{\boldsymbol{G}} = -\Delta_s\boldsymbol{G} \tag{16.3}$$

(6.23). Notice that we expect $\boldsymbol{U}$ to be negative, rather than positive, as it was in (16.2). The three matrices involved in the (16.3) equilibrium formulation ($\boldsymbol{U}$, $\widehat{\boldsymbol{G}}$, $\boldsymbol{\gamma} - \boldsymbol{\beta}\boldsymbol{\beta}^{\mathrm{T}}$) are not independent. In particular, if any two matrices have common principal components (share all their eigenvectors), so will the third (Jones et al. 2007). In other words, if the matrices describing mutation and selection are aligned, $\boldsymbol{G}$ will be aligned as well.

16.4 Simulations Studies of *G*-matrix Stability and Evolution

Several studies have used simulations to circumvent the limitations of analytical work and assess the conditions that promote *G*-matrix stability. These studies also build our intuition for how the *G*-matrix will evolve in response to mutation, drift, and selection (Table 16.1). The usual approach in these studies is to define a plausible set of conditions (genome size, population size, mutation, selection) and then let a set of independent replicate populations evolve to a state of quasi-equilibrium under those conditions. After this burn-in period of several thousand generations, each replicate is followed for an experimental period of an additional few to several thousand generations so that the equilibrium properties of the *G*-matrix can be characterized. The experimental results are then summarized and graphed to evaluate effects on the stability and evolution of the *G*-matrix. An additional four simulation studies, with peak movement (Jones et al. 2007, 2012; Revell 2007), will be discussed in Chapter 17.

Some conventions are used to evaluate and summarize the results of simulated *G*-matrix evolution. To assess the stability of $\boldsymbol{G}$, for example, we can characterize a *G*-matrix by its elements or, equivalently, by its eigenvectors and eigenvalues. This last approach is useful, because—as we shall see—eigenvectors and eigenvalues respond differently to evolutionary forces. In the simplest case of a 2 × 2 *G*-matrix, with three unique elements (G_{11}, G_{22}, G_{12}), we define the following three parameters: *size* (Σ, sum of the eigenvalues), *eccentricity* (ε, the ratio of the smallest eigenvalue to the largest), and *orientation* (φ, the angle of the leading eigenvector to the axis of the first trait, z_1). To assess stability, we compute the per generation change in each of these characteristics and plot their time courses.

Simulations show that—as expected (16.3)—the equilibrium *G*-matrix is a compromise between the characteristic patterns of the mutation and selection matrices (Figure 16.3). When the processes of mutation and selection are aligned, the *G*-matrix adopts that same alignment. Simulations also show that the overall size of the *G*-matrix increases in large populations.

Simulations also show that different aspects of *G*-matrix stability respond differently to evolutionary factors. The orientation of $\boldsymbol{G}$, for example, is stabilized by large population size and strong correlations in mutation and selection, especially when these correlations are aligned. This stabilizing effect can be seen in the bottom rows of Figure 16.3. These same factors also have a stabilizing effect on the shape (eccentricity) of the *G*-matrix

Table 16.1 Parameters and procedures used in simulation studies of *G*-matrix stability and evolution with stationary adaptive landscapes. Over 800 parameter combinations were explored in these studies. Bürger and Lande (1994) simulated the evolution of 400,000 genetic variances. The four studies summarized here and in Table 17.1 simulated the evolution of over 700 million *G*-matrices. The scale of parameters is relative to a standardized environmental variance ($E_{ii} = 1$).

Parameter	Symbol	Bürger and Lande 1994	Jones et al. 2003	Jones et al. 2004	Guillaume and Whitlock 2007	Jones et al. 2014
Number of traits		1	2	2	2	2
Number of loci	n	50	50	50	100	20
Mutational variance	α^2	0.05	0.05	0.05	0.05	0.05
Mutational correlation	r_μ	no	−0.9–0.90	−0.9–0.90	0–0.90	−0.9–0.90
Mutation rate per locus	μ	0.00002–0.002	0.0002	0.0002	0.0002	0.0005
Constant mutation process	yes	yes	yes	no	yes	no
Strength of stabilizing selection	ω	9–999	9–49	9	50	49
Selectional correlation	r_ω	no	0–0.90	0–0.90	0–0.90	0–0.90
Effective population size	N_e	11–683	342–2731	342–2731	1000	64–4096
Migration	m	no	no	no	yes	no
Number of replicate runs per combination		10–100	20	50–200	100	20
Number of combinations		16	145	509	100	40
Burn-in period		10^3–10^4	10,000	10,000	20,000	5,000
Generations per run		10^2–10^3	2–4x10^3	10,000	4,000	5,000
Number of *G* or *G*-matrices		0.4 mil	8.7 mil	636 mil	40 mil	4.0 mil

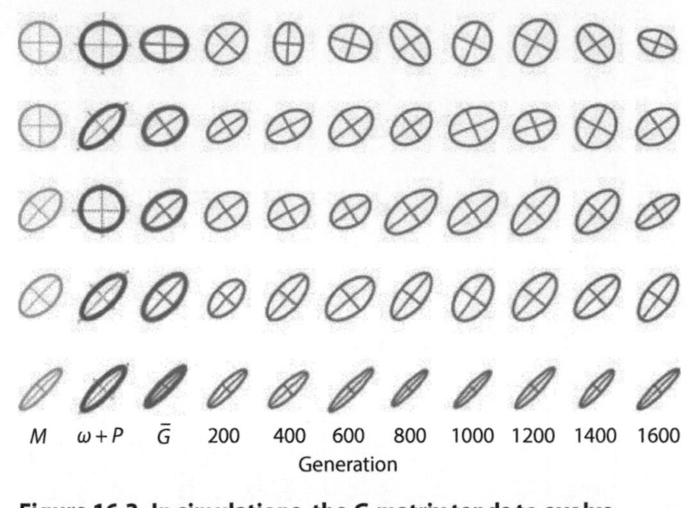

Figure 16.3 In simulations, the *G*-matrix tends to evolve towards alignment with the *M*-matrix and the adaptive landscape. Each row shows the results of a single 2000 generation simulation run. The first three ellipses in each row show the 95% confidence ellipses for mutation (***M***), the adaptive landscape ($\boldsymbol{\Omega} = \boldsymbol{\omega} + \boldsymbol{P}$), and the average *G*-matrix (shown on different scales). The last eight ellipses in each row show the *G*-matrix every 200 generations during the run. From top to bottom the values of r_μ and r_ω are: (a) $r_\mu = r_\omega = 0$; (b) $r_\mu = 0$, $r_\omega = 0.75$; (c) $r_\mu = 0.5$, $r_\omega = 0$; (d) $r_\mu = 0.5$, $r_\omega = 0.75$; (e) $r_\mu = r_\omega = 0.90$.

From Arnold et al. (2008) with permission.

(Figure 16.4, right-hand column). In contrast, this same array of mutation and selection conditions has relatively little effect on the stability of *G*-matrix size (Figure 16.4, left-hand column). In summary, the most stable *G*-matrices will occur in large populations and in traits under strong correlational selection that is aligned with the pattern of pleiotropic mutation.

16.5 The Effect of an Evolving Mutation Process on the Dynamic Properties of the *G*-matrix

The capacity for evolution of the genetic distribution has been limited in the simulation studies that we have reviewed up to this point in this chapter. In particular, we have held the mutational process constant and asked how the cloud of genetic values, described by the *G*-matrix, responds to constant or changing selection regimes. In actual genetic systems, the mutational process itself might be affected by selection, so that it evolves in parallel with the *G*-matrix and has a second order effect on *G*-evolution. In such actual systems, epistasis provides variation in pleiotropic effects that in turn feeds the evolution of ***G***.

In this section we will simulate a poor man's version of epistasis. We will allow individual variation in the mutational correlation between two traits, without specifying the epistasis that could produce such variation in actual genetic systems. As we shall see, despite its simplicity, this kind of variation in mutation allows the mutational process

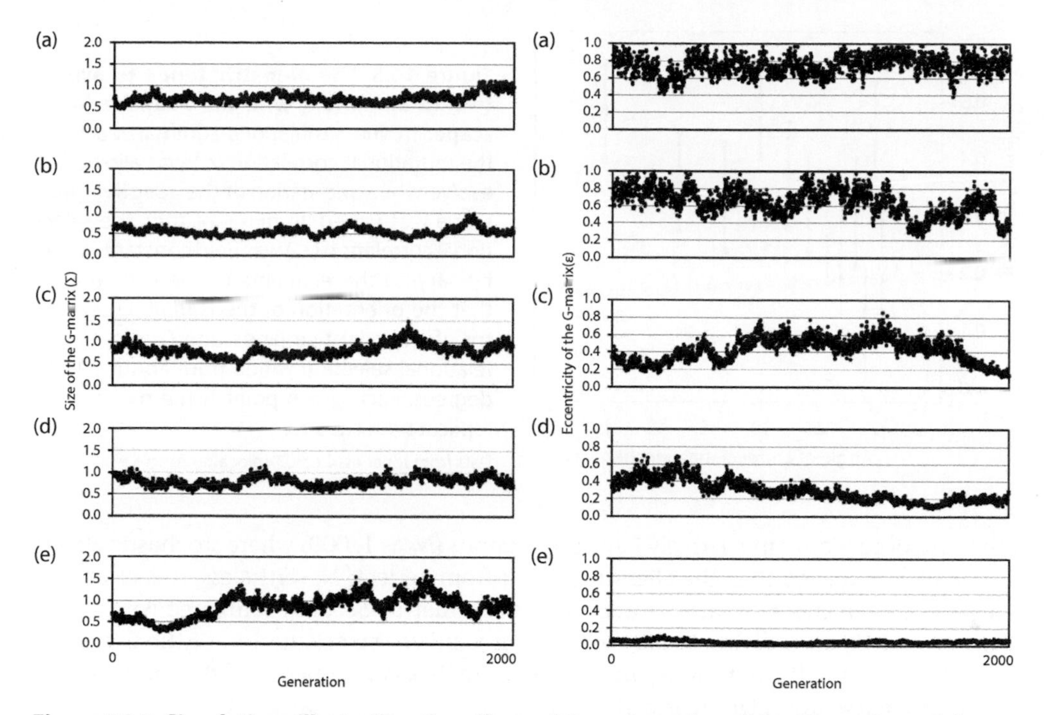

Figure 16.4 Simulations illustrating the effects of the orientation of the *M*-matrix and the adaptive landscape on the size and eccentricity of the *G*-matrix. (Left panel) Time series for the size of the *G*-matrix (sum of eigenvalues) when (a) $r_\mu = r_\omega = 0$; (b) $r_\mu = 0$, $r_\omega = 0.75$; (c) $r_\mu = 0.5$, $r_\omega = 0$; (d) $r_\mu = 0.5$, $r_\omega = 0.75$; (e) $r_\mu = r_\omega = 0.90$. (Right panel) Time series for the eccentricity of the *G*-matrix (smaller eigenvalue divided by larger eigenvalue) with parameters (a) through (e) as in left panel.

From Jones et al. (2003) with permission.

to respond to selection and evolve. But how exactly will this mutational evolution affect the evolution and stability of the *G*-matrix? Before turning to that question, we must first take a smaller step and ask how strongly does selection act on mutational variation in the guise of r_μ, the mutational correlation?

Jones et al. (2007) were able to answer this question under special, simplifying circumstances. For weak stabilizing selection ($\omega \gg 1$) that acts equally on two traits ($\omega = \omega_{11} = \omega_{22}$) in an equilibrated genetic system in which loci are mutationally equivalent, $G_{11} = G_{22}$, and the *directional selection gradient for* r_μ is

$$\frac{\partial ln\overline{W}}{\partial r_\mu} \approx \frac{G}{\omega + E}\left(\frac{1}{\sqrt{1 - r_\omega}\sqrt{1 - r_\mu}} - \frac{1}{\sqrt{1 + r_\omega}\sqrt{1 + r_\mu}}\right) \quad (16.4)$$

(Jones et al. 2007). The term in brackets is less than 10, as long as both of the correlations are ≤ 0.9. So, with weak selection ($\omega + E \approx \Omega \geq 100$), the force of directional selection on r_μ will be less than 0.1. In other words, selection on the mutational correlation is generally a weak force, stronger when the *M*-matrix is aligned with the AL, but disappearing entirely when $\boldsymbol{M}$ and the AL are in anti-alignment (Jones et al. 2007).

Simulations of evolving *G*-matrices when the mutational process is allowed to evolve yield two important results (Jones et al. 2007). First, the *M*-matrix tends to evolve towards alignment with the AL (Figure 16.5). As might be expected from the weak force involved,

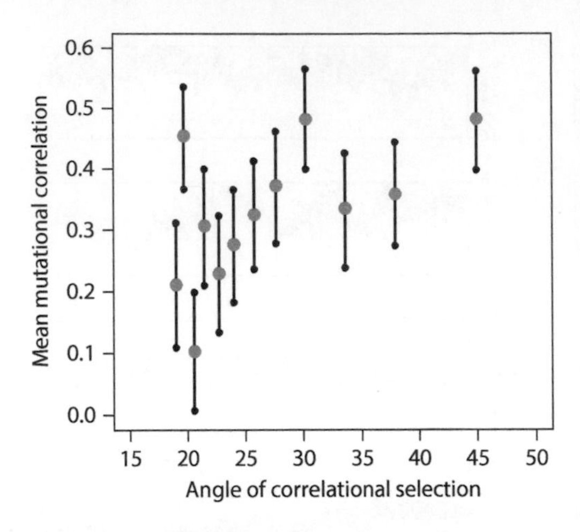

Figure 16.5 The *M*-matrix tends to evolve toward alignment with the adaptive landscape. In the simulations summarized here, the mutational correlation (r_μ) was allowed to evolve while orientation of the adaptive landscape was varied. In different runs the selectional correlation (r_ω) was held constant at 0.90 by varying the elements of the ω-matrix, so that the orientation of the leading eigenvector of the adaptive landscape (angle of correlational selection varied from about 19–45 degrees. Each green point is the mean of 50 replicate runs (± s.e.).

Data from Jones et al. (2007), plot after Arnold et al. (2008).

the trend occurs only in relatively large populations ($N_e > 1,000$) where stochastic effects are smaller, and even then the alignment is far from perfect. Nevertheless, it is clear that the AL exerts an effect on the orientation of the *M*-matrix. Second, compared with a constant *M*-matrix, an evolving *M*-matrix tends to stabilize the orientation of the *G*-matrix. The effect is dramatic, not subtle, and is undoubtedly a consequence of the tendency for $\boldsymbol{M}$ to evolve towards alignment with the AL.

These results imply that the mutational process evolves in response to selection in nature. Natural genetic systems should present a richer field of variation in the mutational process than was created in the computer simulations that we have just discussed. As a consequence, when gene effects are epistatic as well as pleiotropic, the mutational system itself should evolve. These considerations mean that the roles typically assigned to mutation, inheritance, and selection in textbooks are over simplified. Mutation and inheritance do not simply supply the raw variation that allows selection to act, they also respond to selection. The overall process works on two time levels. In the short term, mutation dribbles variation into the *G*-matrix in a manner that can be treated as a constant flow. But in the longer term, selection shapes both the mutational process and the *G*-matrix.

Jones et al. 2014 conducted a more rigorous test of the proposition that selection can shape the *M*-matrix by simulating a model in which epistasis is included so that $\boldsymbol{M}$ can evolve. Their simulations included an adjustable gauge which allowed them to change the amount of pleiotropy from run to run. That gauge was the variance in epistatic effects. This gauge is an element in the *multilinear model of epistasis* developed by (Hansen and Wagner 2001).

Earlier studies used the multilinear model to evaluate the effects on epistasis on the evolution of a single phenotypic trait (Hermisson et al. 2003; Carter et al. 2005; Le Rouzic et al. 2013). In this single trait version of the model, an individual's genotypic value for the quantitative trait z is given by

$$x = \xi_0 + \sum_i y^{(i)} + \sum_i \sum_{j:j>i} \varepsilon^{(i,j)} y^{(i)} y^{(j)} \tag{16.5}$$

where ξ_0 is the value of an arbitrary reference genotype, $y^{(i)}$ is the reference effect on an individual's genotype at locus i, and $\varepsilon^{(i,j)}$ is an epistatic coefficient that determines the nature of the interaction between locus i and locus j. The value of ξ_0 is assumed to be zero because we are employing a model with a stationary intermediate optimum with a value of zero. The two allelic values for the diploid individual in this model are summed to obtain the genotype's reference value, hence the first summation term on the right side of the equation.

In the *multiple trait version of the multilinear model*, an individual's genotypic value is

$$_{a}x = {}_{a}\xi_0 + \sum_i {}_{a}y^{(i)} + \sum_i \sum_{j:j>i} \sum_b \sum_c {}_{abc}\varepsilon^{(i,j)}\, {}_{b}y^{(i)}\, {}_{c}y^{(j)}, \qquad (16.6)$$

where $_{a}x$ is an individual's genotypic value for trait a, $_{a}\xi_0$ is the value of the reference genotype (zero), $_{a}y^{(i)}$ is the individual's reference genotypic value for trait a at locus i, and $_{abc}\varepsilon^{(i,j)}$ gives the epistatic effect on trait a of the interaction between the effects of locus i on trait b and locus j on trait c.

Using (16.6), Jones et al. (2014) calculated an individual's genotypic value across all loci for both traits, taking into account all possible pairwise epistatic interactions. To simulate environmental variance, a value for each individual was drawn independently from a normal distribution with a mean of zero and a variance of one for each trait. The phenotypic value for each individual is the sum of these genotypic and environmental values. Because genotypic and phenotypic variances change during the simulations, environmental variance was used as the scale-setting parameter in the study (i.e., $E = 1$).

The use of the multiple trait multilinear model for the genome means that hundreds of epistatic coefficients need to be specified in the two-trait, 20 locus implementation used by Jones et al. (2014). They solved this problem by drawing coefficients at random from a normal distribution with a mean of zero and a variance of σ_{ε}^2. Consequently, this variance is used as a gauge to adjust the level of epistasis in a particular run.

It is important to remember that in this study the M-matrix can evolve in response to selection acting on genetic variation provided by epistasis. In Figure 16.6 we see the evolved responses of $\boldsymbol{M}$ and $\boldsymbol{G}$ to three configurations of the adaptive landscape. In the top panel the configuration of the ω-matrix strongly tilts the landscape in the direction of trait 2, graphed on the y-axis. In the middle panel the tilt is less and in the bottom panel correlational selection is absent ($\omega_{12} = 0$). In response to these regimes of selection, the M-matrix and the G-matrix evolve into *triple alignment* with the principal axis of the ω-matrix. Note that the ellipse diagrams in the right-hand column are not cartoons, they are the averages of 20 replicate simulation runs.

16.6 Effects of Migration on Trait Means and the *G*-matrix

Up to this point we have considered single populations, independent of all others. Before we consider how migration might affect the G-matrices of two or more populations, let us consider the effect of migration on the trait means of a single trait in two populations that exchange migrants. Let there be two populations, a and b, with different trait optima, θ_a and θ_b, for Gaussian selection surfaces with the same width parameter, ω. For simplicity we will assume that the two populations have the same phenotypic variance, P, and the same additive genetic variance, G. In the absence of migration, the equilibrium difference in trait means, D, equals the difference in optima,

$$D = \bar{z}_a - \bar{z}_b = \theta_a - \theta_b, \qquad (16.7)$$

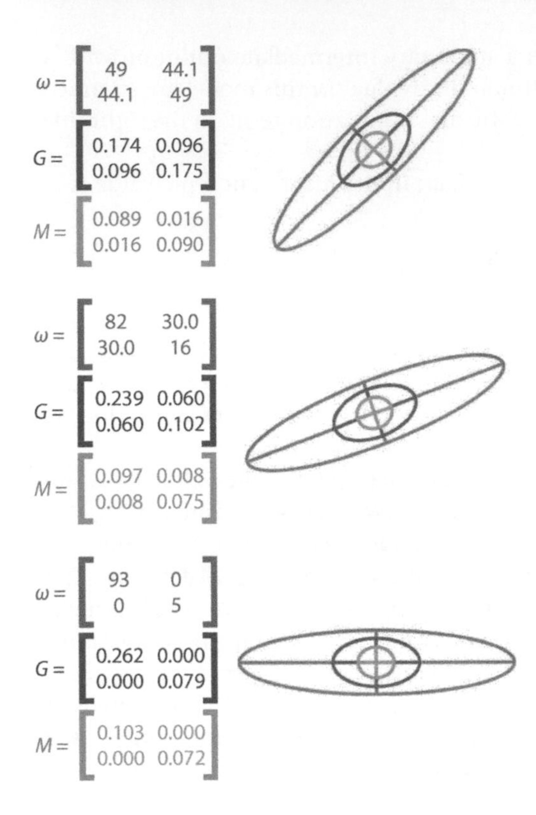

Figure 16.6 Triple alignment of selection, genetic variation, and mutation is promoted by epistasis. In each of three sets of simulation runs, the ω-matrix was held at three different values (orange matrices in the first vertical panel, orange ellipse in the second vertical panel). *G*-matrices (blue matrices, blue ellipses) and *M*-matrices (green matrices, green ellipses) were allowed to evolve in response to those specified selection surfaces. As the angle of the selection surface ($\boldsymbol{\omega_{max}}$) is rotated, the angles of ***G*** and ***M*** evolve in response resulting in triple alignment of the three matrices.

From Jones et al. (2014) with permission.

but with migration in both directions between the two populations, the equilibrium difference in trait means becomes

$$D^* = \bar{z}_a - \bar{z}_b = (\theta_a - \theta_b)\left[\frac{G}{G(1-\hat{m}) + \Omega\hat{m}}\right], \tag{16.8}$$

where $\Omega = \omega + P$, and $\hat{m}$ is the total amount of mixing between the two populations (the proportion of population *a* made up of migrants from population *b* plus the proportion of population *b* made up of migrants from population *a*) (Hendry et al. 2007). Expression (16.8) is for a model in which mixing occurs before selection each generation. If mixing occurs after selection, the numerator of the expression in brackets becomes $G(1-\hat{m})$ (Hendry et al. 2007). In other words, migration displaces each trait mean off of its optimum in the direction of the other population's optimum. Although this result is for a single trait, we will see below that it is qualitatively correct for two traits.

We now consider how the details of gene flow between populations might affect the configuration of the *G*-matrix. We have now entered a region that lacks analytical results and use simulations to find our way (Guillaume and Whitlock 2007). To make things tractable we will assume that migration is one way, from a mainland population to a series of island populations (Figure 16.7). This convention allows the mainland population, unaffected by migration, to equilibrate in mutation-drift-selection balance. Only the island populations are affected by migration and each one is affected independently (i.e., no migration among islands). Other parameters in the simulations are detailed in Table 16.1.

The case of *G*-matrix evolution in the case of strong migration from the mainland to the islands is shown in Figure 16.8. First, consider Figure 16.8a. Here the adaptive landscape

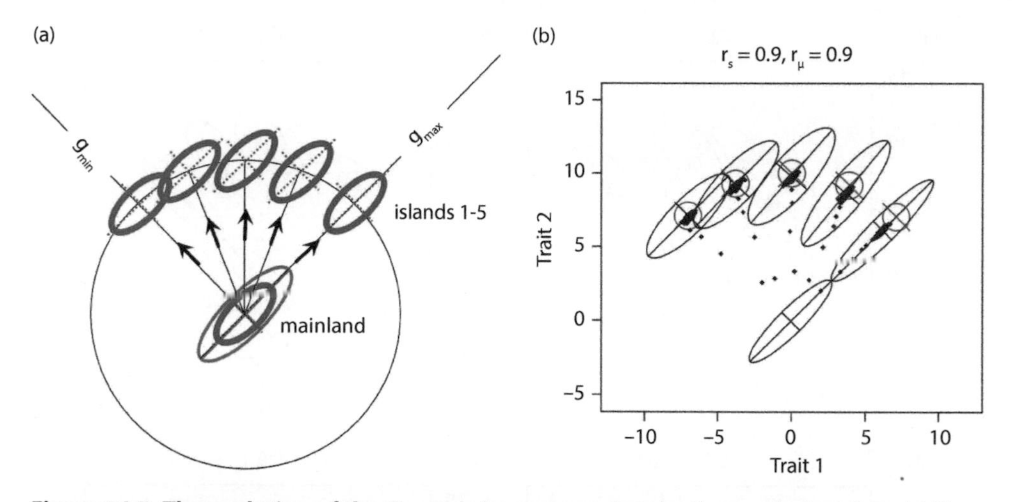

Figure 16.7 The evolution of the *G*-matrix in response to selection and migration. (a) The selection and migration scheme. Selection surfaces are shown as red ellipses. The *G*-matrix is shown as a blue ellipse. The trait optimum (intersection of the dotted red lines) for the mainland population is located at the centre of the circle (shown in black). The *G*-matrix has evolved to equilibrium with the configuration and position of the selection surface. The trait optimum for an island population is at one of five positions on the circle. The solid straight lines connect the optimum of the mainland with the optimum of each island population. This line is the *expected line of divergence* (*LoD*) for the bivariate island mean. The black arrows show the directional selection that acts on the island trait mean as it begins its evolutionary trajectory. (b) The evolution of the *G*-matrix in response to the selection-migration scheme shown on the left. The mainland and island *G*-matrices are shown as black ellipses. The position of the trait optimum is shown by a red cross inside a red circle. Correlational selection denoted r_s is the same as r_ω. The evolutionary trajectory of the bivariate trait mean for one simulation run for each island configuration is represented by black dots, shown at intervals of 10 generations.

From Guillaume and Whitlock (2007) with permission.

is circular while the mutation ellipse is cigar-shaped and oriented at a 45 deg angle. Island 5 lies on the mainland $\boldsymbol{g}_{max}$ which coincides with its *LoD*. The trait mean for 5 evolves rapidly along these coinciding lines and equilibrates close to the 5 optimum but pulled slightly towards the mainland optimum. In contrast, island 1 lies on the mainland $\boldsymbol{g}_{min}$ which coincides with its *LoD* (expected line of divergence) (Figure 16.7). Compared with island 5, island 1 has relatively little genetic variance along its trajectory and indeed it evolves slowly along its path. At equilibrium, the island 1 trait mean is pulled further towards the mainland optimum than is island 5. The evolution of islands 2–4 is initially rapid and biased towards gmax, but then each island mean takes a slow approach towards its optimum, as it evolves along $\boldsymbol{g}_{min}$, equilibrating at a position away from its optimum and biased towards $\boldsymbol{g}_{min}$. In Figure 16.8d, we see a similar pattern in the evolutionary trajectory of the island means: rapid initial evolution along $\boldsymbol{g}_{max}$, a slow approach to the island optimum along $\boldsymbol{g}_{min}$, with equilibration away from the island optimum that is exaggerated along a -45 deg angle ($\boldsymbol{\omega}_{max}$).

Turning to the *G*-matrix and comparing Figure 16.8a with 16.8d, we see that the negatively inclined cigar-shaped landscape in Figure 16.8d has reduced the size of $\boldsymbol{G}$ and changed its orientation from alignment with $\boldsymbol{M}$ to a compromise between alignment with $\boldsymbol{\omega}_{max}$ (island 1) and alignment with $\boldsymbol{g}_{max}$ (island 5).

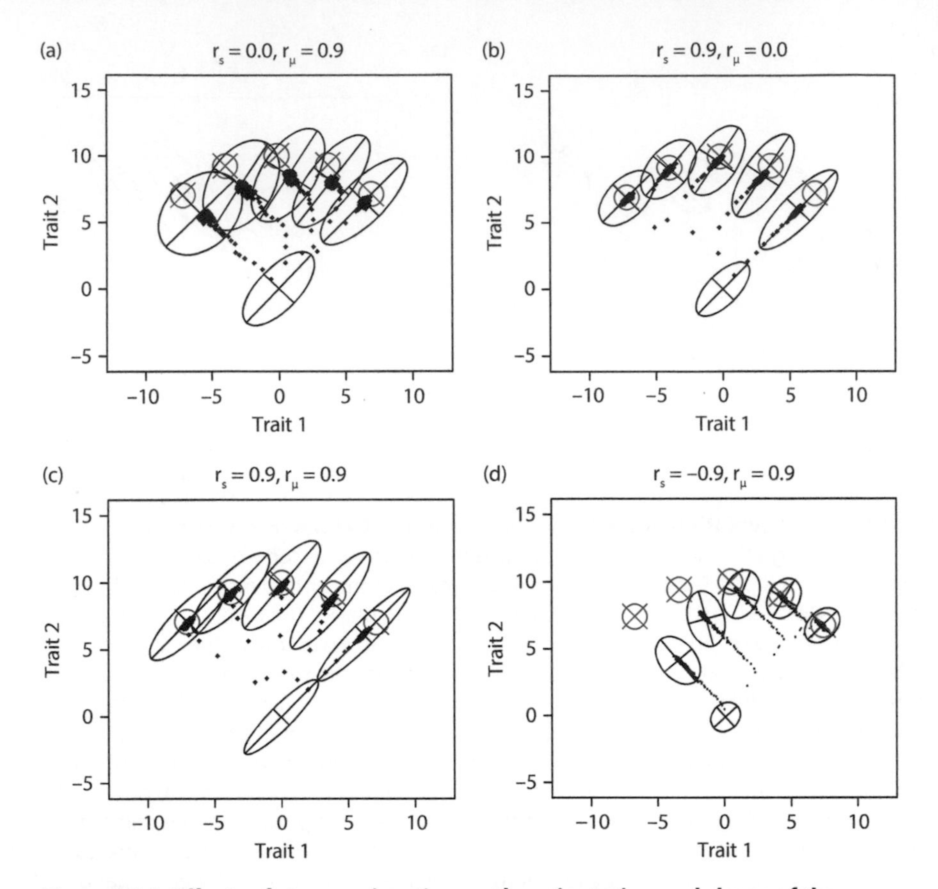

Figure 16.8 Effects of strong migration on the orientation and shape of the *G*-matrix for two traits. Correlational selection denoted r_s is the same as r_ω. (a) No correlational selection, strong mutational correlation. (b) Strong correlational selection, no mutational correlation. (c) Strong correlational selection and strong mutational correlation. (d) Strong negative correlational selection, strong positive mutational correlation.

From Guillaume and Whitlock (2007) with permission.

In summary, these and other simulation runs by Guillaume and Whitlock (2007) show that magnitude and orientation of migration can affect the size, eccentricity, and orientation of the *G*-matrix. Furthermore, in the plausible case that $\boldsymbol{M}$ is aligned with the adaptive landscape (Figure 16.8c), an especially eccentric (and hence stable) *G*-matrix evolves if the optima of two populations are displaced along $\boldsymbol{\omega_{max}}$.

16.7 Comparative Studies of the *G*-matrix

The centrality of the *G*-matrix in predicting responses to drift and selection has inspired several dozen comparative studies to empirically assess its constancy in evolutionary time. These studies are important because they tell us how the *G*-matrices of actual genomes vary under selection regimes in nature. At the same time, the data are limited. Because of the labour required in estimation, a total of only a few hundred *G*-matrices have been

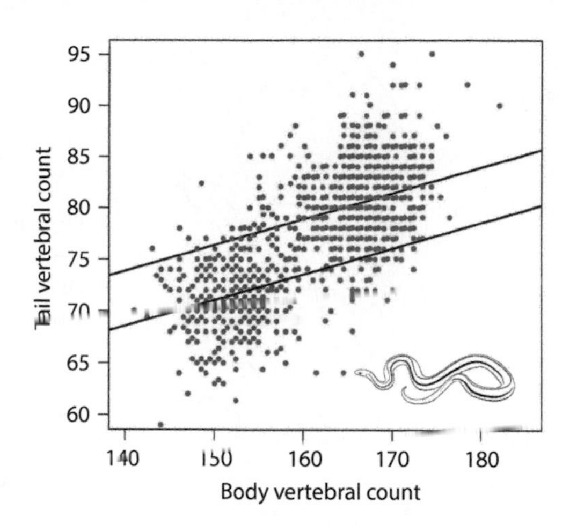

Figure 16.9 Numbers of tail and body vertebrae in samples of newborn females from two California populations of the garter snake *T. elegans*. Females from the inland population are shown as light blue dots (n = 690), females from the coastal population are shown as dark blue dots (n = 474). Least squares regression lines the regression of tail count on body count are shown for both populations.

Data are from the study reported by (Arnold and Phillips (1999) and Phillips and Arnold (1999).

compared, usually two at a time. In contrast, for every pair of actual *G*-matrices that have been compared, at least a million virtual pairs have been assessed in simulation studies (Tables 16.1 and 17.1). Nevertheless, despite its limitations, the direct empirical approach provides information about natural systems that no amount of simulation can reveal.

The basic data in an empirical comparison of *G*-matrices are measurements of phenotypic values and estimates of genetic matrices in two or more populations. Figure 16.9 shows the dispersion clouds for vertebral counts in two different populations of *Thamnophis elegans*. Although these are phenotypic rather than genetic clouds, our attention is attracted to the similarity in the size and orientation of these clouds.

Our goal then in comparing two populations is to see whether features of multivariate similarity commonly prevail in comparisons between populations and species. We also want to characterize those features of similarity.

16.7.1 *Comparing matrices with the Flury hierarchy*

A variety of approaches have been proposed for comparison of *G*-matrices (Steppan et al. 2002; Teplitsky et al. 2014). We will focus on just one comparison technique, the Flury hierarchy (Flury 1988), because it directly tackles the complex multivariate nature of *G*-matrix comparison. The application of Flury's comparison hierarchy to *G*-matrices is complicated by the fact that genetic covariances are estimated from the same sets of relatives, which introduces an extra element of sampling covariance. Phillips and Arnold (1999) describe a bootstrapping solution to this complication.

The *Flury hierarchy* uses comparisons of eigenvectors and eigenvalues to assess similarity between two or more *G*-matrices (Flury 1988). The comparisons are arranged in a hierarchy that permits a succession of statistical tests using the principles of maximum likelihood (Figure 16.10). At the first level in the hierarchy, we can ask if all the eigenvectors and eigenvalues are identical (the matrices are equal). Secondly, if the eigenvectors are the same, we can ask if all the eigenvalues differ by the same constant of proportionality (proportional matrices). Thirdly, if the matrices are neither equal nor proportional, we can ask if the matrices have *common principal components* (i.e., the same eigenvectors) even though their eigenvalues are non-proportional (the matrices are said to be CPC). Finally, the matrices might differ in both their eigenvectors and eigenvalues (matrices unrelated). For the sake of simplicity we have skipped over some ground at the third step.

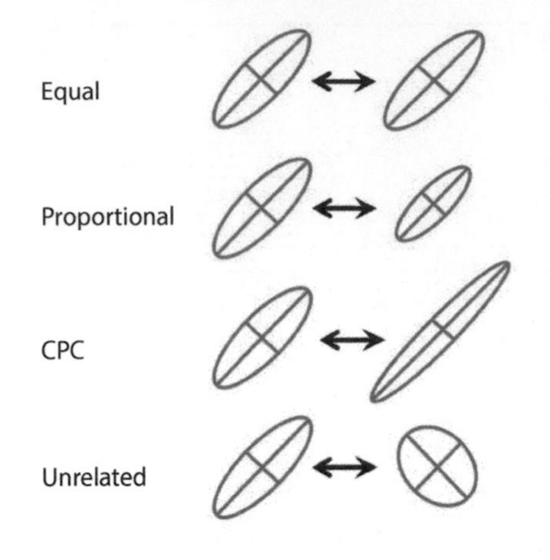

Figure 16.10 The Flury hierarchy for comparing *G*-matrices is a nested series of hypotheses that are tested by comparing eigenvalues and eigenvectors. The illustrated hierarchy is for the comparison of bivariate *G*-matrices, represented by their 95% confidence ellipses. CPC refers to common principal components, the sharing of eigenvectors, even though eigenvalues may differ.

From Arnold et al. (2008).

Some but not all of the eigenvectors might be identical, so that the matrices might have 1, 2, 3, $\ldots m - 1$ principal components in common (i.e., partial CPC), where m is the number of traits. To use this hierarchy in a hypothesis-testing framework, one typically starts at the top and moves down a step at a time, as each hypothesis in turn is rejected (Phillips and Arnold 1999).

A case study of *G*-matrix comparison in the garter snake *T. elegans* illustrates the insights that the Flury hierarchy can provide (Phillips and Arnold 1999, Arnold and Phillips 1999). This study sampled two conspecific populations, one from a coastal site and one from an inland site in California. The phylogenetic separation time of these two populations is approximately 2 million years. Six meristic traits were scored, representing scale counts on the head, around the body, and along the length of the body and tail. Two traits (vent and sub) correspond, respectively, to body and tail vertebral counts. Due to sexual dimorphism, separate matrices were estimated for males and females. Analyses were based on the counts of wild-caught gravid females and their sets of full-sib offspring. Details of the analysis are provided by Arnold and Phillips (1999). An expanded analysis with three geographic sites and special focus on the interpretation of male and female matrices is provided by Barker et al. (2010).

The results of matrix comparisons and statistical tests using the Flury hierarchy are shown in Table 16.2. The results of *E*-matrix comparisons are surprisingly simple, in all cases the highest level of the hierarchy, matrix equality, could not be rejected. Furthermore, all of the *E*-matrices were nearly diagonal in form with small off-diagonal elements. Because $\boldsymbol{P} = \boldsymbol{G} + \boldsymbol{E}$ (6.4), the equality of *E*-matrices implies that any structural differences we might observe in *P*-matrices, reflect differences in *G*-matrices. Turning to *P*-matrix comparisons, pairwise tests generally produced full or partial CPC, depending on mode of analysis (covariance components or individual datapoints). In other words, *P*-matrices tended to share all of or a subset of their eigenvectors. *G*-matrices showed a similar pattern but one that tilted a little more strongly towards full CPC. When a partial CPC result was obtained for *P*- or *G*-matrices, eigenvectors reflecting the vent and sub axes were generally included. In other words, conservation of eigenvectors applied especially to body and tail vertebral count axes.

The similarity of *G*-matrix ellipses based on vertebral counts across two populations and two sexes is evident in Figure 16.11. The similarity of orientation is particularly striking. The leading eigenvectors share a common angle, which is similar to the axis of differentiation in trait means.

Table 16.2 Results of hierarchial comparison of *P* -, *G*-, and *E*-matrices from male and female offspring from coastal and inland populations of *Thamnophis elegans*. Two estimates of the *P*-matrix were used: an estimate from the within- and among-litter components of variance and covariance (cov. comp.) and an estimate using each individual offspring as a datapoint (individual). Two estimates of the *G*-matrix were compared: an estimate from regression of litter means on maternal values (regression) and from the variances and covariances of litter means (family-mean). The first six columns show the results for all possible pairwise comparisons. The last column shows results of simultaneous comparison of all four matrices. Entries give the highest point in the hierarchy at which the listed null hypothesis could not be rejected. [1]CPC(4) when coastal males are excluded. [2]Full CPC when coastal males are excluded. From Arnold and Phillips (1999).

	Coastal male vs coastal female	Coastal male vs inland male	Coastal male vs inland female	Coastal female vs Inland male	Coastal female vs Inland female	Inland male vs Inland female	All together
Phenotypic:							
Cov. comp.	Equal	Full CPC	Full CPC	Full CPC	Full CPC	Full CPC	Full CPC
Individual	Full CPC	CPC(2)	CPC(4)	CPC(4)	CPC(3)	CPC(4)	CPC(2)[1]
Genetic:							
Regression	CPC(2)	CPC(1)	Full CPC	Full CPC	Full CPC	Full CPC	CPC(2)[2]
Family-mean	Full CPC	CPC(3)	CPC(4)	Full CPC	Full CPC	Full CPC	CPC(4)[2]
Environmental	Equal	Equal	Equal	Equal	Equal	Equal	Equal

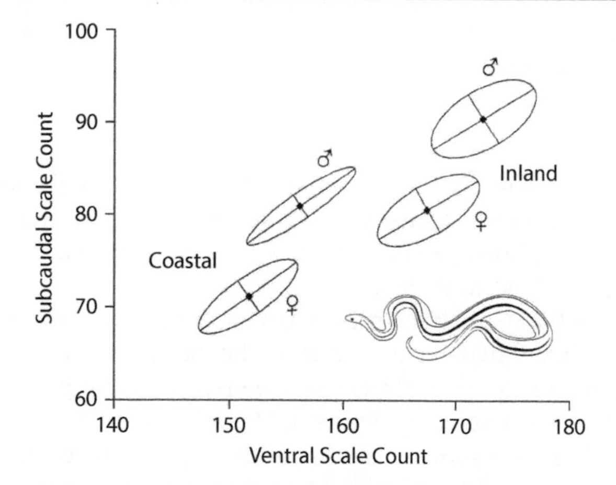

Figure 16.11 Comparison of G-matrix ellipses of males and females of the coastal and inland populations of *Thamnophis elegans* showing the common principal component result. The figure shows projections of the first two principal components onto the tail (subcaudal) and body (ventral scale count) axes.

From Arnold and Phillips (1999).

The theme of common principal components evident in the *T. elegans* example is also evident in a larger sample of studies (Figure 16.12). In the majority of studies some if not all eigenvectors of ***G*** are conserved during the divergence of species and conspecific populations (Arnold et al. 2008). Recall that by the nature of the hierarchy, the only statistical

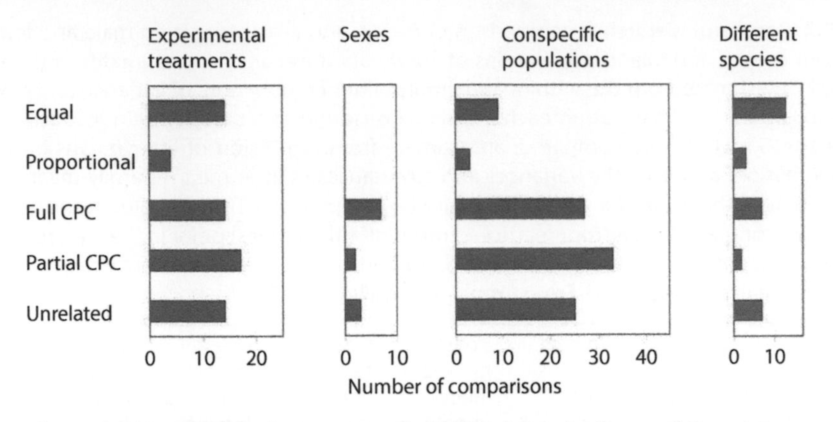

Figure 16.12 Graphical summary of empirical comparisons of *G*-matrices. Only the 31 studies that made comparisons using the Flury hierarchy are included here. From left to right, the four panels summarize results from four kinds of studies: conspecific populations exposed to different environmental treatments (n = 63 pairwise comparisons), males versus females from the same population (n = 12), conspecific populations sampled from nature (n = 97), and different species sampled from nature (n = 32). Outcomes of statistical tests are classified into the categories described in Figure 16.10. Full CPC means that matrices had all principal components (eigenvectors) in common; partial CPC means that at least one but not all principal components were in common. Some studies compared multiple pairs of matrices, and in such cases, all the outcomes are tabulated.

From Arnold et al. (2008).

outcome in which no eigenvectors are shared is the case of unrelated matrices. Applying this realization to the results shown in Figure 16.12, we find that from left to right the proportion of comparisons yielding partial or complete similarity in eigenvectors is, respectively, 78%, 74%, 75%, and 78%.

The empirical result that *G*-matrix eigenvectors are commonly stable, directs our attention to simulation studies that have examined the causes of such stability. The main result, reviewed earlier in this chapter, is that eigenvector stability is promoted by large population size, strong eigenvector structure in *M*- and *G*-matrices (i.e., correlations both in pleiotropic mutations and genetic values) and in the AL (i.e., correlational selection). Eigenvector stability is especially promoted by a three-way alignment of the *M*-, *G*-, and Ω-matrices. As we shall see in the next chapter, movement of the adaptive peak in the same direction as this three-way alignment adds an extra element of stability. A simple way to visualize the conditions most conducive to the kind of structural stability of ***G*** that we observe in empirical comparisons is to imagine that the ridges of the AL are stable on a very long timescale (for tens of thousands to millions of generations) so that ***G***, and even ***M***, evolve towards alignment with that stable structure.

16.7.2 *Phylogenetic comparisons of matrices*

If *G*-matrix stability reflects the underlying stability of the adaptive landscape, we might expect the structural stability of the *G*-matrix to gradually decay as time unfolds and evolutionary change in the AL becomes progressively more inevitable. If we depict comparisons of ***G*** on an evolutionary tree, we would expect to see more instances of

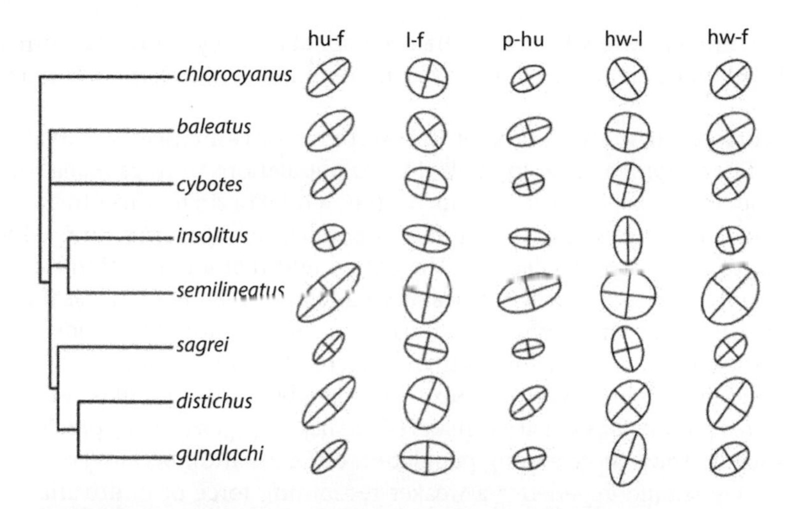

Figure 16.13 Phylogenetic comparison of morphometric *P*-matrix ellipses from eight species of *Anolis* lizards. The phylogeny is shown in the first column. Columns 2–6 show 95% confidence ellipses for pairs of morphometric traits. The five illustrated pairs of trait measurements are: humerus vs. femur (hu-f), lamellae 4th vs. femur (l-f), pelvis vs. humerus (p-f), head width vs. lamellae 4th (hw-l), and head width vs. femur (hw-f). After Kolbe et al. (2011).

equal and proportional matrices on pairs of short branches and more instances of unequal matrices on pairs of long branches. Morgan et al. (2001) found exactly this kind of phylogenetic pattern in a comparison of life-history trait *G*-matrices for *Daphnia* populations inhabiting ponds and lakes in Oregon. Steppan (1997) found a similar phylogenetic picture for the *P*-matrices of morphometric traits in sigmodontine mice. The results of a phylogenetic study of morphometric *P*-matrices in *Anolis* lizards (Kolbe et al. 2011) are shown in Figure 16.13. In this case, however, the phylogeny has relatively little structure and portrays common descent from a common distant ancestor. Nevertheless, the diagram illustrates the conclusion from simulation studies that cigar-shaped matrices (the hu-f column compared with all the others) tend to be more stable than circular matrices with nearly equal eigenvalues.

16.8 Evidence That Selection Shapes the *G*-matrix

Some comparative studies provide evidence that selection plays a role in shaping the *G*-matrix. Usually studies of this kind do not test an a priori hypotheses about the particular ways in which *G*-matrices will differ. Instead, two or more *G*-matrices are sampled from two or more environments that are known to or thought to differ in selection regimes. The test for an effect of selection consists of testing whether differences among *G*-matrices is greater between environments than within environments. Only a few tests of this kind have been conducted and tests for matrix differences vary in stringency (Barker et al. 2010):

- Jernigan et al. (1994) compared genetic correlation matrices for sensory structures in amphipods (*Gammarus minus*) sampled for a surface and a cave population in each of two areas, i.e., in a total of four populations. Focusing on pairs of traits most likely to respond differently to selection in the two habitats (i.e., eye-antennal correlations), they did find that genetic correlations were more similar within habitats than between clades.

- In another revealing study, Cano et al. (2004) compared *G*-matrices for larval morphometric and developmental timing traits in two populations of frogs (*Rana temporaria*) reared under three experimental treatments (i.e., a total of six matrices in a 2 × 3 factorial design). The authors detected significant population differences in ***G*** which they attributed to differences in selection. Their argument that a population difference in selection caused the differences in ***G*** was bolstered by a $Q_{ST} - F_{ST}$ analysis that showed the phenotypic traits had experienced a history of directional selection (i.e., more differentiation in phenotypic means than expected under neutrality).
- Barker et al. (2010) compared within-sex (*G*-) and between-sex (*B*-) matrices for meristic traits in three populations of garter snakes (*Thamnophis elegans*). They predicted that *B*-matrices would show more among-population differentiation because, on theoretical grounds, they should experience a weaker restraining force of multivariate stabilizing selection. *B*-matrices were indeed more variable than *G*-matrices, supporting the conjecture of weaker selection on *B*-matrices.

Although these and a few other published tests provide evidence that selection shapes the *G*-matrix, additional empirical work along these lines is clearly needed.

16.9 Tests for Alignment Between the *G*-matrix and the Adaptive Landscape

Another line of evidence that selection shapes the *G*-matrix comes from tests for alignment of eigenvectors. Tests of this kind go to the heart of the argument that the *G*-matrices should evolve towards alignment with the AL. Only a few tests of this proposition have been conducted, and they give conflicting results:

- Brodie's (1992, 1993) garter snakes (*Thamnophis ordinoides*) are a particularly compelling case, because pleiotropy seems unlikely to be the cause of the genetic correlation that he observed between escape behaviour and colouration pattern. Instead, selection must have generated the correlation by promoting linkage disequilibrium, unassisted by pleiotropic mutation. The observation of correlational selection with the same sign as the genetic correlation strengthens the argument and makes a strong case for evolved alignment between ***G*** and the AL.
- In contrast to Brodie's result, Blows et al. (2004) found a variety of structural differences between the *G*-matrix for male cuticular hydrocarbons in *Drosophila serrata* and the selection surface generated by female mating preferences. The case is complicated, as the authors acknowledge, by directional selection that springs from mating preference, and which may or may not be balanced by unmeasured viability selection.
- On the other hand, Hunt et al. (2007) found substantial similarity between the *G*-matrix for male vocalization variables in the cricket (*Teleogryllus commodus*) and the selection surface generated by female preferences for those variables (Brooks et al. 2005).
- Finally, Hohenlohe and Arnold (2008) demonstrated substantial alignment between *G*-matrices based on meristic traits in garter snakes (*Thamnophis elegans*) and two different selection surfaces.

In other words, although the evidence is not entirely in favour of alignment between ***G*** and the AL, it is predominantly so.

16.10 Conclusions

- In the absence of selection, the *G*-matrix is a compromise between the input of variation from mutation and the loss of variation by drift.
- In a large population experiencing selection, the equilibrium *G*-matrix is a compromise between mutation and selection.
- Simulations allow us to study the stability and evolution of the *G*-matrix under general conditions of mutation, drift, and selection.
- Simulations with a stationary adaptive landscape and a constant *M*-matrix show that the *G*-matrix tends to evolve towards alignment with the *M*-matrix and the adaptive landscape.
- The effect of different forces on the stability of the *G*-matrix is best understood by viewing the *G*-matrix in terms of its eigenvalues and eigenvectors.
- The orientation of the *G*-matrix (angle of its leading eigenvector) is stabilized by large population size and strong among-trait correlations in mutation and stabilizing selection.
- An evolving mutation process adds an additional increment of stability to the orientation of the *G*-matrix.
- When epistasis is present, the *M*-matrix and the *G*-matrix evolve towards a triple alignment with the adaptive landscape.
- If the *M*-matrix is aligned with the adaptive landscape, an especially stable *G*-matrix evolves if optima of two populations connected by migration are displaced along a selective line of least resistance.
- Empirical comparisons of the *G*-matrices of related populations often show alignment (common principal components). This observation in conjunction with simulation results suggest that these related *G*-matrices are shaped by adaptive landscapes of similar configuration.

Literature Cited

Arnold, S. J., R. Bürger, P. A. Hohenlohe, B. C. Ajie, and A. G. Jones. 2008. Understanding the evolution and stability of the G-matrix. Evolution 62:2451–2461.

Arnold, S. J., and P. C. Phillips. 1999. Hierarchical comparison of genetic variance-covariance matrices. II. Coastal-inland divergence in the garter snake, *Thamnophis elegans*. Evolution 53:1516–1527.

Barker, B. S., P. C. Phillips, and S. J. Arnold. 2010. A test of the conjecture that G-matrices are more stable than B-matrices. Evolution 64:2601–2613.

Blows, M. W., S. F. Chenoweth, and E. Hine. 2004. Orientation of the genetic variance-covariance matrix and the fitness surface for multiple male sexually selected traits. Am. Nat. 163:329–340.

Brodie, E. D. I. 1992. Correlational selection for color pattern and antipredator behavior in the garter snake *Thamnophis ordinoides*. Evolution 46:1284–1298.

Brodie, E. D. I. 1993. Homogeneity of the genetic variance-covariance matrix for antipredator traits in two natural populations of the garter snake *Thamnophis ordinoides*. Evolution 47:844–854.

Brooks, R., J. Hunt, M. W. Blows, M. J. Smith, L. F. B. Re, and M. D. Jennions. 2005. experimental evidence for multivariate stabilizing sexual selection. Evolution 59:871–880.

Bürger, R., and R. Lande. 1994. On the distribution of the mean and variance of a quantitative trait under mutation-selection-drift balance. Genetics 138:901–912.

Cano, J. M., A. Laurila, J. U. Palo, and J. Merilä. 2004. Population differentiation in G matrix structure due to natural selection in *Rana temporaria* . Evolution 2013–2020.

Carter, A. J. R., J. Hermisson, and T. F. Hansen. 2005. The role of epistatic gene interactions in the response to selection and the evolution of evolvability. Theor. Popul. Biol. 68:179–196.

Flury, B. D. 1988. Common Principal Components and Related Multivariate Models. John Wiley and Sons, New York.

Guillaume, F., and M. C. Whitlock. 2007. Effects of migration on the genetic covariance matrix. Evolution 61:2398–2409.

Hansen, T. F., and G. P. Wagner. 2001. Modeling genetic architecture: a multilinear theory of gene interaction. Theor. Popul. Biol. 59:61–86.

Hendry, A. P., T. Day, and E. B. Taylor. 2007. Population mixing and the adaptive divergence of quantitative traits in discrete populations: a theoretical framework for empirical tests. Evolution 55:459–466.

Hermisson, J., T. F. Hansen, and G. P. Wagner. 2003. Epistasis in polygenic traits and the evolution of genetic architecture under stabilizing selection. Am. Nat. 161:708–734.

Hohenlohe, P. A., and S. J. Arnold. 2008. MIPoD: A hypothesis-testing framework for microevolutionary inference from patterns of divergence. Am. Nat. 171:366–385.

Hunt, J., M. W. Blows, F. Zajitschek, M. D. Jennions, and R. Brooks. 2007. Reconciling strong stabilizing selection with the maintenance of genetic variation in a natural population of Black Feld Crickets (*Teleogryllus commodus*). Genetics 177:875–880.

Jernigan, R. W., D. C. Culver, and D. W. Fong. 1994. The dual role of selection and evolutionary history as reflected in genetic correlations. Evolution 587–596.

Jones, A. G., S. J. Arnold, and R. Bürger. 2003. Stability of the G-matrix in a population experiencing pleiotropic mutation, stabilizing selection, and genetic drift. Evolution 57:1747–1760.

Jones, A. G., S. J. Arnold, and R. Bürger. 2004. Evolution and stability of the G-matrix on a landscape with a moving optimum. Evolution 58:1639–1654.

Jones, A. G., S. J. Arnold, and R. Bürger. 2007. The mutation matrix and the evolution of evolvability. Evolution 61:727–745.

Jones, A. G., R. Bürger, and S. J. Arnold. 2014. Epistasis and natural selection shape the mutational architecture of complex traits. Nat. Commun. 5:3709–3719.

Jones, A. G., R. Bürger, S. J. Arnold, P. A Hohenlohe, and J. C. Uyeda. 20012. The effects of stochastic and episodic movement of the optimum on the G-matrix and the response of the trait mean to selection. Journal of Evolutionary Biology 25:2210–2231.

Kolbe, J. J., L. J. Revell, B. Szekely, E. D. Brodie III, and J. B. Losos. 2011. Convergent evolution of phenotypic integration and its alignment with morphological diversification in Caribbean *Anolis* ecomorphs. Evolution 65:3608–3624.

Lande, R. 1979. Quantitative genetic analysis of multivariate evolution, applied to brain:body size allometry. Evolution 33:402–416.

Lande, R. 1980. Genetic variation and phenotypic evolution during allopatric speciation. Am. Nat. 116:463–479.

Le Rouzic, A., J. M. Álvarez-Castro, and T. F. Hansen. 2013. The evolution of canalization and evolvability in stable and fluctuating environments. Evol. Biol. 40:317–340.

Lynch, M., and W. G. Hill. 1986. Phenotypic evolution by neutral mutation. Evolution 40: 915–935.

Morgan, K. K., J. Hicks, K. Spitze, L. Latta, M. E. Pfrender, C. S. Weaver, M. Ottone, and M. Lynch. 2001. Patterns of genetic architecture for life-history traits and molecular markers in a subdivided species. Evolution 55:1753–1761.

Phillips, P. C., and S. J. Arnold. 1999. Hierarchical comparison of genetic variance-covariance matrices. I. Using the Flury hierarchy. Evolution 53:1506–1515.

Phillips, P. C., M. C. Whitlock, and K. Fowler. 2001. Inbreeding changes the shape of the genetic covariance matrix in *Drosophila melanogaster*. Genetics 158:1137–1145.

Revell, L. J. 2007. The G matrix under fluctuating correlational mutation and selection. Evolution 61:1857–1872.

Steppan, S. J. 1997. Phylogenetic analysis of phenotypic covariance structure. I. Contrasting results from matrix correlation and common principal component analyses. Evolution 51:571–586.

Steppan, S. J., P. C. Phillips, and D. Houle. 2002. Comparative quantitative genetics: evolution of the G matrix. Trends Ecol. Evol. 17:320–327.

Teplitsky, M. R. Robinson, and J. Merilä. 2014. Evolutionary potential and constraints in wild populations. Pp. 190–208 *in* A. Charmaniter, D. Garant, and L. E. B. Kruuk. Quantitative Genetics in the Wild, Oxford University Press, Oxford.

Wright, S. 1951. The genetical structure of populations. Ann Eugen. 15:323–354.

CHAPTER 17

Evolution of the *G*-matrix on Dynamic Adaptive Landscapes

Overview—Experimental studies show that the *G*-matrix can be shaped by directional selection, so it is natural to ask whether the directional selection associated with peak movement will have appreciable, long-lasting effects on $\boldsymbol{G}$. Simulation studies have explored the evolution of a two-trait *G*-matrix under a variety of peak movement scenarios. For example, when the adaptive peak moves at a constant, moderate rate, genetic variation is enhanced, not diminished, by peak movement. Furthermore, peak movement along a genetic line of least resistance promotes stability of the *G*-matrix. Thus, the greatest stability of the *G*-matrix can be expected when the *M*-matrix, the *G*-matrix, and the adaptive landscape are in three-way alignment, and the adaptive peak moves in that direction as well. Under these circumstances the *G*-matrix evolves towards an especially stable cigar shape. Simulation studies have also shown that both peak movement and correlational selection can promote the evolution of modular structure in the *P*- and *G*-matrices. Finally, we note that studies of *G*-matrix response to realistic selection regimes provide a bridge to genomic study of adaptation, a point that we will explore in the final chapter.

17.1 Effect of Steady, Linear Peak Movement on the Bivariate Mean

Steady linear motion of the optimum is the easiest dynamic pattern to characterize, so it constitutes a good starting point for the discussions that follow. In particular, consider the directional selection on the trait mean that is induced by steady movement of the peak of a Gaussian AL that does not change in configuration. In any particular generation, t, the strength of directional selection will be proportional to distance from the phenotypic trait mean to the peak and inversely proportional to the multivariate curvature of the AL,

$$\boldsymbol{\beta}_t = \boldsymbol{\Omega}^{-1}\left(\boldsymbol{\theta}_t - \bar{\boldsymbol{z}}_t\right). \tag{17.1}$$

With the peak moving at a constant steady per generation rate, $\Delta\boldsymbol{\theta}$, the distance to the peak will eventually stabilize at a constant value, an *equilibrium lag*,

$$\boldsymbol{L} \equiv \boldsymbol{\theta}_t - \bar{\boldsymbol{z}}_t = \boldsymbol{\Omega}\boldsymbol{G}^{-1}\Delta\boldsymbol{\theta} \tag{17.2}$$

(Lynch and Lande 1993; Jones et al. 2004). If $\Delta\boldsymbol{\theta}$ is small, so is $\boldsymbol{L}$, so that the phenotypic mean will appear to be virtually on top of the optimum as it assiduously tracks the peak.

Evolutionary Quantitative Genetics. Stevan J. Arnold, Oxford University Press.
DOI: 10.1093/oso/9780192859389.003.0018

Although the message from (17.2) about the magnitude of lag is straightforward, the direction of lag is less intuitive. For example, we can expect the direction of lag to be out of alignment with the direction of peak movement, when eigenvectors of $\boldsymbol{G}$ and the AL are diametrically opposed, a phenomenon called the *flying kite effect*.

A population of finite size that tracks a steadily moving peak will show variation in $\boldsymbol{G}$ and $\boldsymbol{P}$, as well as in its phenotypic trait mean, but we can nevertheless characterize the *expected value of lag*,

$$E_t\left[\boldsymbol{\theta}_t - \bar{\mathbf{z}}_t\right] = \bar{\boldsymbol{\Omega}}\,\bar{\boldsymbol{G}}^{-1}\Delta\boldsymbol{\theta}, \tag{17.3}$$

where bars denote averages and $\bar{\boldsymbol{\Omega}} = \boldsymbol{\omega} + \bar{\boldsymbol{P}}$ (Jones et al. 2004). This last expression will be useful in an upcoming discussion in which we anticipate the behaviour of the bivariate trait mean and the G-matrix in simulation studies with a moving peak.

17.2 Controversy Over the Net Selection Gradient

Lande (1979) argued that we might be able to estimate the net force of directional selection over some elapsed number of generations without knowing the details of peak movement. Consider our standard expression for response to selection (10.1) summed over some number of elapsed generations, T,

$$\sum_{t=0}^{T-1}\Delta\bar{\mathbf{z}}_t = \sum_{t=0}^{T-1}\boldsymbol{G}_t\boldsymbol{\beta}_t. \tag{17.4a}$$

Lande (1979) argued that $\boldsymbol{G}$ might be relatively stable so that we might take its average value and rearrange our expression in the following way,

$$\sum_{t=0}^{T-1}\Delta\bar{\mathbf{z}}_t = \bar{\boldsymbol{G}}\sum_{t=0}^{T-1}\boldsymbol{\beta}_t \tag{17.4b}$$

$$\boldsymbol{net\beta} \equiv \sum_{t=0}^{T-1}\boldsymbol{\beta}_t = \bar{\boldsymbol{G}}^{-1}\sum_{t=0}^{T-1}\Delta\bar{\mathbf{z}}_t, \tag{17.5}$$

where $\boldsymbol{net\beta}$ is Lande's *net selection gradient*, the sum of the forces of directional selection that have acted during the time interval, T. The significance of this last expression is that to estimate $\boldsymbol{net\beta}$ we apparently only need know the net change in the multivariate mean and the average value of $\boldsymbol{G}$. For example, consider the case of two populations or species, a and b, that have descended from a common ancestor. We estimate the net change in mean as

$$\sum_{t=0}^{T-1}\Delta\bar{\mathbf{z}}_t = \Delta\bar{\mathbf{z}}_T = \left|\bar{\mathbf{z}}_a - \bar{\mathbf{z}}_b\right|, \tag{17.6}$$

where T is the total number of generations separating the two taxa, which might be unknown. Then, to use expression (17.5), we would also need an estimate of the average value of $\boldsymbol{G}$.

A more complicated expression takes temporal variation in $\mathbf{G}$ and $\boldsymbol{\beta}$ into account,

$$\boldsymbol{net\beta} = \bar{\mathbf{G}}^{-1}\left[\Delta\bar{\mathbf{z}}_T - \sum_{t=0}^{T-1}\left(\mathbf{G}_t - \bar{\mathbf{G}}\right)\boldsymbol{\beta}_t\right], \tag{17.7}$$

where the summation term inside the square brackets is proportional to the covariance between $\mathbf{G}$ and the directional selection gradient in the same generation (Turelli 1988). Do we need this more complicated expression? Turelli argued that we do, because large fluctuations in $\boldsymbol{\beta}$ might produce a large summation term even though the fluctuations in $\mathbf{G}$ might be small. On the other hand, one could argue (R. Lande pers. comm.) that $\boldsymbol{\beta}$ in the preceding generation, not $\boldsymbol{\beta}$ in the same generation, should affect $\mathbf{G}$, so the summation term should be large only if the fluctuations in $\boldsymbol{\beta}$ were autocorrelated (Jones et al. 2004). The simulations reported in an upcoming section will help to settle this argument. The argument is important, because if Turelli is right, we have no means of estimating $\boldsymbol{net\beta}$. We shall refer to the troublesome summation term in (17.7) as the *Turelli effect*. We wish to know if this effect is small enough to be ignored. We get one kind of answer to this question in Section 17.4.

17.3 Simulation Studies of the *G*-matrix Evolving on Dynamic Adaptive Landscapes

In the following sections we review the results of four simulation studies in which the adaptive landscape changes as a function of time (Table 17.1). Most studies examine the effects of peak movement on the stability and evolution of the *G*-matrix, but one study looks at the effects of varying the orientation of the adaptive landscape. Conventions for choice of parameters and implementing the simulations are described in Chapter 16.

17.3.1 *Effect of steady, linear peak movement on the G-matrix*

Understanding the effects of a moving peak on the configuration and stability of the *G*-matrix is important for a couple of reasons. In the first place, movement of the peak adds a feature of directional selection that is either lacking or much reduced in magnitude when the peak is stationary.

Secondly, most profound instances of adaptation are thought to represent in situ temporal change in environment or a change in selection resulting from migration into new territory. Both situations can be modelled as deterministic or stochastic trends in the position of the peak of the AL. In other words, movement of the peak represents a mode of selection that poses a special and realistic challenge to the capacity of the population to adapt. We can also expect this mode of selection to have potentially serious effects on the *G*-matrix, especially on its stability. We will first review the effects of a steadily moving peak, as revealed in simulation studies with constant mutation processes, and then turn to cases in which the peak moves episodically or stochastically.

Jones et al. (2004) used computer simulations to study the effects of a steadily moving peak on the configuration and stability of the *G*-matrix. These simulations used the same conventions reviewed in the last chapter for the case of a stationary peak. The experimental condition were created by varying population size (N_e), mutational correlation (r_μ), the intensity of stabilizing selection (ω_{ii}), and the angle of the AL (r_ω). The new feature is movement of the peak ($\boldsymbol{\theta}$) in one of three directions in bivariate trait space (stationary,•; to the

Table 17.1 Parameters and procedures used in simulation studies of *G*-matrix stability and evolution with a dynamic adaptive landscape. These four studies used a total of 294 parameter combinations to simulate the evolution of nearly 22 million *G*-matrices. The scale of many parameters is relative to a standardized variance (E_{ii} = 1, first three studies or $P_{ii} \approx 1$, in the last study).

Parameter	Symbol	Jones et al. 2004	Revell 2007	Jones et al. 2012	Melo and Marroig 2015
Number of traits		2	2	2	10
Number of loci	n	50	50	50	500
Mutational variance	α^2	0.05	0.05	0.05	0.02
Mutational correlation	r_μ	0–0.9	−0.5–0.9	0–0.9	1.0
Mutation rate per locus	μ	0.0002	0.0002	0.0002	0.0005
Constant mutation process	yes	yes	no	yes	yes
Strength of stab. selection	ω	49	49	9	10
Selectional correlation	r_ω	−0.9–0.9	0–0.9	−0.9–0.9	0 or 0.8
Effective population size	N_e	342–2731	1000	1366	5,000
Type of dynamic AL		Steady peak movement	Random walk of AL orientation	Episodic & stochastic peak movement	Steady peak movement, 20 rates
Migration	m	no	no	no	no
Number of replicate runs per combination		20	20	20	50
Number of parameter combinations		235	23	16	20
Burn-in period (generations)		10,000	10,000	10,000	20,000
Generations per run		2,000	2,000	6,000	10,000
Number of G-matrices		9.4 mil	0.92 mil	1.92 mil	10 mil

right, ➔; upwards at 45°, ➚; downward at 45°, ➘) at a modest rate ($0.01\sqrt{E} \approx 0.008\sqrt{P}$ per generation), consistent with ecological observations (Kinnison and Hendry 2001). As before, the *G*-matrix that evolved in response to a particular set of these conditions was characterized in terms of its average size (Σ), eccentricity (ε), and angle (φ). Finally, the stability of the *G*-matrix was described in terms of the average per-generation change in those same three characteristics.

Snapshots of the simulated *G*-matrices at 300 generation intervals (Figure 17.1) illustrate several features of lag, including the flying kite effect. Notice that in the four upper panels, the bivariate mean closely tracks the moving (➔) peak, but the peak is either slightly above or below the direction of peak movement, depending on the orientation of the AL and pleiotropic mutation. When the correlation of pleiotropic mutation is especially strong (r_μ = 0.9), so that the mutation process is considerably out of alignment with the AL (r_ω = 0), we see the mean lagging both high above and behind the moving peak, a strong flying kite effect (second panel from the bottom in Figure 17.1). This is exactly the effect

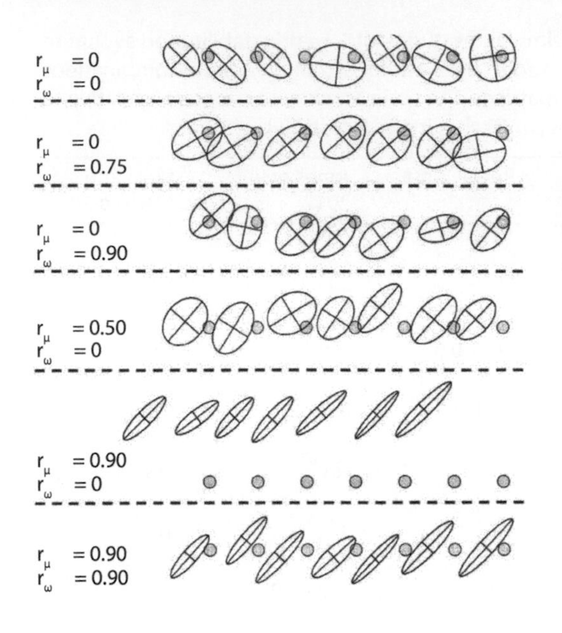

Figure 17.1 Evolution and stability of the *G*-matrix when the optimum is steadily moving. In these simulations, the optimum is steadily moving to the right ($\Delta\theta_1 = 0.007\sqrt{P}$), so that the optimum value for trait 1 is constantly changing, but the optimum value for trait 2 is constant. The *G*-matrix is represented by its 95% confidence ellipse, shown at 300 generation intervals. The position of the optimum in the same generation as the *G*-matrix snapshot is shown with the small, filled circle. Parameter values are: $N_e = 342, \omega_{11} = \omega_{22} = 49$.

From Jones et al. (2004).

we anticipated from (17.2). The effect is potentially serious because the means of both traits are pulled far away from their optima. The resulting maladaptation persists as long as the peak keeps moving.

Movement of the peak affects both the configuration of the *G*-matrix and its stability properties in ways that could not be anticipated from available analytical theory (Jones et al. 2004). Simulations show that the *G*-matrix tends to enlarge when the peak is moving, an *enhancement effect*. The genetic variances of both traits can increase by as much as 20%, but the direction of peak motion is not very important. This effect has been found in both univariate (see Chapter 14) and bivariate simulations. Persistent directional selection created by the moving peak apparently causes a subset of new mutations to sweep to fixation. These sweeps recur in a steady stream, enhancing the genetic variance of both traits. In contrast to these effects on *G*-matrix size, movement of the peak has little effect on the shape and orientation of the *G*-matrix (Jones et al. 2004).

Peak movement also affects the stability of the *G*-matrix. The *G*-matrix is especially stable in a situation of three-way alignment, when the principal axes of the *M*-matrix and the AL are aligned and the optimum is moving in the same direction (Figure 17.2). In other words, evolution along SLLRs is especially conducive to *G*-matrix stability. In contrast, peak movement perpendicular to the long axis of the *G*-matrix makes ***G*** less stable than it is when the optimum is stationary (Jones et al. 2004).

Aside from these special effects, the general features of *G*-matrix evolution and stability, revealed with simulations of a stationary optimum (Chapter 16), also hold in the case of a steadily moving peak (Jones et al. 2004). The mutational process (r_μ) and the AL (r_ω) affect the size and orientation stability of the *G*-matrix, but not the stability of genetic variances and *G*-matrix shape (ε). The stability conferring properties of mutation and selection are most pronounced when those processes are aligned. Large population size promotes stability of *G*-matrix shape, size, and orientation. As expected from theoretical considerations, selection and mutation tend to create a *G*-matrix in their own images, producing a compromise if they differ in shape and orientation.

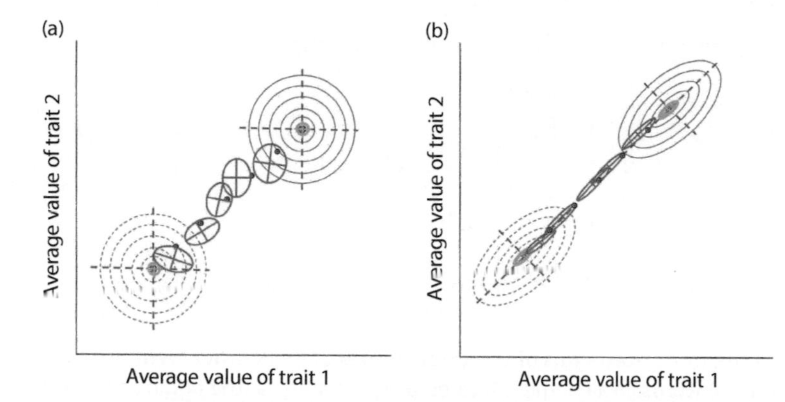

Figure 17.2 Contrasting conditions for *G*-matrix stability. The configuration of the AL is shown in red contours. The location of the adaptive peak (red-filled small circles) is moving up and to the right. Snapshots of the *G*-matrix, represented by 95% confidence ellipses (blue), are shown at 300 generation intervals. The configuration of the *M*-matrix is shown with green ellipses. (a) The absence of correlational selection and correlated mutation results causes instability in the orientation of the *G*-matrix. (b) Three-way alignment of mutation, the AL and peak movement produces a very stable *G*-matrix. Parameter values are: $N_e = 342, \omega_{11} = \omega_{22} = 49, N_e = 342, r_\mu = 0$ or $0.9, r_\omega = 0$ or 0.9.
Simulation data from Jones et al. (2004).

17.3.2 *Effect of episodic and stochastic peak movement on the G-matrix*

Modelling temporal change in the environment with a steadily moving peak, as in the preceding section, is obviously a tremendous simplification. Although we might capture deterministic trends with such a model, we have completely ignored stochastic variation in the environment. In this section we will consider a simulation study with stochastic variation (Jones et al. 2012) that also revealed an important *invariance principle*, the revelation that we do not need to simulate a large sample of stochastic paths of peak movement to understand how the evolution and stability of $\boldsymbol{G}$ are affected. Our task is simplified because large subsets of possible paths will have the same consequences for $\boldsymbol{G}$ and its stability. To see how this simplification arises we need to consider the details of the Jones et al (2012) simulations. But first, we need to review the motivation for such simulations.

We can imagine that stochastic peak movement might be especially likely to destabilize the *G*-matrix. In particular, such stochastic movement might induce such chaotic behaviour in $\boldsymbol{G}$ that our estimates of $\boldsymbol{G}$ would be useless in evolutionary inference. Existing theory does not settle this claim, but we can use simulations to gain some traction on the problem. In particular, Jones et al. (2012) investigated the consequences of three modes of peak movement: steady, episodic, and stochastic (Figure 17.3). Although just three modes appears to be a sparse sample of possibilities, this study revealed a useful principle of invariance and hence simplification.

Examples of *G*-matrix dynamics in response to two of these modes of peak movement are shown in Figures 17.4 and 17.5. In these two figures, all parameters are the same except episodic movement of the peak in Figure 17.4 replaces steady movement of the peak in Figure 17.1. The effect of that episodic movement is apparent in the saw-tooth

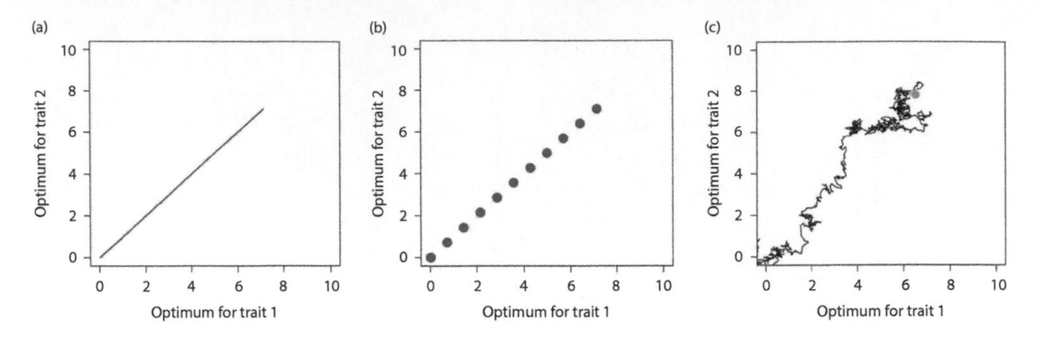

Figure 17.3 Schematic portrayals of the three modes of peak movement used in the Jones et al. (2012) simulations. In all of these figures, the peak starts at the origin and moves up and to the right. (a) The peak moves every generation for 1000 generations in a constant direction at a constant rate ($\Delta\theta_1 = \Delta\theta_2 = 0.0071$). (b) The peak jumps to a new position every 100 generations ($\Delta\theta_1 = \Delta\theta_2 = 0.071$). (c) The peak moves every generation ($\Delta\theta_1 = \Delta\theta_2 = 0.0071$), but peak movement also includes an element of Brownian motion stochasticity each generation (see Section 14.2 for the model). In the illustrated simulation, the stochasticity variance parameter (σ^2_θ) was 0.01. The orange dot shows the final position of the optimum at generation 1000.
After Jones et al. (2012).

pattern produced in overall *G*-matrix size (Σ, upper panel) and directional selection (β_1, bottom panel; Figure 17.4). Remarkably, however, the time courses of *G*-matrix shape and orientation (middle two panels) appear unaffected by the change in mode of peak movement (the *invariance principle* that we mentioned earlier). These results are typical of many additional simulations. The total amount and direction of movement matters for *G*-matrix stability, but not the intervening pattern of peak movement (Jones et al. 2012). In other words, the stability properties of the *G*-matrix are not affected by the details of how the peak moves.

The invariance principle, revealed when the mode of peak movement is varied, is important because of its implications for evolutionary inference. The principle indicates that the direction and total extent of peak movement is important for *G*-matrix stability but not the details of movement. Without this principle, one might consider detailed analysis of responses to various realistic peak trajectories. Instead, one can use readily available information of direction and extent of peak movement to assess how *G*-matrix stability might have been affected.

As expected from previous studies (Jones et al. 2004), the orientation of the *G*-matrix evolves in response to the orientation of the AL and the mutation matrix, as well as the direction of peak movement (Jones et al. 2012). This same pattern of evolutionary response was seen with all three modes of peak movement. Once again, simulations reveal unexpected robustness in the evolution of the *G*-matrix.

17.3.3 *Prospects for retrospective selection analysis using* ***net*β**

Simulations of a moving optimum also inform the controversy over whether it is reasonable to estimate the net selection gradient using Lande's (1979) expression (17.5). The same runs used to assess the stability of the *G*-matrix can also be used to determine the magnitude of the Turelli effect and assess the accuracy of Lande's expression. In the Jones et al. (2004) study, the bivariate mean drifts about, as it tracks the steadily moving peak,

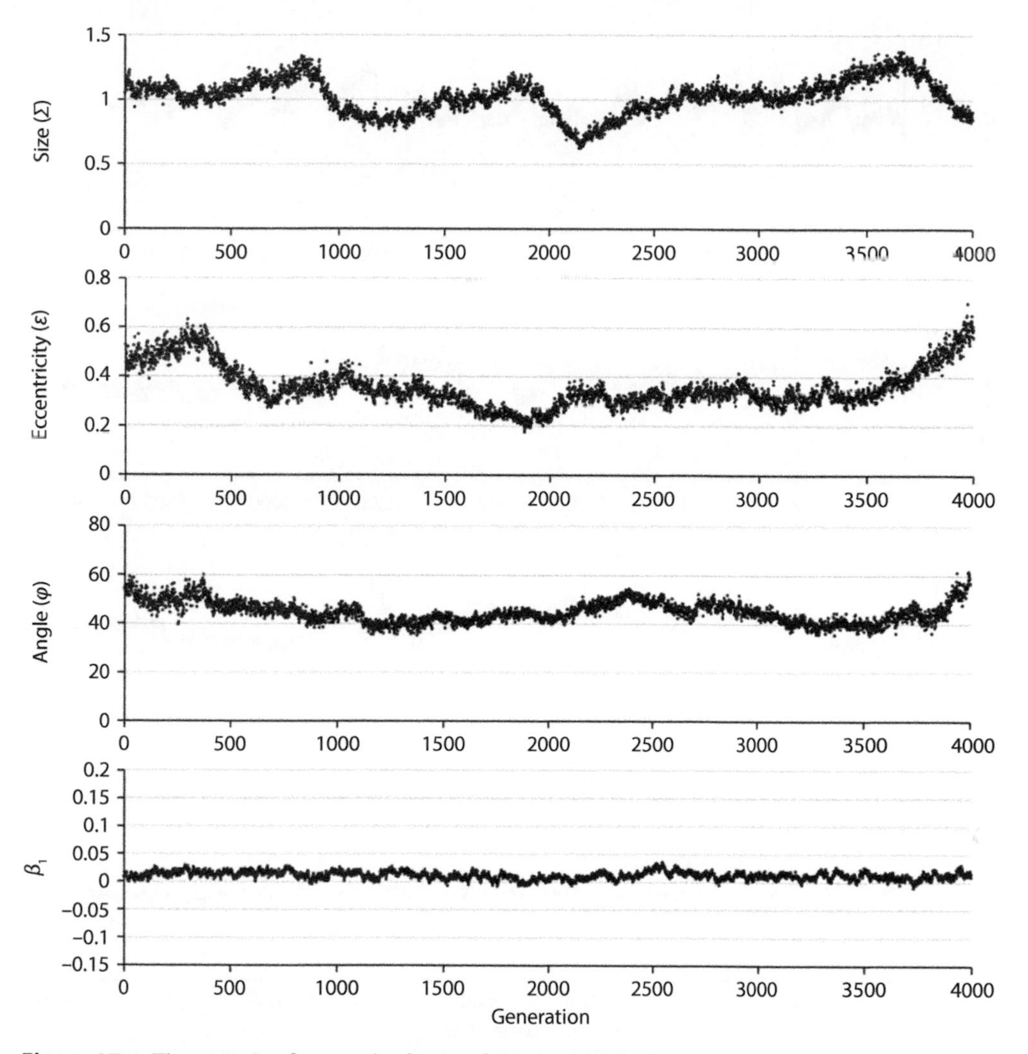

Figure 17.4 Times series from a single simulation run showing responses of some aspects of the G-matrix to a steadily moving optimum, in which the optimum is moving in a positive direction for both traits (↗). In this example, $\omega_{11} = \omega_{22} = 9$, $r_\omega = 0$, $r_\mu = 0.5$, $\Delta\theta_1 = \Delta\theta_2 = 0.007\sqrt{P}$, $N_e = 1024$.

From Jones et al. (2012).

because population size is finite. This drift adds an element of stochasticity to the directional selection experienced by the population. At the end of a run, the authors calculated the magnitude of the Turelli effect (Section 11.6) using known values of $\boldsymbol{\beta}_t$, $\mathbf{G}_t$, and $\bar{\mathbf{G}}$. Calculations of this kind show that the Turelli effect is usually small and negligible, constituting about 6% of ***netβ*** when populations are small (N_e = 342) and 1–2% of ***netβ*** when populations are an order of magnitude larger. Jones et al. (2004) also showed that the estimate of $\bar{\mathbf{G}}$ has a potentially large role in introducing error into an estimate of ***netβ***. In general, circumstances that allow good estimation of $\mathbf{G}$ are also the most favourable for estimating ***netβ***.

The effects of episodic and stochastic peak movement on the estimation of ***netβ*** were assessed by Jones et al. (2012). In particular, they studied the case of stochastic peak

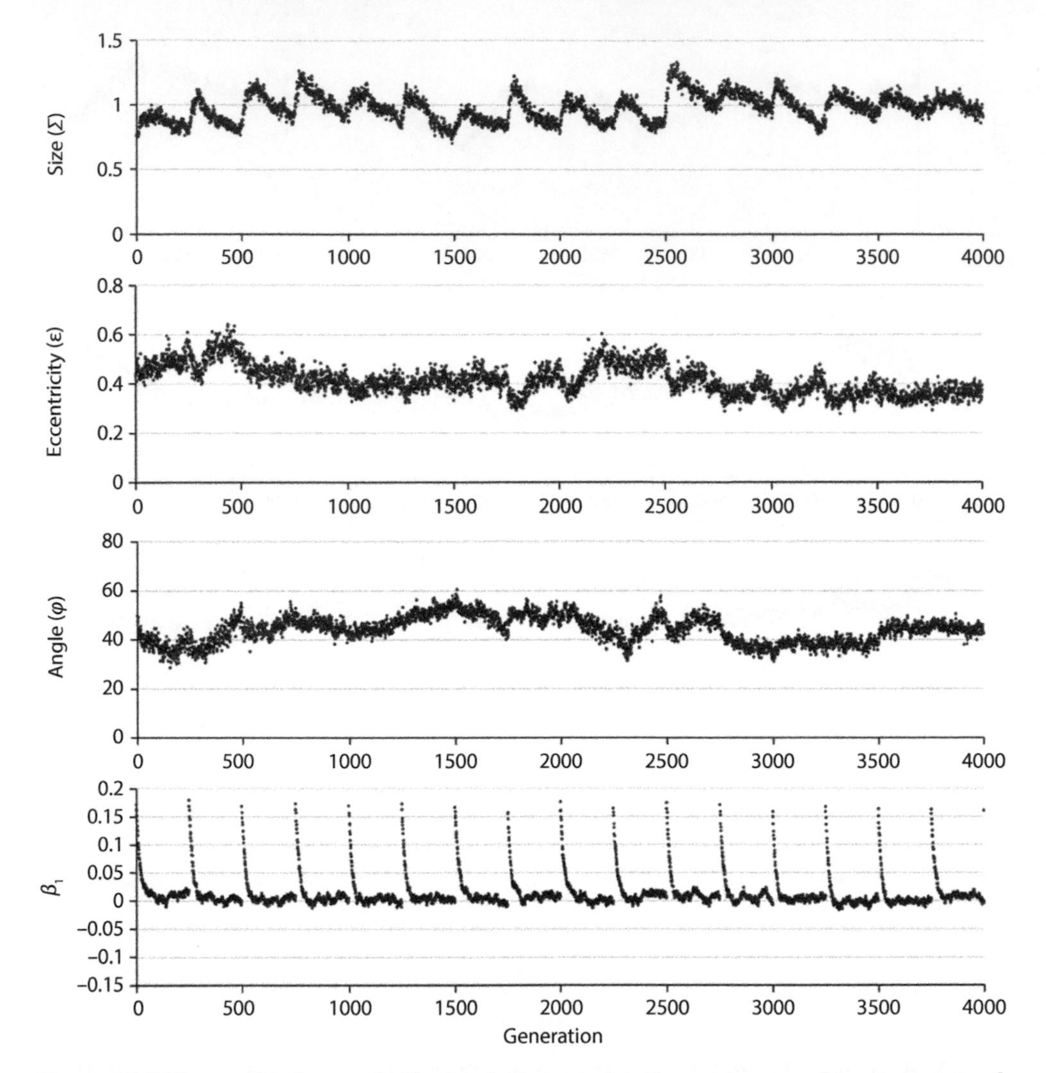

Figure 17.5 Time series from a single simulation run showing responses of some aspects of the *G*-matrix to an optimum that moves episodically every 250 generations in the same direction and at the same long-term average rate as in Figure 17.4. All other parameter values are identical to those used for Figure 17.4.

From Jones et al. (2012).

movement, the mode that is probably the most likely to produce inaccuracies in estimation. On the one hand, results were encouraging because the Turelli effect was small and the angle of ***netβ*** was estimated with accuracy and without bias. A message of caution was delivered as well. Stochastic peak movement caused the magnitude of ***netβ*** to be systematically underestimated, apparently because directional selection skews the distribution of genetic values in a way that retards the response to selection. Although researchers are likely to be more interested in the angle of ***netβ*** than in its absolute magnitude, because the magnitude of the elapsed time interval is usually unknown, these simulation results remind us that departures from the assumption of normal trait distribution can occur and have consequences.

17.3.4 *Effect of Brownian motion in the angle of the AL on the G-matrix*

In the preceding sections, we considered the effects of an AL that changes its location ($\boldsymbol{\theta}$) without varying its configuration. Here we consider the opposite situation in which the AL fluctuates in configuration without changing its location. This kind of variation in selection regime might correspond to some types of environmental fluctuations that lack an overall temporal trend. Revell (2007) used computer simulations to study the effects of Brownian motion in the angle of a bivariate Gaussian AL with a stationary optimum. In particular, Revell held ω_{11} and ω_{22} constant at 49, set $r_\omega = 0$ in the initial generation and then imposed Brownian motion in r_ω each generation thereafter, for 2000 generations after burn-in, by taking draws from a normal distribution with a mean of 0 and a variance, $\sigma^2\,(r_\omega)$, of 0.0001, 0.001, 0.01, or 0.1, depending on the experimental run.

Rapid change in the angle of the AL, i.e., large $\sigma^2\,(r_\omega)$ tended to produce G-matrices that were smaller and less eccentric. The same rapid change caused the orientation of $\boldsymbol{G}$ to be less stable, but size and shape stability were relatively unaffected. One problem in evaluating these results is that we do not know what constitutes ecologically realistic values for $\sigma^2\,(r_\omega)$. Revell (2007) used a broad range of values, so it's likely that the range brackets reality, but we do not know where to focus within that range.

17.4 The Evolution of Modularity in *G* and *P*

In Chapter 7 we discussed the meaning and significance of trait modularity. We also developed the idea that modularity might reflect the structure of the γ-matrix or its Gaussian analogue, the ω-matrix. In a similar vein, Melo and Marroig (2015) used a simulation study to evaluate the roles of peak movement and correlational selection in creating modular patterns in continuous traits. Using an individual-based simulation approach similar to that of Jones et al. (2004, 2012), they used a Gaussian selection surface (ω-matrix) to simulate multivariate stabilizing selection and moved the selection peak at a constant rate in different directions. Parameter values were generally similar to Jones et al. (2012), but with a couple of important differences. To model the creation and maintenance of modular structure, they used a 10×10 ω-matrix of the following form

$$\omega = \begin{bmatrix} \boldsymbol{Wi} & \boldsymbol{Bt} \\ \boldsymbol{Bt} & \boldsymbol{Wi} \end{bmatrix},$$

$$\boldsymbol{Wi} = \begin{bmatrix} 10 & 8 & 8 & 8 & 8 \\ 8 & 10 & 8 & 8 & 8 \\ 8 & 8 & 10 & 8 & 8 \\ 8 & 8 & 8 & 10 & 8 \\ 8 & 8 & 8 & 8 & 10 \end{bmatrix},$$

$$\boldsymbol{Bt} = \begin{bmatrix} 0 & 0 & 0 & 0 & 0 \\ 0 & 0 & 0 & 0 & 0 \\ 0 & 0 & 0 & 0 & 0 \\ 0 & 0 & 0 & 0 & 0 \\ 0 & 0 & 0 & 0 & 0 \end{bmatrix},$$

where $\boldsymbol{Wi}$ and $\boldsymbol{Bt}$ are, respectively, the within- and between-module matrices. The first five traits constitute a module, as do the second set of five traits. The patterns of positive correlational selection (with ω_{ii} = 10 and ω_{ij} = 8) should promote positive correlations within the two modules but not between modules.

The characterization of mutation was another major difference with other simulation studies. Melo and Marroig (2015) followed Wagner (1989) in using a *B*-matrix to characterize the relationship between additive genetic values and phenotypic value. They also constrained the *B*-matrix so that when a locus affected more than one trait, it affected them in the same direction and by the same amount.

In their simulations Melo and Marroig (2015) examined the effects of population size, correlational selection, and peak movement on the evolutionary creation and maintenance of modularity, assessed by tracking covariances and correlations in the output *P*-matrix. They used the ratio of average within- to between-module correlations (*average ratio*, AVG) as a measure of modularity. In a typical simulation run, a burn-in period with no selection (10,000 generations) was followed by an additional 10,000 burn-in generations with uncorrelated stabilizing selection (diagonal ω-matrix) to achieve mutation-drift-stabilizing selection equilibrium. The effects of peak movement were then tracked over the next 20,000 experimental generations. Although the modularity results in this study are expressed in terms of the AVG ratio of the *P*-matrix, they apply equally to the *G*-matrix, because the two matrices are proportional.

Directional selection via peak movement was imposed in different directions in the two-trait modules. The peak moved towards higher values for traits 1 – 5 and towards lower values for traits 6 – 10. The rate of peak movement ranged from $0.0001\sqrt{P}$ to $0.004\sqrt{P}$, which at the upper value is comparable to the value of $0.007\sqrt{P}$ used by Jones et al. (2004, 2012).

The effect of the rate of peak movement on degree of modularity is shown in Figure 17.6. These simulations show the degree of modularity that was established after 10,000 generations of peak movement at different rates. High levels of modularity were achieved by relatively modest rates of peak movement.

Although peak movement is relatively efficient at establishing modularity, a modest degree of modularity can also be established by correlational selection (Figure 17.7). After about 2000 generations of correlational selection, a stable but low level of modularity is established. In contrast, in the time course for populations subjected to peak movement of 0.004 (not shown), modularity reached high levels after 10,000 generations (Figure 17.6) and was still climbing when the simulations were terminated.

Once modularity is established by peak movement, it can be maintained by correlational selection after the peak stops moving. Figure 17.8 shows the time course of modularity under three different selection regimes after peak movement ceases. Correlational selection maintains the same level of modularity that was established by peak movement. With stabilizing selection alone (no correlational selection), modularity decays rapidly at first and then slowly for the next 9000 generations. With no selection and N_e = 5000 (as in the other two groups), drift leads to rapid decay in modularity.

Melo and Marroig (2015) also showed that drift dampens the ability of peak movement to establish modularity. In other words, modularity evolves faster in populations of large effective size.

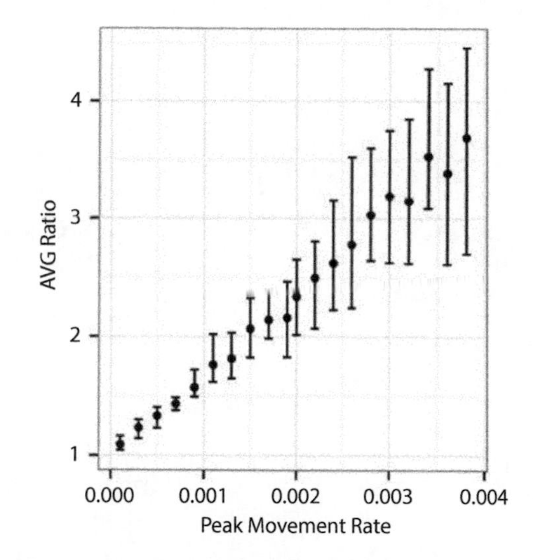

Figure 17.6 The effect of rate of peak movement on trait modularity. Faster rates of peak movement establish more pronounced modularity. Each bar shows the 95% confidence limits for the mean of 10 simulations after 10,000 generations. The mean is indicated by a black dot. Faster rates of movement result in smaller between-module correlation and hence larger sampling variance in the AVG ratio. The peak moved at rates varying from 0.0001 to 0.004, which are expressed in units of within-lineage environmental variation. Parameter values in the simulations were: $N_e = 5000$, per-locus mutation rate = 5×10^{-4}, per-locus mutational variance = 0.02, ω_{ii} = 10 and ω_{ij} = 8 within modules, $\omega_{ii} = \omega_{ij} = 0$ between modules, 500 loci, 10 traits).

From Melo and Marroig (2015) with permission.

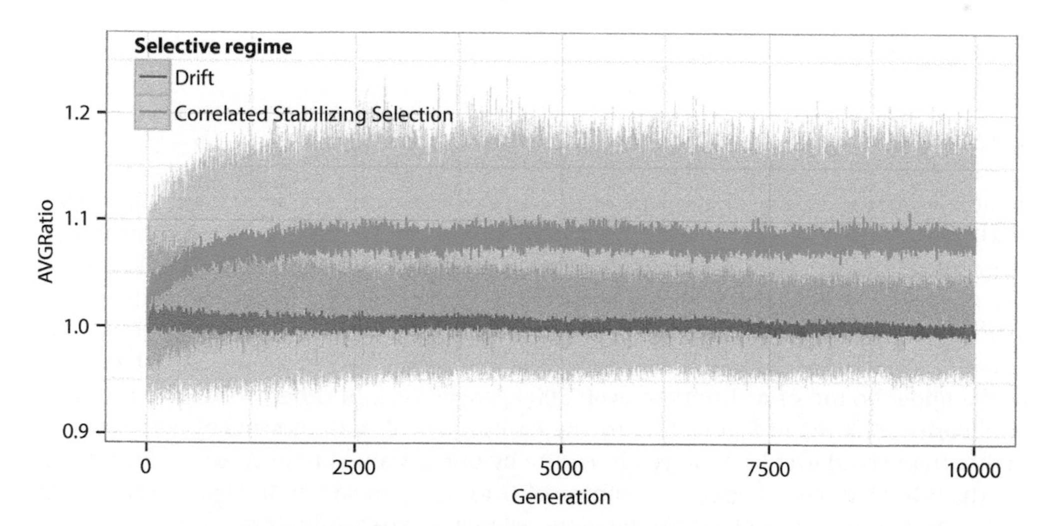

Figure 17.7 The establishment of modularity by correlational selection. At generation zero (after a 20,000 generation burn-in period), correlational selection was established in the lineages shown in green, but not in the lineages shown in red. Gray bars show the 95% confidence limit for the mean of 50 replicates. Means are shown in colours. Parameter values as in Figure 17.6.

From Melo and Marroig (2015) with permission.

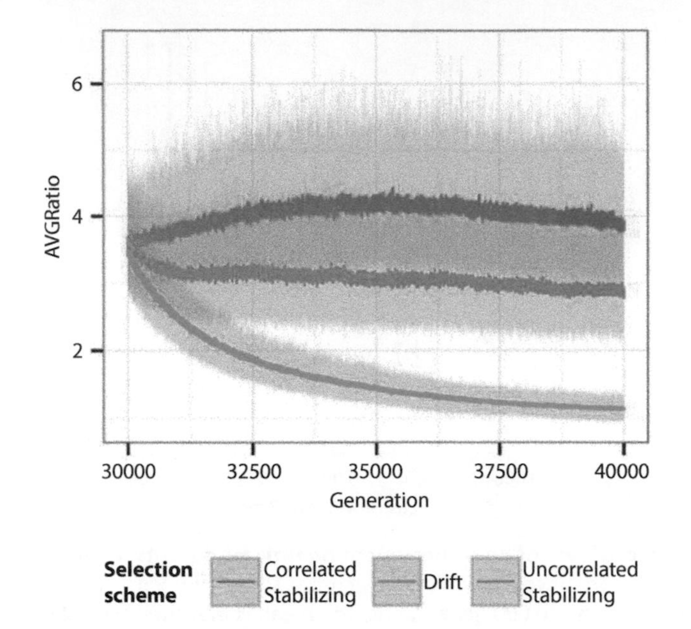

Figure 17.8 Maintenance of modularity by correlational selection and loss of modularity under regimes of drift and uncorrelated selection. During the 20,000 generation burn-in period, modularity was established in all lineages by directional selection. Beginning at generation 30,000 directional selection ceased. In addition all stabilizing and correlational selection ceased in the 100 drift lineages (shown in green), only stabilizing selection was maintained in the 100 uncorrelated stabilizing selection lineages (blue), whereas correlated and stabilizing selection was maintained in the 1000 correlated selection lineages (red). All other parameter values as in Figure 17.6 and graphic conventions as in Figure 17.7.

From Melo and Marroig (2015) with permission.

17.5 Discussion

Although the values of peak movement used in the simulations reviewed here ($0.004\sqrt{P}$-$0.007\sqrt{P}$) are modest and close to the geometric mean rate suggested by Kinnison and Hendry's (2001) survey of microevolutionary studies ($0.006\sqrt{P}$), these rates could not be sustained for many generations in nature. At a rate of $0.004\sqrt{P}$, the trait mean would move $4\sqrt{P}$ in 1000 generations, $40\sqrt{P}$ in 10,000 generations, and $400\sqrt{P}$ in 100,000 generations. These considerations suggest that a simulation run of 1000 generations might be closer to the upper bound of reality than even 5000 generations. In other words, the levels of modularity achieved in Figure 17.6 by the fastest rates of peak movement are probably higher than could ever be achieved in nature by one sustained bout of peak movement. On the other hand the degrees of response of $\boldsymbol{G}$ to selection shown in Figures 17.1, 17.4, and 17.5 are probably well within the range we might expect in nature.

17.6 Conclusions

- Movement of the adaptive peak in certain directions can produce sustained maladaptation (flying kite effect). In particular, when the mutation matrix is considerably out of alignment with the AL (e.g., $r_\mu = 0.9$ but $r_\omega = 0$), the trait mean can lag appreciably and perpetually behind the moving peak.

- The G-matrix tends to enlarge in the direction of peak movement (enhancement effect).
- Alignment of the mutation matrix with the AL produces a cigar-shaped G-matrix that is inherently stable.
- Movement of the adaptive peak movement along the principal axis of a cigar-shaped G-matrix amplifies the stability of the G-matrix.
- The overall magnitude and direction of peak movement affects the stability of the G-matrix, but the details and rate of peak movement are relatively unimportant (invariance principle).
- Retrospective analysis of selection (Lande's ***net*$\boldsymbol{\beta}$**) can be successful if the average G-matrix is known with some accuracy. Simulations indicate that the covariance between $\boldsymbol{G}$ and $\boldsymbol{\beta}$ (the Turelli effect) is small.
- Rapid change in the angle of the AL tends to produce a smaller, rounder G-matrix.
- Peak movement is more effective at establishing modularity of the G-matrix than is correlational selection.
- Once modularity has been established, it can be maintained by correlational selection.
- Large effective population size enhances the evolution of modularity.

Literature Cited

Jones, A. G., S. J. Arnold, and R. Bürger. 2004. Evolution and stability of the G-matrix on a landscape with a moving optimum. Evolution 58:1639–1654.

Jones, A. G., R. Bürger, S. J. Arnold, P. A. Hohenlohe, and J. C. Uyeda. 2012. The effects of stochastic and episodic movement of the optimum on the evolution of the G-matrix and the response of the trait mean to selection. J. Evol. Biol. 25:2210–2231.

Kinnison, M. T., and A. P. Hendry. 2001. The pace of modern life II: From rates of contemporary microevolution to pattern and process. Pp. 145–164 *in* A. P. Hendry and M. T. Kinnison, eds. Microevolution: Rate, Pattern, Process. Springer, Dordrecht.

Lande, R. 1979. Quantitative genetic analysis of multivariate evolution, applied to brain: body size allometry. Evolution 33:402–416.

Lynch, M., and R. Lande. 1993. Evolution and extinction in response to environmental change. Pp. 234–250 *in* P. Kareiva, J. Kingsolver, and R. Huey eds. Biotic Interactions and Global Change. Sinauer Associates, Inc., Sunderland.

Melo, D., and G. Marroig. 2015. Directional selection can drive the evolution of modularity in complex traits. Proc. Natl. Acad. Sci. USA 112:470–475.

Revell, L. J. 2007. The G matrix under fluctuating correlational mutation and selection. Evolution 61:1857–1872.

Turelli, M. 1988. Phenotypic evolution, constant covariances, and the maintenance of additive variance. Evolution 42:1342–1347.

Wagner, G. P. 1989. Multivariate mutation-selection balance with constrained pleiotropic effects. Genetics 122:223–234.

CHAPTER 18

Evolution Along Selective Lines of Least Resistance

Overview—In Chapter 4 we introduced the idea of a selective line of least resistance (SLLR), the principal axis of a selection surface. Trait variation in this direction causes the smallest loss in fitness and hence the least resistance to selection. In Chapters 12–14 we showed that successful models of adaptive radiation allow the adaptive peak to move according to a stochastic process (e.g., Brownian motion). In this chapter we connect these two ideas by formulating the supposition of *SLLR peak movement*, an argument that the adaptive peak tends to move in the direction of the SLLR of the adaptive landscape. After examining the basis for this supposition, we explore the ability of SLLR peak movement to explain allometric evolution and the evolutionary maintenance of bilateral symmetry.

Although stochastic movement of the peak is undoubtedly a component in most adaptive radiations, we can also expect systematic trends in peak movement both from first principles and from empirical experience. From first principles, it seems likely that the same biomechanical laws that create principal axes in selection surfaces will encourage the peak to move in that same multivariate direction, i.e., SLLR movement. We will refer to this argument as the *extrapolation principle*. The biomechanical laws that concern us are not only persistent through time in an individual lineage, they are unavoidable over a wide range of ecological conditions and consequently common to all the lineages in an adaptive radiation. Furthermore, evolution may commonly occur along the main axes of triple alignment (i.e., alignment of the matrices that describe mutation, inheritance, and selection) (Chapter 17).

In the spirit of this extrapolation principle, one simple but powerful expectation is that bilaterally symmetrical traits will retain symmetry over evolutionary time (Section 18.2). This expectation is equivalent to saying that the adaptive landscape for right and left sides is a ridge-shaped surface with a peak that tends to move along the long axis of the ridge (i.e., at a 45° angle). Notice that deviation from this symmetry ridge, means that the right and left sides of the organism will differ in size or shape, a configuration that will generally be opposed by selection (Palmer and Strobeck 1986, 1992; Palmer 1996). Similar arguments for SLLR movement apply in models that produce scaling laws and allometric rules (Section 18.3). Direct empirical tests are scarce but do provide some evidence for evolution along selective lines of least resistance.

Evolutionary Quantitative Genetics. Stevan J. Arnold, Oxford University Press. © Stevan Arnold (2023).
DOI: 10.1093/oso/9780192859389.003.0019

18.1 Peak Controller Processes With Linear Trends

Evolution in multivariate space commonly proceeds along recognizable, recurrent trajectories. We shall argue that this phenomenon can arise by SLLR movement of adaptive peaks as peaks move along restricted sets of tracts in phenotypic space. Those restricted tracts correspond to the axes of the AL that are most permissive with respect to selection. Our argument takes the following form. Correlational selection is a common mode of selection in which certain combinations of traits enhance fitness (Chapter 4). Correlational selection in turn is often a reflection of energetic or biomechanical principles that are broadly applicable, not only to the population in question but to all organisms with similar combinations of morphology and ecology (Chapter 7). In other words, the multiple trait relationship to fitness that is captured by correlational selection can be extrapolated beyond the bounds of variation realized in a particular population. The simplest way to extrapolate a particular selection regime is to change its position and in particular the position of its optimum, without changing its overall configuration. SLLR movement performs this operation of extrapolation and remains faithful to the principles that underlie multivariate stabilizing selection general, and to the principles underlying correlational selection in particular. SLLR movement is an especially likely direction for peak movement because it is the direction least likely to disrupt the operation of functional complexes of traits (Chapter 7).

We can construct a more detailed, quantitative version of this argument using concepts developed in earlier chapters. When the adaptive landscape is Gaussian, the SLLR is the first principal component (leading eigenvector) of the Ω-matrix, which is closely approximated by the ω-matrix, if stabilizing selection is weak (Chapter 4). Consequently, we will use the leading eigenvector of ω-matrix, $\boldsymbol{\omega_{max}}$, as our best estimate of the SLLR. Our proposition is then that the broad applicability of the principle that produced $\boldsymbol{\omega_{max}}$ in a particular population will tend to produce the same configuration of selection and hence similar individual selection surfaces in related populations with similar ecology. Consequently, $\boldsymbol{\omega_{max}}$ will be conserved across those related populations and through time. As a consequence of the principle of extrapolation, adaptive peaks will tend to move along the SLLR, $\boldsymbol{\omega_{max}}$, and to a lesser extent along other axes of the AL as an adaptive radiation unfolds. One particularly strong test of this prediction would be to test for alignment between the main axis of the cloud of multivariate population means, $\boldsymbol{d_{max}}$, and $\boldsymbol{\omega_{max}}$.

An illustration of our argument is shown in Figure 18.1, which is based on the analysis of evolution of vertebral numbers in natricine snakes (Hohenlohe and Arnold 2008) that we reviewed in Chapter 9. In that test case, $\boldsymbol{d_{max}}$ was aligned with $\boldsymbol{\omega_{max}}$ based on field growth rate, which we might expect if the same biomechanical principles that translate vertebral numbers into growth rate also govern the movement of adaptive peaks. Notice that conservation of individual adaptive landscapes, and in particular conservation of their principal axes, is a key element in the argument that $\boldsymbol{d_{max}}$ coincides with the major axis of SLLR movement.

Figure 18.1 does not represent a multi-peaked adaptive landscape, but rather the superimposed landscapes of a few dozen individual species. Multi-peaked landscapes are sometimes used to portray a global vision of the trait space occupied by many species (Dawkins 1996; Schluter 2000). A disadvantage of such multi-peaked landscapes is that their complexity rules out straightforward analytical treatment. In contrast, Figure 18.1 portrays the location of multiple local landscapes that are experienced simultaneously but privately by individual species. The stochastic process of moving peaks that produced

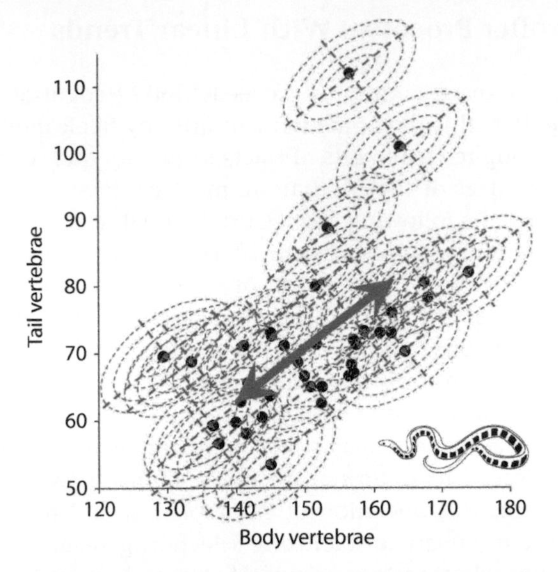

Figure 18.1 An interpretation of bivariate evolution of vertebral numbers in natricine snakes. The data for this example are plotted in Figure 9.3. Here we superimpose identical hypothetical adaptive landscapes for each population (red dotted contour plots), assuming that the landscapes are homogeneous in shape and size and that population means are located close to their optima. The long axes of the selection surfaces represent a common $\boldsymbol{\omega}_{max}$. The principal axis of dispersion ($\boldsymbol{d}_{max}$) is shown as the long double-headed black arrow, which coincides with the major axis of SSLR movement of the adaptive peaks. The minor axis of dispersion ($\boldsymbol{d}_{min}$) is shown as the short double-headed black arrow, which coincides with the minor axis of SSLR movement of the adaptive peaks.

the dispersion of individual landscapes portrayed in Figure 18.1 can be mathematically described and explored. The distinction between these two visions of adaptive landscapes is discussed by Arnold et al. (2001).

Alignment tests comparing $\boldsymbol{\omega}_{max}$ or $\boldsymbol{\Omega}_{max}$ with $\boldsymbol{d}_{max}$ are scarce. Nevertheless, a wide variety of observations bear on proposition of evolution along SLLR, as we shall see.

18.2 A Model for Evolution in Response to SLLR Peak Movement

In this section, we construct a model of multivariate trait means chasing SLLR movement of their adaptive peaks by building on the work of Hansen et al. (2008). In Section 14.3, we introduced their model for univariate trait response to Brownian motion of an adaptive peak. We now consider the multivariate version of that model and make some key assumptions about the matrix that specifies multivariate Brownian motion of the adaptive peak.

The response to selection by the vector of trait means at generation $t+1$ is

$$\Delta\bar{\mathbf{z}}_{t+1} = \boldsymbol{G}\Omega^{-1}\left(\boldsymbol{\theta_t} - \bar{\mathbf{z}}_t\right), \tag{18.1}$$

in which we recognize our standard form for directional selection that is a function of distance from a Gaussian adaptive peak. The column vector $\boldsymbol{\theta}_t$ specifies the peak for each trait at generation t, but these peaks are constantly changing position. The peak of the

Gaussian adaptive landscape at any given generation is the vector of positions in the previous generation plus a draw from a multivariate normal distribution with a mean vector of zeros and a dispersion matrix given by $\boldsymbol{Br}$, the Brownian motion orientation matrix,

$$\boldsymbol{\theta}_t = \boldsymbol{\theta}_{t-1} + N(\mathbf{0}, \boldsymbol{Br}). \tag{18.2a}$$

For example, in the two-trait case at generation t, we have

$$\boldsymbol{\theta}_t = \begin{pmatrix} \theta_{t1} \\ \theta_{t2} \end{pmatrix}, \tag{18.2b}$$

$$\boldsymbol{Br} = \begin{pmatrix} \sigma_{\theta 1}^2 & \sigma_{\theta 12} \\ \sigma_{\theta 12} & \sigma_{\theta 2}^2 \end{pmatrix}, \tag{18.2c}$$

where $\sigma_{\theta 1}^2$ and $\sigma_{\theta 2}^2$ are, respectively, the variances in the positions of peaks of traits 1 and 2, and $\sigma_{\theta 12}$ is the covariance in the positions of the two peaks. In other words, the response to selection equation for the trait means (18.1) is coupled to Brownian motion of the optimum, which is described by (18.2).

At any given generation, $t + 1$, the lineage traits means are multivariate normally distributed with a variance-covariance matrix given by

$$\boldsymbol{D}(\bar{\mathbf{z}}_{t+1}) = \left(\boldsymbol{Br} + N_e^{-1}\boldsymbol{G}\right)\boldsymbol{a}^{-1}\left[1 - \exp\{-2t\boldsymbol{a}\} + t\boldsymbol{Br}\left[1 - 2(1 - \exp\{-t\boldsymbol{a}\})(t\boldsymbol{a})^{-1}\right]\right], \tag{18.3}$$

where

$$\boldsymbol{a} = \boldsymbol{G}\Omega^{-1}$$

(Hansen et al. 2008). We refer to the leading eigenvector of the dispersion matrix $\boldsymbol{D}(\bar{\mathbf{z}}_{t+1})$ as $\boldsymbol{d_{max}}$. The among-lineage dispersion matrix for the positions of the optima after t elapsed generations, $\boldsymbol{D}(\boldsymbol{\theta}_t)$, is proportional to the Brownian motion matrix, $\boldsymbol{Br}$,

$$\boldsymbol{D}(\boldsymbol{\theta}_t) = t\boldsymbol{Br}. \tag{18.4}$$

Consequently, the leading eigenvectors of $\boldsymbol{D}(\boldsymbol{\theta}_t)$ and $\boldsymbol{Br}$ are the same; we will refer to that eigenvector as $\boldsymbol{Br_{max}}$. The among-lineage dispersion matrix for the vector of traits means (18.3) rapidly converges on this matrix (18.4). As $t \to \infty$,

$$\boldsymbol{D}(\bar{\mathbf{z}}_t) = t\boldsymbol{Br}, \tag{18.5}$$

just we might expect from the univariate case described in Section 14.2.

This model for multiple traits evolving in response to Brownian motion of their optima becomes a model for SLLR peak movement, if we assume that the Brownian motion matrix, $\boldsymbol{Br}$, has the same eigenvectors as the curvature matrix for a Gaussian adaptive landscape, $\boldsymbol{\Omega}$. For the purposes of illustration, we will satisfy this condition by making a stronger assumption, viz., that the two matrices are proportional,

$$\boldsymbol{Br} = c\,\Omega, \tag{18.5}$$

where c is a constant scalar. In other words, we have assumed that the eigenvectors are the same and the eigenvalues of the two matrices are proportional. Consequently,

$\boldsymbol{Br}_{max} = \boldsymbol{\Omega}_{max}$, the leading eigenvector of $\boldsymbol{Br}$ is the selective line of least resistance for the adaptive landscape.

In the bivariate case of SLLR peak movement, it will be useful to refer to correlation in the nonlinear selection imposed by the Gaussian adaptive landscape,

$$r_{\Omega} = \Omega_{ij} / \sqrt{\Omega_{ii}\Omega_{jj}}$$

and *correlation in the Brownian motion of the peak,*

$$r_{Br} = \sigma_{\theta ij} / \sqrt{\sigma^2_{\theta i}\sigma^2_{\theta j}}.$$

Simulations of this model of evolution in response to SLLR peak movement are shown in Figure 18.2. These illustrated cases incorporate the kind of strong alignment that we expect between the *G*-matrix and the adaptive landscape under regimes of correlational selection (Chapter 16 and 17). In each of the three cases in Figure 18.2, we see that the evolved patterns in dispersions of trait means and optima closely match the bivariate pattern of peak movement, as expected from expressions 18.4 and 18.5. Indeed, the trait means have closely tracked the movements of their optima and nearly match the positions of their optima after 1000 generations. It is this phenomenon of close tracking that makes the *Br*-matrix relevant to the final outcome and the *G*-matrix generally irrelevant.

We can, however, concoct cases in which the *G*-matrix is relevant and manifestly does affect the shape of the radiation in the short term. For example, if we make the genetic covariance negative in the simulation illustrated in Figure 18.2*c* and *d*, we create a situation in which the *G*- and *Br*-matrices are in maximum misalignment (i.e., anti-alignment). We explored an analogous scenario in Chapter 17 in which the *G*-matrix was misaligned with the direction of peak movement, resulting in the 'flying kite effect', a condition of permanent lag as the trait mean tracks the moving optimum. Likewise, in the case of anti-alignment of the *G*- and *Br*-matrices, the trait mean tends to lag behind its optimum so that only the central region, along the long axis of the orange, dotted ellipse in Figure 18.2d is over-represented by the bivariate trait means and the regions at the ends of the long axis are under-represented. However, as time proceeds beyond 1000 generations, the ends of the ellipse fill in and the orange 95% confidence ellipse does accurately portray the dispersion of bivariate trait means. Even in this contrived case, it is *Br*-matrix and time that govern the size and shape of the adaptive radiation and its expansion, rather than the *G*-matrix.

Figure 18.2 also illustrates the role of correlational selection and correlated peak movement in shaping adaptive radiation under SLLR movement. As correlational selection increases from $r_{\Omega} = 0.5$ to 0.9 and 0.98 (from top to bottom in Figure 18.2), r_{Br} shows the same trend, and the adaptive radiation becomes increasingly cigar-shaped.

18.3 Bilateral Symmetry as an Instance of a SLLR

In certain instances, a statistical relationship between pairs of traits is apparent within populations and is consistent across populations and species. Such instances may represent response to SLLR peak movement. Cases of bilaterally symmetrical traits are probably the most common situation of this kind. In many organisms with a preferred direction of whole-body movement, the sizes or counts of homologous structures on the right and

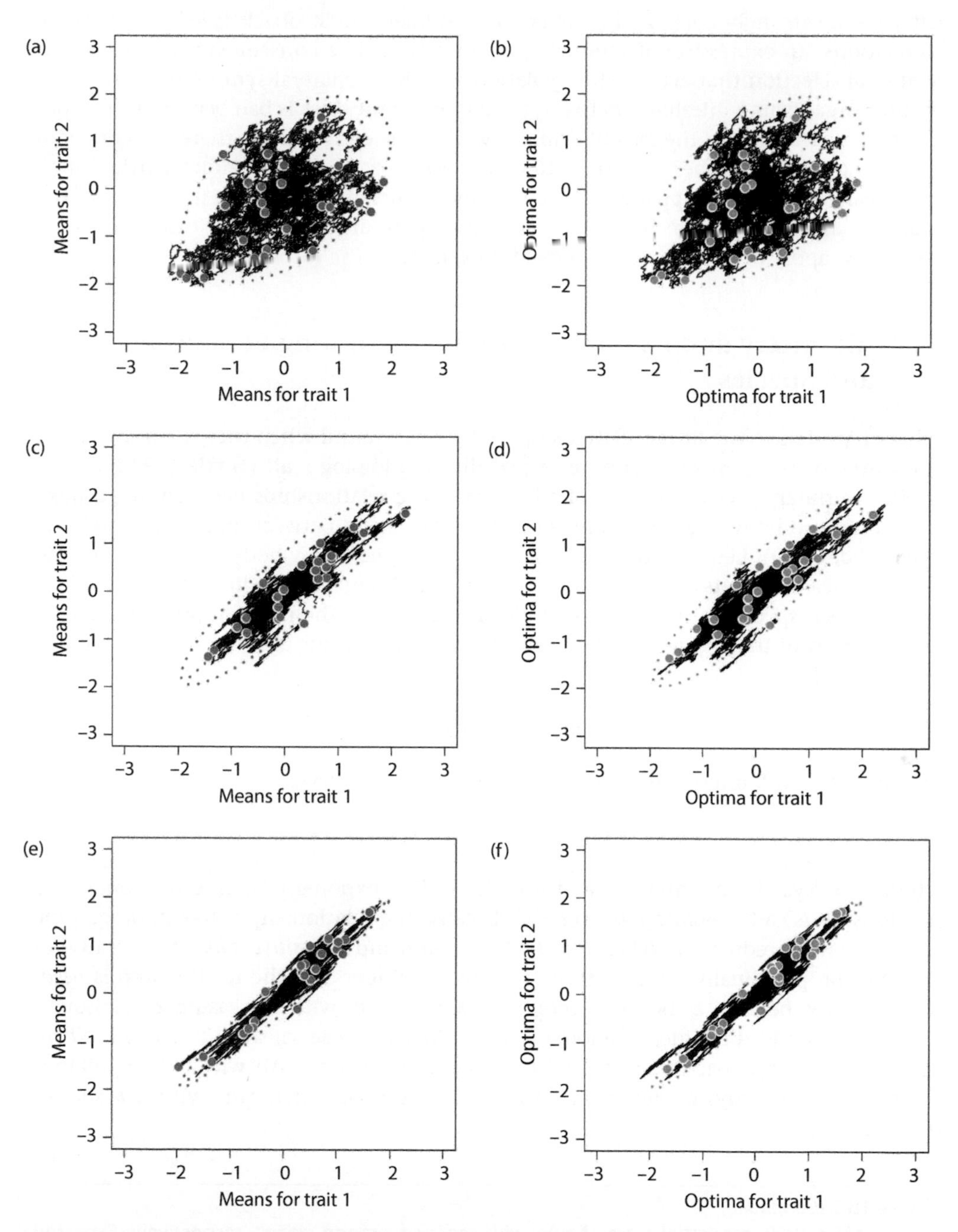

Figure 18.2 Simulations of evolution along selective lines of least resistance. Simulated evolution of 25 replicate bivariate trait means in response to Brownian motion of their bivariate optima along the leading eigenvector of their adaptive landscapes for 1000 generations. Effective population size is 500. Cases of modest (r = 0.5), strong (r = 0.9), and very strong correlation (r = 0.98) in inheritance, selection, and peak movement are illustrated. Evolutionary trajectories of trait means and optima are shown with black lines. The positions of replicate trait means and

(*Continued*)

left sizes are strongly correlated (Palmer 1996; Palmer and Strobeck 1986, 1992). These correlations, an expression of symmetry, are undoubtedly a consequence of strong correlational selection that acts in all populations in which bilateral symmetry is favoured. In most organisms a difference between the left and right sides is bad because it impedes forward progress along the axis of symmetry. The axis of symmetry therefore coincides with a SLLR that describes a major feature of selection within populations. Furthermore, SLLR peak movement is apparently common and occurs mainly along a major axis of the adaptive landscape (Figure 18.3). The same principles, apparent in the case of bilateral symmetry, apply in wider circumstances, for example in the case of ordinary allometry.

18.4 Allometry and Its Causes Within and Among Populations and Species

Allometry refers to the linear relationship that is often found when the sizes of two body parts are plotted against one another, especially on a log-log scale (Huxley 1932; Gayon 2000). A primary focus in studies of allometry is the relationships between linear measurements and body size, for example on the relationship between brain size and body size (Lande 1979). Let $\bar{z}_1$ be the mean of one measurement, say body size, and $\bar{z}_2$ be the mean of measurements of some other body part, e.g., brain size. Finding a linear relationship between species means, i.e., when plotting the log of the mean of one trait, $log\ \bar{z}_2$, as a function of the log of the other trait mean, $log\ \bar{z}_1$, means that the relationship can be portrayed as

$$log\ \bar{z}_2 = a + b\ log\ \bar{z}_1 \tag{18.1}$$

or, alternatively, but less conveniently, as an exponential curve,

$$\bar{z}_2 = A\ \bar{z}_1^b, \tag{18.2}$$

where $a = log\ A$ is the intercept and b is the scaling exponent (allometric slope). The relationship is called *isometry* when $b = 1$, because the relationship is like photographic enlargement or reduction. When $b > 1$, the relationship is *positive allometry*, because $\bar{z}_2$ becomes proportionally larger with increasing $\bar{z}_1$. When < 1, the relationship is *negative allometry*, because $\bar{z}_2$ becomes proportionally smaller with increasing $\bar{z}_1$. Although we have introduced the idea of allometry with plots of species means (*interspecific allometry*), the same approach can be used to describe relative growth within a population (*longitudinal allometry*) or geographic variation of population means within a species (Cock 1966).

Figure 18.2 (*Continued*)
optima after 1000 generations are shown with red and orange circles, respectively. Expected outcomes after 1000 generations are shown with 95% confidence ellipses based on expression (18.4) for optima (orange, dotted ellipses) and lineage trait means (blue, dotted ellipses). The leading and trailing eigenvectors of these ellipses are shown with dotted lines. In these simulations, the leading eigenvectors of the *G*-matrix, Ω-matrix, and *Br*-matrix are identical. (a) Evolution of 25 replicate trait means with $\boldsymbol{G}$ = (0.4 0.2, 0.2 0.4), and $\boldsymbol{\Omega}$ = (10 5, 5 10). (b) Brownian motion of 25 replicate bivariate optima with $\boldsymbol{Br}$ = (0.001 0.0005, 0.0005 0.001). (c) Evolution of 25 replicate trait means with $\boldsymbol{G}$ = (0.4 0.36, 0.36 0.4), and $\boldsymbol{\Omega}$ = (10 9, 9 10). (d) Brownian motion of 25 replicate bivariate optima with $\boldsymbol{Br}$ = (0.001 0.0009, 0.0009 0.001). (e) Evolution of 25 replicate trait means with $\boldsymbol{G}$ = (0.4 0.392, 0.392 0.4), and $\boldsymbol{\Omega}$ = (10 9.8, 9.8 10). (f) Browinan motion of 25 replicate bivariate optima with $\boldsymbol{Br}$ = (0.001 0.00098, 0.00098 0.001).

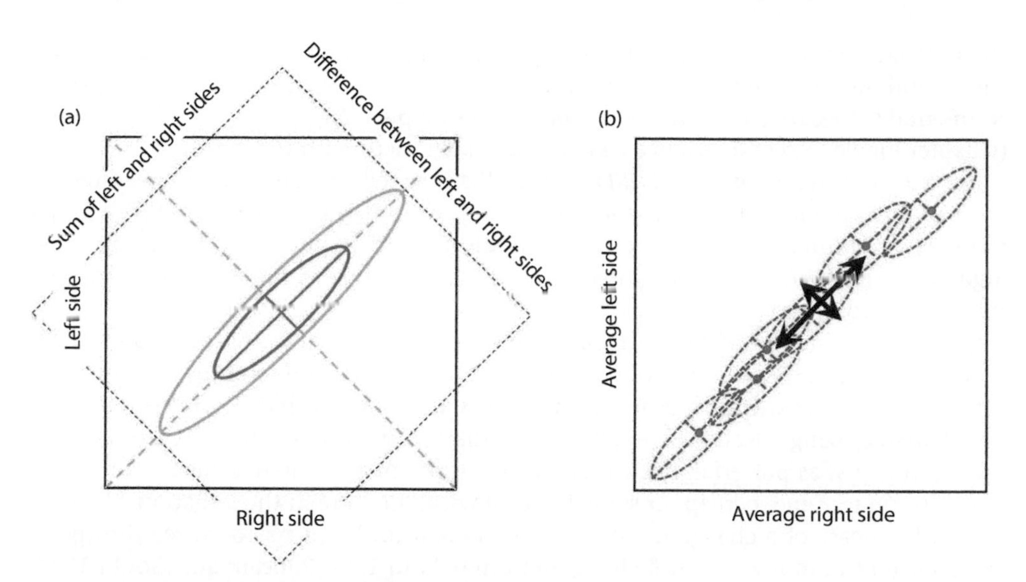

Figure 18.3 Conceptual diagram illustrating selection for bilateral symmetry and the evolution of bilateral traits in the direction of SLLR movement. (a) Selection on bilateral traits can be visualized in two related frames of reference. In the *first frame of reference* (solid box), the trait axes are the scores on the left and right sides of the organism. Selection favoring bilateral symmetry is a cigar-shaped ISS (orange ellipse) in which the major axis is equality of left- and right-side scores, shown as a dashed orange line from the lower-left to the upper-right corner. The minor axis of the selection surface is a dashed orange line from the upper-left to the lower-right corner. Stabilizing selection is much stronger along this axis than along the major axis. We expect that this correlational selection surface will promote correlation between left and right sides and produce a trait distribution before selection (shown as a blue ellipse) that is aligned with the selection surface. In the *second frame of reference* (dotted box), the axes are rotated 90 deg. One of the new axes is the sum or average of left and right sides. Stabilizing selection is relatively weak along this axis. The orthogonal axis is the differences between left and right sides. Stabilizing selection is strong on this axis and favors zero difference between left and right sides. (b) The expected directions of evolution of trait averages in replicate populations is shown as a long double-headed black arrows, which represent the major direction of SLLR movement. Each evolving trait mean (blue dot) tracks the peak of its adaptive landscape, shown as a dashed, red ellipse. These peaks tend to move along the major axis of the adaptive landscape, which is equivalent to equality of left and right side averages, and to a much lesser extent along the minor axis of the adaptive landscape, which is equivalent to difference between left- and right-side averages.

When ontogenetic trait means (representing growth within a population), population means, and species means for the same pair of traits are included in one plot, a regular pattern in allometric coefficients is often found. Usually, the allometric slope is less for growth and higher for species means, with geographic variation falling in between (Lande 1979). Various explanations have been offered for this regularity (Pagel and Harvey 1988, 1989), but a consideration of how pairs of traits equilibrate under regimes of stabilizing selection offers a new perspective.

Allometry in a plot of individuals of the same age (*static allometry*) will usually reflect pleiotropy and is subject to the set of proximate and ultimate shaping forces considered in Chapters 1–8 (bivariate mutation, inheritance, selection, and drift). The main axis of the bivariate cloud of genetic values (g_{max}) is a compromise between the main axes of the AL and the cloud of mutational values (Chapters 16–17). When we add bivariate environmental values to genetic values to obtain phenotypic values in a static allometry plot (Chapter 6), we expect to achieve less not more alignment with the AL.

In contrast, lines of *interspecific allometry* are likely to reflect regularities in the direction of peak movement and, in well-conserved cases, will present a clearer picture of ω_{max} than will the principal axis of phenotypic variation within populations. According to this argument, we should not be surprised if the allometric slope, *b*, increases as we move to higher taxonomic levels.

Comparative studies of allometry sometimes reveal striking constancy in both *a* and *b* during radiations (see examples below). Situations of this kind lend themselves to the interpretation that the allometric relationship is an SLLR that has been maintained by bivariate stabilizing selection throughout the radiation. In other situations, the SLLR interpretation is just as powerful, but either *a* or *b* has jumped to a new value at some point in evolutionary time, perhaps at some key ecological juncture in the radiation.

A striking case of a change in allometric exponent for brain size occurred during the emergence of homonids, with *b* changing from 0.33 to 1.73 (Pilbeam and Gould 1974). The increase in *b* in the homonid line was viewed as evidence for a new selective regime, but the fact of an overall brain-body relationship, present throughout the vertebrate radiation, was taken as evidence of an underlying law, summarized by Pilbeam & Gould's remark that 'a larger brain is needed to satisfy the functional demands of a larger body'. While this statement is undoubtedly true, it does not help us understand how allometry might be produced and maintained by ordinary evolutionary forces and processes.

A path forward in the quest to understand allometry was discovered by physiologists. In the early days of allometric studies, physiologists searched for laws that might explain the evolution of metabolic rate and other measures of whole organism performance. Those studies have continued to the present time. They have helped us understand that allometry is an expression of inheritance (especially pleiotropy), and that it is powerfully shaped by selection, especially by correlational selection.

Studies of the relationship between metabolic rate and body size reveal tight relationships that are maintained over the full range of vertebrate body sizes with characteristic allometric coefficients. Perhaps more importantly, the relationship hints at a functional relationship that holds with law-like regularity (Benedict 1938; Kleiber 1947). In general, let a phenotypic trait mean, *Y*, be an exponential function of average body mass, *M*, (18.2), so that

$$Y = Y_0\, M^b, \tag{18.3}$$

Where $\log Y_0$ is the intercept on a log-log plot, and *b* is the scaling exponent. In other words, we wish to understand evolutionary regularities in the values of Y_o and *b*.

Recently investigators have been increasingly successful in finding energetic and biomechanical explanations for particular scaling exponents (West et al. 1997, 1999; Enquist et al. 1998, 1999). In particular, it is useful to recognize internal network properties of the organism as well as the standard geometric descriptors of organismal size and shape, because of their different scaling properties (Table 18.1). In particular the scaling exponent for network variables is in multiples of 1/4, not in multiples of 1/3 as might be expected from geometric considerations (West et al. 1999). The one-quarter power arises from the scaling of networks of tissues involved in uptake and delivery of nutrients (Savage et al. 2008).

Table 18.1 The scaling of network variables (a, l, v) and geometric variables (L, A, V) as a function of body mass (M). The network variables describe the length of vascular or diffusion pathways (l), area of energy or water collecting networks (a), and volume of vascular or cellular networks (v). The conventional geometric (Euclidian) variables are length (L), area (A), and volume (V). From West et al. (1999) with permission.

Variable	Geometric	Network
Length	$L \propto A^{1/2} \propto V^{1/3} \propto M^{1/3}$	$l \propto a^{1/3} \propto v^{1/4} \propto M^{1/4}$
Area	$A \propto l^2 \propto V^{2/3} \propto M^{2/3}$	$a \propto l^3 \propto v^{3/4} \propto M^{3/4}$
Volume	$V \propto L^3 \propto M$	$v \propto l^4 \propto M$

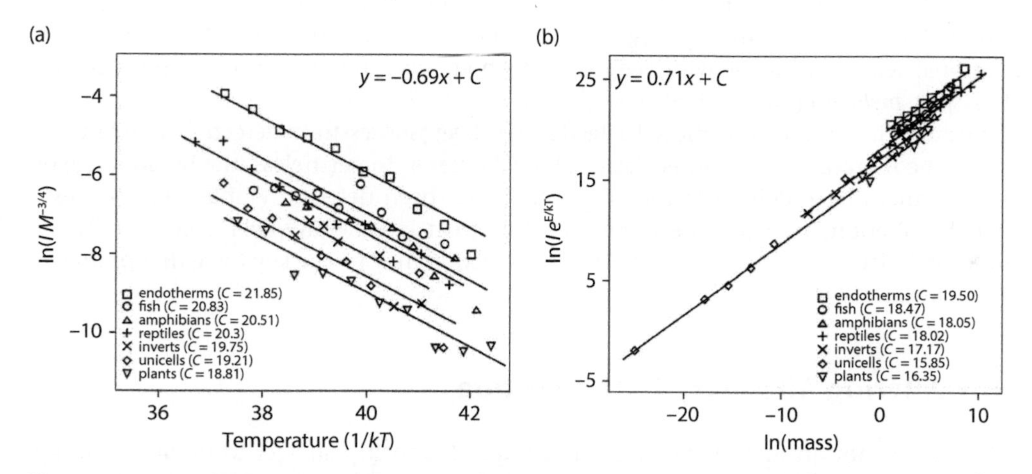

Figure 18.4 Metabolic rate as a function of temperature and mass in various organisms. (a) Mass-corrected metabolic rate, ln $\left(IM^{-3/4}\right)$ as a function of temperature, $1/kT$, measured in degrees Kelvin (K). (b) Temperature-corrected metabolic rate, ln $\left(I\, e^{E/kT}\right)$, measured in watts, as a function of body mass, ln(M), measured in grams.
From Brown et al. (2004) with permission.

Gillooly et al. (2001) argue that metabolic rate is governed by two underlying, interacting aspects: the temperature dependence of biochemical processes (Boltzmann's constant, k) and a scaling of biological rates with body size (a quarter power allometric relationship). Their argument is encapsulated in an equation that relates metabolic rate (B) to body mass (M) and temperature (T),

$$B = I\, M^{3/4}\ exp\,\{-E_i/kT\}\,, \tag{18.4}$$

where I is a constant, and E_i is the average activation energy for the rate-limiting enzyme-catalysed biochemical reactions of metabolism. The universalities of Boltzmann's constant and the three-quarters scaling law constrain, respectively, the evolution of metabolic rate as a function of temperature (Figure 18.4a) and as a function of body mass (Figure 18.4b). Because these constraints are persistent, evolution occurs along lines with a common slope and with elevations that are characteristic of particular clades.

These results suggest that biochemical and scaling laws constrain peak movement to lines with a particular slope (Figure 18.4) for hundreds of millions of years. In contrast, the elevation of these evolutionary trajectories appears to jump on a shorter but still lengthy timescale.

18.5 Correlational Selection, the Missing Ingredient in Allometry Studies

The missing ingredient in past explanations of allometry is the idea that selection can act on combinations of traits. In the case of wing size allometry in birds, we have evidence that correlational selection shapes the allometric relationship. Interspecific studies of wing length allometry in birds reveal an allometric coefficient in the range 0.35–0.37 (Nudds 2007). In line with these results, Schluter and Nychka (1994) detected positive correlational selection for the wing length versus body mass relationship in song sparrows (*Melospiza melodia*) (Figure 18.5).

Instances in which correlational selection has been successfully detected are relatively rare in the literature, for reasons discussed in Chapter 4. Nevertheless, the broad range of circumstances in which correlational selection has been detected (Table 18.2) suggests that the phenomenon may be more prevalent than is generally acknowledged. These successes bolster optimism that correlational selection may be the key force that produces and maintains allometry.

18.6 Altering Allometry With Selection

Allometry is not an immutable law, it is a trait relationship subject to modification by ordinary evolutionary forces. Selection experiments have shown that allometric relationships can be quickly changed by selection. For example, Weber (1990 imposed selection

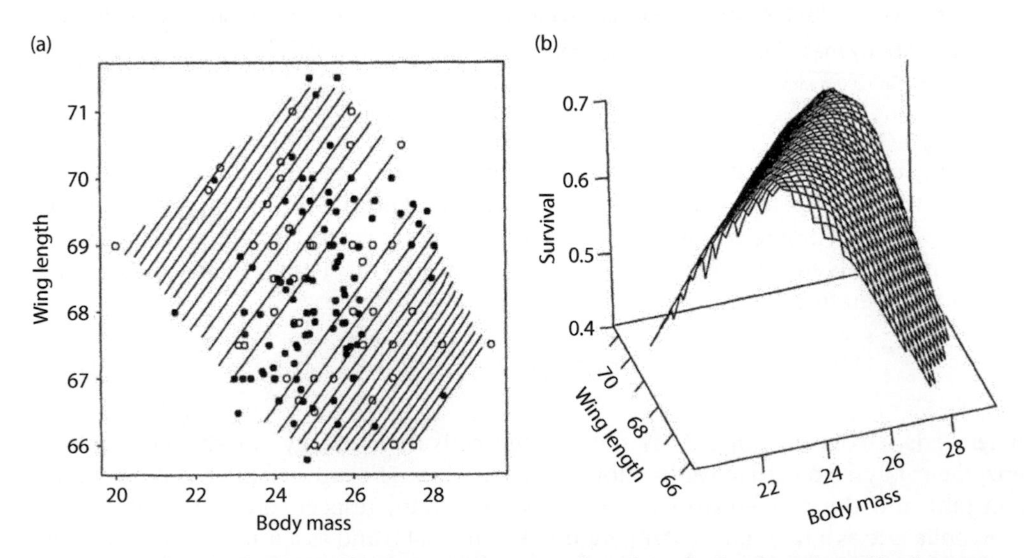

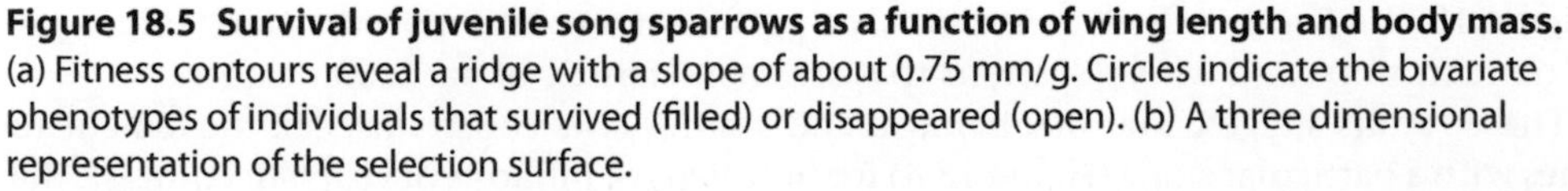
Figure 18.5 Survival of juvenile song sparrows as a function of wing length and body mass. (a) Fitness contours reveal a ridge with a slope of about 0.75 mm/g. Circles indicate the bivariate phenotypes of individuals that survived (filled) or disappeared (open). (b) A three dimensional representation of the selection surface.

From Schluter and Nychka (1994) with permission.

Table 18.2 Statistically significant instances of correlational selection acting on morphological traits.

Taxon	Trait 1	Trait 2	Fitness measure	Reference
Aquarius remigis	hind leg size	body length	male mating status	Fairbairn and Preziosi 1996
Libellula luctosa	white patch size	wing length	female breeding success	Moore 1990
Cypripedium acaule	flower height	labellum length	pollination success	O'Connell and Johnston 1998
Thamnophis elegans	tail vertebrae	body vertebrae	growth rate	Arnold 1988
Thamnophis radix	tail vertebrae	body vertebrae	crawling speed	Arnold and Bennett 1988
Anolis sagrei	hind leg length	sprint sensitivity	survival	Calsbeek and Irschick 2007

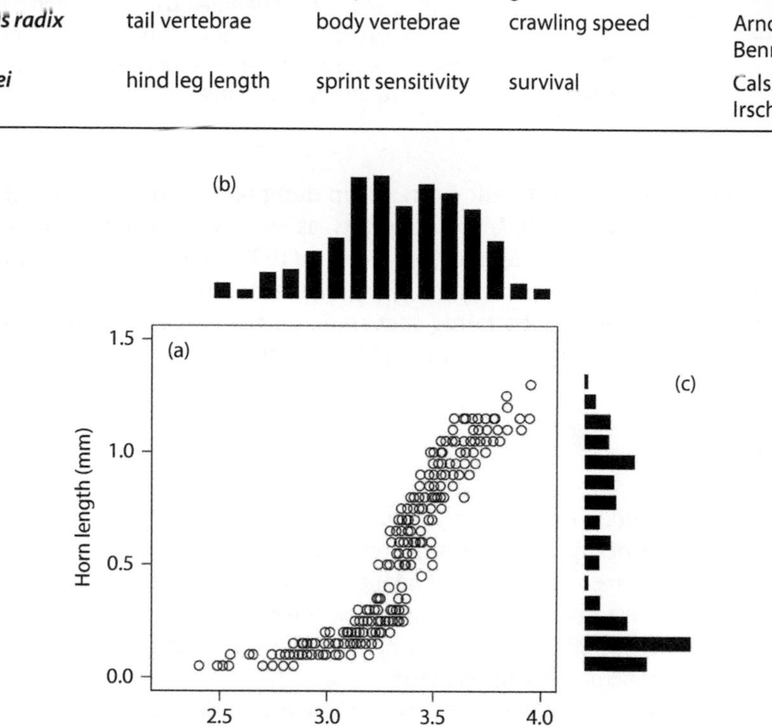

Figure 18.6 Horn length as a function of body size (prothorax width) in a base population of dung beetles (*Onthophagus acuminatus*). (a) The original sigmoidal bivariate relationship. (b) The almost unimodal distribution of body size. (c) The strongly bimodal distribution of horn size. After Emlen (1996).

on bivariate aspects of wing shape in *Drosophila melanogaster* and achieved substantial differentiation in allometry in 16 generations (Figure 11.6). Weber was able to move bivariate means apart by an average of nearly 15 phenotypic standard deviations. Selection for divergent allometry of eye stalks in stalk-eyed flies (*Cyrtodiopsis dalmani*) produced pronounced differences in 10 generations (Wilkinson 1993). Likewise, Emlen (1996) was

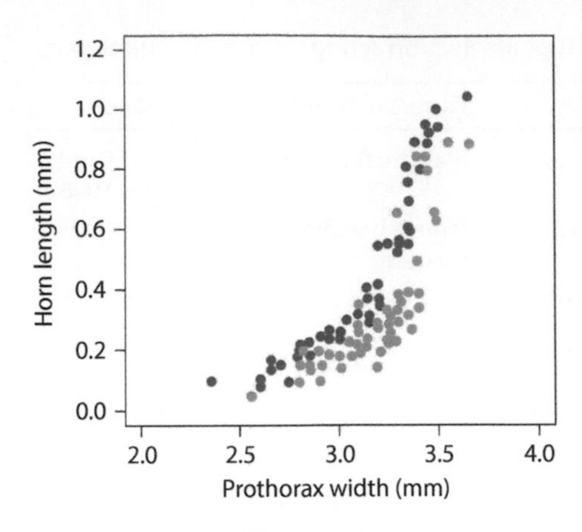

Figure 18.7 Dung beetle horn length as a function of body size after seven generations of bidirectional selection. Selection for increased (blue circles denote up line) and decreased (green circles denote down line) residual horn length caused a lateral divergence in the sigmoidal relationship. After Emlen (1996).

able to alter the allometry of cephalic horn size in dung beetles in just seven generations of selection (Figures 18.6 and 18.7). These results, as well as comparable experiments by Robertson (1962) and Cock (1966), support Weber's (1990) view that the bivariate developmental patterns in these systems can be changed in virtually any direction by selection. In other words, the genetic constraints that underlie these systems can be shaped by selection. They do not constitute insurmountable hurdles for the evolutionary process.

18.7 Conclusions

- Why are some regions of phenotypic space densely occupied, while other regions are rarely or never occupied? In this chapter we argue that densely occupied regions often correspond to recurrent paths of adaptive peak movement, while rarely occupied regions represent regions of adaptive incompatibility.
- The idea of recurrent paths of peak movement focuses our attention on the forces and constraints that control peak movement.
- The set of selective constraints that produce a selective line of least resistance (SLLR) (also known as $\boldsymbol{\Omega}_{max}$) may predispose the adaptive peak to move in that same direction. When the constituent species in a radiation have a common topography for their ALs (and a common $\boldsymbol{\Omega}_{max}$ in particular), peak movement along that SLLR will produce alignment of the trait radiation with that common AL feature (i.e., alignment of $\boldsymbol{d}_{max}$ with the common $\boldsymbol{\Omega}_{max}$).
- A stochastic model of trait evolution with SLLR peak movement reveals that the matrix that describes Brownian motion of the adaptive peak also predicts the size and shape of the adaptive radiation, as well as its rate of expansion.
- SLLR peak movement may be the key to understanding the evolution of bilateral symmetry and many kinds of allometry, including isometric series.
- Some conserved features of allometry may reflect underlying biochemical constraints and laws of physics. Thus, the conserved relationship between metabolic rate and temperature is a reflection of Boltzmann's constant. The universality of this constant, constrains the path of peak movement during the evolution of metabolic rate.
- Correlational selection is often a missing ingredient in studies of allometry. When correlational selection has been measured, it provides evidence that peak movement coincides with $\boldsymbol{\omega}_{max}$.

Literature Cited

Arnold, S. J. 1988. Quantitative genetics and selection in natural populations: Microevolution of vertebral numbers in the garter snake *Thamnophis elegans*. pp. 619–636 *in* B. S. Weir, E. J. Eisen, M. M. Goodman, and G. Namkoong, eds. Proceedings of the Second International Conference on Quantitative Genetics. Sinauer Associates, Inc., Sunderland.

Arnold, S. J., and A. F. Bennett. 1988. Behavioural variation in natural populations. V. Morphological correlates of locomotion in the garter snake (*Thamnophis radix*). Biol. J. Linn. Soc. 34:175–190.

Arnold, S. J., M. E. Pfrender, and A. G. Jones. 2001. The adaptive landscape as a conceptual bridge between micro-and macroevolution. Genetica 112:9–32.

Benedict, F. G. 1938. Vital energetics: A study in comparative basal metabolism. Carnegie Inst. Washington Publication No. 503.

Brown, J. H., J. F. Gillooly, A. P. Allen, V. M. Savage, and G. B. West. 2004. Toward a metabolic theory of ecology. Ecology 85:1771–1789.

Calsbeek, R., and D. J. Irschick. 2007. The quick and the dead: Correlational selection on morphology, performance, and habitat use in island lizards. Evolution 61:2493–2503.

Cock, A. G. 1966. Genetical aspects of metrical growth and form in animals. Q. Rev. Biol. 41: 131–190.

Dawkins, R. C. 1996. Climbing Mount Improbable. Norton, New York.

Emlen, D. J. 1996. Artificial selection on horn length-body size allometry in the horned beetle *Onthophagus acuminatus* (Coleoptera: Scarabaeidae). Evolution 50:1219.

Enquist, B. J., J. H. Brown, and G. B. West. 1998. Allometric scaling of plant energetics and population density. Nature 395:163–165.

Enquist, B. J., G. B. West, E. L. Charnov, and J. H. Brown. 1999. Allometric scaling of production and life-history variation in vascular plants. Nature 401:907–911.

Fairbairn, D. J., and R. F. Preziosi. 1996. Sexual selection and the evolution of sexual size dimorphism in the water strider, *Aquarius remigis*. Evolution 50:1549.

Gayon, J. 2000. History of the concept of allometry. Am. Zool. 40:748–758.

Gillooly, J. F., J. H. Brown, G. B. West, V. M. Savage, and E. L. Charnov. 2001. Effects of size and temperature on metabolic rate. Science 293:2248–2251.

Hansen, T. F., J. Pienaar, and S. H. Orzack. 2008. A comparative method for studying adaptation to a randomly evolving environment. Evolution 62:1965–1977.

Hohenlohe, P. A., and S. J. Arnold. 2008. MIPoD: a hypothesis-testing framework for microevolutionary inference from patterns of divergence. Am. Nat. 171:366–385.

Huxley, J. S. 1932. Problems of Relative Growth. Lincoln MacVeagh-The Dial Press, New York.

Kleiber, M. 1947. Body size and metabolic rate. Physiol. Rev. 27:511–541.

Lande, R. 1979. Quantitative genetic analysis of multivariate evolution, applied to brain: bodysize allometry. Evolution 33:402–416.

Moore, A. J. 1990. The evolution of sexual dimorphism by sexual selection: The separate effects of intrasexual selection and intersexual selection. Evolution 44:315–331.

Nudds, R. L. 2007. Wing-bone length allometry in birds. J. Avian Biol. 38: 515–519.

O'Connell, L. M., and M. O. Johnston. 1998. Male and female pollination success in a deceptive orchid, a selection study. Ecology 79:1246.

Pagel, M. D., and P. H. Harvey. 1988. The taxon-level problem in the evolution of brain size: Facts and artifacts. Am. Nat. 132:344–359.

Pagel, M. D., and P. H. Harvey. 1989. Taxonomic differences in the scaling of brain on body weight among mammals. Science 244:1589–1593.

Palmer, A. R. 1996. Waltzing with asymmetry. BioScience 46:518–532.

Palmer, A. R., and C. Strobeck. 1986. Fluctuating asymmetry: Measurement, analysis, patterns. Annu. Rev. Ecol. Syst. 17:391–421.

Palmer, A. R., and C. Strobeck. 1992. Fluctuating asymmetry as a measure of developmental stability: Implications for non-normal distributions and power of statistical tests. Acta Zool Fenn. 191:57–72.

Pilbeam, D., and S. J. Gould. 1974. Size and scaling in human evolution. Science 186:892–901.

Robertson, F. W. 1962. Changing the relative size of the body parts of *Drosophila* by selection. Genet. Res. 3:169–180.

Savage, V. M., E. J. Deeds, and W. Fontana. 2008. Sizing up allometric scaling theory. PLOS Comput. Biol. 4:e1000171.

Schluter, D. 2000. The Ecology Of Adaptive Radiation. Oxford University Press, Oxford.

Schluter, D., and D. Nychka. 1994. Exploring fitness surfaces. Am. Nat. 143:597–616.

Weber, K. E. 1990. Selection on wing allometry in *Drosophila melanogaster*. Genetics 126:975–989.

West, G. B., J. H. Brown, and B. J. Enquist. 1997. A general model for the origin of allometric scaling laws in biology. Science 276:122–126.

West, G. B., J. H. Brown, and B. J. Enquist. 1999. The fourth dimension of life: Fractal geometry and allometric scaling of organisms. Science 284:1677–1679.

Wilkinson, G. S. 1993. Artificial sexual selection alters allometry in the stalk-eyed fly *Cyrtodiopsis dalmanni* (Diptera: Diopsidae). Genet. Res. 62:213–222.

CHAPTER 19

Speciation and Extinction of Lineages

Overview—This chapter focuses on the roles of speciation and extinction, as well as within-lineage evolutionary process (mutation, inheritance, selection, population size), in shaping adaptive radiations. We begin by pointing out that on theoretical grounds trait evolution associated with speciation is likely to make a smaller contribution than within-lineage processes. We then introduce a theoretical framework that allows us to assess the effects of trait evolution on sexual isolation, speciation, and radiation. Using a sequence of models, we show that the efficacy of the Fisher-Lande process depends on relaxing selection on female mating preferences or allowing random movement of natural selection optima. Finally, we turn to traits that influence resource utilization and competition between species. We show that random movement of the natural selection optima of such traits can drive ecological radiation, isolation, and speciation.

19.1 The Contributions of Speciation and Extinction to Adaptive Radiation

Up to this point we have ignored both speciation and extinction of lineages and the effects they might have on adaptive radiations. We now consider the possibility that trait evolution affects speciation and extinction rates and hence the production and pruning of branches on phylogenetic trees. This possibility has been extensively considered in methods to detect such effects on phylogenetic trees by both categorical and quantitative traits (Maddison et al. 2007; FitzJohn 2010; Louca and Pennell 2020). Our focus, however, is on how quantitative traits affect the pattern of radiation via speciation and extinction. Consequently, we turn to Slatkin (1981) which provides a modelling framework that will enable us to consider those effects.

Slatkin (1981) used a diffusion approximation of a Markov process to model the evolution of a trait mean. The model is in continuous time, rather than in discrete time, as in the earlier chapters of this book. Following his approach, we will assume that each species in a set of species can be characterized by a single value for the mean of a single phenotypic trait ($\bar{z}$), and that the entire ensemble of species at time t can be described by the distribution of trait means, $n(\bar{z}, t)$, which gives the number of species with the trait value $\bar{z}$ at time t. We proceed by modelling the changes in $n(\bar{z}, t)$ due to within-lineage process (e.g., via drift and/or selection), speciation, and extinction.

Evolutionary Quantitative Genetics. Stevan J. Arnold, Oxford University Press. © Stevan Arnold (2023).
DOI: 10.1093/oso/9780192859389.003.0020

To model *ordinary, within-lineage evolution*, we assume that during some short time interval, Δt, a species with mean $\bar{z}$ evolves by within-lineage processes to $\bar{z} + \Delta\bar{z}$, where $\Delta\bar{z}$ is a random variable with a distribution that is a function of Δt. To derive a diffusion equation for $n(\bar{z}, t)$, Slatkin (1981) assumes that mean and variance of $\Delta\bar{z}$ are of the same order of magnitude as Δt and that the third and higher moments of $n(\bar{z}, t)$ are negligible. Let the mean and variance of $\Delta\bar{z}$ be

$$E(\Delta\bar{z}) = M_o(\bar{z})\,\Delta t \tag{19.1a}$$

$$Var(\Delta\bar{z}) = V_o(\bar{z})\,\Delta t, \tag{19.1b}$$

where $E(\ldots)$ denotes an expectation and $Var(\ldots)$ denotes a variance. $M_o(\bar{z})$ is the expected change in the trait mean in each species each generation. If $M_o(\bar{z}) \neq 0$, then some directional force such as selection is causing all the species to evolve in the same way. Stochastic influences on $\Delta\bar{z}$ are reflected in $V_o(\Delta\bar{z})$. In the case of genetic drift of the trait mean (8.3), the per-generation input of variance is

$$V_o(\Delta\bar{z}) = \frac{G}{N_e}. \tag{19.1c}$$

To model speciation, we assume that in the time interval Δt, each species with trait mean $\bar{z}$ produces a new species with probability $s(\bar{z})\,\Delta t$. We assume that the trait mean of the new species is $\bar{z}+\Delta y$, where Δy is a random variable with a known distribution. Making the same assumptions as in the case of within-lineage evolution, let the mean and variance of Δy be

$$E(\Delta y) = M_s(\bar{z})\,\Delta t \tag{19.1d}$$

$$Var(\Delta y) = V_s(\bar{z})\,\Delta t. \tag{19.1e}$$

If $M_s(\bar{z}) \neq 0$, the change in trait mean that occurs at speciation is biased in a particular direction. $V_s(\bar{z})$ reflects stochastic change in the trait mean at speciation. The case of trait evolution affecting speciation is modelled by letting the speciation rate, $s(\bar{z})$, be a function of $\bar{z}$.

To model extinction, we assume that a species with trait mean $\bar{z}$ goes extinct in the time interval Δt with probability $e(\bar{z})\,\Delta t$. The trait mean can affect extinction by letting $e(\bar{z})$ be a function of the trait mean.

We now combine the influence of the three processes we have considered on the distribution of trait means. Letting the time interval go to its limit of zero, we obtain a single differential equation,

$$\frac{\partial n(\bar{z},t)}{\partial t} = \frac{1}{2}\frac{\partial^2}{\partial \bar{z}^2}\left[(V_o(\bar{z}) + s(\bar{z})\,V_s(\bar{z}))\;n(\bar{z},t)\right] - \frac{\partial}{\partial \bar{z}}\left[(M_o(\bar{z}) + s(\bar{z})\,M_s(\bar{z}))\;n(\bar{z},t)\right] + (s(\bar{z}) - e(\bar{z}))\;n(\bar{z},t). \tag{19.2}$$

This expression predicts future values of $n(\bar{z}, t)$ from values at some particular time. The model implies that ensemble of species are monophyletic descendants from ancestors with a trait value of $\bar{z}_0$ at $t = 0$. See Slatkin (1981) for the justification of the expression's derivation and other details. We will now use the three parts of this framework to model particular cases in which trait evolution affects speciation and extinction.

19.1.1 *Trait diversification with constant rates of speciation and extinction that are not affected by trait values*

Assume that the trait mean does not show directional change due to speciation ($M_s = 0$) or ordinary evolution ($M_o = 0$). Assume that the variance contributions from speciation and ordinary evolution, V_s and V_o, are both constant, as are the speciation and extinction rates, $s(\bar{z}) = s$ and $e(\bar{z}) = e$. It this case, (19.2) becomes

$$\frac{\partial n}{\partial t} = \frac{1}{2}(V_o + sV_s)\frac{\partial^2 n}{\partial \bar{z}^2} + (s - e)\,n. \tag{19.3}$$

With the initial condition a Dirac delta function,$n(x, 0) = \delta(\bar{z} - \bar{z}_0)$, the expression above has the solution

$$n(\bar{z}, t) = \frac{1}{\sqrt{2\pi Vt}}\, exp\{(s - e_0)\,t\} - \frac{(\bar{z} - \bar{z}_0)^2}{2Vt}, \tag{19.4}$$

where $V = V_o + sV_s$. In other words, the expected number of species increases or decreases exponentially at a rate equal to the difference between the constant gain from speciation and the constant loss from extinction $(s - e)$. Furthermore, $\bar{z}$ is normally distributed with a mean of $\bar{z}_0$ and a variance of Vt. If, for example, the ordinary evolutionary process is drift of the trait mean, then the variance among species means that characterizes the radiation at time t is

$$Var(\bar{z}_t) = Vt = V_o t + sV_s t = \frac{Gt}{N_e} + sV_s t. \tag{19.5a}$$

We would like to know the relative sizes of the two contributions to Vt. Slatkin (1981) points out that if we use Stanley's (1979) estimate of approximately 25 speciation events per million years, $s = 0.4$ per million years. For speciation and ordinary evolution to make equal contributions to the among-species trait variance in some time interval, $V_o t = sV_s t$, so $V_o = sV_s$. Using the last term in (19.5a) and substituting the empirical modal value 0.5 for G with $P = 1$, we obtain $V_s = 500{,}000/(0.4\,N_e)$. Suppose that the instantaneous contribution of speciation to the trait variance among species is equal to the within-species phenotypic variance ($V_s = P = 1$), then $N_e = 1{,}250{,}000$. Consequently, it is difficult to imagine that a single speciation event could produce such a large shift in a trait mean and the case for speciation to be an equal partner with within-lineage process in producing trait radiation sinks on this observation. Speciation plays at best a minor role in radiations.

If we change the within-lineage process to the trait mean tracking an adaptive peak that moves by Brownian motion, Section 14.2, then (19.1c) becomes $V_o(\bar{z}) = \sigma_\theta^2$, which is the per-generation variance in the position of the peak, θ. If we let $\sigma_\theta^2 = 0.00055$ (Arnold and Houck 2016), the overall contribution of within-lineage processes becomes $V_o t = \sigma_\theta^2 t = 0.00055 * 500{,}000 = 275$. In other words, this alternative version of Brownian motion creates an even larger value for the change needed by speciation, 37 times the within-species standard deviation. We again conclude that speciation is likely to play a minor role in adaptive radiation.

This analysis does not dispute the point that speciation plays an important role in adaptive radiation. Our claim is that speciation is unlikely to make an important contribution to the total variation in species trait means across an adaptive radiation unless some special event (e.g., a major jump in the position of the adaptive peak) is routinely conflated with speciation.

19.1.2 *Trait diversification with a constant rate of speciation and an extinction rate that depends on trait value*

Slatkin(1981) points out that the solution for the triple process is much more difficult if speciation or extinction rates depend on $\bar{z}$ but gives a partial solution under the following set of assumptions. Suppose $\bar{z} > 0$, V_S, V_O, and s are constants and $e(\bar{z}) = b/\bar{z}$, where b is a positive constant that represents the intensity of selection imposed by extinction. Consequently, lineages with larger trait means are less likely to go extinct. These assumptions change (19.1) to

$$\frac{\partial n}{\partial t} = \frac{1}{2} V \frac{\partial^2}{\partial \bar{z}^2} + \left(s - \frac{b}{\bar{z}} \right) n \tag{19.6}$$

for $\bar{z} > 0$, where $V = V_O + sV_S$. Slatkin (1981) was not able to find a solution for this expression but did show that the solution depends on the ratio b/V, which is the ratio of the intensity of selection produced by extinction to the among-species variance in trait means generated by ordinary within-lineage processes and speciation. This result implies that the contribution of extinction to the extent of the radiation depends on the intensity of extinction selection whether it acts through among-species trait variance produced by ordinary evolution or by lineage birth and death processes.

19.2 The Connections Between Trait Evolution, Sexual Isolation, and Speciation

The plan in this section—and the ones that follow—is to illustrate the mathematical steps from trait evolution to adaptive radiation. The lack of a step-by-step charting of this logical pathway has led to conceptual confusion and hesitancy in the literature. In particular, models of sexual selection typically stop short of making concrete connections to speciation. To take the missing step, we need to recognize that sexual isolation is the key concept that connects sexual selection to speciation.

In each subsection, we will follow the plan of attack used by Uyeda et al. (2009), which includes isolation and proceeds in the following sequence: (1) introduce equations for trait evolution under a particular stochastic model of process and illustrate examples of evolutionary trajectories that flow from those equations, (2) introduce and apply a formula for sexual isolation so that the evolution of isolation and its connection to speciation can be examined, and finally (3), where possible, introduce and apply a formula that characterizes the time course and extent of adaptive radiation. See de Queiroz (2007) for a synthetic discussion of species concepts.

Our goal is to illustrate the steps from equations for process and then to formulas for pattern, rather than to explore the effects of the many parameters of process on adaptive radiation. In each section we present and illustrate the simplest case of the multivariate theory, two traits in each sex, from which the general multivariate case can be easily derived. In the example of coevolution between the sexes, this simplest multivariate version consists of two sexually selected ornament traits in males and two preference traits in female. The empirical rationale for exploring multivariate sexual selection models is discussed in Section 19.2c.

We closely follow Lande's (1981) model for evolution by sexual selection and subsequent developments of that theory. For example, using Lande's (1981) results,

Arnold et al. (1996) produced an explicit model of sexual isolation based on the phenotypic values of male and female traits in two populations. Using that model, Uyeda et al. (2009) explored the connection between isolation and speciation by focusing on stochastic evolution of sexual isolation between populations of finite size. Hohenlohe and Arnold (2010) used that same model of sexual isolation to determine how traits participate in evolutionary trajectories that lead to speciation. Later in this section we extend these analyses to include Brownian motion of male and female trait optima, which can also facilitate the evolution of sexual isolation. Our expectation of what to expect from that extension springs from Arnold and Houck's (2016) model and simulation of bivariate sexual radiation.

A roadmap for the sections that follow is provided in Table 19.1, which also summarizes the parameters used in simulations. A principal feature of the model sequence is that one key feature is changed in each subsequent model, while retaining the same values for

Table 19.1 Results and values for parameters in models and simulations of sexual and ecological isolation and speciation. [1]Denotes the widths of a Gaussian stabilizing selection function (infinite width means no stabilizing selection, 50 means weak stabilizing selection) in bivariate trait space. [2]*JI* denotes joint sexual isolation, *EI* denotes ecological isolation, *ED* denotes ecological distance. [3]The permanent or initial position of the stabilizing selection optimum in bivariate trait space. [4]Mean or variance in *JI*, *EI*, or *ED* based on 100 replicate pairs of lineages after 10,000 generations. [5]Variance in *JI*, *EI*, or *ED* based on 10,000 replicate pairs of lineages after 10,000 generations. [6]Width of the cloud of bivariate trait means in units of $\sqrt{P}$ after 10,000 generations. [7]Percent of 100 replicates that achieve a *JI* or *EI* level of 1.6 or higher, a level characteristic of sympatric species.

	Type of isolation and speciation						
	Sexual				Ecological		
Chapter section	19.2.1	19.2.2	19.2.4	19.2.5	19.3.1	19.3.2	19.3.3
Parameter (symbol)							
Genetic variance, male (G_{ii})	0.6	0.6	0.6	0.6	0.6	0.6	0.6
Genetic variance, female (H_{ii})	0.6	0.6	0.6	0.6	0.6	NA	NA
Genetic covariance (B_{ij})	0.48	0.48	0.48	0.48	0.48	NA	NA
Population size (N_e)	1000	1000	1000	1000	1000	1000	1000
Position of optimum ($\boldsymbol{\theta}$)[3]	0, 0	0, 0	0, 0	0, 0	0, 0	0, 0	0, 0
Width of nat. selection (ω)[1]	50, ∞	50, ∞	50,50	50,50	50,50	50,50	50,50
Width of mate choice (ν)[1]	5	5	5	5	5	NA	NA
Sexual selection on males	yes	yes	yes	yes	no	NA	NA
Pheno. variance, male (P)	1	1	1	1	1	1	1
Pheno. variance, female (Q)	1	1	1	1	1	NA	NA
No. traits in each sex	1	2	2	2	2	NA	NA
No. traits in each lineage	2	4	4	4	4	2	2
Measure of isolation[2]	*JI*	*JI*	*JI*	*JI*	*EI*	*ED*	*ED*
Brownian motion of $\boldsymbol{\theta}$	no	no	no	yes	yes	yes	yes
Trait-mediated interaction	no	no	no	no	no	no	yes
Extent of trait radiation[6]	20	25	0.8	6	10	10+	10+
Mean isolation[4]		1.61	0.012	0.75	1.24	4.02	4.32
Variance in isolation[4]		0.30	0.0001	0.25	0.34	4.68	5.18
Speciation percentage[7]	44.5[5]	77	0	5	38	NA	NA

the core set of parameters. This feature allows us to assess the effects of key features on overall aspects of radiation and speciation. For example, in making the transition from the model in Section 19.2.1 to the model in Section 19.2.2, only the number of traits is changed. This feature allows us to attribute a large increase in speciation rate to an increase in trait number.

19.2.1 *Sexual speciation and radiation arising from genetic drift of a single ornament and preference*

In Chapter 13 we introduced Lande's (1981) model for the Fisher-Lande process in which male and female traits interact and affect each other's evolution. In that model, a sex-limited female trait (mating preference) exerts sexual selection on a sex-limited male trait (the ornament). The ornament also experiences stabilizing natural selection, but the preference is selectively neutral and evolves as a correlated response to natural and sexual selection on the ornament. We argued that a stable, walk-towards equilibrium was the most plausible of the two evolutionary outcomes discussed by Lande (1981). Lande also showed that the walk-towards equilibrium was a line, and that rapid diversification could occur by genetic drift along this line. However, although this potential for rapid diversification by drift suggests speciation, it does not actually demonstrate speciation or quantify the path to speciation.

To verify the implication that speciation can arise by drift and to predict rates of speciation under Lande's (1981) model, or any of its many derivatives (Mead and Arnold 2004), we need a model that connects sexual selection to sexual isolation hence to speciation. Arnold et al. 1996 made those connections using Lande's (1981) model. Their *triple Gaussian model* of sexual isolation was used by Uyeda et al 2009 in their simulation study of speciation by drift. In this section we summarize Uyeda et. al's (2009) model and results.

In the scenario we have in mind, two populations, A and B, have just become geographically isolated. Gene flow has ceased between the two populations, and we now follow divergence in an ornament and a preference that results from drift along a line of equilibrium. Each generation we evaluate sexual isolation by computing an index of isolation that can be related to speciation.

Let the ornament trait of males, z, be normally distributed within populations with means, $\bar{z}_A$ and $\bar{z}_B$, and with variance P for each trait in each population. Likewise, let the preference trait of females, y, be normally distributed within populations with means, $\bar{y}_A$ and $\bar{y}_B$, and variance Q for each trait in each population. The additive genetic variances of male and female traits are, respectively, G and H, are the same in both populations. The additive genetic covariance between z and y is likewise the same in both populations and designated B.

As we discussed in an earlier chapter, the deterministic change in trait means each generation due to selection is

$$\Delta\bar{z}_A = \frac{1}{2}G\beta_A \tag{19.7a}$$

$$\Delta\bar{y}_A = \frac{1}{2}B\beta_A \tag{19.7b}$$

$$\Delta\bar{z}_B = \frac{1}{2}G\beta_B \tag{19.7c}$$

$$\Delta\bar{y}_B = \frac{1}{2}B\beta_B, \tag{19.7d}$$

where the A and B subscripts denote populations. The two selection gradients are

$$\beta_A = \frac{S_A}{P} = \frac{\bar{y}_A/\alpha - (1 + 1/\alpha)\,\bar{z}_A + \theta}{\omega_A} \tag{19.8a}$$

$$\beta_B = \frac{S_B}{P} = \frac{\bar{y}_B/\alpha - (1 + 1/\alpha)\,\bar{z}_B + \theta}{\omega_B}, \tag{19.8b}$$

where $\alpha = \nu/\omega$, ν is the width of the Gaussian female mate choice function, and $\omega = \omega_A = \omega_B$ is the width of the Gaussian functions that describe natural selection on the male trait (Lande 1981).

Setting (19.7) to zero and solving for the deterministic evolutionary equilibrium, we obtain

$$\bar{y} = (\alpha + 1)\,\bar{z} + \alpha\theta, \tag{19.9}$$

a line of equilibria. Once the bivariate mean is pushed to this line by selection, the total selection gradients (19.8) take a value of zero, and the mean tends to drift along the line of equilibrium (Lande 1981).

In general (e.g., away from the line), the trait means evolve by selection and drift in populations of finite size, N_e. Adding the stochastic contributions of genetic drift each generation, we obtain the full model for change in the trait means,

$$\Delta\bar{z}_A = \frac{1}{2}G\beta_A + N(0,\ G/N_e) \tag{19.10}$$

$$\Delta\bar{y}_A = \frac{1}{2}B\beta_A + N(0, H/N_e) \tag{19.11}$$

$$\Delta\bar{z}_B = \frac{1}{2}G\beta_B + N(0, G/N_e) \tag{19.12}$$

$$\Delta\bar{y}_B = \frac{1}{2}B\beta_B + N(0, H/N_e)\,. \tag{19.13}$$

The variance-covariance matrix that describes dispersion of replicate lineage trait means at generation t, as they differentiate by drift along the line of equilibria, is

$$\boldsymbol{D}(t) = \frac{H\left(1 - r_g^2\right)t}{N_e(\alpha + 1 - B/G)^2}\begin{pmatrix} 1 & \alpha + 1 \\ \alpha + 1 & (\alpha + 1)^2 \end{pmatrix}, \tag{19.14}$$

where $r_g = B/\sqrt{GH}$ (Lande 1981).

To assess sexual isolation, we need to specify the probability of mating when sexual partners are drawn from the same and from different populations. We will use the triple Gaussian model of sexual isolation, which assumes that three elements are Gaussian: male ornaments and female preferences are normally distributed, and the female preference function is Gaussian in shape (Arnold et al. 1996). Focusing for the moment on a single population, recall that the frequency of males with phenotype z is

$$p(z) = \frac{1}{\sqrt{2\pi P}} exp\left\{-\frac{(z-\bar{z})^2}{2P}\right\}, \tag{19.15a}$$

with a mean of $\bar{z}$ and a variance of P. The female mate preference values, y, are likewise normally distributed but with mean $\bar{y}$ and variance Q,

$$q(y) = \frac{1}{\sqrt{2\pi Q}} exp\left\{-\frac{(y-\bar{y})^2}{2Q}\right\}. \tag{19.15b}$$

In the case of absolute mating preferences, the probability that a female with phenotype y will mate given an encounter with a male with phenotype z is a Gaussian phenotype matching function,

$$\psi(z|y) \propto exp\left\{-(z-y)^2/2v\right\}. \tag{19.16}$$

Averaging over the population of females, the overall probability of mating for a male with ornament value z is

$$\psi(z) = b\ exp\left\{-(z-\bar{y})^2/2(Q+v)\right\}, \tag{19.17}$$

where b is a positive constant equal to or less than one.

Consequently, the average probability of mating between a randomly chosen male and a randomly chosen female is

$$\pi = c\ exp\left\{-(\bar{y}-\bar{z})^2/2\Sigma\right\}, \tag{19.18}$$

where c is a positive constant equal to or less than one, and $\Sigma = v + Q + P$ (Arnold et al. 1996).

We use this last expression to specify the probabilities that a randomly drawn female from one population (first subscript) will mate with a randomly drawn male from another population (second subscript),

$$\pi_{AA} = c\ exp\left\{-(\bar{y}_A - \bar{z}_A)^2/2\Sigma\right\} \tag{19.19a}$$

$$\pi_{AB} = c\ exp\left\{-(\bar{y}_A - \bar{z}_B)^2/2\Sigma\right\} \tag{19.19b}$$

$$\pi_{BB} = c\ exp\left\{-(\bar{y}_B - \bar{z}_B)^2/2\Sigma\right\} \tag{19.19c}$$

$$\pi_{BA} = c\ exp\left\{-(\bar{y}_B - \bar{z}_A)^2/2\Sigma\right\}, \tag{19.19d}$$

where c is a constant less than 1 but greater than 0, and the variation parameter Σ is assumed to be the same in both populations.

These same four probabilities can be estimated in an experimental assay of sexual isolation (see Section 19.2.3). In such an assay, mating partners are drawn at random from two different populations, four kinds of sexual encounter are staged and replicated, mating success is scored for each kind of encounter, and the four probabilities of mating are estimated directly from those data. One common and useful measure of sexual isolation that summarizes such data is called the *index of joint isolation*,

$$JI = \pi_{AA} + \pi_{BB} - \pi_{AB} - \pi_{BA} \tag{19.20}$$

(Bateman 1949; Merrell 1950; Malogolowkin-Cohen et al. 1965). This index effectively ranges from 0 (when the probabilities of within- and between-population matings are equal) to 2 (when all within but no between-population encounters are successful).

Under simplifying assumptions ($\bar{z}_A = \bar{y}_A$, $\bar{z}_B = \bar{y}_B$, $\pi_{AA} = \pi_{BB}$, $\pi_{AB} = \pi_{BA}$), the key variables affecting the distribution of JI is the distribution of the probability $\pi_{AB} = \pi_{BA}$. Uyeda et al. 2009 show that under these conditions the probability that π_{AB} takes the value x at generation t is, for $0 \leq x \leq 1$,

$$f_{\pi_{AB}}(x) = \frac{\Sigma x^{\Sigma/2D_z(t)}}{x\sqrt{-2\pi D_z(t)\,\Sigma \ln(x)}}, \tag{19.21}$$

where Σ is the variation parameter and $D_z(t)$ is the variance of $\bar{z}$ among replicate populations at generation t as approximated by the first element of the $\boldsymbol{D}(t)$ matrix (19.14) (Uyeda et al. 2009). The expected value of JI at generation t is

$$E[JI(t)] = 2\left(1 - E[\pi_{AB}(t)]\right), \tag{19.22a}$$

where $E[\pi_{AB}(t)]$ is the expected value of π_{AB} at generation t,

$$E[\pi_{AB}] = \int_0^1 \frac{\Sigma^2 x^{\Sigma^2/2D_z(t)}}{\sqrt{-2\pi D_z(t)\,\Sigma^2 \ln(x)}}\,dx \tag{19.22b}$$

(Uyeda et al. 2009, Appendix).

To model the evolution of JI, we consider a representative pair of populations, A and B, that have evolved to the line of stable equilibrium in Lande's (1981) model and are now drifting along it. To proceed, it will help to make some simplifying assumptions. First, we note (and confirm by simulation) that the difference between preference and ornament means within a population will be negligible relative to the expected between-population difference in means. Under this assumption, from (19.19), $\pi_{AA} = \pi_{BB} = 1$ and $\pi_{AB} = \pi_{BA}$. The distribution of $\pi_{AB} = \pi_{BA}$ among replicate pairs of populations is then a function of the corresponding distribution of pairs of ornament means (see illustrated pairs in Figure 19.1), which we can determine using (19.22).

In summary, the following parameter values were used in the simulations: $G_{ii} = 0.6$, $H_{ii} = 0.6$, $B_{ii} = 0.48$, $N_e = 1000$, $\theta = 0$, $\omega = 50$, $\nu = 5$, $\alpha = \nu/\omega = 0.1$, $P = 1$, $Q = 1$, $\Sigma = \nu + Q + P = 7$. Using (19.14) and these parameter values, the expected 95% confidence intervals for ornament and preference means at generation 10,000 are both about $\pm 1.96 * 5\sqrt{P} = \pm 10\sqrt{P}$.

Uyeda et al (2009) provide detailed justifications for these choices in parameter, but since we will use them in the series of models that follow, we will briefly summarize their arguments here. Uyeda et al. (2009) explored the effects of a range of parameter values about the core, central values listed above. Those core values represent moderately small effective population size ($N_e = 1000$), relatively weak stabilizing natural selection ($\omega = 50$; Kingsolver et al. 2001; Estes and Arnold 2007), relatively strong sexual selection ($\nu = 5$), and a moderately strong genetic correlation between the sexes ($r_g = B/\sqrt{GH} = 0.8$), arising from linkage disequilibrium generated by sexual selection on ornaments (Lande 1981).

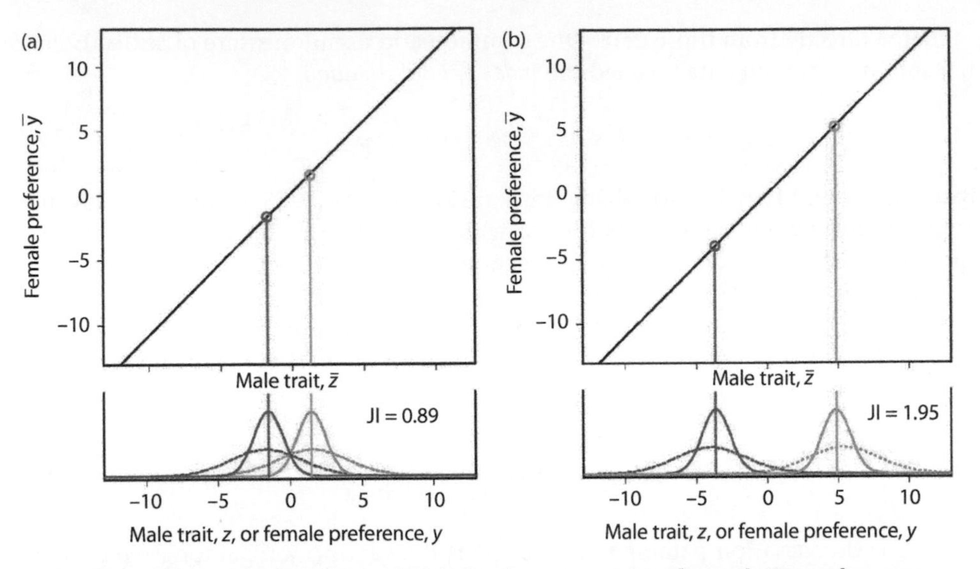

Figure 19.1 Two examples of sexual isolation between pairs of populations whose mean values for ornaments and preferences have diverged along the line of evolutionary equilibrium. The upper panels show the positions of male and female trait means on the line of equilibrium. Population *A* is shown in purple and population *B* is shown in green The lower panels show the distributions of male ornaments ($p(z)$, solid curves) and female preferences ($\psi(z)$, dotted curves). (a) Modest trait divergence between the two populations resulting in a modest level of sexual isolation ($JI = 0.89$). (b) Profound trait divergence resulting in a nearly maximal value for sexual isolation ($JI = 1.95$). Parameter values for this these examples are $P = Q = 1$, and $v = 5$; see column 19.2.1 in Table 19.1.

From Uyeda et al. (2009).

Pursuing this course, we find that the function giving the expected value of *JI* as a function of time is not simple, but it can be evaluated numerically or the process can be simulated (Uyeda et al. 2009). Simulations reveal that the evolutionary trajectory of *JI* for any pair of populations is often chaotic, as we might expect from the overwhelming influence of drift. Instead of progressing steadily towards increasing isolation, individual trajectories often alternate between episodes in which isolation either waxes or wanes (Figure 19.2). On the other hand, when we follow the histories of many replicate pairs of populations, we do see a trend towards increasing isolation (Figure 19.3). The distribution of *JI* is characteristically bimodal during intermediate stages, until finally the mode becomes strong sexual isolation ($JI \approx 2$), typically after some thousands of generations.

Uyeda et al. (2009) conclude that the Fisher-Lande process, coupled with assumptions of no selection on female preferences and moderate population size, can lead to rapid speciation.

19.2.2 *Sexual speciation and radiation arising from genetic drift of multiple ornaments and preferences with selection only on ornaments*

In this section we extend the model just considered in Section 19.2.1 to the case of two ornament and two preference traits. Our rationale for this multivariate extension is that from a variety of perspectives it seems likely that mate choice is based on multiple traits (Hohenlohe and Arnold 2010; Arnold and Houck 2016). The extension to a total of four

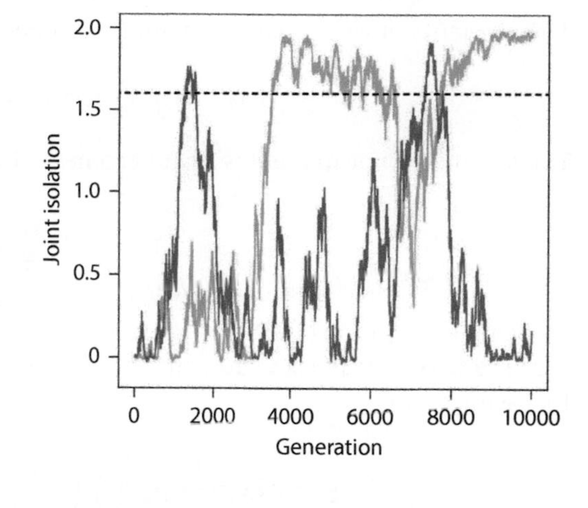

Figure 19.2 Two pairs of simulated evolutionary trajectories for sexual isolation. The dotted line corresponds to $JI = 1.6$ which represents a substantial level of isolation (see text). In each pair of populations (purple and green) joint isolation (JI) waxes and wanes as trait means of each population drift away from and towards each other. Parameter values in both examples are $G = H = 0.6$, $\omega = 50$, $v = 5$, and $N_e = 1000$.

From Uyeda et al. (2009).

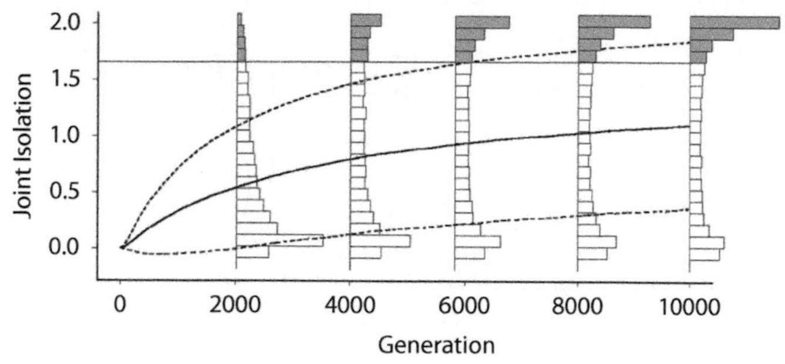

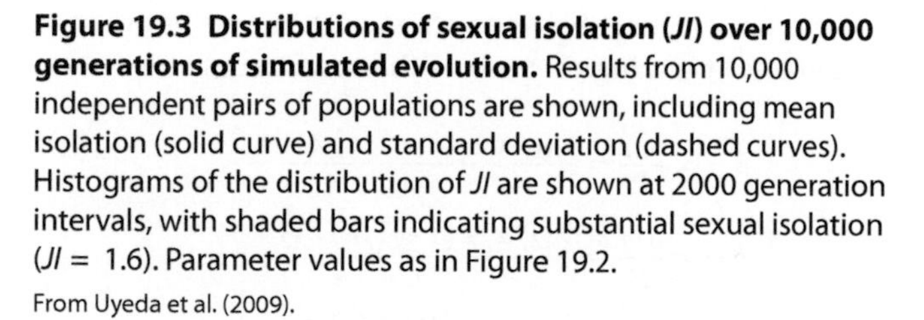

Figure 19.3 Distributions of sexual isolation (JI) over 10,000 generations of simulated evolution. Results from 10,000 independent pairs of populations are shown, including mean isolation (solid curve) and standard deviation (dashed curves). Histograms of the distribution of JI are shown at 2000 generation intervals, with shaded bars indicating substantial sexual isolation ($JI = 1.6$). Parameter values as in Figure 19.2.

From Uyeda et al. (2009).

traits plus the assumption that selection acts only on ornaments means that the evolutionary equilibrium will be a plane rather than a line. We wish to determine whether this increase in dimensionality will increase the rate of evolution of joint isolation and increase the extent of sexual radiation.

In the bivariate case, $\boldsymbol{G}$, $\boldsymbol{B}$, and $\boldsymbol{H}$ are 2×2 matrices. In the simulations that follow, bivariate inheritance is modelled with the following versions of those matrices:

$$\boldsymbol{G} = \begin{pmatrix} G_{11} & G_{12} \\ G_{12} & G_{22} \end{pmatrix} = \begin{pmatrix} 0.6 & 0 \\ 0 & 0.6 \end{pmatrix} = \boldsymbol{H} \tag{19.23a}$$

$$\boldsymbol{B} = \begin{pmatrix} B_{11} & B_{12} \\ B_{12} & B_{22} \end{pmatrix} = \begin{pmatrix} 0.48 & 0 \\ 0 & 0.48 \end{pmatrix}. \tag{19.23b}$$

The elements of these matrices define the genetic correlation between the sexes,

$$r_g = B_{11}/\sqrt{G_{11}H_{11}} = B_{22}/\sqrt{G_{22}H_{22}}. \tag{19.23c}$$

The key matrix for modelling drift becomes a 4 × 4 matrix,

$$\boldsymbol{V} = \frac{1}{N_e}\begin{pmatrix} \boldsymbol{G} & \boldsymbol{B} \\ \boldsymbol{B} & \boldsymbol{H} \end{pmatrix}, \tag{19.23d}$$

compare with Lande (1981) expression (17).

In the two traits per sex case, our deterministic response to selection equations take the following form:

$$\Delta\bar{\boldsymbol{z}}_{\boldsymbol{A}} = \frac{1}{2}\boldsymbol{G}\boldsymbol{\beta}_{\boldsymbol{A}} = \frac{1}{2}\begin{pmatrix} G_{11} & G_{12} \\ G_{12} & G_{22} \end{pmatrix}\begin{pmatrix} \beta_{A1} \\ \beta_{A2} \end{pmatrix} \tag{19.24a}$$

$$\Delta\bar{\boldsymbol{y}}_{\boldsymbol{A}} = \frac{1}{2}\boldsymbol{B}\boldsymbol{\beta}_{\boldsymbol{A}} = \frac{1}{2}\begin{pmatrix} B_{11} & B_{12} \\ B_{12} & B_{22} \end{pmatrix}\begin{pmatrix} \beta_{A1} \\ \beta_{A2} \end{pmatrix} \tag{19.24b}$$

$$\Delta\bar{\boldsymbol{z}}_{\boldsymbol{B}} = \frac{1}{2}\boldsymbol{G}\boldsymbol{\beta}_{\boldsymbol{B}} = \frac{1}{2}\begin{pmatrix} G_{11} & G_{12} \\ G_{12} & G_{22} \end{pmatrix}\begin{pmatrix} \beta_{B1} \\ \beta_{B2} \end{pmatrix} \tag{19.24c}$$

$$\Delta\bar{\boldsymbol{y}}_{\boldsymbol{B}} = \frac{1}{2}\boldsymbol{B}\boldsymbol{\beta}_{\boldsymbol{B}} = \frac{1}{2}\begin{pmatrix} B_{11} & B_{12} \\ B_{12} & B_{22} \end{pmatrix}\begin{pmatrix} \beta_{B1} \\ \beta_{B2} \end{pmatrix}. \tag{19.24d}$$

Assuming no correlational selection between traits within and between sexes, the total directional selection gradients acting on each of the ornaments are:

$$\beta_{A1} = \frac{S_{A1}}{P} = \frac{\bar{y}_{A1}/\alpha - (1 + 1/\alpha)\,\bar{z}_{A1} + \theta}{\omega_A} \tag{19.25a}$$

$$\beta_{A2} = \frac{S_{A2}}{P} = \frac{\bar{y}_{A2}/\alpha - (1 + 1/\alpha)\,\bar{z}_{A2} + \theta}{\omega_A} \tag{19.25b}$$

$$\beta_{B1} = \frac{S_{B1}}{P} = \frac{\bar{y}_{B1}/\alpha - (1 + 1/\alpha)\,\bar{z}_{B1} + \theta}{\omega_B} \tag{19.25c}$$

$$\beta_{B2} = \frac{S_{B2}}{P} = \frac{\bar{y}_{B2}/\alpha - (1 + 1/\alpha)\,\bar{z}_{B2} + \theta}{\omega_B}, \tag{19.25d}$$

where S_{Ai} and S_{Bi} denote, respectively, the selection differentials for ornaments in populations A and B, $\alpha = \nu/\omega$, where ν is the strength of female preference for a particular male ornament, ω is the width of the Gaussian natural selection function acting on ornaments, and θ is the scalar optimum of the Gaussian natural selection function acting on ornaments.

Setting these selection gradients equal to zero, we find that the conditions for equilibrium (i.e., no deterministic evolution) are:

$$\bar{y}_{A1} = (\alpha + 1)\,\bar{z}_{A1} + \alpha\theta \tag{19.26a}$$

$$\bar{y}_{A2} = (\alpha + 1)\,\bar{z}_{A2} + \alpha\theta \tag{19.26b}$$

$$\bar{y}_{B1} = (\alpha + 1)\,\bar{z}_{B1} + \alpha\theta \tag{19.26c}$$

$$\bar{y}_{B2} = (\alpha + 1)\,\bar{z}_{B2} + \alpha\theta, \tag{19.26d}$$

which are four lines in trait space. In contrast, the coevolution of $\bar{y}_{A2}$ as a function of $\bar{y}_{A1}$ or $\bar{z}_{A2}$ as a function of $\bar{z}_{A1}$ are not linearly constrained, and likewise in population B. In these trait dimensions, the evolutionary equilibrium is a plane.

The combined deterministic and stochastic changes in trait means from generation to generation are:

$$\Delta\bar{\mathbf{z}}_{\mathbf{A}} = \frac{1}{2}\mathbf{G}\boldsymbol{\beta}_{\mathbf{A}} + N\left[0, G/N_e\right] = \frac{1}{2}\begin{pmatrix} G_{11} & G_{12} \\ G_{12} & G_{22} \end{pmatrix}\begin{pmatrix} \beta_{A1} \\ \beta_{A2} \end{pmatrix} + N\left[\begin{pmatrix} 0 \\ 0 \end{pmatrix}, \frac{1}{N_e}\begin{pmatrix} G_{11} & G_{12} \\ G_{12} & G_{22} \end{pmatrix}\right] \tag{19.27a}$$

$$\Delta\bar{\mathbf{y}}_{\mathbf{A}} = \frac{1}{2}\mathbf{B}\boldsymbol{\beta}_{\mathbf{A}} + N\left[0, H/N_e\right] = \frac{1}{2}\begin{pmatrix} B_{11} & B_{12} \\ B_{12} & B_{22} \end{pmatrix}\begin{pmatrix} \beta_{A1} \\ \beta_{A2} \end{pmatrix} + N\left[\begin{pmatrix} 0 \\ 0 \end{pmatrix}, \frac{1}{N_e}\begin{pmatrix} H_{11} & H_{12} \\ H_{12} & H_{22} \end{pmatrix}\right] \tag{19.27b}$$

$$\Delta\bar{\mathbf{z}}_{\mathbf{B}} = \frac{1}{2}\mathbf{G}\boldsymbol{\beta}_{\mathbf{B}} + N\left[0, G/N_e\right] = \frac{1}{2}\begin{pmatrix} G_{11} & G_{12} \\ G_{12} & G_{22} \end{pmatrix}\begin{pmatrix} \beta_{B1} \\ \beta_{B2} \end{pmatrix} + N\left[\begin{pmatrix} 0 \\ 0 \end{pmatrix}, \frac{1}{N_e}\begin{pmatrix} G_{11} & G_{12} \\ G_{12} & G_{22} \end{pmatrix}\right] \tag{19.27c}$$

$$\Delta\bar{\mathbf{y}}_{\mathbf{B}} = \frac{1}{2}\mathbf{B}\boldsymbol{\beta}_{\mathbf{B}} + N\left[0, H/N_e\right] = \frac{1}{2}\begin{pmatrix} B_{11} & B_{12} \\ B_{12} & B_{22} \end{pmatrix}\begin{pmatrix} \beta_{B1} \\ \beta_{B2} \end{pmatrix} + N\left[\begin{pmatrix} 0 \\ 0 \end{pmatrix}, \frac{1}{N_e}\begin{pmatrix} H_{11} & H_{12} \\ H_{12} & H_{22} \end{pmatrix}\right]. \tag{19.27d}$$

This model is a straight-forward bivariate extension of the model studied by Uyeda et al. 2009 and—not surprisingly—simulations revealed that the dynamic system is a drift-selection balance. In each population, the ornament means are drawn towards their natural selection optima (which may be separated in a trait space plane) by stabilizing selection and are pulled away from their optima by sexual selection on the ornaments.

We can specify the expected effects of stabilizing selection on ornaments in the absence of sexual selection and use those expectations as a benchmark. With no sexual selection, the variance among lineages in ornament mean is expected to equilibrate at

$$Var\left(\bar{z}_{\infty}\right) = \frac{\Omega}{2N_e}, \tag{19.28}$$

where $\Omega = \omega_z + P$ (13.11). We use this expression to gauge the observed extent of the sexual radiation, which should exceed (19.28). Substituting in parameter values used in the simulations ($\omega = 50, P = 1, N_e = 1000$), we expect the among-lineage ornament means to equilibrate at a value of $Var\left(\bar{z}\right) = 0.0255$. The expected standard deviation is $\sqrt{0.0255} \approx 0.16$ and the expected 95% confidence limits are $\pm\,(1.96 * 0.16) = 0.31$, which is less than one-third of a within-population phenotypic standard deviation. In other words, in the absence of sexual selection, drift-selection balance would produce an exceedingly modest radiation of ornaments.

Turning to the specification of joint isolation, define the following Euclidean distances between vectors of preference and ornament means:

$$d_{AA} = \sqrt{(\bar{y}_{A1} - \bar{z}_{A1})^2 + (\bar{y}_{A2} - \bar{z}_{A2})^2} \tag{19.29a}$$

$$d_{AB} = \sqrt{(\bar{y}_{A1} - \bar{z}_{B1})^2 + (\bar{y}_{A2} - \bar{z}_{B2})^2} \tag{19.29b}$$

$$d_{BB} = \sqrt{(\bar{y}_{B1} - \bar{z}_{B1})^2 + (\bar{y}_{B2} - \bar{z}_{B2})^2} \tag{19.29c}$$

$$d_{BA} = \sqrt{(\bar{y}_{B1} - \bar{z}_{A1})^2 + (\bar{y}_{B2} - \bar{z}_{A2})^2}. \tag{19.29d}$$

Our expressions for the probabilities of mating within and between populations become:

$$\pi_{AA} = c\ exp\left\{-(d_{AA})^2/2\Sigma\right\} \tag{19.30a}$$

$$\pi_{AB} = c\ exp\left\{-(d_{AB})^2/2\Sigma\right\} \tag{19.30b}$$

$$\pi_{BB} = c\ exp\left\{-(d_{BB})^2/2\Sigma\right\} \tag{19.30c}$$

$$\pi_{BA} = c\ exp\left\{-(d_{BA})^2/2\Sigma\right\}, \tag{19.30d}$$

where, for simplicity, $\Sigma = v + Q + P$ in both populations. Using our expression from the previous section, the index of joint isolation at any given generation is, as before,

$$JI = \pi_{AA} + \pi_{BB} - \pi_{AB} - \pi_{BA}.$$

We made a number of simplifying assumptions in the simulations, many of which are bivariate extensions of assumptions used in the last section: the 2×2 G- and H-matrices are diagonal with genetic variances of 0.6 and zero genetic covariance. The 2×2 B-matrix is likewise diagonal so that genetic covariance between ornaments and preferences with the same numeric subscript is 0.48. This assumption means that genetic correlation between the sexes is 0.8 when trait pairs have the same numeric subscript and zero when they do not. In the specification of drift, mate choice, and selection, the following variables are the same in both populations and regardless of trait numeric subscripts: $N_e = 1000$, $\theta = 0$, $\omega = 50$, $v = 5$, $\alpha = v/\omega = 0.1$, $P = 1$, $Q = 1, \Sigma = v + Q + P = 7$.

In the typical simulation example shown in Figure 19.4, the ornament and preference means begin their evolution at the trait optima, at the centre of the figure. The A population is shown in purple, the B population is shown in green. For the next 10,000 generations the ornament and preference means of each population pair remain close to each other, drawn close by the mate choice parameter v. We show each pair as a dumbbell with a black line connecting the two means. Arnold and Houck (2016) referred to the movement of these dumbbells as a *phenotypic tango* in which dance partners dance about the optimum, systematically drawn to the optimum by stabilizing selection and randomly moving away by drift. For animations of their figures, comparable to Figure 19.4, see https://phenotypicevolution.com/?p=221.

From (19.28) we see that in the absence of sexual selection, the ornament radiation in Figure 19.4 should have equilibrated with a radius of about $0.31\sqrt{P} = 0.31$, which is about

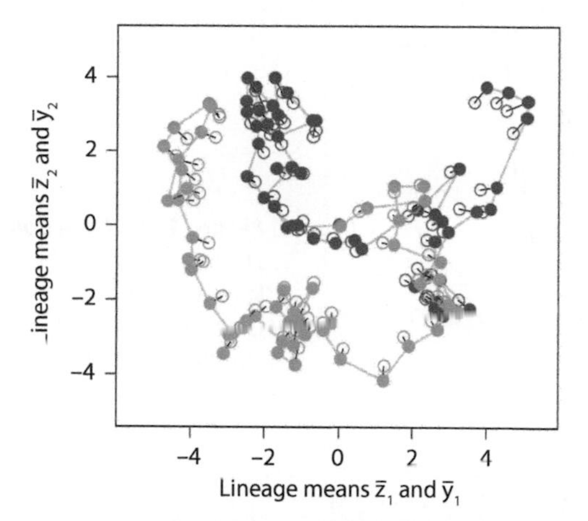

Figure 19.4 Joint evolution of two ornament traits and two mating preference traits undergoing the Fisher-Lande process in a finite population with natural selection on ornaments but not on preferences. Ornament means are shown in open circles and preference means are shown in solid. Ornament and preference means in the same generation are connected by solid black lines, but are often not visible because the means of the two sexes are close together. Means are plotted every 200 generations and preference means are connected by grey lines. One population is shown in green, the other in purple. The simulation begins with the means of both populations at the origin and ends with the means far apart in the upper left and upper right corners of the plot. The trajectories result from 10,000 generations of interactions between drift (N_e = 1000) and selection. The genetic variance of male and female traits is 0.6. Traits are genetically uncorrelated within sexes. Genetic correlations between the sexes are 0.6 for traits with the same subscript and 0 otherwise. For both ornaments, $\omega = 50$ (with no correlational selection), and $P = 1$; for both preferences, $v = 5$ (with no correlational preferences), $Q = 1$, and $\alpha = 0.1$. The simulation begins with all traits at the values of the optimum of the ornaments traits (x-axis value = 0, y-axis value = 0), at the centre of the figure.

one tenth of the radius we observe (about 4.0). In other words, most of the radiation that we observe in Figure 19.4 can be attributed to amplification of ornament evolution by sexual selection.

A typical time course for the evolution of sexual isolation is shown in Figure 19.5. As we saw in the univariate model of this same process (Figure 19.2), joint isolation waxes and wanes during the first few thousand generations, as the trait means of the two populations vacillate between moving away from each other during the first 3000 generations, towards each other from generation 3000 to 5000, and then away from each other. Eventually however, in the last 3000 generations the trend is towards greater and greater separation, as *JI* moves towards its asymptotic value of 2.0. As we pointed out in the last section, a *JI* value of 1.6 (shown as a dashed line in Figure 19.5) is characteristic of sympatric, fully isolated species and so can be taken as a empirical benchmark for sexual speciation.

The *haystack plot* of divergence for pairs of bivariate ornament means (Figure 19.6a) leaves off the symbols that were used in Figure 19.4 to indicate the positions of bivariate ornament and preference means for each lineage. At the scale used in Figure 19.6a those symbols and the distance between them would be barely visible. Notice too that the extent of the radiation of replicates after 10,000 spans 20 within-population phenotypic standard deviations ($\pm 20\sqrt{P}$), which is appreciably larger than the span achieved by a single pair of populations after 5000 generations (Figure 19.4).

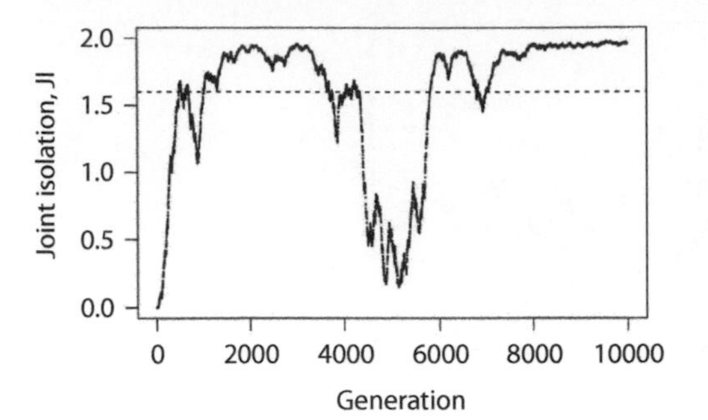

Figure 19.5 Time course for joint isolation, *JI*. The horizontal dashed line shows the lower level of *JI* that is characteristic of sympatric, isolated species. *JI* values computed from the simulation shown in Figure 19.4.

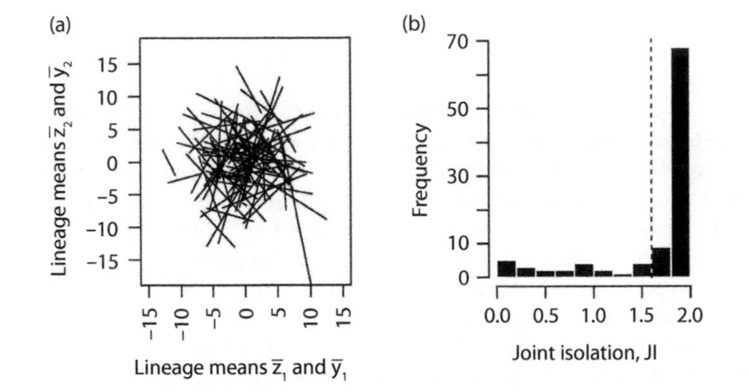

Figure 19.6 Joint evolution of two ornament traits and two preference traits in replicate pairs of populations undergoing the Fisher-Lande process in a finite populations with natural selection on ornaments but not on preferences. Results are shown after 10,000 generations of simulation using parameters listed in Figure 19.4. (a) Haystack plot showing divergence between ornament means in 100 replicate pairs of populations. Each line connects the ornament means of a replicate pair of populations. (b) Histogram of joint isolation measures for the 100 pairs of populations. The dashed line shows the expected level of reproductive isolation ($JI = 1.6$) expected in sympatric pairs; 77% of pairs have achieved that level of isolation or higher. This distribution has a mean of 1.61 and a variance of 0.30.

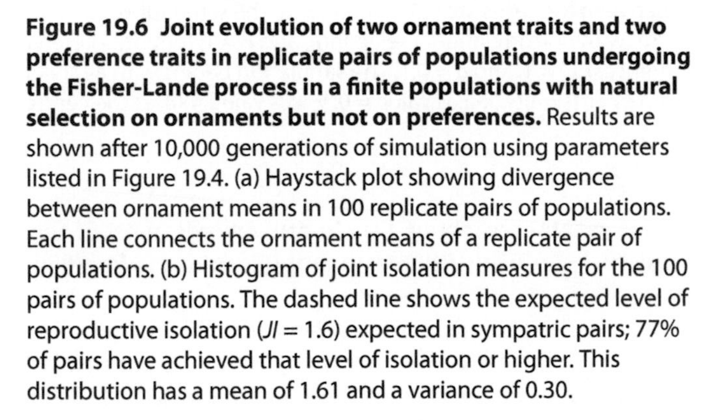

In a simulation with 100 pairs of populations (Figure 19.6), we see that 77% of replicate pairs achieved a joint isolation level of 1.6 or higher after 10,000 generations.

Comparing the present model with the previous one (Section 19.2b), we see a good case for the argument that increasing the number of traits under sexual selection will increase the rate of sexual speciation. With just one trait in each sex, about 45% of lineages achieved the speciation cut-off point in 10,000 generations (Table 19.1). Using the same parameter values, but increasing to two traits per sex, the average increased to 77% after 10,000 generations. The argument also makes sense on intuitive grounds. Increasing the dimensions of the dance floor from a line to a plane means many more opportunities for trait divergence. That increase in opportunities translates into an increased rate of evolution of sexual isolation.

19.2.3 *Dimensionality of sexual isolation and sexual speciation*

Courtship and other forms of premating communication between mating partners usually involve multiple sensory modalities. Consequently, the breakdown in sexual communication that is reflected in sexual isolation is likely to involve multiple phenotypic traits in both sexes. Although their results may appear to be a magic trick, Hohenlohe and Arnold (2010) provide evidence that sexual isolation is based on two to four trait dimensions in *Drosophila*, cichlid fishes and plethodontid salamanders. Because the operations for estimating and interpreting sexual isolation may be unfamiliar to some readers, we will begin with a concrete, empirical example.

An example of a survey of sexual isolation is shown in Table 19.2. In such a survey the idea is to assess sexual isolation in the laboratory between sexual partners drawn from allopatric populations. The sexual isolation that is assessed gives an indication of whether sexual barriers exist that might impede mating should the populations come into sympatry. In this particular example, joint isolation was assessed among nine allopatric populations of salamanders in the *Desmognathus ochrophaeus* complex in the Appalachian Mountains of eastern USA (Tilley et al. 1990). Joint isolation ranged from 0.20 ± 0.15 s.e. (SI × JK) to 1.50 ± 0.12 s.e. (MR × HP) and averaged 0.72 ± 0.31 sd (McCullagh and Nelder 1989; Arnold et al. 1996). In other words, sexual isolation covered the gamut from undetectable to values we might expect from sympatric, isolated species.

In a survey such as this, isolation is assessed, but the phenotypic traits responsible for isolation are usually not identified (but see Verrell and Arnold 1989). On intuitive grounds, however, we can see that the survey in conjunction with the triple Gaussian model (Arnold et al. 1996) provides insight into the average trait values that produced the pairwise values of isolation we see in Table 19.2. A pair of populations, such as SI × JK, with a low value of *JI*, apparently has trait means that are close together in trait space, whereas a pair such as MR × HP, with a high value for *JI* apparently has trait means that are much further apart. Since these pairs represent the *JI* extremes in the sample, the trait means of all the other pairs must lie in between the extremes. Encouraged by this insight, we can imagine estimating the trait differences of all the pairs in Table 19.1 and from these estimating the trait values themselves. The hypothetical values are not actual traits

Table 19.2 A survey of sexual isolation among nine allopatric populations of salamanders in the *Desmognathus ochrophaeus* complex. Each entry reports the number of successful matings (k_{ij}, numerator) in a number of trials (n_{ij}, denominator). In each trial, a single female was housed with a male overnight. The population identity of females is given at the left, the population identity of males is given at the top. From Arnold et al. (1996).

	MR	MM	RB	SI	HP	UN	JK	WA	WR
MR	106/180	17/30	1/30	0/30	3/30	6/30	0/0	5/30	0/0
MM	5/30	124/210	1/30	2/30	2/30	10/30	0/30	7/30	0/0
RB	1/30	12/30	150/240	15/30	45/90	13/30	0/0	13/30	0/0
SI	1/30	1/30	4/30	163/270	5/60	2/30	20/30	14/30	10/30
HP	0/30	7/30	17/90	9/60	162/300	9/30	3/30	8/30	0/0
UN	2/30	17/30	8/30	3/30	6/30	147/240	0/30	10/30	7/30
JK	0/0	0/30	0/0	18/30	1/30	6/30	68/120	0/0	0/0
WA	1/30	8/30	7/30	9/30	4/30	7/30	0/0	113/210	0/30
WR	0/0	0/0	0/0	12/30	0/0	6/30	0/0	4/30	52/90

but rather 'latent trait' values that we infer from the survey table. This argument is more than a thought experiment, however, because we can use it as the logic behind a maximum likelihood approach to estimate the values of the latent traits. Furthermore, we can use maximum likelihood to help us decide whether a single dimension of latent traits produces the values in the table, or whether we need to invoke two or more dimensions of latent trait variation.

Following the argument just developed, Hohenlohe and Arnold (2010) devised a maximum likelihood approach and used it both to estimate the latent trait means that produced Table 19.2 and to decide whether the means were arrayed in two or more latent trait dimensions, rather than just one. Moreover, they used this approach to analyse a total of nine surveys of sexual isolation, representing studies of *Drosophila* and fish, as well as salamanders.

Turning to the details of their approach, recall that the when mate choice is based on a single pair of traits, the probability of mating between a randomly chosen female from population i and a randomly chosen male from population j is

$$\pi_{ij} = c_{ij}\, exp\left\{-\left(\bar{z}_j - \bar{y}_i\right)^2/2\Sigma_{ij}\right\}. \tag{19.31}$$

Extending this model to the case of multivariate mate choice based on a vector of d traits in each sex, the absolute preference function becomes multivariate Gaussian with a $d \times d$ interaction (matching) matrix $\boldsymbol{N}$ (upper case nu), analogous to v in the univariate case,

$$\boldsymbol{N} = \begin{pmatrix} v_{11} & v_{12} & \cdots & v_{1d} \\ v_{21} & v_{22} & \cdots & v_{2d} \\ \vdots & \vdots & \ddots & \vdots \\ v_{d1} & v_{d2} & \cdots & v_{dd} \end{pmatrix}, \tag{19.32a}$$

where columns and rows denote male and female traits, respectively. Let the indices p and q denote particular traits in females and females, respectively. Then v_{pq} denotes the female tolerance for mating when the trait difference is between her preference value y_p and his ornament value z_q. That is to say, y_p is the phenotypic value of the female's most preferred mate. The probability that she will mate with a male falls off as a Gaussian function with an optimum of y_p as the male's phenotype, z_q, deviates from that optimum. When $p \neq q$ the trait match is between the female's most preferred value of a particular male trait and the male's phenotypic values for some other trait. We will assume that the matrix $\boldsymbol{N}$ is the same in all populations.

The probability of mating between an individual male from population j with phenotype vector $\mathbf{z}_j$ and an individual female from population i with phenotype vector $\mathbf{y}_i$ is

$$\psi(\mathbf{z}|\mathbf{y}) \propto exp\left\{-\frac{1}{2}\left(\mathbf{z}_j - \mathbf{y}_i\right)^{\mathrm{T}}\boldsymbol{N}^{-1}\left(\mathbf{z}_j - \mathbf{y}_i\right)\right\}, \tag{19.32b}$$

which is the multivariate analogue of equation 2 in Arnold et al. (1996). We assume that the traits have multivariate normal distributions within populations with means $\bar{\mathbf{y}}_i$ and $\bar{\mathbf{z}}_j$, and phenotypic variance-covariance matrices $\boldsymbol{Q}_i$ and $\boldsymbol{P}_j$. The probability of mating between a randomly chosen male from population j and a randomly chosen female from population i is

$$\pi_{ij} = c_{ij}\, exp\left\{-\frac{1}{2}\left(\bar{\mathbf{z}}_j - \bar{\mathbf{y}}_i\right)^{\mathrm{T}} \Sigma_{ij}^{-1} \left(\bar{\mathbf{z}}_j - \bar{\mathbf{y}}_i\right)\right\}, \tag{19.33}$$

where $\bar{\mathbf{z}}_j$ and $\bar{\mathbf{y}}_i$ are, respectively, the column vectors of trait means in males of population j and of trait means in females of population i, and $\mathbf{\Sigma}_{ij} = \mathbf{Q}_i + v_{ij} + \mathbf{P}_j$.

To solve for the trait axes latent in the set of π_{ij} values, we assume that the covariance term Σ_{ij} is constant across all comparisons of population i with population j. In particular, we assume that the matrices $\mathbf{Q}_i$, $\mathbf{N}$, and $\mathbf{P}_j$ are constant and consequently the matrix $\mathbf{\Sigma}_{ij}$ is constant across all populations. Because $\mathbf{\Sigma}_{ij}$ is a variance-covariance matrix, common to all j population combinations, it can be decomposed into a set of orthogonal eigenvectors with non-negative eigenvalues. We use those eigenvectors and eigenvalues to scale and rotate $\mathbf{y}_i$ and $\mathbf{z}_j$ into a new set of d latent traits. In this rotated space, (19.33) simplifies to

$$\pi_{ij} = c_{ij}\, exp\left\{-\frac{1}{2}\left(\bar{\mathbf{z}}_j - \bar{\mathbf{y}}_i\right)^{\mathrm{T}}\left(\bar{\mathbf{z}}_j - \bar{\mathbf{y}}_i\right)\right\}. \tag{19.34}$$

The analyses that follow were conducted in this rotated space.

To take the maximum likelihood approach, we need an expression that relates the π_{ij} values, and hence latent trait means, to the probability of observing the data. Let the probability of observing k_{ij} matings in n_{ij} trials be a binomial function of π_{ij} (McCullagh and Nelder 1989, pp. 439–450). The log likelihood that the entire set of population mean trait vectors ($\mathbf{Z}$, $\mathbf{Y}$) will produce the observed table of π_{ij} values (e.g., Table 19.2) is then

$$\ln L\,(\mathbf{Z},\mathbf{Y}) = \sum_{i,j}\left[ln\begin{pmatrix} n_{ij} \\ k_{ij}\end{pmatrix} + k_{ij}\, ln\pi_{ij} + \left(n_{ij} - k_{ij}\right)\ln\left(1 - \pi_{ij}\right)\right]. \tag{19.35}$$

This expression was used to solve for the values of male and female traits, $\mathbf{Z}$ and $\mathbf{Y}$, that maximize the likelihood of the π_{ij} observations. Furthermore, we can solve for the likelihood of the observations when $\mathbf{Z}$ and $\mathbf{Y}$ are 1, 2, 3, ... n-element vectors and use the corresponding cAIC values to compare the fits of these models that differ in trait dimensions. In this way, Hohenlohe and Arnold (2010) determined the best-fitting model for trait dimensionality in nine different data sets.

The surprising take-away message is that only a few orthogonal trait dimensions were required to account for each table of mating results. In *Desmognathus* salamanders (Table 19.2) three trait dimensions gave the best fit, in *Drosophila* two trait dimensions were usually optimal (two was best in six data sets, four was the best in a seventh set), and likewise, two trait dimensions gave the best fit in a cichlid fish data set. To simplify interpretation, in each dataset, Hohenlohe and Arnold (2010) rotated the latent trait axes so that they were orthogonal and ordered them from most to least variance explained, so they were analogous to principal components. This analogy makes it clear that each latent trait is a linear combination of traits rather than a single trait. In other words, each latent trait represented an ensemble of correlated traits. Such an ensemble is likely to represent a functional complex of traits (Chapter 7) with all of its associated attributes (e.g., trait interactions molded by correlational selection).

The details of the best-fitting trait values in *Desmognathus* are shown in Figure 19.7. The figure shows the first two orthogonal dimensions of variation in population trait means that account for most of the variation in the probability of between-population matings.

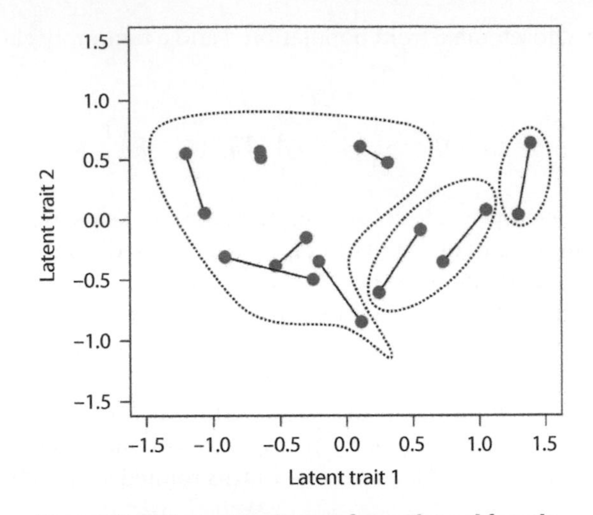

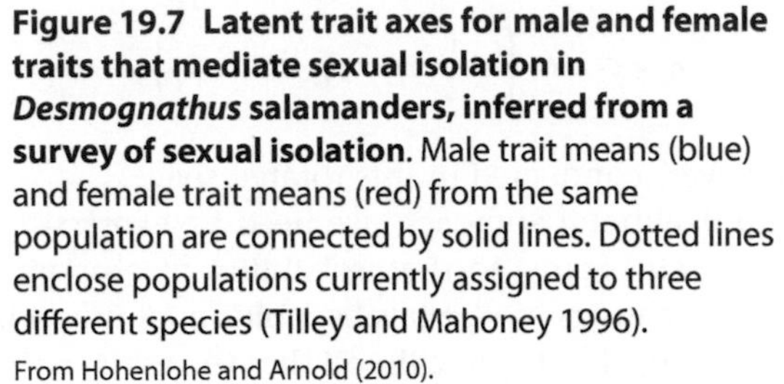

Figure 19.7 Latent trait axes for male and female traits that mediate sexual isolation in *Desmognathus* salamanders, inferred from a survey of sexual isolation. Male trait means (blue) and female trait means (red) from the same population are connected by solid lines. Dotted lines enclose populations currently assigned to three different species (Tilley and Mahoney 1996).
From Hohenlohe and Arnold (2010).

Sexual isolation between particular pairs of populations is a function of the separation between the male and female means. Female means showed more divergence than male means (red and blue dots in Figure 19.6), a pattern that was observed in seven out of nine datasets. This pattern of greater female divergence is predicted by Lande's (1981) model with no selection on preferences, if stabilizing selection on male traits is strong relative to female preferences (ω small relative to v) or by a model with selection on female preferences, if stabilizing selection on preferences is weaker than stabilizing selection on ornaments (Arnold and Houck 2016).

Although inference about the traits that mediate sexual isolation is usually difficult from surveys alone, in the case of *Desmognathus,* observations of mating trials within and among populations provide hints that guide interpretation and suggest experiments. Verrell and Arnold (1989) observed mating trials within and between the same set of nine populations (Table 19.1) that were surveyed for sexual isolation by Arnold et al. (1996). In each mating trial, they scored progress through three stages of courtship (no contact, pursuit, persuasion), as well as the duration of the persuasion stage, which can include an attempt at sperm transfer. These stages of courtship are easily recognized during observations. The characteristic transition from pursuit to persuasion is shown in Figure 19.8 for *Plethodon metcalfi,* a species related to *Desmognathus* and with similar courtship. Verrell and Arnold (1989) found that courtship tended to break down at particular stages or transitions in courtship that implicated particular trait complexes in sexual isolation (e.g., pheromone delivery and perception, male displays during the pursuit phase).

Hohenlohe and Arnold (2010) looked for correlations between latent isolation traits and the behavioural indicators of isolation scored by Verrell and Arnold (1989)

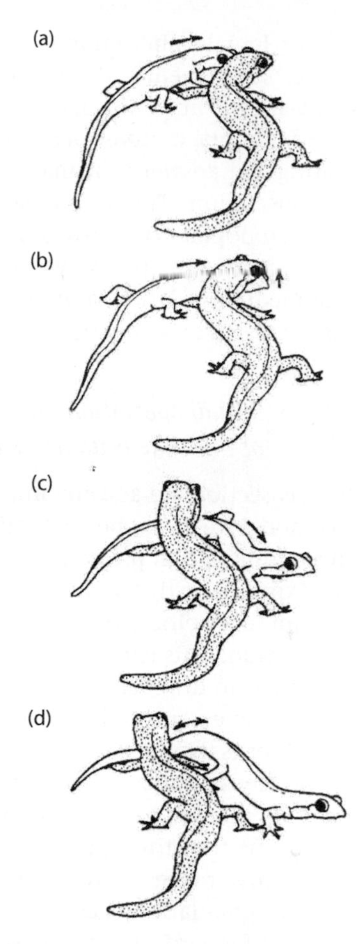

Figure 19.8 The transition from the pursuit to persuasion phase of courtship in the salamander *Plethodon metcalfi*. (a) After pursuing the female (stippled), the male (unstippled) touches her cheek with his snout. (b) The male crawls under her chin, lifting his head as he does so. (c) The male's body contacts her chin as he slides forward. (d) The male stops, while arching and undulating his tail under her chin. If the female steps astride the male's tail, the persuasion phase of courtship begins.
From Arnold (1976) with permission.

Table 19.3 Correlations between latent traits and phases of courtship in *Desmognathus*. The first three courtship measures are the proportions of mating trials between populations that terminated at the indicated courtship phase, given that the trial progressed to the previous phase. Time to persuasion is the mean time that elapsed before the persuasion phase began. Table entries show Spearman rank correlation coefficients (P values) between pairwise male-female distances along latent trait axes in the best-fit model and measures of the behavioural basis of sexual isolation.
* indicates significance at the 0.05 level after correcting for false discovery rate. Behavioural data from Verrell & Arnold (1989), table from Hohenlohe and Arnold (2010).

Latent trait	No activity	Pursuit	Persuasion	Time to persuasion
1	0.307 (0.093)	0.569 (0.0008)*	0.403 (0.025)	−0.066 (0.782)
2	0.302 (0.099)	0.410 (0.022)	0.190 (0.306)	0.496 (0.028)
3	−0.088 (0.637)	0.089 (0.634)	0.099 (0.595)	−0.081 (0.734)

with some success. A tabular summary of the correlation analysis is shown in Table 19.3. Sexual communication is the essence of plethodontid courtship (Arnold et al. 2017). Consequently, breakdown at particular phases implicates particular male and female traits as well as particular sensory modalities in sexual isolation. Thus, breakdown after the pursuit phase implicates incompatibility in pheromone communication (male pheromones and/or female pheromone perception) which occurs primarily during the persuasion phase. Consequently, the pattern of behavioural correlations with latent trait 1 may reflect a breakdown in chemical communication.

In summary, a survey of sexual isolation in conjunction with a model of trait-mediated mating can answer fundamental questions about the dimensionality of sexual isolation and speciation. The additional element of observing interactions during mating trials between populations provides an empirical path to identifying the traits that confer isolation. Although an observational study by itself may not conclusively identify the traits that mediate isolation, as in the *Desmognathus* case, such studies generate hypotheses that can be tested experimentally.

19.2.4 *Sexual speciation and radiation arising from drift-selection balance of multiple ornaments and sexual preferences*

In this section we add the important element of stabilizing selection on preferences to the model we developed in Section 19.2.2. This addition is important because it addresses the argument that preferences are unlikely to be selectively neutral (Pomiankowski 1987). We expect that this addition of stabilizing selection will collapse the evolutionary equilibrium to a point, from a line (Lande 1981, Uyeda et al. 2009) or from a plane, as in the last section. This expectation of a point equilibrium raises the primary question for this section. Can drift-selection balance about a point equilibrium support a sexual radiation of realistic extent and promote sexual speciation? As we shall see, the answer to both questions is 'No!'

The scenario that we have in mind for this mode is the case of two sister populations, *A* and *B*, that have recently become geographically separated. The recent origin of allopatry means that their phenotypic trait means, including those for male ornaments and female preferences, have not diverged. The fact of allopatry means that gene flow between the two populations has ceased. In our model and simulations, we follow the two populations forward in time to examine any divergence in ornaments and preferences that might occur and to plot the time course of evolving sexual isolation.

By including selection on female preferences, our expression for combined deterministic response to selection and genetic drift for two ornaments, z_1 and z_2, and two preferences, y and y_2, in population *A* change from the responses given by (12.21) to

$$\begin{pmatrix} \Delta\bar{\mathbf{z}} \\ \Delta\bar{\mathbf{y}} \end{pmatrix} = \begin{pmatrix} \Delta\bar{z}_1 \\ \Delta\bar{z}_2 \\ \Delta\bar{y}_1 \\ \Delta\bar{y}_2 \end{pmatrix} = \frac{1}{2}\begin{pmatrix} \mathbf{G} & \mathbf{B} \\ \mathbf{B} & \mathbf{H} \end{pmatrix}\begin{pmatrix} \boldsymbol{\beta}_z \\ \boldsymbol{\beta}_y \end{pmatrix} + N\left[0, \frac{1}{2N_e}\begin{pmatrix} \mathbf{G} & \mathbf{B} \\ \mathbf{B} & \mathbf{H} \end{pmatrix}\right] \tag{19.36}$$

and similarly in population *B* ($\boldsymbol{\beta}_z$ and $\boldsymbol{\beta}_y$ are each two-element column vectors). See Arnold and Houck (2016) (appendix) for a derivation. In our simulations of this model, we let

$$\boldsymbol{G} = \boldsymbol{H} = \begin{pmatrix} 0.6 & 0 \\ 0 & 0.6 \end{pmatrix} \tag{19.37a}$$

and

$$\boldsymbol{B} = \begin{pmatrix} 0.48 & 0 \\ 0 & 0.48 \end{pmatrix}. \tag{19.37b}$$

In other words, there is no phenotypic or genetic covariance between the two ornaments, or between the two preferences, but there is a genetic correlation of 0.8 ($= 0.48/\sqrt{0.6 * 0.6}$) between z_1 and y_1 and likewise between z_2 and y_2.

Gaussian stabilizing natural selection acts on the ornaments and has optima at

$$\boldsymbol{\theta_z} = (\theta_{z1}\ \theta_{z2})^{\mathrm{T}} \tag{19.38a}$$

and width parameters

$$\boldsymbol{\omega_z} = \begin{pmatrix} \omega_{z11} & 0 \\ 0 & \omega_{z22} \end{pmatrix}. \tag{19.38b}$$

Similarly Gaussian stabilizing selection acts on the preferences and has optima at

$$\boldsymbol{\theta_y} = \left(\theta_{y1}\ \theta_{y2}\right)^{\mathrm{T}} \tag{19.38c}$$

and width parameters

$$\boldsymbol{\omega_y} = \begin{pmatrix} \omega_{y11} & 0 \\ 0 & \omega_{y22} \end{pmatrix}. \tag{19.38d}$$

Only natural selection acts on preferences in females, so the total directional selection gradient for the *ith* preference trait is

$$\beta_{yi} = Q^{-1}S_{yi} = \left(\omega_{yii} + Q\right)^{-1}\left(\theta_y - \bar{y}_i\right). \tag{19.39a}$$

Sexual selection acts after natural selection in males, so the total, combined directional selection gradient for the *ith* ornament is

$$\beta_{zi} = P^{-1}S_{zi} \approx \omega_{zii}^{-1}\left[\frac{\bar{y}_i^*}{\alpha_{ii}} - \left(1 + \frac{1}{\alpha_{ii}}\right)\bar{z}_i + \theta_z\right] \tag{19.39b}$$

where $\bar{y}_i^*$ denotes the mean of the *ith* preference after natural selection, which is either

$$\bar{y}_1^* = \left(\bar{y}_1\ \omega_{y11}\right) / \left(\omega_{y11} + Q_{11}\right) \tag{19.40a}$$

or

$$\bar{y}_2^* = \left(\bar{y}_2\ \omega_{y22}\right) / \left(\omega_{y22} + Q_{22}\right). \tag{19.40b}$$

In the absence of sexual selection, after many generations, the drift-selection balance process that we have specified will achieve an equilibrium in which ornament and preference lineage means each has a multivariate normal distribution with specified variance-covariance matrices. After many generations, as $t \to \infty$, these matrices are approximately

$$Var\left[\bar{\mathbf{z}}\left(t\right)\right] = \left(\boldsymbol{\omega_z} + \boldsymbol{P}\right)/2N_e \tag{19.41a}$$

$$Var\left[\bar{\mathbf{y}}\left(t\right)\right] = \left(\boldsymbol{\omega_y} + \boldsymbol{Q}\right)/2N_e \tag{19.41b}$$

(Arnold and Houck 2016). In the simulations, $\boldsymbol{P}$ and $\boldsymbol{Q}$ are both identity matrices. We will use these expressions in a plot of simulation results to show the extent of adaptive radiation expected in the absence of sexual selection.

To calculate sexual isolation (*JI*), we use the model and expressions developed in Section 19.2.2. In summary, we used the following values for parameters in the simulations: $G_{ii} = 0.6$, $B_{ii} = 0.48$, $N_e = 1000$, $\theta = \theta_{zi} = \theta_{yi} = 0$, $\omega = \omega_{zii} = \omega_{yii} = 50$, $\nu = \nu_{ii} = 5$, $\alpha = \nu/\omega = 0.1$, $\tau = 1$, $P_{ii} = Q_{ii} = 1, \Sigma = \nu + Q + P = 7$.

In a typical simulation run, the dumbbells, representing the ornament-to-preference means in each population, dance about the joint optimum value of x = 0, y = 0 on the plot axes (Figure 19.9). Stabilizing selection draws the means toward the joint optimum each generation, but finite parental sampling causes them to drift away. In the absence of sexual selection, we would expect the trait means to have reached an equilibrium after 10,000 generations in which the variance of trait means in either ornament would be

$$Var\left[\bar{z}\left(t\right)\right] = \left(\omega_z + P\right)/2N_e = 51/2000 = 0.0255, \tag{19.42}$$

and the standard deviation of trait means would be $\sqrt{0.0255} \approx 0.16$, and similarly for each of the preference trait means. The 95% confidence limits would be $\pm 1.96 * 0.16 \approx \pm 0.31$. Almost all the dispersion in Figure 19.9 is inside these limits. In any case, the dispersion is an order of magnitude less than we saw in an otherwise comparable simulation in which preferences were not constrained by stabilizing selection (Figure 19.4).

The take-away lesson from this model, with stabilizing selection on both ornaments and preferences, is that the radiations that are produced in both traits are very much smaller than iconic cases in nature (Arnold and Houck 2016). To produce a match to the diversification we see in birds of paradise, for example, we need to invoke either extremely weak stabilizing selection on preferences (e.g., $\omega \geq 10,000$) or unrealistically small population size (< 500).

Turning to the prospects for sexual isolation evolving in a model with stabilizing selection on both ornaments and preferences, we encounter another disappointing result (Figure 19.10). When the ornament and preference means of two populations are anchored to the same values for intermediate optima, the means do not diverge enough to produce substantial sexual isolation.

The results of running 100 replicates of the simulation example in Figure 19.10 are shown in Figure 19.11. Although the pattern of the haystack plot shown in Figure 19.11a

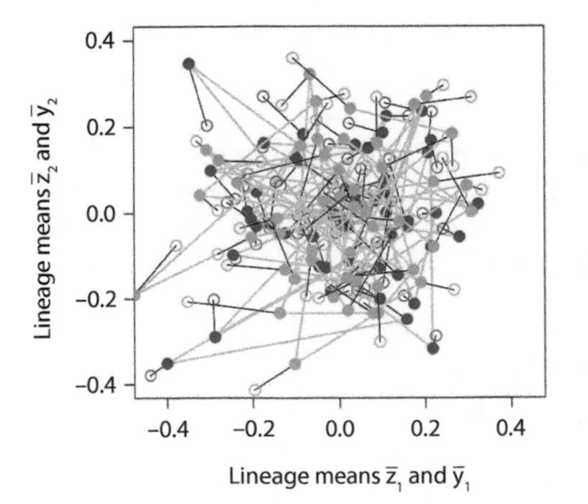

Figure 19.9 Joint evolution of two ornaments and two preferences in two populations undergoing the Fisher-Lande process with drift and stabilizing selection on ornaments and preferences. Same graphic conventions as in Figure 19.4, with same parameter values (column 19.2.5 in Table 19.1). The width parameter for Gaussian stabilizing selection on each of four traits is $\omega = 50$.

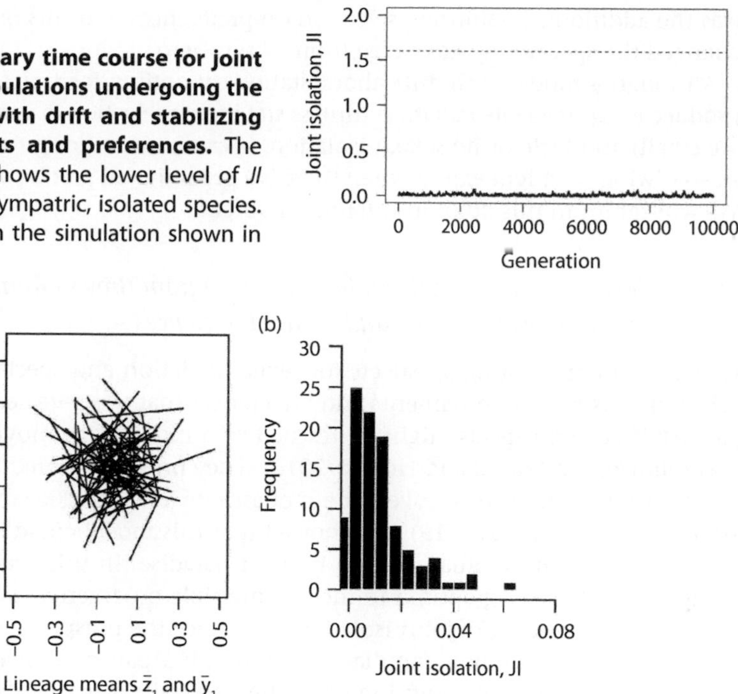

Figure 19.10 Evolutionary time course for joint isolation, *JI*, in two populations undergoing the Fisher-Lande process with drift and stabilizing selection on ornaments and preferences. The horizontal dashed line shows the lower level of *JI* that is characteristic of sympatric, isolated species. *JI* values computed from the simulation shown in Figure 19.9.

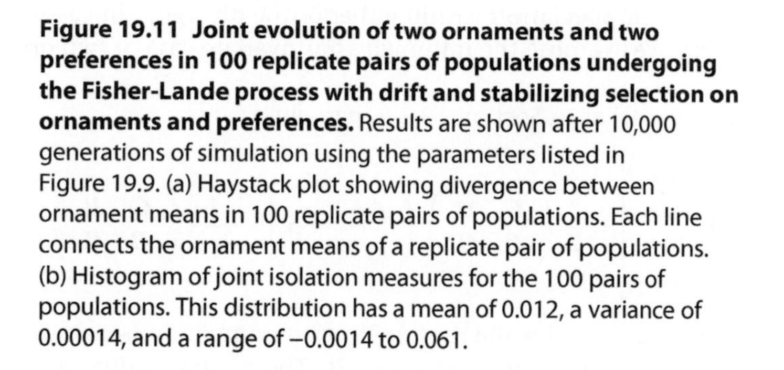

Figure 19.11 Joint evolution of two ornaments and two preferences in 100 replicate pairs of populations undergoing the Fisher-Lande process with drift and stabilizing selection on ornaments and preferences. Results are shown after 10,000 generations of simulation using the parameters listed in Figure 19.9. (a) Haystack plot showing divergence between ornament means in 100 replicate pairs of populations. Each line connects the ornament means of a replicate pair of populations. (b) Histogram of joint isolation measures for the 100 pairs of populations. This distribution has a mean of 0.012, a variance of 0.00014, and a range of −0.0014 to 0.061.

resembles the plot of coevolution when the preference traits are not under stabilizing selection, notice the great difference in scale. At generation 10,000 the lineages in Figure 19.11 have long since reached drift-stabilizing selection balance. The bounds of the radiation in ornament means (and preference means) are about $\pm 0.4\sqrt{P}$ and are stationary. In contrast, with no selection on preferences, the same bounds are about 50 times larger ($\pm\, 20\sqrt{P}$) (Figure 19.6) and will continue to endlessly expand. Stabilizing selection has tied both bivariate means to their optima and greatly limited the radiation in ornaments and preferences.

Stabilizing selection has also greatly restrained the evolution of sexual isolation. After 10,000 generations the ornament means are close to the preference means of other populations, and vice versa, so the opportunity for sexual isolation is very limited. Figure 19.11b illustrates the constraint on sexual isolation. In all of the 100 pairs, joint isolation was less than 0.061, with a mode close to zero. None of the 100 pairs are sexually isolated. The impact of stabilizing selection is vividly illustrated by comparing the results in Sections 19.2.2 and 19.2.4 (Table 19.1). The one change between the two models was the addition of stabilizing selection on preferences, but this one selective constraint changed the speciation percentage from 77% to zero.

Although a model with drift about stationary optima for traits in both sexes fails to produce an appreciable radiation (unless stabilizing selection is very weak or populations are small) and little or no sexual isolation, it can produce long-term stasis. This kind of stasis is what we might expect when there is a premium on precise or intricate interaction. We will return to this point in the final chapter.

19.2.5 *Sexual speciation and radiation arising from Brownian motion of the optima of multiple ornaments and sexual preferences*

Given the disappointing prospects for sexual radiation and speciation when stabilizing selection acts on both ornaments and preferences that was revealed in Section 19.2.4, we can ask if both prospects might be rescued by a model with moving optima. This logic was followed by Arnold and Houck (2016). They briefly explored the consequences of a preference optimum that moved at a modest rate ($\sigma_\theta^2 = 0.00055$) with moderately weak stabilizing selection ($\omega_y = 19$) and showed that this model could account for the sexual radiation in plumage ornaments in birds of paradise. In this section, we continue the exploration of moving optima to include multiple ornaments and preferences. We also evaluate the evolution of joint isolation and hence the prospects for sexual speciation.

Our scenario is the now familiar one in which a pair of allopatric sister populations undergoes independent evolution of sex-limited traits (ornaments and preferences). For simplicity, we assume no speciation within either of the diverging lineages and no gene flow between them. We examine the minimally multivariate case of two ornaments and two preferences.

We use the same model as in Section 19.2.4, namely

$$\begin{pmatrix} \Delta\bar{\mathbf{z}} \\ \Delta\bar{\mathbf{y}} \end{pmatrix} = \begin{pmatrix} \Delta\bar{z}_1 \\ \Delta\bar{z}_2 \\ \Delta\bar{y}_1 \\ \Delta\bar{y}_2 \end{pmatrix} = \frac{1}{2}\begin{pmatrix} \mathbf{G} & \mathbf{B} \\ \mathbf{B} & \mathbf{H} \end{pmatrix}\begin{pmatrix} \boldsymbol{\beta}_z \\ \boldsymbol{\beta}_y \end{pmatrix} + N\left[0, \frac{1}{2N_e}\begin{pmatrix} \mathbf{G} & \mathbf{B} \\ \mathbf{B} & \mathbf{H} \end{pmatrix}\right], \tag{19.43}$$

but we add a coupled process for the movement of ornament and/or preference optima and evaluate the evolutionary time course of *JI*. The coupled equation that imposes Brownian motion change in the position of the optima each generation (Hansen et al. 2008) takes the form of a draw from a multivariate normal distribution with a 4 × 4 variance-covariance matrix, $\boldsymbol{Br}$, in which all covariance terms take a value of zero in the simulations, namely

$$\boldsymbol{Br} = Var(\boldsymbol{\theta}_z\ \boldsymbol{\theta}_y)^{\mathrm{T}} = \begin{pmatrix} \sigma_{\theta z1}^2 & 0 & 0 & 0 \\ 0 & \sigma_{\theta z2}^2 & 0 & 0 \\ 0 & 0 & \sigma_{\theta y1}^2 & 0 \\ 0 & 0 & 0 & \sigma_{\theta y2}^2 \end{pmatrix}. \tag{19.44}$$

Consequently, the vector of optimum change each generation takes the simplified form

$$\begin{pmatrix} \Delta\boldsymbol{\theta}_z \\ \Delta\boldsymbol{\theta}_y \end{pmatrix} = \begin{pmatrix} \Delta\theta_{z1} \\ \Delta\theta_{z2} \\ \Delta\theta_{y1} \\ \Delta\theta_{y2} \end{pmatrix} = \begin{pmatrix} N\left(0, \sigma_{\theta z1}^2\right) \\ N\left(0, \sigma_{\theta z2}^2\right) \\ N\left(0, \sigma_{\theta y1}^2\right) \\ N\left(0, \sigma_{\theta y2}^2\right) \end{pmatrix}. \tag{19.45}$$

Notice that while we used a simplified version of ***Br*** in (19.44), this matrix can be used to specify a much more complicated Brownian motion peak-controller process. We will return to this point in the final chapter.

In Chapter 13 we reported the Hansen et al. (2008) exact univariate result for the distribution of the means with BM motion of the optimum. Starting with zero means at generation zero, at generation t the lineage means are normally distributed about a zero mean with a variance that is approximately the same as the among replicate variance in the position of the optimum,

$$Var\left[\bar{\mathbf{z}}\,(t)\right] = Var\left[\bar{\mathbf{y}}\,(t)\right] = \sigma_\theta^2 t, \tag{19.46}$$

where $\sigma_\theta^2 = \sigma_{\theta z1}^2 = \sigma_{\theta z2}^2 = \sigma_{\theta y1}^2 = \sigma_{\theta y2}^2$ (Arnold and Houck 2016). We used these expressions to gauge how much differentiation to expect in simulations that last 10,000 generations.

To calculate sexual isolation (*JI*), we used the model and expressions developed in Section 19.2.2. In particular, we used the following values for parameters in the simulations: $G_{ii} = 0.6$, $B_{ii} = 0.48$, $N_e = 1000$, $\theta = \theta_{zi} = \theta_{yi} = 0$, $\omega = \omega_{zii} = \omega_{yii} = 50$, $\nu = \nu_{ii} = 5$, $\alpha = \nu/\omega = 0.1$, $\tau = 1$, $P = 1$, $Q = 1, \Sigma = \nu + Q + P = 7$, $\sigma_\theta^2 = 0.00055$.

We report three sets of simulation results in the following order: (1) both ornament and preference optima move, (2) only the preference optima moves, and (3) only the ornament optima moves. In each set we show plots of optimum movement, evolutionary trajectories of trait means ($\bar{z}_2$versus $\bar{z}_1$ superimposed on $\bar{y}_2$versus $\bar{y}_1$), and the evolutionary time course of *JI*.

Starting with the case in which both male and female trait optima move (Figure 19.12), we see that the evolutionary trajectories of the trait means mirror the movement of the preference optima but are pulled off the preference optima in the direction of the ornament optima. Notice that the within-population trait means are so close together that the black lines connecting them are barely visible in Figure 19.12a and Figure 19.12b.

How extensive is the radiation in the sample simulation shown in Figure 19.12? In the absence of sexual selection, after 10,000 generations we would expect the among-lineage variance of ornament or preference means to be

$$Var\left[\bar{\mathbf{z}}\,(t)\right] = Var\left[\bar{\mathbf{y}}\,(t)\right] = \sigma_\theta^2 t = \; 0.00055 * 10,000 = 5.5, \tag{19.47}$$

with 95% confidence limits around the mean of means equal to $\pm 1.96 * \sqrt{5.5} = \pm 4.60$ where, as usual, the units are within-population trait standard deviation, ($\sqrt{P} = \sqrt{Q}$). As we expect, the lineage trait means in Figure 19.12 are inside these limits.

A typical simulation of the case in which only the preference optimum moves by Brownian motion is shown in Figure 19.13. As in the previous case, the evolutionary trajectories of the trait means mirror the motion of the preference optimum (shown in orange) but are anchored near the ornament optimum at the origin. In the current case, however, the ornament and preference means are not pulled towards a moving ornament optimum. As in the preceding case, the position of the trait means after 10,000 generations is well inside the confidence limits we would expect in the absence of sexual selection.

A typical simulation of the case in which only the ornament optimum moves by Brownian motion is shown in Figure 19.14. In this case, however, the evolutionary trajectories of the ornament means are anchored near the preference optimum at the origin throughout the simulation. As time goes on, the ornament optimum (yellow) wanders farther and farther from its starting position (at the origin). As in the preceding case, the position

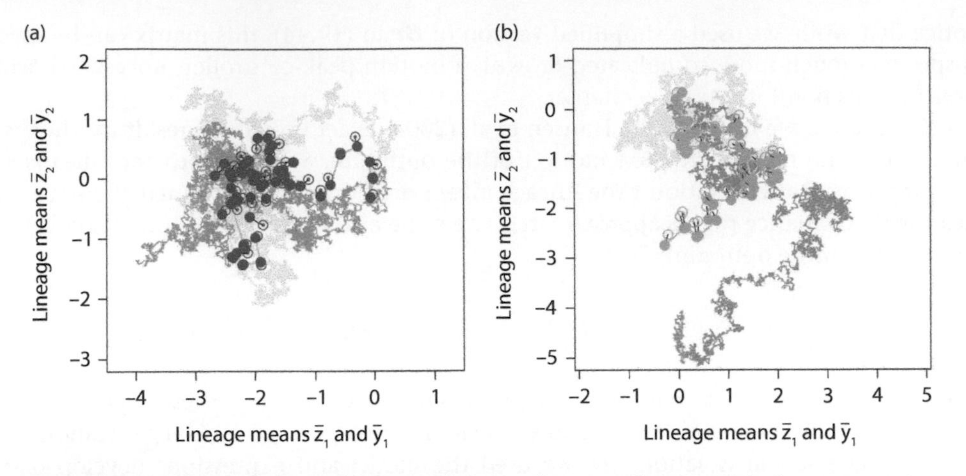

Figure 19.12 The evolution of two ornament and two preference traits in response to Brownian motion of both the natural selection optima of ornaments and preferences and with sexual selection on the ornaments. Movement of optima generation by generation is shown over 10,000 generations of the simulation: ornament optima are shown in yellow, preference optima shown in orange. Trait means are shown in two separate plots, rather than superimposed as in Figure 19.4. Graphical conventions as in Figure 19.4, but because of the larger scale, ornament and preference means in the same generation (connected by black line) appear to be closer together. Parameters as in Figure 19.4, except for Brownian motion of optima with per-generation variance of $\sigma_\theta{}^2 = 0.00055$ for both traits in each sex in each population. The simulation begins with the values for the bivariate optima of ornaments and preferences at the centre of the figure (x-axis value = 0, y-axis value = 0). (a) Evolution in population *A* in response to Brownian motion of its optima. Trait and preference means (open and solid purple circles) are shown every 200 generations. (b) Evolution in population *B* in response to independent Brownian motion of its optima. Trait and preference means (open and solid green circles) are shown every 200 generations.

of the trait means after 10,000 generations is well inside the confidence limits we would expect in the absence of sexual selection.

In summary, under three different versions of the moving optimum model (both preference and ornament optima move, only preference optimum moves, or only ornament optima moves), we see very much the same picture of trait diversification. When both optima move, trait trajectories are a compromise between the moving positions of the two optima. When just one optimum moves, trait trajectories are pulled towards the stationary position of the other optimum. We can ask whether this pull is sufficient to impede the chasing of the other optimum and consequently place a limit on trait radiation. Indeed, it is. When the length of simulation runs was increased to 100,000 generations with the ornament optimum stationary and the preference optimum moving, the bivariate means remained anchored to within a few $\sqrt{P}$ of the ornament optimum.

Turning to the evolutionary time courses of joint isolation under these same three models of optimum movement, we see three very different pictures of stochastic change in *JI* (Figure 19.15): waxing and waning in one case (Figure 19.15a), steady increase in another (Figure 19.15b), and consistent low values in the third (Figure 19.15c). In all three examples, the pairs of populations fall short of *JI* = 1.6, but the cases in which both ornament and preferences optima move (Figure 19.15a) or just preference optima move

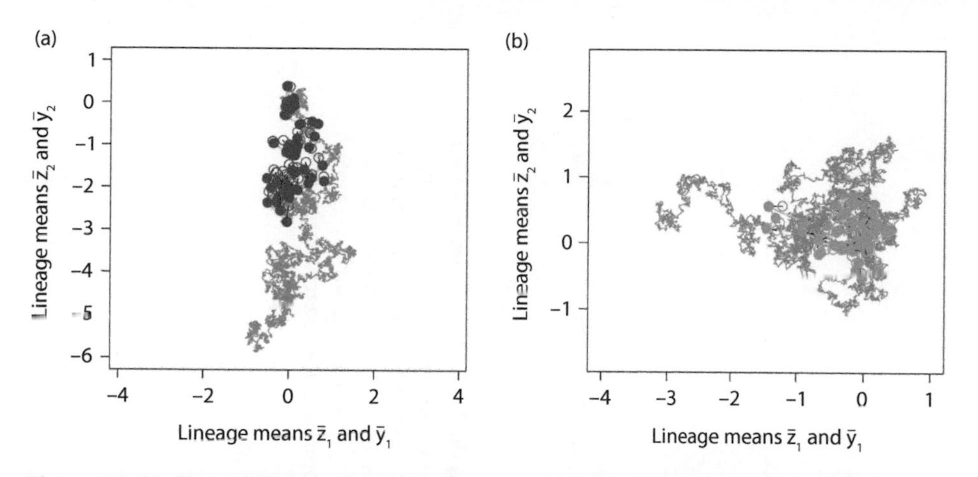

Figure 19.13 The evolution of two ornament and two preference traits in response to Brownian motion of their natural selection optima for preferences (but not ornaments) and with sexual selection on the ornaments. Graphic and simulation parameters as in Figure 19.12. Brownian motion of natural selection optima for preferences with per-generation variance of $\sigma_\theta{}^2 = 0.00055$ for both preference traits in each population. (a) Evolution in population *A* in response to Brownian motion of its preference optima. (b) Evolution in population *B* in response to Brownian motion of its preference optima.

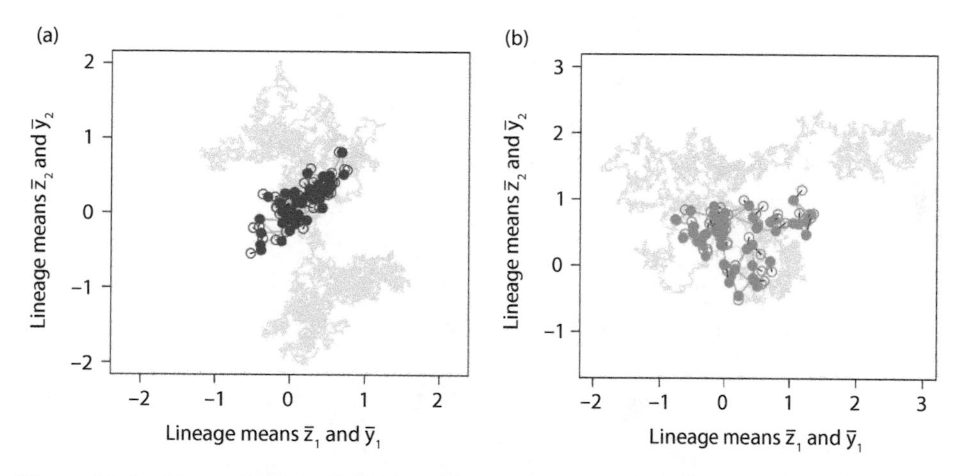

Figure 19.14 The evolution of two ornament and two preference traits in response to Brownian motion of their natural selection optima for ornaments (but not preferences) and with sexual selection on the ornaments. Graphic and simulation parameters as in Figure 19.12. Brownian motion of natural selection optima for ornaments with per-generation variance of $\sigma_\theta{}^2 = 0.00055$ for both ornament traits in each population. (a) Evolution in population *A* in response to Brownian motion of its ornament optima. (b) Evolution in population *B* in response to Brownian motion of its ornament optima.

(Figure 19.15b) *JI* achieves values of 1 or higher. In contrast, when just the ornament optima move, *JI* is substantially smaller.

The results of simulating three sets of 100 replicates of the Fisher-Lande process with moving optima are shown in Figure 19.16. Compared with movement of just ornament or preference optima, we see that movement of both preference and ornament

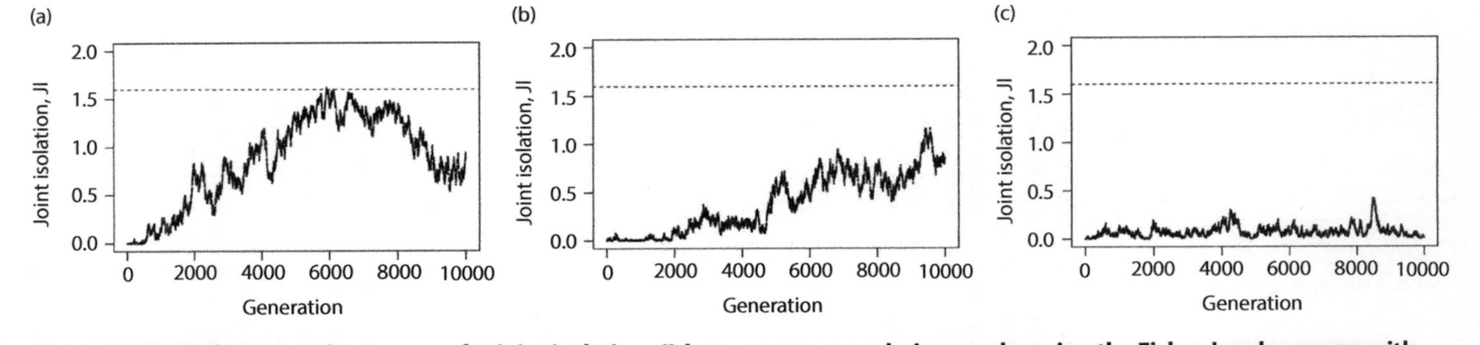

Figure 19.15 Evolutionary time courses for joint isolation, *JI*, between two populations undergoing the Fisher-Lande process with drift and stabilizing selection on ornaments and/or preferences. Optima of ornaments and/or preferences move by Brownian motion each generation. *JI* values computed from the simulation shown in Figure 19.12–14. The horizontal dashed line shows the lower level *JI* that is characteristic of sympatric, isolated species. (a) Evolution of *JI* when both ornament and preference optima undergo Brownian motion. *JI* values computed from the simulation shown in Figure 19.12. (b) Evolution of *JI* when only the preference optima undergo Brownian motion. *JI* values computed from the simulation shown in Figure 19.13. (c) Evolution of *JI* when only the ornament optima undergo Brownian motion. *JI* values computed from the simulation shown in Figure 19.14.

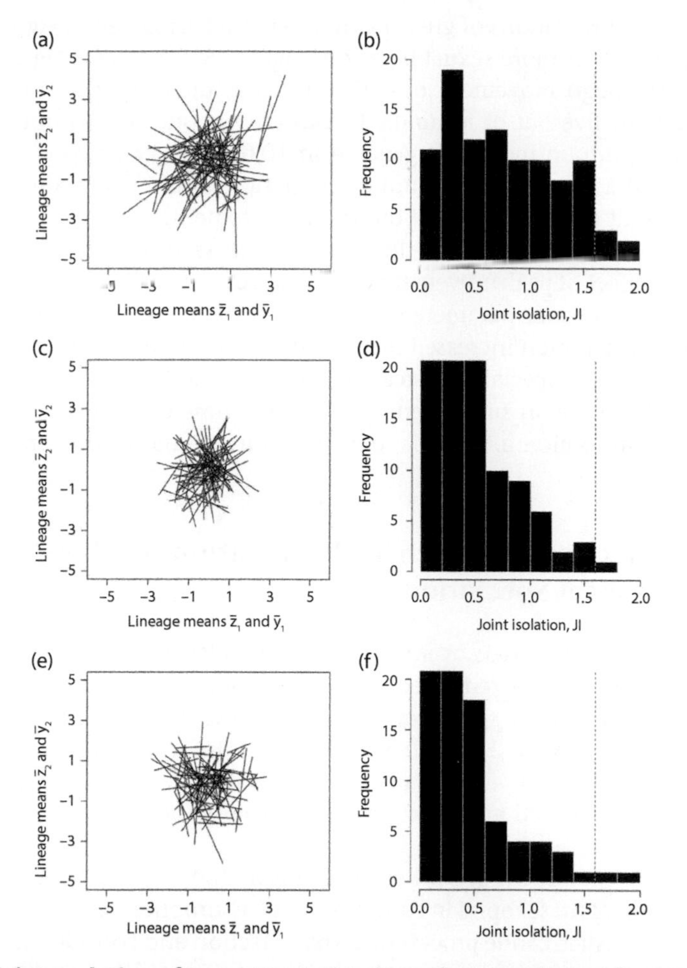

Figure 19.16 Joint evolution of two ornament traits and two mating preference traits undergoing the Fisher-Lande process with natural selection optima moving by Brownian motion. Results are shown after 10,000 generations of simulation using parameters listed in Figure 19.12. (First row) Both ornament and preference optima undergo Brownian motion: (a) Haystack plot showing divergence between ornament means in 100 replicate pairs of populations. Each line connects the ornament means of a replicate pair of populations. (b) Histogram of joint isolation measures for the 100 pairs of populations (two pairs with slightly negative *JI* are omitted from the plot). The dashed line shows the expected level of reproductive isolation (*JI* = 1.6) expected in sympatric pairs of species; 5% of pairs have this level of isolation or higher. This distribution has a mean of 0.75, a variance of 0.25 and a range of −0.0049 to 1.925. (Second row) Only preference optima undergoes Brownian motion: (c) Haystack plot showing divergence between ornament means in 100 replicate pairs of populations, (d) Histogram of joint isolation measures for the 100 pairs of populations. The dashed line shows the expected level of reproductive isolation (*JI* = 1.6) expected in sympatric pairs of species; 1% of pairs have this level of isolation or higher. This distribution has a mean of 0.49, a variance of 0.16, and a range of 0.0002 to 1.71. (Third row) Only ornament optima undergoes Brownian motion: (e) Haystack plot showing divergence between ornament means in 100 replicate pairs of populations, (f) Histogram of joint isolation measures for the 100 pairs of populations. The dashed line shows the expected level of reproductive isolation (*JI* = 1.6) expected in sympatric pairs of species; 2% of pairs achieved this level of isolation. This distribution has a mean of 0.42, a variance of 0.15, and a range of 0.0026 to 1.82.

optima produces trait radiations of greater extent (Figure 19.16a versus Figure 19.16c and Figure 19.16e), as well as more sexual isolation (Figure 19.16b versus Figure 19.16d and Figure 19.16f). Although movement of both optima resulted in speciation-level differentiation of traits in five out of a hundred cases (versus one or two out of a hundred cases, when one optimum moves) at generation 10,000, we can expect differentiation to continue so that a substantial amount of speciation would occur after, say, 100,000 generations under all three scenarios of optimum movement.

Stepping back from the details of the three cases in this section, it is important to recognize the dramatic effect of peak movement on the rate of trait radiation and speciation. In this section we used the same parameters as in Section 19.2d but added Brownian motion of trait optima. This addition increased the extent of trait radiation by more than sevenfold, while increasing the speciation percentage from zero to 8% after 10,000 generations (Table 19.1). Furthermore, in subsequent generations these differences will continue to increase as Brownian motion moves adaptive peaks further and further apart.

19.3 The Connections Between Trait Evolution, Ecological Distance, and Ecological Speciation

In the absence of process models over the last few decades, the conceptual foundations of ecological speciation have remained on shaky grounds. Although recent accounts and discussions of ecological speciation have ably described the empirical landscape (Rundle and Nosil 2005; Nosil 2012), they are based on informal, verbal models that fail to provide conceptual clarity and a path forward. The solution to this predicament—which has probably impeded empirical progress—would appear to be concrete, formal models of process.

In this section we explore models for ecological isolation and speciation in an attempt to anchor discussion of these topics in concrete sets of assumptions and consequences. We turn away from the Fisher-Lande process of sexual selection and consider the evolution of traits that affect habitat and resource use. In the first of these models (Section 19.3.1), we consider two habitat traits that are expressed in both sexes but may be sexually dimorphic. These traits affect encounters between males and females in different populations and hence ecological isolation between populations. In the second model (Section 19.3.2), we consider two resource utilization traits that are identically expressed in males and females. Divergence in these traits between populations leads to ecological distance between them. In the third model (19.3.3), we consider divergence in two traits that mediate competition between coexisting species. Our goal is to determine how trait divergence affects the evolution of ecological distance.

19.3.1 *Ecological speciation arising from habitat divergence that results in reproductive isolation*

Our first model is the same as in Section 19.2.5, except that both sexes experience stabilizing natural selection towards intermediate optima—but *no sexual selection*—and those optima undergo Brownian motion. Our model is of habitat traits in males and females that are not sex-limited. Unlike Section 19.2.5, we explore only the case in which the optima of both male and female traits undergo Brownian motion. Furthermore, the habitat traits of males and females track the *same* moving optimum.

In our scenario, the traits are indicators of habitat use in both males and females. Geographic separation has recently interrupted gene flow between two populations, *A* and *B*. The male and female traits are now free to diverge in the two populations in the absence of migration between them. The selective mechanism that promotes trait divergence is independent Brownian motion of the trait optima in the two populations.

We imagine that lineage divergence in the habitat traits affect the probability of encounter in the hypothetical situation of sympatry. In other words, although the populations continue to diverge in allopatry without gene flow, each generation we calculate a hypothetical index of encounter between A males and B female, and between B males and A females. This index is a function of the habitat trait values in the two populations. The more that population trait means diverge, the less likely are the populations to encounter one another and interbreed. As usual we use the time course of this index in simulations to gauge progress towards ecological speciation.

To visualize the scenario we have in mind, consider the hypothetical case of two populations of a hypothetical bird, MacArthur's warbler. One pair of traits (z_1, y_1) affects the tendency of both males and females to forage in oak trees versus pine trees. The second pair of traits (z_2, y_2) affects the tendency of both males and females to forage low versus high in the vertical dimension trees. In each population the two habitat traits evolve in response to an intermediate optimum in the oak-pine and low-high trait spaces. For simplicity we will assume that these intermediate optima are the same for males and females.

Consider the details of optimum movement. Initially, the habitat optima are the same in the two allopatric populations, but as time unfolds, the optima undergo independent Brownian motion in each population. As the two populations chase optima that stochastically move further and further apart, we expect the habitat traits to diverge to a point at which males of one population would no longer encounter females of the other population should they exist in sympatry. We assess these probabilities of encounter using the triple Gaussian model developed by Arnold et al. (1996) and implemented in Section 19.2 to assess joint isolation, *JI*. In this section, we use that model to compute an index of ecological isolation. The primary question we wish to address is whether this model of the evolution of ecological isolation describes an efficient and rapid process for achieving ecological speciation.

Our full model for change in means from generation to generation is the same as in Section 19.2.5, but the traits are not sex-limited.

$$\begin{pmatrix} \Delta\bar{\boldsymbol{z}} \\ \Delta\bar{\boldsymbol{y}} \end{pmatrix} = \begin{pmatrix} \Delta\bar{z}_1 \\ \Delta\bar{z}_2 \\ \Delta\bar{y}_1 \\ \Delta\bar{y}_2 \end{pmatrix} = \begin{pmatrix} \boldsymbol{G} & \boldsymbol{B} \\ \boldsymbol{B} & \boldsymbol{H} \end{pmatrix} \begin{pmatrix} \boldsymbol{\beta}_z \\ \boldsymbol{\beta}_y \end{pmatrix} + N\left[0, \frac{1}{2N_e}\begin{pmatrix} \boldsymbol{G} & \boldsymbol{B} \\ \boldsymbol{B} & \boldsymbol{H} \end{pmatrix}\right], \tag{19.48}$$

where $\boldsymbol{\beta}_z$ and $\boldsymbol{\beta}_y$ are column vectors with two elements and the three inheritance matrices are each 2 × 2. The first expression on the right gives the responses to selection and the second gives the contributions of genetic drift.

Each generation the changes in the positions of the four optima are drawn from a multivariate normal distribution with a variance-covariance matrix in which all covariance terms take a value of zero, as described in Section 19.2e.

In our simulations of this model, we will let

$$\boldsymbol{G} = \boldsymbol{H} = \begin{pmatrix} 0.6 & 0 \\ 0 & 0.6 \end{pmatrix} \tag{19.49a}$$

and

$$\boldsymbol{B} = \begin{pmatrix} 0.48 & 0 \\ 0 & 0.48 \end{pmatrix}. \tag{19.49b}$$

Only Gaussian stabilizing natural selection acts on the four habitat traits with directional selection gradients given by

$$\beta_{zi} = P^{-1} S_{zi} = (\omega_{zii} + P)^{-1} (\theta_i - \bar{z}_i) \tag{19.50a}$$

$$\beta_{yi} = Q^{-1} S_{yi} = (\omega_{yii} + Q)^{-1} (\theta_i - \bar{y}_i). \tag{19.50b}$$

Probabilities of encounters between males and females within and between populations are governed by overlap in normal distributions of the habitat traits $\boldsymbol{z}$ and $\boldsymbol{y}$, as specified by the following expressions

$$\pi_{AA} = c\ exp\left\{-(\bar{y}_A - \bar{z}_A)^2/2\Sigma\right\} \tag{19.51a}$$

$$\pi_{AB} = c\ exp\left\{-(\bar{y}_A - \bar{z}_B)^2/2\Sigma\right\} \tag{19.51b}$$

$$\pi_{BB} = c \exp\left\{-(\bar{y}_B - \bar{z}_B)^2/2\Sigma\right\} \tag{19.51c}$$

$$\pi_{BA} = c\ exp\left\{-(\bar{y}_B - \bar{z}_A)^2/2\Sigma\right\}, \tag{19.51d}$$

where the variation parameter $\Sigma = P + Q + \nu$.

By analogy with the model and expressions developed in Section 19.2b for sexual isolation, our *index of ecological isolation* is

$$EI = \pi_{AA} + \pi_{BB} - \pi_{AB} - \pi_{BA}. \tag{19.52}$$

Within each lineage, Lande's (1980) model (evolution of sexual dimorphism via natural selection alone) is in play (see his Figure 1, top row). In other words, the traits do not influence mate choice, just encounter. Each sex tends to evolve towards its own trait optimum, delayed by any genetic correlation between the sexes that might be specified. Letting those optima move via Brownian motion, we have a stochastic, perpetual model for the evolution of *EI*.

In summary, we used the following values for parameters in the simulations: $G_{ii} = 0.6$, $B_{ii} = 0.48$, $N_e = 1000$, $\theta = \theta_{zi} = \theta_{yi} = 0$, $\omega = \omega_{zii} = \omega_{yii} = 50$, $\nu = \nu_{ii} = 5$, $\alpha = \nu/\omega = 0.1$, $P = 1$, $Q = 1, \Sigma = \nu + Q + P = 7$, $\sigma_\theta^2 = 0.00055$.

A typical simulation trait evolution in two allopatric lineages is shown in Figure 19.17. Although the male and female trait means are usually some distance apart (shown as black lines), the trait pairs (dumbbells) closely track the moving optima in both lineages. In our stochastic model, there is no guarantee that the bivariate optima of both lineages will move away from the origin position at the centre of the figures (x = 0, y = 0). Indeed, in the illustrated simulation, one optimum moves far away from the original position (Figure 19.17a), while the other does not (Figure 19.17b). Nevertheless, although *EI* waxes and wanes during the first 8000 generations of the simulation, the overall trend is for steady increase in isolation, so that in the last 1000 generations the threshold of $EI \geq 1.6$

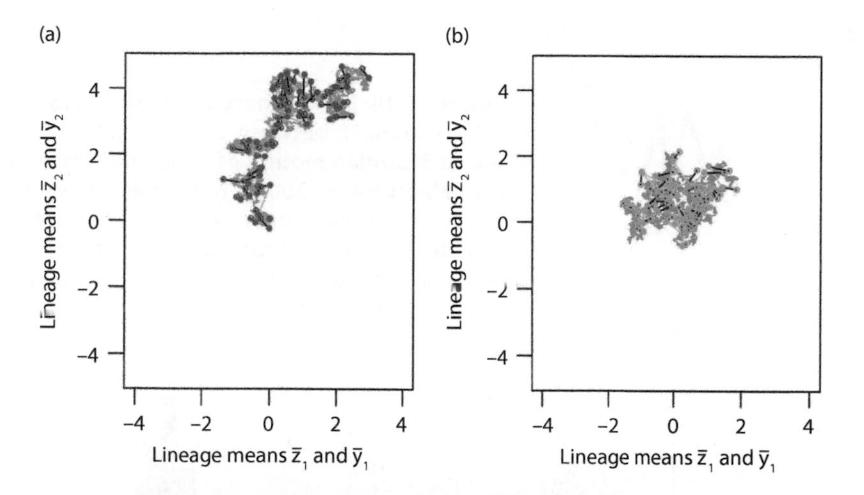

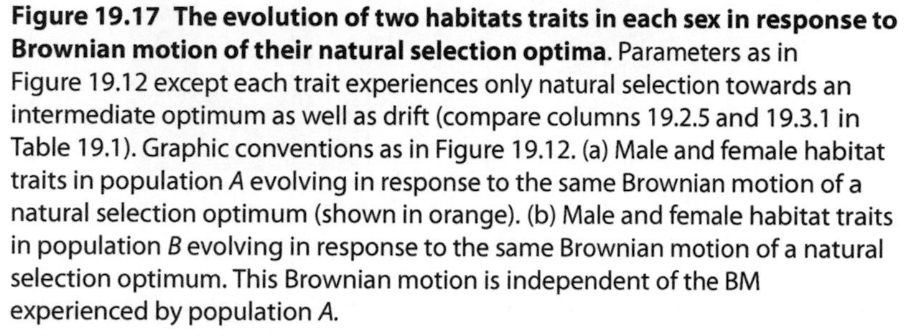

Figure 19.17 The evolution of two habitats traits in each sex in response to Brownian motion of their natural selection optima. Parameters as in Figure 19.12 except each trait experiences only natural selection towards an intermediate optimum as well as drift (compare columns 19.2.5 and 19.3.1 in Table 19.1). Graphic conventions as in Figure 19.12. (a) Male and female habitat traits in population *A* evolving in response to the same Brownian motion of a natural selection optimum (shown in orange). (b) Male and female habitat traits in population *B* evolving in response to the same Brownian motion of a natural selection optimum. This Brownian motion is independent of the BM experienced by population *A*.

is achieved and maintained (Figure 19.18). In other words, in this particular simulation, ecological speciation was achieved in 9000 generations.

Returning to the hypothetical example of MacArthur's warblers that evolve differences in tree habitat (oak versus pine trees, on the z_1, y_1 axis) and foraging height (low versus high, on the z_2, y_2 axis). Each of the two populations begins its simulated evolution at an optimum that is situated in the middle of each trait axis, corresponding to 50–50 use of oaks versus pines and foraging at intermediate height. As the simulation unfolds, the bivariate optima move and gradually diverge). One population evolves towards the upper-right corner of the plot (Figure 19.17a) which represents foraging high in pine trees, while the trajectory of the other population circles about the ancestral values of intermediacy in each trait (Figure 19.17b). Despite the apparent modesty of this divergence between populations, it is sufficient to lead a value of joint isolation characteristic of sympatric species after about 9000 generations (Figure 19.18).

In Section 19.2.5 we calculated the expected variance among lineage means, assuming no sexual selection, and obtained a value of ± 4.60 for the 95% confidence limit of the bivariate mean after 10,000 generations. The two simulated lineages in Figure 19.17 match that expectation.

The simulation shown in Figures 19.17 and 19.18 is typical of simulations in which the male and female optima of the two lineages undergo independent BM. With independent BM, *EI* increased through time towards an asymptote of 2. However, in other simulation runs, male and female traits were allowed to track the same moving optimum in *both* populations. With such completely correlated peak movement, *EI* did not accumulate and grow beyond an average value of zero. Non-zero correlation between peak movement in the two lineages would produce *EI* evolution in between these two extremes.

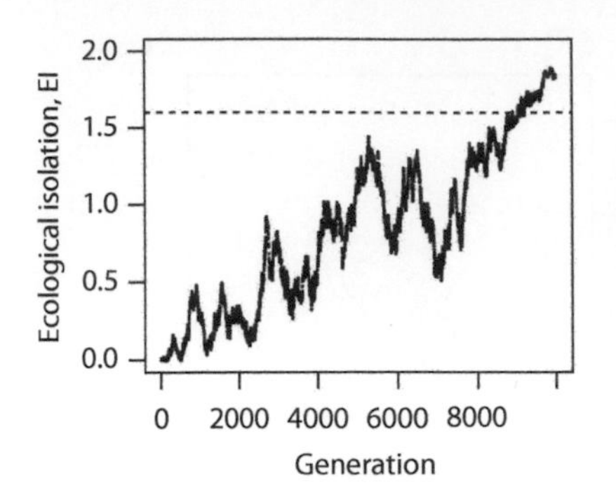

Figure 19.18 Evolutionary time courses for ecological isolation, *EI*, between two populations responding to Brownian motion of male and female optima for habitat traits. Optima of the traits move independently by Brownian motion each generation in each population. The horizontal dashed line shows the lower level of *EI* that is characteristic of joint isolation in sympatric, isolated species. *EI* values computed from the simulation shown in Figure 19.17.

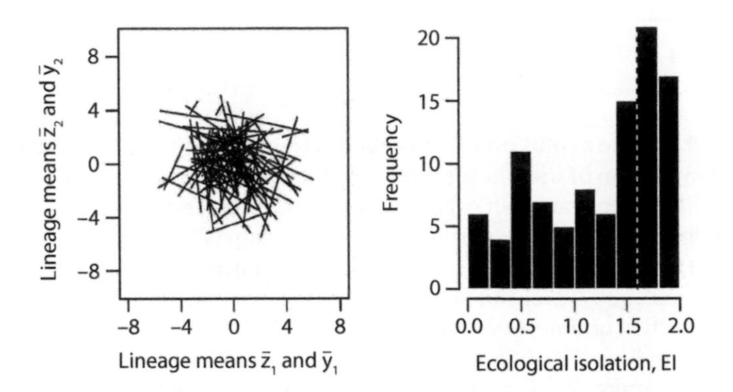

Figure 19.19 Joint evolution of two habitat traits each in males and females in 100 replicate pairs of populations with Brownian motion of natural selection optima. Results are shown after 10,000 generations of simulation using the parameters listed in Figure 19.12. (a) Haystack plot showing divergence between habitat trait means in 100 replicate pairs of populations. Each line connects the habitat trait means of a replicate pair of populations. (b) Histogram of ecological isolation measures for the 100 pairs of populations. The vertical dashed line shows the expected level of reproductive isolation (*EI* = 1.6) expected in sympatric pairs of species; 38% of pairs have this level of isolation or higher. This distribution has a mean of 1.24, a variance of 0.34, and a range of 0.03 to 1.99.

The results of running 100 replicates of the ecological isolation process with moving optima are shown in Figure 19.19. The haystack plot of differentiated pairs of habitat traits at generation 10,000 is shown in Figure 19.19a. We see that the trait radiation spans about 10 within-population phenotypic standard deviations ($10\sqrt{P}$). This amount of trait differentiation translates into substantial levels of ecological isolation between some pairs of populations. In Figure 19.19b we see that 38% of population pairs attained the level of isolation we could expect in pairs of sympatric species.

We also have the surprising result that sexual selection can retard trait radiation and speciation when trait optima are perpetually moving. In the models we have been considering, trait means constantly chase their moving natural selection optima. In this section we have removed sexual selection, so that evolving traits respond to only one master, ever-changing natural selection. The effect, comparing this section with Section 19.2.5, was to increase the rate of trait radiation (from six to 10) and the rate of speciation (from 8% to 38%). In other words, the effect of sexual selection was to bind pairs of traits together (ornaments bound to preferences) and this selective constraint impeded both trait radiation and speciation.

19.3.2 *Speciation arising from ecological niche divergence*

In this section we consider ecological speciation that arises from population divergence in important characteristics of the ecological niche. We have in mind traits that affect resource utilization and so affect competition between two species. In contrast to the case of habitat traits in Section 19.3a, these niche traits do not affect the probabilities of encounter between mates from different populations. In the next two chapters we will explore trait-mediation of species interactions in detail. For now, we ignore the details of interactions and argue that as niche traits diverge, interaction declines, and the populations evolve towards ecological isolation and speciation.

Our focus is on the tempo of progress towards ecological isolation and speciation. We carry forward the theme, developed in the last few sections, that Brownian motion of adaptive peaks drives divergence.

Our model is the same as in the last section, but instead of four traits in each population, we have only two. Instead of modelling sexual dimorphism which might arise by drift within populations, we assume sexual monomorphism in each of two traits in each population. For example, in a hypothetical snake population A, z_{A1} is the tendency to feed on lizards, whereas z_{A2} is the tendency to feed on anurans (frogs and toads). The two traits have the same meaning in population B. Both traits are under Gaussian stabilizing selection towards an intermediate optimum in both populations.

The two populations are sisters, recently separated into allopatry. The niche trait optima of the two populations move according to our standard model for Brownian motion. Initially the trait optima for the two populations are the same, but independent Brownian motion moves the optima apart. As the trait means of the two populations chase their separate moving peaks, they tend to move apart in niche space. We gauge the separate in niche space with an index of ecological distance (defined below). We wish to know how rapidly this model of niche divergence can take sister populations along the path to ecological speciation.

Our equations for change in trait means from generation to generation are

$$\Delta\bar{z}_{A1} = G_{11}\beta_{A1} + N[0, G/2N_e] \quad (19.53a)$$

$$\Delta\bar{z}_{A2} = G_{22}\beta_{A2} + N[0, G/2N_e] \quad (19.53b)$$

$$\Delta\bar{z}_{B1} = G_{11}\beta_{B1} + N[0, G/2N_e] \quad (19.53c)$$

$$\Delta\bar{z}_{B2} = G_{22}\beta_{B2} + N[0, G/2N_e], \quad (19.53d)$$

where the selection gradients are

$$\beta_{zAi} = P^{-1}S_{zAi} = (\omega_{zAii} + P)^{-1}(\theta_{zAi} - \bar{z}_{Ai}) \quad (19.54a)$$

$$\beta_{zBi} = P^{-1}S_{zBi} = (\omega_{zBii} + P)^{-1}(\theta_{zBi} - \bar{z}_{Bi}). \quad (19.54b)$$

Results in Section 19.2e were used to define a coupled set of equations that specified the stochastic per-generation changes in positions of the optima.

We define our *index of ecological distance, ED,* to be the Euclidean distance between the vectors of population means, $\bar{\mathbf{z}}_{\mathbf{A}}$ and $\bar{\mathbf{z}}_{\mathbf{B}}$, each consisting of q trait elements,

$$ED = d(\bar{\mathbf{z}}_{\mathbf{A}}, \bar{\mathbf{z}}_{\mathbf{B}}) = \sqrt{\sum_{i=1}^{q} (\bar{z}_{Bi} - \bar{z}_{Ai})^2}. \quad (19.55)$$

In simulations of this model we used the following values for parameters:$G = G_{11} = G_{22} = 0.6, N_e = 1000,\ \theta = \theta_{zAi} = \theta_{zBi} = 0,\ \omega = \omega_{zAi} = \omega_{zBi} = 50, P = 1,\ \sigma_\theta^2 = 0.00055$.

The univariate version of this model for trait evolution was described in Section 13.2. Here we expand the model to the bivariate case and use it to analyse the evolution of isolation between two independent lineages. As in the last section, we can use the analytical expression from Section 13.2 for the expected variance among replicate lineage means after 10,000 generations of independent evolution, namely ± 4.60 for the 95% confidence limit of the bivariate mean.

A typical simulation of two independently evolving lineages is shown in Figure 19.20. In the absence of sexual selection or any other systematic force on the trait means, they closely track their moving optima, as expected. After 10,000 generations of peak tracking, the trait means are inside the 95% confidence limits (± 4.60) that we expect for divergence.

Returning to our hypothetical example of lizard- and frog-eating in snakes, our plots represent lizard-eating that ranges from absent to frequent on the x-axis and frog-eating that ranges from absent to frequent on the y-axis (Figure 19.20). Our two populations begin their evolution at an optimum that is intermediate both traits, at the centre of the plot. Over the course of 10,000 generations the optimum of one population moves by Brownian motion towards the lower-right corner of the plot (Figure 19.20b), towards more frequent lizard-eating and less frequent frog-eating. Over the same period, the bivariate optimum of the other populations circles erratically about its initial position (Figure 19.20a).

Using our hypothetical example of two resource utilization traits, we can make the concept of moving trait optima more concrete. Let changes in the abundance of lizards be the basis for the Brownian motion in the lizard-eating trait peak, and likewise change

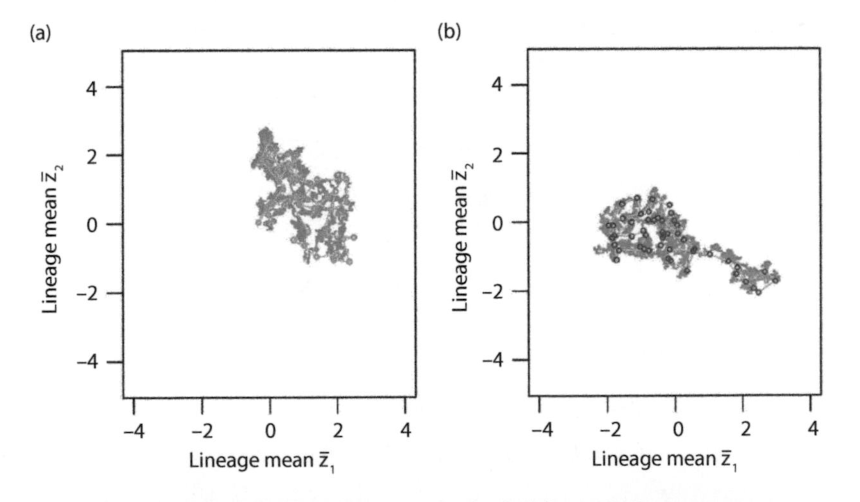

Figure 19.20 The evolution of two ecological niche traits in response to Brownian motion of their natural selection optima. Parameters as in Figure 19.12 except each trait experiences only natural selection towards an intermediate optimum and drift (compare columns 19.2.5 and 19.3.2 in Table 19.1). Graphic conventions as in Figure 19.12. (a) Two niche traits in population *A* evolving in response to the same Brownian motion of a natural selection optimum (shown in orange). (b) Two niche traits in population *B* evolving in response to bivariate Brownian motion of a natural selection optimum. This Brownian motion is independent of the BM experienced by population *A*.

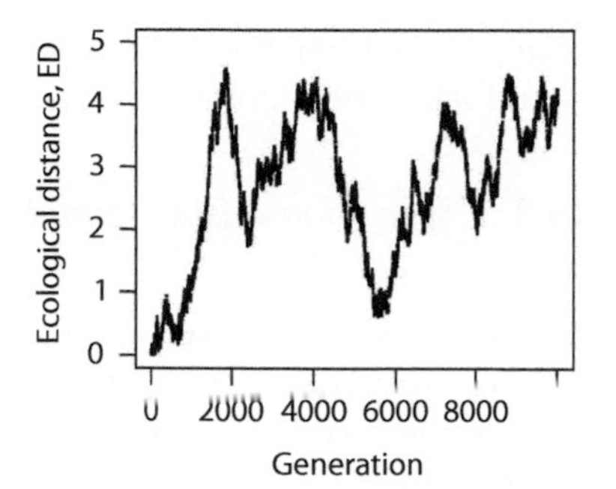

Figure 19.21 Evolutionary time course for ecological distance, *ED*, between two populations responding to Brownian motion of the optima for two ecological traits. Optima of the traits move independently by Brownian motion each generation in each population. *ED* values computed from the simulation shown in Figure 19.20.

in frog abundance is the basis of movement of the frog-eating peak. This proposal has some basis in ecological reality because the species abundance of lizard-eating snakes is statistically correlated with the lizard species abundance. Likewise, the species abundance of frog-eating snakes is correlated with frog species abundance (Arnold 1972). Although these empirical results probably reflect broad rules for community assembly, they may also reflect the causal basis of perpetual evolution in resource utilization traits.

Turning to the corresponding evolutionary trajectory of ecological distance (Figure 19.21), we see an erratic trajectory over the first 6000 generations, with distance increasing initially and then decreasing. Nevertheless, over the entire course of 10,000 generations the overall trend is towards increasing distance between population means.

The results of running 100 replicates of the ecological differentiation process with moving optima are shown in Figure 19.22. The haystack plot of differentiated pairs of habitat traits at generation 10,000 is shown in Figure 19.22a. We see that the trait radiation spans about 10 within-population phenotypic standard deviations ($10\sqrt{P}$). This amount of trait differentiation translates into substantial ecological distance between some pairs

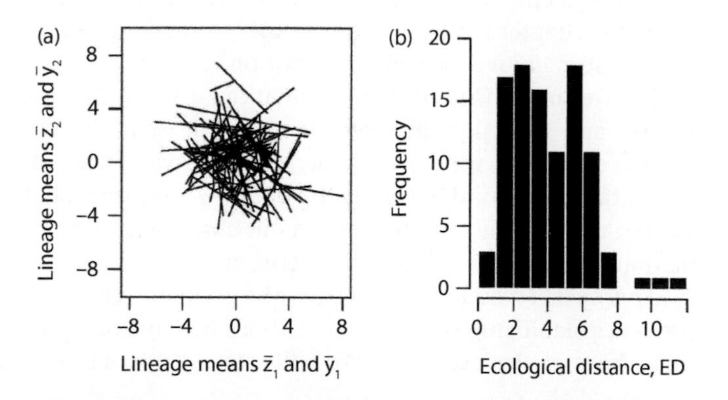

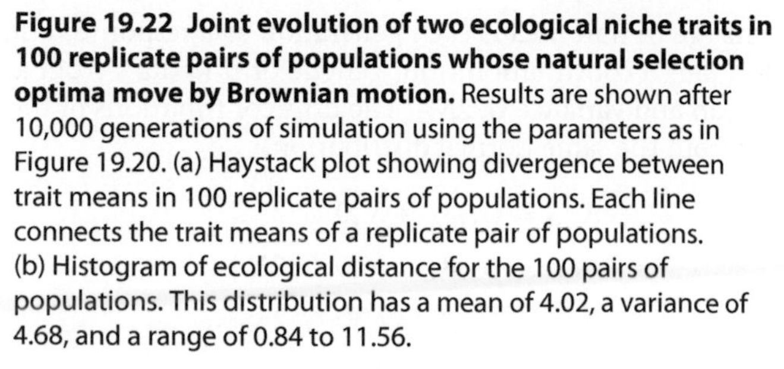
Figure 19.22 Joint evolution of two ecological niche traits in 100 replicate pairs of populations whose natural selection optima move by Brownian motion. Results are shown after 10,000 generations of simulation using the parameters as in Figure 19.20. (a) Haystack plot showing divergence between trait means in 100 replicate pairs of populations. Each line connects the trait means of a replicate pair of populations. (b) Histogram of ecological distance for the 100 pairs of populations. This distribution has a mean of 4.02, a variance of 4.68, and a range of 0.84 to 11.56.

of populations (Figure 19.22b). Unfortunately, we lack a gauge of ecological distance that would tell us how much distance to expect before and after speciation.

19.3.3 *Speciation arising from divergence in traits that mediate an ecological interaction*

Up to this point we have modelled the evolution of traits that reflect ecological niches without specifying how the traits affect ecological interactions. In this section we briefly explore the evolution of traits that mediate ecological interactions and determine how that evolution can lead to ecological speciation. The coevolution of interaction-mediating traits will be a major topic in the next two chapters, but for now we focus on one question: How does the fact of interaction mediation affect the evolutionary time course of ecological distance? To pursue this question, we build on Nuismer's (2017) model for a matching interaction between two competing species.

In Nuismer's (2017) model of a competitive matching interaction, the strength of competition in an interaction between two individuals is a quadratic function of the difference in individual trait values. Consider the example of two competing species, A and B. Competition between these two species is mediated by two traits z_{j1} and z_{j2}, where j denotes species (i.e., A or B). Strength of the interaction is at a maximum when trait values are matched ($z_{A1} = z_{B1}$ and $z_{A2} = z_{B2}$) and falls off as a negative exponential function as trait values differ. For example, the strength of interaction for species A in a competitive interaction with species B mediated by trait 1 is

$$p(z_{A1}, z_{B1}) = exp\left\{-\alpha(z_{A1} - z_{B1})^2\right\}, \tag{19.56}$$

where α is a positive constant (see below). Nuismer (2017) uses this characterization of a competitive interaction to derive expressions for effects of the interaction on fitness and hence on selection gradients. We will examine those derivations in detail in the next chapter. The goal in this chapter is to use those results to model the evolution of trait divergence when trait optima move by Brownian motion.

In the scenario that we have in mind, two populations (A and B) have just become reproductively isolated and gene flow between them has ceased. The two populations can now be considered coexisting species with identical trait values. We now follow the process of trait differentiation forward in time using our now familiar model for Brownian motion of adaptive peaks. The only new element in our characterization of peak chasing is the fact that the traits in question mediate competition.

In addition to the *interaction selection* arising from the competitive interaction, each of the four traits experience stabilizing selection towards an intermediate optimum, which takes a form familiar from earlier examples. We will refer to this selection as *functional selection.*

We model the change in trait means each generation as a response to interaction and functional selection plus a contribution from genetic drift (a draw from a normal distribution with zero mean and variance $G/2N_e$). The drift contributions of the four traits are independent draws from the same normal distribution.

$$\Delta\bar{z}_{A1} = G_{11}\beta_{A1} + N[0, G/2N_e] \tag{19.57a}$$

$$\Delta\bar{z}_{A2} = G_{22}\beta_{A2} + N[0, G/2N_e] \tag{19.57b}$$

$$\Delta\bar{z}_{B1} = G_{11}\beta_{B1} + N[0, G/2N_e] \tag{19.57c}$$

$$\Delta\bar{z}_{B2} = G_{22}\beta_{B2} + N[0, G/2N_e]. \tag{19.57d}$$

The four traits are under stabilizing selection and so directional selection pulls their means back towards their functional optima. Those optima ($\theta_{zA1}, \theta_{zA2}, \theta_{zB1}, \theta_{zB2}$) change slightly from the position in the preceding generation by values given by independent draws from a normal distribution with a mean of zero and a variance of σ_θ^2, as described in Section 19.2.5.

Nuismer (2017, Chapter 13) derives expressions for selection gradients for a slightly simpler version of this model (a single trait and no Brownian motion of the optima). Applying and extending those results, the total directional selection gradients are composed of the following functional and interaction gradients:

$$\beta_{A1} = \beta_{zFA1} + \beta_{zIA1} \tag{19.58a}$$

$$\beta_{A2} = \beta_{zFA2} + \beta_{zIA2} \tag{19.58b}$$

$$\beta_{B1} = \beta_{zFB1} + \beta_{zIB1} \tag{19.58c}$$

$$\beta_{B2} = \beta_{zFB2} + \beta_{zIB2}. \tag{19.58d}$$

The functional selection gradients are

$$\beta_{zFA1} = (\omega_{zA11} + P)^{-1} (\theta_{zA1} - \bar{z}_{A1}) \tag{19.59a}$$

$$\beta_{zFA2} = (\omega_{zA22} + P)^{-1} (\theta_{zA2} - \bar{z}_{A2}) \tag{19.59b}$$

$$\beta_{zFB1} = (\omega_{zB11} + P)^{-1} (\theta_{zB1} - \bar{z}_{B1}) \tag{19.59c}$$

$$\beta_{zFB2} = (\omega_{zB22} + P)^{-1} (\theta_{zB2} - \bar{z}_{B2}). \tag{19.59d}$$

The interaction selection gradients are

$$\beta_{zIA1} = k (\bar{z}_{A1} - \bar{z}_{B1}) \tag{19.60a}$$

$$\beta_{zIA2} = k (\bar{z}_{A2} - \bar{z}_{B2}) \tag{19.60b}$$

$$\beta_{zIB1} = k (\bar{z}_{B1} - \bar{z}_{A1}) \tag{19.60c}$$

$$\beta_{zIBi} = k (\bar{z}_{B2} - \bar{z}_{A2}), \tag{19.60d}$$

where $k = \alpha s / (1 - s)$. The constants α and s are Nuismer's (2017) conversion factors that convert trait values into fitness; we will discuss them in detail in the next chapter. For now, the important point is that k is a positive constant. As a consequence, when $\bar{z}_{A1} > \bar{z}_{B1}$, the selection gradient β_{zIA1} is also positive, indicating that selection favours larger values of $\bar{z}_{A1}$ and the selection gradient β_{zIB1} favours smaller values of $\bar{z}_{B1}$. In other words, the four interaction selection gradients push the means of interacting traits in opposite directions, exaggerating the difference between the two species.

We use our expression for an index of ecological distance, developed in the last section,

$$ED = d(\bar{\mathbf{z}}_{\mathbf{A}}, \bar{\mathbf{z}}_{\mathbf{B}}) = \sqrt{\sum_{i=1}^{q} (\bar{z}_{Bi} - \bar{z}_{Ai})^2}, \tag{19.61}$$

to assess ecological distance between the two vectors of species means.

Because this model includes an extra force of selection (the interaction selection gradients) driving the species means apart, we expect that ecological distance will evolve more rapidly than it did in the previous section (Figure 19.22) without that force.

In simulations of this model we used the following values for parameters: $G = G_{11} = G_{22} = 0.6$, $N_e = 1000$, initial $\theta = \theta_{zAi} = \theta_{zBi} = 0$, $\omega = \omega_{zAi} = \omega_{zBi} = 50$, $P = 1$, $\sigma_\theta^2 = 0.00055$, $k = 0.000033$. In our implementation of Nuismer's (2017) model of trait matching, we used a value of $as/(1-s)$ corresponding to a case of strong divergence (Nuismer 2017, p. 350). In this case, $S = 0.00002 = asG/(1-s)$, so with $G = 0.6$, $as/(1-s) = 0.000033$.

The results of a typical simulation run are shown in Figure 19.23. In both lineages the trait means closely track their moving optima despite the fact that the interactions between species tend to push the trait means apart. Our standard expression for the 95% confidence limits of the bivariate mean after 10,000 generations (i.e., ± 4.60) does not account for the between species interaction, but nevertheless the two trajectories remain within those limits.

The evolutionary time course for ecological isolation between the two lineages shown in Figure 19.24 shows a striking absence of profound isolation (Figure 19.24a), reflecting the fact that the bivariate means of the two lineages barely diverge over the course of 10,000 generations. In other replicates of the process, however, the bivariate means do tend to diverge and produce a trend towards increasing ecological isolation (e.g., Figure 19.24b).

The results of running 100 replicates of the ecological differentiation process with moving optima are shown in Figure 19.25. The haystack plot of differentiated pairs of habitat traits at generation 10,000 is shown in Figure 19.25a. We see that the trait radiation spans more than 10 within-population phenotypic standard deviations ($> 10\sqrt{P}$).

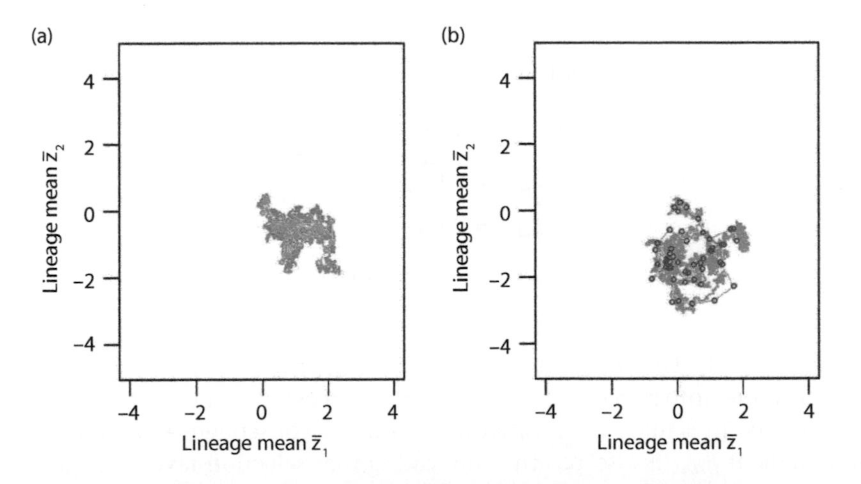

Figure 19.23 The evolution of two traits that mediate competition between two species in response to Brownian motion of their natural selection optima. Parameters as in Figure 19.9, except each trait experiences only natural selection towards an intermediate optimum and drift (compare columns 19.2.5 and 19.3.3). Graphic conventions as in Figure 19.12. (a) Two competition-mediating traits in population *A* evolving in response to Brownian motion of a bivariate natural selection optimum (shown in orange). (b) Two competition-mediating traits in population *B* evolving in response to Brownian motion of a bivariate natural selection optimum. This Brownian motion is independent of the BM experienced by population *A*.

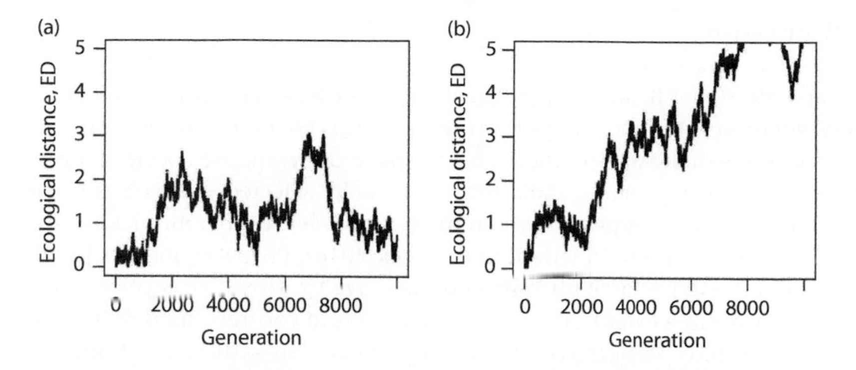

Figure 19.24 Evolutionary time course for ecological distance, *ED*, between two populations responding to Brownian motion of the optima for two competition-mediating traits. Optima of the traits move independently by Brownian motion each generation in each population. (a) *ED* values computed from the simulation shown in Figure 19.23. (b) *ED* values computed from a replicate simulation showing a strikingly different outcome. Simulation parameters as in Figure 19.23.

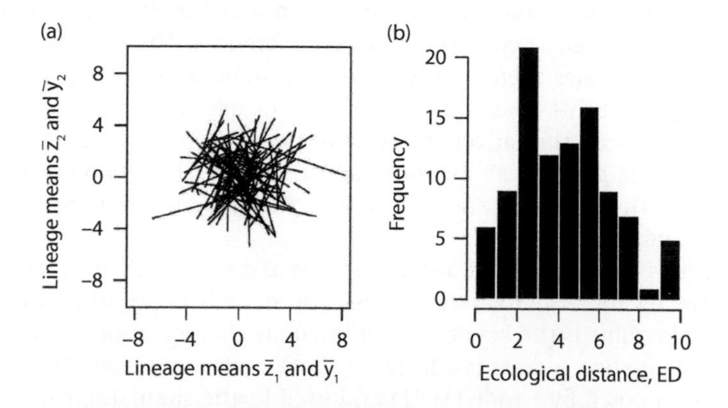

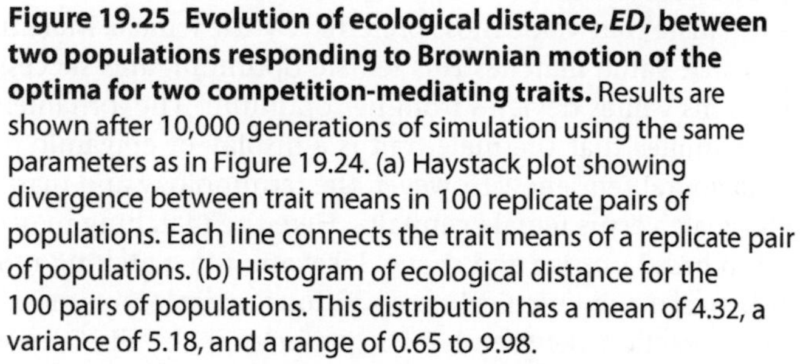

Figure 19.25 Evolution of ecological distance, *ED*, between two populations responding to Brownian motion of the optima for two competition-mediating traits. Results are shown after 10,000 generations of simulation using the same parameters as in Figure 19.24. (a) Haystack plot showing divergence between trait means in 100 replicate pairs of populations. Each line connects the trait means of a replicate pair of populations. (b) Histogram of ecological distance for the 100 pairs of populations. This distribution has a mean of 4.32, a variance of 5.18, and a range of 0.65 to 9.98.

This amount of trait differentiation translates into substantial ecological distance between some pairs of populations (19.25b). We expected that the interaction selection differential in the present model would produce more trait differentiation than models without trait-mediated competition, but we see about the same level of differentiation as in Section 19.3.2. The effect of interaction on selection is controlled by the key variable k, which was held at a single value (0.000033) in the present simulations. It would be interesting to determine how much trait differentiation could be produced if k were throttled up to higher levels.

19.4 Discussion

The demonstration that Brownian motion of an adaptive peak is a feasible model for speciation and radiation is the most important result in this chapter. Models with this feature can produce levels of sexual isolation characteristic of sympatric species within 10,000 generations. The corresponding radiations in sexually selected male traits can span 10 or more within-population phenotypic trait standard deviations, mimicking the extent of sexual radiations in nature (Arnold and Houck 2016). Likewise, models in which the optima of habitat and resource utilization traits move by Brownian motion can produce radiation of comparable extent on the same timescale. In contrast, models in which sexually selected traits have stationary optima can produce only radiations of limited extent, unless stabilizing selection on female mating preferences is extremely weak.

This success of models with peak movement encourages us to examine the causes of that movement. In the case of resource utilization traits, change in the abundance of prey and other resources may produce peak movement. In the case of female mating preferences, the causes of peak movement are less obvious. The concept of an intermediate natural selection optimum for preferences is used to represent some of the costs and conflicts associated with preferences. When we use a convex Gaussian function to represent fitness as a function of preference value (equivalent to a convex function representing costs), the model imposes the highest costs on extreme preferences. Those highest costs might be incurred because females spend excessive time searching or waiting for males with extreme phenotypes. In this case, a moving optimum may correspond to a change in environmental conditions that affect the costs of mating with extreme and hence rare male phenotypes (e.g., predator abundance, habitat cover). These two examples suggest that in many cases the ultimate causes of peak movement may be temporal change in environmental conditions.

Although we have used ornament and preference as a shorthand for male and female traits that evolve by the Fisher-Lande process, the models apply to a wide variety of trait categories. Matching is the key feature of the mathematical function that is used to describe how male mating success is affected by both male and female traits. In the particular function proposed by Lande (1981) and used in the simulations of this chapter, the female trait is the male trait value most preferred by the female. Male mating success is highest when his trait value matches this female optimum, and success falls off as a Gaussian function as his values deviates from her optimum. The 'ornament versus preference' terminology implies that the male trait is a display or epigamic trait, while the female trait is a behavioural tendency. However, the terminology and mating model also apply if we consider male versus female genitalia. Here the matching may correspond to a physical interaction based on size and shape. Likewise, the matching model may refer to the interaction between the chemical compositions of male pheromone and female receptors or to the interaction between habitat preferences. In all of these cases, if the matching function applies, the evolutionary consequences of the response to selection model follow.

Similar matching functions can be used to represent trait-mediated sexual interactions between mating partners and trait-mediated ecological inactions between different species. In both cases a Gaussian function can be used in which fitness falls off with an increasingly poor match between the traits of interacting individuals (Lande 1981; McPeek 2017; Nuismer 2017). Alternatively, in some kinds of ecological interaction, fitness may increase with an increasingly poor match, as when a defensive trait in a prey species enables it to evade its predator. Details aside, these matching functions enable us to model the effects of trait-mediated interaction on the coevolutionary process. In this chapter we briefly considered the effects of such interactions on coevolving, competing species. In the next two chapters we will take up this topic in much greater detail.

One of our most surprising results in this chapter is that sexual selection can impede a trait radiation propelled by Brownian motion of natural selection optima. In Section 19.2.5, we allowed the optima of two sexually interacting traits (ornaments and preferences) to move by independent processes of Brownian motion. Although this Fisher-Lande process resulted in a substantial radiation, the rate of radiation increased when we dropped the feature interaction (i.e., eliminated sexual selection) (see column 19.3.1 versus column 19.2.5 in Table 19.1). In one case (Section 19.2.5) trait interaction linked two processes of Brownian motion, while in the other (Section 19.3.1), the Brownian motion processes were unlinked. This result suggests that other interesting phenomena will be discovered as we explore the consequences of Brownian motion of multiple trait peaks.

19.5 Conclusions

- Ordinary evolution within lineages (e.g., evolutionary change resulting from drift and or selection), speciation, and extinction can make complementary or opposing contributions to trait evolution in a radiation.
- In a model with constant rates of speciation and extinction that are not affected by trait values, the number of species with a particular trait value increases or decreases exponentially at a rate equal to the difference between a constant gain from speciation and a constant loss from extinction.
- Furthermore, at any particular time, the variance among species trait means in a radiation (i.e., the size of the radiation) reflects the stochastic input from ordinary evolution and from speciation. However, applying the model to actual data reveals that the contribution from speciation is likely to be orders of magnitude smaller than the contribution from ordinary evolution within lineages.
- If rates of speciation and extinction do depend on trait values, the contribution of extinction to the size of a radiation depends on the magnitude of selection that it exerts on within- and among-lineage processes.
- The effects of trait evolution on sexual speciation and radiation are mediated by sexual isolation. Quantitative genetic models have illuminated each of the following conceptual steps: trait evolution $\rightarrow$ sexual isolation $\rightarrow$ sexual speciation $\rightarrow$ sexual radiation.
- Models of the Fisher-Lande process (coevolution of male ornaments and female mating preferences) with no selection on preferences can rapidly produce extensive radiation of ornaments as well as profound sexual isolation between populations even in lineages of appreciable effective size.
- When stabilizing natural selection acts of preferences as well as ornaments, the ability of the Fisher-Lande process to produce sexual isolation and radiation is sharply curtailed. The scope of radiation is so dramatically reduced that models of this type will fail to account for extensive radiation, although they may account for long-term stasis within ensembles of interacting traits.
- Brownian motion of natural selection optima partially counteracts the restraining effects of stabilizing selection on ornaments and preferences and can result in ever-expanding trait radiations, persistent sexual isolation, and sexual isolation.

- Brownian motion of natural selection optima can also drive with- and among-species differentiation of ecological traits, resulting in ecological isolation and speciation.
- When evolving traits also mediate ecological interactions (e.g., competition), the effects of Brownian motion of peaks may be enhanced or depressed. More theoretical work is needed.

Literature Cited

Arnold, S. J. 1972. Species densities of predators and their prey. Am. Nat. 106:220–236.

Arnold, S. J. 1976. Sexual behavior, sexual interference and sexual defense in the salamanders *Ambystoma maculatum, Ambystoma tigrinum* and *Plethodon jordani*. Z. Für Tierpsychol. 42: 247–300.

Arnold, S. J., and L. D. Houck. 2016. Can the Fisher-Lande process account for birds of paradise and other sexual radiations? Am. Nat. 187:717–735.

Arnold, S. J., K. M. Kiemnec-Tyburczy, and L. D. Houck. 2017. The evolution of courtship behavior in plethodontid salamanders, contrasting patterns of stasis and diversification. Herpetologica 73:190.

Arnold, S. J., P. A. Verrell, and S. G. Tilley. 1996. The evolution of asymmetry in sexual isolation: A model and a test case. Evolution 1024–1033.

Bateman, A. J. 1949. Analysis of data on sexual isolation. Evolution 3:174–177.

De Queiroz, K. 2007. Species concepts and species delimitation. Syst. Biol. 56:879–886.

Estes, S., and S. J. Arnold. 2007. Resolving the paradox of stasis: models with stabilizing selection explain evolutionary divergence on all timescales. Am. Nat. 169:227–244.

FitzJohn, R. G. 2010. Quantitative traits and diversification. Syst. Biol. 59:619–633.

Hansen, T. F., J. Pienaar, and S. H. Orzack. 2008. A comparative method for studying adaptation to a randomly evolving environment. Evolution 62:1965–1977.

Hohenlohe, P. A., and S. J. Arnold. 2010. Dimensionality of mate choice, sexual isolation, and speciation. Proc. Natl. Acad. Sci. 107:16583–16588.

Kingsolver, J. G., H. E. Hoekstra, J. M. Hoekstra, D. Berrigan, S. N. Vignieri, C. E. Hill, A. Hoang, P. Gibert, and P. Beerli. 2001. The strength of phenotypic selection in natural populations. Am. Nat. 157:245–261.

Lande, R. 1980. Sexual dimorphism, sexual selection, and adaptation in polygenic characters. Evolution 34:292–305.

Lande, R. 1981. Models of speciation by sexual selection on polygenic traits. Proc. Natl. Acad. Sci. USA 78:3721–3725.

Louca, S., and M. W. Pennell. 2020. Extant timetrees are consistent with a myriad of diversification histories. Nature 580:502–505.

Maddison, W. P., P. E. Midford, and S. P. Otto. 2007. Estimating a binary character's effect on speciation and extinction. Syst. Biol. 56:701–710.

Malogolowkin-Cohen, C. A., A. Solima Simmons, and H. Levene. 1965. A study of sexual isolation between certain strains of *Drosophila paulistorum*. Evolution 19:95–103.

McCullagh, P., and J. Nelder. 1989. Generalized Linear Models. Chapman and Hall, London.

McPeek, M. A. 2017. Evolutionary Community Ecology. Princeton University Press, Princeton.

Mead, L. S., and S. J. Arnold. 2004. Quantitative genetic models of sexual selection. Trends Ecol. Evol. 19:264–271.

Merrell, D. J. 1950. Measurement of sexual isolation and selective mating. Evolution 4:326–331.

Nosil, P. 2012. Ecological Speciation. Oxford University Press, Oxford.

Nuismer, S. 2017. Introduction to Coevolutionary Theory. MacMillan Higher Education, New York.

Pomiankowski, A. 1987. The costs of choice in sexual selection. J. Theor. Biol. 128:195–218.

Rundle, H. D., and P. Nosil. 2005. Ecological speciation. Ecol. Lett. 8:336–352.

Slatkin, M. 1981. A diffusion model of species selection. Paleobiology 7:421–425.

Stanley, S. M. 1979. Macroevolution: Pattern and Process. W. H. Freeman, San Francisco.

Tilley, S. G., and M. J. Mahoney. 1996. Patterns of genetic differentiation in salamanders of the *Desmognathus ochrophaeus* complex (Amphibia: Plethodontidae). Herpetol. Monogr. 10:1–42.

Tilley, S. G., P. A. Verrell, and S. J. Arnold. 1990. Correspondence between sexual isolation and allozyme differentiation: A test in the salamander *Desmognathus ochrophaeus*. Proc. Natl. Acad. Sci. USA 87:2715–2719.

Uyeda, J. C., S. J. Arnold, P. A. Hohenlohe, and L. S. Mead. 2009. Drift promotes speciation by sexual selection. Evolution 63:583–594.

Verrell, P. A., and S. J. Arnold. 1989. Behavioural observations of sexual isolation among allopatric populations of the mountain dusky salamander, *Desmognathus ochrophaeus*. Evolution 43: 745–755.

CHAPTER 20

Coevolution of Species with Trait-based Interactions

Overview—Models for the coevolution of traits in interacting species produce theoretical expectations with the potential to guide empirical research on adaptive radiation. The nature of the interspecies interaction (e.g., antagonistic versus mutualistic, a basis in trait differences versus matching, etc.) profoundly affects coevolutionary trajectories and outcome. Mutualistic interaction produces a coevolutionary outcome with a single, stable outcome. In contrast, antagonistic interactions (e.g., between predators and prey) can result in an unstable arms race. Interactions based on trait matching exaggerate curvature of coevolutionary trajectories and delay approach to a stable equilibrium. In contrast, interactions based on trait differences produce less curvature in trajectories and a more rapid approach to equilibrium. The interactions that specify key parameters for coevolution also shape geographic mosaics and community structure. For example, the strength of interspecific interactions affects equilibrium rates of interaction which in turn affect the degree of connectivity and modularity in the community interaction network. In other words, the models reviewed in this chapter successfully fuse evolutionary and ecological visions as they illuminate processes of coevolution and their effects on adaptive radiation. They also lay the groundwork for empirical studies that will place bounds on key parameter values.

20.1 Coevolution of a Plant and its Pollinator

Kiester et al. (1984) present a general model for the coevolution of traits in two species with a mutualistic interaction based on a plant floral trait and a matching trait of its pollinator. The model is closely patterned after Lande's (1981) model for the within-species coevolution of a male sexually selected trait and a female preference trait. Kiester et al. (1984) provide details of the new model's derivation. The model is cast in continuous time to account for the possibility that the life cycles of the plant and pollinator are not temporally aligned. Kiester et al. (1984) use varieties of their model to illuminate particular instances of coevolution (e.g., orchids and orchid bees, figs and fig wasps, yuccas and yucca moths). Our focus, however, will be on the speed and extent of adaptive radiation in coevolving species.

Evolutionary Quantitative Genetics. Stevan J. Arnold, Oxford University Press.
DOI: 10.1093/oso/9780192859389.003.0021

Let the generation time of the plant be τ_x and the generation time of the pollinator be τ_y. Then our standard response to selection equations (11.2) become, in continuous time,

$$\frac{d\bar{x}}{dt} = \frac{G_x S_x}{\tau_x P_x} \tag{20.1a}$$

$$\frac{d\bar{y}}{dt} = \eta \frac{G_y S_y}{\tau_y P_y}, \tag{20.1b}$$

where x is a normally distributed plant floral trait and y is a normally distributed pollinator trait that interacts with the floral trait, and the constant parameter η depends on the system of inheritance and the expression of the pollinator trait. The phenotypic variances of the traits are P_x and P_y and their additive genetic variances are G_x and G_y. When there is no sexual dimorphism in a diploid species, the eta parameter, $\eta = 1$. When pollination is sex-limited, $\eta = 1/2$. When the pollinator is haplodiploid (as in bees), $= 1/3$, when pollinators are males, and $\eta = 2/3$ when pollinators are females. Notice that the two equations (20.1) are not linked by a genetic covariance between the sexes, as they are in (13.5).

Following Kiester et al. (1984), we now sketch the derivation of the two selection differentials, S_x and S_y. The model assumes that the plant-pollinator interaction affects a single component of fitness in each species, so we only need to specify a single shift in the trait mean in each species. Kiester et al. (1984) explore three models of pollinator preference for the floral trait, but for simplicity we will consider only the simplest of these (absolute preference). Coevolutionary results are similar for all three models.

Let the relative fitness for plants with phenotype x be proportional to the average number of pollinator visits,

$$w_x(x) = \int p(y)\, \psi(x|y)\, dy, \tag{20.2}$$

where $p(y)$ is the frequency of pollinator phenotype y and $\psi(x|y)$ is an interaction function that gives the fitness contribution to the plant population from plants with phenotype x interacting with pollinators having phenotype y. We will assume that the pollinator preference function is absolute, such that each individual pollinator's preference is characterized by a Gaussian function with an optimum at y and a constant-valued variance-like width of v^2,

$$\psi(x|y) \propto \exp\left\{-\frac{(x-y)^2}{2v^2}\right\}. \tag{20.3}$$

In other words, a pollinator with phenotype y is most likely to pollinate a flower if its phenotype matches the pollinator's preference, $x = y$, and that probability falls off as the flower trait deviates from that optimum. Consequently, the selection differential for the floral trait is, averaging over the two trait distributions,

$$S_x = \int p(y) \int x p(x)\, \psi(x|y)\, dx dy - \bar{x} = P_x \frac{\bar{y} - \bar{x}}{v^2 + P_x}. \tag{20.4}$$

By analogous steps, the selection differential for the pollinator trait is

$$S_y = P_y \frac{-(\bar{y} - \bar{x})}{P_y + (v^2 + P_x)(2v^2 + P_x)/P_x}. \tag{20.5}$$

Substituting these expressions for the selection differentials into (20.1) and setting the response equations equal to zero, we find that the *deterministic equilibrium* is a line given by

$$\bar{y} = \bar{x}. \tag{20.6}$$

To solve for the approach to equilibrium, we write the response equations (20.1) in matrix form as

$$\frac{d}{dt}\begin{pmatrix} \bar{x} \\ \bar{y} \end{pmatrix} = \begin{pmatrix} -c_x & c_x \\ c_y & -c_y \end{pmatrix}\begin{pmatrix} \bar{x} \\ \bar{y} \end{pmatrix} = \boldsymbol{M}\begin{pmatrix} \bar{x} \\ \bar{y} \end{pmatrix}, \tag{20.7}$$

where

$$c_x = G_x/\tau_x \left(v^2 + P_x\right)$$

$$c_y = \eta\, G_y/\tau_y \left[P_y + \left(v^2 + P_x\right)\left(2v^2 + P_x\right)/P_x\right].$$

With the equations in this form, it is clear that the coevolution of the species means occurs along lines of constant slope,

$$\frac{d\bar{y}}{d\bar{x}} = -\frac{c_y}{c_x}. \tag{20.8}$$

The eigenvalues of the matrix $\boldsymbol{M}$, the first term on the right in 20.7, are $\lambda_0 = 0$, corresponding to no evolution along the line of equilibrium and $\lambda = -c_x - c_y$, corresponding to evolution along lines of constant slope (20.8). In other words the trait means evolve along lines of motion toward the line of equilibrium at the rate $e^{\lambda t}$. The line of equilibrium is stable. The dynamics are much like those depicted in Figure 13.8, except coevolution of means is along lines with negative slope (Figure 20.1).

In a population of infinite size, the trait means will stop evolving once they reach the line of equilibrium. However, in a population of finite size, the bivariate mean can drift off the line and then be pushed back towards it by selection, so that net effect is drift up and down the line, as in Lande's (1981) model of evolution by sexual selection.

Now consider the case in which the effective population sizes of the plant, N_{ex}, and of the pollinator, N_{ey}, are large enough to produce stable values of G but small enough that appreciable evolution happens by drift. The input sampling variance of the vector of species means per unit time is

$$\boldsymbol{V} = \begin{pmatrix} G_x/\tau_x N_{ex} & 0 \\ 0 & G_y/\tau_y N_{ey} \end{pmatrix}. \tag{20.9}$$

Replicate pairs of populations have a joint distribution of trait means that is approximately Gaussian with an expectation governed by (20.7). The stochastic dispersion of replicate pairs of coevolving means as they drift along the line of equilibrium converges on a variance-covariance matrix

$$\boldsymbol{D}(t) \simeq \frac{\left(\dfrac{c_y^2 G_x}{\tau_x N_{ex}}\right) + \left(\dfrac{c_x^2 G_y}{\tau_y N_{ey}}\right)}{\left(c_x + c_y\right)^2}\boldsymbol{I}t \tag{20.10}$$

plus a constant matrix, where $\boldsymbol{I}$ is a 2×2 identity matrix.

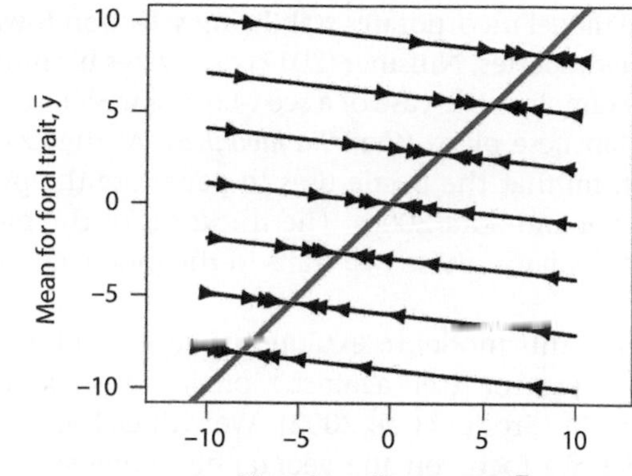

Figure 20.1 Coevolution of a floral trait and a matching pollinator trait according to the model of Kiester et al.1984. The line of stable equilibria is shown in red. Coevolutionary trajectories, lines of motion (black), are shown from 14 starting points with arrowheads every 10 generations. Parameter values are $G_x = G_y = 0.4$, $P = P_x = P_y = 1$, $\tau_x = \tau_y = 1$, $\eta = 1$, $v^2 = 4$.

These results tell us that the values of the floral and pollinator trait means will rapidly evolve towards the line of equilibrium at a negative exponential rate. For example, using (20.7) and taking reasonable but simplifying values for parameters ($G_x = G_y = 0.4$, $P = P_x = P_y = 1$, $\tau_x = \tau_y = 1$, $\eta = 1$, $v^2 = 4$), the instantaneous rate of evolution towards the line of equilibrium is $e^{\lambda t} = 0.9151$, where $\lambda = -0.0887$ and $t = 1$. After 100 time intervals, the rate is 0.0001 as the bivariate mean slowly approaches the line of equilibrium.

Once the line of equilibrium has been reached, differentiation by drift along the line is initially exceedingly slow. In addition to the parameter values in the last paragraph, let $N_{ex} = N_{ey} = 500$, so that we might expect dispersion by drift to be relatively rapid. Using our expression for $\boldsymbol{D}(t)$, after 100 generations the variance among trait means is only $0.066\sqrt{P}$ (about 6.6% of a within-species phenotypic standard deviation) and after 10,000 generations it is a modest $6.6\sqrt{P}$. On a very long timescale, however, the accumulation of among species variance in floral or pollinator traits can be appreciable: $66\sqrt{P}$ after 100,000 generations, $659\sqrt{P}$ after a million generations.

Although the model does not incorporate stabilizing selection on either the floral or the pollinator trait, we can easily deduce the consequences of such selection (Section 13.6). Stabilizing selection on each of the two traits converts the equilibrium line into an equilibrium point. Finite population size adds a drift component to the response equation each generation (13.30), and consequently the populations will equilibrate in drift-selection balance, forming a stochastic cloud about the point equilibrium.

20.2 Coevolution of Two Interacting Species at a Single Locality

Nuismer (2017, Chapter 3) presents a more general model than the one we considered in the preceding section for the coevolution of traits in two interacting species. First, the function that models the species interactions captures a broader range of ecological possibilities. Second, the model incorporates stabilizing selection toward an intermediate optimum in each of the two species. Nuismer (2017) motivates his model with two biological examples. The first example is the case of a seed-boring weevil (*Curculio cameliae*) that feeds on the seeds of a Japanese plant (*Camelia japonica*). At the crux of the interaction is the length of the rostrum that the beetle uses to penetrate the protective outer layer (pericarp) of the seed (Toju and Sota 2006). The thickness of the protective layer is the coevolving plant trait. We will use these two traits in the model description immediately below.

Nuismer (2017) also uses this model to explore the case of a newt (*Taricha granulosa*) that uses a neurotoxin to protect itself against a predatory snake (*Thamnophis sirtalis*) that is resistant to the toxin (Brodie et al. 2002). We will elaborate on this second case in Section 20.3, in which we focus on the geographic dimension of an evolving trait interaction.

To make the model more concrete, consider two normally distributed interacting traits: an insect trait (e.g., rostrum length), x, and a plant trait (e.g., pericarp thickness), y. The response to selection equations are

$$\Delta\bar{x} = G_x\beta_x \tag{20.11a}$$

$$\Delta\bar{y} = G_y\beta_y, \tag{20.11b}$$

where both of the selection gradients are functions of an interaction between the traits of the two species. As we shall see, each selection gradient is composed of two parts: one arising from the interaction with the other species and one arising from stabilizing selection that is not part of the interaction.

Suppose that the interaction between the insect and plant can be captured by a function that describes the probability of insect success in the interaction between the insect trait and the plant trait. This probability is a function of the *difference* in the values of x and y,

$$p(x,y) = \frac{1}{1 + exp\{\alpha(y - x)\}}, \tag{20.12}$$

where α describes the sensitivity of insect success to the difference in trait values (Figure 20.2). Suppose that while success for the insect, X, means an increment in insect fitness, it also means failure for the plant, Y, and hence a decrement in plant fitness. We model those aspects with two interaction fitness functions, one for the insect and one for the plant,

$$W_{XI}(x,y) = 1 + s_X\, p(x,y) \tag{20.13a}$$

$$W_{YI}(x,y) = 1 - s_Y\, p(x,y), \tag{20.13b}$$

where s_X represents the change in insect fitness from an successful interaction and s_Y is the change in plant fitness. Notice the corresponding difference in signs in (20.13).

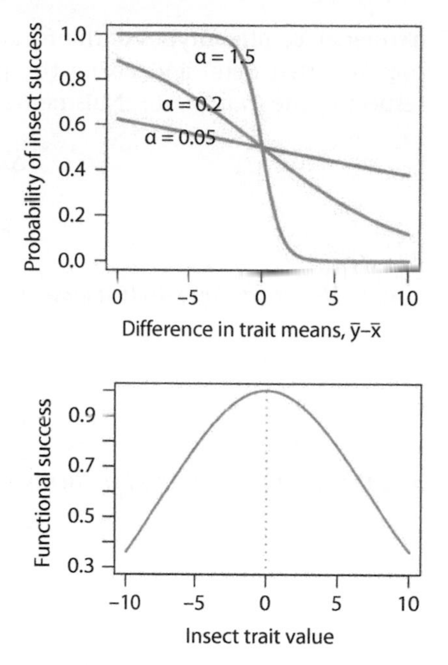

Figure 20.2 Interaction success as a function of difference in insect and plant trait means. Three examples are shown for $\alpha = 0.005, 0.2$, and 1.5. After Nuismer 2017.

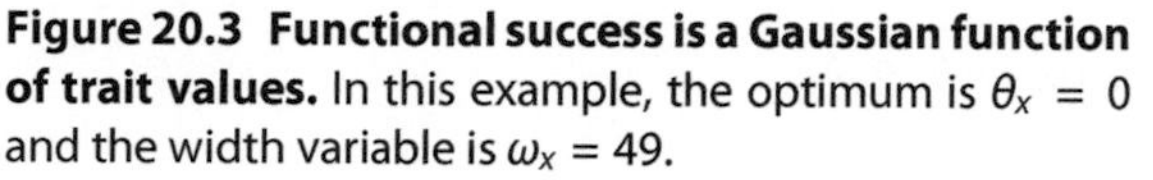

Figure 20.3 Functional success is a Gaussian function of trait values. In this example, the optimum is $\theta_X = 0$ and the width variable is $\omega_X = 49$.

Next we incorporate stabilizing selection. The corresponding functional components of fitness for the insect and the plant are:

$$W_{XF} = exp\left\{\frac{-(x-\theta_x)^2}{2\omega_x}\right\} \tag{20.14a}$$

$$W_{YF} = exp\left\{\frac{-(y-\theta_y)^2}{2\omega_y}\right\}, \tag{20.14b}$$

where θ_x and θ_y are, respectively, the insect and plant trait optima, and ω_x and ω_y are the widths of the corresponding insect and plant Gaussian ISS functions (3.8) (Note that in Nuismer's (2017) account these equations are parameterized with a gamma variable equal to $1/(2\omega)$, but this gamma is not the same as the γ of Lande and Arnold (1983)). The first of these functions (20.14a) is plotted in Figure 20.3. This stabilizing selection can be envisioned as arising from functional (e.g., biomechanical) considerations. It does not reflect interaction between species.

Assuming that the two fitness components act independently in each species, the total fitness consequences for the insect and plant from a single insect-plant interaction are products of the two components:

$$W_X(x,y) = W_{XF}W_{XI}(x,y) = exp\left\{\frac{-(x-\theta_x)^2}{2\omega_x}\right\}[1 + s_X\, p(x,y)] \tag{20.15a}$$

$$W_Y(x,y) = W_{YF}W_{YI}(x,y) = exp\left\{\frac{-(y-\theta_y)^2}{2\omega_y}\right\}[1 - s_Y p(x,y)]\,. \tag{20.15b}$$

Using these expressions, we derive equations for the population mean fitnesses ($\bar{W}_X$, $\bar{W}_Y$) by averaging over the frequency of phenotypes of the interacting species and over the frequency of phenotypes of the focal species (the weevil). Evaluating those expressions (not reported here) and taking the partial derivative of population mean fitness with respect to the trait means, Nuismer (2017) obtains the coevolutionary equations,

$$\Delta\bar{x} = G_X\left(\beta_{xI} + \beta_{xF}\right) \tag{20.16a}$$

$$\Delta\bar{y} = G_Y\left(\beta_{yI} + \beta_{yF}\right), \tag{20.16b}$$

where the *interaction selection gradients* are

$$\beta_{xI} = \alpha\left(s_X/2\left(2 + s_X\right)\right)$$

$$\beta_{yI} = \alpha\left(s_Y/2\left(2 - s_Y\right)\right)$$

and the *functional selection gradients* are

$$\beta_{xF} = \left(\theta_x - \bar{x}\right)/2\omega_x$$

$$\beta_{yF} = \left(\theta_y - \bar{y}\right)/2\omega_y.$$

A graphic summary of these equations is shown in Figure 20.4.

Evaluation of expressions (20.16) reveals that the trait means evolve towards a unique, stable equilibrium,

$$\hat{x} = \theta_x + \beta_{xI}\omega_x \tag{20.17a}$$

$$\hat{y} = \theta_y + \beta_{yI}\omega_y. \tag{20.17b}$$

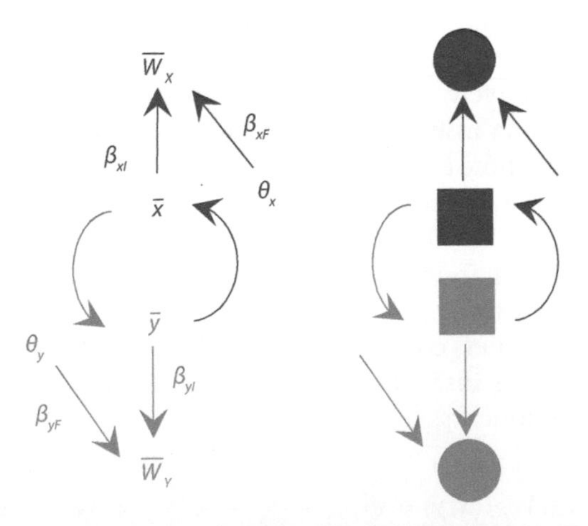

Figure 20.4 Path diagram and network depiction of the model for two interacting species. The plant species is shown in green and the insect species is shown in blue. (Left) Path diagram showing how the selection gradients (β_{xI}, β_{xF}, β_{yI}, β_{yF}) translate the effects of an intermediate optimum (θ_x, θ_y) and the species interaction between the trait means of the two species ($\bar{x}$, $\bar{y}$) to effects on the population mean fitness of the two species ($\overline{W}_X$, $\overline{W}_Y$). The curved arrows are meant to show that interaction selection gradients in each species (β_{xI}, β_{yI}) are functions of the trait means of both species. (Right) Network diagram, a symbolic version of the path diagram.

These expressions show that at equilibrium both species are pulled off of their functional adaptive peaks by the trait interaction (Figure 20.5). The extent of the equilibrium departure from the optimum depends on the strength of the interaction (α, s_X or s_Y) and the weakness of stabilizing selection (ω_x or ω_y). Weak stabilizing selection (larger ω) can cause one species to depart further from its optimum than the other species (Figure 20.5b).

Although coevolution proceeds along straight lines to the equilibrium if all parameter values are symmetrical (i.e., take the same values in the two species, Figure 20.5a), asymmetry in the key variables of the response to selection equations (20.16) can cause the trajectories to be curved. Figure 20.5b shows a case of curved trajectories caused by stabilizing selection that is much weaker on the insect trait than on the plant trait. An almost identical picture of the dynamics with curved trajectories can be produced by asymmetry in genetic variance (*e.g.*, $G_y = 0.4$, $G_x = 0.2$).

Nuismer (2017) points out that the directional selection (β_{xI}, β_{yI}) component of his model specifies a coevolutionary arms race. The strength of the two interaction selection gradients and the amount of genetic variances for the two traits determine the rate of trait exaggeration without specifying limits on that exaggeration. Countervailing stabilizing selection brings the arms race under control. The functional selection gradients reflect distance of the trait means from their optima and the weakness of stabilizing selection (ω_x, ω_y). At a certain distance the optima (20.17), the two kinds of selection gradients exactly balance and the arms races is brought to a halt.

Nuismer (2017) also considers the issue of how the arms race is affected by the interaction between the two species. For example, suppose we change the signs of the fitness effects, so that both species benefit from the interaction. Despite this fundamental change in the interaction, the arms race is unchecked in the absence of stabilizing selection.

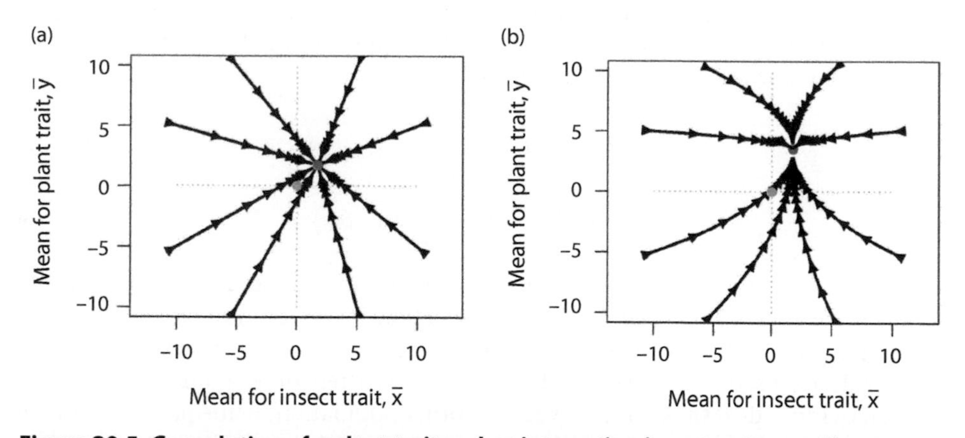

Figure 20.5 Coevolution of a plant trait and an interacting insect trait according to a model by Nuismer 2017. The location of the joint intermediate optima for the two traits is shown as a solid orange dot at the intersection of dotted orange lines. The coevolutionary equilibrium ($\hat{x}$, $\hat{y}$) is shown as a solid blue dot. Coevolutionary trajectories (black lines) are shown from eight starting positions. (a) Straight coevolutionary trajectories arise from symmetrical variables for the two species ($G_X = G_Y = 0.4$, $s_X = s_Y = 0.1$, $\alpha = 1.5$, $\theta_x = \theta_y = 0$, $\omega_x = \omega_y = 49$). Arrowheads are shown every 50 generations for 450 generations. The blue dot at the evolutionary equilibrium is hidden by arrowheads. (b) Curved coevolutionary trajectories are due to asymmetrical stabilizing selection ($\omega_x = 49$, $\omega_y = 99$) with all other parameters as in the preceeding case. Arrowheads positioned every 50 generations for 600 generations.

The primary difference at equilibrium with stabilizing selection is that the trait mean of one species exceeds the functional optimum, while the trait mean of the other species is smaller than the functional optimum. In the sections that follow we wish to determine whether the nature of species interaction can control an arms race or whether—as in the present case—that role must be surrendered to stabilizing selection.

20.3 Coevolution of Two Interacting Species at Multiple Localities

In this section we follow Nuismer's (2017, Chapter 9) account of his model for a trait interaction that evolves at two or more localities. This model is built upon the one we explored in the preceding section, but here we will illustrate the extended model using the case of the toxin resistance in a snake (*Thamnophis*) that coevolves with toxin level in its salamander prey (*Taricha*) (Brodie et al. 2002). A central objective of the model is to model the geographic aspect of the interaction. Toxicity of the newt varies geographically and so does the toxin resistance of the snake. This kind of *geographic mosaic* is found in many other pairs of coevolving species (Thompson 2005), so we need to take such a mosaic into account in thinking about coevolving interactions.

Consider a normally distributed snake trait (resistance), x, that interacts and coevolves with a normally distributed newt trait (toxicity), y. As in the preceding section, two variables (s_X and s_Y) capture the crux of the interaction. When a snake successfully ingests a newt, the interaction reduces newt fitness by an amount s_Y and increases snake fitness by an amount s_X. Using these variables and an interaction function that depends on *trait differences* (20.12), we model the fitness consequences of the interaction for the snake and the newt (see 20.15). As before, we derive expressions for the selection gradients in each species that arise from the interaction (β_{xI}, β_{yI}) and functional constraints (β_{xF}, β_{yF}). The response to selection equations for resistance evolution in the snake and toxicity evolution in the newt take the following form

$$\Delta\bar{x} = \bar{x}(t) - \bar{x}(t-1) = G_x(\beta_{xI} + \beta_{xF}) \tag{20.18a}$$

$$\Delta\bar{y} = \bar{y}(t) - \bar{y}(t-1) = G_y(\beta_{yI} + \beta_{yF}), \tag{20.18b}$$

where $\bar{x}(t-1)$ and $\bar{y}(t-1)$ are the trait means in generation $t-1$ and $\bar{x}(t)$ and $\bar{y}(t)$ are the trait means in generation t, after response to selection but before migration. To capture the geographic dimension, we assume that the two species co-occur at multiple localities, denoted with a subscript i. At each of these localities, after the response to selection, a proportion of individual snakes m_x moves to another population, while proportion $1-m_x$ does not, and likewise m_y specifies migration between newt populations. We denote the trait means at each locality, after selection response and migration, as

$$\bar{x}_i^*(t) = (1-m_x)\ \bar{x}_i(t) + (m_x)\ \bar{x}_i(t) \tag{20.19a}$$

$$\bar{y}_i^*(t) = (1-m_y)\ \bar{y}_i(t) + (m_y)\ \bar{y}_i(t). \tag{20.19b}$$

While these equations could be used to model coevolution under a regime that includes migration as well as selection, it will be easier to see the implications of the model if we change to variables that describe the geographic trait averages and differences in each species. For simplicity, we shift to just two localities and define the trait averages (μ_i) and differences (δ_i), where the subscript denotes predator or prey, as

$$\mu_x = \frac{\bar{x}_1 + \bar{x}_2}{2} \tag{20.20a}$$

$$\mu_y = \frac{\bar{y}_1 + \bar{y}_2}{2} \tag{20.20b}$$

$$\delta_x = \bar{x}_1 - \bar{x}_2 \tag{20.21a}$$

$$\delta_y = \bar{y}_1 - \bar{y}_2. \tag{20.21b}$$

From these definitions, it follows that the geographic average of the interaction selection gradients are

$$\bar{\beta}_{xI} = (\beta_{x1I} + \beta_{x2I})/2 \tag{20.22a}$$

$$\bar{\beta}_{yI} = (\beta_{y1I} + \beta_{y2I})/2, \tag{20.22b}$$

while the geographic differences in interaction selection gradients are

$$\delta_{\beta_{xI}} = (\beta_{x1I} - \beta_{x2I}) \tag{20.23a}$$

$$\delta_{\beta_{yI}} = (\beta_{y1I} - \beta_{y2I}). \tag{20.23b}$$

Finally, assuming that migration between populations is relatively small, Nuismer (2017) derive the following approximate expressions for the changes in geographic trait averages across generations

$$\Delta\mu_x = G_x\left(\bar{\beta}_{xI} - \beta_{xF}\right) \tag{20.24a}$$

$$\Delta\mu_y = G_y\left(\bar{\beta}_{yI} - \beta_{yF}\right), \tag{20.24b}$$

while the changes in geographic trait differences across generations are

$$\Delta\delta_x = G_x\,\delta_{\beta_{xI}} - \delta_x\left(\frac{G_x}{\omega_x} + 2m_x\right) \tag{20.25a}$$

$$\Delta\delta_y = G_y\,\delta_{\beta_{yI}} - \delta_y\left(\frac{G_y}{\omega_y} + 2m_y\right). \tag{20.25b}$$

Analysis of these dynamic equations shows that the trait averages and differences evolve towards a stable equilibrium. Setting expressions for the response in means (20.24) equal to zero, we see that at equilibrium the trait means are pulled off of their functional optima by interaction:

$$\hat{\mu}_x = \theta_x + \bar{\beta}_{xI}\omega_x \tag{20.26a}$$

$$\hat{\mu}_y = \theta_y + \bar{\beta}_{yI}\omega_y. \tag{20.26b}$$

Likewise, setting expressions for the differences in trait means (20.25) equal to zero, we see that at equilibrium the geographic trait differences are:

$$\hat{\delta}_X = \frac{G_X \, \delta_{\beta_{xI}}}{(G_X/\omega_X) + m_X} \tag{20.27a}$$

$$\hat{\delta}_Y = \frac{G_Y \, \delta_{\beta_{yI}}}{(G_Y/\omega_Y) + m_Y}. \tag{20.27b}$$

These expressions reveal that at equilibrium the geographic trait differences are exaggerated by geographic differences in directional selection arising from the interaction ($\delta_{\beta_{xI}}$ and $\delta_{\beta_{yI}}$) and diminished by functional stabilizing selection and gene flow.

Nuismer (2017) draws four lessons from his model. First, the bigger the geographic difference in coevolutionary selection (i.e., larger $\bar{\beta}_{xI}$ and $\bar{\beta}_{yI}$), the larger the geographic difference in trait means in each species. Spatial variation in coevolutionary selection drives geographic variation in the interacting trait means. Second, *coevolutionary hot spots* (Thompson 1994, 2005, 2013) with the strongest coevolutionary selection should coincide with locations where the interacting traits are most exaggerated. Coming back to our snake-newt example, locations in which newts are most toxic and snakes are most resistant should represent hot spots with the strongest coevolutionary selection. Third, congruent coevolutionary selection (larger $\bar{\beta}_{xI}$ geographically matched with larger $\bar{\beta}_{yI}$) should produce geographic congruence in the interacting traits. In other words, geographic congruence in interaction selection gradients should promote a positive correlation between newt toxicity and snake resistance across localities. Finally, gene flow should smooth out the geographic differentiation in traits that is being driven by coevolutionary selection.

20.4 Local Adaptation Between Interacting Species in a Geographic Array

Many pairwise interactions between species are enduring relationships that persist through evolutionary time and across geography. Local adaptation is a key aspect of the geography dimension. In particular, we wish to know whether sympatric players in the interaction are more closely matched in their phenotypes than pairs of players chosen at random across the geographic extent of the interaction.

In this section we focus on Nuismer's (2017, Chapter 10) account of theory and tests for local adaptation which are built on the models considered in Section 20.3. One of the interesting results that arises in the present account is that the nature of local adaptation depends on whether the species interaction is based on phenotypic differences or phenotypic matching. We will begin by considering an interaction based on trait differences, the case of toxin resistance of snakes coevolving with the toxicity of their newt prey, and consider the issue of local adaptation.

Imagine a fully factorial experimental design in which the performance of every population of snakes is assayed in tests against every population of newts. For each pair of populations, we assay the average probability, $\bar{p}_{i,j}$, that a snake from population i successfully ingests a newt from population j. To assess local adaptation of snake populations, A_X, we calculate the difference between the average probability of successful ingestion by snakes interacting with sympatric populations compared with success in all pairwise combinations of snake and newt populations. In the case of a 2 × 2 array of snake and newt populations, our measure of local adaptation in snakes is

$$A_x = \frac{\bar{p}_{1,1} + \bar{p}_{2,2}}{2} - \frac{\bar{p}_{1,1} + \bar{p}_{2,2} + \bar{p}_{1,2} + \bar{p}_{2,1}}{4}. \quad (20.28)$$

This measure ranges from zero (if all the probabilities are equal) to 0.5 (if snakes are completely successful with sympatric newts and completely unsuccessful with allopatric newts).

The measure of local adaptation depends on whether the interaction is a function of trait differences or matching. In the case of trait differences, the probability that a snake is successful in an interaction with a newt is

$$p(x, y) = \frac{1}{1 + exp\{\alpha(y - x)\}}. \quad (20.29)$$

The average probability of success of snakes from population i interacting with newts from population j is

$$\bar{p}_{i,j} = \frac{1}{2} + \frac{\alpha}{4}(\bar{x}_i - \bar{y}), \quad (20.30)$$

where $\bar{x}_i$ and $\bar{y}_j$ are respectively average toxin resistance of snakes from population i and the average toxicity of newts from population j. Substituting (20.30) into (20.28) we find that

$$A_x \approx 0. \quad (20.31)$$

The surprising result that we should not expect local adaptation makes sense when we realize that a snake population with a certain level of resistance can successfully interact with newt populations with a wide range of toxicities, and vice versa for a newt population. By its nature, a difference interaction does not promote specificity in coevolutionary outcome.

In contrast, we get a very different result if the species interaction depends on *trait matching*. Under this form of interaction, the probability that an avian nest parasite (a cuckoo) has a successful interaction with its host (a warbler) is

$$p(x, y) = exp\left\{-\alpha(x - y)^2\right\}, \quad (20.32)$$

where x is the colour of parasite's egg and y is the colour of the host's egg. The average probability of success of a parasite from population i interacting with hosts from population j is

$$\bar{p}_{i,j} = 1 - \alpha(\bar{x}_i - \bar{y}_j)^2 - \alpha(P_{x,i} + P_{y,j}), \quad (20.33)$$

where $P_{x,i}$ and $P_{y,j}$ are respectively the phenotypic variance in mimic egg colour at locality i and the phenotypic variance in host egg colour at locality j.

Substituting (20.33) into expression (20.28), we find that the measure of local adaptation is

$$A_x = 2\alpha\, Cov(\bar{x}, \bar{y}), \quad (20.34)$$

where $Cov(\bar{x}, \bar{y})$ is the covariance between the average egg colours of parasites and hosts across geographic locations. The covariance term is a signature of local adaptation. We expect it to be positive when we use (20.34) to assess local adaptation in the mimic and negative when we use it to assess local adaptation in the host. More broadly, an interaction based on trait matching rather than trait difference can promote a geographic mosaic with tight local adaptation.

Up to this point we have focused on analysing the results of a reciprocal, factorial design in which, for example, newt toxicity and snake resistance are assessed across an array of localities and the results are summarized by the local adaptation measure A_x (expressions 20.28, 20.31, 20.34). Nuismer (2017) goes beyond that analysis to derive expressions for equilibrium geographic differences in the two trait means ($\hat{\delta}_x$ and $\hat{\delta}_y$). Those complicated expressions are not reported here, but we will consider the conclusions that follow from them. In particular, Nuismer found that the relative values of the interaction selection gradients (β_{xI} and β_{yI}) strongly affect the degree of local adaptation and which species evolves local adaptation. Returning to the cuckoo and warbler host example, if selection on egg colour of the cuckoo is under stronger selection than the egg colour of the host ($\beta_{xI} \gg \beta_{yI}$), then local adaptation will tend to evolve in the cuckoo. Conversely, $\beta_{xI} \ll \beta_{yI}$ promotes local adaptation in the warbler, but in that case particular parameter combinations are required as well. Nuismer's (2017) numerical analysis also revealed interesting, counterintuitive roles for gene flow in promoting local adaptation.

20.5 Coevolution of Multiple Traits Among Multiple Interacting Species

Up until this point we have considered only a single trait of one species interacting with a single trait of another species. In this section we expand our focus to multiple traits, as we follow Nuismer's (2017, Chapter 11) account of his model for coevolution in parasite-host interactions with multiple hosts and in predator-prey interactions with multiple prey. We will also briefly touch on coevolution in multiple species engaged in mutualistic interactions.

20.5.1 *Coevolution of a parasite interacting with two hosts*

Nuismer (2017, Chapter 11) motivates the following model with the empirical result that an avian nest parasite (a cuckoo) with two hosts (two species of warblers) lays eggs that are phenotypically intermediate between the egg phenotypes of the avian hosts (Drobniak et al. 2014). In the model, we imagine that two traits of the hosts interact with two traits of the parasite.

Suppose that the success of the nest parasite (*X*) depends on matching its egg colour (x_1) to the egg colour (y) of one host (*Y*) and its egg size (x_2) to the egg size (z) of the other host (*Z*) (Figure 20.6). More specifically, let the matching functions take the following forms,

$$p(x_1, y) = exp\left\{-\alpha(x_1 - y)^2\right\} \quad (20.35a)$$

$$p(x_2, z) = exp\left\{-\alpha(x_2 - z)^2\right\}. \quad (20.35b)$$

Notice that both effects on fitness (20.35) have negative signs. A plot of this matching function is shown in Figure 20.7.

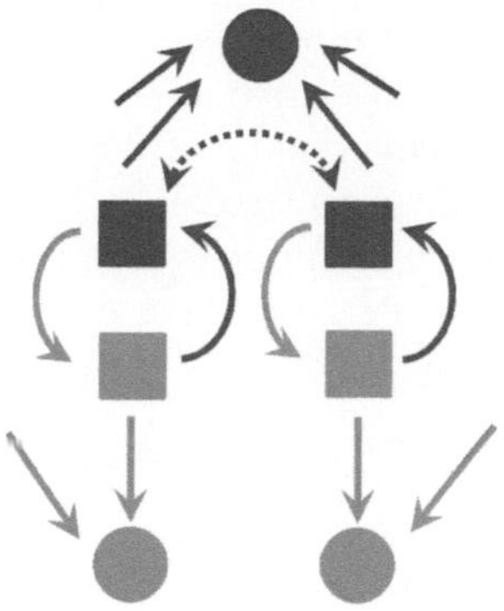

Figure 20.6 Network diagram of a parasite *X* with two traits (blue boxes) interacting with the traits of two host species *Y* and *Z* (green boxes). Population mean fitnesses are shown as circles. The genetic correlation between the two parasite traits is shown as a double-headed dotted arrow. Other conventions as in Figure 20.4.

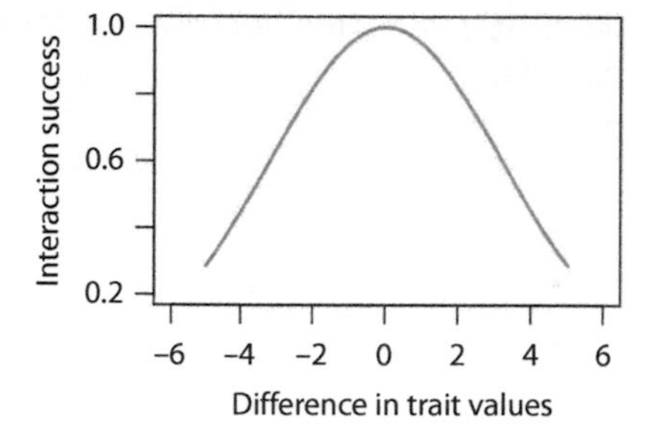

Figure 20.7 Interaction success is a matching, quadratic function of the difference in trait values, (20.35). In this example, $\alpha = 0.05$.

Assuming that stabilizing selection acts towards an intermediate optimum for each of the four traits, we obtain the following expressions for individual fitness in each of the three species as a function of trait values and interactions:

$$W_X = exp\left\{\frac{-(x_1-\theta_{x,1})^2}{2\omega_{x,1}}\right\} exp\left\{\frac{-(x_2-\theta_{x,2})^2}{2\omega_{x,2}}\right\}$$

$$[1 + f_Y\, s_{XY}\, p\,(x_1, y) + f_Z\, s_{XZ}\, p\,(x_2, z)] \tag{20.36a}$$

$$W_Y = exp\left\{\frac{-(y-\theta_y)^2}{2\omega_y}\right\}(1 - s_Y\, p\,(x_1, y)) \tag{20.36b}$$

$$W_Z = \exp\left\{\frac{-(z-\theta_z)^2}{2\omega_z}\right\}(1 - s_Z\, p\,(x_1, z))\,, \tag{20.36c}$$

where f_Y and f_Z are the frequencies of the two host species (assumed to be constant). The fitness benefits to the parasite from interactions with the two hosts are s_{XY} and s_{XZ}. The fitness costs to the two hosts from parasitism are s_y and s_z. Notice the positive effect of the interactions on parasite fitness in (20.36a) and the negative effects on the host fitnesses in (20.36b, c).

Averaging these individual fitness expressions over the trait distributions and differentiating with respect to each trait mean, we obtain the approximate response to selection equations for the two host traits:

$$\Delta\bar{y} \approx G_y\ (\beta_{yI} + \beta_{yF}) \tag{20.37a}$$

$$\Delta \bar{z} \approx G_z \left(\beta_{zI} + \beta_{zF}\right) \tag{20.37b}$$

where

$$\beta_{yI} = 2\alpha \left(s_Y / \left(1 - s_Y\right) \left(\bar{y} - \bar{x}_1\right)\right)$$

$$\beta_{yF} = \left(\theta_y - \bar{y}\right) / 2\omega_y$$

$$\beta_{zI} = 2\alpha \left(s_Z / \left(1 - s_Z\right) \left(\bar{z} - \bar{x}_2\right)\right)$$

$$\beta_{zF} = \left(\theta_z - \bar{z}\right) / 2\omega_z.$$

Similarly we obtain the response to selection equations for the two egg traits of the parasite, which include the possibility of genetic correlation between the two traits, namely:

$$\Delta \bar{x}_1 \approx G_{x,11} \left(\beta_{x,1I} + \beta_{x,1F}\right) + G_{x,12} \left(\beta_{x,2I} + \beta_{x,2F}\right) \tag{20.37c}$$

$$\Delta \bar{x}_2 \approx G_{x,22} \left(\beta_{x,2I} + \beta_{x,2F}\right) + G_{x,12} \left(\beta_{x,1I} + \beta_{x,1F}\right) \tag{20.37d}$$

where

$$\beta_{x,1I} = s_{x,1} \left(\bar{y} - \bar{x}_1\right)$$

$$\beta_{x,1F} = \left(\theta_{x,1} - \bar{x}_1\right) / 2\omega_{x,1}$$

$$\beta_{x,2I} = s_{x,2} \left(\bar{z} - \bar{x}_2\right)$$

$$\beta_{x,2F} = \left(\theta_{x,2} - \bar{x}_2\right) / 2\omega_{x,2},$$

with $s_{x,1} = \left(2\alpha f_Y s_{XY}\right) / \left(1 + s_{XY} f_Y + s_{XZ} f_z\right)$ and $s_{x,2} = \left(2\alpha f_Z s_{XZ}\right) / \left(1 + s_{XY} f_Y + s_{XZ} f_z\right)$.

Analysing the stability properties of these response equations, Nuismer (2017) found a surprising diversity of possible evolutionary outcomes. In particular, three outcomes are possible: (1) convergence on a stable equilibrium point defined by expressions (20.37a–d), (2) an unstable coevolutionary runaway, or (3) coevolutionary oscillations. The third possibility is unstable in the sense that the magnitude of the oscillations increases through time.

Cases in which the evolutionary outcome is stable have a simple bottom line. The equilibrium values for average egg colours of the parasite and host ($\bar{x}_1$ and $\bar{y}$) are both compromises between the functional optima for x_1 and y. Likewise, the equilibrium values for average egg sizes of parasite and host ($\bar{x}_2$ and $\bar{z}$) are both compromises between the functional optima for x_2 and z. In other words, the model retrieves the essential compromise in parasite traits observed by (Drobniak et al. 2014), and it also reveals that the host traits are pulled off of their optima. Instead of illustrating evolution towards these compromised outcomes, we will focus on other aspects of the coevolutionary trajectories in the stable case.

A case of stable outcome is depicted in Figure 20.8, which illustrates the effects of a genetic correlation between parasite traits. In this illustrated case, values of the functional optima are the same for all traits, so there is no appearance of compromise in the outcome. In this example, the host trait means and interacting parasite trait means evolve along curved trajectories towards their joint optima (Figure 20.8a). A strong genetic correlation between the two parasite traits exaggerates period of slow approach to the stable equilibrium (Figure 20.8b). The genetic correlation has the same delaying effect on the approach to the optimum, when we focus on the joint evolution of the two parasite traits (Figure 20.8c and d), but now the slow approach is along the genetic line of greatest resistance, g_{min} (Figure 20.8d).

Nuismer (2017, Figure 11.2) showed how the stability of the outcome depends on the values of the interaction coefficients (s_y, s_z, α), stabilizing selection on host traits, and the genetic correlation between the two parasite traits. In general, a stable outcome is

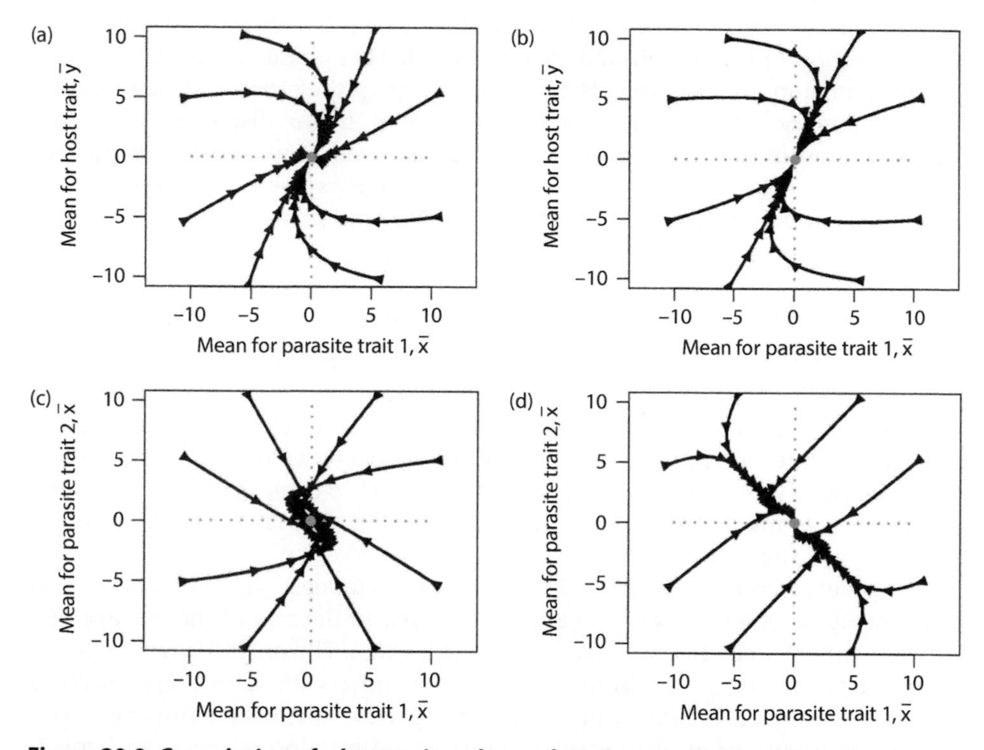

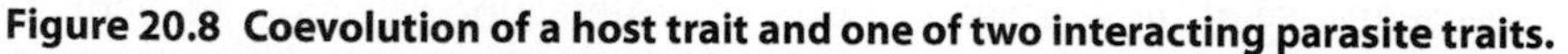

Figure 20.8 Coevolution of a host trait and one of two interacting parasite traits. Graphic conventions as in Figure 20.5. Arrowheads every 50 generations for 500 generations. Parameter values used in calculations: $G_{x,11} = G_{x,22} = 0.4$, $G_y = G_z = 0.4$, $s_y = s_z = s_{x,1} = s_{x,2} = 0.01$, $\theta_y = \theta_z = \theta_{x,1} = \theta_{x,2} = 0$, $\omega_y = \omega_z = \omega_{x,1} = \omega_{x,1} = 49$. Companion plots of the mean of host trait, $\bar{z}$, as a function of the mean of the parasite trait, $\bar{x}_2$, were identical to the plots shown here. (a) Coevolutionary trajectories of the mean of host trait, $\bar{y}$, as a function of the mean of the parasite trait, $\bar{x}_1$, when the genetic correlation between the two parasite traits is zero ($G_{x,12} = 0$). (b) Coevolutionary trajectories of the mean of host trait, $\bar{y}$, as a function of the mean of the parasite trait, $\bar{x}_1$, when the genetic correlation between the two parasite traits is 0.9 ($G_{x,12} = 0.36$). (c) Coevolutionary trajectories of the mean of parasite trait 2, $\bar{x}_2$, as a function of the mean of the parasite trait 1, $\bar{x}_1$, when the genetic correlation between the two parasite traits is zero ($G_{x,12} = 0$). (d) Coevolutionary trajectories of the mean of parasite trait 2, $\bar{x}_2$, as a function of the mean of the parasite trait 1, $\bar{x}_1$, when the genetic correlation between the two parasite traits is 0.8 ($G_{x,12} = 0.32$).

encouraged by modest values of the interaction coefficients, strong stabilizing selection on host traits, and no genetic correlation between parasite traits.

Nuismer (2017, Chapter 11) also examined the factors that encourage the evolution of specialization in host-parasite systems. In the stable case, specialization tends to evolve when the benefits to the parasite differ widely between the two hosts ($s_{x,1} \neq s_{x,2}$). Not surprisingly, the parasite tends to specialize on the host that is more frequently encountered and yields the larger benefit. The parasite also tends to evolve towards specialization on the host that it harms the least (e.g., toward host species Y if $s_y < s_z$). Specialization on host Y evolves because under this fitness condition, Y is less likely to make an evolutionary escape from the parasite by evolving an extremely divergent trait value. Stabilizing selection of the traits of the parasite and the hosts also affects the outcome of evolution towards specialization.

Nuismer (2017, Chapter 11) used numerical analysis to characterize the evolution of host specialization in the two unstable cases. In the runaway case, one of the host traits can evolve more rapidly away from the parasite trait than the other host trait, resulting in the evolutionary escape of one or both host species. In the case of coevolutionary oscillation in trait means, the parasite may switch between hosts at different points in the trait cycles. In other words, whether the coevolutionary system is stable or unstable, extreme specialization on a single host species is a likely outcome.

20.5.2 *Coevolution of species in mutualistic interaction*

In this section we consider mutualistic interactions, a category of species interactions that is widespread in nature. A common empirical observation in systems of species involved in mutualistic interactions is that generalization is more common than extreme specialization. For example, often a set of many pollinators services a wide array of plant species. Pursuing this observation, Nuismer (2017) uses his model to compute the expected level of specialization in mutualistic systems and test the prediction that it should be less than in systems of parasites and hosts.

Following Nuismer's (2017, Chapter 11) account of his models, we find that the coevolutionary outcome of mutualistic interactions is similar to the case of stable outcome in host-parasite interactions that we examined in Figure 20.8. The only difference from the model in the previous section is in the expressions for fitness effects (expressions 20.35), because in the mutualistic case all the interacting species experience positive effects on fitness. As a consequence of these positive effects, the interaction selection gradients for the two mutualistic species (β_{yI} and β_{zI}) are positive. Nuismer (2017) finds that the resulting evolutionary response equations converge on a single, stable equilibrium, unlike the model of host-parasite coevolution, in which runaway coevolution was a possibility. Furthermore, numerical analysis of the model shows that specialization is much less likely to evolve with the mutualistic model than with the host-parasite model.

20.5.3 *Coevolution of a predator interacting with two prey species*

In this section we return to the interaction structure depicted in Figure 20.6 but consider interaction functions based on phenotypic differences rather than phenotypic matching. As we shall see, difference interactions lead to much less curvature in coevolutionary trajectories that we observed with matching interactions. The following results are from Nuismer's (2017, Chapter 11) account of his model.

Nuismer (2017) illustrates his model using the insight that success in an interaction between predator and prey often depends on a difference in sprint speed (Bro-Jorgenson 2013) or a difference in body size (Sinclair et al. 2003). Using these *difference interactions*, Nuismer (2017) goes on to anchor his model in case of the cheetah (X) whose predatory success with a small antelope (impala) (Y) depends on a difference in speed (x_1 versus y), while its success with a large antelope (topi) (Z) depends on a difference in body size (x_2 versus z)). In other words, each of two traits (x_1 and x_2) of a single predator species (X) interacts with a single trait (y or z) in each of two prey species (Y and Z) (Figure 20.6).

In parallel with expressions (20.37), we specify the probability of predator success functions for trait x_1 interacting with trait y and for trait x_2 interacting with trait z:

$$p(x_1, y) = 1/(1 + exp\{-\alpha(x_1 - y)\}) \tag{20.38a}$$

$$p(x_2, z) = 1/(1 + exp\{-\alpha(x_2 - z)\}). \tag{20.38b}$$

Notice that both effects on fitness have negative signs.

We need to specify the amount that fitness of the predator is increased by successful predation on each prey (s_{XY}, s_{XZ}) and the reduction in fitness suffered by prey by those interactions (s_Y, s_Z). Using those constants and expression 20.38, we derive expressions for the average fitness of the three species:

$$\bar{W}_X = \frac{\alpha}{4}\left(s_{XY}f_Y(\bar{x}_1 - \bar{y}) + s_{XZ}f_Z(\bar{x}_2 - \bar{z})\right) + \frac{\kappa_X\left(1 - (\bar{x}_1 - \theta_{x,1})^2 + P_{x,1}\right)}{\omega_{x,1}} - \frac{(\bar{x}_2 - \theta_{x,2})^2 + P_{x,2}}{\omega_{x,2}} \tag{20.39a}$$

$$\bar{W}_Y = \frac{\alpha}{4}\left(s_Y(\bar{y} - \bar{x}_1)\right) + \kappa_Y\left(1 - ((\bar{y} - \theta_y)^2 + P_y\right)/\omega_y \tag{20.39b}$$

$$\bar{W}_Z = \frac{\alpha}{4}\left(s_Z(\bar{z} - \bar{x}_2)\right) + \kappa_Z\left(1 - ((\bar{z} - \theta_z)^2 + P_z\right)/\omega_z), \tag{20.39c}$$

where $\kappa_X = 1 + (1/2)\,s_{XY}f_Y + (1/2)\,s_{XZ}f_Z$, $\kappa_Y = (2 - s_Y)/2$, and $\kappa_Z = (2 - s_Z)/2$.

Differentiating each of these expressions with respect to the trait means, we obtain the interaction and functional selection gradients for each trait:

$$\beta_{x,1I} = (\alpha f_Y\, s_{XY})/2\,(2 + s_{XY}f_Y + s_{XZ}f_Z) \tag{20.40a}$$

$$\beta_{x,2I} = (\alpha f_z\, s_{XZ})/2\,(2 + s_{XY}f_Y + s_{XZ}f_Z) \tag{20.40b}$$

$$\beta_{YI} = \alpha s_Y/2\,(2 - s_Y) \tag{20.40c}$$

$$\beta_{ZI} = \alpha s_Z/2\,(2 - s_Z) \tag{20.40d}$$

$$\beta_{x,1F} = (\theta_{x,1} - \bar{x}_1)/2\omega_{x,1} \tag{20.40e}$$

$$\beta_{x,2F} = (\theta_{x,2} - \bar{x}_2)/2\omega_{x,2} \tag{20.40f}$$

$$\beta_{yF} = (\theta_y - \bar{y})/2\omega_y \tag{20.40g}$$

$$\beta_{zF} = (\theta_z - \bar{z})/2\omega_z. \tag{20.40h}$$

Substituting into the four-trait version of the multivariate response to selection equation we obtain the following approximate expressions for evolutionary change

$$\Delta\bar{y} \approx G_y\left(\beta_{yI} + \beta_{yF}\right) \tag{20.41a}$$

$$\Delta\bar{z} \approx G_z\left(\beta_{zI} + \beta_{zF}\right) \tag{20.41b}$$

$$\Delta\bar{x}_1 \approx G_{x,11}\left(\beta_{x,1I} + \beta_{x,zF}\right) + G_{x,12}\left(\beta_{x,2I} + \beta_{x,2F}\right) \tag{20.41c}$$

$$\Delta\bar{x}_2 \approx G_{x,22}\left(\beta_{x,2I} + \beta_{x,2F}\right) + G_{x,12}\left(\beta_{x,1I} + \beta_{x,1F}\right). \tag{20.41d}$$

Setting these expressions (20.41) equal to zero and solving for the equilibrium values, we obtain

$$\hat{x}_1 = \theta_{x,1} + \beta_{x,1I}\omega_{x,1} \tag{20.42a}$$

$$\hat{x}_2 = \theta_{x,2} + \beta_{x,2I}\omega_{x,2} \tag{20.42b}$$

$$\hat{y} = \theta_y + \beta_{yI}\omega_y \tag{20.42c}$$

$$\hat{z} = \theta_z + \beta_{zI}\omega_z. \tag{20.42d}$$

Nuismer's (2017) stability analysis of (20.41) revealed that the coevolution of these three interacting species does not result in the unstable outcomes that we observed in the case of a parasite and two hosts. Instead the three traits evolve towards a stable equilibrium point defined by (20.42).

We see from expressions (20.42) that at equilibrium all four trait means are pulled off of their functional optima by amounts that depend on the weakness of stabilizing selection and the strength of the interaction selection gradient.

The difference interactions in this model do not cause curvature in the coevolutionary trajectories of prey traits plotted as a function of predator traits (Figure 20.9a) or in the trajectories of prey traits plotted against each other (Figure 20.9c), as they do in the case of matching interactions (Figure 20.8a, b), so long as the genetic correlation between the two predator traits is zero. However, a nonzero genetic correlation between predator traits can induce profound curvature in coevolutionary trajectories (Figure 20.9e, b, d, f). The effect of the genetic correlation is especially profound on the trajectories of the prey traits plotted against each other (Figure 20.9 b, f). In these cases, the initial part of the trajectory is relatively rapid along $\boldsymbol{g}_{max}$ but then much slower along $\boldsymbol{g}_{min}$. In the case of a positive genetic correlation, $\boldsymbol{g}_{max}$ is tilted at a positive 45 degree angle, while a negative genetic correlation tilts $\boldsymbol{g}_{max}$ at a negative 45 degree angle. But these complications in curvature are relatively simple compared with the complexities of curvature in the case of matching interactions (Figure 20.8).

Nuismer (2017, Chapter 11) used numerical analysis to characterize the evolution of specialization in this model with difference interactions, just as he did in the corresponding model with matching interactions. As in that case, asymmetries in fitness rewards and stabilizing selection drive the predator towards specialization on one of the two prey species. In the case of the cheetah, it tends to specialize on the prey with the highest fitness reward using the predator trait constrained by the weakest stabilizing selection. Parameters of the prey also affect the specialization outcome. The prey most

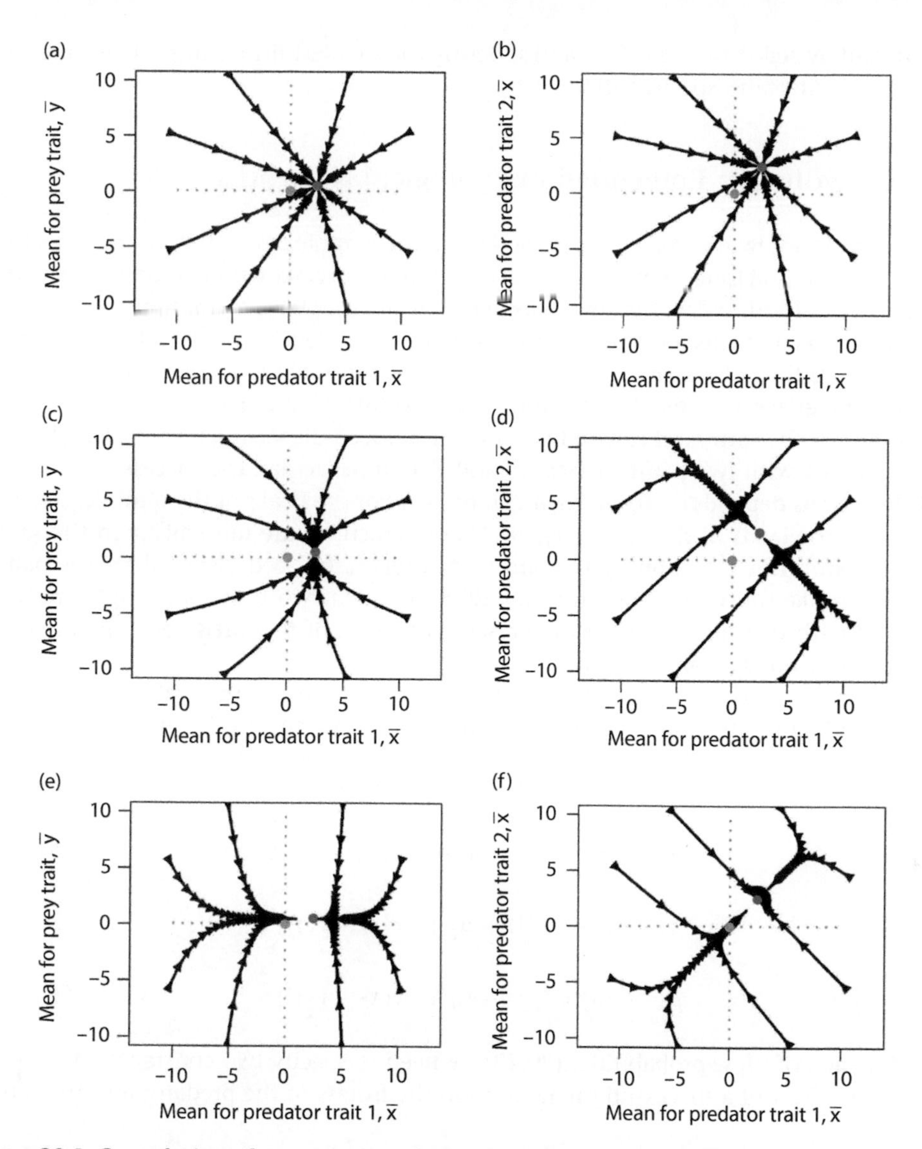

Figure 20.9 Coevolution of a prey trait and one of two interacting predator traits. Graphic conventions as in Figure 20.5. The genetic correlation between the two predator traits is zero ($G_{x,12} = 0$) in the first row of three graphs, 0.8 ($G_{x,12} = 0.32$) in the second row, and 0.8 ($G_{x,12} = -0.32$) in the third row. Arrowheads every 50 generations for 500 generations in the top row. Arrowheads every 50 generations for 1000 generations in the middle and bottom rows. Parameter values used in all calculations: $G_{x,11} = G_{x,22} = 0.4$, $G_y = G_z = 0.4$, $s_y = s_z = 0.01$, $s_{x,1} = s_{x,2} = 0.05$, $\theta_y = \theta_z = \theta_{x,1} = \theta_{x,2} = 0$, $\omega_y = \omega_z = \omega_{x,1} = \omega_{x,1} = 49$. Companion plots of the mean of prey trait, $\bar{z}$, as a function of the mean of the predator trait, $\bar{x}_2$, were identical to the plots shown here. (a) Coevolutionary trajectories of the mean of prey trait, $\bar{y}$, as a function of the mean of the predator trait, $\bar{x}_1$, when the genetic correlation between the two predator traits is zero. (b) Coevolutionary trajectories of the mean of predator trait 2, $\bar{x}_2$, as a function of the mean of predator trait 1, $\bar{x}_1$, when the genetic correlation between the two parasite traits is zero ($G_{x,12} = 0$). (c) Coevolutionary trajectories of the mean of prey trait, $\bar{y}$, as a function of the mean of the predator trait, $\bar{x}_1$, when the genetic correlation between the two predator traits is 0.8. (d) Coevolutionary trajectories of the mean of predator trait 2, $\bar{x}_2$, as a function of the mean of predator trait 1, $\bar{x}_1$, when the genetic correlation between the two parasite traits is 0.8 ($G_{x,12} = 0.32$). (e) Coevolutionary trajectories of the mean of prey trait, $\bar{y}$, as a function of the mean of the predator trait, $\bar{x}_1$, when the genetic correlation between the two parasite traits is −0.8. (f) Coevolutionary trajectories of the mean of predator trait 2, $\bar{x}_2$, as a function of the mean of predator trait 1, $\bar{x}_1$, when the genetic correlation between the two parasite traits is −0.8 ($G_{x,12} = -0.32$).

constrained by stabilizing selection and suffering the smallest fitness impact becomes the target of the cheetah's specialization.

20.6 Coevolution Embedded in Ecological Networks

In this section we take a step closer to modelling the complexity of the interactions that are found in natural food webs by allowing each trait to interact with two traits in another species. In addition, we will follow Nuismer's (2017, Chapter 12) account by considering four interacting species, two in each of two trophic levels (Figure 20.10). Nuismer (2017) stresses the point that Figure 20.8 depicts the minimal model needed to study the fundamental effects of ecological networks on coevolution and vice versa.

Consider the hypothetical case of two fruit-eating, seed-dispersing bird species (X_1 and X_2) interacting with two plant species (Y_1 and Y_2). In particular, the success or failure of the interactions depends on the *difference* in the sizes of the beaks of the birds (x_1, x_2) and the sizes of the fruits of the plants (y_1, y_2). The interactions are mutualistic in the sense that successful fruit ingestion by the bird means successful seed dispersal for the plant. However, unlike the simple case in Figure 20.5 in which two traits of a single predator interacted with two prey traits, in the present case each of two predator traits interacts with *both* prey traits (Figure 20.10).

Modelling four pairwise interactions as differences between the traits of birds and plants, we obtain four expressions for the probability of success in the interactions:

$$p(X_1, Y_1) = 1/\left[1 + exp\left\{-\alpha\left(x_1 - y_1\right)\right\}\right] \tag{20.43a}$$

$$p(X_1, Y_2) = 1/\left[1 + exp\left\{-\alpha\left(x_1 - y_2\right)\right\}\right] \tag{20.43b}$$

$$p(X_2, Y_1) = 1/\left[1 + exp\left\{-\alpha\left(x_2 - y_1\right)\right\}\right] \tag{20.43c}$$

$$p(X_2, Y_2) = 1/\left[1 + exp\left\{-\alpha\left(x_2 - y_2\right)\right\}\right]. \tag{20.43d}$$

Next, for each of these probabilities (20.43) we need to specify two constants, one representing the effect of a successful interaction on the fitness of the predator and the other

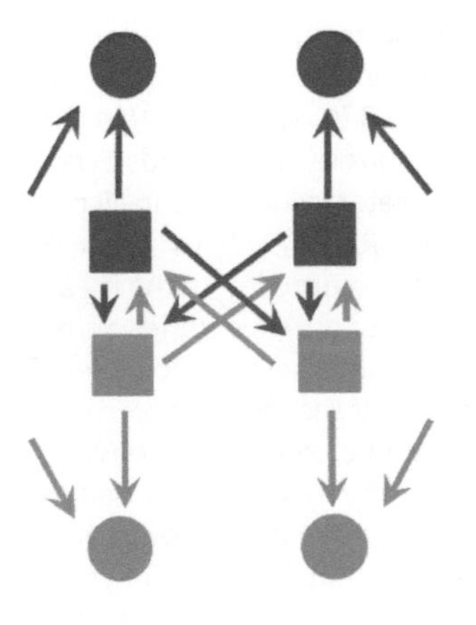

Figure 20.10 Network diagram of the bill traits of two bird species (blue) interacting with the fruit traits of two plant species (green). Population mean fitnesses are shown as circles; trait means are shown as boxes. Other conventions as in Figure 20.4.

the fitness effect on the prey. Then, assuming that each trait is under stabilizing selection towards a function optimum, we specify the individual fitness of each species as function of interactions among traits and stabilizing selection on individual traits. Averaging these fitness functions over trait distributions, we obtain expressions for average fitness in each of the four species and solve for the selection gradients acting on the four traits.

Using those selection gradient expressions (shown below) and the univariate version of our standard expression for response to selection (since we model the evolution of a single trait in each species) we obtain the following expressions for response to selection:

$$\Delta\bar{x}_1 = G_{x_1}\left(\beta_{x_1 I} + \beta_{x_1 F}\right) \tag{20.44a}$$

$$\Delta\bar{x}_2 = G_{x_2}\left(\beta_{x_2 I} + \beta_{x_2 F}\right) \tag{20.44b}$$

$$\Delta\bar{y}_1 = G_{y_1}\left(\beta_{y_1 I} + \beta_{y_1 F}\right) \tag{20.44c}$$

$$\Delta\bar{y}_2 = G_{y_2}\left(\beta_{y_2 I} + \beta_{y_2 F}\right), \tag{20.44d}$$

where the interaction and functional selection gradients are:

$$\beta_{x,1I} = s_{x,1} \tag{20.45a}$$

$$\beta_{x,2I} = s_{x,2} \tag{20.45b}$$

$$\beta_{y,1I} = -s_{y,1} \tag{20.45c}$$

$$\beta_{y,2I} = -s_{y,2} \tag{20.45d}$$

$$\beta_{x,1F} = \left(\theta_{x_1} - \bar{x}_1\right)/2\omega_{x_1} \tag{20.45e}$$

$$\beta_{x,2F} = \left(\theta_{x_2} - \bar{x}_2\right)/2\omega_{x_2} \tag{20.45f}$$

$$\beta_{y,1F} = \left(\theta_{y_1} - \bar{y}_1\right)/2\omega_{y_1} \tag{20.45g}$$

$$\beta_{y,2F} = \left(\theta_{y_2} - \bar{y}_2\right)/2\omega_{y_2}. \tag{20.45h}$$

Notice that the interaction selection gradients have contrasting signs in the two species.

Nuismer's (2017, Chapter 12) analysis of expressions (20.44) revealed that the trait means evolve towards a stable equilibrium point given by:

$$\hat{x}_1 = \theta_{x_1} + \beta_{x,1I}\omega_{x_1} \tag{20.46a}$$

$$\hat{x}_2 = \theta_{x_2} + \beta_{x,2I}\omega_{x_2} \tag{20.46b}$$

$$\hat{y}_1 = \theta_{y_1} - \beta_{y,1I}\omega_{y_1} \tag{20.46c}$$

$$\hat{y}_2 = \theta_{y_2} - \beta_{y,2I}\omega_{y_2}. \tag{20.46d}$$

Notice that the bird and plant trait means deviate in opposite directions from their optima and that this is a consequence of the bird-plant difference in the signs of the interaction selection gradients (20.45a–d). In other words, the plant species evolve fruits that are smaller than optimal, while the bird species evolve bills that are larger than optimal.

This model is similar to the one described in Section 20.2, except that we have two birds each with one trait (bill size) interacting with two plants each with one trait (fruit size). As a consequence, we have more complex expressions for the interaction selection gradients. However, if we simplify by treating these gradients as constants, then the coevolutionary dynamics are similar to those in Section 20.2. For example, if we use symmetrical values of parameters (e.g., $S_{x_1} = S_{y_1} = S_{x_2} = S_{y_2}$, , $G_{x_1} = G_{y_1} = G_{x_2} = G_{y_2}$), then bivariate means evolve on straight line trajectories towards their equilibria, as in Figure 20.3a. However, asymmetrical values of interaction coefficients (e.g., $S_{x_1} \neq S_{y_1}$) change the position of the equilibrium point. Asymmetrical values of genetic variance ($G_{x_1} \neq G_{y_1}$) or stabilizing selection ($\omega_{x_1} \neq \omega_{y_1}$) produces curved trajectories on the approaches to equilibria (see Figure 20.5b, for the case of asymmetrical stabilizing selection).

Nuismer (2017, Chapter 12) also uses his model to characterize the interaction network of the four mutualistic, interacting species at evolutionary equilibrium. The first step in this analysis is to average the success probabilities (20.43) over the trait distributions to obtain expressions for expected (average) rates of interaction between all pairs of species. These expressions are then evaluated at the evolutionary equilibrium by substituting the equilibrium trait means (20.46) into the expressions for average rates of interaction. Nuismer uses his expressions for average rates of interaction to compute *interaction efficiency* (Nuismer et al. 2013), a measure of connectivity in the community network. This efficiency measure predicts the probability that a randomly chosen bird and plant will interact successfully in the sense that the fruit is eaten and seeds are dispersed,

$$
\begin{aligned}
I &= \frac{1}{2} + \frac{\alpha}{4}\left(\bar{\beta}_{xI}\omega_x + \bar{\beta}_{yI}\omega_y + \bar{\theta}_x - \bar{\theta}_y\right) \\
&= \frac{1}{2} + \frac{\alpha}{4}\left[\left(\bar{\theta}_x + \bar{\beta}_x\omega_x\right) - \left(\bar{\theta}_y - \bar{\beta}_{yI}\omega_y\right)\right] \\
&= \frac{1}{2} + \frac{\alpha}{4}\left(\hat{x} - \hat{y}\right),
\end{aligned}
\tag{20.47}
$$

where $\bar{\beta}_{xI} = (\beta_{x,1I} + \beta_{x,2I})/2$, $\bar{\beta}_{yI} = (\beta_{y,1I} + \beta_{y,2I})/2$, $\omega_x = (\omega_{x_1} + \omega_{x_2})/2$, $\omega_y = (\omega_{y_1} + \omega_{y_2})/2$, $\bar{\theta}_x = (\theta_{x_1} + \theta_{x_2})/2$, $\bar{\theta}_y = (\theta_{y_1} + \theta_{y_2})/2$, $\hat{x} = (\hat{x}_1 + \hat{x}_2)/2$, and $\hat{y} = (\hat{y}_1 + \hat{y}_2)/2$. The third version of this expression shows that it reaches it maximum value if the average values of the bird traits at equilibrium is greater than the average value of the plant traits at equilibrium. The significance of this expression is that it connects a measure of community structure to the selection parameters that drive coevolution in the community.

Nuismer (2017) points out that the overall message of expression (20.47) is that communities experiencing strong coevolutionary selection (large $\bar{\beta}_{xI}$ and $\bar{\beta}_{yI}$) evolve the highest rates of species interaction and the highest levels of connectivity. Those highest rates and levels are manifested by the equilibrium trait values of higher trophic level species being more extreme than those of the lower tropic level species. Functional stabilizing selection constrains trait evolution and works against the evolution of those high rates and levels of interaction. Nuismer (2017) also shows that asymmetries in interaction and functional selection gradients can bias evolution towards nested network structures.

Nuismer (2017, Chapter 12) uses the model (diagrammed in Figure 20.10) to model antagonistic coevolution in a community of four interacting species. The change from mutualistic to antagonistic interactions is accomplished by changing the signs of the interaction coefficients that specify the fitness consequences to interacting species. The consequence of those changes is that trait values evolve in an arms race towards a stable equilibrium in which the traits values of all four species are larger than their functional equilibria. One sign in the expression for interaction efficiency (20.47) is changed as well, but this change makes it harder to make predictions about connectivity in communities with antagonistic interactions.

Finally, Nuismer (2017, Chapter 12) uses his model (Figure 20.10) to explore mutualistic coevolution in a community of four species with matching rather than difference interactions. As in the difference case, matching interactions leads to equilibria with high levels of interactions that in some cases promote modular structure of the interaction network.

20.7 Effects of Stochastic Coevolution on Adaptive Radiations

In this section we introduce a stochastic dimension to the coevolution of traits in two interacting species. Up until this point, we have considered only the deterministic coevolution of two or more traits. In particular, those results ignored the contribution of errors of parental sampling that occur in populations of finite size. The stochastic contribution of that sampling each generation causes trait means to drift, which can in turn produce differences among replicate lineages exposed to the same parameters of inheritance and selection.

We follow Nuismer's (2017, Chapter 13) account of drift in coevolving traits with a focus on the general properties of the process. In particular, we will consider whether species interaction can take the place of stabilizing selection in constraining an adaptive radiation based on drift. We will also add stabilizing selection to Nuismer's (2017) model and consider the coevolutionary balance that it can produce in conjunction with drift.

A crucial part of Nuismer's (2017) analysis is to follow trait evolution through the process of speciation and afterwards when the diverging trait values becomes the crux of competitive interaction between the two sister species. More generally, Nusimer models a succession of speciation events as competitive interactions spread through an adaptive radiation. Our initial focus, however, is on a particular speciation event at time τ_1 and the stochastic trait evolution that unfolds afterwards in the two sister species (Figure 20.11).

Consider a trait in a pair of sister species (1 and 2) that speciate at time τ_1 and then diverge by genetic drift until τ_2 when another speciation event occurs (Figure 20.11). Employing our model of stochastic divergence by drift among replicate lineages (Section

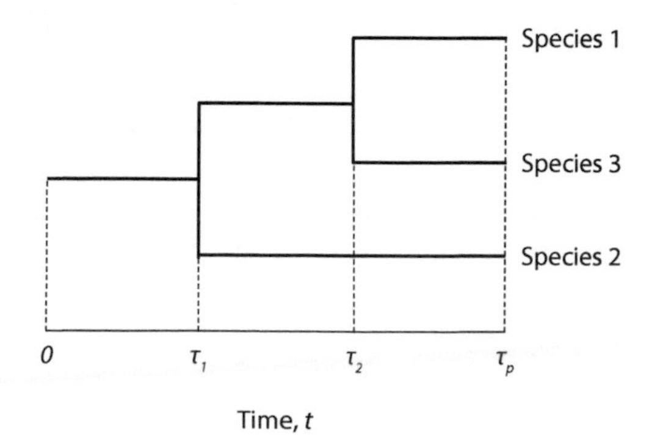

Figure 20.11 Phylogeny of three competing species. Speciation events occur at times τ_1 and τ_2. The present time is denoted τ_p. After Nuismer (2017).

8.5), we do not expect the value of the trait mean to change from generation, but each generation the variance about that expectation increases by an amount equal to G/N_e. Setting the ancestral trait value at zero and with no variance about that value at time 0, we follow the lineage forward in time. At time τ_1, the instant after speciation, the two sister species have the same expected value of the trait mean (0), and the variance about those expectations is proportional to the length of the elapsed interval. In other words,

$$E\left(\bar{z}_{1,\tau_1}\right) = 0 \tag{20.48a}$$

$$E\left(\bar{z}_{2,\tau_1}\right) = 0 \tag{20.48b}$$

$$D_{1,\tau_1} = \frac{1}{N_e} G\tau_1 \tag{20.48c}$$

$$D_{2,\tau_1} = \frac{1}{N_e} G\tau_1 \tag{20.48d}$$

$$D_{12,\tau_1} = \frac{1}{N_e} G\tau_1. \tag{20.48e}$$

The dispersion denoted D_{12,τ_1} is the covariance between the means of the two sister species, which initially has the same value as the within-lineage variances, D_{1,τ_1} and D_{2,τ_1}. This process of drift producing increasing variance about the expected value of the trait is depicted in Figure 8.2.

We now model the ecological interaction that occurs between the two sister species after speciation. We follow Nuismer (2017) in assuming direct interference competition that depends on the difference in trait values and ignore competition within each species. Our model for individual fitness within each species is:

$$W_1 = 1 - s\frac{1}{1 + exp\left\{-\alpha\left(z_2 - z_1\right)\right\}} \tag{20.49a}$$

$$W_2 = 1 - s\frac{1}{1 + exp\left\{-\alpha\left(z_1 - z_2\right)\right\}}, \tag{20.49b}$$

where s measures the fitness cost of losing an interaction and α converts the difference in trait values into fitness units. Notice that effects on fitness have negative signs in both species.

Averaging these functions over the two distributions of trait values, we obtain approximate expressions for population mean fitness in each species:

$$\bar{W}_1 \approx 1 - \frac{s}{4}\left(2 + \alpha\left(\bar{z}_2 - \bar{z}_1\right)\right) \tag{20.50a}$$

$$\bar{W}_2 \approx 1 - \frac{s}{4}\left(2 + \alpha\left(\bar{z}_1 - \bar{z}_2\right)\right). \tag{20.50b}$$

As before, we differentiate these expressions with respect to the trait mean to obtain the interaction selection gradient,

$$\beta_I = \alpha s/\left(4 - 2s\right), \tag{20.51}$$

which takes the same value in both species. See Nuismer (2017) for the details of this somewhat surprising result.

We substitute into our standard response to selection equation and add a per-generation contribution due to drift of the mean, which is a normally distributed random variable with a mean of zero and a variance of G/N_e. The change in the trait mean each generation in each species is

$$\Delta\bar{z}_1 = G\beta_I + N(0, G/N_e) \tag{20.52a}$$

$$\Delta\bar{z}_2 = G\beta_I + N(0, G/N_e)\,. \tag{20.52b}$$

Next we follow the process of trait divergence during the period of interspecies competition from time τ_1 to τ_2. Summing up the changes in trait mean given by (20.52) over that interval, and after a little algebra, we obtain the following expressions for the expected trait mean, variance about that expectation for two sets of replicate lineages, and the covariance between those two sets at time τ_2:

$$E\left(\bar{z}_{1,\tau_2}\right) = G\beta_I\left(\tau_2 - \tau_1\right) \tag{20.53a}$$

$$E\left(\bar{z}_{2,\tau_2}\right) = G\beta_I\left(\tau_2 - \tau_1\right) \tag{20.53b}$$

$$D_{1,\tau_2} = \frac{1}{N_e}G\tau_2 \tag{20.53c}$$

$$D_{2,\tau_2} = \frac{1}{N_e}G\tau_2 \tag{20.53d}$$

$$D_{12,\tau_2} = \frac{1}{N_e}G\tau_1\,. \tag{20.53e}$$

These expressions describe steady change in the trait means of the two species under the force of interaction selection gradients of constant value over the time interval after speciation, from τ_1 to τ_2. For the moment we specified those gradients as having the same value in both species, although the picture would be similar if the values were different. Around each of those evolving means is a steadily expanding 95% confidence interval specified by increasing variances and constant covariance given by the last three expressions. We discussed the single trait version of this model in Section 8.3 and depicted it in Figure 8.2.

Nuismer (2017) uses the model elaborated up to this point to analyse how coevolution shapes the pattern of interaction among competing species. In particular, he shows how the model might lead to asymmetries in competitive ability and hence to the probability that competition will have a consistent winner and loser. Nuismer (2017) also derives an expression for an *index of asymmetry* in competitive ability and shows that the equilibrium value of this index is

$$\hat{A}_{ij} = \frac{\alpha^2}{2}\frac{G}{N_e}\left(\tau_p - \tau_{ij}\right), \tag{20.54}$$

where τ_{ij} is the time when species i and j speciated, and the time interval in parentheses is the elapsed time since speciation. In other words, the probability of a consistent winner and loser increases steadily after speciation as a consequence of the drift in the traits of the two sister species.

The model under consideration has severe limitations as a model for adaptive radiations, because directional selection on the trait means is modelled as a constant. Using

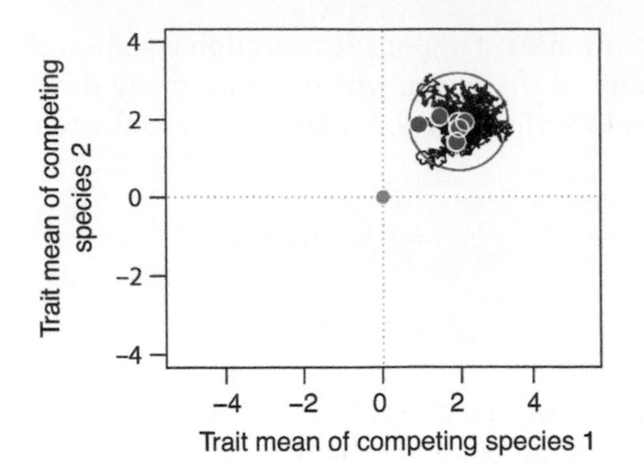

Figure 20.12 Stochastic coevolution of the means of the interacting traits of a pair of competing species in drift-selection balance. Same parameter values as in Figure 20.4a. Stochasticity is a consequence of genetic drift of the mean. Effective population size is 100 in each species. Five independent trajectories are shown, all beginning at the expected equilibrium ($\hat{x} = 1.93$, $\hat{y} = 1.93$), shown as a solid blue dot (almost totally obscured by the red dots), surrounded by the 95% confidence ellipse (blue circle) for coevolutionary outcome under drift-selection balance at time infinity (Section 13.4). The final position of each trajectory after 500 generations is shown as a solid red dot. The position of the functional optimum is shown as a solid orange dot.

the benchmark of empirical values for trait divergence, the model produces too much differentiation, especially on long timescales (Section 10.1). For this reason, the modelling literature (and this book) have moved beyond perpetual, constant directional selection to consider stabilizing selection and stochastic peak movement (e.g., Chapter 14). It is, however, straightforward to incorporate stabilizing selection into the stochastic drift model specified early in this section so that evolution is bounded rather than ever expanding, as in (20.53).

To incorporate functional stabilizing selection, we use the trait difference model in Section 20.2, i.e., expression (20.12), and add a random drift term, as in expression (20.52). Sample trajectories using this combined model are shown in Figure 20.12. The change from a parasite-host interaction to competition has changed the position of the expected coevolutionary equilibrium (compare with Figure 20.5a). More importantly, the addition of drift in the means has created a stochastic cloud of trajectory endpoints at any particular time around the deterministic equilibrium point. The 95% confidence limits that describe that stochastic cloud are a function of the weakness of stabilizing selection ($\Omega = \omega + P$) and effective population size, Section 13.4. We see in Figure 20.12 that the expected span of the radiation is about 1 within-population phenotypic standard deviations away from the equilibrium point. In other words, we have specified a process that produces radiation of very modest size. Furthermore, no feasible combination of parameters will enable this model to produce a radiation that matches even the modal much less the upper values of empirical size estimates (Chapter 12).

20.8 Summary and Assessment

The approach taken in this chapter has illuminated many features of coevolution. At the same time it is important to realize that the formulas we have used to characterize interspecies interaction are not explicit functions of the abundance of the two species.

That simplification is convenient, but it means that the formulations of interaction are divorced from a large ecological literature on interaction. In the next chapter we review theoretical studies of trait coevolution that incorporate abundance and so offer a stronger bridge to empirical validation.

20.9 Conclusions

- Species interactions can prolong the evolution of the bivariate trait mean of two interacting species towards a stable equilibrium.
- Transit of the trait mean to the equilibrium can be additionally delayed by genetic correlations between traits within one or both species.
- The nature of species interactions affects coevolutionary outcome as well as the trajectory. In particular, interactions based on trait differences between the two species lead to a different outcome than do interactions based on trait matching. An interaction based on trait matching leads to more curvature in coevolutionary trajectories than an interaction based on trait differences and hence to more delay in reaching a stable equilibrium, as well as increased fitness costs.
- Coevolutionary trajectories can converge on a stable equilibrium point, runaway from an unstable point, or oscillate with increasing amplitude. A stable outcome is probably the most likely of these possibilities because it demands less restrictive assumptions about parameter values.
- Interactions between species can result in a stable outcome in which trait means equilibrate at some distance from an optimum defined by functional considerations.
- Geographic mosaics in which trait values vary geographically in a pair of interacting species are common in nature. The extent of such geographic variation is exaggerated by geographic differences in the strength of the interaction and diminished by functional stabilizing selection and gene flow.
- The nature of interaction (trait difference versus trait matching) affects the scope of local adaptation between a pair of interacting species. Tight local adaptation can occur if with a trait matching interaction but not with a trait difference interaction.
- The sign of species interactions affects coevolutionary process and outcome. When interactions are mutualistic (all signs positive) coevolution results in a single, stable deterministic outcome. In contrast, antagonistic interactions (with different signs) can result in unstable, runaway coevolution. Furthermore, resource specialization is less likely to evolve if interactions are mutualistic rather than antagonistic.
- The nature of selection generated by species interactions can affect community structure. Communities with stronger coevolutionary selection (arising from species interactions) evolve higher rates of interaction and higher levels of connectivity and modularity.
- Coevolution can also shape the equilibrium probability that competition will have a consistent winner and loser.
- Finite population size creates a stochastic envelope of possibilities around the determinist trajectories and outcomes of the coevolutionary process. However, if the coevolutionary process leads to a stable equilibrium point, the stochastic cloud around this point is modest in size and inadequate as an explanation for the size of radiations observed in nature.

Literature Cited

Bro-Jorgensen, J. 2013. Evolution of sprint speed in African savannah herbivores in relation to predation. Evolution 67:3371–3376.

Brodie, E. D., B. J. Ridenhour, and E. D. Brodie III. 2002. The evolutionary response of predators to dangerous prey: Hotspots and coldspots in the geographic mosaic of coevolution between garter snakes and newts. Evolution 56:2067–2082.

Drobniak, S. M., A. Dyrcz, J. Sudyka, and M. Cichon. 2014. Continuous variation rather than specialization in the egg phenotypes of cuckoos (*Cuculus canorus*) parasitizing two sympatric reed warbler species. PLoS ONE 9:e106650.

Kiester, A. R., R. Lande, and D. W. Schemske. 1984. Models of coevolution and speciation in plants and their pollinators. Am. Nat. 124:220–243.

Lande, R. 1981. Models of speciation by sexual selection on polygenic traits. Proc. Natl. Acad. Sci. USA 78:3721–3725.

Lande, R. and S. J. Arnold 1983. The measurement of selection on correlated characters. Evolution 37: 1210–1226.

Nuismer, S. 2017. Introduction to Coevolutionary Theory. MacMillan Higher Education, New York.

Nuismer, S. I., P. Jordano, and J. Bascompre. 2013. Coevolution and the architecture of mutualistic interactions. Evolution 67: 338–354.

Sinclair, A. R., S. Mduma, and J. S. Brashares. 2003. Patterns of predation in a diverse predator-prey system. Nature 425: 288–290.

Thompson, J. N. 1994. The Coevolutionary Process. University of Chicago Press, Chicago.

Thompson, J. N. 2005. The Geographic Mosaic of Coevolution. University of Chicago Press, Chicago.

Thompson, J. N. 2013. Relentless Evolution. University of Chicago Press, Chicago.

Toju, H., and T. Sota. 2006. Imbalance of predator and prey armament: geographic clines in phenotypic interface and natural selection. Am. Nat. 167:105–117.

CHAPTER 21

Coevolution of Species with Density-dependent Interactions

Overview—In this chapter we explore trait coevolution in a model that allows the adaptive landscape to evolve as a consequence of the changing abundances of interacting species. As in the last chapter, we recognize that ecological interactions are mediated by phenotypic traits, and we use mathematical characterizations of those interactions to fold ecology into our treatment of trait coevolution. The primary new feature in this chapter is the recognition that fitness and population growth rate are equivalent currencies. We use that equivalence to build a formal bridge between population dynamics and trait coevolution. Using that bridge we follow McPeek's (2017a, b) treatment of the case of two interacting species (resource and consumer), in which interaction is mediated by a single trait in each species. Using McPeek's equations for demographic dynamics and trait coevolution, we study his use of simulations to draw general lessons about coevolution, focusing on five take-away messages from this test case. (1) The two species can often reach a joint demographic and evolutionary equilibrium in which species coexist and their population sizes and the values of trait means are stable. (2) Ecological interactions push this equilibration point downslope from adaptive peaks. (3) The approach to this equilibrium can be slow and circuitous. In other words, interaction induces temporary as well as permanent maladaptation. Coevolution does not rescue species from this predicament. (4) Because interaction affects the abundance of each species, their adaptive landscapes change through time. (5) Temporal change in the location and configuration of adaptive peaks can cause trait means to shift from one peak to another.

The perspectives on trait coevolution developed by Nuismer (2017) and McPeek (2017a, b), which form a common foundation for this chapter and Chapter 20, rests on two premises. (1) Ecological interactions are mediated by phenotypic traits. This statement seems obvious enough but it carries the less obvious implication that the consequences of ecological interactions can be deduced by using trait-based functions to assess the strength of interaction. Furthermore, empirical evidence supports the view that phenotypic traits mediate ecological interactions (Ohgushi et al. 2012; Hendry 2016). (2) The mathematical form of the interaction, couched in terms of the trait values of interacting parties, can take various forms such as the matching and difference functions used by Nusimer (2017) or the uni- or bi-directional functions used by McPeek (2017b). The diversity of these interaction functions enlarges our concept of selection arising from ISSs (Chapters 3, 4) to include asymptotic as well as unimodal selection surfaces.

The new perspective in this chapter formalizes coevolution using two sets of equations. One set describes change in the abundance of interacting species, the other set describes

Evolutionary Quantitative Genetics. Stevan J. Arnold, Oxford University Press. © Stevan Arnold (2023).
DOI: 10.1093/oso/9780192859389.003.0022

evolution of the traits that mediate those species interactions. Abrams (2000), Holt and Barfield (2012), and others have used this approach with profit, but we will focus on McPeek's (2017a, b) formulation, which articulates a simple but powerful connection between the two equation sets.

21.1 The Connection Between Fitness and Population Growth Rate

Fitness is the currency of evolutionary biology, just as population growth rate is the currency of ecology. In this chapter we detail the equivalence of these currencies and use that equivalence to model the coevolution of traits in interacting species. The equivalence of the two currencies was elaborated by Lande (1982, 2007) and Charlesworth (1994) in their models of life history evolution and portrayed by McPeek (2017a, b) (Figure 21.1).

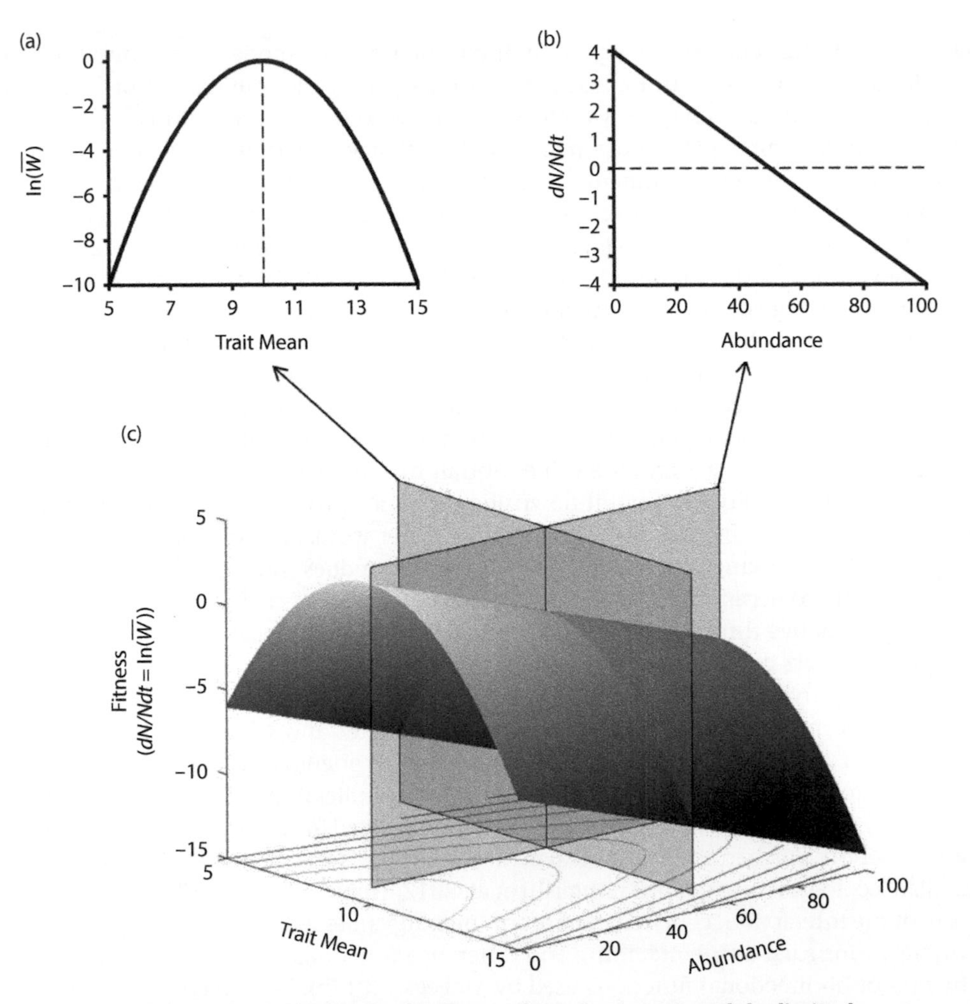

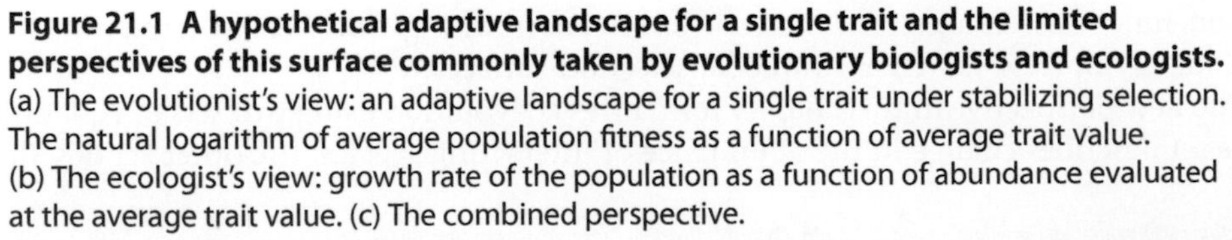

Figure 21.1 A hypothetical adaptive landscape for a single trait and the limited perspectives of this surface commonly taken by evolutionary biologists and ecologists. (a) The evolutionist's view: an adaptive landscape for a single trait under stabilizing selection. The natural logarithm of average population fitness as a function of average trait value. (b) The ecologist's view: growth rate of the population as a function of abundance evaluated at the average trait value. (c) The combined perspective.

From McPeek (2017a) with permission.

The incorporation of population growth rate into our models of trait coevolution will mean that species evolve on adaptive landscapes in which the shape of the surface as well as the position of optima changes through time. R. C. Lewontin surmised that the changing abundance of interacting species would cause the adaptive landscape to behave like an *environmental trampoline* (Gould 1997), but—as we shall see—McPeek's evolving landscapes are less chaotic than Lewontin's metaphor implies.

Our account of the currency equivalence follows McPeek's (2017b) treatment, which is based on the approach of Lande (1982, 2007) and Charlesworth (1994). In the quantitative genetic models pioneered by Lande (1976, 1979), fitness of a particular phenotype is a lifetime count of newly conceived progeny. Likewise in ecology, population growth is the net infusion of newly conceived individuals into the population. Consequently, the equivalence of the two currencies rests on a firm intuitive foundation. Indeed, the equivalence can be formally established with just a few algebraic steps.

For simplicity, we assume at first that all individuals in a population have the same phenotype. Let N_i be the number of individuals of the *ith* species and let lnW_i be the absolute fitness of one of the identical individuals in that species, then the rate of population growth is

$$\frac{dN_i}{dt} = N_i lnW_i. \tag{21.1}$$

Relaxing the assumption of identical phenotypes and letting $n_i(z_i)$ be the number of individuals with phenotype z_i, we can also express the number of individuals in the population as an integration over phenotypes,

$$N_i = \int n_i(z_i)\, dz_i. \tag{21.2}$$

The proportion of individuals with a particular phenotype is a simple ratio,

$$p_i(z_i) = \frac{n_i(z_i)}{N_i}. \tag{21.3}$$

Letting $lnW_i(z_i)$ be the absolute fitness of an individual with phenotype z_i, we can express population growth as an integration over phenotypes,

$$\frac{dN_i}{dt} = \int n_i(z_i)\, lnW_i(z_i)\, dz_i = N_i \int p_i(z_i)\, lnW_i(z_i)\, dz_i. \tag{21.4}$$

Notice that in these few steps, we have derived an expression for population growth, a fundamental ecological process, in terms of the fitness of individuals, a fundamental evolutionary property.

21.2 A Model for Species Interaction and Coevolution With Changing Abundance

We now turn to the fitness consequences of interaction between two species, *i* and *j*, again following McPeek's (2017a, b) account. We will assume that the interaction is mediated by a single phenotypic trait in each species, z_i and z_j. Define $f_{ij}(z_i, z_j)$ as the interaction function that produces fitness consequences in each species. For the moment, let us assume that all fitness is mediated through this function. Then we can express the absolute fitness of an individual in species *i* as a sum of interaction contributions from each of the *p* species ($j = 1, 2, \cdots p$) with which it interacts,

$$lnW_i(z_i) = \sum_j N_j \int p_j(z_j) f_{ij}(z_i, z_j)\, dz_j = \sum_j N_j f_{ij}(z_i, \bar{z}_j)\,. \tag{21.5}$$

The first expression on the right gives absolute fitness in terms of an integration over phenotypic variability in each of the species ($j = 1, 2, \cdots p$), while the second expression approximates each of those integrations by simply using the average phenotypic in each species ($j = 1, 2, \cdots p$).

We are now in a position to connect our ecological model of species interaction with our standard response to selection equation for trait evolution. In continuous time, the rate of change in the mean of the trait of the *ith* species is a product of genetic variance for the trait, G_i, and a directional selection gradient for that trait evaluated at the trait mean,

$$\frac{d\bar{z}_i}{dt} = G_i \left.\frac{\partial lnW_i(z_i)}{\partial z_i}\right|_{z_i=\bar{z}_i} = G_i \left(\sum_j N_j \left.\frac{\partial f_{ij}(z_i, \bar{z}_j)}{\partial z_i}\right|_{z_i=\bar{z}_i} \right). \tag{21.6}$$

In the first expression on the right, we have expressed the directional selection gradient as a partial derivative of individual fitness with respect to trait value, evaluated at the mean value of the trait (12.1). In the second expression on the right, the selection gradient is equivalently expressed as a sum of fitness effects from the species that interact with species *i*.

With these expressions (21.5, 21.6), we have completed our bridge between ecology and evolution. Focusing on two interacting species, we have an equation (21.4) based on (21.5) for population growth, and an equation (21.6) based on (21.5) for trait evolution arising from their ecological interaction. For an analogous pair of equations, without the explicit link to quantitative genetic models of trait evolution, see Holt and Barfield (2012). Next we make these formulations operational by focusing on a single pair of species in which fitness is a function of deviation from a trait optimum as well as the interaction between species.

We will begin with the meaning of interaction, encapsulated in the function $f_{ij}(z_i, z_j)$. Following McPeek's (2017b, Chapter 2) account, consider a simple two-level community consisting of a single resource species, *R*, and a single consumer species, *N*. We follow McPeek and assume that trait value of the resource species affects the fecundity of the consumer species, whereas the trait value of the consumer species affects the mortality of the resource species. These interaction effects on individual fitness and growth rate of the two species are expressed in the following two equations:

$$lnW_N(z_N) = \frac{dN_{z_N}}{Ndt} = \frac{ba(\bar{z}_R, z_N)R}{1 + a(\bar{z}_R, z_N)hR} - f(z_N) - gN \tag{21.7a}$$

$$lnW_R(z_R) = \frac{dR_{z_R}}{Rdt} = c(z_R) - dR - \frac{a(z_R, \bar{z}_N)N}{1 + a(z_R, \bar{z}_N)hR}. \tag{21.7b}$$

See Table 21.1 for the definitions of the parameters in these equations, which are based on the Rosenzweig-MacArthur consumer-resource model (Rosenzweig and MacArthur 1963)

Table 21.1 Definitions of variables and parameters used by McPeek (2017a, b). In those cases in which a different symbol is used in this chapter (shown in the first column), the equivalent McPeek symbol is shown in parentheses in the second column.

Symbol	Meaning
$lnW_N\,(z_N)$	$\log_e$ absolute fitness of an individual consumer with phenotype z_N
$lnW_R\,(z_R)$	$\log_e$ absolute fitness of an individual resource with phenotype z_R
N	Number of consumer individuals
R	Number of resource individuals
N_{z_N}	Number of consumer individuals with phenotype z_N
N_{z_R}	Number of resource individuals with phenotype z_R
b	Conversion efficiency for a particular consumer phenotype feeding on a particular resource phenotype
$a\,(\bar{z}_R, z_N)$	Attack coefficient for consumer phenotype z_N feeding on resource phenotype $\bar{z}_R$; the rate at which a consumer kills a resource (Holling 1959)
a_0	Maximum value of attack success
ω_a	Width parameter for the Gaussian attack function ($\beta^2/2$)
$f(z_n)$	Intrinsic deathrate of the consumer with phenotype z_N
f_0	Minimum value of the consumer's intrinsic death rate
γ_f	Stabilizing selection coefficient for the consumer's intrinsic death rate (θ)
θ_f	Value of z_R that minimizes consumer's intrinsic death rate ($\bar{z}_N^f$)
g	Density-dependent rate of increase in the consumer's death rate
$c\,(z_R)$	Intrinsic birth rate of resource species with phenotype z_R
c_0	Maximum value of the resource intrinsic birth rate
γ_c	Stabilizing selection coefficient for the resource's intrinsic birth rate (γ)
θ_c	Value of z_N that minimizes resource's intrinsic death rate ($\bar{z}_R^c$)
d	Density-dependent rate of decrease in the birth rate of the resource
h	Handling time for a particular consumer phenotype feeding on a particular resource phenotype

and Holling's (1959) saturating functional response expression. In (21.7a), the fitness of the consumer is a function of its trait value (z_N), as well as the average trait value of the resource ($\bar{z}_R$). In particular, we see that the individual fitness function of the consumer species consists of three terms. The first term represents the fitness gain from feeding on the resource. The second term represents loss in fitness from intrinsic causes of death. The third term represents loss in fitness from density-dependent causes of death. Likewise in (21.7b), the fitness of the resource species is a function of its trait value (z_R) as well as the average trait value of the consumer ($\bar{z}_N$) and consists of three terms. The first term represents the fitness gain from the resource's intrinsic birth rate. The second term represents fitness loss from density-dependent causes of death. The third term represents fitness loss from interactions with the consumer.

Evaluating these expressions (21.7) at the trait mean of each focal species, we obtain the following expressions for *per capita growth rate* in each species,

$$\frac{dN}{Ndt} = \frac{ba\,(\bar{z}_R - \bar{z}_N)\,R}{1 + a\,(\bar{z}_R - \bar{z}_N)\,hR} - f(\bar{z}_N) - gN \tag{21.8a}$$

$$\frac{dR}{Rdt} = c\,(\bar{z}_R) - dR - \frac{ba\,(\bar{z}_R - \bar{z}_N)\,N}{1 + a\,(\bar{z}_R - \bar{z}_N)\,hR}. \tag{21.8b}$$

In the preceding paragraphs we discussed the meaning of each of the terms on the right sides of these expressions. Because of the equivalence of population growth rate and fitness, the expressions for fitness evaluated at the traits means (derived from 21.7) describe the adaptive landscape in each species for each trait.

We derive an explicit pair of response to selection equations by substituting our new expressions for fitness (derived from 21.7) into expression (21.6) and obtain:

$$\frac{d\bar{z}_N}{dt} = G_{z_N}\left(\beta_{z_N b} + \beta_{z_N d}\right) \tag{21.9a}$$

$$\beta_{zb} = \left.\frac{\partial\left(\frac{ba(\bar{z}_R - \bar{z}_N)R}{1 + a(\bar{z}_R - \bar{z}_N)hR}\right)}{\partial z_N}\right|_{z_N=\bar{z}_N} = \left.\frac{bR\left(\frac{\partial a(\bar{z}_R - z_N)}{\partial z_N}\right)}{\left(1 + a\left(\bar{z}_R - z_N\right)hR\right)^2}\right|_{z_N=\bar{z}_N} \tag{21.9b}$$

$$\beta_{z_N d} = \frac{\partial\left(-f(z_N) - gN\right)}{\partial z_N}\Big|_{z_N=\bar{z}_N} = -\frac{\partial f(z_N)}{\partial z_N}\Big|_{z_N=\bar{z}_N}. \tag{21.9c}$$

In these expressions we have defined two directional selection gradients, $\beta_{z_N b}$ and $\beta_{z_N d}$. The first of these, $\beta_{z_N b}$, is the *birth selection gradient*, which reflects the contribution of trait values to the birth component of fitness in the consumer species. The first expressions on the right in (21.9b and 21.9c) give the general expression of the partial derivative, while the second expressions give the evaluation of each derivative. Likewise, $\beta_{z_N d}$ is the *death selection gradient*, which reflects the contribution of trait values to the death component of fitness in the consumer species.

Turning to the response to selection expression for the resource species trait, we see the same two components of the directional selection gradient, the birth selection gradient, $\beta_{z_R b}$, and the death selection gradient, $\beta_{z_R d}$,

$$\frac{d\bar{z}_R}{dt} = G_{z_R}\left(\beta_{z_R b} + \beta_{z_R d}\right) \tag{21.9d}$$

$$\beta_{z_R b} = \left.\frac{\partial\left(c\left(z_R\right) - dR\right)}{\partial z_R}\right|_{z_R=\bar{z}_R} = \left.\frac{\partial c\left(z_R\right)}{\partial z_R}\right|_{z_R=\bar{z}_R} \tag{21.9e}$$

$$\beta_{z_R d} = \left.\frac{\partial\left(-\frac{a(z_R - \bar{z}_N)N}{(1 + a(\bar{z}_R - \bar{z}_N)hR)}\right)}{\partial z_R}\right|_{z_R=\bar{z}_R} = \left.\frac{N\left(\frac{\partial a(z_R - \bar{z}_N)N}{\partial z_R}\right)}{\left(1 + a\left(\bar{z}_R - \bar{z}_N\right)hR\right)}\right|_{z_R=\bar{z}_R}. \tag{21.9f}$$

As before with the consumer species, we have given general and specific forms for these components of the selection gradient. However, we need to formulate specific models for the $a(z_R, z_N)$, $c(z_R)$, and $f(z_N)$ functions before we can fully evaluate our selection gradient components.

The attentive reader will have noticed that in making the transition from the individual selection surfaces (21.7) to the adaptive landscape implicit in (21.8) and (21.9), we have skipped the step of averaging the ISS over the phenotypic trait distributions, a step we used, for example, in Chapter 20. Instead we have merely substituted in the trait means for trait values in (21.7). In using this approximation, McPeek (2017b) follows a procedure that is justified and described by Iwasa et al. (1991, Appendix), which assumes that within-population phenotypic variation (P) is so small relative to the width of the ISS that it can be ignored.

Beginning with the evaluation of the *birthrate and deathrate functions*, $c(z_R)$ and $f(z_N)$, we assume that trait value in a species directly affects individual fitness in that same

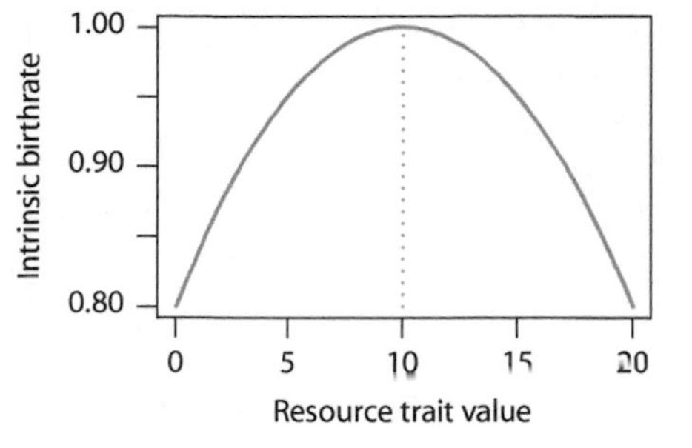

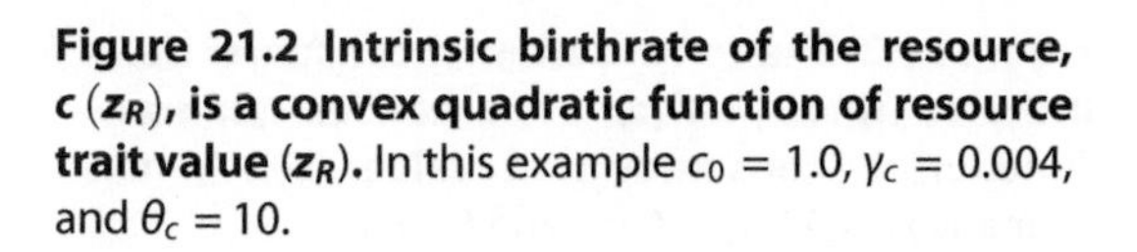

Figure 21.2 Intrinsic birthrate of the resource, $c\,(z_R)$, is a convex quadratic function of resource trait value (z_R). In this example $c_0 = 1.0$, $\gamma_c = 0.004$, and $\theta_c = 10$.

species. We will assume that stabilizing selection acts on the traits in each species and has the effect of anchoring trait evolution in the neighbourhood of an optimum value. These effects are comparable to functional stabilizing selection in Nuismer's (2017) models (Chapter 19). In the resource species we will model the effect with a quadratic selection model, in which the birthrate contribution to individual fitness is given by

$$c\,(z_R) = c_0 \left(1 - \gamma_c (z_R - \theta_c)^2\right), \tag{21.10}$$

where c_0 is the maximum value of the resource species intrinsic birthrate, γ_c is a quadratic selection coefficient, and θ_c is the value of the trait z_R that yields the maximum value of the intrinsic birthrate. Notice that γ_c is a parameter that always takes a positive value, consequently the function is stabilizing (Figure 21.2).

Similarly, in the consumer species the deathrate contribution to individual fitness is given by

$$-f(z_N) = -\, f_0 \left(1 + \gamma_f (z_N - \theta_f)^2\right), \tag{21.11}$$

where f_0 is the maximum value of intrinsic deathrate in the resource species, γ_f is a quadratic selection coefficient, and θ_f is the value of the trait z_C that yields the minimum value of the intrinsic deathrate. Notice that γ_f is a parameter that always takes a positive value, while the function is preceded by a positive sign in the expression, so the function is convex (Figure 21.3).

The contribution of species interaction to individual fitness in both species is given by a Gaussian function

$$a\,(z_R, z_N) = a_0\, exp \left\{ \frac{-(z_N - z_R)^2}{2\omega_a} \right\}, \tag{21.12}$$

where a_0 is the maximum value of attack success and ω_a is the width parameter for the Gaussian function (analogous to a variance). This function is always convex and takes a maximum value when trait values match (Figure 21.4).

McPeek (2017b) uses the example of the bill size of a seed-eating bird to explicate his model for the case of two interacting species and the functional forms shown in Figures 21.2–4. In the plant species, stabilizing selection pulls seed size towards an intermediate optimum that is not related to the interaction with the seed predator (Figure 21.2).

In the bird species, stabilizing selection pulls bill size towards an intermediate optimum that minimizes fitness losses but is unrelated to interaction with the seeds of the plant species (Figure 21.3).

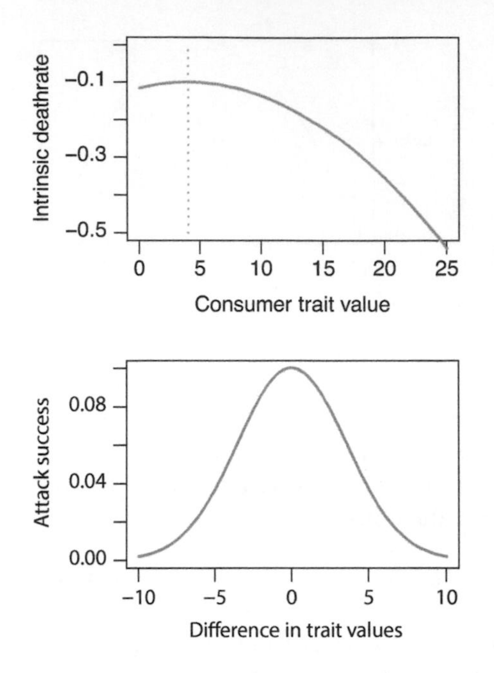

Figure 21.3 The contribution of consumer deathrate to fitness, $-f(z_N)$, is a convex quadratic function of consumer trait value (z_N). In this example $f_0 = 0.1$, $\gamma_f = 0.01$, and $\theta_f = 4$.

Figure 21.4 Attack success, given by $a(z_R, z_N)$, is a matching Gaussian function of the difference in trait values, $(z_R - z_N)$. In this example $a_0 = 0.1$ and $\omega_a = 12.5$.

The success of the bird in attacks on seeds is maximized when the bill size of the bird is matched by a particular seed size (Figure 21.4). High attack success increases bird fitness but decreases plant fitness.

Substituting (21.10–12) into our general expressions for population growth (21.8), we obtain the following specific expressions for population growth:

$$\frac{dN}{dt} = N\left[\frac{ba_0\, exp\left\{\frac{-(\bar{z}_N-\bar{z}_R)^2}{2\omega_a}\right\}R}{1+a_0\, exp\left\{\frac{-(\bar{z}_N-\bar{z}_R)^2}{2\omega_a}\right\}hR} - f_0\left(1+\gamma_f(\bar{z}_N-\theta_f)^2\right) - gN\right] \tag{21.13a}$$

$$\frac{dR}{dt} = R\left[c_0\left(1-\gamma_c(\bar{z}_R-\theta_c)^2\right) - dR - \frac{a_0\, exp\left\{\frac{-(\bar{z}_N-\bar{z}_R)^2}{2\omega_a}\right\}N}{1+a_0\, exp\left\{\frac{-(\bar{z}_N-\bar{z}_R)^2}{2\omega_a}\right\}hR}\right]. \tag{21.13b}$$

Likewise, substituting (21.10–12) into our general expressions for the four directional selection gradients (21.9), we obtain the following specific expressions for the evolutionary response of the consumer trait mean to selection:

$$\frac{d\bar{z}_N}{dt} = G_{z_N}\left(\beta_{z_Nb} + \beta_{z_Nd}\right), \tag{21.14a}$$

where the birth selection gradient is

$$\beta_{z_{Nb}} = -\frac{2ba_0\left(\bar{z}_N-\bar{z}_R\right)\, exp\left\{\frac{-(\bar{z}_N-\bar{z}_R)^2}{2\omega_a}\right\}R}{2\omega_a\left(1+a_0\, exp\left\{\frac{-(\bar{z}_N-\bar{z}_R)^2}{2\omega_a}\right\}hR\right)^2}, \tag{21.14b}$$

and the death selection gradient is

$$\beta_{zNd} = -f_0\gamma_f\left(\bar{z}_N-\theta_f\right). \tag{21.14c}$$

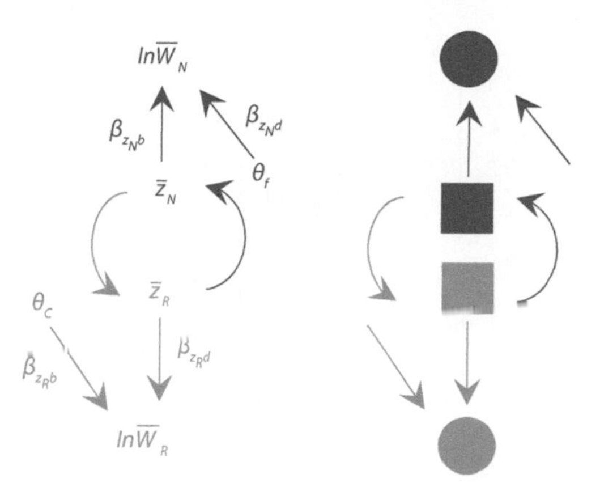

Figure 21.5 Path diagram and network depictions of McPeek's fitness model for two interacting species. The resource species is shown in green and the consumer species is shown in blue. (Left) Path diagram showing how the selection gradients (β_{zNb}, β_{zNd}, β_{zRb}, β_{zRd}) translate the effects of an intermediate optimum (θ_f, θ_c) and the species interaction between the trait means of the two species ($\bar{z}_N$, $\bar{z}_R$) into effects on the population mean fitness of the two species ($ln\overline{W}_N, ln\overline{W}_R$). The curved arrows remind us that the interactions selection gradients in each species (β_{zNb}, β_{zRd}) are functions of the trait means of both species. (Right) Network diagram, a symbolic version of the path diagram.

Likewise the evolutionary response of the resource trait mean to selection is

$$\frac{d\bar{z}_R}{d} = G_{z_R}\left(\beta_{z_Rb} + \beta_{z_Rd}\right), \tag{21.14d}$$

where the birth selection gradient is given by

$$\beta_{zRb} = -c_0\gamma_c\left(\bar{z}_R - \theta_c\right), \tag{21.14e}$$

and the death selection gradient is given by

$$\beta_{zRd} = -\frac{2\left(\bar{z}_R - \bar{z}_N\right)a_0\, exp\left\{\frac{-(\bar{z}_N-\bar{z}_R)^2}{2\omega_a}\right\}N}{2\omega_a\left(1 + a_0\, exp\left\{\frac{-(\bar{z}_N-\bar{z}_R)^2}{2\omega_a}\right\}hR\right)}. \tag{21.14f}$$

The role of these four selection gradients in response to selection (Figure 21.5) is directly analogous to their roles Nuismer's (2017) model (Chapter 20, Figure 20.4). However, in McPeek's (2017a, b) formulation, both of the interaction selection gradients (β_{zNb}, β_{zRd}) are functions of population abundances (N and R), and so change through time (21.13).

21.3 An Example of Trait Coevolution With Changing Abundance

The complexity of our four dynamic equations (21.13, 21.14) frustrates an analytical solution to the issues of evolutionary equilibrium and stability, but we can simulate the coevolution process to gain insight. An example of the dynamic behaviour of the equations is shown in Figure 21.6 and discussed at length by McPeek (2017a, b). In this example, the initial position of the trait means is at their functional equilibria ($\theta_f = 4$, $\theta_c = 10$), with the consumer at a very low abundance ($N = 0.01$) and the resource species at a considerably higher abundance ($R = 50$). In the absence of species interaction, the interaction selection gradients vanish ($\beta_{z_{Nb}} = \beta_{zRd} = 0$) and the remaining two selection gradients (β_{zNd} and β_{zRb}) govern the coevolution of the trait means. We can see from the composition of those selection gradients that, regardless of starting point, the means would evolve towards a stable equilibrium point at which $\bar{z}_N = \theta_f$ and $\bar{z}_R = \theta_c$.

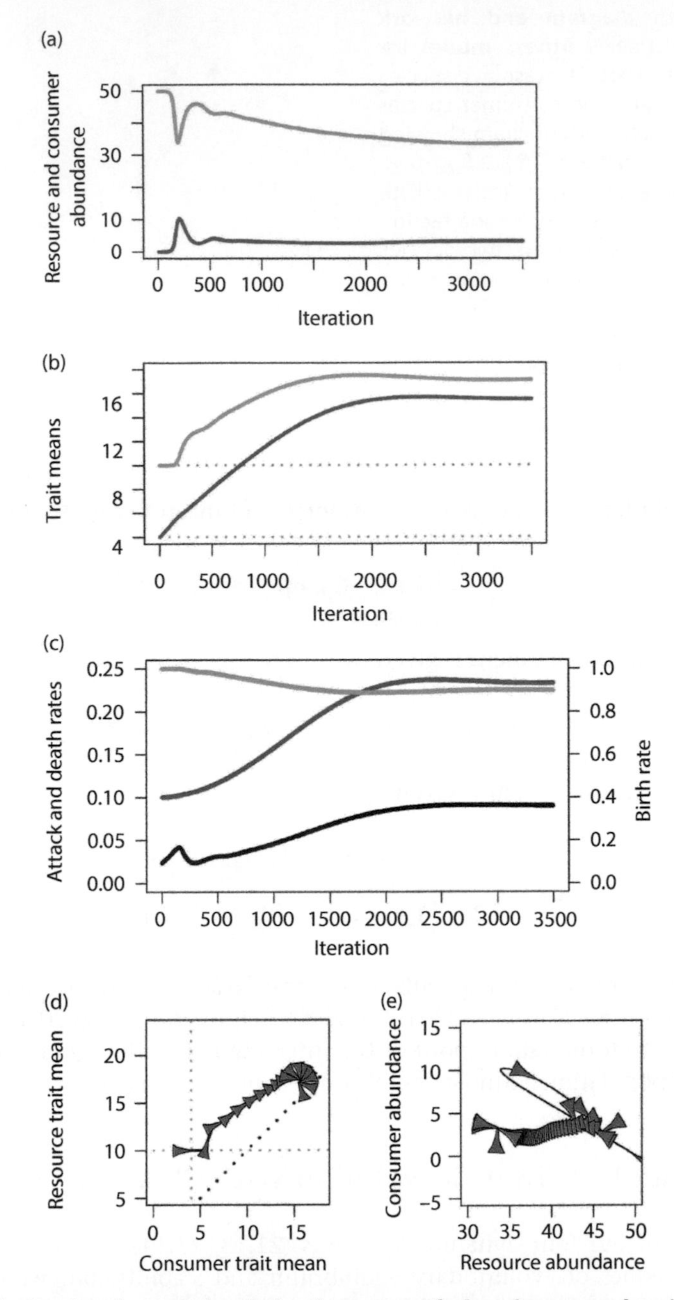

Figure 21.6 A simulated example of the dynamics of abundances and trait means during coevolution of a resource and consumer species. This figure shows different views of a single simulation lasting for 3500 iterations with the following fixed values for parameters: c_0 = 1.0, d = 0.02, a_0 = 0.1, b = 0.1, h = 0.1, f_0 = 0.1, g = 0.0, ω_a = 12.5, γ_c = 0.002, γ_f = 0.01, θ_c = 10, θ_f = 4, $G_{z_R} = G_{z_N}$ = 0.2 and the following initial values of variables: $\bar{z}_R$ = 10, R = 50.0, $\bar{z}_N$ = 4.0, N = 0.01. (a) Time course of resource abundance, R (shown in green) and consumer abundance, N (shown in blue). (b) Time course of resource and consumer trait means, $\bar{z}_R$ (shown in green) and $\bar{z}_N$ (shown in blue). The position of the two functional optima are shown with orange dotted lines. (c) Time courses

(*Continued*)

In the presence of interaction, however, the means do not evolve to this equilibrium point but instead to a distant point (Figure 21.6b, d). Species interaction causes this departure from a simple equilibrium. Focusing on the actual trajectories of the trait means under a regime of interaction (Figure 21.6b), we see an initial period of rapid evolution over the first 1500 iterations followed by a final period of slow evolution for the next 2000 iterations during a slow approach to an equilibrium point. The final approach is especially remarkable because the joint trajectory slowly spirals towards the equilibrium point (Figure 21.6d). The slow period of coevolution represents a steady approach to an adaptive ridge defined by the attack success function, $a(\bar{z}_R - \bar{z}_N)$, when $\bar{z}_R = \bar{z}_N$. However, the joint mean never reaches this ridge and instead equilibrates at a point at which the resource trait mean is slightly larger than the consumer trait mean ($\bar{z}_N = 15.5$, $\bar{z}_R = 17.1$).

The abundances of the two species (N and R) change throughout the coevolution of the trait means and influence their trajectories. The temporal dynamics of N and R are especially rapid and complicated during the first 500 generations (Figure 21.6a). It is during this period that the consumer species grows from very small to low population size, while the abundance of the resources species rapidly declines, increases, and then steadily declines. In a bivariate plot of these dynamics (Figure 21.6e), we see that the abundances execute a loop before beginning a slow approach towards an abundance equilibrium at $N = 3.3$, $R = 33.6$.

Because we lack a formal stability analysis, we resort to simulations to explore the stability of the abundance equilibrium point revealed in Figure 21.6e. A limited exploration by local perturbation (using interactive software on M. A. McPeek's website) reveals some characteristics of this equilibrium point. If we begin simulations using the endpoint values $N = 3.3$ and $R = 33.6$, but perturb the start points from $\bar{z}_N = 15.5$ and $\bar{z}_R = 17.1$ by adding a constant, k, to both values, we return to the equilibrium points revealed in Figure 21.6d and Figure 21.6e, if k is sufficiently small (less than about 5). However, if $k \geq 5$ the consumer species goes extinct. In other words, the equilibria we see in Figures 21.6d and e appear to be relatively fragile local equilibrium, rather than robust regional equilibrium.

The functions incorporating attraction to functional optima and trait interaction (21.10–12) are functions of trait values just as they are in Nuismer's (2017) models (Chapter 20), but their dynamics are more complicated (Figure 21.6c). The complication arises from the fact that the evolution of trait means is a function of species abundances. In the example here, over 2000 iterations are required for these three functions to equilibrate, mirroring the equilibration time for the trait means.

Figure 21.6 (*Continued*)
of attack rate, $a(\bar{z}_R, \bar{z}_N)$ (shown in black, scale on left), intrinsic resource birthrate, $c(\bar{z}_R)$ (shown in green, scale on right), and intrinsic consumer deathrate, $f(\bar{z}_N)$ (shown in blue, scale on left). (d) Consumer trait mean ($\bar{z}_N$) as a function of resource trait mean ($\bar{z}_R$) during coevolution. Arrowheads are shown every 200 iterations. Orange dotted lines show optimum values for consumer deathrate (θ_c, hortizontal line) and resource birthrate (θ_f, vertical line). The dotted black line show the optimum combinations of trait means specified by attack success, $a(\bar{z}_R, \bar{z}_N)$, i.e., $\bar{z}_R = \bar{z}_N$. (e) Consumer abundance (N) as a function of resource abundance (R) during coevolution. Arrowheads are shown every 75 iterations.
After McPeek (2017a, b).

21.4 Dynamics of the Adaptive Landscape and Its Components

One of the most striking features of McPeek's (2017a, b) formulation of trait evolution is that the adaptive landscape changes its shape as well as the position of its peak(s) over time. Although the AL shape dynamic is unapparent in Figure 21.6, we focus on that dynamic now because it goes a long ways towards illuminating the causes of trait coevolution (Figure 21.7). Recall that the McPeek formulation recognizes two components of fitness (birth and death) in each of the interacting species. In the resource species, birth fitness is the first term and death fitness is the second and third terms in (21.7b). Likewise, in the consumer species, birth fitness is the first term and death fitness is the last two terms in (21.7a). Notice that the death landscape is constant in the consumer species because g takes a value of zero in the example we are considering. The other overarching consideration to keep in mind is that evolutionary equilibria that are eventually attained in each trait represent balance between the birth and death selection gradients. These gradients are tangents to the red curves at the points of the solid red dots in Figure 21.7b versus c and in Figure 21.7e versus f. These slopes face uphill and are of equal magnitude but opposite sign.

Consider the evolution of the resource trait mean ($\bar{z}_R$) as a first example (Figure 21.7a–c). In each figure the succession of open red circles shows the trajectory of the trait mean in relation to an evolving landscape. In the top figure (Figure 21.7a), we see that the bivariate mean takes a straight path towards its final position at about $\bar{z}_R = 17.2$. Throughout this trajectory the total fitness remains constant with a value close to zero as the adaptive landscape changes shape, rapidly at first and then decelerating to its final configuration. The dynamics on the birth landscape (Figure 21.7b) are different with the bivariate mean taking a curving trajectory towards its final position, downslope from the functional optimum (10) at about $\bar{z}_R = 17.2$. The birth landscape retains its shape and position while shifting upwards due to the density-dependent term dR. Meanwhile on the death landscape, the bivariate mean takes a curving trajectory until it equilibrates at a suboptimal value for death fitness (Figure 21.7c). At equilibrium the birth selection gradient is negative (slopes towards lower trait values) and is exactly balanced by the death selection gradient (slopes towards higher trait values).

The landscape dynamics of evolving consumer trait ($\bar{z}_N$) are simpler (Figure 21.7d–f) than the evolving resource trait ($\bar{z}_R$) because the death landscape is constant (Figure 21.7f). On the overall landscapes (Figure 21.7d), total fitness remains essentially constant at a value of zero, as the trait mean steadily moves toward higher values, eventually equilibrating at about $\bar{z}_N = 15.5$. The trajectory on the birth landscape is similar, except that the bivariate mean takes an upward, curving trajectory, until the mean equilibrates downslope from the birth optimum (Figure 21.7e), yielding a positive birth selection gradient. Meanwhile on the death landscape (Figure 21.6 f), the trait mean is pulled downslope, away from its functional optimum, yielding a negative death selection gradient that balances the positive interaction selection gradient in Figure 21.7e.

Notice that while the adaptive landscapes and their components constantly change during coevolution, they do not change erratically or rhythmically as is implied by Lewontin's trampoline metaphor. Instead, although the configuration of the total landscape constantly changes, it is always a simple sum of its two components (birth and death). Furthermore, throughout our simulation, total fitness in both species never departs far from zero, as shifting adaptive landscapes drive coevolution to an evolutionary equilibrium.

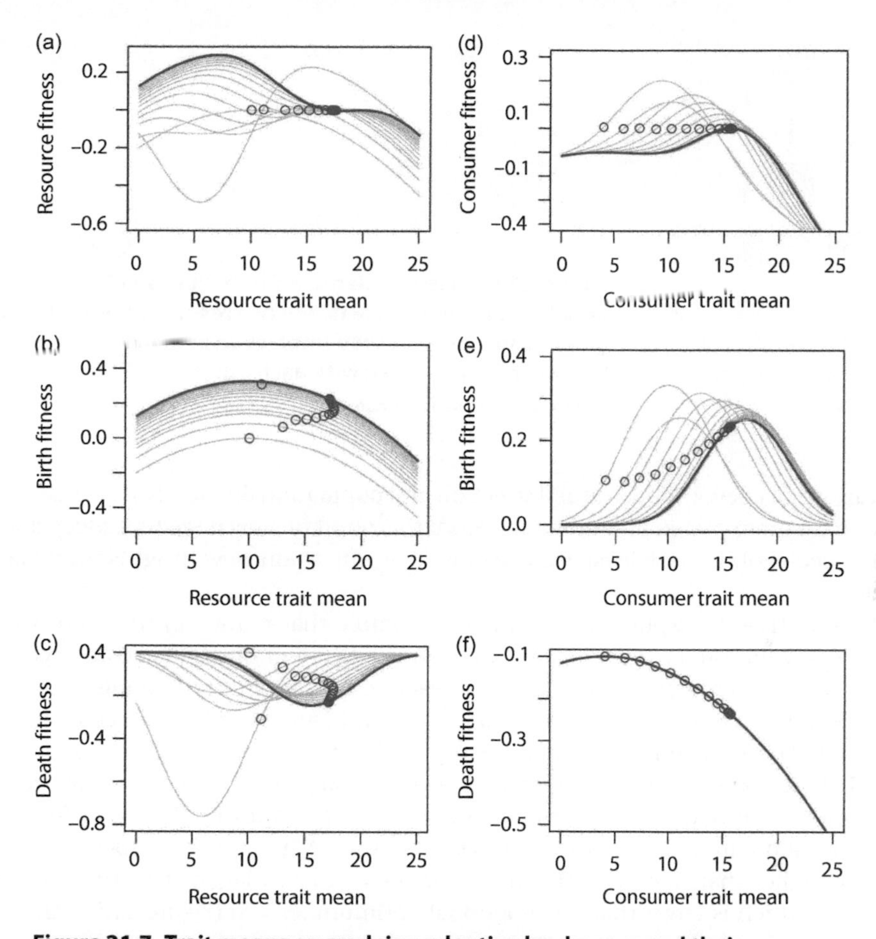

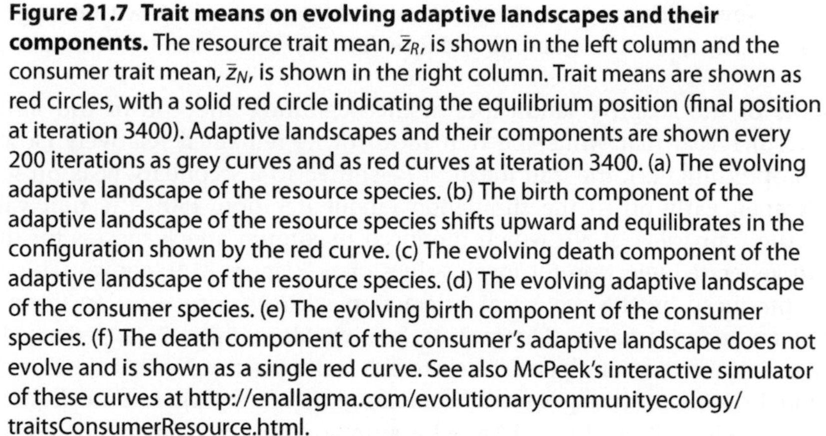

Figure 21.7 Trait means on evolving adaptive landscapes and their components. The resource trait mean, $\bar{z}_R$, is shown in the left column and the consumer trait mean, $\bar{z}_N$, is shown in the right column. Trait means are shown as red circles, with a solid red circle indicating the equilibrium position (final position at iteration 3400). Adaptive landscapes and their components are shown every 200 iterations as grey curves and as red curves at iteration 3400. (a) The evolving adaptive landscape of the resource species. (b) The birth component of the adaptive landscape of the resource species shifts upward and equilibrates in the configuration shown by the red curve. (c) The evolving death component of the adaptive landscape of the resource species. (d) The evolving adaptive landscape of the consumer species. (e) The evolving birth component of the consumer species. (f) The death component of the consumer's adaptive landscape does not evolve and is shown as a single red curve. See also McPeek's interactive simulator of these curves at http://enallagma.com/evolutionarycommunityecology/traitsConsumerResource.html.

21.5 Peak Shift Induced by Eco-evolutionary Dynamics

One of the most important, overarching messages from McPeek's (2017a) formulation is that both the trait-based and abundance-based perspectives in Figure 21.1 must be simultaneously embraced to completely understand the ecological and evolutionary

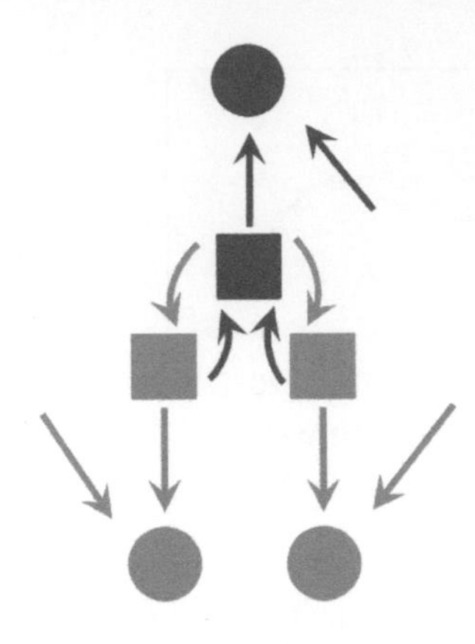

Figure 21.8 Network diagram for McPeek's (2017a) model of apparent competition between two resource species (shown in green) showing that they interact with the consumer species (shown in blue) but not with each other.

Graphic conventions as in Figure 21.5.

implications of coevolution. Natural selection and population dynamics are not separable, although we often pretend that they are. From this point of view it seems inescapable that a formal eco-evolutionary perspective will give us important new insights into adaptive radiation.

McPeek (2017a, b) explores an instructive example that reinforces the point that his formal eco-evolutionary theory leads to insights that are unobtainable using ecological or evolutionary perspectives alone. In a three-species community module, a consumer (N_1) relies on two resources species (R_1 and R_2) with identical trait parameter values; see McPeek's (2017a) account of the rich history of theoretical work on this ecological scenario. This scenario is sometimes called *apparent competition* because the two resource species interact only with the consumer and not with one another (Figure 21.8).

McPeek's simulation begins with both resources species having trait values (ca. 20) considerably higher than their functional optima (θ_c = 12) and larger than the consumer's trait value, which is larger than its functional optimum (θ_f = 5) (Figure 21.9). After 4000 iterations, eco-evolutionary dynamics produced a stable equilibrium with divergent trait values ($\bar{z}_{R_1}$ = 2.3, $\bar{z}_{R_2}$ = 15.8) and abundance (ca. 15, 45) in the two resource species (R_1, R_2).

Snapshots of the adaptive landscapes at the beginning and end of the simulation (Figure 21.10) reveal that while the trait mean of R_2 remained relatively close to its functional optimum (12), the trait mean of R_1 shifted to a secondary peak on its adaptive landscape). Early on in the simulation (about iteration 450), the trait values of R_1 match the trait value of N_1, which produced maximum attack success for the consumer and generated selection for lower values of $\bar{z}_{R_1}$ (Figure 21.9, above). This intense selection, produced by the ecological interaction with N_1, drove $\bar{z}_{R_1}$ to lower values and across an adaptive valley (Figure 21.9, below). Meanwhile, R_2 experienced much weaker selection because of a poor match to the trait mean of N_1. $\bar{z}_{R_2}$ remains close to its functional optimum throughout the simulation, equilibrating at a much lower trait value than $\bar{z}_{R_1}$. See McPeek (2017a, b) and his animation website (http://enallagma.com/evolutionarycommunityecology/traitsApparentCompetition.html) for more detail and a fuller account of the eco-evolutionary dynamics in this example.

The take-home message from this example is that the peak shift of R_1 was accomplished by a novel mechanism, eco-evolutionary dynamics. R_1 was not driven by drift to a secondary peak on a stationary landscape, as in the model of Lande (1985, 1986) discussed in Chapter 12 (Section 12.4). Instead the shift to a secondary peak happened

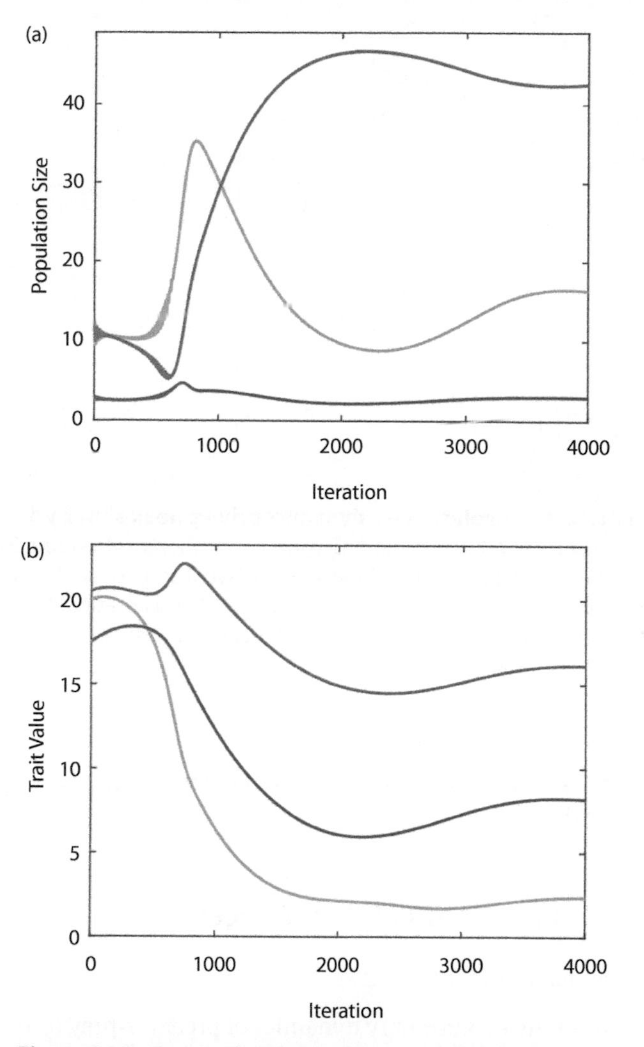

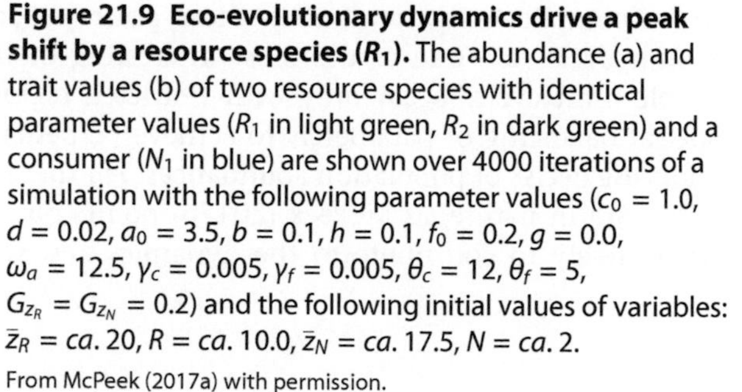

Figure 21.9 Eco-evolutionary dynamics drive a peak shift by a resource species (R_1). The abundance (a) and trait values (b) of two resource species with identical parameter values (R_1 in light green, R_2 in dark green) and a consumer (N_1 in blue) are shown over 4000 iterations of a simulation with the following parameter values ($c_0 = 1.0$, $d = 0.02$, $a_0 = 3.5$, $b = 0.1$, $h = 0.1$, $f_0 = 0.2$, $g = 0.0$, $\omega_a = 12.5$, $\gamma_c = 0.005$, $\gamma_f = 0.005$, $\theta_c = 12$, $\theta_f = 5$, $G_{z_R} = G_{z_N} = 0.2$) and the following initial values of variables: $\bar{z}_R = ca.\ 20$, $R = ca.\ 10.0$, $\bar{z}_N = ca.\ 17.5$, $N = ca.\ 2$.

From McPeek (2017a) with permission.

on a continuously changing landscape and was driven by selection arising from the ecological interaction with a consumer, N_1. Furthermore, the dynamics of the interaction with the consumer rapidly produced a stable pattern of trait divergence between two resource species that were initially identical. The lesson for community assembly is that

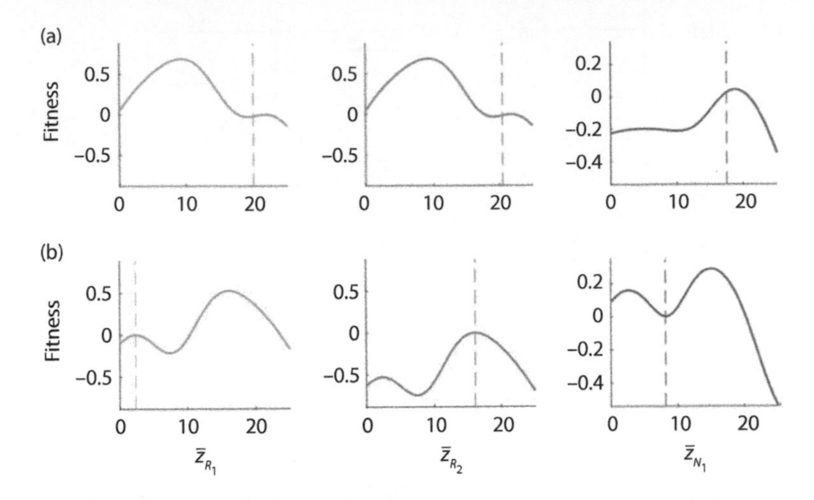

Figure 21.10 Eco-evolutionary dynamics drive a peak shift by a resource species (R_1), continued. Graphic conventions as in Figure 21.7. The adaptive landscapes of the three species (two resource species, R_1 and R_2, and a consumer, N_1, are shown at iteration 0 (a) and iteration 4000 (b). Vertical dashed lines show the positions of mean trait values.

From McPeek (2017a) with permission.

evolutionary dynamics, rather than ecological interaction alone, can play a large role in generating stable coexistence.

21.6 Other Important Results From McPeek

21.6.1 *The prospect for stable limit cycles*

In other explorations of the evolutionary dynamics of predator-prey interactions, McPeek (2017b) discovered more complicated patterns than those portrayed in Figures 21.6–7 and 21.9–10. Stable limit cycles are one of the most interesting of those alternative patterns. However, stable limit cycles occur only with restricted regions of parameter values and require special balancing of parameters to achieve perpetual coevolution of traits coupled with enduring cycles of population abundances. For these reasons, cyclical patterns are likely to be rare in nature, as McPeek (2017b) points out. In other words, stable limit cycles are unlikely to contribute to the dynamics and scope of adaptive radiations.

21.6.2 *Coevolution in mutualistic species arrays*

Although McPeek (2017b) does not simulate the coevolution of traits in arrays of species with mutualistic interactions, he does analyse the conditions necessary for coexistence. Because mutualistic species have positive effects on each other's abundance, they create new opportunities for coexistence (McPeek et al. 2021). It is not a big step to imagine that mutualistic interactions also promote speciation and increase the extent of adaptive radiation.

21.6.3 *The effect of background ecology on coevolution*

McPeek (2017b) points out that his expressions for functional selection gradients (21.14c, e) enable us to predict the ecological circumstances that will pull traits toward their functional peaks in evolutionary time. For example, higher productivity (larger c_0) and more benign conditions for the consumer (lower f_0) will promote steeper directional selection gradients that will tend to constrain trait means near their functional optima. In this way, the parameters of McPeek's (2017b) formulation provide a roadmap of how background ecological conditions will affect selection on traits and hence their coevolution.

21.7 The Strengths of McPeek's Approach

As we pointed out earlier, the overarching strength of McPeek's (2017a, b) approach is that it simultaneously models changes in species abundance and evolution of trait means. The effects of interaction change with abundance and those effects are taken into account in modelling selection on interacting traits. The incorporation of abundance into models of trait evolution has two important ramifications that we have not stressed, implications for ecological opportunity and an explicit ecological model for peak shifts.

A demographic perspective on trait evolution also provides an operational definition for *ecological opportunity* (McPeek 2017b). Ecological opportunity has many definitions and connotations, but its most fundamental characteristic is that the opportunity provides conditions for population growth, $dN/dt > 0$. This crucial benchmark for assessing establishment and persistence is wholly absent in accounts that do not incorporate abundance.

Secondly, McPeek's (2017a, b) approach provides an explicit deterministic model for peak shift rather than a purely statistical description with limited biological connection. Before we continue, we must realize that this perspective has changed our vocabulary from that used in the earlier chapters of this book. In Chapter 11, we used 'peak shift' to denote stochastic movement of the trait mean from one adaptive peak to another as a consequence of drift. The mean shifted from one peak to another but the peaks did not move; the landscape was static. In Chapter 14, we used 'peak movement' to denote stochastic movement of a single adaptive peak with unchanging curvature. The trait mean chased these moving peaks, as described by the standard model for directional selection. We considered several alternative models for peak movement, but these were purely statistical models that evoked stochastic processes such as Brownian motion and Levy processes. In contrast, in McPeek's formulation, peak movement arises from a deterministic process of species interactions mediated by trait values. These deterministic processes cause the adaptive landscape to change shape and as a consequence the trait mean can shift from one peak to another (Figure 21.10). Our point here is that we have a richer model for peak movement than we had in Chapter 14, one that is couched in ecological terms. Those ecological terms include abundance and interaction and hence provide a bridge to a variety of ecological processes and connections.

21.8 The Limitations of McPeek's Approach

As we appreciate the strength of McPeek's (2017a, b) approach, it is also important to keep in mind some of its limitations. For example, in our account we have only considered the consequences of interaction in community modules consisting of two or three species.

McPeek (2017b) enlarges his framework to incorporate an arbitrary number of incorporate multiple species, but expressions become correspondingly large and cumbersome. Analysis using this enlarged framework is possible using simulations, but those simulations involve a very large number of parameters and variables. The upshot of this situation is that it may be difficult to extract generalizations about community modules of realistic size.

Neither Nuismer's (2017) nor McPeek's (2017a, b) formulations incorporate stochasticity. The most fundamental and relevant type is stochasticity in vital rates (Lande et al. 2017). In addition, stochasticity arising from population size, which affects the trajectory of the trait mean (Chapters 8–9), as well as the stability and evolution of the *G*-matrix (Chapters 14–16), was not included by Nuismer and McPeek (but see the end of Chapter 20). Incorporating drift into McPeek's formulation will be straightforward because effective size is related to population abundance. Alternatively, stochasticity could be introduced by stochastic movement of trait optima. The expected consequence of either of these incorporations will be the creation of stochastic clouds around points or lines of evolutionary equilibria (Figure 20.12), hence enlarging the size of adaptive radiations. Although the effect is in the direction of enlargement, it will not be appreciable in the case of drift, although it might substantial in the case of Brownian motion of peaks.

Although McPeek (2017b) extends his formulation to include multiple interacting species, he limits his analysis to a single trait in each species. From simulations of Nuismer's (2017) model, which includes multiple, genetically correlated traits, we know that the inclusion of multiple traits generally produces curved evolutionary trajectories, which like those produced by species interaction, delay the approach to an evolutionary equilibrium and subject the population to maladaptation. Consequently, we can expect that an extension of McPeek's formation to multiple traits will produce trajectories that experience additional delay in reaching evolutionary equilibria.

21.9 Conclusions

- Ecological interactions are mediated by phenotypic traits.
- Mathematical characterization of these ecological interactions can take a variety of function forms, such as trait matching or trait difference.
- Fitness and population growth rate are equivalent currencies.
- The equivalence of these currencies provides a bridge that connects trait coevolution with population dynamics.
- The bridge that connects coevolution and population dynamics takes the form of coupled sets of equations. One set models species abundance through time, the other set models the coevolution of interacting traits.
- In the case of two interacting species, in which the interaction is mediated by a single trait in each species, the demographic and evolutionary behaviour of the two sets of equations can be studied by computer simulation.
- In a simple version of this case, McPeek (2017a, b) recognizes two components of fitness (birth and death) in each of two species (a resource species and a consumer species). Trait-mediated interaction affects the birth-related fitness of the consumer and the death-related fitness of the resource (e.g., predation increases the fecundity of the predator, but lowers the survival of the prey). In each species the trait selection mediated through the complementary component of fitness (death in the consumer,

birth in the resource) is modelled as stabilizing selection towards an intermediate optimum. This stabilizing selection is intrinsic (functional) and not influenced by the ecological interaction.

- In his two-species, one-trait test case, McPeek (2017a, b) uses trait matching to model the ecological interaction. In this model, the success of the predator in attacking the prey is a function of the difference in the trait values of the predator and prey. Success is highest when the trait values are the same (match), but success falls off as prey's trait value becomes less than or greater than the predator's trait value. Higher success means increased fitness for the predator but lower fitness for the prey.
- One common outcome of two-species, one-trait test cases is demographic stability in which both species maintain stable population sizes and evolutionary equilibrium in which trait values in both species are stable through time.
- At such stable evolutionary equilibria, trait means typically equilibrate downslope from adaptive peaks as a consequence of ecological interaction.
- Interaction between the species also causes a slow, curved approach to the evolutionary equilibrium in bivariate trait space. In other words, interaction induces prolonged temporary as well as permanent maladaptation. Likewise, interaction causes loops in demographic trajectories and a curved final approach to the bivariate demographic equilibrium. Both features are expressions of maladaptation.
- Evolution of adaptive landscapes is a general feature of McPeek's (2017a, b) formulation. Because mean fitness of each species is a function of the abundance of the other species, the shapes and peak locations of adaptive landscapes change through time, although they attain a stable final configuration at evolutionary and demographic equilibrium.
- The dynamic nature of the adaptive landscape means that trait means can shift from one peak to another because of interaction-mediated change in the landscape itself rather than because of drift or other conventional modes of peak shift.

Literature Cited

Abrams, P. A. 2000. The evolution of predator-prey interactions: theory and evidence. Annu. Rev. Ecol. Syst. 31:79–105.

Charlesworth, B. 1994. Evolution in Age-Structured Populations. Cambridge University Press, Cambridge.

Gould, S. J. 1997. Review: Self-help for a hedgehog stuck on a molehill. Evolution 51:1010–1023.

Hendry, A. P. 2016. Eco-evolutionary Dynamics. Princeton University Press, Princeton.

Holling, C. S. 1959. The components of predation as revealed by a study of small mammal predation of the European Pine sawfly. Can. Entomol. 8:293–332.

Holt, R. D., and M. Barfield. 2012. Trait-mediated effects, density dependence and the dynamic stability of ecological systems. Pp. 89–106 *in* T. Ohgushi, O. Schmitz, and R. D. Holt, eds. Trait-Mediated Indirect Interactions: Ecological and Evolutionary Perspectives. Cambridge University Press, Cambridge.

Iwasa, Y., A. Pomiankowski, and S. Nee. 1991. The evolution of costly mate preferences ii. the 'Handicap' principle. Evolution 45:1431.

Lande, R. 1976. Natural selection and random genetic drift in phenotypic evolution. Evolution 30:314–334.

Lande, R. 1979. Quantitative genetic analysis of multivariate evolution, applied to brain: body size allometry. Evolution 33:402–416.

Lande, R. 1982. A quantitative genetic theory of life history evolution. Ecology 63:607–615.

Lande, R. 1985. Expected time for random genetic drift of a population between stable phenotypic states. Proc. Natl. Acad. Sci. USA 82:7641–7645.

Lande, R. 1986. The dynamics of peak shifts and the pattern of morphological evolution. Paleobiology 12:343–354.

Lande, R. 2007. Expected relative fitness and the adaptive topography of fluctuating selection. Evolution 61:1835–1846.

Lande, R., S. Engen, and B.-E. Sæther. 2017. Evolution of stochastic demography with life history tradeoffs in density-dependent age-structured populations. Proc. Natl. Acad. Sci. USA 114:11582–11590.

McPeek, M. A. 2017a. The Ecological Dynamics of Natural Selection: Traits and the Coevolution of Community Structure. Am. Nat. 189:E91–E117.

McPeek, M. A. 2017b. Evolutionary Community Ecology. Princeton University Press, Princeton.

McPeek, S. J., J. L. Bronstein, and M. A. McPeek. 2021. The evolution of resource provisioning in pollination mutualisms. Am. Nat. 198:441–459.

Nuismer, S. 2017. Introduction to Coevolutionary Theory. MacMillan Higher Education, New York.

Ohgushi, T., O. Schmitz, and R. D. Holt. 2012. Ohgushi, T., Schmitz, O. and Holt, R. D. eds., 2012. Trait-mediated indirect interactions: Ecological and evolutionary perspectives. Cambridge University Press, Cambridge.

Rosenzweig, M. L., and R. H. MacArthur. 1963. Graphical representation and stability conditions of predator-prey interactions. Am. Nat. 97:209–223.

CHAPTER 22

From Evolutionary Process to Pattern—a Synthesis

Overview—In earlier chapters, we established the feasibility of accounting for fundamental processes of evolution and adaptive radiation with models of moving peaks. In this chapter, we review those arguments and examine their implications. A model of the trait mean chasing a perpetually moving adaptive peak provides a strong basis for understanding adaptive evolution and radiation. Both the foundations and the predictions of this model are supported by empirical evidence. Furthermore, adaptive peak models bridge between the domains of evolution and ecology, as well as between micro- and macroevolution. Peak movement is the common denominator of adaptive peak models that successfully account for data on trait divergence on all timescales, as well as iconic adaptive radiations. Although stochastic models of peak movement can predict the broad features of adaptive radiation, the study of limits on radiation remains a growing point in theoretical work. Regions of trait space densely populated by species means may often represent regions of recurrent peak movement. In addition to shaping adaptive radiations, the adaptive landscape shapes patterns of mutation, inheritance, and peak movement.

Most if not all traits are elements of functional complexes. Interactions between traits in such complexes can enhance or curtail adaptive radiation. The ubiquity of trait interactions stresses the importance of multiple-trait evolutionary theory. The structure of that multivariate theory is encapsulated in ten kinds of matrices. Chasing of separate peaks by different lineages can lead to relatively rapid sexual and ecological speciation. Trait-mediated interactions can profoundly affect the coevolution of interacting species. When the feature of peak movement is added to coevolutionary models, maladaptation can range from mild to severe and may be permanent.

22.1 The Trait Mean Chases Its Adaptive Peak

The overarching thesis of this book is that a model of the trait mean chasing a perpetually moving adaptive peak provides a strong basis for understanding adaptive evolution and radiation. First, surveys of selection coefficients in natural populations suggest that populations are generally close to their adaptive peaks (Estes and Arnold 2007). Second, the observation of bounded evolution on all timescales, as well as surveys of selection coefficients, suggests that downward curvature of the peak (stabilizing selection) is common if not universal. Finally, as we point out below, peak movement is the common denominator of models that successfully account for substantial adaptive radiation.

Evolutionary Quantitative Genetics. Stevan J. Arnold, Oxford University Press. © Stevan Arnold (2023).
DOI: 10.1093/oso/9780192859389.003.0023

If we adopt the perspective of a perpetually moving peak, several empirical results and consequences snap into perspective. (1) Trait means are close to their adaptive peaks but not coincident because peaks are always moving, forcing trait means to chase them. (2) Genetic variation is generally available, especially in directions favoured by selection because peak movement helps create genetic variation. (3) Peak movement exacts a cost in the demographic growth of populations that chase their adaptive peak. (4) On long evolutionary timescales, peak movement is the primary determinant of the broad character of adaptive radiations rather than within-lineage processes (e.g., drift away from and selective attraction to the peak).

22.2 The Evolutionary Ecology of the Adaptive Peak

The evolutionary currency that describes the topology of the adaptive landscape is also the fundamental currency of ecology (Chapter 21). Average fitness of a population is a function of population trait averages according to the concept of an adaptive landscape for phenotypic traits, developed by Simpson (1944, 1953) and especially by Lande (1976, 1979). As Lande (1982) and McPeek (2017) have emphasized, the currency that measures the height of the adaptive landscape is equivalent to population growth, the currency of population ecology. This currency equivalency means that we can simultaneously model trait evolution and the population growth of interacting species. The currency equivalency also means that the adaptive landscape of a particular species changes through time as a consequence of interaction with other species. Although coevolving, interacting species may often coexist in stable evolutionary equilibria, periods of maladaptation and depressed populations growth are inevitable consequences of ecological interaction.

22.3 The Empirical Necessity and Appeal of Peak Movement

Movement of adaptive peaks is required to account for evolutionary patterns in large samples of divergence data, as well as in iconic examples of adaptive radiation (Chapters 12, 13, and 19). This conclusion is based on several results. In the first place, fitting models to data from actual radiations and requiring realistic values for parameters rules out several candidates as sufficient explanations: (1) genetic drift of the trait mean, (2) stabilizing selection–drift balance about a stationary peak, (3) shifting between peaks by drift of the trait mean, and (4) early burst of diversification followed by diminished radiation. In all of these insufficient modes, the adaptive peak is stationary.

Secondly, peak movement is the common denominator of successful models. Indeed, the appeal of peak movement as a model for adaptive evolution and radiation rests on its success in accounting for observed evolutionary pattern without requiring unrealistic parameter values. Furthermore, models of peak movement reflect temporal stochasticity and trends in selection-generating features of the environment. This feature can be used to identify the environmental drivers of adaptive radiation.

22.4 Stochastic Peak Movement and Adaptive Radiation

Stochastic models of peak movement provide a link between within-lineage process and among-lineage pattern (Chapter 14). In these models the adaptive peak (intermediate optimum) continuously moves according to some stochastic process. The analytically tractable candidate processes for movement are Brownian motion or white noise movement of the peak, combined with stabilizing selection centred on the peak (OU). Although many model comparisons have been made using generic versions of BM and OU and tree-based data (e.g., Butler and King 2004; Harmon et al. 2010; Beaulieu et al. 2012), we focus instead on process-specific BM and OU models (Chapter 14). The requirement of model specificity means that we can cross-check parameter values with empirical results (Uyeda and Harmon 2014). If, for example, success of a process-specific BM model requires out of bound values for population size or rate of peak movement, the basis for failure is visible. In contrast, such points of failure may not visible if the model is generic. A further advantage of the process-specific models is the link they provide between within-lineage processes and among-lineage results. In other words, these models link the domains of micro- and macro-evolution.

The most powerful tests of process-specific models use thousands of data points representing a huge range of timescales (Estes and Arnold 2007; Uyeda et al. 2011; Landis et al. 2013; Landis and Schraiber 2017). At the short end of the timescale, trait divergence is assessed over a few generations. At the long end, trait means diverge over tens or hundreds of millions of years. These test cases show that Brownian motion of a peak does an excellent job of accounting for the data, although superimposing rare jumps in the position of the peak can improve the fit. At the same time, certain models are ruled out by these tests (e.g., drift of the trait mean, selection-mutation balance).

Brownian motion of an adaptive peak is an open-ended process that produces an ever-expanding adaptive radiation. It seems likely that real adaptive radiations are limited by some process that curtails the perpetual expansion of trait differences. In a two-layered OU process model, curtailment is accomplished by over-arching control of peak movement (Reitan et al. 2012). However, achieving control of peak movement, while permitting enough movement to generate a substantial radiation, requires OU control of the peak that is orders of magnitude weaker than the stabilizing selection that has been documented to control the trait mean in natural populations. A single-layered OU process can generate substantial radiation by assuming very weak stabilizing selection; a double-layered OU process achieves the same goal by placing the assumption of very weak control at a higher level. Nevertheless, multiple-layered OU process models do have explanatory potential. For example, linear trends or time series that represent environmental variables, such as short-term weather patterns or long-term climate change, can be incorporated with three-layer models. The take-away message from these models and test cases is that exploration of peak-controller processes that can bound and otherwise constrain adaptive radiation may be a profitable growing point in future work.

22.5 Peak Controller Processes

A variety of peak-controller processes will undoubtedly be required to account for the diversity of evolutionary modes in nature. We have already pointed out that while BM is a good candidate for a peak-controller process, stasis in some traits is so extreme that a virtually stationary peak, or one with very modest movement, is probably the best model. Stochastic, rare jumps in the peak create another mode of evolution that may be relevant to some traits and in some ecological circumstances. If peak jumps are rare, their effects will only be observed when we sample evolution over long stretches of time. In the rest of this section we will consider this effect of timescale.

Uyeda et al. (2011) discuss two possible interpretations for the observation that trait divergence is bounded on short timescales (< 1 Myr) but expansive on longer timescales. In Simpson's (1944) perspective, evolution is bounded as lineages diverge within an adaptive zone. In a clade of burrowing mammals such as gophers, for example, lineages typically speciate and diversify within defined morphological limits on size and on structures used for excavation (e.g., teeth and claws). A systematist working on gophers sees a radiation, but for the rest of us a gopher is a gopher; we see a clade in which the species are instantly recognizable as gophers. The environmental exigencies that make gophers so recognizable probably reflect the biomechanical demands that soil excavation places on functional complexes. Horns for male combat, for instance, are out of the question. The point here is that actual factors must bound evolution within the gopher adaptive zone even if they are not explicitly articulated in the evolutionary model. On the other hand, escape from an adaptive zone is explicitly modelled, for example, as rare jumps in the position of the adaptive peak. These rare jumps can accumulate on a timescale >1 Myrs and produce a major departure from the gopher body plan and the constrained gopher way of life.

Ephemeral divergence (Futuyma 1987, 2010) is another way of accounting for escape from bounded evolution on long timescales. According to this perspective, the lineages that make up a species are continuously tracking local and regional variation in the positions of their adaptive peaks. This tracking is reflected in geographic variation in morphology and other aspects of the phenotype. However, geographic variation in peak position prevents local adaptations from spreading across the range of the species. Because adaptation is always local, evolution is bounded. Range-wide escape from a bound of this kind might be rare and typically observable on timescales >1 Myr. Such rare escapes might arise from global change in the environment and hence in peak position. Alternatively, the range of a species might contract, resulting in a biased sample of peak positions. Lineage means respond to this biased sample of peaks and the species as a whole escapes from its bounds (Gould 2002; Futuyma 2010).

Although these two perspectives on escape from persistent, bounded evolution dramatically differ in detail, both invoke rare events in pattern of peak movement. Once again, our focus on peak-controller processes is encouraged.

22.6 Evolution Along Selective Lines of Least Resistance

It seems unlikely that evolution is equally probable in all directions of trait space (Chapter 18). Counter to the idea that evolution is equi-probable in all directions, we argue that densely occupied regions of trait space often represent regions of recurrent peak movement, while rarely occupied regions are zones of adaptive incompatibility. Evolution in response to peak movement along selective lines of least resistance (the major axes of the adaptive landscape) may represent one important category of recurrent peak movement (i.e., SLLR peak movement). The recurrent evolutionary trajectories caused by this mode of peak movement help explain allometry in morphology and physiology.

Implementation of a data analysis framework for a Brownian motion model of SLLR peak movement (Section 18.2) could follow the approach used by Hohenlohe and Arnold (2008) in which a framework for testing a model of multivariate trait evolution is based on Brownian motion of the trait mean (Section 9.4). That framework used estimates of the G-matrix, a time-calibrated phylogeny, multivariate trait means in tip taxa, effective population size, and the curvature of the selection surface (ω-matrix) to diagnosis the role of selection in shaping an adaptive radiation. A future implementation of the SLLR model would depend on these same ingredients but would emphasize the role of the ω-matrix, rather than the G-matrix, in predicting the rate and direction of peak movement and hence the size and shape of the adaptive radiation, as well as its rate of expansion.

22.7 The Adaptive Landscape Shapes Patterns of Mutation, Inheritance, and Peak Movement

The configuration and movement of adaptive peaks profoundly shapes the pattern and stability of mutation and inheritance (Chapters 16 and 17). In conventional portrayals of the evolutionary paradigm, variation and selection are independent variables. The independence of roles is captured in the oft-repeated phrase 'selection acts on variation to produce evolutionary change'. Although this phrase is accurate in the short term, it does not capture the long-term effects of selection on variation and covariation.

Analytical theory and computer simulations show that selection can shape both mutation and inheritance on a timescale of thousands of generations. A single generation of selection arising from an adaptive peak can cause slight contraction in genetic variances and may slightly torque the main axis of the G-matrix. If the configuration of the adaptive peak is stable on an evolutionary timescale, repeated contraction and torqueing will promote alignment between $\boldsymbol{G}$ and the adaptive landscape (i.e., produce common principal components). If stability of the adaptive landscape persists for thousands of years, the same selective process will promote alignment of the M-matrix with both $\boldsymbol{G}$ and the landscape (Ω-matrix). This long-term selective response of the M-matrix is facilitated by epistasis. As a colleague once remarked, 'Selection creates mutation and inheritance matrices in its own image'. Computer simulations and some empirical observations support the plausibility and reality of this three-way alignment.

An existing regime of three-way alignment changes our expectations about likely directions of peak movement. Consider, for example a ridge-shaped adaptive landscape that is aligned with ridge-shaped M- and G-matrices. The ridge in the landscape is a selective line of least resistance. We say 'least resistance' because movement along the ridge incurs the least cost in fitness, whereas movement perpendicular to the ridge incurs the highest cost. The fitness ridge of the landscape reflects the fact that particular combinations of the values of different traits produce the highest fitness. Those combinations correspond to the location of the ridge. This consideration suggests that peak movement might be biased in the direction of the main axis of the adaptive landscape. Finally, computer simulations repeatedly reveal a surprising result. Peak movement tends to enhance genetic variation (i.e., elongate the G-matrix) in the direction of peak movement. In other words, we might expect four-way alignment between the adaptive landscape, $\boldsymbol{M}$, $\boldsymbol{G}$, and peak movement.

22.8 Modules of Interacting Traits and Their Evolution

Multivariate selection is the key to understanding functional complexes of traits and their evolution (Chapter 7). In that chapter we argued that the modular structure of phenotypic traits may be reflected in the modular structure of multivariate selection (i.e., in the modular structure of γ- and ω-matrices). The multivariate selection characterized by these matrices builds genetic correlation among traits in functional complexes, profoundly affecting their evolution. Trait modularity can also be expressed in separate interacting individuals (e.g., a plant and its pollinator). We expect those modularities to be reflected in the interaction matrix, $\boldsymbol{N}$. Our particular focus in the rest of this section is on the modular organization of interacting traits and the puzzling fact that the same evolutionary model that explains exuberant sexual radiations can also account for extreme stasis.

Before discussing the reasons that one model can toggle between these two extremes of expression, it will be helpful to consider two examples drawn from studies of vertebrate courtship behaviour. In the first case study, the evolution of courtship behaviour in plethodontid salamanders provides an instructive example of extreme stasis (Arnold et al. 2017). In the second case study, the evolution of display behaviour in birds of paradise provides an informative example of rapid diversification (Scholes 2008). Can one model account for both observations?

In plethodontids and their relative *Rhyacotriton*, the basic mechanism for coordinating sexual partners during sperm transfer at the end of courtship has remained unchanged for more than 100 million years (Arnold et al. 2017; Doten et al. 2017). As we mentioned in an earlier chapter (Section 18.2c), plethodontid courtship is organized into a sequence of five major modules that culminate in a tail-straddling walk and sperm transfer. A useful method for recognizing such modules and diagramming them was invented by Edwin Scholes (2008). An application of his method to plethodontid courtship is shown in Figure 22.1. For example, in the top left corner of Figure 22.1 we see a diagram of the simplest behavioural path to insemination in *Pseudotriton*. In an actual sequence, TSW might revert to HC and then resume TSW. Ignoring loops such as this that occur in actual sequences, the idealized behavioural sequence shown in the diagram consists of the following steps: orient, follow, nudge, swing, slide, under, forward, vent, deposit, lift-off, flex, sigmoid, backup, and extend. Descriptions of each of these steps, as well as a phylogenetic analysis, are provided by Arnold et al (2017). The caption for Figure 22.1 describes the translation of idealized sequences into modules and the graphical conventions for recognizing sequence variants in each module.

This graphical method is designed to highlight sequence differences among species or, in the case of Figure 22.1, among genera. Furthermore, by comparing modules among genera we can determine whether particular modules are more prone to adaptive radiation than others. For example, the first module (AP, approach) is both simple and largely invariant across the entire radiation. In contrast, the second module (HC, head contact) is highly variable across the radiation. The behavioural variety among genera in this module consists of alternative tactile displays, some of which deliver pheromones to the female from a gland on the male's chin (Wilburn et al. 2017). The chemical constituents of the male pheromone (a protein cocktail) are known to shorten the duration of courtship. Indeed, the function of the entire HC module seems to be female persuasion. The final three modules (TSW, SD, POS) follow one another in rapid succession and culminate in sperm transfer (Figure 22.2). Like the approach module, these three modules largely invariant across the radiation. But, unlike the approach module, these modules are highly complex and involve intricate coordination between sexual partners (video of SD and POS, https://www.youtube.com/watch?v=_UIpOcmejvI). Arnold et al (2017) argue that successful sperm transfer places a fitness premium on intricate coordination and drives long-term evolutionary stasis in the final modules of courtship.

In contrast, intricate coordination between sexual partners is largely absent from the courtship displays of birds of paradise. Furthermore, this radiation is renowned for its extraordinary diversity in male plumage and courtship display. Scholes (2008) has shown that plumage and displays in the genus *Parotia* are organized into complex modules and phases (Figure 22.3). Male *Parotia* build and maintain a display arena with a cleared court and perches for visiting females. When a female visits his arena, the male executes a series

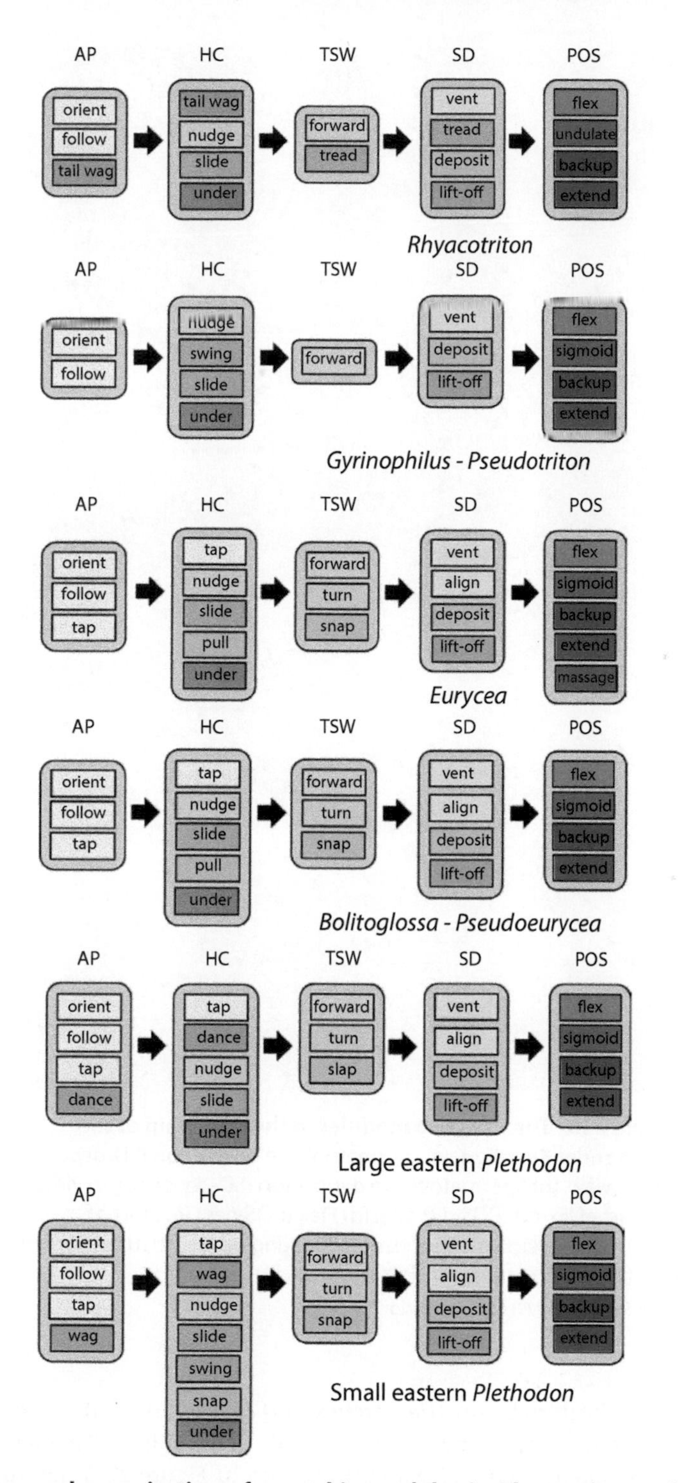

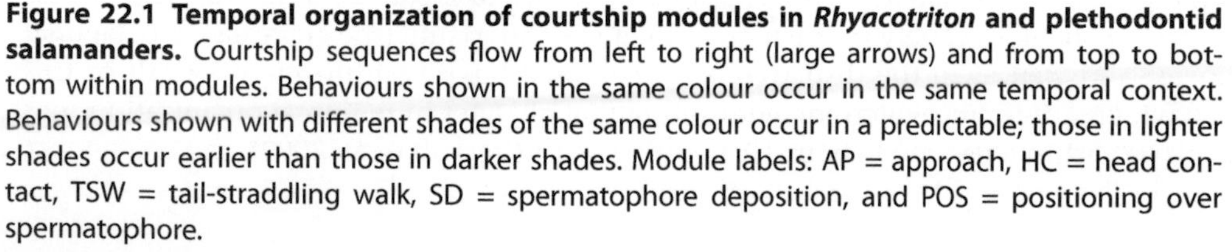

Figure 22.1 Temporal organization of courtship modules in *Rhyacotriton* and plethodontid salamanders. Courtship sequences flow from left to right (large arrows) and from top to bottom within modules. Behaviours shown in the same colour occur in the same temporal context. Behaviours shown with different shades of the same colour occur in a predictable; those in lighter shades occur earlier than those in darker shades. Module labels: AP = approach, HC = head contact, TSW = tail-straddling walk, SD = spermatophore deposition, and POS = positioning over spermatophore.

From Arnold et al. (2017) with permission.

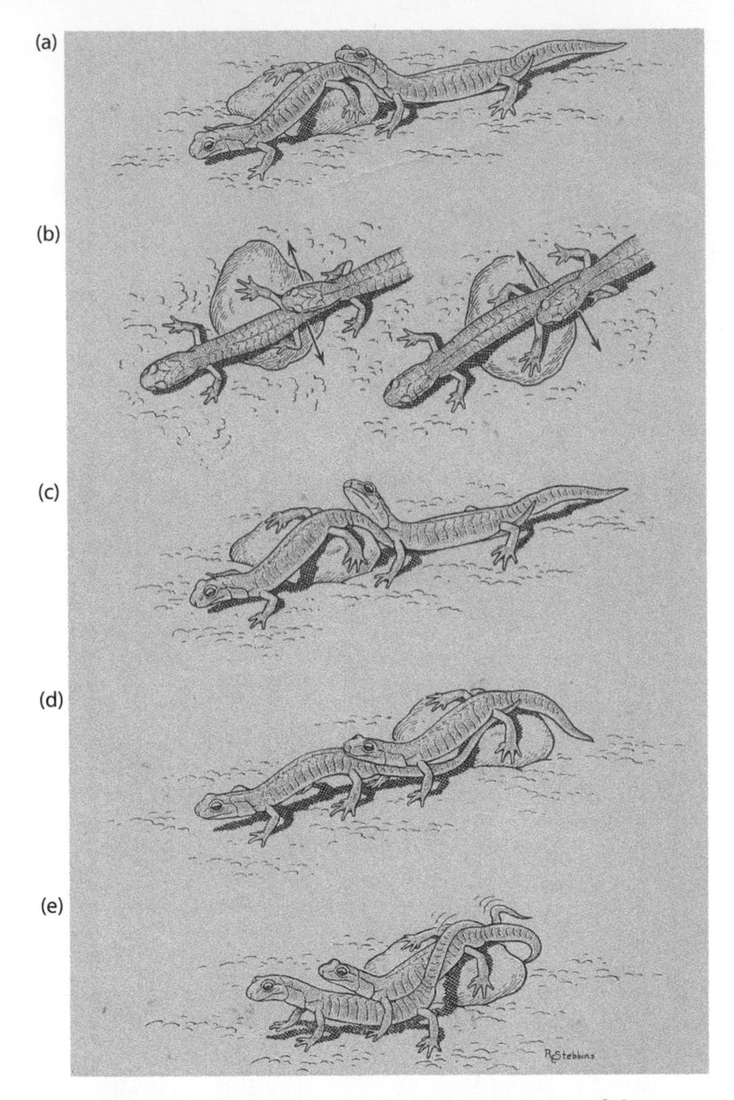

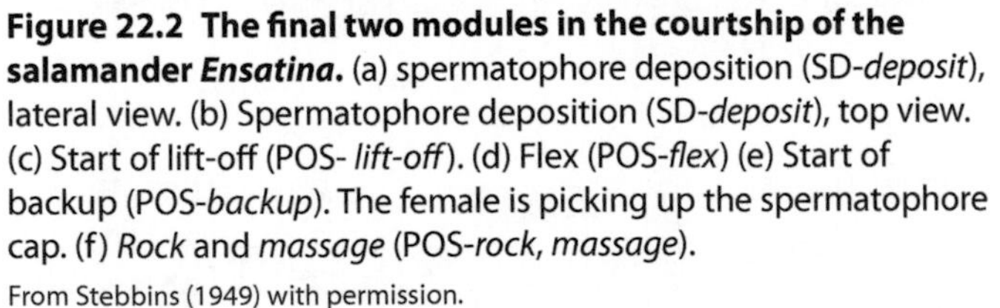

Figure 22.2 The final two modules in the courtship of the salamander *Ensatina*. (a) spermatophore deposition (SD-*deposit*), lateral view. (b) Spermatophore deposition (SD-*deposit*), top view. (c) Start of lift-off (POS- *lift-off*). (d) Flex (POS-*flex*) (e) Start of backup (POS-*backup*). The female is picking up the spermatophore cap. (f) *Rock* and *massage* (POS-*rock, massage*).

From Stebbins (1949) with permission.

of display modules, beginning on the perches and culminating in the ballerina dance (BD) on the cleared court, below the perching female (courtship of *Parotia lawesii*, https://www.youtube.com/watch?v=GEDVSSjamm0). We see in Figure 22.3 that diversity across the genus includes variety in the major modules of courtship (the grey-shaded octagonal boxes) and well as in the display phases (coloured boxes) within modules. All of this behavioural diversity, which includes substantial changes in temporal organization, has evolved within a period of about 10.4 million years (Irestedt et al. 2009).

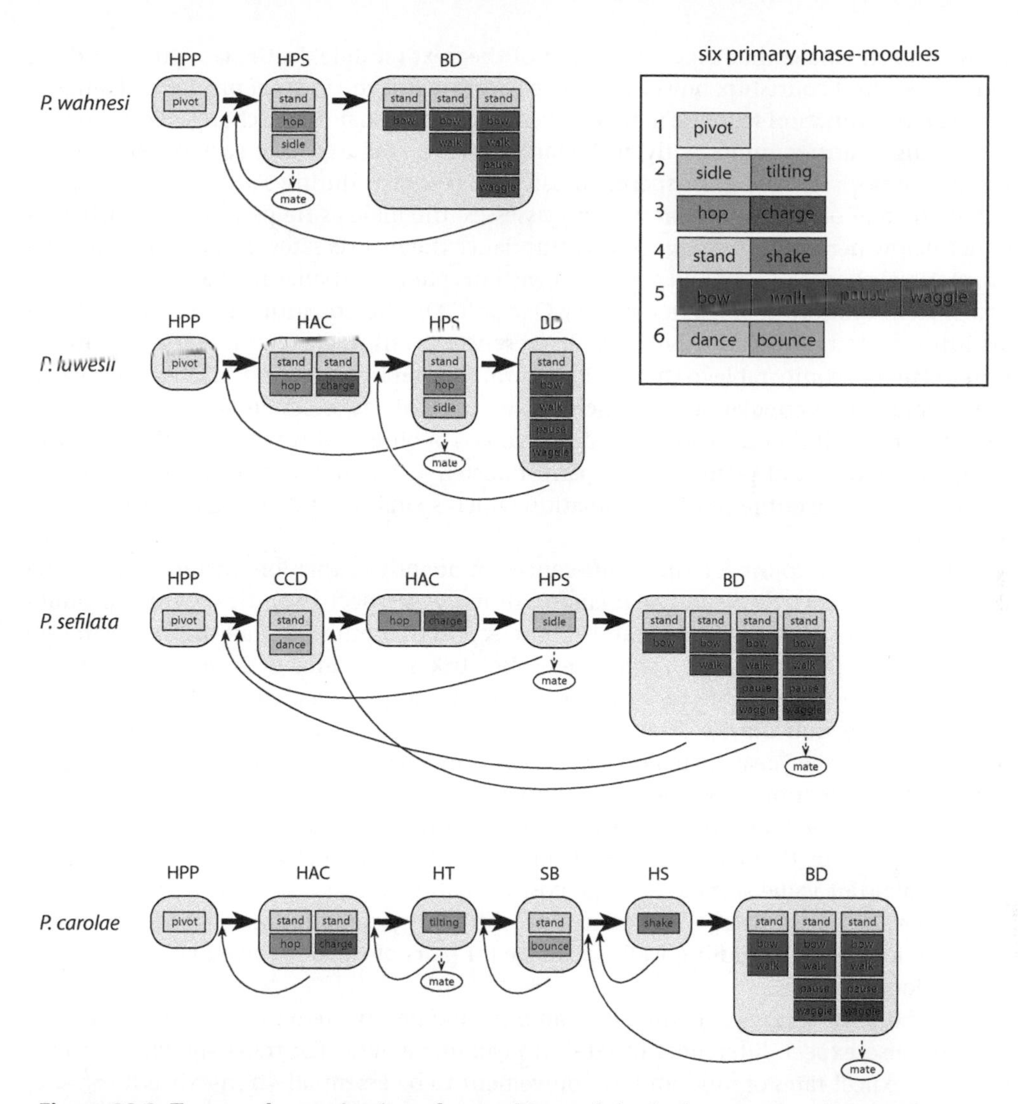

Figure 22.3 Temporal organization of courtship modules in four species of birds of paradise in the genus *Parotia*. Display modules are shown as labelled, grey octagonal boxes (HPP = horizontal perch pivot, HPS = horizontal perch slide, HAC = hops-across-court, CCD = court clearing dance, HT = head tilting, SB = swaying bounce, HS = hop and shake, and BD = ballerina dance). Courtship sequences flow from left to right (large arrows) and from top to bottom within modules. Colored rectangles denote phases within modules. Phases are of six major kinds (inset box). Alternative versions of phase sequences are shown as side-by-side columns within modules.

From Scholes (2008) with permission.

What accounts for the contrast between the rapid diversification of courtship choreography in *Parotia* over a period of 10 million years and the extreme stasis of sperm transfer behaviours in plethodontids over a period of 100 million years? In *Parotia*, as in plethodontids, the female governs the temporal flow of courtship from module to module. In both cases, female control consists of subtle movements, usually directed towards rather than away from the male, that trigger the onset of the next module. In the last three modules of plethodontid courtship, however, female control takes the form of precise and lengthy physical coordination with the male's movements. The female's role changes from spectator to tango partner. Importantly, if the tango partners fail to follow each other's moves, the dance does not result in sperm transfer. In contrast, during *Parotia* courtship, the female merely observes, and presumably assesses, the male as he performs the ballerina dance below her in the clearing. During this dance the male creates rhythmically moving geometrical shapes (circles and crescents) with his plumage (ballerina dance of *P. sefilata*, https://www.youtube.com/watch?v=wTwOxcOqlCA). The constituent courtship phases are intricate, but they require only female presence, not intimate coordination. Coordination in *Parotia*, comparable to that during sperm transfer in plethodontids, happens when the female solicits copulation by crouching and the male balances on her back to achieve cloacal contact. It is in the coordinated cloacal kiss of birds that we see evolutionary stasis comparable to that of plethodontid sperm transfer. As we might expect, the cloacal kiss is static across the entire passerine radiation which spans about 75 million years (Ericson et al. 2003).

How can we account for these differences in adaptive radiation within and among courtship modules in birds and salamanders from the perspective of the moving optimum model? First, let us focus on the stabilizing selection that pulls the trait mean towards an intermediate optimum (peak), and which also strikes a balance with drift of the mean away from the peak. In the case of interacting sexual partners, we model multivariate stabilizing selection on the interacting traits with the *N*-matrix with elements v_{ij}. We imagine a Gaussian peak in male- and female-trait space with sexual success falling off as a Gaussian function as we move away from a perfect functional match between male- and female-trait values. The *N*-matrix defines the premium on such a match. A small value of v_{ij} means that any departure from a perfect match between male-trait value z_j and female-trait value y_i incurs a large cost in mating success. According to our model the values of v_{ij} (analogous to a variance) would be very small for male and female traits involved in the tail-straddling walk, but large for pairs of traits involved in looser sexual interaction.

Aside from these expected differences among modules in strengths of interaction selection, we also expect differences in rates of peak movement. For traits showing extreme stasis we expect rates of random peak movement to be essentially non-existent ($\sigma_\theta^2 \approx 0$) but very much larger in the case of male display traits that do not require coordinated interaction with the female ($\sigma_\theta^2 \geq 0.00055$). But, despite the inference of strong interaction selection on traits involved in the tail-straddling walk and essentially no peak movement, we do see rare evolutionary innovation in this module. For example, the *Ensatina* lineage has deleted pheromone delivery from the TSW module (Arnold et al. 2017). Such trait change in an otherwise static module may reflect drift during an episode of small population size or temporary relaxation of interaction selection. Coming back to the main topic of this section, change in two parameters (v_{ij} and σ_θ^2) enables the model to toggle between extreme stasis and exuberant radiation.

22.9 The Structure of Multivariate Evolutionary Theory: Ten Matrices

We have assiduously presented and discussed the multiple-trait version of evolutionary quantitative genetic theory for several reasons. (1) The individuals that concern us are composed of multiple phenotypic traits and consequently the evolution of their

Table 22.1 Ten matrices that specify the structure of multivariate evolutionary theory. All of these matrices have *p* rows and *p* columns, where *p* is the number of traits. Abbreviations: individual selection surface (ISS), adaptive landscape (AL), vector of directional selection gradients (β).

Category	Short name	Long name	Symbol	Describes	of	Text description
Mutation	*M*-matrix	Mutation var-cov matrix	$\boldsymbol{M}$	mutational effects	locus	Chap 6
Inheritance	*G*-matrix	Additive genetic var-cov matrix	$\boldsymbol{G}$	additive genetic values	individ.	Chap 6
Variation	*E*-matrix	Environmental var-cov matrix	$\boldsymbol{E}$	environ. values	individ.	Chap 6
Variation	*P*-matrix	Phenotypic var-cov matrix	$\boldsymbol{P}$	phenotypic values	individ.	Chap 6
Selection	γ-matrix	Quadratic selection gradient matrix	$\boldsymbol{\gamma}$	interactive effects on fitness	ISS	Chap 4
Selection	AL shape matrix	Quadratic curvature matrix	$\boldsymbol{\gamma} - \boldsymbol{\beta}\boldsymbol{\beta}^T$	curvature	AL	Chap 4
Selection	ω-matrix	Gaussian selection gradient matrix	$\boldsymbol{\omega}$	interactive effects on fitness	ISS	Chap 4
Selection	Ω-matrix	Gaussian curvature matrix	$\boldsymbol{\Omega}$	curvature	AL	Chap 4
Selection	*N*-matrix	Interaction matrix	$\boldsymbol{N}$	trait matching	separate individ.	Chap 18
Selection	*Br*-matrix	Brownian motion controller matrix	$\boldsymbol{Br}$	movement	adaptive peak	Chap 13, 18

populations is multivariate. (2) The consideration of even two phenotypic traits introduces new and fundamental issues that do not arise in the univariate case. In contrast, the transition from two to three or more traits is much less profound. (3) These new, fundamental issues are of two kinds (covariance and interaction) that are unfamiliar to many students and even some professional evolutionary biologists. (4) These fundamental multivariate issues are handled by arranging coefficients of covariation and interaction in matrices (Table 22.1). (5) Mathematical treatment of equations using these matrices reveals new evolutionary phenomena that are important but unapparent in the univariate case. Examples of these multivariate revelations are provided below:

- $\boldsymbol{M}$—Mutational covariances (the off-diagonal terms in $\boldsymbol{M}$) describe pleiotropic effects of mutation at a locus on pairs of phenotypic traits, in particular the covariance between traits induced by mutation at a particular locus.
- $\boldsymbol{G}$—Genetic covariances (off-diagonal terms in $\boldsymbol{G}$) can cause maladaptive curvature in the evolutionary trajectories of the trait mean.
- $\boldsymbol{E}$—Environmental covariances (off-diagonal terms in $\boldsymbol{E}$) describe environmental contribution to the phenotypic covariance between traits.
- $\boldsymbol{P}$—Phenotypic covariances (off-diagonal terms in $\boldsymbol{P}$) can induce indirect effects of selection (shifts in mean, variance and covariance) and so disguise the actual target of selection.
- $\boldsymbol{\gamma}$ and $\boldsymbol{\omega}$—Correlational selection (off-diagonal terms in both matrices) can increase genetic correlation between traits and shape functional complexes.

- $\gamma - \beta\beta^T$ and Ω—Orientation of the adaptive landscape (off-diagonal terms in both matrices) affects evolutionary trajectories of the trait mean and promotes multivariate alignment with $\boldsymbol{M}$ and $\boldsymbol{G}$.
- $\boldsymbol{N}$—Matching between different traits of sexual partners (diagonal and off-diagonal terms in $\boldsymbol{N}$) mediates the evolutionary consequences of sexual interactions. A similar matrix mediates the evolutionary consequences of ecological interactions between individuals of different species (e.g., predator and prey).
- $\boldsymbol{Br}$—Rate and direction of Brownian peak movement (off-diagonal terms in $\boldsymbol{Br}$) in trait space affect the multivariate shape of adaptive radiations, producing—for example—radiations that are shaped like tilted cigars.

22.10 Chasing of Separate Peaks by Different Lineages Can Lead to Sexual and Ecological Speciation

Peak movement is the crucial missing ingredient in earlier treatments of speciation (Chapter 19). In many treatments of speciation, selection that leads to trait divergence is characterized in vague terms. An appeal to 'disruptive selection' (Rundle and Nosil 2005), for example, is insufficient because of its vagueness. 'Ecologically divergent selection' (Nosil 2012) is closer to the mark, but again, we need an explicit model of selection. Vague characterizations of selection also fail to supply an empirical agenda. If we are to assess selection, we need hypotheses about the specific kinds of selection we expect to be important to speciation. In our comparison of the ability of specific selective mechanisms to continuously drive trait differentiation before and after speciation (Chapter 19), peak movement emerged as the leading candidate driver of divergence.

We also need to follow the lead of Dobzhansky (1937) and his students in recognizing that new species arise by evolving reproductive isolation. Critics of the biological species concept (with isolation as its mantra) complain that the concept is not operational (Cracraft 1983). This criticism misses the point that isolation can be measured in contemporary populations and consequently evolutionary progress towards speciation can be assessed. Furthermore, the most powerful analytical theory of speciation is built using isolation as its centrepiece (Chapter 19).

The ascendancy of the moving optimum model as a selective mechanism that drives sexual speciation is an outcome of model comparisons. The problem facing rival selection models is that the realistic versions of the Fisher-Lande process for sexual selection that lack peak movement produce a single, stable evolutionary equilibrium. In that realistic version, stabilizing selection acts on both ornaments and preferences, a condition that sharply curtails among-population differentiation. Adding the feature of perpetual, random movement of ornament and preference optima allows traits in separate populations to continuously diverge. This continuous differentiation can lead to sexual isolation and speciation within a few thousand generations.

The moving peak model also emerges as a candidate driver of ecological differentiation and speciation. If the optima of traits that characterize ecological niches and interactions move by a continuous random process it is not hard to imagine that ecological differentiation and speciation will follow suit. This intuition was verified in Chapter 19. As in the case of sexual speciation, modest rates of peak movement can lead to ecological speciation in a few thousand generations. If traits mediate ecological interactions, the path to speciation can be amplified.

22.11 Coevolution on Static and Evolving Adaptive Landscapes

We have discussed the impact of peak movement on the evolution and radiation of lineages without reference to their ecological interactions. Not surprisingly, trait-mediated interactions can profoundly affect the growth and evolution of interacting species. An important contemporary perspective recognizes that evolution of the traits that mediate interactions is anchored by functional constraints. These constraints define adaptive peaks that would be occupied in the absence of interaction. Ecological interactions between coevolving species cause temporary as well as permanent departure from these functional peaks. Trait-based models of ecological interaction between competing species can generate selection that drives species apart and reduces the intensity of interaction. Finally, when the feature of peak movement is added to coevolutionary models, maladaptation can range from mild to severe and may be permanent.

22.12 Conclusions

- A model of the trait mean chasing a perpetually moving adaptive peak provides a strong basis for understanding adaptive evolution and radiation. This model is supported by empirical evidence that buttresses both its foundations and its predictions.
- Adaptive peak models bridge between the domains of evolution and ecology. Recently these models have illuminated the coevolution of interacting species.
- Peak movement is the common denominator of models that successfully account for data on trait divergence on all timescales, as well as iconic adaptive radiation.
- Although stochastic models of peak movement can predict the broad features of adaptive radiation, the study of limits on radiation remains a growing point in theoretical work.
- Regions of trait space densely populated by species means may often represent regions of recurrent peak movement.
- The adaptive landscape shapes patterns of mutation, inheritance, and peak movement.
- Most if not all traits are elements of functional complexes. Interactions between traits in such complexes can curtain or allow adaptive radiation.
- The ubiquity of trait interactions stresses the importance of multiple-trait evolutionary theory. The structure of that multivariate theory is encapsulated in ten kinds of matrices.
- Chasing of separate peaks by different lineages can lead to sexual and ecological speciation.
- Trait-mediated ecological interactions can profoundly affect the coevolution of interacting species. When the feature of peak movement is added to coevolutionary models, maladaptation can range from mild to severe and may be permanent.

Literature Cited

Arnold, S. J., K. M. Kiemnec-Tyburczy, and L. D. Houck. 2017. The evolution of courtship behavior in plethodontid salamanders, contrasting patterns of stasis and diversification. Herpetologica 73:190–205.

Beaulieu, J. M., D.-C. Jhwueng, C. Boettiger, and B. C. O'Meara. 2012. Modeling stabilizing selection: Expanding the Ornstein-Uhlenbeck model of adaptive evolution. Evolution 66:2369–2383.

Butler, M. A., and A. A. King. 2004. Phylogenetic comparative analysis: A modeling approach for adaptive evolution. Am. Nat. 154:683–695.

Cracraft, J. 1983. Species concepts and speciation analysis. pp. 159–187 *in* R. F. Johnston, ed. Current Ornithology. Springer, New York.

Doten, K., G. W. Bury, M. Rudenko, and S. J. Arnold. 2017. Courtship in the torrent salamander, *Rhyacotriton*, has an ancient and stable history. Herpetol. Conserv. Biol. 12:457–469.

Ericson, P. G. P., M. Irestedt, and U. S. Johansson. 2003. Evolution, biogeography, and patterns of diversification in passerine birds. J. Avian Biol. 34:3–15.

Estes, S., and S. J. Arnold. 2007. Resolving the paradox of stasis: Models with stabilizing selection explain evolutionary divergence on all timescales. Am. Nat. 169:227–244.

Futuyma, D.J. 1987. On the role of species in anagenesis. Am. Nat. 130:465–473.

Futuyma, D. J. 2010. Evolutionary constraint and ecological consequences. Evolution 64:1865–1884.

Gould, S. J. 2002. The Structure of Evolutionary Theory. Harvard University Press, Cambridge.

Harmon, L. J., J. B. Losos, T. Jonathan Davies, R. G. Gillespie, J. L. Gittleman, W. Bryan Jennings, K. H. Kozak, M. A. McPeek, F. Moreno-Roark, T. J. Near, A. Purvis, R. E. Ricklefs, D. Schluter, J. A. Schulte II, O. Seehausen, B. L. Sidlauskas, O. Torres-Carvajal, J. T. Weir, and A. Ø. Mooers. 2010. Early bursts of body size and shape evolution are rare in comparative data. Evolution 64:2385–2396.

Hohenlohe, P. A., and S. J. Arnold. 2008. MIPoD: a hypothesis-testing framework for microevolutionary inference from patterns of divergence. Am. Nat. 171:366–385.

Irestedt, M., K. A. Jønsson, J. Fjeldså, L. Christidis, and P. G. Ericson. 2009. An unexpectedly long history of sexual selection in birds-of-paradise. BMC Evol. Biol. 9:235. doi:10.1186/1471-2148-9-235

Lande, R. 1976. Natural selection and random genetic drift in phenotypic evolution. Evolution 30:314–334.

Lande, R. 1979. Quantitative genetic analysis of multivariate evolution, applied to brain: body size allometry. Evolution 33:402–416.

Lande, R. 1982. A quantitative genetic theory of life history evolution. Ecology 63:607–615.

Landis, M. J., and J. G. Schraiber. 2017. Pulsed evolution shaped modern vertebrate body sizes. Proc. Natl. Acad. Sci. 114:13224–13229.

Landis, M. J., J. G. Schraiber, and M. Liang. 2013. Phylogenetic analysis using Lévy processes: Finding jumps in the evolution of continuous traits. Syst. Biol. 62:193–204.

McPeek, M. A. 2017. The ecological dynamics of natural selection: traits and the coevolution of community structure. Am. Nat. 189:E91–E117.

Nosil, P. 2012. Ecological Speciation. Oxford University Press, Oxford.

Reitan, T., T. Schweder, and J. Henderiks. 2012. Phenotypic evolution studied by layered stochastic differential equations. Ann. Appl. Stat. 6. DOI: 10.1214/12-AOAS559.

Rundle, H. D., and P. Nosil. 2005. Ecological speciation. Ecol. Lett. 8:336–352.

Scholes II, E. 2008. Evolution of the courtship phenotype in the bird of paradise genus *Parotia* (Aves: Paradisaeidae): Homology, phylogeny, and modularity. Biol. J. Linn. Soc. 94:491–504.

Simpson, G. G. 1944. Tempo and Mode in Evolution. Columbia University Press, New York.

Simpson, G. G. 1953. The Major Features of Evolution. Columbia University Press, NewYork.

Stebbins, R. C. 1949. Courtship of the plethodontid salamander *Ensatina eschscholtzii*. Copeia 1949:274–281.

Uyeda, J. C., T. F. Hansen, S. J. Arnold, and J. Pienaar. 2011. The million-year wait for macroevolutionary bursts. Proc. Natl. Acad. Sci. USA108:15908–15913.

Uyeda, J. C., and L. J. Harmon. 2014. A novel Bayesian method for inferring and interpreting the dynamics of adaptive landscapes from phylogenetic comparative data. Syst. Biol. 63:902–918.

Wilburn, D. B., S. J. Arnold, L. D. Houck, P. W. Feldhoff, and R. C. Feldhoff. 2017. Gene duplication, co-option, structural evolution, and phenotypic tango in the courtship pheromones of plethodontid salamanders. Herpetologica 73:206–219.

Glossary of Notation and Symbols

Symbol, name **[Chapter transition]**

z, phenotypic trait value **[1]**
$\bar{z}$, average phenotypic trait value
$p(z)$, phenotypic trait distribution before selection
$w(z)$, relative fitness as a function of trait value
$\bar{w}$, average relative fitness
$W(z)$, absolute fitness as a function of trait value
$\bar{W}$, average absolute fitness
$p^*(z)$, phenotypic trait distribution after selection
$\int p^*(z)\,dz$, integral, area under the curve, $p^*(z)$
P, phenotypic trait variance before selection
$\bar{z}^*$, average phenotypic trait value after selection
s, directional selection differential
s', variance-standardized directional selection differential
i, selection intensity
z_{1-p}, phenotypic trait value used as a truncation selection point
$Cov(x, y)$, covariance between random variables x and y
$Cov(w, z)$, covariance between relative fitness and phenotypic trait value
P^*, phenotypic trait variance after selection
C, nonlinear selection differential
z, vector of phenotypic trait values for an individual **[2]**
m, the number of traits
$\bar{\mathbf{z}}$, vector of phenotypic trait means for the population
$p(\mathbf{z})$, multivariate distribution of vector of phenotypic trait values before selection
$\boldsymbol{P}$, phenotypic trait variance-covariance matrix before selection
$\boldsymbol{P}^{-1}$, the matrix inverse of $\boldsymbol{P}$
$\left|\boldsymbol{P}^{-1}\right|$, the determinant of $\boldsymbol{P}^{-1}$
$p^*(\mathbf{z})$, multivariate distribution of vector of phenotypic trait values after selection
$w(\mathbf{z})$, relative fitness as a function of the vector of phenotypic trait values
$W(\mathbf{z})$, absolute fitness as a function of the vector of phenotypic trait values
$\boldsymbol{s}$, vector of directional selection differentials
$\boldsymbol{\beta}$, vector of directional selection gradients
$\boldsymbol{C}$, matrix of nonlinear selection differentials
$\boldsymbol{P}^*$, phenotypic trait variance-covariance matrix after selection

$\boldsymbol{s}^{\mathrm{T}}$, transpose of $\boldsymbol{s}$, the elements of $\boldsymbol{s}$ are arranged in column, $\boldsymbol{s}^{\mathrm{T}}$ has the same elements arranged in a row

$\tilde{z}$, the deviation of a phenotypic trait value from its mean.

γ, matrix of nonlinear selection gradients

$\boldsymbol{\Lambda}$, canonical form of $\boldsymbol{\gamma}$ with the eigenvalues of γ, λ_i, on its main diagonal

$\mathbf{M}$, matrix with columns that are the eigenvectors of $\boldsymbol{\gamma}$

$\partial w(z)/\partial z$, first derivative of $w(z)$ with respect to z **[3]**

$\partial^2 w(z)/\partial z^2$, second derivative of $w(z)$ with respect to z

α, elevation of a linear or quadratic approximation to the individual selection surface (ISS)

ε, deviation of an individual trait value from a linear or quadratic approximation to the ISS

$ln\bar{W}$, natural logarithm of average absolute fitness in the population

θ, optimum of a Gaussian individual selection surface or the peak of a Gaussian adaptive landscape

ω, variance-like width of a Gaussian individual selection surface

Ω, variance-like width of a Gaussian adaptive landscape

$\partial\, ln\bar{W}/\partial\bar{z}$, first derivative of the adaptive landscape evaluated at the trait mean, slope

$\partial^2\, ln\bar{W}/\partial\bar{z}^2$, second derivative of the adaptive landscape evaluated at the trait mean, curvature

θ_t, position of the optimum at time t in the Chevin et al. (2015) model

x_t, environmental variable at time t that governs the movement of the optimum

ϵ_t, stochastic contribution to the position of the optimum at time t

σ_ϵ^2, the variance of the stochastic variable ϵ_t

β', variance-standardized directional selection gradient

β_μ, mean-standardized directional selection gradient

$\beta_{i,j}$, the directional selection gradient for the *ith* trait in the *jth* temporal period

$\partial w(\mathbf{z})/\partial z_1$, partial derivative of $w(\mathbf{z})$ with respect to z_1 **[4]**

$\partial^2 w(\mathbf{z})/\partial z_1^2$, partial second derivative of $w(\mathbf{z})$ with respect to z_1

$\partial^2 w(\mathbf{z})/\partial z_1 \partial z_2$, partial second derivative $w(\mathbf{z})$ with respect to z_1 and z_2

$\boldsymbol{m_i}$, the *ith* eigenvector of γ, *ith* column of $\mathbf{M}$, corresponding to the *ith* eigenvalue, λ_i

$\mathbf{z}_0$, stationary point on a fitted quadratic approximation to the individual selection surface; a fitness maximum, minimum, or saddle point

w_0, value of relative fitness at the stationary point

$\boldsymbol{\gamma_{min}}$, eigenvector of $\boldsymbol{\gamma}$ with the smallest eigenvalue (least curvature of the ISS), selective line of least resistance

$\partial ln\bar{W}/\partial\bar{\mathbf{z}}$, vector of first derivatives of the adaptive landscape evaluated at the vector of trait means

$\partial^2 ln\bar{W}/\partial\bar{\mathbf{z}}^2$, matrix of second derivatives of the adaptive landscape evaluated at the vector of trait means

$\boldsymbol{\omega}$, matrix of Gaussian nonlinear selection gradients, ω-matrix

$\boldsymbol{\theta}$, vector of optima for a Gaussian ISS

$\boldsymbol{\Omega}$, matrix of Gaussian nonlinear curvature coefficients for a Gaussian adaptive landscape, $\boldsymbol{\Omega}$- matrix

$\boldsymbol{\omega_{max}}$, the eigenvector of $\boldsymbol{\omega}$ with the largest eigenvalue, selective line of least resistance

$\boldsymbol{\Omega}_{max}$, the eigenvector of $\boldsymbol{\omega}$ with the largest eigenvalue, selective line of least resistance
$\boldsymbol{\Lambda}_{\omega}$, canonical form of ω
$\mathbf{M}_{\omega}$, eigenvectors of $\boldsymbol{\Lambda}_{\omega}$
Λ_{Ω}, canonical form of $\boldsymbol{\Omega}$
$\mathbf{M}_{\Omega}$, eigenvectors of $\boldsymbol{\Lambda}_{\Omega}$
$f(\mathbf{z})$, projection pursuit approximation of the true ISS in the formulation of Schluter and Nychka (1994)
f_i, single-variable regression coefficient (ridge function)
$\boldsymbol{a}_i$, projection that identify a direction in a cross-section of $f(\mathbf{z})$
p, the number of dimensions in the projection pursuit approximation surface
x, additive genetic or breeding trait value of an individual **[5]**
e, environmental trait value of an individual
$\bar{x}$, mean additive genetic value for a trait in a population
$\bar{e}$, mean environmental value for for a trait in a population
G, additive genetic variance of a trait within a population
E, environmental variance of a trait within a population
h^2, heritability of a trait within a population
y, dominance value of a trait for an individual
i, epistasis value of a trait for an individual
i_{AA}, contribution of additive by additive interactions across all loci
i_{AD}, contribution of additive by dominance interactions across all loci
i_{DD}, contribution of dominance by dominance interactions across all loci
r, coefficient of relationship, the probability that relatives X and Y have the same allele through identity by descent
u, coefficient of consanguinity, the probability that two alleles at a locus drawn at random, one each from relative X and relative Y, will be identical by descent
G_D, genetic variance arising from dominance, variance of dominance deviations
G_{AA}, genetic variance arising from additive by additive epistasis
G_{AD}, genetic variance arising from additive by dominance epistasis
a, trait value of a single locus homozygous genotype
d, trait value of a single locus heterozygote
α_1, average effect of substituting one allele (A_1) at a locus for another (A_2)
α_2, the average effect of substituting one allele (A_2) at a locus for another (A_1)
α, the average effect of allelic substitution at a locus
p, the number of fixed effects in the animal model of Henderson (1984)
q, the number of individuals in the pedigree in the animal model
n, the number of individuals with phenotypic scores in the animal model
$\mathbf{z}$, column vector of phenotypic values of all individuals in a sample
z_i, phenotypic trait value of the *ith* individual in a sample
μ_j, contribution of fixed effect j to z_i
u_k, contribution of random effect k to z_i
e_i, contribution of a residual effect to z_i
$\boldsymbol{b}$, fixed, usually unknown column vector of length p, giving the contributions of the fixed effect to $\mathbf{z}$

$\boldsymbol{u}$, random, unknown column vector of additive values of length q, with zero means, giving the contribution of the random effect to $\mathbf{z}$

$\boldsymbol{e}$, random, unknown column vector of length n, with zero means, giving the residual contribution to $\mathbf{z}$

$\boldsymbol{X}$, known n x p design matrix (with zero and one entries) which assigns the fixed effects to individuals

$\boldsymbol{Z}$, known n x q design matrix (with zero and one entries) which assigns the random effects to individuals

$\boldsymbol{F}$, q x q symmetric variance-covariance matrix matrix that is usually nonsingular (has a matrix inverse), $Var(\boldsymbol{u})$

$\boldsymbol{R}$, n x n diagonal matrix with each individual's error on the diagonal and zeros elsewhere, $Var\ (\boldsymbol{e})$

n, number of freely recombining loci affecting the trait in Kimura's (1965) model

μ, mutation rate per haploid locus per generation

α^2, variance of mutant allele effects (i.e., variance of average effects, α) at a locus

ζ, total genomic mutation rate

U, per generation input from mutation to genetic variance in a trait

M, variance in mutational effects summed across all loci

$\Delta_s G$, reduction in genetic variance each generation due to stabilizing selection

$\hat{G}$, equilibrium genetic variance at mutation-selection balance

$\boldsymbol{G}$, additive genetic variance-covariance matrix for m traits before selection **[6]**

$\mathbf{z}$, vector of phenotypic values for m traits

$\boldsymbol{x}$, vector of additive genetic (breeding) values for m traits

$\boldsymbol{e}$, vector of environmental values for m traits

$\bar{\mathbf{z}}$, vector phenotypic means for m traits

$\bar{\boldsymbol{x}}$, vector of additive genetic means for m traits

$\bar{\boldsymbol{e}}$, vector of environmental means for m traits

$\boldsymbol{P}$, phenotypic variance-covariance matrix before selection

$\boldsymbol{E}$, environmental variance-covariance matrix before selection

r_g, additive genetic correlation between two traits

z_{oi}, phenotypic value of the *ith* trait in an offspring of a mother exerting maternal effects in the Kirkpatric and Lande (1989) model

x_{0i}, additive genetic value of the *ith* trait in an offspring of a mother exerting maternal effects

e_{oi}, environmental value of the *ith* trait in an offspring of a mother exerting maternal effects

m_{ij}, maternal effect of the *jth* maternal trait on the *ith* offspring trait

$\boldsymbol{m}$, matrix of maternal effect coefficients with elements m_{ij}

z^*, phenotypic value of the *jth* maternal trait after selection

$\boldsymbol{C_{az}}$, matrix of covariances between additive genetic and phenotypic values

$\boldsymbol{M}$, mutational effects matrix, a variance-covariance matrix for per locus effects of mutation on m traits in the multivariate version of Kimura's (1965) model

r_μ, mutational correlation, the correlation between per locus effects of mutation on two traits

$\boldsymbol{U}$, the total genomic mutational input matrix for$m$ traits

$\widehat{\boldsymbol{G}}_{ii}$, the equilibrium contribution of the *ith* locus to the *G*-matrix at mutation-selection balance

$\Delta_S\boldsymbol{G}$, the change in the *G*-matrix within a generation due to multivariate Gaussian selection

f, performance in the Arnold (1983) model **[7]**

$\beta_{f_j z_i}$, linear performance gradient for the *ith* phenotypic trait via the *jth* measure of performance

ε_f, residual performance not explained by performance gradients

ε_w, residual fitness not explained by fitness gradients

$\beta_{w|j}$, fitness gradient for the *jth* measure of performance

$\gamma_{fz_i^2}$, nonlinear performance gradient for the *ith* phenotypic trait

$\gamma_{fz_iz_j}$, correlational performance gradients for the *ith* and *jth* phenotypic traits

$\boldsymbol{\beta_f}$, column vector of linear performance gradients for *m* phenotypic traits

$\boldsymbol{\beta_w}$, column vector of linear fitness gradients for a set of performance measures

$\boldsymbol{\gamma_f}$, matrix of nonlinear performance gradients for *m* phenotypic traits

$\boldsymbol{\gamma_w}$, matrix of nonlinear fitness gradients for a set of performance measures

$\boldsymbol{m_i}$, *ith* eigenvector of the γ –matrix

N_e, the effective size of a population **[8]**

n, number of demes in Wright's (1939) island model

$\bar{N}$, average effective size of the demes in Wright' island model

F_{ST}, among deme component of variance in gene frequency

$Var(\vartheta)$, variance in the eventual contributions of demes to the effective size of the whole population in the Whitlock and Barton (1997) model of effective size of a metapopulation

N_i, the effective size of the *ith* deme

F_{STi}, the average of pair-wise F_{ST}s for the ith deme

$\boldsymbol{m}$, n x n migration matrix in which the element m_{ij} represents the per-generation migration rate from deme *j* to deme *i* (the proportion of individuals in deme *i* newly arrived from deme *j*)

$Var(\bar{z}_0)$, variance in mean trait values at generation 0 of replicate lineages evolving by a stochastic process (e.g., drift) in Lande's (1976) model

$\Phi(\bar{z}_t)$, probability distribution of mean trait values at generation *t* of replicate lineages evolving by a stochastic process (e.g., drift)

$D(t)$, variance in mean trait values at generation *t* of replicate lineages evolving by a stochastic process (e.g., drift)

$\boldsymbol{T}$, phylogeny of a set of *r* taxa represented by an *r* x *r* shared ancestry matrix in which the off-diagonal *ijth* element is the time (in generations) from the root of the tree to the most recent common ancestor of taxon *i* and taxon *j*, and the diagonal elements are the time from the root to extant taxon *i* in the Hansen and Martins (1996) model

$\boldsymbol{A}$, an *r* x *r* matrix analogue of $D(t)$, in which the off-diagonal, *ijth* element of is the expected covariance in trait mean values between two taxa, *i* and *j*, with a shared coancestry equal to the *ijth* element of $\boldsymbol{T}$. Similarly, the diagonal, *iith* element of $\boldsymbol{A}$ is the expected variance in mean trait values after an elapsed time given by the *iith* element of $\boldsymbol{T}$

$\Phi(\bar{\mathbf{z}})$, probability distribution of a vector of mean trait values of replicate lineages at the tips of a phylogeny specified by $\boldsymbol{A}$ and evolving by a stochastic process (e.g., drift)

$\hat{G}$, equilibrium within-population genetic variance for a populations differentiating by the Lande (1992) model for mutation-migration-drift balance in a metapopulation

$\hat{G}_a$, equilibrium among-population genetic variance corresponding to $\hat{G}$

Q_{ST}, among-population genetic variance expressed as a fraction of the total genetic variance for a trait

$Var(\bar{\mathbf{z}}_1)$, variance in the vector of trait means at generation 1 of replicate lineages evolving by a stochastic process (e.g., drift) **[9]**

$\Phi(\bar{\mathbf{z}}_t)$, distribution of replicate lineage means for m traits at generation t evolving by a stochastic process (e.g., drift)

$\boldsymbol{D}(t)$, variance-covariance (dispersion) matrix for the vector trait means at generation t of replicate lineages evolving by a stochastic process (e.g., drift), D-matrix

$\Phi(\bar{\bar{\mathbf{z}}})$, probability distribution of the stacked column vector of mean trait values of replicate lineages, $\bar{\bar{\mathbf{z}}}$, at the tips of a phylogeny specified by $\boldsymbol{A}$ and evolving by a stochastic process (e.g., drift) in the account by Hohenlohe and Arnold (2008)

$\otimes$, Kronecker product, a rule for multiplying two matrices

$\boldsymbol{p}_{max}$, leading eigenvector of the within-locality phenotypic matrix in the study by Hunt (2007)

$\boldsymbol{g}_{max}$, leading eigenvector of the G –matrix

ω_{max}, leading eigenvector of the ω – matrix

γ_{min}, trailing eigenvector of the γ – matrix (direction of weakest curvature of ISS)

$\boldsymbol{d}_{max}$, leading eigenvector of the D –matrix

$\boldsymbol{d}_{max^*}$, phyologeny-corrected version analogue of $\boldsymbol{d}_{max}$

R-matrix, a trait dispersion matrix in Houle et al. (2017) equivalent to the observed D-matrix

$\widehat{\boldsymbol{G}}$, equilibrium within-population additive genetic variance-covariance matrix in the model of Kremer et al. (1997)

$\widehat{\boldsymbol{G}_a}$, equilibrium among-population additive genetic variance-covariance matrix

$\boldsymbol{Q}_{ST}$, matrix of among-population components of additive genetic variance and covariance corrected for genetic covariance among m traits

$\Delta\bar{z}$, deterministic response to selection, change of the trait mean after one generation of directional selection **[10]**

$\Delta\bar{z}_t$, change of the trait mean after t generations of directional selection

I_A, square of the additive genetic coefficient of variation

CV_A, additive genetic coefficient of variation

$\beta(x)$, directional selection gradient at spatial position x in the model of Slatkin (1978)

L, characteristic length, the minimum length over which the geographic pattern in the trait mean will respond to geographic differences in selection and not be swamped by gene flow among populations

l, standard deviation of dispersal distance

$\partial^2\bar{z}/\partial x^2$, curvature of the geographic variation surface that relates the trait mean to spatial position, curvature of geographic variation

$\bar{z}_{ii}$, phenotypic trait mean of population i grown in its native environment

$\bar{z}_{icg}$, phenotypic trait mean of population i grown in a common environment (cg)

$\bar{g}_i$, genotypic trait mean of population i

$\bar{e}_{cg}$, average environmental contribution of the common garden to the phenotypic trait mean

$\bar{e}_i$, average contribution of the environment of population i to the phenotypic trait mean

$\bar{z}_{ij}$, phenotypic trait mean of population i grown in the environment of population j

δ_{ij}, interaction effect resulting from rearing genotypes from population i in environment j

b, the trait truncation point corresponding to a level of minimum selective mortality in Lande's (1976) model

P_1, one of two parental generations in a cross

P_2, the other parental generation

F_1, first filial generation, first generation of progeny in a cross between two parental populations (P_1 and P_2)

F_2, second filial generation, second generation of progeny produced by crossing individuals of the F_1

BC1, a backcross generation produced by crossing the F_1 and P_1

BC2, a backcross generation produced by crossing the F_1 and P_2

δ_i, difference in average effects at locus i on a phenotypic trait in two parental populations in the model of Lande (1981)

σ_δ^2, variance in δ_i across all loci

$mean\,(P_{i2})$, average effect of locus i on a phenotypic trait in population j

$\bar{\delta}$, mean value of δ_i average across all relevant loci

σ_S^2, segregational variance, the extra genetic variance appearing in the F_2 progeny beyond that in the F_1 progeny

$Var\,(F_k)$, trait phenotypic variance in the *kth* filial generation of a Mendelian cross

$mean\,(P_i)$, trait mean of the *ith* parental generation

n, the actual number of factors (loci) contributing to the mean trait difference between samples of the two parental populations raised in a common environment under the Wright (1968) model

n_E, the effective number of factors with equal magnitudes of effect and producing the same pattern of segregation as observed in a Mendelian cross

$Var\,(n_E)$, sampling variance of n_E

$Var\left(\sigma_S^2\right)$, sampling variance of σ_S^2

QTL, quantitative trait locus

b, an unknown parameter that represents the phenotypic effect of a single allele substitution at a putative QTL in the Lander and Botstein (1989) model

g_i, an indicator variable that denotes the number of alleles from P_2 (0 or 1) at a marker locus

ε, normally distributed error contribution (with a mean of 0 and variance σ^2) to the phenotypic trait value of an individual in a backcross progeny

$\prod_i x_i$, product of variables x_i, where i =1,2,3, . . .

$L\left(a, b, \sigma^2\right)$, probability of the observed data, likelihood

$L\left(\hat{a}, \hat{b}, \hat{\sigma}^2\right)$, the likelihood that maximizes $L\left(a, b, \sigma^2\right)$, where $\hat{a}, \hat{b}$, and $\hat{\sigma}^2$ are the parameter values that give this maximum

LOD, log10 of the ratio of the unconstrained, $L\left(\hat{a}, \hat{b}, \hat{\sigma}^2\right)$, and constrained likelihoods, $L\left(\bar{z}_1, 0, \sigma_{B1}^2\right)$; evidence for a QTL

$\Delta\bar{\mathbf{z}}$, change in the vector of phenotypic trait mean from one generation to the next **[11]**

I, desired gains index

PC 1, first principal component in a principal component analysis, the eigenvector with the largest eigenvalue

$\boldsymbol{net\beta}$, net selection gradient

I, selection index

$\Delta ln\bar{W}$, change in the natural log of mean fitness arising from response to selection in the trait mean **[12]**

$Var(\bar{z}_t)$, variance in trait means among replicate lineages evolving according to a stochastic process (e.g., OU) after *t* generations

$Var(\infty)$, equilibrium variance in trait means among replicate lineages evolving according to a stochastic process (e.g., OU)

$\Phi(\bar{z}_\infty)$, equilibrium probability distribution of trait means among replicate lineages evolving according to a stochastic process (e.g., OU)

$\bar{z}_t$, phenotypic trait mean at generation *t*

T, the expected time for the trait mean, initially situated between two peaks, to evolve to either of the peaks (Lande 1985)

$\bar{W}_a$, height of the first of two adaptive peaks

$\bar{W}_v$, height of the valley between the two peaks

$E(\bar{z}_t)$, expected value of phenotypic trait mean at generation *t*

$\partial ln\bar{W}/\partial\,\bar{\mathbf{z}}$, direction of steepest uphill slope on the adaptive landscape from a point given by the vector of trait means **[13]**

$\Delta\bar{\mathbf{z}}^2$, generalized phenotypic distance

$\boldsymbol{\Omega_{min}}$, trailing eigenvector of the Ω-matrix

$\boldsymbol{\Omega_{max}}$, leading eigenvector of the Ω-matrix

G_m, additive genetic variance of a sex-limited male trait (Lande 1980)

G_f, additive genetic variance of a sex-limited female trait

B, additive genetic covariance between a sex-limited male trait and a sex-limited female trait

β_m, total directional selection gradient for a male trait, *z*

β_f, total directional selection gradient for a female trait, *y*

θ_m, natural selection optimum for a sex-limited male trait

θ_f, natural selection optimum for a sex-limited female trait

Ω_m, width of Gaussian adaptive landscape for a sex-limited male trait

Ω_f, width of Gaussian adaptive landscape for a sex-limited female trait

ω_m, width of a Gaussian ISS for a sex-limited male trait

ω_f, width of a Gaussian ISS for a sex-limited female trait

c_m, constant directional sexual selection gradient acting on a sex-limited male trait

c_f, constant directional sexual selection gradient acting on a sex-limited female trait

a, average of the mean of a sex-limited, male trait and the mean of a sex-limited female trait; sexual average

d, difference between the mean of a sex-limited, male trait and the mean of a sex-limited female trait; sexual dimorphism

Δa, change in a from one generation to the next

Δd, change in d from one generation to the next

$\bar{\mathbf{z}}_t$, vector of phenotypic trait means at generation t, in multivariate extension of models by Lande (1976) and Hansen and Martin (1996)

$Var(\bar{\mathbf{z}}_t)$, variance-covariance matrix for the vector of phenotypic trait means at generation t

$Var(\bar{\mathbf{z}}_\infty)$, equilibrium variance-covariance matrix for the vector of phenotypic trait means at generation t

$\Phi(\bar{\mathbf{z}}_\infty)$, equilibrium probability distribution for the vector of phenotypic trait means

P_m, phenotypic variance for a sex-limited male trait, before selection (Lande 1981)

P_m^*, phenotypic variance for a sex-limited male trait, after selection

$\bar{z}^*$, phenotypic male trait (ornament) mean after natural selection

$\bar{z}^{**}$, phenotypic male trait (ornament) mean after natural and sexual selection

$\psi(z|y)$, tendency of the female with preference value y to mate with a male with ornament value z

$\psi(z)$, overall probability of females mating as a function of male ornament value, z

y, normally distributed, female preference trait

$\bar{y}$, mean female preference before selection

τ^2, phenotypic variance in female preference trait

v^2, 'variance' of the Gaussian function $\psi(z|y)$

λ, eigenvalue of the matrix in Lande's (1981) re-parameterized response to selection equations

H, additive genetic variance for female mating preference

$N[\boldsymbol{m}, \boldsymbol{V}]$, a draw of a vector from a multivariate normal distribution with mean vector, $\boldsymbol{m}$, and variance-covariance matrix, $\boldsymbol{V}$

$\boldsymbol{\beta}_{\mathbf{z}}$, two-element vector of total (natural and sexual) selection gradients for the two male ornament traits (Arnold and Houck 2016)

$\boldsymbol{\beta}_{\mathbf{y}}$, two-element vector of natural selection gradients for the two female preference traits

$\boldsymbol{B}$, 2×2 between-sex additive genetic covariance matrix

$\boldsymbol{G}$, 2×2 additive genetic variance-covariance matrix for male ornament traits

$\boldsymbol{H}$, 2×2 additive genetic variance-covariance matrix for female preference traits

α, OU selection parameter estimated by OUwie (Beaulieu et al. 2012)

σ^2, Brownian motion parameter estimated by OUwie

θ_t, position of the optimum of a Gaussian ISS at generation t (Hansen et al. 2008) **[14]**

σ_θ^2, variance in the position of the optimum, θ_t

$N(m, V)$, a draw from a normal distribution with a mean of m and a variance of V

$Var[\theta_t]$, variance among replicate lineages in the position of the optimum at generation t

$E(\bar{z}_t)$, expected value for the trait mean of a lineage at generation t (Lynch and Lande 1993)

$E(\bar{z}_\infty)$, expected equilibrium value for the trait mean of a lineage at generation t

$Var(\bar{z}_\infty)$, expected equilibrium variance among lineages for the trait mean

L, distance between the optimum at generation t and the trait mean at generation t (lag)

X_i, partially measured variable in visible layer i (Reitan et al. 2012)

$dX_i(t)$, small change in X_i at time t

W_i, white noise variable in visible layer i

$W_i(t)$, white noise variable at time t

$dW_i(t)$, small change in W_i at time t

α_1, strength of pulling force towards the optimum in layer 1, viz., X_2

α_2, strength of pulling force towards the optimum in layer 2

μ_0, optimum in layer 2

σ_i, standard deviation of W_i

t_i, characteristic time of the pulling force towards the optimum in layer i

$E(X_1)$, expected value of X_1 as time goes to infinity

$Var(X_1)$, variance in the means of X_1 in replicate lineages as time goes to infinity

$\theta(t)$, optimum of trait z at time t

Ψ, optimum of $\theta(t)$

Υ, variance-like width of a Gaussian function that governs change in $\theta(t)$

λ, constant representing the rate of occurrence displacement in a Poisson model for the occurrence of displacement in the position of a trait optimum (Uyeda et al 2011)

σ_θ^2, variance in white noise contributions to the position of the optimum

d, magnitude of a displacement in the optimum

σ_d^2, variance in d

n, number of optimum displacement events

$p_n(t)$, probability of observing n displacement events in some time interval of length t

$K[\bar{z}(t)]$, excess kurtosis in the distribution of replicate lineage trait means at generation t (Landis and Schraiber 2017)

$\hat{G}(N)$, expected value of G at equilibrium under drift-mutation balance (Lynch and Hill 1986) **[15]**

$Var\left[\hat{G}(N)\right]$, expected equilibrium variance in G among replicate populations under drift-mutation balance

$\hat{G}(SHC)$, Stochastic house-of-cards approximation for the expected value of G at equilibrium under mutation-drift-selection

$Var(G)_B$, expected equilibrium variance in G among replicate populations under drift-mutation-selection balance

F, inbreeding coefficient for a population **[16]**

$\widehat{\boldsymbol{G}}$, equilibrium $\boldsymbol{G}$ under mutation-drift balance

$\boldsymbol{U}$, matrix of genetic variance and covariance inputs to the G-matrix each generation from mutation

$\Delta_s\boldsymbol{G}$, change in the G-matrix within a generation due to multivariate selection

Σ, size of a G-matrix, sum of its eigenvalues (Jones et al. 2003)

ε, eccentricity of a G-matrix, the ratio of the smallest eigenvalue to the largest

φ, orientation of a G-matrix, the angle of the leading eigenvector to the axis of the first trait, z_1

$\partial ln\bar{W}/\partial r_\mu$, directional selection gradient for the mutational correlation, r_μ

ξ_0, value of an arbitrary reference genotype in the multilinear model of epistasis (Hansen and Wagner 2001)

$y^{(i)}$, the reference effect on an individual's genotype at locus i

$\varepsilon^{(i,j)}$, epistatic coefficient that determines the nature of the interaction between locus i and locus j

${}_ax$, individual's genotypic value for trait a in the multivariate multilinear model of epistasis (Jones et al. 2014)

${}_a\xi_0$, value of the reference genotype (zero)

${}_ay^{(i)}$, individual's reference genotypic value for trait a at locus i

${}_{abc}\varepsilon^{(i,j)}$, epistatic effect on trait a of the interaction between the effects of locus i on trait b and locus j on trait c

σ^2_ε, variance of epistatic coefficients

D, difference in trait means between two populations with no migration between them (Hendry et al. 2007)

D^*, difference in trait means between two populations, taking migration into account

LoD, expected line of divergence in trait means between an island and a mainland population (Guillaume and Whitlock 2007)

$\boldsymbol{\beta}_t$, vector of directional selection gradients at generation t (Jones et al. 2004) **[17]**

$\bar{\mathbf{z}}_t$, vector of phenotypic trait means at generation t.

$\boldsymbol{\theta}_t$, vector of trait optima (peaks) at generation t

$\Delta\boldsymbol{\theta}$, vector of rates of peak movement

$\Delta\theta_i$, rate of peak movement for phenotypic trait mean $\bar{z}_i$

$\boldsymbol{L}$, vector of equilibrium lags, distances of trait means from their adaptive peaks

$E_t\left[\boldsymbol{\theta}_t - \bar{\mathbf{z}}_t\right]$, expected lag in a finite population at generation t in a set of replicate lineages

•, a stationary adaptive peak

→, peak movement towards larger values of $\bar{z}_1$ and an unchanging value of $\bar{z}_1$

↗, peak movement towards larger values of $\bar{z}_1$ and $\bar{z}_2$

↘, peak movement towards smaller values of $\bar{z}_1$ and $\bar{z}_2$

σ^2_θ, stochastic variance in position of an adaptive peak

r_ω, correlation in ω-values, selectional correlation

$\sigma^2\left(r_\omega\right)$, stochastic variance in r_ω (Revell 2007)

Wi, matrix of within-module stabilizing and correlational selection coefficients, ω_{ii} and ω_{ij} (Melo and Marroig 2015)

Bt, matrix of between-module stabilizing and correlational selection coefficients, ω_{ii} and ω_{ij}

Br, Brownian motion orientation matric with elements giving stochastic variances and covariances for positions of the adaptive peak **[18]**

$\sigma^2_{\theta i}$, stochastic variance in position of an adaptive peak in direction of trait mean $\bar{z}_i$

$\sigma_{\theta ij}$, stochastic covariance in position of an adaptive peak in directions of trait means $\bar{z}_i$ and $\bar{z}_j$

$\boldsymbol{D}\left(\bar{\mathbf{z}}_{t+1}\right)$, variance-covariance matrix for a vector of trait means at generation t

$\boldsymbol{D}\left(\boldsymbol{\theta}_t\right)$, variance-covariance matrix for a vector of adaptive peak positions at generation t

$\boldsymbol{Br}_{\boldsymbol{max}}$, leading eigenvector of ***Br***

r_{Ω}, correlation in nonlinear selection imposed by the Gaussian adaptive landscape

r_{Br}, correlation in Brownian motion of an adaptive peak

a, elevation (intercept) in an expression for allometry

b, allometric slope (scaling exponent)

Y_0, intercept in an allometric plot

B, average metabolic rate in an allometric plot

M, average body mass in an allometric plot

I, intercept in an allometric plot

k, Boltzmann's constant

T, temperature

E_i, average activation energy for the rate-limiting enzyme-catalysed biochemical reactions of metabolism

$n(\bar{z},t)$, distribution of the number of species with trait mean $\bar{z}$, at time t in the Slatkin (1981) model **[19]**

$E(\Delta\bar{z})$, the expected value of $\Delta\bar{z}$

$Var(\Delta\bar{z})$, the among species variance in $\Delta\bar{z}$

$M_o(\bar{z})$, expected per generation change in $\bar{z}$ due to ordinary within-lineage evolution

$V_o(\Delta\bar{z})$, input to variance among species in $\Delta\bar{z}$ arising from stochastic processes within-lineages

$s(\bar{z})\,\Delta t$, probability that a species with trait mean $\bar{z}$ produces a new species in the time interval Δt

Δy, random variable representing the amount that the new species trait means deviates from its parental species at speciation.

$M_s(\bar{z})\,\Delta t$, expected change in the trait mean that occurs at speciation.

$V_s(\bar{z})\,\Delta t$, variance in the change in the trait mean at speciation

$e(\bar{z})\,\Delta t$, probability that a species with trait mean $\bar{z}$ goes extinct in time interval Δt

$\partial n(\bar{z},t)$, small change in $n(\bar{z},t)$

∂t, small interval of time

$\delta(\bar{z}-\bar{z}_0)$, Dirack delta function giving the initial value of the trait mean

s, constant value for the speciation rate, $s(\bar{z})\,\Delta t$

e, constant rate for the extinction rate, $e(\bar{z})\,\Delta t$

n, constant number of species with trait mean $\bar{z}$

$Var(\bar{z}_t)$, variance among species in trait mean at time t, arising from within-lineage processes, speciation, and extinction.

π_{AB}, probability of mating between a randomly drawn female from population A and a randomly drawn male from population B (Arnold et al. 1996)

JI, Index of Joint Isolation between two populations

$\boldsymbol{D}(t)$, 2×2 variance-covariance matrix giving the dispersion of replicate lineage trait means (ornament and preference) at generation t during drift along the line of equilibria in the Lande (1981) model

$D_z(t)$, first element of $\boldsymbol{D}(t)$, variance among replicate lineages in ornament means at generation t

$f_{\pi_{AB}}(x)$, probability that π_{AB} takes the value x at generation t (Uyeda et al. 2009)

$E[JI(t)]$, the expected value of JI at generation t

$E[\pi_{AB}]$, the expected value of π_{AB} at generation t

d_{AB}, Euclidean distance between the preference means of females from population A and the ornament means of males from population B

$\boldsymbol{N}$, interaction (matching) matrix, rows represent female traits and columns represent male traits, elements are Gaussian coefficients, such that the ν_{pq} element is the width of a Gaussian function that gives the probability of mating with a male when the trait difference is between her preference value, y_p, and his ornament value, z_q (Hohenlohe and Arnold 2010)

$\mathbf{z}_j$, vector of hypothetical male traits in population j with vector of means, $\bar{\mathbf{z}}_i$, and variance-covariance matrix, $\boldsymbol{P}_j$

$\boldsymbol{y}_i$, vector of hypothetical female traits in population i with vector of means, $\bar{\boldsymbol{y}}_i$, and variance-covariance matrix, $\boldsymbol{Q}_i$

$\boldsymbol{Z}$, a column vector of hypothetical male trait observations that will account for the π_{ij} observations

$\boldsymbol{Y}$, a column vector of hypothetical female trait observations that will account for the $\boldsymbol{Y}$ observations

$\ln L(\boldsymbol{Z}, \boldsymbol{Y})$, the log likelihood that the $\boldsymbol{Z}$ and $\boldsymbol{Y}$ trait vectors will produce the observed table of π_{ij} values

$\boldsymbol{\theta}_{\mathbf{z}}$, vector of optima for multiple male ornaments (Arnold and Houck 2016)

$\boldsymbol{\omega}_{\mathbf{z}}$, matrix of Gaussian stabilizing selection coefficients for multiple male ornaments

$\boldsymbol{\theta}_{\boldsymbol{y}}$, vector of optima for multiple female preferences

$\boldsymbol{\omega}_{\boldsymbol{y}}$, matrix of Gaussian stabilizing selection coefficients for multiple female preferences

S_{yi}, directional natural selection differential for the *ith* female preference

S_{zi}, directional natural selection differential for the *ith* male ornament

$\bar{y}_i^*$, mean of the *ith* female preference after natural selection

$Var[\bar{\mathbf{z}}(t)]$, the equilibrium variance among replicates in the vector of ornaments means in the absence of sexual selection, $t = 10,000$ generations

β_{zi}, directional selection gradient for the *ith* male habitat trait

β_{yi}, directional selection gradient for the *ith* female habitat trait

π_{AB}, probability that a female from population A encounters a male from population B as a function of their habitat traits

EI, Index of Ecological Isolation, a function of π_{ij} values

z_{ij}, phenotypic trait value for the tendency to utilize ecological resources i and j

β_{zij}, directional selection gradient for z_{ij}

θ_{zij}, peak of a Gaussian landscape for z_{ij}

$p(z_{Ai}, z_{Bi})$, strength of an interaction between species A and B mediated by the values of the *ith* phenotypic trait z_{ij}

β_{Ai}, total directional selection gradient for the *ith* trait in species A

β_{zFAi}, functional selection gradient for the *ith* trait in species A

β_{zIAi}, interaction selection gradient for the *ith* trait in species A

k, a function that converts trait values into fitness

$d\bar{x}$, a small change in the mean of phenotypic trait, x, in the Kiester et al. (1984) model

[20]

dt, a small interval of time

τ_x, generation time of a species with trait x

η, a parameter denoting sex-limitation of the trait and diploidy versus haplodiploidy of a species

$\psi(x|y)$, a matching Gaussian function (with width v) that gives strength of the interaction between phenotype x in one species and phenotype y in the other species

$w_x(x)$, relative fitness for plants with floral phenotype x

S_x, directional selection differential for a floral trait, x

S_y, directional selection differential for a pollinator trait, y

$\boldsymbol{M}$, matrix of selection and inheritance parameters that governs the evolution of the trait means, $\bar{x}$ and $\bar{y}$

$\Delta\bar{x}$, change in the mean of an insect trait, x, from one generation to the next (Nuismer 2017)

$\Delta\bar{y}$, change in the mean of an plant trait, y, from one generation to the next

$p(x, y)$, strength of the difference interaction between the insect trait and the plant trait

α, constant that translates interaction success into fitness

$W_{XI}(x, y)$, change in insect fitness as a consequence of interaction with the plant

$W_{YI}(x, y)$, change in plant fitness as a consequence of interaction with the insect

W_{XF}, functional component fitness of the insect as a function of deviation from its trait optimum, θ_x

W_{YF}, functional component fitness of the plant as a function of deviation from its trait optimum, θ_y

$W_X(x, y)$, fitness of the insect as a function of stabilizing selection (with width ω_x) on trait value x, and interaction with a plant with trait value y

$W_Y(x, y)$, fitness of the plant as a function of stabilizing selection (with width ω_y) on trait value y, and interaction with an insect with trait value x

β_{xI}, interaction selection gradient for the insect trait x

β_{yI}, interaction selection gradient for the plant trait y

β_{xF}, functional selection gradient for insect trait x

β_{yF}, functional selection gradient for plant trait y

$\hat{x}$, equilibrium value for $\bar{x}$

$\hat{y}$, equilibrium value for $\bar{y}$

m_x, proportion of individuals that moves from its natal population to another population after selection

$\bar{x}_i^*(t)$, trait mean at locality i after selection response and migration

μ_x, average of trait means at two localities

δ_x, difference between trait means at two localities

$\bar{\beta}_{xI}$, average of interaction selection gradients among localities

$\delta_{\beta_{xI}}$, difference between interaction selection gradients at two localities

$\Delta\mu_x$, change in geographic trait average from one generation to the next

$\Delta\delta_x$, change in geographic trait differences from one generation to the next

$\bar{p}_{i,j}$, average probability that a predator from population i successfully ingests a prey from population j

A_x, measure of local adaptation, a function of $\bar{p}_{i,j}$ values

$s_{x,1}$, sensitivity coefficient that is a function of a constant that translates costs and benefits of interaction into fitness (α), the frequencies of the two host species (f_Y and f_Z), and the fitness benefits to the parasites from the interaction (s_{XY} and s_{XZ}).

$\bar{W}_X$, the average fitness of a predator species X

$\bar{W}_Y$, the average fitness of a prey species, Y

κ_X, a sensitivity coefficient that puts interaction effects and effects of stabilizing selection into register (the same currency)

I, interaction efficiency, a measure of connectivity in the community network

$E(\bar{z}_{i,\tau_1})$, the expected value of the trait mean of species i at the instant of speciation

D_{i,τ_1}, within-lineage variance of the trait in lineage i at the instant of speciation

D_{12,τ_1}, the covariance between the trait means of two sister species at the instant of speciation

W_i, fitness of an individual of species i as a function of a trait-mediated interaction with another species

$\bar{W}_i$, average fitness of an individual of species i as a function of a trait-mediated interaction with another species, and the distributions of trait values in both species

D_{i,τ_2}, within-lineage variance of the trait in lineage i at time τ_2

$\hat{A}_{ij}$, equilibrium value for an index of asymmetry in competitive ability, a function of drift and elapsed time

N_i, number of individuals of the *ith* species

lnW_i, $\log_e$ absolute fitness of an individual of the *ith* species

$n_i(z_i)$, number of individuals with phenotype z_i

$p_i(z_i)$, proportion of individuals with phenotype z_i

$lnW_N(z_N)$, $\log_e$ absolute fitness of an individual consumer with phenotype z_N (McPeek 2017) **[21]**

$lnW_R(z_R)$, $\log_e$ absolute fitness of an individual resource with phenotype z_R

N, number of consumer individuals

R, number of resource individuals

N_{z_N}, number of consumer individuals with phenotype z_N

N_{z_R}, number of resource individuals with phenotype z_R

b, conversion efficiency for a particular consumer phenotype feeding on a particular resource phenotype

$a(\bar{z}_R, z_N)$, attack coefficient for consumer phenotype z_N feeding on resource phenotype $\bar{z}_R$; the rate at which a consumer kills a resource (Holling 1959)

a_0, maximum value of attack success

ω_a, width parameter for the Gaussian attack function

$f(z_n)$, intrinsic death rate of the consumer with phenotype z_N

f_0, minimum value of the consumer's intrinsic death rate

γ_f, stabilizing selection coefficient for the consumer's intrinsic death rate

θ_f, value of z_R that minimizes consumer's intrinsic death rate

g, density-dependent rate of increase in the consumer's death rate

$c(z_R)$, intrinsic birth rate of resource species with phenotype z_R

c_0, maximum value of the resource intrinsic birth rate

γ_c, stabilizing selection coefficient for the intrinsic birth rate of the resource species

θ_c, value of z_N that minimizes the death rate of the resource species

d, density-dependent rate of decrease in the birth rate of the resource

h, handling time for a particular consumer phenotype feeding on a particular resource phenotype

dN/Ndt, per capita growth rate of the consumer species

dR/Rdt, per capita growth rate of the resource species

$d\bar{z}_N/dt$, evolutionary change in $\bar{z}_N$, continuous time version of $\Delta\bar{z}_N$

G_{z_N}, additive genetic variance for the phenotypic trait of the consumer species, z_N

$\beta_{z_N b}$, the birth selection gradient for z_N, reflects the contribution of trait values to the birth component of fitness in the consumer species

$\beta_{z_N d}$, death selection gradient for z_N, reflects the contribution of trait values to the death component of fitness in the consumer species

dN/dt, population growth rate of the consumer species

dR/dt, population growth rate of the resource species

Subject Index

Notes
Tables and figures are indicated by an italic *t* or *f* following the page number.